Biomedical Imaging
The Chemistry of Labels, Probes and Contrast Agents

RSC Drug Discovery Series

Editor-in-Chief:
Professor David Thurston, *London School of Pharmacy, UK*

Series Editors:
Dr David Fox, *Pfizer Global Research and Development, Sandwich, UK*
Professor Salvatore Guccione, *University of Catania, Italy*
Professor Ana Martinez, *Instituto de Quimica Medica-CSIC, Spain*
Dr David Rotella, *Montclair State University, USA*

Advisor to the Board:
Professor Robin Ganellin, *University College London, UK*

Titles in the Series:

How to obtain future titles on publication:
A standing order plan is available for this series. A standing order will bring delivery of each new volume immediately on publication.

For further information please contact:
Book Sales Department, Royal Society of Chemistry, Thomas Graham House, Science Park, Milton Road, Cambridge, CB4 0WF, UK
Telephone: +44 (0)1223 420066, Fax: +44 (0)1223 420247, Email: books@rsc.org
Visit our website at http://www.rsc.org/Shop/Books/

Biomedical Imaging
The Chemistry of Labels, Probes and Contrast Agents

Edited by

Martin Braddock
AstraZeneca, Loughborough, UK

RSC Publishing

RSC Drug Discovery Series No. 15

ISBN: 978-1-84973-014-3
ISSN: 2041-3203

A catalogue record for this book is available from the British Library

Published by The Royal Society of Chemistry,
Thomas Graham House, Science Park, Milton Road,
Cambridge CB4 0WF, UK

Registered Charity Number 207890

For further information see our web site at www.rsc.org

Preface

The concept of medical imaging is one of the cornerstones of modern medicine. Although its origins can be found in 19th century photography, the field only properly emerged in 1895 following W. C. Röntgen's discovery of X-rays. Since then, insights from across physics and chemistry have devised many more modalities, such as magnetic resonance imaging (MRI), optical imaging (including fluorescence), X-ray imaging (including X-ray Computed Tomography, CT), gamma imaging (including Single Photon Emission Computed Tomography, SPECT), positron emission tomography (PET) and ultrasound techniques.

In this exemplary new book a distinguished group of experts from both industry and academia present a comprehensive review on how medical imaging is being used in screening, diagnosis, patient management, clinical research and to assist in the development of new therapeutic drugs.

Biomedical Imaging: The Chemistry of Labels, Probes and Contrast Agents begins with a comprehensive introduction to endogenous and exogenous contrast in medical imaging. The book is then broken down into four sections. Section one presents a review of some of the more important advances in recent years such as the development of radiotracers and radio-pharmaceuticals as biomedical imaging tools, recent developments in imaging agents for selected brain targets that are of clinical relevance in psychiatry and neurology and of pharmacological interest in drug discovery and development and the synthesis of radiopharmaceuticals for application in SPECT imaging. Section two focuses on the design and synthesis of contrast agents, MRI and X-ray modalities. Topics covered include the synthesis and applications of MRI contrast agents, synthetic methods used for the preparation of DTPA and DOTA derivative ligands, MRI contrast agents based on metallofullerenes, applications of MRI in radiotherapy treatment and the use of autoradiography in the pharmaceutical discovery and development of

RSC Drug Discovery Series No. 15
Biomedical Imaging: The Chemistry of Labels, Probes and Contrast Agents
Edited by Martin Braddock
© Royal Society of Chemistry 2012
Published by the Royal Society of Chemistry, www.rsc.org

xenobiotics. Section three concentrates on optical imaging techniques and the value of fluorescence optical imaging in pharmacological research and drug development. There are also chapters on fluorescence lifetime imaging applied to microviscosity mapping and fluorescence modification studies in cells and the design and use of contrast agents for ultrasound imaging. The final section focuses on physical techniques and application, with a review of recent advances in brain imaging that provide opportunities to develop biomarkers for diseases of the central nervous system (CNS) and current progress and future prospects of using MRI to assist in the drug discovery and development process. The final chapter brings the book to a close peering into the future of MRI contrast agents.

This book will be essential reading for medicinal and physical scientists working in both industry and academia in the fields of chemistry, physics, radiology, biochemistry and pharmaceutical sciences.

Contents

RSC Drug Discovery Series No. 15
Biomedical Imaging: The Chemistry of Labels, Probes and Contrast Agents
Edited by Martin Braddock
© Royal Society of Chemistry 2012
Published by the Royal Society of Chemistry, www.rsc.org

**Chapter 4 Design and Synthesis of Radiopharmaceuticals for
SPECT Imaging   144**
David Hubers and Peter J. H. Scott

CHAPTER 1

Medical Imaging: Overview and the Importance of Contrast

JOHN C WATERTON, PhD CChem CSci FRSC(UK)[a,b]

[a] Personalised Healthcare & Biomarkers, AstraZeneca, Alderley Park, Macclesfield, Cheshire, SK10 4TG UK; [b] Biomedical Imaging Institute, University of Manchester, Stopford Building, Oxford Road, Manchester, M13 9PT UK

1.1   Introduction

The concept of medical imaging—using a **device** to capture **images** which have **medical utility** from living humans—is one of the cornerstones of modern medicine. Although its origins can be found in 19th century photography, the field emerged properly following W. C. Röntgen's discovery, in 1895, that X-rays could image the skeleton inside a living human, an achievement for which he was awarded, in 1901, the first Nobel Prize for Physiology or Medicine. Since then, insights from across physics and chemistry have been used to devise many more imaging modalities that can be used in living humans. Some of these other modalities, such as Magnetic Resonance Imaging (MRI), or X-ray Computed Tomography (CT), have themselves also been associated with Nobel prizes, and are reportedly considered by physicians to be among the most important medical advances of the 20th century.[1] While such extraordinarily complex (and expensive) 3D imaging techniques have become essential tools in the diagnosis and management of many conditions, including cancer and cardiovascular diseases, traditional inexpensive medical imaging techniques such as planar X-ray images are still routinely used, for example to diagnose and

RSC Drug Discovery Series No. 15
Biomedical Imaging: The Chemistry of Labels, Probes and Contrast Agents
Edited by Martin Braddock
© Royal Society of Chemistry 2012
Published by the Royal Society of Chemistry, www.rsc.org

treat fractures. In developed countries, almost everyone is imaged at least once by ultrasound when a foetus, and is usually imaged again throughout life for screening, diagnosis, and treatment monitoring. In addition to its critical role in patient care, medical imaging has also revolutionised our understanding of human function and physiology, perhaps most notably in the brain, and in our understanding of psychiatric illness. Before brain imaging, psychiatrists had only a vague and ill-defined concept of the "mind", but after two or three decades of brain imaging research, neuroscientists are now beginning to be able to describe exactly how the activation and wiring of specific structures in the brain makes a mind work or malfunction.[2]

There are several distinct medical imaging modalities, some more familiar than others. Each relies on a different physical principle. The most important are listed in Table 1.1, and mapped in Figure 1.1 according to the signal they detect.

The over-arching requirement for any useful medical imaging technology is **contrast**. We need to exploit some physical principle that permits one structure in the body to report a different signal than another. So, for example, soft X-rays are stopped more by higher atomic number (Z) nuclei than by low Z nuclei, so that the skeleton (containing calcium and phosphorus in hydroxylapatite) has a higher signal than the brain (composed largely of water), which in turn has a higher signal than the air-filled lungs. A hypothetical medical imaging technique based on the absorption of, say, neutrinos, is most unlikely to be useful, because all tissues of the body are essentially transparent to neutrinos. On the other hand, medical imaging in the terahertz range is almost equally useless, because water absorbs terahertz radiation very strongly: the radiation cannot penetrate past the outermost layers of the skin, and deep organs are obscured.

Contrast may be classified as **endogenous** or **exogenous**. Familiar examples of endogenous contrast are the difference between signal from bone and soft tissue in X-ray imaging, or the difference between signal from grey and white matter in MRI (due to differences in the nuclear magnetic relaxation of water protons in the respective brain tissues). Exogenous contrast, on the other hand, is created by administering a foreign substance, usually called a tracer or a contrast agent. Many are listed in the MICAD database.[3] It is exogenous contrast that is of particular interest to the medicinal chemist, since tracers and contrast agents only exist because of chemists' creativity and ingenuity. These substances are discussed in detail in subsequent chapters. Fewer than 100 tracers and contrast agents have ever been approved by regulatory authorities for use in human healthcare, but they cover a wide variety of chemistries, including small inorganics, small organic molecules, chelates, tagged peptides, tagged proteins *e.g.* monoclonal antibodies, noble gases, nanoparticles and microbubbles. Many more have been used investigationally in humans or animals. Section 1.2 in this chapter describes the most important medical imaging modalities, and how they achieve contrast.

Once we have a new medical imaging modality, based on a physical principle which creates endogenous contrast, or with some chemistry to provide

Table 1.1 Significant Imaging Modalities.

Names and synonyms	Physical principle	Endogenous contrast	Examples of exogenous contrast chemistries approved for human use[a] [additional investigational chemistries in square brackets]
• Ultrasound, sonography • Echocardiography	Transmission and reflection of sound waves	Yes	Sometimes — Microbubbles
• Magnetic Resonance Imaging (MRI), Magnetic Resonance Tomography (MRT), Nuclear Magnetic Resonance (NMR) Imaging • Magnetic Resonance Spectroscopy (MRS), Magnetic Resonance Spectroscopic Imaging (MRSI) • Functional MRI (fMRI)	Nuclear magnetic resonance and relaxation	Yes	Sometimes — Gadolinium chelates, manganese chelates, iron nanoparticles, [other paramagnetic substances such as nitroxyls or O_2], [small molecules containing ^{19}F or ^{17}O], [hyperpolarised noble gases], [hyperpolarised small molecules containing ^{13}C], [small diamagnetic or paramagnetic compounds containing exchangeable protons]
• Optical imaging • Fluorescence imaging • Endoscopy	Excitation of valence electron: absorption, fluorescence	Yes	Sometimes — Substances which absorb or fluoresce in the visible or NIR
• X-Ray imaging, X-radiography (planar) • X-Ray computed tomography (CT) (tomographic) • Fluoroscopy • Dual-energy X-ray absorptiometry (DEXA)	Photoelectron absorption, Compton scattering	Yes	Sometimes — Organoiodines, $BaSO_4$, [other substances containing heavy atoms]
• Gamma-camera, scintigraphy (planar) • Single Photon Emission Computed Tomography (SPECT) (tomographic)	Radioactive decay with emission of gamma ray	Almost none[b]	Always — Substances containing gamma-emitting isotopes such as ^{99m}Tc, ^{111}In, ^{201}Tl, ^{133}Xe, ^{67}Ga, ^{123}I, ^{131}I, ^{51}Cr, ^{169}Yb, ^{81m}Kr
• Positron Emission Tomography (PET) (tomographic)	Radioactive decay with emission and annihilation of positron	No	Always — Substances containing positron-emitting isotopes e.g. ^{18}F, ^{82}Rb, ^{13}N, [^{11}C, ^{15}O, ^{64}Cu, ^{76}Br, ^{89}Zr, ^{124}I]

[a]Source: regulatory agency websites (FDA, EMA), accessed 2011.
[b]There is a very weak signal from endogenous ^{40}K.

exogenous contrast, there are two further conditions which must be fulfilled before it can come into practical use. Firstly, society demands that the benefits of imaging with the new modality must exceed the costs and risks. Many clever medical imaging techniques have been devised but never came into widespread use, because they lacked a commercially viable application. Secondly, it is essential to have a regulatory and legal framework within which medical imaging can be performed while ensuring acceptable levels of patient safety. Section 1.3 discusses the various uses of medical imaging, using the "biomarker" concept, while Section 1.4 outlines some regulatory and economic considerations.

1.2   Medical Imaging Modalities

1.2.1   Some General Ideas

1.2.1.1   *Formats: 2-D planar, 2-D tomographic, 3-D and 4-D*

Medical images can be created and displayed in a number of different formats. The oldest is a simple 2-D planar image typified, say, by a chest X-ray. Here the signal represents X-ray absorption, a silhouette, summed and projected through the body. Gamma scintigraphy is also a planar projection technique. More modern techniques like CT, MRI, ultrasound, SPECT, and PET can produce full 3-D data sets which can be rendered for display, although usually 2-D tomographic sections (slices) through the body are extracted for viewing. Time, of course, is also a dimension, and to understand the functioning, say, of the heart, 4-D data may be acquired from ultrasound, MRI or CT.

1.2.1.2   *Molecular or Functional Imaging vs. Anatomic Imaging*

This term "molecular imaging" is sometimes used to describe modalities that rely on signal from a specific molecule, labelled perhaps (in the case of PET or SPECT) with a radioisotope. These modalities can provide functional or physiological information such as receptor occupancy or enzyme flux. In contrast, imaging modalities such as ultrasound or MRI, sometimes seem to provide predominantly anatomic information such as organ size and shape. However this is not a rigid distinction, and in reality all modalities offer a mixture of anatomic and functional information.

1.2.1.3   *Dynamic Scans*

The term "dynamic" imaging implies repetitive imaging before and during the (usually) intravenous administration or inhalation of a tracer or contrast agent. Examples include dynamic contrast-enhanced (DCE) MRI, CT or ultrasound, and dynamic PET. From these time-dependent data, maps of uptake, distribution and clearance of the agent can be made. The pharmacokinetic parameters obtained from compartmental modelling of these maps are frequently useful in quantitative imaging biomarker studies (see Section 1.3).

1.2.2 Imaging and the Electromagnetic Spectrum

Most medical imaging employs electrical and magnetic fields, or electro-magnetic radiation. The remainder use pressure waves or sound. Modalities can be classified according to the frequency at which they operate. Electromagnetic fields and electromagnetic radiation penetrate well into the body at very low frequencies (below, say, 200 MHz), or at very high frequencies (wavelengths below, say, 0.1 nm), so most useful medical imaging techniques use either the low-energy or high-energy window (Figure 1.1). In addition, electromagnetic

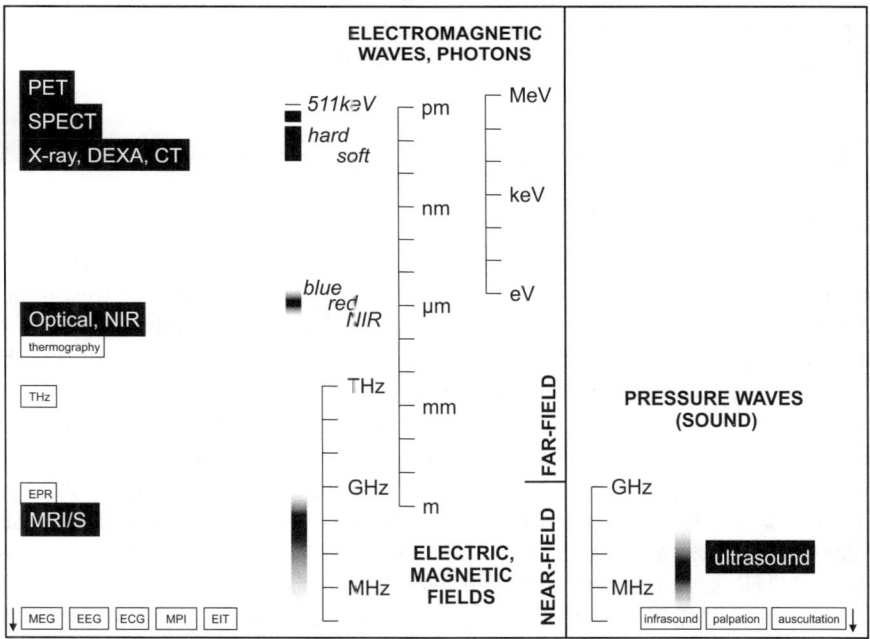

Figure 1.1 Imaging modalities mapped onto the electromagnetic spectrum (and its equivalent for pressure and sound waves). In white text in black boxes are the six most medically significant modalities. In smaller black font in white boxes are shown some other investigational or lesser-used imaging modalities, together with related modalities which similarly use a device to detect a signal (electromagnetic or pressure) from a patient, but do not necessarily produce an image. Abbreviations: PET: Positron Emission Tomography; SPECT: Single Photon Emission Computed Tomography; DEXA: Dual Energy X-ray Absorptiometry; NIR: Near Infra-Red; EPR: Electron Paramagnetic Resonance; MRI/S: Magnetic Resonance Imaging and *in vivo* Magnetic Resonance Spectroscopy; MEG: Magnetoencephalography; EEG: Electroencephalography; ECG: Electrocardiography; MPI: Magnetic Particle Imaging; EIT: Electrical Impedance Tomography.

radiation can penetrate tissue to some extent at visible and near-infrared frequencies, and these also can be useful.

1.2.3   Radio Frequencies and Below

In the low-energy window, the electromagnetic wavelength exceeds the size of the human body, so we are using not waves, or photons, but time-varying magnetic and electrical fields described by the equations for near-field effects. In addition to MRI, investigational or specialist techniques in this low-frequency range include Electrical Impedance Tomography (EIT),[4] and Magnetic Particle Imaging (MPI).[5] MPI uses iron nanoparticles to create contrast. These techniques are potentially attractive in that they do not use ionising radiation, but have not yet come into widespread use. Other important modalities (albeit not necessarily imaging modalities) in this window include magnetoencephalography (MEG), electrocardiography (ECG), and electroencephalography (EEG), together with related electrophysiological techniques.

1.2.4   Magnetic Resonance

1.2.4.1   Acquiring MRI

The most commonly used technique in the low-frequency window is Magnetic Resonance Imaging (MRI). MRI is a form of Nuclear Magnetic Resonance (NMR). It relies on the insight that, since the NMR Larmor frequency is proportional to the magnetic field, if the field is caused to vary across the sample (body), then nuclei at different positions will resonate at different frequencies. Some of the earliest experiments were performed literally by varying the field, and scanning point by point through the sample, but this technique is very inefficient: much faster and more efficient pulse sequences are used today. Many are variations of the two-dimensional MR techniques familiar to chemists. Just as in other forms of 2-D MR, pulse sequences start with excitation of the NMR signal, followed by an evolution period in which a phase shift accumulates, and an acquisition period. However, unlike conventional 2-D NMR, the parameters obtained are not spectral parameters such as chemical shifts or scalar couplings, but are spatial positions encoded by gradients in the magnetic field. Hundreds of pulse sequences have been developed for MRI, each weighting the signal according to different combinations of contrast mechanisms, and optimised for a specific body location.

An MRI system has a lot in common with the chemist's NMR spectrometer. The most obvious difference is that most MRI magnets have horizontal rather than vertical access to allow patients to be scanned while lying down, and of course the bore size is much larger, almost a metre, to allow human beings to be scanned. In addition, the B_0 magnetic field strengths are much smaller. Typical MRI magnets in hospitals operate at 1.5 Tesla (63 MHz for ^{1}H) or 3 T (126 MHz), and experimental systems for human use have been built at fields as high as 9.4 T (400 MHz). However, reasonably good images can be obtained at fields as low as 0.5 T (21 MHz) and useful images have been obtained at fields of 0.1 T (4 MHz) and below. MRI systems are quite widely available in research centres and larger hospitals in the developed world, and mobile systems on lorries extend availability further. There are also many horizontal-bore animal systems in research centres: these typically operate at field strengths between 4.7 T (200 MHz for ^{1}H) and 11.7 T (500 MHz for ^{1}H).

MRI has very few problems in detecting signal from deep tissues, and is ideal for studies of the brain, neck, spine, limbs, joints, and pelvis. Motion artefact is one of the more important practical limitations. Since 2-D tomographic slices or 3-D volumes typically take several seconds or minutes to acquire, cardiac, respiratory, or peristaltic motion can degrade the images, although there are many experimental approaches available to overcome these problems, and very good images or the heart, lungs, kidney, and liver can be obtained if suitable pulse sequences are employed.

The spatial resolution of MRI is limited more by practical considerations than by the inherent limitations of the technique. Typically, images are composed of a matrix of between 128×128 and 256×256 voxels (volume elements) in plane, usually with fewer voxels in the third dimension. Larger matrix sizes require impractically long imaging times. Consequently, better resolution is typically obtained in the hand than in the chest.

1.2.4.2 Endogenous Contrast in MRI

The strength of MRI is the great variety of contrast mechanisms, both endogenous and exogenous. Perhaps the most obvious is the water proton density, high in most tissues, but low in the air-filled lung, and in bone. In soft tissues, water concentrations are generally rather similar, and other contrast mechanisms have more discriminating power. Most important are the longitudinal and transverse relaxation times, respectively T_1 and T_2. T_1 is the time constant for spins to return to thermal equilibrium following excitation. It can thus be associated with the enthalpy of the spin system. The main relaxation mechanism for water protons in tissue is dipolar relaxation induced by fluctuating magnetic fields at frequencies close to the Larmor frequency: the most important source of these fluctuating fields is protons on tumbling proteins. Thus T_1 depends to some degree on the local concentration of macromolecules in the tissue. Parts of the body almost devoid of macromolecules, such as urine in the bladder, or cerebrospinal fluid in the ventricles of the brain, have long T_1 values similar to pure water, around three seconds. At the other extreme tissues which consist of densely packed macromolecules, *e.g.* fibrous tissues such as ligaments, in which almost all of the water is bound, show extremely short T_1 values of a few milliseconds or less. More typical T_1 values for water protons in tissue are between about half a second and two seconds, somewhat dependent on B_0. T_2 is the time constant for the loss of phase coherence: it can thus be associated with the entropy of the spin system. T_2 is affected by the same dipolar processes as T_1, but it can also be associated with exchange processes, including the exchange of water protons into the hydration sphere of macro-molecules or on to the hydroxyl or amine groups of the macromolecules themselves. Typically in tissues, exchange processes make T_2 1-2 orders of magnitude lower than T_1, although, as for T_1, body parts almost devoid of macromolecules can have very long T_2 values.

Nearly always, MRI measures water protons. Water is the most abundant molecule in tissues: the water proton is present at concentrations around 80 M. Also, protons have the highest gyromagnetic ratio of any stable isotope, pro-viding a large NMR signal. While MRI easily distinguishes regions with little or no water, such as the air-filled lungs, from areas of high water content such as muscle, the real power of the technique lies not in distinguishing different water concentrations, but in distinguishing areas with different water proton relaxa-tion times, or with other biophysical differences such as flow, diffusion or chemical exchange which can be measured by NMR. The only other spin routinely imaged in medical MRI is the triglyceride methylene moiety, whose concentration can exceed 10 M in fatty tissues. Other endogenous substances that can be measured by ^{1}H MRI include the choline trimethylammonium moiety, and the methyl groups in lactate and in *N*-acetylaspartate, albeit at coarse resolutions. Sodium ions, present at 0.15 M in extracellular fluid can be imaged using ^{23}Na MRI. Measurements in living humans can also be made using some other endogenous isotopes including ^{13}C (*e.g.* for triglyceride and glycogen) and ^{31}P (*e.g.* phosphocreatine, adenosine triphosphate, and $HPO_4{}^{2-}$). Unfortunately, the organic molecules of most interest to the chemist are gen-erally present in the body in such low concentrations that their NMR signal can seldom be detected at all, let alone used for the construction of an image.

Another relaxation time of interest is T_2*. This is equivalent to the reciprocal line width in conventional high-resolution NMR. The T_2* relaxation time incorporates not only the loss of phase coherence due to T_2 relaxation, but also loss of phase coherence due to local magnetic field inhomogeneities. Typical magnetic field inhomogeneities in the body occur where domains exhibiting different magnetic susceptibility abut: for example, air-water interfaces in the lung, or the erythrocyte cell membrane. Arterial erythrocytes contain oxyhae-moglobin, which has diamagnetic low-spin Fe^{II}, while venous erythrocytes, having given up their O_2, contain deoxyhaemoglobin, which has paramagnetic high-spin Fe^{II}. The high concentration (20 mM) of deoxyhaemoglobin in the venous erythrocyte makes the intracellular magnetic susceptibility quite different from the surrounding blood plasma, and creates local inhomogeneities, reducing T_2*. (This is the basis of the blood-oxygen-level dependence or BOLD effect, used to create maps of brain activations during sensory, motor, or cognitive tasks.).

Other important sources of endogenous contrast in MRI include coherent motion (*e.g.* blood flow) and molecular diffusion: water diffusion is faster and less tortuous when cell packing decreases or when cells die. Chemical exchange can also create contrast, just as in high-resolution NMR. This includes both chemical exchange with macromolecules where the macromolecule signal itself is broad, maybe many kHz wide and NMR-invisible, and chemical exchange with, say, slowly exchanging NH protons resonating a few hundred Hz away.

In the presence of exchange, saturation transfer and attenuation of the water proton signal can be achieved by saturating spins at the frequency of the exchanging partner.

1.2.4.3 Exogenous Contrast in MRI

The unpaired electron has a much higher gyromagnetic ratio than the proton, and paramagnetic substances can also induce dipolar relaxation of tissue water protons. A common approach is to increase dipolar relaxation, and reduce T_1, by administration of a paramagnetic substance. Most effective are organic radicals, or metal ions with relatively long electron T_1 values. Such compounds are rare in the body but can be introduced in the form of contrast agents. Typically these are manganese or gadolinium ions (chosen because of the half filled d or f orbital and consequent long electron T_1), which have been chelated to ensure rapid clearance, usually *via* the kidney, and avoid toxicity from the release of free metal ion into the body. Nine gadolinium chelates and one manganese chelate have been approved for parenteral use in human healthcare, and many more have been used experimentally. The first, and one of the most widely used, is Gd-DTPA^{2-} (gadopentetate). Administered intravenously, it enhances the blood signal allowing measurements of perfusion and flow, and leaks slowly across capillary endothelia (although is excluded from healthy brain by the blood-brain barrier). In diseases where endothelial permeability increases, such as inflammation and neoplasia, the contrast agent leaks rapidly into the extra-vascular extracellular space, enhancing the signal from the pathologic tissue. It does not bind to albumin and does not enter cells. Other agents, such as gado-fosveset, are highly protein-bound, which lengthens the residence time in blood. Some agents, such as mangafodipir and gadoxetate, are taken up into normal liver cells but not liver tumours, assisting in the diagnosis of liver cancer.

It is important to appreciate that contrast in MRI is indirect: in other words we do not detect the gadolinium (or manganese) directly; rather we detect the effect of the gadolinium on the relaxation of water. To quantitate gadolinium concentration in tissues we need to measure the change in T_1:

$$\frac{1}{T_1} = \frac{1}{T_{1,0}} + r_1 \times [Gd.X] \tag{1}$$

where $T_{1,0}$ is the T_1 in the absence of contrast agent, and r_1 with dimensions $\sec^{-1} M^{-1}$, is the relaxivity, *i.e.* the efficacy of a specific contrast agent Gd.X in reducing T_1.

Although dipolar relaxation caused by such contrast agents affects T_2 as well as T_1, in most cases it is the effect on T_1 that is most easily detected.

An alternative contrast mechanism is to affect T_2^*. This can be achieved using iron oxide nanoparticles, of which two have been approved for human use, and several more have been used experimentally. These are super-paramagnetic, and create enormous local magnetic field inhomogeneities in the blood after they are injected or, around the cells in which they reside if they are taken up into macrophages. The clearance is highly dependent on the nano-particle size and coating, and they may be taken up by lymph nodes, Kupffer cells in the liver, or macrophages in other sites of inflammation.

Another substance sometimes used as a contrast agent is the O_2 molecule, which is also paramagnetic. Dissolved O_2 reduces plasma T_1 in normoxic tissue, while in hypoxic tissue O_2 converts deoxyhaemoglobin to oxyhaemoglobin, increasing T_2*. O_2 may have a function in mapping tumour hypoxia.[6]

Other investigational endogenous substances that can be measured by MRI include noble gases. Helium (^{3}He NMR) or xenon (^{129}Xe NMR) when hyperpolarised, can produce high quality images of the airspaces of the lung.[7] Measurements in living humans have also been made using exogenous sub-stances (mainly drugs)[8] containing other isotopes, especially ^{19}F, but also ^{1}H, ^{7}Li, ^{13}C, ^{15}N, and ^{17}O. Some of these, notably ^{13}C, can also be hyperpolarised.

Contrast agents based on chemical exchange saturation transfer (CEST) between exchangeable protons in the contrast agent and tissue water have been proposed. These could be small diamagnetic organic molecules,[9] where the exchangeable protons are chemically shifted from ^{1}H$_2$O, or paramagnetic organometallics,[10] where the exchangeable protons exhibit a large contact or pseudo contact shift. While no agent specifically developed for CEST contrast has been approved for use in healthcare, there is an existing approved CT contrast agent, iopamidol, which exhibits CEST activity.[11]

1.2.5 Microwaves

The electromagnetic spectrum, between radio frequencies and the near infrared, is of little use to the medical imager. The THz region, relatively unexplored by spectroscopists, has the attraction of employing non-ionising radiation. Unfortunately, above 1 GHz, tissue water absorbs very strongly, making the imaging of deep tissues almost impossible. Thus medical imaging using rota-tional spectroscopy in the microwave region, very high field NMR, or EPR (except at very low field,[12] where it may have a role in mapping tissue oxyge-nation if suitable tracers can be developed), are unlikely to be very useful.

1.2.6 Optical Imaging

The human body is not, of course, transparent at optical wavelengths, because of absorption and scattering. However photons in the visible red and near-infrared ranges can penetrate around 20 mm into tissue, as anybody who has shone a torch through their hand in a darkened room will have observed. For studies in mice and rats, this is sufficient to achieve almost full-body penetra-tion. Many medically important sites in the human body can potentially be accessed for optical imaging. The retina, of course, is easily seen at the back of the eye and is routinely imaged by opticians, particularly when screening for early degenerative changes because of diabetes. The blood supply to the retina can be measured more selectively using the technique of retinal angiography, where fluorescein is injected intravenously, imparting fluorescence to the blood. Endoscopy is the name for techniques where a camera is introduced through some natural or surgeon-created orifice. The gastrointestinal (GI) tract can be accessed for considerable distances from either end, and indeed capsule endoscopy,[13] using a swallowed camera designed to transit the GI tract and eventually be recovered in the faeces. The lungs can be partially accessed by bronchoscopy, as can the bladder and the uterus in a similar fashion. Fibre optics can even be introduced into the arteries to allow imaging of the inside of an arterial wall *e.g.* the carotid or coronary artery where atherosclerotic plaque often develops. While the adult brain is not generally accessible for optical imaging, in the newborn the bones of the skull are not completely fused leaving small windows, the fontanelles, through which optical signals can be detected.

Optical imaging can use many of the techniques familiar to optical spectroscopists including absorption, scattering, single photon or two photon imaging, fluorescence, and bioluminescence. Bioluminescence imaging uses mice or other organisms genetically modified to express luciferase: following the injection of luciferin, in the presence of ATP, a visible photon is emitted. For obvious reasons, this technique has not been implemented in humans. Mice can also be genetically modified to express fluorescent proteins such as green fluorescent protein (GFP). Apart from genetic modification, optical imaging can employ both endogenous and exogenous contrast. One important source of endogenous contrast is haem: oxyhaemoglobin and deoxyhaemoglobin have different spectra, so the oxygenation of venous blood can be measured. Regarding exogenous contrast, there is of course an enormous range of readily-available coloured compounds: while many have been used experimentally only a handful are approved for human use. These include fluorescein and indocyanine green which are used to highlight blood vessels or lymphatics.

Optical imaging is attractive because it does not use ionising radiation. In addition, although exogenous contrast requires the introduction of foreign substances to the body, these are often given at very low concentrations: techniques for detecting fluorescence are sensitive and, unlike radioisotope decay, the signal source is not destroyed after the signal is measured, so signal averaging is feasible.

In the absence of scattering, the resolution of optical imaging is limited only by the wavelength (sub-micron), as evident in microscopy, but for *in vivo* applications scattering makes resolution quite poor, and quantitation difficult. Many hospitals have installed specialist equipment for niche applications such as endoscopy. Over the past decade, dedicated *in vivo* optical imaging systems for mice and rats have become much more widespread. Compared with humans, depth penetration in rodents is less problematic, a much wider range of coloured tracers can be given, and genetic modification can be employed.

1.2.7   Ultraviolet

The ultraviolet region is another region in which tissue is opaque and thus of little use to the medical imager. Beyond the ultraviolet is the X-ray region, which of course has been of enormous value to the medical imager for over 100 years.

1.2.8 X-Ray

X-Ray imaging as introduced by Röntgen is a planar technique: essentially the image is a projection or silhouette of X-ray absorption through the body. X-Ray imaging was revitalised and revolutionised in the 1970s through the introduction of tomographic imaging (CT). The key insight is that X-ray projections taken from different angles can be used to infer the underlying internal 3-D structure using a back-projection algorithm. This allows 2-D tomographic slices or full 3-D volumes to be acquired. Continuing developments in CT technology have increased both the spatial resolution and also the speed, so that complete volumes can be obtained in a period of time that is short in comparison with the cardiac cycle, essentially freezing cardiac and respiratory motion.

At the energies used in biomedical imaging, X-rays interact with tissues by photoelectric (PE) absorption (ejection of an inner-shell electron) and by Compton scattering (CS) (ejection of an outer-shell electron and scattering of a lower-energy X-ray). Both PE and CS occur less efficiently at higher energies but PE declines much more steeply. Hence PE predominates with lower-energy (soft) X-rays while CS predominates with hard X-rays. CS is approximately independent of Z, while PE absorption depends approximately on Z^3, providing good endogenous contrast. Tissue absorbs more strongly than air ($Z \sim 0$), and fatty tissues, predominantly triglyceride ($Z \sim 6$), absorb slightly less strongly than other tissues which consist mainly of water ($Z \sim 8$). Calcium phosphate ($Z \sim 15, 20$) absorbs very strongly, allowing not only the skeleton to be detected, but also calcifications associated with disease *e.g.* in breast tumours or in atherosclerotic plaque in the coronary artery. Exogenous contrast is achieved by introducing molecules rich in high-Z atoms such as iodine or barium, xenon, or historically, thorium. Many organoiodines have been introduced for human use over the past century, although only around ten are still commonly used. As with MRI contrast agents, when administered intravenously, they increase the blood signal allowing measurements of perfusion and flow, and, in inflammation and neoplasia, leak across capillary endothelia. While most are cleared renally, some agents, such as iodipamide, are taken up into the liver and pass into the bile. Other iodine or barium containing agents are used to opacify the gastrointestinal tract or other body spaces or cavities. Potentially any drug containing a high-Z atom given in reasonably high doses is potentially detectable, such as platinum containing anti-cancer drugs or the anti-arrhythmic drug amiodarone, which contains two iodine atoms.[14] The gadolinium-containing agents used as MRI contrast media can also be detected by CT,[15] so radiologists need to take care when interpreting CT scans acquired immediately after contrast-enhanced MRI scans. An alternative source of contrast in CT is to reduce Z. Although no longer used, in the early days of X-ray imaging, air was sometimes introduced into the ventricles of the brain, the so-called pneumoencephalography technique, to create some contrast within the head.

Planar X-rays are somewhat difficult to quantitate because the signals from overlapping structures are combined in projection, but CT quantification is relatively straightforward, the increase in X-ray absorption (measured in Hounsfield units) due to the contrast agent being directly related to its concentration. For quantitation of endogenous contrast, where several different elements contribute significantly to tissue composition, dual-energy (DEXA or DXA) measurements are useful. They acquire images at two different monochromatic wavelengths, with different relative contributions of PE ($\sim Z^3$) and CS ($\sim Z^0$), allowing one, with some assumptions, to solve for concentration of one specific atom or molecule, such as hydroxylapatite. DEXA scanning is very widely used to measure bone mineral density and screen for osteoporosis.

CT is somewhat more widely available than MRI, and somewhat less expensive. CT generally has better spatial and temporal resolution than MRI, and is less sensitive to motion artefact, although its full technical potential is not usually achieved in the clinic because of the need to minimise exposure to ionising radiation. CT is often the method of choice where calcified tissue is involved, or in the presence of physiologic motion (*e.g.* in the heart, lungs and abdomen); however contrast in CT is less versatile than MRI, making MRI preferable for many soft tissues, especially the brain. CT systems for small animals have also become available in research centres in recent years.

1.2.9 Gamma Rays and Nuclear Medicine

At higher energies, both PE and CS decline, and tissue becomes more transparent, so endogenous contrast is less useful. However radioisotopes that decay to produce high-energy gamma rays, or positrons, can be used: this is the domain of nuclear medicine.

1.2.10 Single Photon Emission Computed Tomography

A considerable number of isotopes decay to produce gamma rays, although only a few are commonly used, notably ^{99m}Tc. This isotope has a conveniently short half life of about six hours and can be chelated and incorporated in a wide variety of tracers, many of which have been approved for human use. It can also be conveniently supplied on-site from a generator (*i.e.* a supply of ^{99}Mo, which slowly decays to ^{99m}Tc). Supply is, however, vulnerable as ^{99}Mo is manufactured in only a few sites worldwide. In the past, gamma emitting isotopes were detected using the gamma camera, a 2-D planar device. However just as X-radiography has been partly replaced by CT, so the gamma camera is being increasingly replaced by its tomographic equivalent, SPECT. The SPECT tracer emits its gamma ray in a random direction: in order to detect the direction from which it originated, collimation is used. Images are reconstructed using back-projection algorithms as for CT.

SPECT has more than 30 tracers available for human use. In addition to small molecules modified with iodine or technetium tags, antibodies and white blood cells can be labelled. Unlike other imaging modalities, SPECT has many approved tracers which have specific molecular targets. These include ^{111}In-labelled pentetreotide used in the localisation of neuroendocrine tumours bearing somatostatin receptors, ^{99m}Tc-labelled nofetumomab and arcitumomab used in cancer diagnosis, $[^{131}$I] - iodide used to evaluate thyroid function and associated with thyroid malignancies, and ^{99m}Tc-medronate, a bone-imaging agent. Other SPECT tracers are used in evaluating cardiac function, inflammation and infections. Indeed, of all the substances ever approved by the US Food and Drug Administration (FDA) for any diagnostic imaging modality, more than half are scintigraphy/SPECT agents.

It can be seen that resolution in SPECT is very much a function of the collimation: a longer collimator can identify the direction of origin precisely, but many gamma rays will be "wasted", undesirably necessitating a higher injected dose. Gamma scintigraphy was (and is) widely used in hospital nuclear medicine departments and is gradually being replaced by SPECT. Animal SPECT systems have also been developed in recent years. In practical terms, while SPECT has the worst spatial resolution in humans of any major medical imaging modality, and is relatively difficult to quantitate, resolution in animals is quite good (because pinhole collimators can be used with higher doses in animals).

1.2.11 Positron Emission Tomography

Certain radioisotopes with a deficit of neutrons decay by emitting a positron, for example ^{18}F to ^{18}O. The ejected positron cannot travel very far before meeting an electron and annihilating. The annihilation produces gamma radiation with properties of interest to the imager. Two photons are produced with a precise energy of 511 keV. Moreover the angle between the trajectories of these photons is almost exactly 180°. Positron emission tomography detectors are designed to detect pairs of 511 keV photons arriving almost simultaneously at different detectors (coincidences). This specificity makes PET a very sensitive technique, since collimation is not needed, and much random noise is excluded because only coincidences are accepted. Images are reconstructed using back-projection algorithms as for CT and SPECT. PET appeals to the chemist because positron emitting ^{11}C, halogens ^{18}F, ^{76}Br, and ^{124}I, and a number of metals, are available. However many of the most interesting PET isotopes have very short half lives and must be produced on site in a cyclotron, for example ^{15}O with a half-life of two minutes. The short half lives also create severe constraints in choice of synthetic routes as discussed in subsequent chapters. For this reason PET is the most costly of all the major medical imaging techniques, and the least widely available. Only one tracer is at all widely used, and that is $[^{18}$F]2-deoxy-2-fluoro-D-glucose (fludeoxyglucose, FDG). FDG is a glucose mimic, which is taken up into cells, phosphorylated and trapped. Many tumours get their energy from glycolysis alone (the

Warburg effect) which is inefficient, requiring large amounts of glucose, so they take up FDG avidly, making it ideal for monitoring tumours.[16] Many PET scanners use only FDG, which, having an ^{18}F half-life of 110 min, can be manufactured remotely and shipped daily. Many hundreds of ^{11}C and ^{18}F labelled molecules have been designed and elegant radiosyntheses devised, providing tracers for assessment of receptor occupancy and biochemical pathways in the brain and other organs, and many have been taken into man on an investigational basis. However very few tracers are available in more than a handful of centres, because of the cost and complexity of radiosynthesis in hot-cells of products intended for human use.

PET is easy to quantitate. Its resolution is fundamentally limited to around 1 mm even in the most favourable situations, because of the path length of the positron before it annihilates (it is the origin of the gamma ray, not the positron, which is detected). This spatial resolution is of little concern in humans, but does fundamentally limit the resolution of the small animal scanners that have been developed in recent years.

PET and SPECT detect only the gamma ray, and cannot identify the chemical form in which the disintegration occurred. Thus it can be difficult to establish whether the signal comes from the injected tracer, or a metabolite.

1.2.12 Ultrasound

Unlike all other forms of medical imaging, ultrasound relies not on electromagnetic radiation, but on pressure waves, *i.e.* sound. Boundaries between domains with different tissue properties reflect ultrasound, so it is echogenicity that creates endogenous contrast. Unfortunately, very echogenic structures such as bones, or media that do not transmit ultrasound well such as air in the lungs, conceal all structures behind them, creating shadows. The ultrasound probe is typically handheld and acoustically coupled through the skin, using a gel to transmit the ultrasound efficiently. Lower frequencies of 1 to 10 MHz penetrate deeply into tissue, so can be used for example in the liver or for the foetus, while higher frequencies provide better resolution but do not penetrate as deeply, so can be used in the skin or the eye, or in small animals. Like optical imaging, ultrasound can be used endoscopically, for example for transoesophageal imaging of the heart or transrectal ultrasound of the prostate, or even within the human coronary artery. Ultrasound images are typically recorded continuously and interpreted during the scanning process by the sonographer. They can be displayed as 1-D (the so called A- or amplitude- mode), or 2-D images, or reconstructed into 3-D blocks. A-mode images can be acquired continuously to measure cardiovascular motion: this is the so-called M- or motion mode. The Doppler effect is used to measure blood flow. Most ultrasound investigations rely on endogenous contrast but specific ultrasound contrast agents have been introduced and approved for human use. These consist of gas-filled micron-sized bubbles (containing *e.g.* SF_6), which are highly echogenic, remain inside blood vessels, and can be used to measure blood flow.

Ultrasound is generally regarded as very safe: it is used in screening in the unborn. It is inexpensive, and widely available. Because of the subjective use of a handheld probe it is somewhat difficult to quantitate, and although micro-bubble contrast agents can be visualised, they are also difficult to quantitate. High-frequency ($\sim$40 MHz) animal ultrasound equipment has also become available in recent years: both spatial resolution ($<$0.05 mm) and time resolution ($<$10 ms) compare very favourably with other modalities.

1.2.13   Multimodal Techniques

In much the same way as coupled techniques such as LC-MS have been explored in analytical chemistry, multi-modality approaches are increasingly favoured in medical imaging. The most common and useful approach is where a pre-dominantly molecular imaging technique such as PET or SPECT is coupled to a predominantly structural approach such as X-ray CT. CT provides the struc-tural locator to identify with which structure the SPECT or PET signal is associated. This is particularly useful in cancer where CT can localise the tumour and PET can identify its metabolic activity. In addition, PET and SPECT image intensities can be distorted because gamma rays are somewhat attenuated by tissues: by simultaneously acquiring a CT scan this attenuation can be corrected providing much more quantitatively accurate PET and SPECT images. PET-CT is now widely used in hospitals and in fact most systems sold today come with CT. The combination of PET with MRI is also under active investigation.

Multimodal techniques allow multimodal contrast media. Some contrast agents are incidentally multimodal: microbubbles create T_2* contrast by a susceptibility mechanism, and gadolinium contrast agents can be detected in both MRI and CT. Optical and MRI contrast media can be radiolabelled, and radioiodine is available both in SPECT and PET. Microbubbles and nano-particles can be decorated with chromophores or radioisotopes, or genuinely multimodal agents can be designed *de novo*. Examples are given in subsequent chapters.

1.3   How is Medical Imaging Used?

Medical imaging is used in humans in screening, in diagnosis, in patient management and monitoring. It is also widely used in clinical research, parti-cularly to assist the development of therapeutic drugs in man.

In the past few years medical imagers have begun to regard measurements and assessments from medical images as biomarkers. The insight is to think about the information we obtain from imaging tests in the same way as we think about information from biochemical markers.

A biomarker has been described as "a characteristic that is objectively measured and evaluated as an indicator of normal biological processes, pathogenic processes, or pharmacological responses to a therapeutic inter-vention".[17] The concept of a biomarker extends to include nearly any biome-dical measurement, including measurements of the molecular markers we find in blood or urine, measurements in biopsied tissue, measurements made from physiological tests such as blood pressure or ECG, and measurements made from images. The roles of biomarkers in medicine can be classified as Prognostic; Predictive; Monitoring or Response.

1.3.1 Prognostic or Diagnostic Biomarkers

The characteristic of a prognostic biomarker is that the patient's future health can be better predicted when the biomarker data are added to the clinical data. In asymptomatic individuals this is called screening. Familiar examples of screening using medical imaging include foetal ultrasound, mammography, colonoscopy, chest X-ray for tuberculosis, or retinal imaging. Any medical test used for screening must be inexpensive and very low risk because it will be used widely on healthy people. Screening applications rarely or never use tracers or contrast agents because of the costs and potential risks.

A prognostic biomarker used with symptomatic patients is called a diagnostic. Medical diagnosis is the process of explaining symptoms reported by the patient in terms of a recognised disease entity with an outcome that is at least to some extent predictable, so that appropriate treatments can be considered. Medical imaging is of course most familiar in hospitals in routine diagnosis. In the developed world even small hospitals may well have MRI and CT as well as ultrasound, X-ray, and a gamma camera or perhaps SPECT, and millions of scans are acquired each year. Although scintigraphy and SPECT always use tracers, and MRI and CT commonly use gadolinium and iodine-containing contrast agents particularly in cancer diagnosis, for the chemist this is a challenging sector in which to introduce new molecules because of the regulatory hurdles are discussed in Section 1.4.

1.3.2 Predictive Biomarkers or Companion Diagnostics

The characteristic of a predictive biomarker is that the patient's future health can be better predicted when the biomarker data are added to the clinical data and used to select a specific therapy. For example, certain therapeutic antibodies (*e.g.* tositumomab) may be prescribed only after imaging with a radionuclide labelled antibody shows the presence of the receptor and likelihood of response (in the case of tositumomab, ^{131}I labelled tositumomab).[18] While the use of imaging biomarkers as companion diagnostics is not common today, the growth of interest in personalised healthcare provides real opportunities for new contrast agent and tracer development.

1.3.3 Monitoring Biomarkers

The characteristic of a monitoring biomarker is that the patient's future health can be better predicted when the changes in the biomarker data are added to the clinical data and used to change therapy. FDG PET is used in this way to monitor for recurrence of cancer and cardiac ultrasound used to monitor for cardiac toxicity. Monitoring biomarkers usually evolve from diagnostic biomarkers and are a similarly difficult sector for new contrast agent or tracer development.

1.3.4   Response Biomarkers

A response biomarker is one which is measured before and after an intervention (such as drug treatment) and the change in the biomarker used to evaluate the intervention. In medical research (including drug development), response bio-markers are widely used in clinical trials in humans and in studies in experimental animals. Also, response biomarkers may be used in "trials of therapy" in individual patients. MRI and CT (with or without contrast) and PET are commonly used to provide response biomarkers, ranging from simple measurement of tumour size using CT (an imaging biomarker used to evaluate cancer therapies), to complex *in vivo* neuropharmacology such as the quantitative analysis of D2 dopamine receptor binding in the human brain by using [^{11}C]raclopride.[19] Because of their use, often in small studies of limited scope in research centres, response biomarkers are an ideal setting in which to evaluate investigational PET and SPECT tracers which do not yet have regulatory approval for use in healthcare.

1.4   Regulatory and Cost Issues

No medical imaging test can be performed in humans without regulatory approvals, and the chemist needs some understanding of this process before embarking on development of new molecules to be used in imaging tests and imaging biomarkers. Regulatory approval is complex, and the laws and guidelines vary in different jurisdictions. Broadly, however, regulatory approvals fall into three areas. Firstly the imaging instrument itself needs approval. This is to minimise the risk of harm to patients from, for example, electrical or mechanical malfunction, thermal injury (*e.g.* from microwave heating from radio-frequency pulses used in MRI), or from ionising radiation (*e.g.* in CT, SPECT, or PET).

Secondly, if any exogenous tracer contrast medium is given, the molecule will likely be regarded as a diagnostic drug. The regulator will need to understand the risks associated with this diagnostic drug, *i.e.* toxicity, and is likely to demand evidence that the benefits in diagnostic decision-making exceed the risks. Regulatory approaches differ considerably between the diagnostic molecules used for the different modalities, and between small molecules, nanoparticles, and "biologicals" (*e.g.* antibodies). The iodinated contrast media used in CT are given in similar or higher doses to therapeutic drugs, and have the potential to damage the kidney (nephrotoxicity); some of the gadolinium contrast agents used in MRI have also been associated with kidney damage in severe renally-impaired patients perhaps due to the release of free gadolinium ion from the chelate. New gadolinium and iodine agents therefore are likely to need extensive toxicology. Safety questions have also arisen from time to time in connection with microbubbles, iron nanoparticles, and radioisotope-labelled antibodies. In contrast, in the case of optical, SPECT, and PET, the amount of substance injected is tiny compared with CT or MRI, and risks of toxicity lower: for positron-emitting small molecules there may be little concern beyond the ionising radiation exposure. Regulators also insist that all diagnostic drugs are manufactured to a high and consistent quality (*e.g.* using Good Manufacturing Practice) which creates challenges for quality assurance for molecules incorporating short half-life isotopes. The two key stages in regulatory approval are, firstly, approval to investigate the diagnostic molecule in humans (an Investigational New Drug, IND, in the United States or Investigational Medicinal Product, IMP, in the European Union), and, secondly, approval to promote, sell, and prescribe the diagnostic drug for use in humans for specified purposes described on the label (a New Drug Application, NDA, in the United States or Marketing Authorisation Application, MAA, in the European Union). The development of a diagnostic drug to NDA/MAA can be very costly: maybe hundreds of millions of dollars for a CT or MRI contrast agent, requiring considerable investment from a pharmaceutical company. Such investments will only be made if there is likely to be a return on the investment and the new agent will command a high price or generate large sales. This requires that those who pay for scans must see significant benefits over existing diagnostic tests, to justify the increased costs to the healthcare system. That harsh commercial reality has prevented very many innovative contrast agent molecules from becoming products which can be used in patients. PET tracers are somewhat less expensive to develop than agents for other modalities, and indeed some agents have been taken into human use through purely public-sector funding mechanisms.

The third area of regulatory and ethical control is where humans participate in clinical trials or medical research. In this case, where patients' participation is voluntary and aimed at furthering medical science or medical research with little prospect of benefiting the patient himself or herself, regulators and ethical committees take great pains to ensure that the risk of harm is minimised either in connection with ionising radiation, or in connection with exposure to novel molecules. A particular issue for those developing therapeutic drugs is the so-called double-IND problem: it is very difficult to employ both an investigational therapeutic drug and an investigational diagnostic drug in the same clinical trial. Thus pharmaceutical companies who would like to use the latest medical imaging techniques to evaluate the efficacy of their new drugs, find themselves limited to a very simple palette of about 50 diagnostic agents which have NDA approval, many of which are quite old and rather similar. The thousands of investigational diagnostic agents being developed around the world by definition lack NDA approval, and so are almost impossible to exploit in imaging biomarkers for therapeutic drug development.

1.5 Conclusion

Medical imaging is rich in physical phenomena to provide endogenous contrast, and there are extensive chemical opportunities to provide exogenous contrast. In addition to the scientific challenges in devising novel contrast mechanisms, however, there are formidable commercial and regulatory challenges in making new tracer and contrast agent molecules available for human use in healthcare.

References

1. V. R. Fuchs and H. C. Sox, Jr, *Health Aff.*, 2001, **20**, 30.
2. H. L. Gallagher and C. D. Frith, *Trends Cogn. Sci.*, 2003, **7**, 77.
3. *Molecular Imaging and Contrast Agent Database (MICAD)*. National Center for Biotechnology Information, MD. 2004-2011. http://www.ncbi.nlm.nih.gov/books/NBK5330/.
4. R. H. Bayford, *Annu. Rev. Biomed. Eng.*, 2006, **8**, 63.
5. B. Gleich and J. Weizenecker, *Nature*, 2005, **435**, 1214.
6. J. P. B. O'Connor, J. H. Naish, G. J. M. Parker, J. C. Waterton, Y. Watson, G. C. Jayson, G. A. Buonaccorsi, S. Cheung, D. L. Buckley, D. M. McGrath, C. M. L. West, S. E. Davidson, C. Roberts, S. J. Mills, C. L. Mitchell, L. Hope, N. C. Ton and A. Jackson, *Int. J. Radiat. Oncol. Biol. Phys.*, 2009, **75**, 1209.
7. S. Fain, M. L. Schiebler, D. G. McCormack and G. Parraga, *J. Magn. Reson. Imaging*, 2010, **32**, 6.
8. P. Bachert, *Progr. NMR Spectroscopy*, 1998, **33**, 1.
9. K. M. Ward, A. H. Aletras and R. S. Balaban, *J. Magn. Reson.*, 2000, **143**, 79.
10. S. Aime, A. Barge, D. Delli Castelli, F. Fedeli, A. Mortillaro, F. U. Nielsen and E. Terreno, *Magn. Reson. Med.*, 2002, **47**, 639.
11. J. Keupp, I. Dimitrov, S. Langereis, O. Togao, M. Takahashi and A. D. Sherry, *Proc. Intl. Soc. Mag. Reson. Med.*, 2011, **19**, 828.
12. H. M. Swartz, N. Khan, J. Buckey, R. Comi, L. Gould, O. Grinberg, A. Hartford, H. Hopf, H. Hou, E. Hug, A. Iwasaki, P. Lesniewski, I. Salikhov and T. Walczak, *NMR Biomed.*, 2004, **17**, 335.
13. T. Nakamura and A. Terano, *J. Gastroenterol.*, 2008, **43**, 93.
14. I. S. Goldman, M. L. Winkler, S. E. Raper, M. E. Barker, E. Keung, H. I. Goldberg and T. D. .Boyer, *Am. J. Roentgenol.*, 1985, **144**, 541.
15. T. Albrecht and P. Dawson, *Br. J. Radiol.*, 2000, **73**, 878.
16. H. Young, R. Baum, U. Cremerius, K. Herholz, O. Hoekstra, A. A. Lammertsma, J. Pruim and P. Price, *Eur. J. Cancer*, 1999, **35**, 1773.
17. A. J. Atkinson Jr, W. A. Colburn, V. G. DeGruttola, D. L. DeMets, G. J. Downing, D. F. Hoth, J. A. Oates, C. C. Peck, R. T. Schooley, B. A. Spilker, J. Woodcock and S. L. Zeger, *J. Clin. Pharm. Ther.*, 2001, **69**, 89.
18. *BEXXAR [prescribing information]*., Research Triangle Park, NC, Glaxo-SmithKline; 2005.
19. L. Farde, H. Hall, E. Ehrin and G. Sedvall, *Science*, 1986, **231**, 258.

CHAPTER 2

Biomedical Imaging: Advances in Radiotracer and Radiopharmaceutical Chemistry

ROBERT N HANSON

Matthews Distinguished University Professor, Department of Chemistry and Chemical Biology, Northeastern University, 360 Huntington Avenue, Boston, MA 02115-5000, USA

2.1 Background

Radiotracers and radiopharmaceuticals constitute a class of compounds that have been prepared in a radioactive form for the express purpose of interrogating a particular system or process. Within a general definition one could include those compounds that decay by beta or alpha emission, however, for this brief review, the topic will be limited to those materials labeled with radionuclides that decay by either positron or gamma ray emission. Radiotracers are materials that are radiolabeled versions of a parent substance and are chemically identical or sufficiently similar to that parent substance such that when introduced into a system or process, will mimic the dynamics of that substance. In this focus on biomedical imaging, the distribution of the radiotracer provides specific information regarding localization or metabolism of the parent substance in an organism, ranging from cell to small animal to a human patient. An example of a radiotracer would be a radioactive analog of a therapeutic drug for which information regarding total body distribution and clearance are required for FDA approval. A radiopharmaceutical, on the other

RSC Drug Discovery Series No. 15
Biomedical Imaging: The Chemistry of Labels, Probes and Contrast Agents
Edited by Martin Braddock
© Royal Society of Chemistry 2012
Published by the Royal Society of Chemistry, www.rsc.org

hand, is a radiotracer prepared with specific applications, such as diagnosis or therapy, in mind. In general, such compounds are required to meet specific criteria, such as high specific activity, ease in preparation and purification, enhanced selectivity for a specific biological process, and metabolic stability. As a result, much more consideration goes into the design and development of radiopharmaceuticals than for radiotracers. A number of reviews over the past decade have described various aspects of radiotracer and radiopharmaceutical development using radio-halogens and radio-metals.[1-14] In this review I will present what I consider to be the key principles that one must consider in radiopharmaceutical design. These key principles are illustrated in examples that are representative of the rational development of biomedical imaging agents.

2.1.1 Factors in Radiopharmaceutical Design and Synthesis

The development of biomedical imaging agents is a multidisciplinary task that requires consideration and integration of a variety of factors (Figure 2.1). Most important is a clear definition of the biological target of the imaging agent. Advances in instrumentation and software have dramatically improved data acquisition and image definition for both positron emission tomography (PET) and single photon emission computed tomography (SPECT), making clinical diagnoses more accurate.[15-17] Nevertheless, one still has to select a target that possesses properties that allow the radiopharmaceutical to localize with some selectivity, and have that binding process represent a relevant clinical function.[18,19] Typically, most of the biological targets are functional receptors

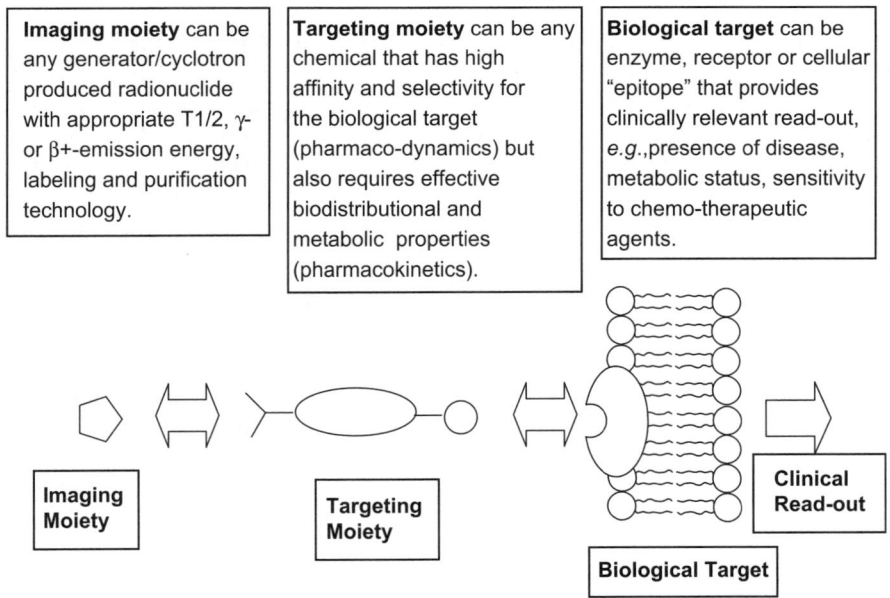

Figure 2.1 Association of radionuclide, targeting moiety, biological target and clinical "read-out".

or enzymes that are present at low levels in normal tissue but are upregulated or over-expressed as the result of pathology. Occasionally there are non-biological processes that can be targeted such as catecholamine storage in pheochromo-cytoma or protein aggregations in Alzheimer's disease. The use of epitopes unique to or overexpressed by specific tissues also represents a targeting mechanism; however, this review will not cover that process. In addition to identifying the molecular target for the radio-imaging agent, its location in the body also plays a factor. Imaging agents that target the central nervous system (CNS) also must penetrate the blood–brain barrier (BBB) which influences the physicochemical properties of the agent. Agents targeting peripheral binding sites do not have to consider these effects. Because imaging agents are usually administered intravenously (IV), such materials do not have to consider properties associated with gastrointestinal absorption. Consideration of cell surface localization *versus* intracellular binding may also play a role as different structural properties may be required to penetrate the cell membrane prior to binding to the molecular target.

Selection of a lead compound as the targeting moiety is a second important factor. Based upon published studies or patents, often pharmaceutically oriented, one can frequently identify small molecules (< 1 Kd) that possess high affinity and selectivity for the specific biological target. Evaluation of the structure-activity relationships (if available) then provides a rationale for selecting a candidate for transformation into a biomedical imaging agent. One should recall that most published studies focus on the development of ther-apeutic pharmaceuticals, which ultimately have different criteria than imaging agents.[20–22] These criteria include efficacy, bioavailability, long-term safety and ease of preparation. As a result, the drugs that emerge successfully from these efforts are not necessarily the most appropriate for (direct) conversion to ima-ging agents. In particular, bioavailability and long-term safety are not issues for radio-diagnostic agents. In terms of synthesis, while overall yields in large-scale pharmaceutical chemistry are very important, for the preparation of a radi-olabeled compound on a very small scale, only the last steps are critically important and primarily focus on efficient introduction of the radionuclide and the clean rapid separation of the radiochemical. As a result, one can analyze the literature and select as candidates, compounds that possess optimized biological and physicochemical properties, but whose syntheses may be more problematic.

Selection of the radionuclide is another important factor because it will ultimately influence the chemistry used in preparing the radiopharmaceutical. Although a number of radio-nuclides are available *via* cyclotron or reactor production for incorporation into imaging agents, only three will be considered in this review – [^{18}F], [^{123}I] and [^{99m}Tc], (Table 1). Because of their overall characteristics, these three radionuclides are incorporated into the majority of the radiopharmaceuticals prepared and evaluated.

These radionuclides are readily available *via* cyclotron or reactor production and accessible to almost all major radiochemistry facilities. Fluorine-18 is the sole positron-emitter in this group and has a 110 min half-life, which is suffi-cient to allow for significant chemical transformations post-isolation from the

Table 2.1 List of clinical relevant radionuclides available for radiopharmaceutical chemistry.

Nuclide	Half-life $T_{1/2}$	Mode of Decay (keV)	Production Method	Radiopharmaceutical Form
^{11}C	20 min	511 ($\beta+$)	Cyclotron	$^{11}CH_3, ^{11}CH_2OH, ^{11}C{=}O, ^{11}CO_2H$
^{18}F	110 min	511 ($\beta+$)	Cyclotron	C(alkyl)-^{18}F, C(aryl)-^{18}F
^{68}Ga	68 min	511 ($\beta+$)	Generator	Chelate as $^{68}Ga^{III}$
^{123}I	13.3 h	159 (γ)	Cyclotron	C(alkyl)-^{123}I, C(alkenyl)-^{123}I, C(aryl)-^{123}I
^{99m}Tc	6 h	140 (γ)	Generator	Chelate as $^{99m}Tc^{I,IV,V}$
^{111}In	2.83 d	171(γ) 245 (γ)	Cyclotron, linear accelerator	Chelate as $^{111}In^{III}$

cyclotron source. Iodine-123 is a single photon emitting radionuclide with a
13 h half-life, which allows it to be produced off-site, shipped and still available
for radio-synthesis. It has gamma ray emission energy of 159 keV and a decay
scheme that generates 10–15 Auger electrons for potential radio-therapeutic
purposes. Technetium-99m is the daughter of molybedenum-99 decay, which
permits its onsite availability with a generator. As such, it is available on
demand which allows it, with its 140 keV gamma emission and 6h half-life, to
be incorporated into many potential imaging agents.

Fluorine-18 and iodine-123 are radio-halogens, and as such, take advantage
of much of the chemistry established for the non-radioactive isotopes. Incorporation of fluorine into biologically active organic compounds has been well
described and proceeds in good yields and stereochemical control for both
aliphatic and aromatic systems.[23–29] C–F bonds are remarkably stable and the
presence of fluorine in biomolecules imparts important pharmacological
properties, such as bioisosterism to C–H bonds and dipolar properties similar
to C–OH groups. However, the synthesis of the radio-fluorinated analog of a
specific biomolecule often presents synthetic challenges. Fluoride ion is a poor
nucleophile and although displacements at primary alkyl sites proceed in good
yields at the no-carrier-added level, displacements at secondary carbons are
much more sluggish and accompanied by elimination reactions. Nucleophilic
displacement by fluoride of leaving groups at sp^2 sites (aryl or alkenyl) is even
more impeded and requires elevated temperatures and electron-withdrawing
groups to facilitate the reaction. Iodine is incorporated into biomolecules much
less frequently than fluorine primarily due to the weaker C–I bond (rapid
deiodination *via* nucleophilic displacement on sp^3 centers and oxidative cleavage on sp^2 centers). The isosteric relationship between iodine and a methyl
group however, does present potential advantages in terms of radio-synthesis
and the ease with which electrophilic iodine species can be generated provides
advantages not available to fluorine.[30–34] The development of the ipso demetallation reaction for the introduction of radio-iodine at the no-carrier-added
level has allowed the preparation and evaluation of many radio-iodinated

biomolecules as potential biomedical imaging agents as well as radiotherapeutics.

Technetium-labeled and other radio-metallated imaging agents that have specific molecular targets present a different set of synthetic challenges. As a metallic ion, one cannot employ typical organic chemistry to prepare C–Tc bonds and generate stable biomolecules. Historically, Tc-99m, as isolated from the generator as pertechnetate, underwent reduction and subsequent chelation in the form of large, anionic complexes that loosely resembled the parent molecule. Fortunately, recent efforts, largely by Alberto, Schibli, and Schubiger,[35–39] have provided a small, stable, low valent $Tc(CO)_3$-core that can be incorporated into biomolecules using trivalent coordination. The ease with which this metallated core can be prepared has allowed for the synthesis and evaluation of many potential radiodiagnostic and radiotherapeutic (using the Re-188 congener) agents.[40–45]

2.2 Recent Examples of Integrated Radiotracer and Radiopharmaceutical Development

An aspect of radiopharmaceutical chemistry and its application to biomedical imaging that is often neglected involves the design and synthesis of radio-chemicals that integrate all of the features previously described. In particular, the synthesis of precursor molecules that can rapidly incorporate specific radionuclides of clinical interest, at sites on the molecule that do not impair the pharmacodynamic or pharmacokinetic properties, has generally been poorly handled. To a large extent this is because synthetic organic chemists, who have the expertise, have not been interested in what may appear to be a mundane resynthesis or minor modification of an existing compound. On the other hand, those investigators who have been interested in modifications often resort to what is available or to compounds that have clinically acceptable (therapeutic) efficacy, without considering the compromises that have gone into that clinical acceptability. For example, direct introduction of fluoride or iodide onto the parent compound or precursor implies the presence of a reactive group in order for the nucleophilic substitution to occur (Figure 2.2).[46–49] These sites, once labeled would also constitute sites for *in vivo* dehalogenation *via* displacement with endogenous nucleophiles. In addition, rapid separation of the unlabeled precursor from the relatively small quantity of labeled product may be difficult unless the properties of the leaving group were substantially different from those of the radiohalide. Alternatively, the halide could be introduced by direct electrophilic substitution (Figure 2.3).

This would require the presence of an aromatic group with electron-donating substituents, such as phenolic or anilinic groups. While electrophilic substitution with radioiodine is well established, frequently positional isomers are formed and the physicochemical properties of the adjacent substituent are significantly modified. Also, introduction of the iodo-group general increases the log$P$ of the product. Electrophilic substitution with radiofluoride is

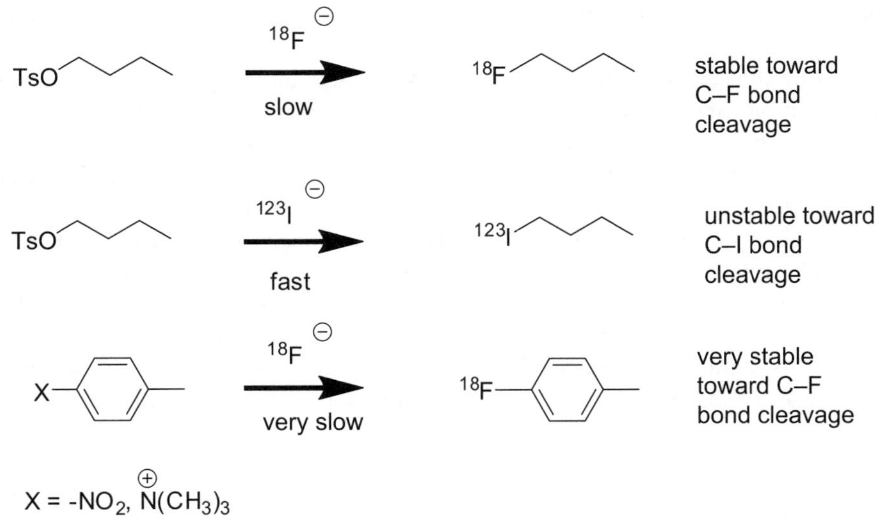

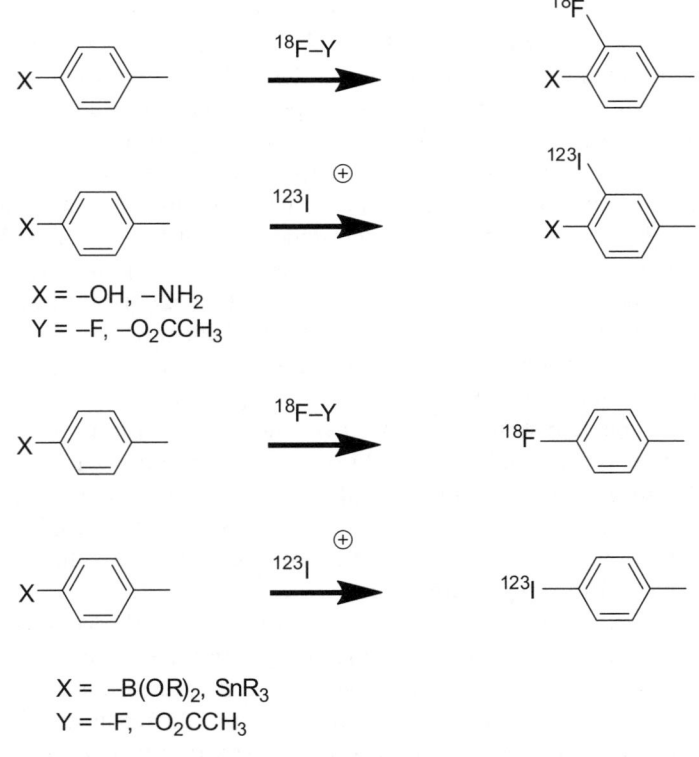

Figure 2.2 Radiohalogenation (^{18}F/^{123}I) *via* nucleophilic displacement.

Figure 2.3 Radiohalogenation (^{18}F/^{123}I) *via* electrophilic substitution.

currently not done for radiotracers requiring very high specific activity. For [^{18}F] radiopharmaceuticals, an alternative to direct radiofluorination involves the use of bioconjugating methods in which a small reagent labeled in high specific activity can subsequently be conjugated to the appropriate precursor. This, of course, implies the presence of a corresponding chemo-reactive group and that modification of this group does not compromise the pharmacodynamic and pharmacokinetic properties of the product. That is usually a difficult task with small molecules for molecular targeting of disease. Nevertheless, a number of approaches for labeling bioactive compounds via [^{18}F] bioconjugating reagents have been described in the recent literature.[7,50–60] This strategy permits the synthesis of specific biomolecules and the labeling moieties to proceed independently and bring together the two components in the final ligation step.

Three examples that illustrate the integration of biological target selection, lead compound identification and specific radionuclide incorporation are the development of bioimaging agents for β-amyloid plaque formation in Alzheimer's disease, for prostate-specific membrane antigen (PSMA) in prostate cancer, and integrin receptors in various cancers. In each case, a clear rationale was developed for the biological target that leads to the selection of a specific class of compounds. Based upon the physicochemical properties or structure-activity relationships, investigators developed strategies that would use the clinical relevant radionuclides. It is not possible to cover all the work done in each of these areas and therefore particular highlights will be presented to emphasize the selection criteria.

2.2.1 β-Amyloid Targeted Agents for Imaging in Alzheimer's Disease

Without question, Alzheimer's disease (AD) is a growing societal problem as the percentage of the population reaching older age increases. It is the most common form of dementia, characterized by diminishment of cognitive function and behavioral impairment. Accurate assessment of this disease would result in the improvement of care for patients and alleviate the burden of caregivers. As a result, there has been significant effort to identify biomarkers that serve as targets for non-invasive imaging agents. Among the potential biological targets, one promising biomarker is the association of β-amyloid plaque formation in the brain with the severity of disease. Observations that organic dyes, such as Congo red and Thioflavin-T, would preferentially stain these aggregations, led investigators to propose the use of β + -labeled radiotracers as potential *in vivo* imaging agents for AD.[61–70] A more successful approach resulted from the choice of Thioflavin-T as the lead compound from which to develop positron (or gamma)-emitting derivatives. One key issue was that for enhancing uptake into the brain, the compounds could not be permanently charged, as in Thioflavin-T. Initially, it was not clear whether the deletion of the quaternary ammonium site on the benzothiazole, while improving passage across the blood-brain barrier (BBB), would adversely affect

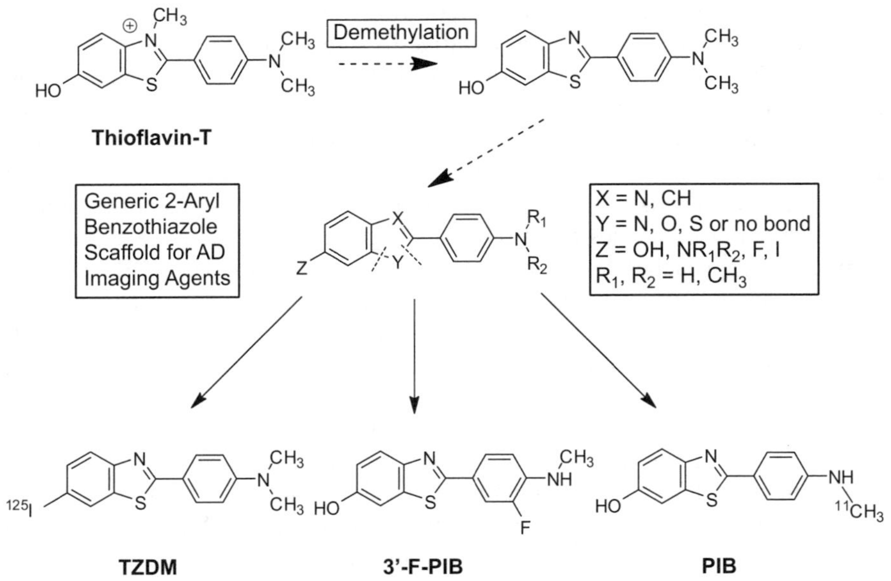

Figure 2.4   Development of first generation AD imaging agents.

the localization with the β-amyloid aggregates.[71,72] Fortunately, both radio-iodinated and [^{11}C]-analogs of N3-demethylated Thioflavin-T were developed,[73–79] and demonstrated selective binding to β-amyloid plaque formation in post-mortem brain slices of AD patients and good brain uptake in rodents. Ultimately, the [^{11}C]-agent, known as (6-OH-BTA-PIB),[62,63] was advanced to clinical trials for imaging AD. However, as successful as this agent is for imaging, the short half-life of ^{11}C (20 min) limits its use in a general clinical setting (Figure 2.4). As a result, efforts continue to identify and prepare either [^{18}F]-, [^{123}I]-, or [^{99m}Tc]-labeled analogs that possess similar selectivity for the β-amyloid aggregates, comparable brain uptake, *in vivo* stability, and better radiosynthetic characteristics.

The effort to develop "second generation" AD imaging agents is best exemplified by the recent work of Kung, *et al.*[80–103] Starting from the core structure of PIB, generated from the structure of Thioflavin-T, this research group has systematically explored the central thiazole moiety, the flanking aryl groups, and peripheral substituents as a means of enhancing the biological and radiochemical properties of the final compounds. In this dissection, either of the peripheral groups X or Y could be the site for the radionuclide. The aryl groups, 1 or 2, could potentially be heterocyclic, although in practice phenyl is preferred. The thiazole component of Thioflavin-T could be considered not only as a heterocyclic unit but also as a constrained alkene, which would give rise to stilbene analogs. In the course of the 8 years of research done by this group, almost every variation which could incorporate [^{18}F] or [$^{123/125}$I] was evaluated (Figure 2.5).

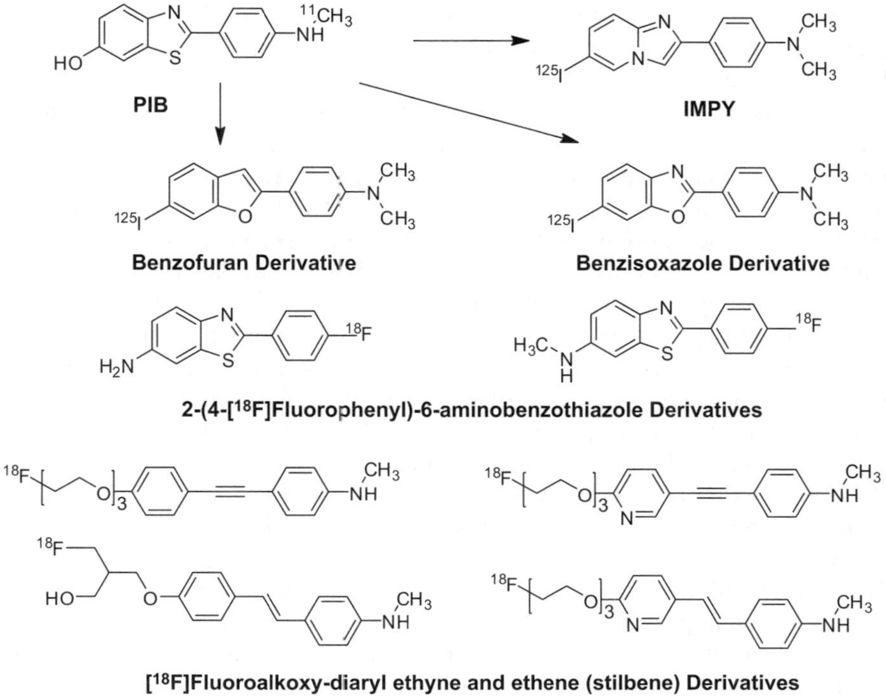

Figure 2.5   Systematic evaluation of structure of β-amyloid targeted agents: second-generation agents.

The initial studies focused more on the central component of the bioimaging agent while maintaining the peripheral groups intact. The thiazolo-moiety was modified by conversion, first to the oxygen analog (benzoxazole) and then to the de*aza* analog (benzofuran).[80,91] In each case, the 6-iodo-substituent and the 2-(4-dimethylaminophenyl) group provided the best activity in the series and the results were consistent with the structure activity relationships generated in the first generation of compounds. Modification of the benzothiazole core to an imidazopyridine core provided an interesting series of derivatives which retained β-amyloid binding capacity and provided [125I]IMPY which has been used in a number of biological studies.[81–83,86,99] Removal of the sulfur heteroatom generated a series of stilbene derivatives that could be labeled with ^{11}C, ^{18}F or [125]I.[88,89,93–95] Optimal activity was still associated with a *para* methylamino group on one phenyl substituent and a hydroxyl or alkoxy group on the other aryl substituent. It is interesting to note, that as in the case of IMPY, the aryl group could also be pyridyl. The physicochemical properties of the molecule could be modulated by variation on the alkoxy substituent, as [18F]hydroxypropoxy and [18F]oligoethylene glycoloxy groups were reasonably well tolerated. In these cases, the synthesis of the compounds was designed to permit introduction of the radiohalogen at the last step, rapidly and in good yields. As the studies proceeded, they evaluated whether true fused or seco-analogs were actually necessary. The two aryl groups were appended to the 2,5-positions of thiophene without a major reduction in biological activity.[88] Synthesis of 1,2,3-triazole derivatives using Huisgen [3 + 2]-cycloaddition (click) chemistry also provided interesting derivatives.[88] Binding to β-amyloid was improved by using an ethynyl spacer generating a series of 1,2-biaryl acetylenes. In these compounds, the orientation of the two groups is not exactly the same as in Thioflavin-T, but is close enough to provide good interaction with the protein. The synthesis of these compounds is relatively simple and permitted an evaluation of substituents.[96,98] Not only the peripheral substituents but the aryl groups could also be easily replaced to permit additional SAR studies.[100–103] More recently, this group returned to the 2-phenylbenzothiazole parent structure to prepare and evaluate the compounds in which the [18F] is incorporated at the *para*-position of the 2-phenyl group rather than at the 6-position of the benzothiazole. This variation also appears to be quite selective for the β-amyloid protein.

The success of this approach is exemplified by the identification of ^{18}F-AV-45 (Florbetapir F-18) as a candidate for clinical imaging of β-amyloid in patients with Alzheimer's disease. Over the past several years, this agent has undergone extensive evaluation in patients with AD.[104–107] The imaging characteristics and absence of serious adverse effects have advanced this agent toward consideration for approval by the FDA. The fast kinetics of localization in the brain and the strong correlation between that localization and pathology are advantages not displayed by competing [18F]amyloid radiotracers. As a result, significant effort has proceeded to develop methods for synthesizing this agent for routine clinical use.[108,109]

Key aspects of these studies are that the work was undertaken with a rational design, required expertise in organic synthesis to prepare the desired precursors and unlabeled compounds, and incorporated structural features that permitted the desired radionuclide to be introduced late in the synthesis, rapidly, in high yield and specific activity, and be readily purified. This was a project that clearly focused on the development of radio-halogenated analogs of the original [11C]PIB. Recent work has also focused on variations of the biaryl core through the synthesis and initial evaluation of styryl, bis-styryl and chalcone derivatives,[110–113] however, these derivatives do not appear superior to previously evaluated agents. In addition, significant efforts have been made to develop [99mTc]-compounds for imaging β-amyloid deposits, but they have generally been unsuccessful.[114–117] The key issue with this research is whether the agent that is eventually approved can enhance the detection, diagnosis and treatment of patients suffering from Alzheimer's disease.

2.2.2 PSMA Targeting for Imaging Prostate Cancer

To date there are no reliable imaging agents for the detection and diagnosis of prostate cancer. Prosta Scint® is an FDA approved labeled antibody, however, it targets an intracellular epitope of the prostate specific membrane antigen PSMA.[118–120] More recently, other investigators have prepared and evaluated radiolabeled derivatives of the J591 monoclonal antibody, which recognizes an extracellular domain of the PSMA, as potential diagnostic and therapeutic agents. Unfortunately, results from these studies have not been sufficiently encouraging to bring the derivatives to advanced clinical trials.[121–124] As a result, investigators have undertaken studies to develop new biomedical imaging agents based on small molecules that target the same protein, but by a different mechanism. Recognition that PSMA is highly similar to *N*-acetylated-α-linked acidic dipeptidase (NAALADase) and glutamate carboxypeptidase II (GCP II) resulted in its characterization as a target for which specific and high affinity inhibitors could be developed.[125,126] Many small molecule inhibitors have been prepared; however, virtually all of the synthetic work was disclosed in the patent literature, which limited transformation of the best candidates into imaging. In 2005, a review was published which proposed the use of some of those compounds as potential imaging agents, from which the following studies resulted.[127]

The most successful approaches to developing biomedical imaging agents for this target are based on peptidomimetic analogs of the substrate *N*-acetyl-aspartyl glutamate in which the internal peptide bond has been replaced with either a urea or phosphonate moiety (Figure 3.6). Structure activity relationships developed for these classes of inhibitors indicated that the C-terminal glutamate moiety is favored and should not be significantly modified while the aspartyl and *N*-acetyl groups could undergo significant changes or substitution without a major diminishment of binding affinity,[128–140] (Figure 2.6). The nature of the steric tolerance demonstrated by the binding pocket for the "aspartyl" substituent suggested that one would be able to introduce ^{99m}Tc-chelating groups as well as the radiohalogens. This can be illustrated by the studies

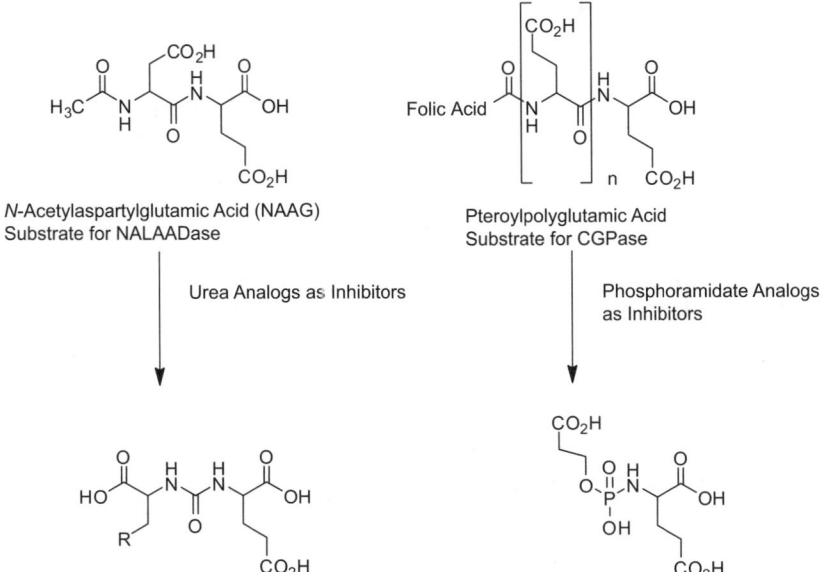

Figure 2.6   Development of urea and phosphoramidate analogs of NAAG and pteroyl polyglutamate substrates for NALAADase and CGPase.

reported by several groups in developing potential prostate cancer imaging agents. All of the research groups were assisted in the design aspects of their work by the publication of X-ray crystal structures of the enzyme and its complexes with inhibitors, and by molecular modeling of putative ligands.[141–148]

The most extensively developed program designed to yield clinically useful PSMA-based prostate cancer imaging agents is based on the urea analogs initially identified by Kozikowski, *et al.*[128] In this approach (Figure 2.7), the terminal glutamate moiety is maintained and appropriate amino acids are coupled through the use of triphosgene (isocyanate intermediate). A major advantage of this scheme is that the chiral starting materials are readily

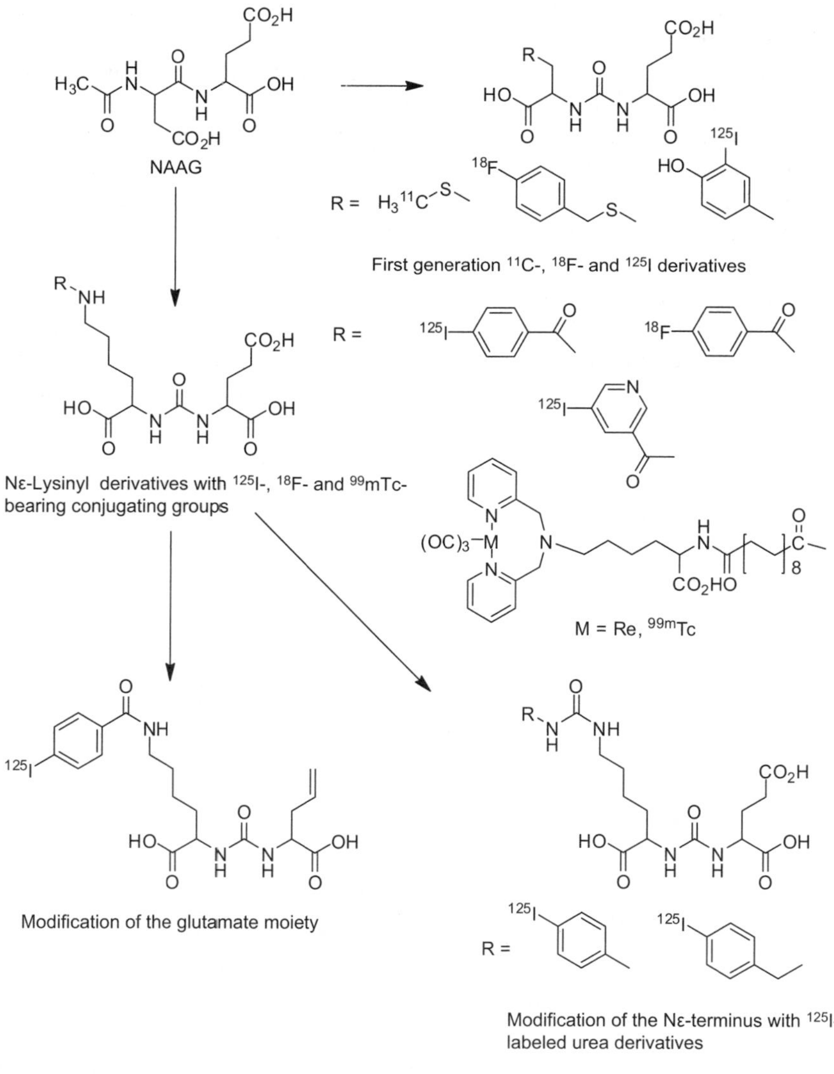

Figure 2.7   Development of urea-based PSMA-targeted imaging agents.

available from the pool of natural amino acids and one needs only to employ the requisite protection-deprotection chemistry to obtain a small library of dipeptidyl analogs bearing the terminal S'-1 glutamate group and an "N-terminal" S-1 site for further functionalization. That functionalization could then incorporate the radionuclide of choice, whether it was radiohalogen (^{18}F/^{123}I) or radiometal (^{99m}Tc). In their initial studies, the S-1 moiety was a tyrosine group that permitted direct radio-iodination of the phenol to give the labeled product. Use of cysteine in the S-1 position provided the ability to use alkylation as the bioconjugation strategy and gave both [^{11}C] and [^{18}F]-labeled materials.[149–151] Use of lysine at the S-1 site allowed the investigators to use acylation chemistry to introduce the radio-halogenated prosthetic group. Conjugation with the activated trialkyl stannyl benzoyl or nicotinyl group followed by radio-iodo destannylation gave the [^{125}I]-labeled species in high yield. Preparation of the activated [^{18}F]fluorobenzoate and ligation to the free amine gave the labeled derivative, also in high yield. Evaluation of the cold material *in vitro* and the labeled material *in vivo* indicated selective localization properties for both radiohalobenzamido derivatives.[152] Variations on this theme using *N*-(halo)-benzyl, *N*-(halo)phenylureido, and *N*-(halo)benzenesulfonamido derivatives also demonstrated reasonable activity as potential imaging agents.[153,154]

This same group has also exploited the structure activity relationships to develop a series of [^{99m}Tc]-labeled ureas as potential PSMA-targeted imaging agents (Figure 2.8). The terminal amino group of the S-1 lysine provides the locus for a variety of ligation strategies. In their recent publication, they describe

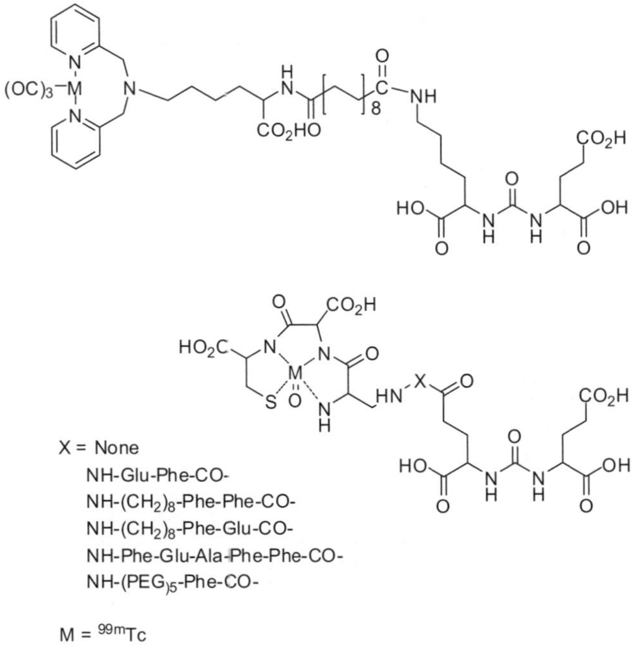

X = None
NH-Glu-Phe-CO-
NH-(CH$_2$)$_8$-Phe-Phe-CO-
NH-(CH$_2$)$_8$-Phe-Glu-CO-
NH-Phe-Glu-Ala-Phe-Phe-CO-
NH-(PEG)$_5$-Phe-CO-

M = ^{99m}Tc

Figure 2.8 ^{99m}Tc labeled PSMA targeted ligands.

the attachment of the technetium-tricarbonyl core to the ligand using the simple bis(picolinyl) moiety or that moiety elaborated with sequentially longer tethers. The *in vitro* studies with the rhenium analogs indicated the need for long rather than short tethers to achieve potent inhibitory activity, and conversion to the [99mTc]-labeled derivatives identified one of the complexes as possessing potential as an imaging agent.[155] The use of a different chelating group on the lysine permits incorporation of [68Ga] for PET imaging while labeling with a near IR (NIR) dye imparts optical imaging properties.[156,157] A variation of this approach was used by Low, *et al.* who prepared the symmetrical Glu-urea-Glu dipeptide. Synthesis of a [99mTc]-chelate, based upon a modified-Phe-Phe-dipeptide linked to an N2S2 core followed by ligation to one of the two equivalent side chain carboxylic acids, gave the precursor. Introduction of the [99mTc] gave the radiochemical rapidly, in high yields and specific activity. *In vivo* imaging studies and biodistributional analyses suggested that this material also demonstrated avidity for PSMA-expressing prostate tumors.[158,159]

A slightly different approach was used by Berkman and his collaborators in developing their version of PSMA-targeted agents (Figure 2.9). In this approach, a phosphoramidate core was used to link the S-1' and S-1 components, rather than the urea moiety, as used in Kozikowski's method. Initial studies, which focused on drug delivery or optical imaging, established the basic chemical methodology and SAR needed to translate this approach for radio-diagnostic purposes.[132,133,135,136,160–162] Ultimately, the results directed the preparation on novel compounds in which either [18F] or [99mTc] could be introduced without significantly modifying the biological integrity of the PSMA inhibitor.[163] The [18F]-labeled derivative indeed demonstrated subnanomolar inhibition *in vitro* and good target to non-target uptake ratios *in vivo*.

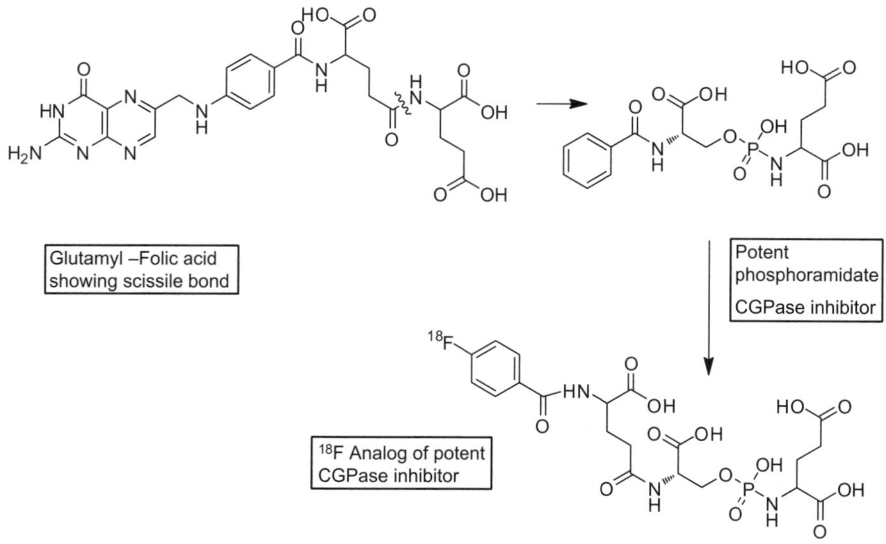

Figure 2.9   Phosphoramidate-based imaging agents.

As in the previous example of AD imaging agents, these studies demonstrated the use of rational drug design, expertise in organic synthesis to prepare the desired precursors and unlabeled compounds, and employed the use of structural features that allowed the desired radionuclide to be introduced late in the synthesis, rapidly, in high yield and specific activity, and be readily purified. This project was designed to use both radio-halogens and radio-metals based on the understanding of the target protein and its mechanism of action.

2.2.3 Integrin Receptor Targeted Agents for Imaging Cancer

One of the physiological processes present in the progression and spread (metastasis) of many cancers involves the establishment of blood vessels to maintain the nutritional demands of the rapidly proliferating cells. These neo-angiogenic tumor vessels typically do not resemble normal vasculature but tend to overexpress receptors found in normal epithelial cells. Among these cell surface receptors that are significantly upregulated in many cancers are the family of integrin receptors, particularly the αvβ3 receptor, which not only is involved in tumor angiogenesis, it alsobinds the exposed arginine-glycine-aspartic (RGD) sequence present in many extracellular matrix proteins, such as vitronectin, fibronectin, laminin, collagen, and osteopontin, among others. This close association between the receptor expression on the cancer cells and their angiogenic behavior has made it a particular target not only for chemotherapeutic drug development but also for developing imaging agents. A number of recent reviews have described this process.[164–178] In this examination of radiopharmaceutical development, particular attention will focus on the translation of the understanding of the RGD-sequence as a receptor-binding component to its development as a bioimaging modality.

The identification of the RGD-sequence as the key recognition and binding component of naturally occurring ligands for the integrin receptors constituted the logical starting point for developing diagnostic imaging agents. There were several fundamental problems associated with this tripeptidyl sequence, such as receptor selectivity, binding affinity, metabolic stability that needed to be overcome before it could be converted to a scaffold for incorporating radionuclides (Figure 2.10). Although evaluation of linear peptidyl analogs and peptidomimetics yielded materials with enhanced receptor selectivity and binding affinity, the more effective approach involved the development of cyclic peptides. In this approach, the requisite RGD motif would be displayed on three vertices and a variety of amino acids, including D-amino acids, could be incorporated into the remaining site(s). Cyclization reduced the conformational mobility of the peptide and, depending upon the nature of the fourth and fifth amino acids, provided a position for introducing other functional groups. This was a critical factor because the parent peptide, linear or cyclic, did not offer functional groups that could readily be labeled without decreasing selectivity, affinity or stability (Figure 2.11). One of the first derivatives of the cyclic RGD peptides used D-tyrosine which was radio-iodinated. The product ([^{123}I]cycloRGDyV) could be imaged *in vivo*, but separation of the labeled from

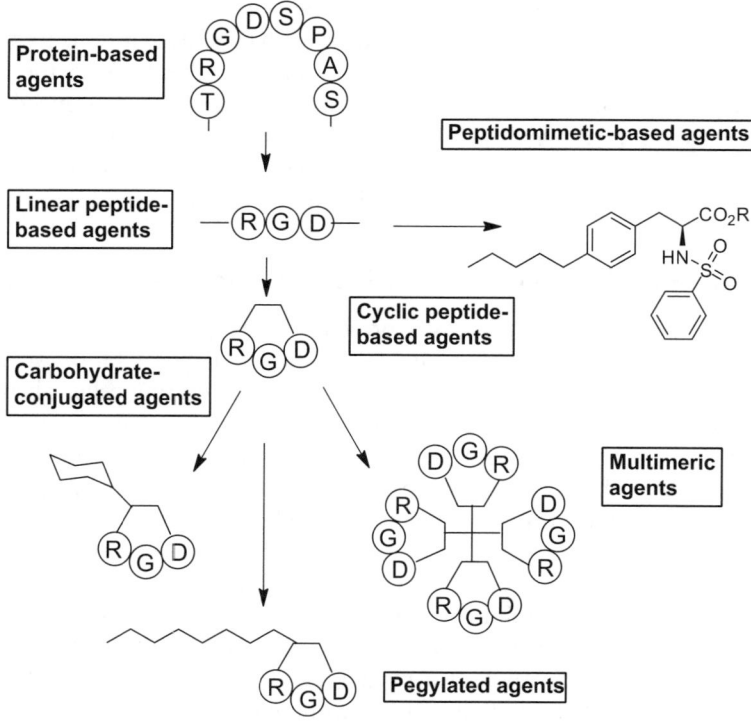

Figure 2.10   Strategies in the development of integrin receptor-targeted molecular imaging agents.

unlabeled material, non-specific binding, and deiodination remained as problems.[173,176] Maintaining the D-phenylalanine/tyrosine and replacing valine with lysine (cRDGf/yK) provided a site for attaching prosthetic imaging groups. Because the pendant lysine side chain is not involved with receptor interactions, radiometal binding components could readily be incorporated into the ligand. This has resulted in the preparation and evaluation of a number of [99mTc]radiopharmaceuticals in which the radionuclide is chelated by HYNIC, bidentate and tridentate moieties.[179–182]

The preparation of [18]-derivatives of cRGDfK proceeded in a manner similar to that used for the [99mTc]-radiopharmaceuticals. The terminal amino group of lysine constituted a convenient site for bioconjugation with [18F]-reagents. The amino group could undergo direct acylation with the [18F]benzoate N-hydroxysuccinimide ester or with 2-[18F]propionate 4-nitrophenyl ester.[183,184] The reactions proceed in good yields under mild conditions, but require pre-synthesis of the labeled reagents and separation of labeled from unlabeled products in order to achieve high specific activity. A similar situation is encountered in the synthesis using 4-[18F]benzaldehyde and the oxyaminoacetyl derivative of cRGDfK.[185] Modification of the lysine side chain to galactose or polyethylene glycol prior to labeling provided alternate methods

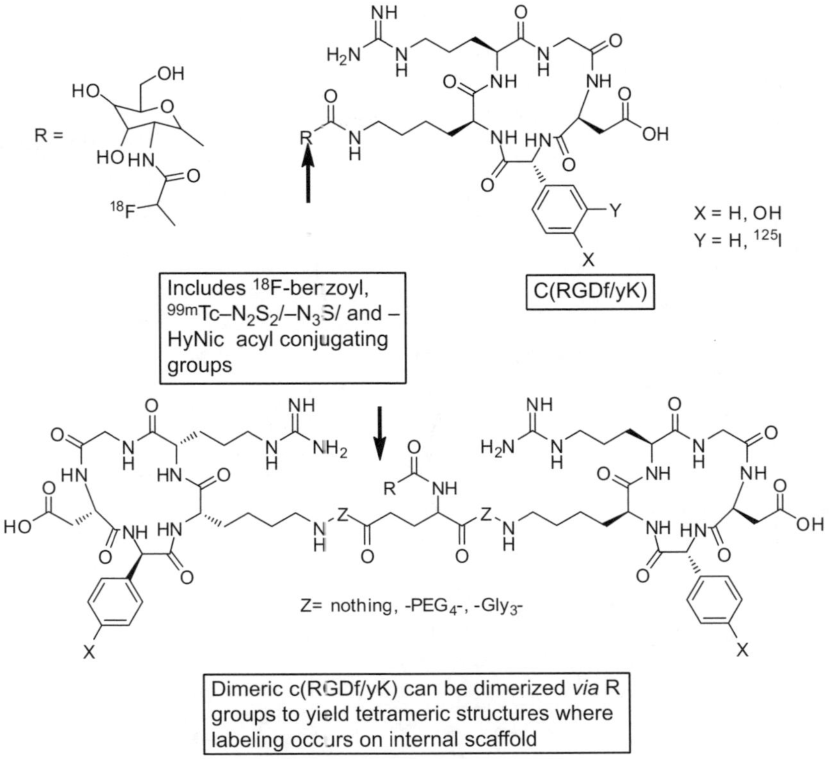

Figure 2.11 Cyclic RGD peptide-based derivatives for molecular imaging agents.

for [^{18}F]-labeling which gave products with greater metabolic stability or improved pharmacokinetics.[186–188] With multimeric cRGD peptides, the [^{18}F]-fluorinating reagents could be conjugated on the tethering moieties.[189,190]

One variation that has been explored in an effort to improve binding and selectivity for integrin receptor expressing cells involves the use of multivalent ligands. In this scenario, the cRDGxX molecules can be linked through the X side chain to a variety of tethers that possess the capacity to chelate radio-nuclides, such as [^{99m}Tc] or [^{18}F]. The strategy is based on the hypothesis that multimeric binding will produce significantly higher affinity than the corresponding monomeric agent will and, if the receptor density is appropriate, higher cellular selectivity compared to normal-expressive cells. Liu, *et al.* have examined this strategy extensively in recent years with the general conclusion that multimeric expression does lead to improved binding up to a point, and then the presence of additional groups does not contribute to enhanced interactions with the receptors.[170,180,191–195]

In virtually all of these instances, the parent ligand is a cyclic peptide that is commercially available and can undergo modification using relatively simple chemical procedures. Incorporation of the radionuclide also employs relatively standard protocols. An alternative approach involves the use of

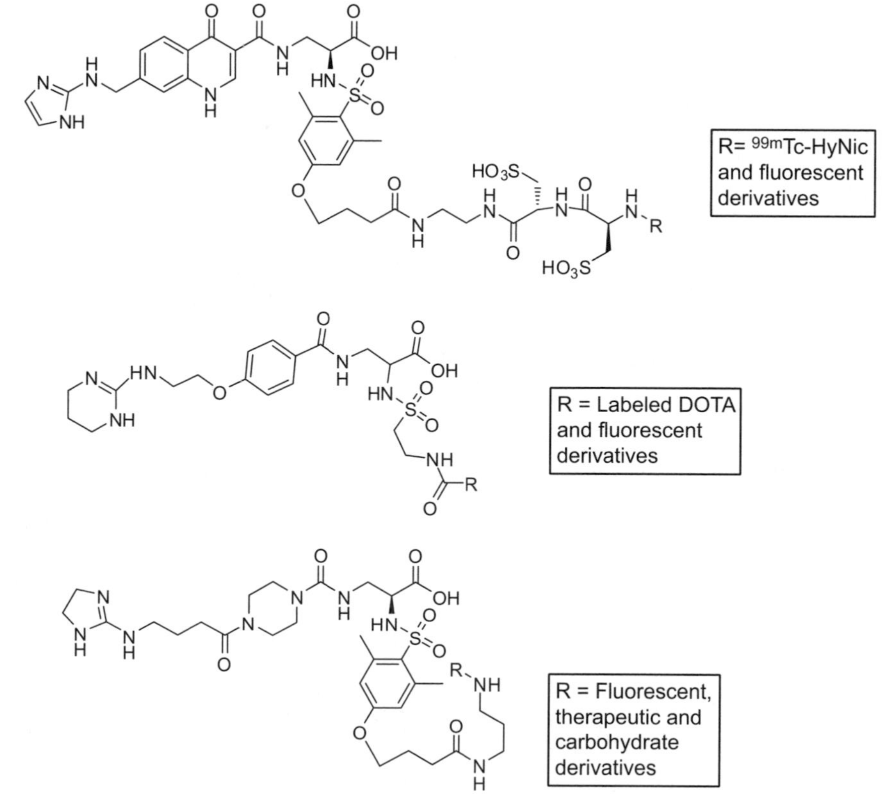

Figure 2.12   Peptidomimetics as integrin receptor imaging agents.

peptidomimetics that use nontraditional functional groups to represent the RGD-motif (Figure 2.12). These compounds have been prepared and evaluated primarily as therapeutic agents by the pharmaceutical industry, but can also constitute leads as imaging agents.[196–198] Most of the work in this area has focused on non-radioactive imaging agents,[199–201] but the strategy can be applied to the radiolabeled compounds as well. Very little has been reported, however, for radio-halogenated radiotracers, but several examples of radio-metallated derivatives have been described.[202–204] It will be interesting to see if these compounds demonstrate better properties than the peptidyl derivatives. They should display significant integrin receptor-subtype selectivity and improved metabolic stability; however, they require greater synthetic expertise in their preparation.

2.3   Conclusions

The previous sections describe rational approaches for the development of molecular imaging agents. The process starts with the selection of a biological target that has clinical significance and leads to the identification of a targeting molecule. The targeting molecule is designed to incorporate features that provide selective binding to the biological target and for introduction of the relevant radionuclide. Competence in organic synthesis is then required to assemble that compound and iteratively modify it to achieve the final properties. In the cases described in this chapter, the ultimate imaging agent has not been prepared, but substantial progress toward that goal has been achieved. Nevertheless, these examples illustrate the process by which advances are made in the field of molecular imaging with radiopharmaceuticals.

Acknowledgments

The author recognizes the support of DOE grant DE-SC0001781.

References

1. F. Dolle, D. Roeda, B. Kuhnast and M. Lasne, *Fluorine and Health.*, 2008, 3–65.
2. T. Mindt, H. Struthers, E. Garcia-Garayoa, D. Desbouis and R. Schibli, *Chimia*, 2007, **61**, 725–731.
3. R. Alberto, In *Bioorganometallics*, ed. G. Jaouen 2006, 97–124.
4. R. Alberto, *Top. Curr.Chem.*, 2005, **252**, 1–44.
5. G. Ferro-Flores, C. Arteage de Murphy and L. Melendez-Alafort, *Curr. Pharm. Anal.*, 2006, **2**, 339–352.
6. M. J. Adam and D. S. Wilbur, *Chem. Soc. Rev.*, 2005, **34**, 153–163.
7. M. Eisenhut and W. Mier, in *Handbook of Nuclear Chemistry*, ed. A. Vertes, Kluwer Academic Publishers, Dordrecht, Netherlands, 2003, **vol. 4**, pp. 257–278.
8. H. J. Wester, in *Handbook of Nuclear Chemistry*, ed. A. Vertes, Kluwer Academic Publishers, Dordrecht, Netherlands, 2003, **vol. 4**, pp. 167–209.
9. S. E. Snyder and M. R. Kilbourn, in *Handbook of Radiopharmaceutical.*, ed. M. J. Welch, and C. S. Redvanly, John Wiley & Sons Ltd., Chichester, UK, 2003, pp. 195–227.
10. M. C. Lasne, C. Perrio, J. Rouden, L. Barre, D. Roeda, F. Dolle and C. Crouzel, *Top. Curr. Chem.*, 2002, **222**, 201–258.
11. F. Dolle, *Curr. Pharm. Des.*, 2005, **11**, 3221–35.
12. N. S. Mason and C. A. Mathis, *Neuroimag. Clin. N. Am.*, 2003, **13**, 671–87.
13. S. Z. Lever, *J. Cell. Biochem.J. Cell. Biochem.*, 2002, **Suppl. 39**, 60–64.
14. T. J. Hoffman, T. P. Quinn and W. A. Volkert, *Nucl. Med. Biol.*, 2001, **28**, 527–539.
15. V. C. Spanoudaki and S. I. Ziegler, *PET and SPECT instrumentation*, in *Molecular Imaging I, Handbook of Experimental Pharmacology 185/I*, ed. W. Semmler and M. Schwaiger, Springer-Verlag, Berlin 2008, pp. 53–74.
16. M. J. Martinez, S. I. Ziegler and T. Beyer, *Recent Res. Cancer.*, 2008, **170**, 1–23.
17. A. H. Maurer, *Health Phys.*, 2008, **95**, 571–576.

18. I. Dijkgraaf, O. C. Boerman, W. J. Oyen, F. H. Corstens and M. Gotthardt, *Anti-Cancer Agents Med. Chem.*, 2007, **7**, 543–551.
19. E. Benedetti, G. Morelli, A. Accardo, R. Mansi, D. Tesauro and L. Aloj, *BioDrugs.*, 2004, **18**, 279–295.
20. M. A. T. Blaskovich, *Curr. Med. Chem.*, 2009, **16**, 2095–2176.
21. S. J. Enna and M. Williams, *J. Pharmacol. Exp. Ther.*, 2009, **329**, 404–411.
22. R. L. M van Montfort and P. Workman, *Trends Biotechnol.*, 2009, **27**, 315–328.
23. W. K. Hagmann, *J. Med. Chem.*, 2008, **51**, 4359–4369.
24. S. Purser, P. R. Moore, S. Swallow and V. Gouveneur, *Chem. Soc. Rev.*, 2008, **37**, 320–330.
25. K. L. Kirk, *Org. Process Res. Dev.*, 2008, **12**, 305–321.
26. C. Jäckel and B. Koksch, *Eur. J. Org. Chem.*, 2005, 4483–4502.
27. D. Cahard, X. Xu, S. Couve-Bonnaire and X. Pannecoucke, *Chem. Soc. Rev.*, 2010, **39**, 558–568.
28. D. O'Hagan, *Chem. Soc. Rev.*, 2008, **37**, 308–319.
29. Aj. Podgorsek, M. Zupan and J. Iskra, *Angew. Chem. Int. Ed. Engl.*, 2009, **48**, 8424–8450.
30. S. Stavber, M. Jereb and M. Zupan, *Synthesis*, 2008, 1487–1513.
31. R. N. Hanson, *New Methods Drug Res.*, 1988, **2**, 79–114.
32. R. M. Baldwin, *Int. J. Appl. Radiat. Isot.*, 1986, **37**, 817–821.
33. G. W. Kabalka and M. L. Yao, *J. Organomet. Chem.*, 2009, **694**, 1638–1641.
34. B. Maziere and C. Loc'h, *Curr. Pharm. Des.*, 2001, **7**, 1931–1943.
35. R. Schibli and P. A. Schubiger, *Eur. J. Nucl. Med. Mol. Imaging.*, 2002, **29**, 1529–1542.
36. R. Alberto, *Top. Curr. Chem.*, 2005, **25**, 1–44.
37. R. Alberto, *J. Nucl. Radiochem. Sci.*, 2005, **6**, 173–176.
38. P. V. Grundler, L. Helm, R. Alberto and A. E. Merbach, *Eur. J. Inorg. Chem.*, 2006, **45**, 10378–10390.
39. R. Alberto, *J. Organomet. Chem.*, 2007, **692**, 1179–1186.
40. G. Ferro-Flores, C. Arteage de Murphy and L. Melendez-Alafort, *Curr. Pharm. Anal.*, 2006, **2**, 339–352.
41. R. Garcia, L. Gano, L. Maria, A. Paulo, I. Santos and H. Spies, *J. Biol. Inorg. Chem.*, 2006, **11**, 769–782.
42. M. B. Mallia, S. Subramanian, A. Mathur, H. D. Sarma, M. Venkatesh and S. Banerjee, *J. Labelled Compd. Radiopharm.*, 2008, **51**, 308–313.
43. H. He, J. E. Morely, E. Silva-Lopez, B. Bottenus, M. Montajano, G. A. Fugate, B. Twamley and P. D. Benny., *Bioconjug. Chem.*, 2009, **20**, 78–86.
44. R. Tavare, R. Torres, Martin De Rosales., P. J. Blower and G. E. D. Mullen, *Bioconjug. Chem.*, 2009, **20**, 2071–2081.
45. J. Yang, H. Guo, F. Gallazzi, M. Berwick, R. S Padilla and Y. Miao, *Bioconjug. Chem.*, 2009, **20**, 1634–1642.
46. J. A. Katzenellenbogen, *J. Fluorine Chem.*, 2001, **109**, 49–54.
47. F. Dolle, *Curr. Pharm. Des.*, 2005, **11**, 3221–3235.
48. R. H. Seevers and R. E. Counsell, *Chem. Rev.*, 1982, **82**, 575–590.

49. J. J. Langone, *Methods Enzymol.*, 1981, **73**, 112–127.
50. T. L. Mindt, C. Muller, F. Stuker, J. F. Salazar, A. R. Hohn, T. Mueggler, M. Rudin and R. Schibli, *Bioconjug. Chem.*, 2009, **20**, 1940–1949.
51. C. Mamat, T. Ramenda and F. R. Wuest, *Mini-Rev. Org. Chem.*, 2009, **6**, 21–34.
52. M. R. Zhang, and K. Suzuki, [18F] *Curr. Top. Med. Chem.*, 2007, **7**, 1817–1828.
53. D. E. Olberg, O. K Hjelstuen, M. Solbakken, J. M. Arukwe, K. Dyrstad and A. Cuthbertson, *J. Labelled Compd. Radiopharm.*, 2009, **52**, 571–575.
54. X. Li, J. M. Link, S. Stekhova, K. J. Yagle, C. Smith, K. A. Krohn and J. F. Tait, *Bioconjug. Chem.*, 2008, **19**, 1684–1688.
55. M. Glaser and E. G. Robins, *J. Labelled Compd. Radiopharm.*, 2009, **52**, 407–414.
56. M. Glaser, E. Arstad, S. K. Luthra and E. G. Robins, *J. Labelled Compd. Radiopharm.*, 2009, **52**, 327–330.
57. M. Erlandsson, H. Hall and B. Laangstroem, *J. Labelled Compd. Radiopharm.*, 2009, **52**, 278–285.
58. K. A. Stephenson, R. Chandra, Z. P. Zhuang, C. Hou, S. Oya, M. P. Kung and H. F. Kung, *Bioconjug. Chem.*, 2007, **18**, 238–246.
59. J. A. H. Inkster, B. Guerin, T. J. Ruth and M. Adam, *J. Labelled Compd. Radiopharm.*, 2008, **51**, 444–452.
60. J. M. Hooker, *Curr. Opin. Chem. Biol.*, 2010, **14**, 105–111.
61. A. Nordberg, *Lancet Neurol.*, 2004, **3**, 519–527.
62. Y. Wang, W. E. Klunk, M. L. Debnath, G. F. Huang, D. P. Holt, S. Li and C. A. Mathis, *J. Mol. Neurosci.*, 2004, **24**, 55–62.
63. C. A. Mathis, B. J. Lopresti and W. E. Klunk, *Nucl. Med. Biol.*, 2007, **34**, 809–822.
64. I. Maezawa, H. S. Hong, R. Liu, C. Y. Wu, R. H. Cheng, M. P. Kung, H. F. Kung, K. S. Lam, S. Oddo, F. M. LaFerla and L. W. Jin, *J. Neurochem.*, 2008, **104**, 457–468.
65. M. P. Kung, Z. P. Zhuang, C. Hou and H. F. Kung, *J. Mol. Neurosci.*, 2004, **24**, 49–53.
66. E. K. Ryu and X. Chen, *Front. Biosci.*, 2008, **13**, 777–789.
67. K. Nagren and J. O. Rinne, *Fluorine and Health*, 2008, 67–84.
68. M. Ono, *Chem. Pharm. Bull.*, 2009, **57**, 1029–1039.
69. K. L. Xiong, Q. W. yang, S. G. Gong and W. G. Zhang, *Nucl. Med. Commun.*, 2010, **31**, 4–11.
70. V. Valotassiou, S. Archimandritis, N. Sifakis, J. Paptriantafyllou and P. Georgoulias, *Curr. Alzheimer Res.*, 2010, **7**, 477–486.
71. D. M. Skovronsky, B. Zhang, M. P. Kung, H. F. Kung, J. Q. Trojanowski and V. M .Y. Lee, *Proc. Natl. Acad. Sci. U. S. A.*, 2000, **97**, 7609–7614.
72. W. E. Klunk, Y. Wang, G. F. Huang, M. L. Debnath, D. P. Holt and C. A. Mathis, *Life Sci.*, 2001, **69**, 1471–1484.
73. Z. P. Zhuang, M. P. Kung, C. Hou, K. Plossl, D. Skovronsky, T. L. Gur, J. Q. Trojanowski, V. M. Y. Lee and H. F. Kung, *Nucl. Med. Biol.*, 2001, **28**, 887–894.

74. H. F. Kung, C. W. Lee, Z. P. Zhuang, M. P. Kung, C. Hou and K. Ploessl, *J. Am. Chem. Soc.*, 2001, **123**, 12740–12741.

75. M. P. Kung, C. Hou, Z. P. Zhuang, D. M. Skovronsky, B. Zhang, T. L. Gur, J. Q. Trojanowski, V. M. Y. Lee and H. F. Kung, *J. Mol. Neurosci.*, 2002, **19**, 7–10.

76. C. A. Mathis, B. J. Bacskai, S. T. Kajdasz, M. E. McLellan, M. P. Frosch, B. T. Hyman, D. I. P. Holt, Y. Wang, G. F. Huang, M. L. Debnath and W. E. Klunk, *Bioorg. Med. Chem. Lett.*, 2002, **12**, 295–298.

77. Y. Wang, W. E. Klunk, G. F. Huang, M. L. Debnath, D. P. Holt and C. A. Mathis, *J. Mol. Neurosci.*, 2002, **19**, 11–16.

78. C. A. Mathis, Y. Wang, D. P. Holt, G. F. Huang, M. L. Debnath and W. E. Klunk, *J. Med. Chem.*, 2003, **46**, 2740–2754.

79. Y. Wang, C. A. Mathis, G. F. Huang, M. L. Debnath, D. P. Holt, L. Shao and W. E. Klunk, *J. Mol. Neurosci.*, 2003, **20**, 255–260.

80. M. Ono, M. P. Kung, C. Hou and H. F. Kung, *Nucl. Med. Biol.*, 2002, **29**, 633–642.

81. M. P. Kung, C. Hou, Z. P. Zhuang, B. Zhang, D. Skovronsky, J. Q. Trojanowski, V. M. Y. Lee and H. F. Kung, *Brain Res.*, 2002, **956**, 202–210.

82. Z. P. Zhuang, K. Pi. Kung, A. Wilson, Ch. W. Lee, K. Ploessl, C. Hou, D. M. Holtzman and H. F. Kung, *J. Med. Chem.*, 2003, **46**, 237–243.

83. M. P. Kung, D. M. Skovronsky, C. Hou, Z. P. Zhuang, T. L. Gur, B. Zhang, J. Q. Trojanowski, V. M. Y. Lee and H. F. Kung, *J. Mol. Neurosci.*, 2003, **20**, 15–23.

84. M. Ono, A. Wilson, J. Nobrega, D. Westaway, P. Verhoeff, Z. P. Zhuang, M. P. Kung and H. F. Kung, *Nucl. Med. Biol.*, 2003, **30**, 565–571.

85. M. P. Kung, Z. P. Zhuang, C. Hou, L. W. Jin and H. F. Kung, *J. Mol. Neurosci.*, 2003, **20**, 249–253.

86. M. P. Kung, C. Hou, Z. P. Zhuang, A. J. Cross, D. L. Maier and H. F. Kung, *Eur. J. Nucl. Med. Mol. Imaging*, 2004, **31**, 1136–1145.

87. M. P. Kung, C. Hou, Z. P. Zhuang, D. Skovronsky and H. F Kung, *Brain Res.*, 2004, **1025**, 98–105.

88. W. Zhang, S. Oya, M. P. Kung, C. Hou, D. L. Maier and H. F. Kung, *J. Med. Chem.*, 2005, **48**, 5980–5988.

89. W. Zhang, S. Oya, M. P. Kung, C. Hou, D. L. Maier and H. F. Kung, *Nucl. Med. Biol.*, 2005, **32**, 799–809.

90. R. Chandra, M. P. Kung and H. F. Kung, *Bioorg. Med. Chem. Lett.*, 2006, **16**, 1350–1352.

91. M. Ono, H. Kawashima, A. Nonaka, T. Kawai, M. Haratake, H. Mori, M. P. Kung, H. F. Kung, H. Saji and M. Nakayama, *J. Med. Chem.*, 2006, **49**, 2725–2730.

92. A. B. Newberg, N. A. Wintering, K. Plossl, J. Hochold, M. G. Stabin, M. Watson, D. Skovronsky, C. M. Clark, M. P. Kung and H. F. Kung, *Int. J. Nucl. Med. Biol.*, 2006, **47**, 748–754.

93. K. A. Stephenson, R. Chandra, Z. P. Zhuang, C. Hou, S. Oya, M. P. Kung and H. F. Kung, *Bioconjug. Chem.*, 2007, **18**, 238–246.

94. W. Zhang, M. P. Kung, S. Oya, C. Hou and H. F. Kung, *Nucl. Med. Biol.*, 2007, **34**, 89–97.
95. W. Qu, M. P. Kung, C. Hou, T. E. Benedum and H. F. Kung, *J. Med. Chem.*, 2007, **50**, 2157–2165.
96. R. Chandra, S. Oya, M. P. Kung, C. Hou, L. W. Jin and H. F. Kung, *J. Med. Chem.*, 2007, **50**, 2415–2423.
97. W. Qu, M. P. Kung, C. Hou, S. Oya and H. F. Kung, *J. Med. Chem.*, 2007, **50**, 3380–3387.
98. W. Qu, M. P. Kung, C. Hou, L. W. Jin and H. F. Kung, *Bioorg. Med. Chem. Lett.*, 2007, **17**, 3581–3584.
99. P. J. Song, S. Bernard, P. Sarradin, J. Vergote, C. Barc, S. Chalon, M. P. Kung, H. F. Kung and D. Guilloteau, *Nucl. Med. Biol.*, 2008, **35**, 197–201.
100. W. Qu, S. R. Choi, C. Hou, Z. Zhuang, S. Oya, W. Zhang, M. P. Kung, R. Manchandra, D. M. Skovronsky and H. F. Kung, *Bioorg. Med. Chem. Lett.*, 2008, **18**, 4823–4827.
101. S. P. Wey, C. C. Weng, K. J. Lin, C. H. Yao, T. C. Yen, H. F. Kung, D. Skoronsky and M. P. Kung, *Nucl. Med. Biol.*, 2009, **36**, 411–417.
102. K. Serdons, K. Van Laere, P. Janssen, H. F. Kung, G. Bormans and A. Verbruggen, *J. Med. Chem.*, 2009, **52**, 7090–7102.
103. S. R. Choi, G. Golding, Z. Zhuang, W. Zhang, N. Lim, F. Hefti, T. R. E. Benedum, M. R. Kilbourn, D. Skovronsky and H. F. Kung, *Int. J. Nucl. Med. Biol.*, 2009, **50**, 1887–1894.
104. K. J. Lin, W. C. Hsu, I. T. Hsiao, S. P. Wey, L .W. Jin, D. Skovronsky, Y. Y. Wai, H. P. Chang, C. W. Lo, C. H. yao, T. C. Yen and M. P. Kung, *Nucl. Med. Bio.*, 2010, **37**, 497–508.
105. D. F. Wong, P. B. Roserberg, Y. Zhou, A. Kumar, V. Raymont, H. T. Ravert, R. F. Dannals, A. Nandi, J. R. Brasic, W. Ye, J. Hilton, C. Lyketsos, H. F. Kung, A. D. Joshi, D. M. Skovronsky and M. J. Pontecorvo, *Int. J. Nucl. Med. Biol.*, 2010, **51**, 913–920.
106. C. M. Clark, J. A. Schneider, B. J. Bedell, T. C. Beach, W. B. Bilker, M. A. Mintun, M. J. Pontecorvo, F. Hefti, A. P. Carpenter, M. L. Flitter, M. J. Krauthammer, H. F. Kung, R. E. Coleman, P. M. Doraiswamy, A. S. Fleisher, M. N. Sabbaghy, C. H. Sadowsky, P. E. M. Reiman, S. P. Zehntner and D. M. Skovronsky, *J. Am. Med. Assoc.*, 2011, **305**, 275–283.
107. N. Okamura and K. Yahai, *IDrugs*, 2010, **13**, 890–899.
108. Y. Lin, L. Zhu, K. Ploessi, S. R. Choi, H. Qiao, X. Sun, S. Li, Z. Zha and H. F. Kung, *Nucl. Med. Bio.*, 2010, **37**, 917–925.
109. C. H. Yao, K. J. Lin, C. C. Weng, I. T. Hsiao, Y. S. Ting, T. R. Jan, D. M. Skovronsky, M. P. Kung and S. P. Wey, *Int. J. Appl. Radiat. Isot.*, 2010, **68**, 2293–2297.
110. M. C. Hong, Y. K. Kim, J. Y. Choi, S. Q. Yang, H. Rhee, Y. H. Ryu, T. H. Choi, G. J. Cheon, G. I. An, H. Y. Kim, Y. Kim, D. J. Kim, J. S. Lee, Y. T. Chang and K. C. Lee, *Bioorg. Med. Chem.*, 2010, **18**, 7724–7730.
111. D. P. Flahery, T. Kiyota. Y. X. Dong, T. Ikezu and J. L. Vennerstrom, *J. Med. Chem.*, 2010, **53**, 7992–7999.

112. M. Ono, H. Watanabe, R. Watanabe, M. Haratake, M. Nakayama and H. Saji, *Bioorg. Med. Chem. Lett.*, 2011, **21**, 117–120.
113. Y. Cheng, M. Ono, H. Kimura, S. Kagawa, R. Nishii, H. Kawashima and H. Saji, *ACS Med. Chem. Lett.*, 2010, **1**, 321–325.
114. K. S. Lin, M. L. Debnath, C. A. Mathis and W. E. Klunk, *Bioorg. Med. Chem. Lett.*, 2009, **19**, 2258–2262.
115. M. Ono, Y. Fuchi, T. Fuchigami, N. Kobashi, H. Kimura, M. Haratake, H. Saji and M. Nakayama, *ACS Med. Chem. Lett.*, 2010, **1**, 443–447.
116. M. Ono, R. Ikeoka, H. Watanabe, H. Kimura, H. Fuchigami, T. Haratake, M. H. Saji and M. Nakayama, *ACS Chem. Neurosci.*, 2010, **1**, 598–607.
117. M. Cui, R. Tang, Z. Li, H. Ren and B. Liu, *Bioorg. Med. Chem. Lett.*, 2011, **21**, 1064–1068.
118. A. A. Mohammed, I. S. Shergill, M. T. Vandal and S. S. Gujral, *Expert Rev. Mol. Diagn.*, 2007, **7**, 345–349.
119. S. S. Chang and W. D. W. Heston, *Urol. Oncol.*, 2002, **7**, 7–12.
120. A. Zaheer, S. Y. Cho and M. G. Pomper, *Int. J. Nucl. Med. Biol.*, 2009, **50**, 1387–1390.
121. S. Vallabhajosula, S. J. Goldsmith, K. A. Hamacher, L. Kostakoglu, S. Konishi, SM. I. Milowski, D. M. Nanus and N. H. Bander, *Int. J. Nucl. Med. Biol.*, 2005, **46**, 850–858.
122. M. I. Milowsky, D. M. Nanus, L. Kostakoglu and C. E. Sheehan, *Am. J. Clin. Oncol.*, 2007, **25**, 540–547.
123. M. J. Morris, N. Pandit-Taskar, C. R. Divgi, S. Bender, J. A. O'Donoghue, A. Nacca, P .Smith-Jones, L. Schwartz, S. Slovin, R. Finn, S. Larson and H. I. Scher, *Clin. Cancer Res.*, 2007, **13**, 2707–2713.
124. J. P. Holland, V. Divilov, N. H. Bander, P. M. Smith-Jones, S. M. Larson and J. S. Lewis, *Int. J. Nucl. Med. Biol.*, 2010, **51**, 1293–1300.
125. R. Luthi-Carter, A. K. Barczak, H. Speno and J. T. Coyle, *J. Pharmacol. Exp. Ther.*, 1998, **286**, 1020–1025.
126. H. Tang, M. Brown, Y. Ye, G. Huang, Y. Zhang, Y. Wang, H. Zhai, X. Chen, T. Y. Shen and M. Tenniswood, *Biochem. Biophys. Res. Commun.*, 2003, **307**, 8–14.
127. J. Zhou, J. H. Neale, M. G. Pomper and A. P. Kozikowski, *Nat. Rev. Drug Discovery.*, 2005, **4**, 1015–1026.
128. A. P. Kozikowski, F. Nan, P. Conti, J. Zhang, E. Ramadan, T. Bzdega, B. Wroblewska, J. H. Neale, S. Pshenichkin and J. T. Wroblewski, *J. Med. Chem.*, 2001, **44**, 298–301.
129. P. F. Jackson, K. L. Tays, K. M. Maclin, Y. S. Ko, W. Li, D. Vitharana, T. Tsukamoto, D. Stoermer, X. C. M. Lu, K. Wozniak and B. S. Slusher, *J. Med. Chem.*, 2001, **44**, 4170–4175.
130. A. J. Oliver, O. Wiest, P. Helquist, M. J. Miller and M. Tenniswood, *Bioorg. Med. Chem.*, 2003, **11**, 4455–4461.
131. A. P. Kozikowski, J. Zhang, F. Nan, P. A. Petukhov, E. Grajkowska, J. T. Wroblewski, T. Yamamoto, T. Bzdega, B. Wroblewska and J. H. Neale, *J. Med. Chem.*, 2004, **47**, 1729–1738.

132. J. P. Mallari, C. J. Choy, Y. Hu, A. R. Martinez, M. Hosaka, Y. Toriyabe, J. Maung, J. E. Blecha, S. F. Pavkovic and C. E. Berkman, *Bioorg. Med. Chem.*, 2004, **12**, 6011–6020.

133. D. W. G. Wone, J. A. Rowley, AI. W. Garofalo and C. E. Berkman, *Bioorg. Med. Chem.*, 2006, **14**, 67–76.

134. P. Majer, B. Hin, D. Stoermer, J. Adams, W. Xu, B. R. Duvall, G. Delahanty, Q. Liu, M. J. Stathis, K. M. Wozniak, B. S. Slusher and T. Tsukamoto, *J. Med. Chem.*, 2006, **49**, 2876–2885.

135. M. O. Anderson, L. Y. Wu, N. M. Santiago, J. M. Moser, J. A. Rowley, E. S. D. Bolstad and C. E. Berkman, *Biorg. Med. Chem.*, 2007, **15**, 6678–6686.

136. L. Y. Wu, M .O. Anderson, Y. Toriyabe, J. Maung, T. Y. Campbell, C. Tajon, M. Kazak, J. Moser and C. E. Berkman, *Bioorg. Med.Chem.*, 2007, **15**, 7434–7443.

137. P. Ding, P. I. Helquist and M. J. Miller, *Org. Biomol. Chem.*, 2007, **5**, 826–831.

138. H. Wang, Y. Byun, C. Barinka, M. Pullambhatla, H. E. C. Bhang, J. J. Fox, J. Lubkowski, R. C. Mease and M. G. Pomper, *Bioorg. Med.Chem. Lett.*, 2010, **20**, 392–397.

139. C. Barinka, Y. Byun, C. L. Dusich, S. R. Banerjee, Y. Chen, M. Castanares, A. P. Kozikowski, R. C. Mease, M. G. Pomper and J. Lubkowski, *J. Med. Chem.*, 2008, **51**, 7737–7743.

140. H. Wang, Y. Byun, C. Barinka, M. Pullambhatia, H. C. Bhang, J. J. Fox, J. Lubowski, R. C. Mease and M. G. Pomper, *Bioorg. Med.Chem. Lett.*, 2010, **20**, 392–397.

141. C. Barinka, K. Hlouchova, M. Rovenska, P. Majer, M. Dauter,, N. Hin, Y. S. Ko, T. Tsukamoto, B. S. Slusher, J. Konvalinka and J. Lubkowski, *J. Mol. Biol.*, 2008, **376**, 1438–1450.

142. S. B. Rong, J. Zhang, J. H. Neale, J. T. Wroblewski, S. Wang and A. P. Kozikowski, *J. Med. Chem.*, 2002, **45**, 4140–4152.

143. M. I. Davis, M. J. Bennett, L. M. Thomas and P. J. Bjorkman, *Proc. Nat. Acad. Sci. U.S.A.*, 2005, **102**, 5981–5986.

144. J. R. Mesters, C. Barinka, W. Li, T. Tsukamoto, P. Majer, B. S. Slusher, J. Konvalinka and R. Hilgenfeld,, *EMBO J.*, 2006, **25**, 1375–1384.

145. C. Barinka, J. Starkova, J. Konvalinka and J. Lubkowski, *Acta. Crystallogr. F: Struct. Biol. Cryst. Commun.*, 2007, **F63**, 150–153.

146. J. R. Mesters, K. Henning and R. Hilgenfeld, *Acta Crystallogr. D: Biolog. Crystallogr.*, 2007, **D63**, 508–513.

147. C. Barinka, M. Rovenska, P. Mlcochova, K. Hlouchova, A. Plechanovova, P. Majer, T. Tsukamoto, B. S. Slusher, J. Konvalinka and J. Lubkowski, *J. Med. Chem.*, 2007, **50**, 3267–3273.

148. P. Mlcochova, A. Plechanovova, C. Barinka, D. Mahadevan, J. W. Saldanha, L. Rulisek and J. Konvalinka, *FEBS J.*, 2007, **274**, 4731–4741.

149. M. G. Pomper, J. L. Musachio, J. Zhang, U. Scheffel, Y. Zhou, J. Hilton, J. A. Maini, R. F. Dannals, D. F. Wong and A. P. Kozikowski, *Mol. Imaging*, 2002, **1**, 96–101.

150. C. A. Foss, R. C. Mease, H. Fan, Y. Wang, H. T. Ravert, R. F. Dannals, R. T. Olszewski, W. D. Heston, A. P. Kozikowski and M. G. Pomper, *Clin. Cancer Res.*, 2005, **11**, 4022–4028.

151. R. C. Mease, C. L. Dusich, C. A. Foss, H. T. Ravert, R. F. Dannals, J. Seidel, A. Prideaux, J. J. Fox, G. Sgouros, A. P. Kozikowski and M. G. Pomper, *Clin. Cancer Res.*, 2008, **14**, 3036–3043.

152. Y. Chen, C. A. Foss, Y. Byun, S. Nimmagadda, M. Pullambhatla, J. J. Fox, M. Castanares, S. E. Lupold, J. W. Babich, R. C. Mease and M. G. Pomper, *J. Med. Chem.*, 2008, **51**, 7933–7943.

153. K. P. Maresca, S. M. Hillier, G. J. Femia, D. Keith, C. Barone, J. L. Joyal, J. C. N. Zimmerman, A. P. Kozikowski, J. A. Barrett, W. C. Eckelman and J. W. Babich, *J. Med. Chem.*, 2009, **52**, 47–357.

154. S. M. Hillier, K. P. Maresca, F. J. Femia, J. C. Marquis, C. A. Foss, N. Nguyen, C. N. Zimmerman, J. A. Barrett, W. C. Eckelman, M. G. Pomper, J. L. Joyal, J. and W. Babich, *Cancer Res.*, 2009, **69**, 6932–6940.

155. S. R. Banerjee, C. A. Foss, M. Castanares, R. C. Mease, Y. Byun, J. J. Fox, J. Hilton, S. E. Lupold, A. P. Kozikowski and M. G. Pomper, *J. Med. Chem.*, 2008, **51**, 4504–4517.

156. S. R. Banerjee, M. Pullambhatia, Y. Byun, S. Nimmagadda, G. Green, J. J. Fox, A. Horti, R. C. Mease and M. G. Pomper, *J. Med. Chem.*, 2010, **53**, 5333–5341.

157. Y. Chen, S. Dhara, S. R. Banerjee, Y. Byun, M. Pullambhatia, R. C. Mease and M. G. Pomper, *Biochem. Biophys. Res. Commun.*, 2009, **390**, 624–629.

158. S. A. Kularatne, K. Wang, H. K. R. Santhapuram and P. S. Low, *Mol. Pharm.*, 2009, **6**, 780–789.

159. S. A. Kularatne, Z. Zhou, J. Yang, C. B. Post and P. S. Low, *Mol. Pharm.*, 2009, **6**, 790–800.

160. T. Liu, L. Y. Wu, M. Kazak and C. E. Berkman, *Prostate*, 2008, **68**, 955–964.

161. T. Liu, L. Y. Wu, J. K. Choi and C. E. Berkman, *Prostate*, 2009, **69**, 585–594.

162. T. Liu, L. Y. Wu, M. R. Hopkins, J. K. Choi and C. E. Berkman, *Bioorg. Med. Chem. Lett.*, 2010, **20**, 7124–7126.

163. S. E. Lapi, H. Wahnishe, D. Pham, L. Y. Wu, J. R. Nedrow-Byers, T. Liu, K. Vejdani, H. F. Van Brocklin, C. E. Berkman and E. F. Jones, *Int. J. Nucl. Med. Biol.*, 2009, **50**, 2042–2048.

164. X. Lu, D. Lu, M. Scully and V. Kakkar, *Perspect. Med. Chem.*, 2008, **2**, 57–73.

165. M. Binder and M. Trepel, *Expert Opin. Drug. Discov.*, 2009, **4**, 229–241.

166. H. M. Sheldrake and L. H. Patterson, *Curr. Cancer Drug Targets*, 2009, **9**, 519–540.

167. P. Clezardin, *Curr. Cancer Drug Targets*, 2009, **9**, 801–806.

168. J. S. Desgrosellier and D. A. Cheresh, *Nat. Rev. Cancer*, 2010, **10**, 9–22.

169. I. Dijkgraaf and O. C. Boerman, *Cancer Biother. Radiopharm.*, 2009, **24**, 637–647.

170. S. Liu, *Mol. Pharm.*, 2006, **3**, 472–487.
171. E. Garanger, D. Boturyn and P. Dumy, *Anti-Cancer Agents Med. Chem.*, 2007, **7**, 552–558.
172. W. Cai and X. Chen, *Int. J. Nucl. Med. Biol.*, 2008, **49** (Suppl. 2), 113S–128S.
173. M. Schottelius, B. Laufer, H. Kessler and H. J. Wester, *Acc. Chem. Res.*, 2009, **42**, 969–980.
174. P. M. Mitrasinovic, *Curr. Radiopharm.*, 2009, **2**, 214–219.
175. S. Liu, *Bioconjug. Chem.*, 2009, **20**, 2199–2213.
176. R. Haubner, H. J. Wester, U. Reuning, R. Senekowitsch-Schmidtke, B. Diefenbach, H. Kessler, G. Stocklin and M. Schwaiger, *Int. J. Nucl. Med. Biol.*, 1999, **40**, 1061–1071.
177. L. Auzzas, F. Zanardi, L. Battistini, P. Carta, G. Rassu, C. Curti and G. Casiraghi, *Curr. Med. Chem.*, 2010, **17**, 1255–1299.
178. T. H. Stollman, T. J. M. Ruers, W. J. G. Oyen and O. C. Boerman, *Methods.*, 2009, **48**, 188–192.
179. S. Liu, W. Y. Hsieh, Y. S. Kim and S. I. Mohammed, *Bioconjug. Chem.*, 2005, **16**, 1580–1588.
180. S. Liu, Z. He, W. Y. Hsieh, Y. S. Kim and Y. Jiang, *Bioconjug. Chem.*, 2006, **17**, 1499–1507.
181. C. Decristoforo, I. Santos, H. J. Pietzsch, J. U. Kuenstler, A. Duatti, C. J. Smith, A. Rey, R. Alberto, E. Von Guggenberg and R. Haubner, *Eur. Int. J. Nucl. Med. Biol. Mol. Imaging.*, 2007, **51**, 33–41.
182. Z. Liu, B. Jia, J. Shi, X. Jin, H. Zhao, F. Li, S. Liu and F. Wang, *Bioconjug. Chem.*, 2010, **21**, 548–555.
183. X. Chen, R. Park, A. H. Shahinian, M. Tohme, V. Khankaldyyan, M. H. Bozorgzadeh, J. R. Bading, R. Moats, W. E. Laug and P. S. Conti, *Nucl. Med. Biol.*, 2004, **31**, 179–189.
184. Y. S. Lee, J. M. Jeong, H. W. Kim, Y. S. Chang, Y. J. Kim, M. K. Hong, G. B. Rai, D. Y. Chi, W. J. Kang, J. H. Kang, D. S. Lee, J. K. Chung, M. C. Lee and Y. G. Suh, *Nucl. Med. Biol.*, 2006, **33**, 677–683.
185. T. Poethko, M. Schottelius, G. Thumshirn, U. Hersel, M. Herz, G. Henriksen, H. Kessler, M. Schwaiger and H. J. Wester, *Int. J. Nucl. Med. Biol.*, 2004, **45**, 892–902.
186. R. Haubner, H. J. Wester, W. A. Weber, C. Mang, S. I. Ziegler, S. L. Goodman, R. Senekowitsch-Schmidtke, H. Kessler and M. Schwaiger, *Cancer Res.*, 2001, **61**, 1781–1785.
187. A. J. Beer, R. Haubner, M. Goebel, S. Luderschmidt, M. E. Spilker, H. J. Wester, W. A. Weber and M. Schwaiger, *Int. J. Nucl. Med. Biol.*, 2005, **46**, 1333–1341.
188. Z. Wu, Z. B. Li, K. Chen, W. Cai, L. He, F. T. Chin, F. Li and X. Chen, *Int. J. Nucl. Med. Biol.*, 2007, **48**, 1536–1544.
189. Z. Liu, F. Wang, S. Liu and X. Chen, *Eur. Int. J. Nucl. Med. Biol. Mol. Imaging.*, 2009, **36**, 1296–1307.
190. Z. Wu, Z. B. Li, W. Cai, L. He, F. T. Chin, F. Li, and X. Chen, *Eur. J. Nucl. Med. Mol. Imaging.*, 2007, **34**, 1823–1831.

191. J. Shi, L. Wang, Y. S. Kim, S. Zhai, B. Jia, F. Wang, and S. Liu, *Eur. J. Nucl. Med. Mol. Imaging.*, 2009, **36**, 1874–1884.

192. J. Shi, Y. S. Kim, S. Chakraborty, B. Jia, F. Wang and S. Liu, *Bioconjug. Chem.*, 2009, **20**, 1559–1568.

193. L. Wang, J. Shi, Y. S. Kim, S. Zhai, B. Jia, H. Zhao, Z. Liu, F. Wang, X. Chen and S. Liu, *Mol. Pharm.*, 2009, **6**, 231–245.

194. J. Shi, l. Wang, Y. S. Kim, S. N. Zhai, Z. Liu, X. Chen and S. Liu, *J. Med. Chem.*, 2008, **51**, 980–7990.

195. J. Wang, Y. S. Kim and S. Liu, *Bioconjug. Chem.*, 2008, **19**, 634–642.

196. K. E. Gottschalk and H. Kessler, *Angew. Chem. Int. Ed. Engl.*, 2002, **41**, 3767–3774.

197. X. Lu, D. Lu, M. Scully and V. Kakkar, *Perspect. Med. Chem.*, 2008, **2**, 57–73.

198. A. Meyer, J. Auernheimer, A. Modlinger and H. Kessler, *Curr. Pharm. Des.*, 2006, **12**, 2723–2747.

199. L. L. Kiessling, J. E. Gestwicki and L. E. Strong, *Angew. Chem. Int. Ed. Engl.*, 2006, **45**, 2348–2368.

200. R. M. Owen, C. B. Carlson, J. Xu, P. Mowery, E. Fasella and L. L. Kiessling, *ChemBioChem.*, 2007, **8**, 68–82.

201. C. B. Carlson, P. Mowery, R. M. Owen, E. C. Dykhuizen and L. L. Kiessling, *ACS Chem. Biol.*, 2007, **2**, 119–127.

202. T. D. Harris, S. Kalogeropoulos, T. Nguyen, G. Dwyer, D. S. Edwards, S. Liu, J. Bartis, C. Ellars, D. Onthank, P. Yalamanchili, S. Heminway, S. Robinson, J. Lazewatsky and J. Barrett, *Bioconjug. Chem.*, 2006, **17**, 1294–1313.

203. I. Dijkgraaf, J. A. Kruijtzer, C. Frielink, A. C. Soede, H. W. Hilbers, W. J. Oyen, F. H. Corstens, R. M. Liskamp and O. C. Boerman, *Nucl. Med. Bio.*, 2006, **33**, 953–961.

204. B. S. Jang, E. Lim, S. Hee, I. S. Park, S. N. Shin, S. N. Danthi, I. S. Hwang, N. Le, S. Yu, J. Xie, K. C. Li, J. A. Carrasquillo and C. H. PaiK, *Nucl. Med. Bio.*, 2007, **34**, 363–370.

CHAPTER 3

Recent Developments in PET and SPECT Radioligands for CNS Imaging

DAVID ALAGILLE,* RONALD M. BALDWIN* AND
GILLES D. TAMAGNAN*

Molecular NeuroImaging, LLC/Institute of Neurological Disorders,
60 Temple Street, Suite 8A, New Haven, CT 95610, USA

3.1 Introduction

Functional radionuclide brain imaging with PET[a] (positron emission tomography) and SPET (single-photon emission tomography; SPECT in the US) provide powerful techniques of high sensitivity to measure *in vivo* neurochemistry by quantitative three-dimensional imaging. Each imaging modality offers distinct advantages. PET radionuclides have short half lives (*e.g.,* ^{11}C, $T_{1/2}$ 20 min, ^{18}F 110 min), and are produced with a local cyclotron. In contrast, SPET radionuclides have longer half lives (*e.g.,* ^{123}I, $T_{1/2}$ 13 h) and can be supplied commercially. Even though the intrinsic resolution of SPET is theoretically better than that of PET, in practical terms, PET instruments generally have higher sensitivity and resolution. The ultimate technical advantage from this standpoint depends on the properties of the radiotracer and the requirements of the imaging studies. Generally, studies in humans and nonhuman primates favor PET imaging, owing to the greater sensitivity and spatial resolution. In small animal studies, however, modern SPET cameras optimized for the purpose approach the theoretical intrinsic resolution of the detectors used. Painstaking studies of the benzodiazepine receptor,[1] and the dopamine

RSC Drug Discovery Series No. 15
Biomedical Imaging: The Chemistry of Labels, Probes and Contrast Agents
Edited by Martin Braddock

transporter,[2] have demonstrated that even in humans, SPET can yield valid quantitative outcome measures for specific pharmacological targets.

This review focuses on ligands for selected targets in the central nervous system (CNS), primarily the brain, which have enjoyed significant attention from pharmacologists and radiotracer imaging scientists because of their clinical relevance. For each target, a brief introduction to the target's function and pharmacology is followed by a discussion of ligands that have been discovered suitable for PET or SPET imaging. The emphasis is on the recent literature.

3.2 Amyloid Plaque

Alzheimer's disease (AD) is a progressive and irreversible neurodegenerative disorder affecting an estimated 35 million people worldwide. AD is characterized by three structural changes: a diffuse neuron loss, intracellular deposit of neurofibrillary tangles (NFT) mainly consisting of hyperphosphorylated tau protein, and extracellular deposit of senile plaque mainly consisting of amyloid beta peptide (Aβ). The etiology of the disease is largely unknown, but robust evidences implicating the amyloid cascade seems to be central in the development of the disease. Amyloid precursor protein (APP) is a ubiquitous 770 amino acid glycosylated transmembrane protein possessing a large extracellular N-terminal domain, a spanning transmembrane domain, and a short C-terminal cytoplasmic domain. While the physiological role of APP is still largely unknown, its role in amyloid plaque formation has been extensively studied. APP is the target of two aspartate proteases (β-secretase and γ-secretase); first the extracellular domain of APP is cleaved by β-secretase releasing a large ectodomain in extracellular fluid and leaving a 99-amino acid C-terminal membrane bound residue at the cell surface. This residue is cleaved by γ-secretase leading to the release of soluble Aβ peptide in extracellular fluid. This cleavage is not specific and three principal forms of Aβ are produced, containing either 38, 40 or 42 amino acids. The catabolism of APP is a normal physiological process and since the resulting Aβ fragments don't seem to have any physiological role it is hypothesized that they are rapidly catabolized in normal brain. However, upon certain, and still unknown conditions, the increase in production of Aβ (from either increased APP cleavage or reduction of Aβ catabolism) will induce peptide fibril formation and subsequent deposit of the insoluble Aβ oligomer hallmark of AD. The amyloid cascade hypothesis of AD is strongly supported by genetic observation. Familial mutation within the presenilin 1 and 2 genes, which are critical for the catalytic activity of γ-secretase, will cause an extremely aggressive and early form of AD. Duplication of the APP locus in extremely rare family mutation of chromosome 21 cause early onset of AD. It should be noted that the load of amyloid deposit in the brain correlates rather poorly with the cognitive deficit in AD, and a general consensus is emerging that soluble Aβ oligomers make up the toxic substance associated with cognitive impairment rather than plaque deposit. Imaging amyloid deposits in living brain has stimulated much research

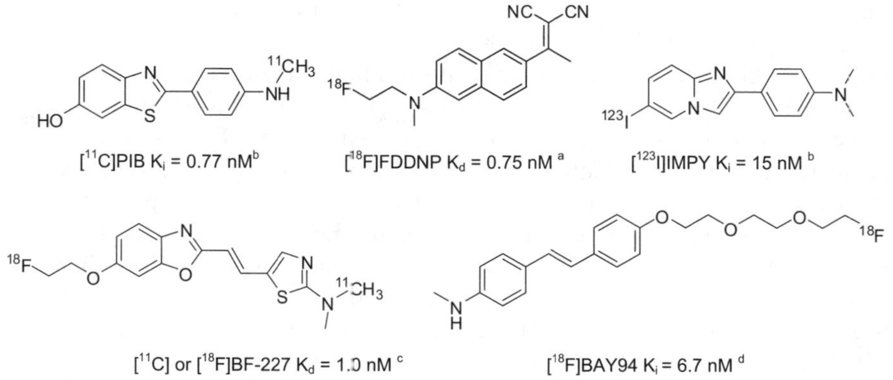

Figure 3.1 Ligands for Amyloid Imaging.
[a]Dissociation constant in homogenates of AD brain. [b]Inhibition constant using [^{125}I]TZDM on synthetic $A\beta_{1-40}$ aggregate. [c]Dissociation constant on synthetic $A\beta_{1-40}$ aggregate. [d]Inhibition constant using [^{125}I]IMPY in homogenates of AD brain.

in the past twenty years, with the goal of providing a reliable diagnostic for AD during the patient's life (until recently only post mortem autopsy was available). Since this goal has been achieved with the discovery of PIB, the new goal will be the early diagnosis of AD. For general reviews on the topic see bibliography.[3–8]

3.2.1 2-(4-([^{11}C]Methyl amino)phenyl)benzo[d]thiazol-6-ol ([^{11}C]PIB)

[^{11}C]PIB is by far the most studied amyloid imaging agent, and numerous reports have been published over the years. The following paragraph is not exhaustive and only represents a fraction of the work achieved with this tracer focusing mainly on human evaluation.

[^{11}C]PIB is synthesized in two steps: first, methylation of the aniline precursor MOM protected at the phenol position using [^{11}C]methyliodide, then deprotection of the phenol in methanolic HCl. The radiochemical yield is in the range of 10–15%, with specific activity averaging 25 GBq μmol^{-1} and radiochemical purity greater than 95%.[9] The first evaluation in humans of this tracer was reported in 2004.[10] The tracer shows fast brain uptake peaking around 4 SUV (standard uptake value) at 5 min post injection (pi), followed by a relatively fast washout. No differences in the TAC were observed between young and old healthy control (21 and 59–77 years of age respectively); in contrast, the AD group showed significantly higher retention of [^{11}C]PIB in cortical regions known to be amyloid-rich (parietal, frontal, temporal, occipital cortices) as well as in striatum, with similar uptake as healthy control in the other parts of the brain (cerebellum, subcortical white matter, pons) consistent with post mortem distribution of amyloid plaque in AD brain. Peripheral metabolism was fast and identical in both groups, with unchanged tracer accounting for 65–68% at 5 min, 30–33% at 12 min, and 7–9% at 60 min; no differences between venous and arterial sampling was observed and only very polar metabolites were detected, unlikely to cross the blood-brain-barrier.[9] Experiments in transgenic mice (both Tg2576 and double Tg PS1/APP strains) failed to demonstrate retention of [11C]PIB compared with wild type, even in elderly mice known to have higher amyloid loads than AD patients. This surprising discrepancy is presumably due to the difference in high affinity binding site of the amyloid deposit. *In vitro* studies revealed less than one binding site per 1000 Aβ molecules in transgenic mice *versus* more than 500 in post mortem AD brain.[11,12] The correlation between *in vivo* imaging with [11C]PIB and postmortem examination was demonstrated in two studies with patients undergoing an [11C]PIB scan followed by postmortem *in vitro* autoradiography, histological fluorescence staining, and immunohistochemistry. A strong correlation with both localization and concentration of the insoluble amyloid load was observed, consistent with the *in vivo* [11C]PIB signal being related to the amount of insoluble Aβ peptide.[13,14] However, in a recent report,[15] brain tissue from a terminal Alzheimer's patient showed no binding of [3H]PIB despite meeting post mortem histopathological criteria for AD, suggesting a selectivity of PIB for a certain polymorphic form of amyloid deposits.

Different kinetic modeling approaches have been reported for this tracer. The first thorough study was published in 2005 and was conducted in healthy controls, mild cognitive impairment (MCI) and AD subjects with a comparison of compartmental analysis and Logan graphical approach.[16] Using the cerebellum as reference region (nonspecific binding) a three-tissue 6-parameter, two-tissue 4-parameter, two-tissue 3-parameter, and one-tissue 2-parameter model were generated; the two-tissue compartment model provided the best fit and the two-tissue 4-parameter was significantly better. Logan analysis correlated well with this model and showed good test-retest reliability and lower inter-subject variability than compartmental analysis, and was therefore the modeling method of choice for [11C]PIB imaging. The use of reference tissue model without blood sampling using 3, 4, and 5 parameters showed a good fit and the 3 parameter was preferred since the 4 and 5 parameters did not provide significant improvement; the model was further refined by simultaneously fitting the model to time-activity curves in all regions of interest (ROI).[17] While the previous studies used the distribution volume ratio as the parameter of interest, binding potentials were calculated and compared using different parametric methods, and the best results were obtained using either receptor parametric mapping or multi-linear reference tissue models (RPM2, MRTM2, both with fixed k'$_2$).[18] Using a voxel principal component analysis (data analysis method extracting patterns) showed better discrimination between AD and healthy subjects than ROI analysis; the first two principal components accounted for 80% of the variability.[19] An automated method of estimation of the standardized uptake value (SUV) ratio was developed and proved better than the manual method to discriminate between AD and healthy controls, with significantly less variation within each group.[20] Partial volume effects and their correction were recently addressed by using partial volume estimate MRI segmentation (taking into account the cortical atrophy in AD subjects), resulting in better classification between AD and healthy subjects.[21] Longitudinal studies of Alzheimer's patients showed no significant differences in [11C]PIB uptake at baseline and two years follow-up scan, independently of the evolution of their mental status (stable AD *vs.* significant decline). However, a significant decrease in regional cerebral metabolic rate for glucose (20%) was observed in cortical regions at follow-up. These results seem to indicate a plateau of amyloid deposit at an early clinical stage of the disease, but a slowly increasing rate of amyloid deposit (not significantly detected in two years) could not be excluded.[22,23] In contrast, cognitive follow-up of PIB-positive and PIB-negative MCI patients over two years showed that 82% of the PIB-positive patients converted to clinical AD and only one patient in the PIB-negative group converted to AD. The conversion rate was faster (less than 1 year) for the PIB-positive patients showing higher [11C]PIB uptake in the anterior cingulate and frontal cortex (p = 0.027 and 0.031 respectively).[24] In a similar study subdividing patients with mild cognitive impairment (MCI) into single-domain amnesic, multi-domain amnesic, and non amnesic, all groups were associated with a high PIB uptake (46%, 83%, 43% respectively); the conversion rate to AD was 38% for the PIB-positive MCI group and 0% in the PIB-negative group.[25] Similar results were observed in non-demented elderly with a significant correlation of higher [11C]PIB retention in cortical regions and change in clinical dementia rating (0 to 0.5) over ten years.[26,27] Many studies have coupled [11C]PIB PET with other imaging techniques (MRI, [18F]FDG), resulting in a better prediction in the conversion from MCI to AD; whereas [11C]PIB provided good discrimination between PIB-positive (likely to evolve to AD) and PIB-negative (relatively spared from AD conversion) the second imaging technique allowed good discrimination in the PIB-positive group between AD conversion or not, since a strong correlation was observed between the AD conversion and the reduction of regional cerebral metabolic rate of glucose or their ventricular expansion.[28–31] [11C]PIB was also used to evaluate the efficacy of potential disease modifying treatment for AD; a phase 2, double-blind, placebo-controlled of bapineuzumab (Aβ amyloid antibody) was recently reported and showed a low but significant reduction of [11C]PIB retention after 78 weeks in the treated group compared to placebo (−0.25 cortex to cerebellum ratio) and baseline (−0.08 cortex to cerebellum ratio), indicating the usefulness of this tracer in drug development for Alzheimer's disease.[32] [11C]PIB was also evaluated in pathologies other than Alzheimer's disease. Evaluation in a very restrained number of patients with atypical dementia syndrome (primary progressive aphasia, PPA, n = 1, posterior cortical atrophy, PCA, n = 1) showed similar tracer retention as the AD group (significantly higher than control) but with higher binding in the occipital cortex for the PCA patients compared to AD, whereas the PPA patient presented with higher accumulation in the left cerebral hemisphere (laterality and symmetry of [11C]PIB was studied in AD and shown to be symmetric bilaterally).[33,34] In frontotemporal lobar degeneration (FTLD), four out of twelve subjects presented a PIB-positive scan (7/7 for the AD group, 1/7 for healthy controls) and further FDG scans indicated likely AD for two of them and a likely FTLD for the remaining two.[35] In variant AD patients (4555-bp deletion encompassing exon 9 of PS1 gene), [¹¹C]PIB distribution was significantly increased for all subjects in caudate (45%) and putamen (41%) compared to healthy age-matched controls; the authors did not include an AD group, but the distribution pattern in variant AD seems different from that described for sporadic AD.[36] Evaluation in Parkinson's disease dementia (PDD) subjects showed increased binding in cortical regions (AD-like) for two out of ten patients compared with healthy subjects, but the eight remaining subjects showed low but significant increase in pons and mesencephalon uptake compared to controls and the AD group; this specific signal was confirmed *in vitro* in PD (substantia nigra) and PDD (brainstem) using fluorescence microscopy.[37] [¹¹C]PIB was also evaluated in cerebral amyloid angiopathy and showed increased binding in cortical regions compared to controls but lower than the AD group except for a small increase in uptake in the lobar regions.[38]

3.2.2   2-(1-(6-((2-[¹⁸F]fluoroethyl)(methyl)amino)-2-naphthyl)ethylidene)malononitrile ([¹⁸F]FDDNP)

[¹⁸F]FDDNP was synthesized by nucleophilic substitution of its tosylate precursor by [¹⁸F]fluoride in 20–25% yield with specific activity of 2000–6000 Ci mmol^{-1}.[39,40] The synthesis was further optimized and automated to satisfy the need of larger quantity for human use.[41,42] [¹⁸F]FDDNP is one of the few tracers able to label both senile plaque (mainly Aβ and neurofibrillary tangles (mainly hyperphosphorylated and microtubule-associated tau protein deposit) as confirmed by autoradiography, wide-field fluorescence microscopy, and immunostaining.[43] As for [¹¹C]PIB, evaluation of this tracer in transgenic mice models of AD (Tg2576) showed no significant differences with wild type *in vivo*. However, contrary to the case of [¹¹C]PIB, autoradiographic studies demonstrated specific binding of [¹⁸F]FDDNP in this mouse strain,[44,45] and *in vivo* imaging in triple transgenic rats showed increased uptake *versus* wild type.[46] In its first human evaluation, this tracer showed fast brain uptake, peaking around 5 min pi, followed by relatively fast washout, despite a relatively high lipophilicity (log P = 3.92). Comparison of relative residence time (RRT) between AD and healthy controls showed significantly higher RRT in the AD group, especially in the hippocampus, amygdala, and entorhinal regions as well as temporal lobes; a good inverse correlation was observed between the RRT values and the mini-mental state exam test score (MMSE).[47,48] Using a Logan graphical method to estimate the relative distribution volume (DVR) with the cerebellum as reference region, a relatively large group of AD (n = 25), MCI (n = 28), and healthy controls (n = 30) were imaged with [¹⁸F]FDDNP, [¹⁸F]FDG, and MRI. The global DVR values (average of all ROI) as well as each ROI permit a clear and significant distinction between the three groups with AD > MCI > healthy control (1.07 control, 1.12 MCI, 1.16 AD, global value, p < 0.001); the best diagnostic accuracy between groups was achieved by [^{18}F]FDDNP scan using the global value as compared to FDG or MRI scan when using the receiver-operating-characteristic analysis.[49] In a similar study using an MRI-derived cortical surface model with a four-dimensional animation technique a strong correlation between the composite cognitive score and the signal intensity in the entorhinal, orbitofrontal, lateral temporal cortices, temporoparietal and perisylvian language area, parietal association cortices, and much of the dorsolateral prefrontal cortex.[50,51] In non-demented patients, three factors (impaired cognitive status, older age and APOE-4 carrier) have been shown to increase [^{18}F]FDDNP binding in all ROI (parietal, temporal cortices, lateral temporal, posterior cingulate, frontal regions) and in the impaired cognitive status group the APOE-4 carrier showed higher medial temporal uptake whereas a normal aging APOE-4 carrier show higher frontal uptake.[52] Different kinetic models were evaluated and the simplified reference tissue model appears to be the model of choice.[53] However, better results (less affected by noise, more precise BP_{ND} estimate) were recently reported using receptor parametric mapping and multi-linear reference tissue model with fixed k'$_2$.[54] Comparison of [^{11}C]PIB and [^{18}F]FDDNP in AD and healthy subjects showed higher uptake (45–76% increase in AD group *vs.* control) for PIB than for FDDNP (11–16% increase in AD group *vs.* control). The distribution patterns of the two tracers were significantly different, especially in the medial temporal cortex where FDDNP showed higher binding increase than PIB, whereas PIB showed higher increase in the neocortical region. To explain this discrepancy, the authors suggest a preferential affinity of [^{18}F]FDDNP for neurofibrillary tangles and a preferential affinity of [^{11}C]PIB for amyloid plaque.[55] A recent *in vitro* binding study comparing [^{3}H]FDDNP and [^{3}H]PIB showed weak binding of [^{3}H]FDDNP to amyloid plaque (K_d = 85 nM), ten-fold lower than that of [^{3}H]PIB, and no labeling of sections containing neurofibrillary tangles by [^{3}H]FDDNP could be observed.[56] These results are very different from earlier reports using [^{18}F]FDDNP,[48] who reported a K_d of 0.75 nM towards neurofibrillary tangles. Re-examination of this tracer affinity should be a priority before further conclusions of *in vivo* studies can be drawn. [^{18}F]FDDNP is quickly metabolized in humans, with only 20% unchanged tracer in plasma after 10 min. The radiolabeled metabolites could be reproduced *in vitro* by incubation of the tracer with human hepatocytes; *in vivo* studies in rats showed that they are capable of penetrating the brain with a uniform distribution.[57] Acute toxicity as evaluated in rats (0–5 mg kg^{-1}) at the highest dose produced mortality of 3 out of 10 rats accompanied with liver damage in those animals.

3.2.3 6-[^{123}I]iodo-2-(4'-dimethylamino)phenyl-imidazo[1,2-*a*]-pyridine ([^{123}I]IMPY)

This tracer was synthesized by iododestannylation of its tributyltin precursor using hydrogen peroxide as oxidizing agent, in radiochemical yield of 20–60%, with specific activity of 2200 Ci mmol^{-1} and radiochemical purity greater than

95%. Evaluation in normal mice showed high brain uptake, reaching 2.9% ID at 2 min pi, followed by fast washout from all brain regions (0.2% ID @ 60 min).[58–61] In transgenic mice (Tg2576), *ex vivo* brain uptake at 4 h pi was 3.3 times higher than in wild type, and autoradiography (thioflavin-S fluorescence) clearly demonstrated the accumulation of [^{123}I]IMPY in amyloid plaque, with a low background even in the white matter.[58] Similar results were observed with double transgenic mice expressing both Tg2576 and PS1 M146L, with increased uptake of 1.79 fold *versus* wild type at 30 min (at the same time the increase was 2.38 for Tg2576 only) and localization of the amyloid-rich part of the brain by autoradiography.[62] The regional brain distribution in this double transgenic model of AD showed higher uptake in all ROI than wild type with the highest ratio (transgenic uptake/wild type uptake) for cortex and hypothalamus (greater than 2), followed by striatum and cerebellum.[63] *In vivo* imaging using micro-PET on transgenic mice was unsuccessful due to the low sensitivity of the pinhole collimator used in this study.[64] *In vitro* evaluation in human AD brain shows a good localization of the amyloid deposit with a low binding to the white matter.[65] Biodistribution and dosimetry in humans showed the highest radiation dose estimates for the gallbladder (0.135 mGy MBq^{-1}) and the lower and upper large intestine wall (0.115 and 0.109 mGy MBq^{-1}, respectively) and the effective dose equivalent (EDE) was estimated at 0.035 mSv MBq^{-1}, with no apparent differences between men, women and AD subjects.[66] [^{123}I]IMPY has been evaluated in human and nonhuman primate studies but no reports have been published to date.

3.2.4   2-(2-(2-Dimethylaminothiazol-5-yl)ethenyl)-6-(2([18F]-fluoro)ethoxy)benzoxazole ([18F]BF227) and 2-(2-(2-N-methyl-N-[11C]methyl-aminothiazol-5-yl)ethenyl)-6-(2-(fluoro)ethoxy)benzoxazole ([11C]BF-227)

[^{11}C]BF227 was synthesized by methylation of its desmethyl precursor using [^{11}C]methyl triflate in 50% yield, with specific activity between 119–128 GBq μmol^{-1} and radiochemical purity exceeding 95%.[67] [^{18}F]BF227 was synthesized by nucleophilic displacement of its tosylate precursor with [^{18}F]fluoride in 17% yield, with specific activity of 42 GBq μmol^{-1} and radiochemical yield superior to 95%.[68,69] In normal mice [^{11}C]BF227 show a fast brain uptake peaking at 7.9% ID g^{-1} at 2 min pi followed by a fast washout (0.64% ID g^{-1} @ 60 min) and both radionuclide demonstrate good staining of the amyloid load in AD brain sections.[70] *Ex vivo* autoradiography in rats infected with Aβ$_{1–40}$ showed good labeling at the site of the lesion and low uptake in the rest of the brain. Acute dosing of 0.1–10 mg kg^{-1} i.v. did not produce any significant change in behavior or post-mortem pathological examination compared to controls.[68] Evaluation in humans showed rapid brain uptake reaching almost 3 SUV at 5 min pi, followed by relatively fast washout from all brain regions; a clear distinction between AD and healthy age-matched controls was revealed by comparing the mean SUV ratio to cerebellum from the frontal, lateral temporal, parietal, temporo-occipital, occipital, anterior and posterior cingulate cortices, with no overlapping between the two groups in the lateral temporal cortex. By comparison, no significant differences were observed between young and elderly healthy subjects.[67] Comparison in AD, MCI, and age-matched controls showed higher binding in the order AD > MCI > control, reaching significance in the lateral temporal, parietal, mean neocortex, frontal and occipital cortex, with a higher specificity and sensitivity than FDG scan.[71] In a similar study applying [11C]BF227 and MRI in healthy, MCI, MCI converted to AD, and AD groups showed significant distinction between AD and MCI converted compared to MCI and healthy groups using [11C]BF227 binding in the neo cortex; better discrimination between MCI and MCI converted was observed with [11C]BF-227 than with MRI.[72] Early results with [11C]BF227 are encouraging but remain to be confirmed by other groups.

3.2.5 *trans*-4-(N-Methylamino)-4'-(2-(2-(2-[18F]fluoroethoxy)-ethoxy)stilbene ([18F]BAY94–9172 or [18F]florbetaben)

[18F]BAY94 was synthesized in two steps, first labeling by nucleophilic displacement of its N-Boc protected tosylate precursor using [18F]fluoride, then deprotection of the Boc with aqueous hydrochloric acid. Overall radiochemical yield was around 30%, with specific activity of 1,300–1,500 Ci mmol^{-1} and radiochemical purity exceeding 99%. Evaluation in wild type mice showed fast brain uptake, reaching 7.7% ID at 2 min pi and followed by a relatively fast washout (1.6% ID @ 60 min); autoradiography in transgenic mice model of AD (Tg2576) following injection of [18F]BAY94 showed distinctive labeling of amyloid deposits, confirmed by co-staining with thioflavin-S; similar results were observed in post mortem AD brain slices.[73] Similarly, fast brain uptake (up to 4 SUV) followed by a relatively fast washout from all brain regions was observed in humans, with slower clearance from the neocortex region of AD patients than of healthy controls. All AD subjects presented extensive cortical uptake especially in frontal and posterior cingulate cortex, whereas fronto-temporal lobar degeneration (FTLD) patients showed low uptake in those cortical regions. Comparison of neocortical SUV ratio to cerebellum in AD, FTLD, and healthy controls showed higher uptake in the AD group (2.02), whereas both FTLD and healthy control had similar lower uptake (around 1.29), leading to clear discrimination between AD and FTLD or healthy control subjects.[74–76] Whole body imaging in humans showed high uptake in the liver, urinary bladder, upper and lower intestine, and spleen with obvious renal and liver excretion. The highest absorbed radiation dose was found in the gall bladder (132 μGy MBq^{-1}), liver (39 μGy MBq^{-1}), and urinary bladder (24 μGy MBq^{-1}), resulting in an EDE of 14.67 μSv MBq^{-1} (30% lower than FDG scan).[77] Plasma analysis showed fast metabolism in both healthy control and AD, with an initial half life around 6 min, and only more polar metabolites detected (the N-desmethyl analogue was detected in small proportion).[78] This tracer is currently in phase 2/phase 3 clinical trial development.

3.3   Metabotropic Glutamate Receptors

Glutamate is the main excitatory neurotransmitter in the central nervous system (CNS) and acts through two classes of receptors. Fast excitatory transmission at glutamate synapses is mediated by the ligand-gated ionotropic receptors (iGluRs), consisting of α-amino-3-hydroxy-5-methyl-4-isoxazole propionic acid (AMPA), kainate (KA), and *N*-methyl-D-aspartate (NMDA) receptors. iGluRs are permeable to both potassium and sodium, and in the case of NMDA receptors and some AMPA receptors, calcium ions. In contrast, metabotropic glutamate receptors (mGluRs) mediate slower responses through G-proteins, coupled to various transduction cascades. Group I mGluRs (mGluR1 and 5) are positively associated with phospholipase C *via* $G_{q/11}$ proteins, leading to phosphoinositide hydrolysis and formation of two intracellular second messengers: inositol tri-phosphate (IP$_3$), which induces intracellular Ca^{2+} release, and diacylglycerol (DAG), which can activate protein kinase C. The mGluR subtypes belonging to Group II (mGluR2 and 3) and group III (mGluR4, 6, 7 and 8) are negatively coupled to adenylate cyclase *via* $G_{i/o}$ proteins and will therefore decrease the intracellular concentration of cAMP. In addition, glutamate can, through the mGluRs, modulate excitatory (AMPA and NMDA) and inhibitory (GABA) signaling pathways as well as other ion channel receptors. Group I mGluR are mainly found postsynaptically whereas group II and III are presynaptical and play an important role in the modulation of neurotransmission. All mGluR possess a large bi-lobed extracellular N-terminal domain (characteristic of the C family of GPCR, including GABAB, pheromone receptor and calcium sensing receptors) containing the glutamate binding site. Pharmacological ligand of mGluR are divided into two groups, the orthosteric ligand which binds compe-titively with glutamate in the bi-lobed N-terminal domain and the allosteric ligand binding within the seven transmembrane domain of the receptor. Orthosteric ligands bind in a much conserved part of the receptor and therefore subtype selective ligandsare difficult to develop, and the few selective compounds are amino acid analogues which greatly limits their use due to poor CNS availability. On the other hand, allosteric ligands have proven extremely useful, being subtype-selective and readily entering the brain. mGluR1 are found in high concentration in the cerebellum, hypothalamus, thalamus, and hippocampus and previously implicated in pain, neurodegeneration and psychiatric disorders including schi-zophrenia. mGluR5s are found in high concentration in the frontal cortex, cau-date, putamen, nucleus accumbens, olfactory tubercle, and hippocampus and have been implicated in a variety of CNS disorders including depression, schi-zophrenia, anxiety, Parkinson's disease, and drug addiction as well as Down syndrome and autism where mGluR5 might become the first pharmacological tool for those diseases. For general reviews on the topic, see bibliography.[79–87]

3.3.1   mGluR1

The field of mGluR1 imaging is only recently emerging, and to date the only reports of mGluR1 tracers are poster presentations with limited information.

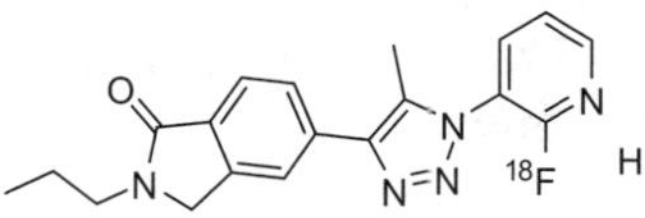

[^{18}F]MK-1312 [a]   [^{18}F]MK-4908 [a]

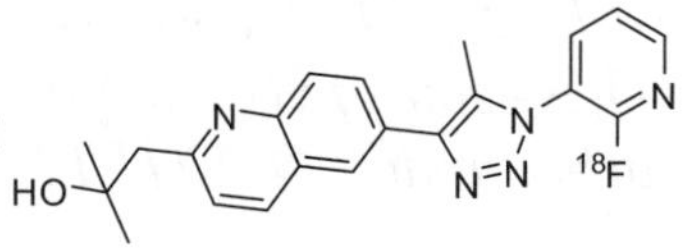

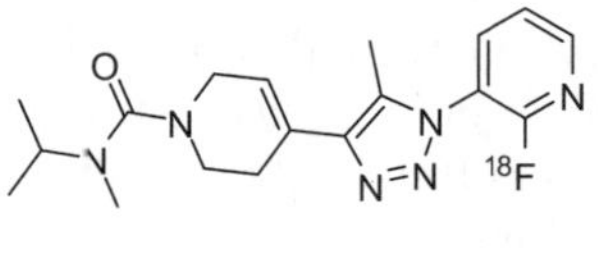

[^{18}F]FTIDC IC$_{50}$ = 5.8 nM [b]

Figure 3.2   mGluR1 Ligands.
[a]No pharmacological data published to date. [b]Half maximal inhibitory concentration using [^{3}H]*L*-quisqualate on CHO cells expressing human mGluR1a.

3.3.1.1   5-(1-(2-[^{18}F]Fluoropyridin-3-yl)-5-methyl-1H-1,2,3-triazol-4-yl)-2-propylisoindolin-1-one ([^{18}F]MK-1312)

This tracer showed fast brain uptake in monkey, reaching 3 SUV at 20 min pi, followed by gradual washout. The regional brain distribution matched the one of mGluR1 with the highest activity in the cerebellum > thalamus > cortex > striatum. Pretreatment with MK-5435 (desfluoro analog) resulted in a significant reduction of signal in all brain regions; however, a structurally different mGluR1 antagonist (MK-7643) failed to displace the signal fully from certain brain regions (thalamus and midbrain). Occupancy estimates of MK-7643 (130 ng ml^{-1} in plasma) using [^{18}F]MK-1312 were 95%, 91%, 62%, and 65% in cortex, cerebellum, striatum, and thalamus, respectively.[88]

3.3.1.2   1-(6-(1-(2-[^{18}F]Fluoropyridin-3-yl)-5-methyl-1H-1,2,3-triazol-4-yl)quinolin-2-yl)-2-methylpropan-2-ol ([^{18}F]MK-4908)

This tracer showed brain uptake peaking at around 2 SUV at 30 min, followed by a slow washout in all brain regions except for the cerebellum. In cerebellum the tracer accumulate over the duration of the scan reaching 4.5 SUV at 90 min. The regional distribution was in accordance with mGluR1 distribution, with the order of tracer accumulation being cerebellum ≫ thalamus > cortex > striatum. Pretreatment with the closely related analog MK-7643 resulted in significant reduction of signal in all brain regions and occupancy estimates of MK-7643 (120 ng ml^{-1} in plasma) were 90%, 90%, 84%, and 84% in cortex, cerebellum, striatum, and thalamus respectively.[88]

3.3.1.3   4-(1-(2-[^{18}F]Fluoropyridin-3-yl)-5-methyl-1H-triazol-4-yl)-N-isopropyl-N-methyl-5,6-dihydropyridine-1(2H)-carboxamide ([^{18}F]FTIDC)

This tracer was synthesized by halogen exchange from its bromo precursor and [^{18}F]fluoride. The tracer showed high accumulation in rat mGluR1-rich regions of the brain (cerebellum > thalamus > hippocampus) similar to immuno-chemical staining. Co-injection with unlabeled ligand greatly reduced tracer accumulation and the specific binding was estimated at 98% in mGluR1-rich regions. Maximum uptake was found in cerebellum at 10 min and was 2–3 fold higher than other brain regions.[89]

3.3.2   Metabotropic Glutamate Type 5 (mGluR5) Receptor

3.3.2.1   3-(6-Methylpyridin-2-yl ethynyl)-cyclohex-2-enone-O-[^{11}C]methyloxime ([^{11}C]ABP688)

This tracer was synthesized by methylation with [^{11}C]methyliodide of the sodium salt of the desmethyl precursor, in 35% yield with specific activity of 232 GBq μmol^{-1} and radiochemical purity exceeding 95%. *Ex vivo* evaluation in rats showed good brain uptake of the tracer with a distribution in accordance with mGluR5 distribution in this species; the highest activity at 30 min was found in the striatum (0.22% ID g) and hippocampus (0.19% ID g). Intermediate uptake was observed in the cortex (0.15% ID g) and the lowest levels were found in the cerebellum (0.03% ID g). Pretreatment with the selective mGluR5 antagonist M-MPEP (1 mg kg^{-1}) resulted in significant decrease in brain uptake to cerebellar levels, and the resulting estimate of specific binding was around 80% in all mGluR5 rich regions. Wild type mice showed similar results in brain uptake and regional distribution of the tracer, whereas mGluR5 knockout mice showed lower brain uptake (cerebellar wild type levels) in all brain regions. [^{11}C]ABP688 was extensively metabolized, with 25% parent compound in plasma after 30 min; all metabolites were more polar than the tracer, and 95% of the activity in the brain was carried by the tracer.[90] Evaluation in healthy humans showed high and rapid brain uptake peaking at 4 min pi (extraction fraction of 0.87), followed by a gradual washout from all brain regions. Quantification using a two-tissue compartment model with arterial input showed high receptor concentration in the anterior cingulate (DV = 5.4), caudate (DV = 5.1), putamen (DV = 4.9), posterior cingulate (DV = 4.5) and the lowest in the brainstem and cerebellum (DV = 1.9), in good agreement with mGluR5 *ex vivo* studies. The tracer was quickly metabolized, with the intact parent compound representing 64%, 44%, 36%, 28, and 25% of the plasma activity at 5, 10, 15, 30, 40, and 60 min pi.[91] A kinetic modeling study in which one-tissue compartment, two-tissue compartment and Logan plots were compared in a bolus or bolus plus constant infusion paradigm, resulted in good fit of the experimental data by the two-tissue compartment and an excellent correlation between Logan plot and the total

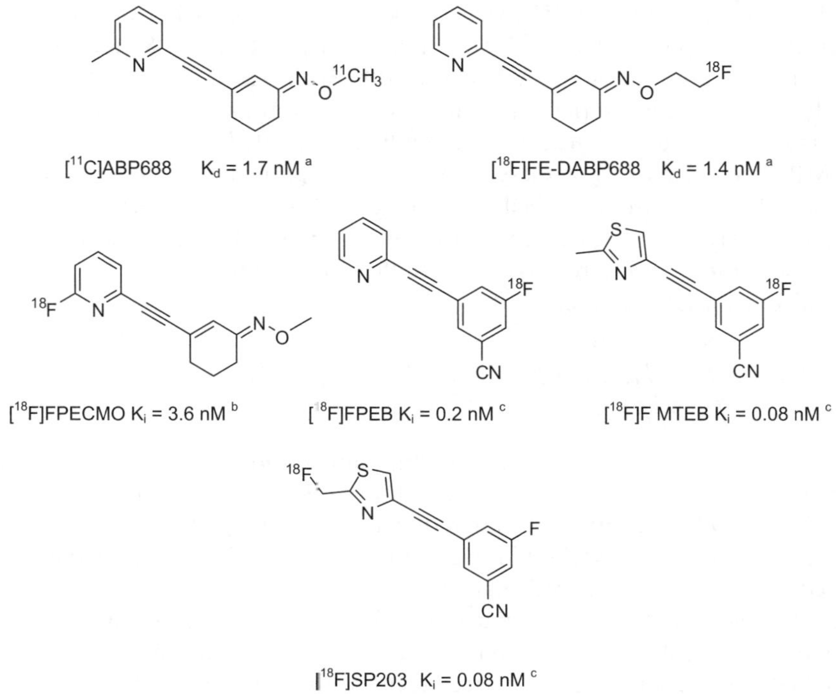

[^{11}C]ABP688 K_d = 1.7 nM [a]

[^{18}F]FE-DABP688 K_d = 1.4 nM [a]

[^{18}F]FPECMO K_i = 3.6 nM [b]

[^{18}F]FPEB K_i = 0.2 nM [c]

[^{18}F]F MTEB K_i = 0.08 nM [c]

[^{18}F]SP203 K_i = 0.08 nM [c]

Figure 3.3 mGluR5 Ligands.
[a]Dissociation constant on whole rat brain homogenate minus cerebellum. [b]Inhibition constant using [^{3}H]M-MPEP on whole rat brain homogenate minus cerebellum. [c]Inhibition constant on rat cortical membrane using [^{3}H]M-PEPy.

distribution volume calculated with two-tissue compartment. The minimum scan duration to obtain stable results was estimated at 45 min.[92] Whole body imaging in humans showed a rapid excretion of the tracer through the hepatobiliary pathway and urinary tract, leading to the highest absorbed dose for the liver (0.0164 mSv MBq^{-1}), gall bladder (0.0813 mSv MBq^{-1}) and kidney (0.0072 mSv MBq^{-1}), with an EDE of 0.0037 mSv MBq^{-1}, well within the limits recommended by international guidelines.[93]

3.3.2.2 3-(Pyridin-2-yl ethynyl)-cyclohex-2-enone-O-[^{18}F]fluoroethyloxime ([^{18}F]FE-DABP688) and 3-((6-[^{18}F]fluoropyridin-2-yl)ethynyl)cyclohex-2-enone O-methyl oxime ([^{18}F]FPECMO)

[^{18}F]FE-DABP688 was synthesized by alkylation with 2-[^{18}F]fluoroethyl tosylate of the sodium salt of the desmethyl precursor in 25% yield, with specific activity of 30 GBq μmol^{-1} and radiochemical purity greater than 98%.[94] [^{18}F]FPECMO was synthesized by nucleophilic displacement with [^{18}F]fluoride of the bromo precursor in 35% yield, a specific activity >240 GBq μmol^{-1} and radiochemical purity >99%.[95] Evaluation of [^{18}F]FE-DABP688 in rats showed fast brain uptake followed by fast washout, with accumulation of tracer in mGluR5-rich regions of the brain. However, the striatum to cerebellum ratio of 2 was relatively low and the specific binding (estimated by pretreatment with M-MPEP at 1 mg kg^{-1}) was less than 50%, well below the values obtained for [^{11}C]ABP688 (striatum/cerebellum around = 6.6, 80% specific signal).[94] [^{18}F]FPECMO was even more disappointing, with only 40% specific signal (from autoradiography results) and insufficient accumulation in the brain to visualize the mGluR5 receptor.[95]

3.3.2.3   3-[^{18}F]Fluoro-5-(2-pyridinylethynyl)benzonitrile ([^{18}F]FPEB)

This tracer was synthesized by nucleophilic substitution of its chloro precursor under microwave irradiation, using [^{18}F]fluoride in 5% yield with specific activity of 1900 mCi μmol^{-1} and radiochemical purity greater than 98%.[96] Evaluation in rats showed rapid brain uptake (1.8% ID g @ 20–30 min), followed by slow washout; regional brain distribution was in agreement with mGluR5 distribution. Highest uptake was found in the striatum, hippocampus, and olfactory bulb and low levels in the cerebellum and cortex. The signal specificity was confirmed by pretreatment with either mGluR5 antagonists MPEP or MTEP (10 mg kg^{-1}, MTEP was more efficient than MPEP) resulting in a decreased uptake in all brain regions, whereas the mGluR1 antagonist YM-298198 did not affect the tracer uptake or distribution. Metabolite analysis showed only 20% unchanged [^{18}F]FPEB in the plasma at 10 min pi. Detected metabolites were polar and deemed unlikely to penetrate the brain.[97] Evaluation in monkey showed rapid brain uptake peaking in the striatum at 4.2 SUV at 55 min, followed by a relatively slow washout. Uptake in the cerebellum was relatively high but the washout was faster than for mGluR5-rich regions. The highest activity was found in the striatum and frontal cortex and the lowest in the white matter. Pretreatment with MTEP (3 mg kg^{-1}) resulted in a significant decrease of uptake in all brain regions and the total/nonspecific activity ratio was of 8.5 at 85 min. [^{18}F]FPEB was relatively stable in blood and up to 62% of the plasma activity was accounted by unchanged radiotracer 90 min pi. *In vitro* liver microsome stability assay indicated an even lower metabolic rate in human than monkey.[96] Discrete species differences in both K_d and B_{max} between human, rhesus, and rat were reported using saturation experiments with [^{18}F]FPEB; low but quantifiable levels of mGluR5 receptors were detected in human and rhesus cerebellum.[98] Biodistribution studies in nonhuman primates revealed hepatobiliary excretion of the tracer with high uptake in the liver and the gallbladder. Estimated absorbed radiation dose in humans showed the highest dose to the upper large intestine (0.18–0.20 mGy MBq^{-1}, male-female) and small intestine (0.16–0.18 mGy MBq^{-1}, male-female). The EDE was estimated at 0.033 and 0.034 mGy MBq^{-1} (male-female), similar to many currently used ^{18}F labeled tracers.[99]

3.3.2.4 3-[^{18}F]Fluoro-5-[(2methyl-1,3-thiazol-4-yl)ethynyl]-benzonitrile ([^{18}F]F-MTEB)

[^{18}F]F-MTEB was synthesized by nucleophilic substitution of the chloro precursor under microwave irradiation by [^{18}F]fluoride in 4% yield with specific activity of 1,350 Ci mmol^{-1} and radiochemical purity greater than 98%. This tracer behaved very similar to [^{18}F]FPEB both *in vitro* (autoradiography) and *in vivo* (brain uptake, distribution, washout, ratio to cerebellum, specific signal); however, [^{18}F]F-MTEB seemed to be metabolically less stable, with only 2% (*versus* 30% for [^{18}F]FPEB) of unchanged parent compound in monkey liver microsomes assay at 30 min.[96,100]

3.3.2.5 3-Fluoro-5-(2-(2-[^{18}F]fluoroethylthiazol-4-yl)ethynyl)-benzonitrile ([^{18}F]SP203)

This tracer was synthesized by nucleophilic displacement of its bromomethyl precursor by [^{18}F]fluoride in 87% yield with specific activity up to 9 Ci µmol^{-1}, and radiochemical purity >99%. Evaluation in monkey showed high and fast brain uptake up to 700% SUV at 10 min pi followed by gradual washout (450% SUV at 180 min). Regional brain distribution was in accordance with mGluR5 distribution (striatum ≈ hippocampus > occipital cortex > cerebellum after 90 min); however a relatively high uptake was reported in the cerebellum, washing out faster than in the other brain regions but higher than striatum and hippocampus before 30 min, and occipital cortex before 90 min. Pretreatment with MPEP (5 mg kg^{-1}) significantly decrease the uptake in all brain regions, and an estimate of 60% specific binding was made for the striatum and hippocampus in the time interval 60–120 min. Only one more polar metabolite was detected in plasma and accounted for nearly all the activity after 25 min. Incubation of [^{18}F]SP203 in human or monkey blood gives very different results with only around 25% of unchanged tracer in monkey plasma after 25 min, while up to 98% of unchanged tracer was detected in human plasma after 90 min; incubation in human or monkey brain homogenates for 90 min afford only 1.4% and 1.8% respectively of radioactive metabolite.[101] Relatively high and increasing mandibular bone uptake was observed, reaching 400% SUV at 180 min, presumably due to accumulation of [^{18}F]fluoride. This hypothesis was investigated thoroughly by LC-MS/MS analysis of SP203 carrier and [^{18}F]SP203, revealing that [^{18}F]fluoride was produced by glutathionylation of the tracer (substitution of the [^{18}F]fluoride by the sulfur of glutathione, presumably catalyzed by glutathione transferase) the glutathione-SP203 conjugate was clearly detected.[102] Evaluation in healthy volunteers showed fast brain uptake, peaking at 580% SUV at 5 min, followed by gradual washout (50% peak reached 90–180 min in all 7 subjects); quantification was best achieved using an unconstrained two-tissue compartment with plasma input and showed highest binding in temporal cortex (DV 25.8); frontal cortex (22.7), cingulate (22.7), putamen (22.3), caudate (21.7), and the lowest in thalamus (15.8), and cerebellum (14.2) with small inter-subject variation.[103]

3.4   Monoamine Transporter Targets

The concept of active reuptake of neurotransmitter from the synaptic cleft to the pre synaptic neuron was introduced in the mid 1950's by J. Axelrod. Since then a plethora of research teams have studied this crucial physiological process and lead the way to the discovery of some of the most prescribed drugs for mental disorders, Prozac and Ritalin. All three monoamine transporters (dopamine transporter DAT, norepinephrine transporter NET, serotonin transporter SERT, containing 620, 617, and 630 residues respectively) share a putative 12 transmembrane domain placing the N and C terminal intracellularly, and characterized by a large extracellular loop between TMD3 and TMD4 possessing potential *N*-glycosylation sites. Human DAT gene is carried by chromosome 5p15.3 and does not seem to have any splice variant. Human NET gene is carried by chromosome 16q12.2 and three splice variants have been characterized differing in their C terminal region. Human SERT gene is carried by chromosome 17q11 and no splice variants have been identified. Co-transport of Na^+ and Cl^- ions is a requirement for substrate uptake of the three transporters. Whereas NET and SERT require one Na^+ and one Cl^-, DAT requires two Na^+ and one Cl^-. It has been postulated that binding of Na^+ and Cl^- produces the conformational change leading to the translocation of the complex transporter-substrate-Na^+-Cl^-. In the brain, all three monoamine transporters are localized in presynaptic neurons with a perisynaptic location, and they are almost exclusively expressed on their substrate neuron (DAT on dopaminergic, NET on noradrenalinergic, SERT on serotoninergic neurons). DAT is highly expressed in striatum, nucleus accumbens, prefrontal cortex, and thalamus (from neurons originating in the substantia nigra and ventral tegmenta); NET is mainly expressed in the hippocampus and cortex (from neurons originating in the brainstem); SERT is more widely expressed in the brain with higher density in the midbrain, thalamus, hypothalamus, cerebral cortex and hippocampus. The critical role of monoamine transporters in the regulation of neurotransmission and transmitter homeostasis has been highlighted by the development of monoamine knockout (KO) animals. DAT-KO mice showed a 300 fold increase in persistence of dopamine in the synaptic junction, revealing that monoamine transport is the main mechanism to terminate neurotransmission (similar results have been reported for NET-KO and SERT-KO mice). Increasing evidence indicates that monoamine transporters regulate and can be regulated by complex interactions with many proteins (protein kinase C, phospatidylinositol 3 kinase, protein phosphatase 2A) revealing the high degree of complexity of those transporters. For general reviews on the topic, see bibliography.[104–123]

3.4.1   Dopamine Transporter (DAT)

The dopamine system and dopamine transporters in particular have been heavily investigated targets for *in vivo* radiotracer imaging. A selection of the field is given here.

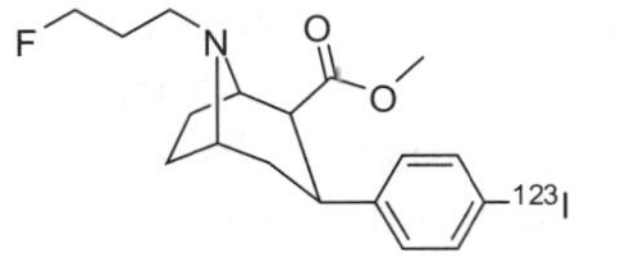

$[^{123}I]$FP-CIT K_i = 3.50 nM [a]

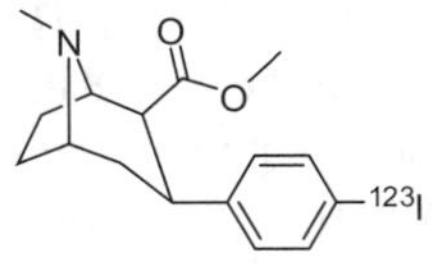

$[^{123}I]β$-CIT K_i = 1.40 nM [a]

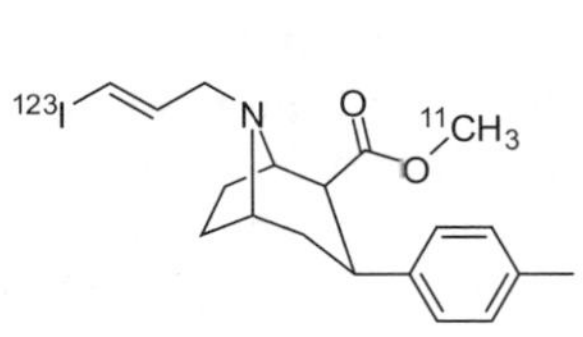

$[^{123}I]$ or $[^{11}C]$PE2I K_d = 0.09 nM [b]

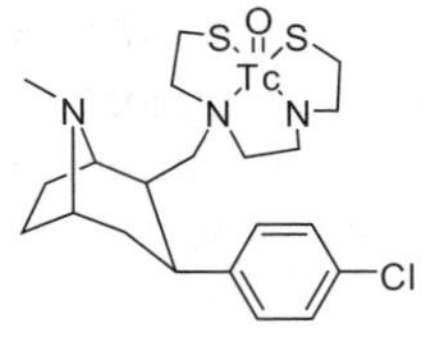

$[^{99m}Tc]$TRODAT-1 K_i = 14.1 nM [d]

K_d = 3.8 nM [c]

Figure 3.4 Dopamine Transporter Ligands.
[a]Inhibition constant using $[^3H]$GBR 12935 on rat forebrain membrane homogenates. [b]Dissociation constant on rat striatal membrane homogenate. [c]Dissociation constant on CHO cells expressing human DAT. [d]Inhibition constant of Re-TRODAT-1 using $[^{125}I]$IPT on rat striatal membrane homogenate.

3.4.1.1 *N-[ω-Fluoropropyl-2β-carbomethoxy-2β-(4-[^{123}I]-iodophenyl)tropane) ([^{123}I]FP-CIT, ioflupane)*

This tracer was synthesized by iododestannylation of its trimethylstannyl precursor using peracetic acid as oxidizing agent in 64% yield, with specific activity exceeding 1,850 Mbq μmol^{-1} and radiochemical purity greater than 98%.[124] Evaluation in baboon showed a relatively slow brain uptake, peaking at 12% ID at around 30 min pi followed by gradual washout from low DAT density brain region (midbrain, occipital cortex, cerebellum washout rate of 16%/h) and a slow washout from high DAT density regions (striatum washout rate of 4.0%/h); with a ratio specific to non specific striatal uptake (defined as (striatum-occipital)/occipital) reaching 6 at 3 h pi, and increasing up to 9 at the end of the scan (5 h). In the periphery major uptake in the liver, intestine, and urinary bladder revealing both renal and hepatobiliary excretion and the highest absorbed radiation dose was estimated for the urinary bladder (0.7 rad mCi^{-1}) and lower large intestine wall (0.9 rad mCi^{-1}).[125] The signal specificity was demonstrated in rats: both pretreatment and displacement with GBR12,909 (5mg kg^{-1}, i.v. selective DAT inhibitor) significantly reduced $[^{123}I]$FP-CIT striatal uptake with no influence on other brain regions, whereas pretreatment and displacement with fluvoxamine (5 mg kg^{-1} i.v., selective SERT inhibitor) resulted in significant decrease in hypothalamus uptake without change in the rest of the brain. Further evaluation in unilateral MPTP-treated monkeys revealed a striatum: occipital ratio of 1 on the side of the lesion and 4 on the untreated side.[126] Similar results were observed in healthy humans, with high uptake peaking at 6.37% ID for the striatum at 30 min pi with no significant washout during the next 8 h; uptake in midbrain and occipital cortex was lower (around 3% ID), peaking at 15 min pi and followed by a relatively fast washout for the first two hours of the scan (washout rate 71% per h) and then stabilized until the end of the scan (8 h).[127] Evaluation of [123I]FP-CIT in Parkinson's disease patients showed decreased uptake in the striatum, with higher reduction in the putamen (57%) than the caudate (29%). In patients with unilateral symptoms the reduction was greater on the contralateral side and the contralateral putamen leads to the best discrimination between PD and healthy group, but a significant reduction on both sides was observed. In the healthy control group a good correlation between age and reduction of striatal uptake was observed and estimated at around 10% per decade. Good correlation between specific/non specific uptake (mean striatum, putamen, and caudate as well as ipsi or contralateral) and duration and severity of the disease (Hoehn and Yahr I-V, UPDRS) was observed. Correlation between striatal uptake ratio and bradykinesia was better than with tremor but both showed relatively low correlation.[128–131] [123I]FP-CIT imaging provided good reliability and reproducibility in both healthy and PD groups.[132] Human biodistribution revealed high initial uptake in liver and lungs and lower levels in the brain and intestine all persisting after 48 h. The radiation dose estimates were the highest for the urinary bladder (0.054 mGy MBq^{-1}), lungs (0.043 mGy MBq^{-1}), lower and upper large intestine (0.042 and 0.038 mGy MBq^{-1}, respectively) with an EDE of 0.024 mSv MBq^{-1}, allowing high-count imaging with radiation exposure within the international safety recommendation.[133,134] Automated and semi-automated quantification methods were developed to evaluate [123I]FP-CIT binding using statistical parametric mapping (SPM), anatomical standardization method, ROI, ordered subset expectation maximization, and filtered back projection methods. All led to good quantification of the uptake and similar or better results than visual examination in the discrimination between PD and healthy subjects in a faster manner, allowing large scale imaging with this ligand.[135–141] While this tracer clearly shows usefulness in the diagnostic and follow-up of Parkinson's disease patients, it was also evaluated in other pathologies. Dementia with Lewy bodies (DLB) is often confused with Alzheimer's disease, which is problematic since those patients are often hypersensitive to antipsychotic medication but respond well to anticholinesterase treatment. However, one of the major differences in those pathologies is the loss of presynaptic dopamine transporters in DLB but not in AD, which can be visualized by use of [123I]FP-CIT. Significant reduction in caudate and putamen binding was observed in DLB subjects compared to AD, with a sensitivity between 77–88% (depending on the studies) and a selectivity between 94–100% for the diagnosis of DLB *versus* AD (definite diagnostic was confirmed at autopsy in one study); however, no differences could be seen among DLB, PD, and PD with dementia, all showing similar reduction in tracer uptake.[142–144] In the case of cerebrovascular disease (CVD), which can evolve to vascular Parkinsonism (VP) and is not easily distinguished from idiopathic Parkinson's disease clinically, the comparison of striatal uptake led to a clear identification of VP *versus* CVD patients evolving towards nigrostriatal dopamine degeneration.[145] The utility of [^{123}I]FP-CIT was further demonstrated in inconclusive or possible Parkinsonism (suspicion of drug-induced parkinsonism, young onset, unresponsive to dopaminergic therapy, atypical symptoms) identifying with 100% accuracy the PD group and around 90% accurate diagnostic was made in the inconclusive parkinsonism (around 10% of non PD were diagnosed with PD at follow-up).[146,147] A recent review on the topic has been published.[148]

Certain prescription drugs can affect the uptake of this DAT tracer, and this issue needs to be taken into account during the design of future studies. In a double-blind, placebo-controlled study using paroxetine (selective serotonin reuptake inhibitor) in healthy subjects, a 10% increase in specific striatal: occipital binding ratio was observed.[149] In rats, pretreatment with the anti-psychotic drug haloperidol showed significant reduction (25%) of striatal uptake.[150] A review on possible drug interaction with this tracer was recently published.[151]

3.4.1.2 2β-Carbomethoxy-3β-(4-[^{123}I]iodophenyl)tropane ([^{123}I]β-CIT)

This tracer is synthesized by iododestannylation of its trimethyl or tributyl tin precursor using either peracetic acid or Chloramine-T as oxidizing agent in 60% yield, with specific activity greater than 5,000 Ci mmol^{-1} and radio-chemical purity exceeding 95%.[152] Evaluation in baboon showed the highest uptake for the striatum peaking at 179 min, followed by the hypothalamus and substantia nigra peaking earlier at 45 min pi; no washout, or extremely low washout from the striatum was observed at the end of the scan (420 min), whereas a more rapid washout was observed in the other brain regions ($T_{1/2}$ = 294 min). Displacement study with non selective DAT-SERT inhibitors (cocaine, CFT, indatraline) all results in decrease binding in both striatum and hypothalamus, in contrast the selective SERT inhibitor citalopram produced selective displacement in hypothalamus with only marginal reduction in striatum.[153,154]

Whole body distribution and dosimetry were evaluated in humans up to 48 h pi; the highest initial uptake was observed in the lungs with subsequent distribution in the liver, intestine, and brain. The highest absorbed dose estimates were found for lungs (0.1 mGy MBq^{-1}), liver (0.087 mGy MBq^{-1}), and large intestine (0.05 mGy MBq^{-1}), with an effective dose equivalent (EDE) estimate of 0.048 mGy MBq^{-1}.[155]

Evaluation in PD and healthy controls revealed much slower kinetics than in baboon. The PD group reached striatal equilibrium in 1 day (peaking around 3 to 4 h, then stable values up to 2 days), the striatal uptake of the control group continued to increase until 8–15 h and stabilized at this levels for up to 2 days; in both cases the activity of the other brain regions washed out at much faster rates. Distinction between the two groups is evident at day 1 (255 min pi, caudate and putamen ratio to occipital, caudate 1.4 fold, putamen 1.8 fold) but became even more pronounced at day two (1330 min pi, caudate 2.4 fold, putamen 4.2 fold) with no overlap between the two groups. The extent of dopaminergic neuron loss between the two groups was estimate between 50–70%, lower than values obtained by post-mortem analysis, presumably due to the less advance stage of the disease of the subjects (Hoehn and Yahr stage 1–3).[156] In a similar study with patients at different stages, 36% reduction was observed for stage I, and 71% at stage V.[157] In patients with unilateral symptoms the striatal uptake was reduced in both sides, with a more pronounced loss in the contralateral (53%) than ipsilateral side (38%) and with a greater loss in the putamen than caudate (1.5 ratio caudate/putamen on the ipsilateral side).[158] Test-retest showed good reproducibility at both 7 h and 24 h for both PD and healthy group.[159,160] A significant correlation was found between striatum/cerebellum ratio (especially on the ipsilateral side) and the stage of the disease as measured by Hoehn-Yahr or UPDRS I-III, but no significant correlation was observed with the tremor rating.[161–163] The influence of current Parkinson therapy (L-Dopa/carbidopa, L-selegiline) in binding and distribution of this tracer was evaluated in drug naïve patient (scan 1) after 4–6 weeks of medication (scan 2) and after withdrawal 1–9 weeks (scan 3) and no significant differences were observed between the three scans. Similarly, the dopamine agonist pergolid did not significantly affect the uptake of $[^{123}I]\beta$-CIT, but a slight trend toward higher striatal uptake (8%) was observed.[164,165] The rate of loss of dopaminergic neurons in PD patients was evaluated and ranged 4.5–12% per year, whereas only 0.8% per year was observed in a healthy control group; the rate loss seems constant over time with no significant differences between the first five and seven years.[166–168] Longitudinal studies of Parkinson therapy incorporating $[^{123}I]\beta$-CIT imaging in addition to clinical assessment with a large cohort of participants have been reported. The CALM-PD-CIT was designed to compare the treatment outcomes of pramipexole *versus* carbidopa/levodopa; in this study a significant lower rates of loss was observed in the pramipexole group than levodopa (reduction in striatal uptake: 7.1% *vs.* 13.5% @ 22 month, 10.9% *vs.* 19.6% @ 34 month, 16.0% *vs.* 25.5% @ 46 month).[169,170] Similar results were observed in the ELIDOPA randomized, double-blind, placebo-controlled trial for levodopa, where significant higher decline in striatal uptake was observed in the levodopa group than placebo (− 7.2%, − 4%, − 6%, − 1.4% at daily dose of 600 mg, 300 mg, 150 mg, placebo) despite improvement of clinical symptoms.[171,172] This tracer was also evaluated in other pathologies. In Tourette's syndrome, a significant increase in striatal V_3'' was observed (mean 37%), in accordance with post mortem observation.[173] In a small cohort of Wilson's disease patients, significant reduction in striatal binding of $[^{123}I]\beta$-CIT compared to control was reported, to a similar extent as the PD group; however, the loss was symmetrical and more pronounced in the caudate than putamen.[174] Discrimination between PD, multiple system atrophy (MSA), progressive supranuclear palsy (PSP), and corticobasal degeneration (CBD) were evaluated; in all cases a significant reduction of striatal tracer uptake was observed, with a pronounced asymmetry in the PD and CBD group and a less pronounced, non-significant, asymmetry in MSA and PSP goup.[175,176] However, applying statistical parametric mapping to the entire brain revealed significant decrease in tracer binding in the midbrain of MSA and PSP patients compared to PD, allowing 91–95% correct classification of PD *versus* PSP or MSA.[177,178] In a small cohort of 13 patients with clinical vascular parkinsonism, no differences in striatal uptake *versus* control were observed in all but one subject, whereas a classical reduction was observed in all PD patients, suggesting the usefulness of this tracer in the diagnosis of VP.[179]

3.4.1.3 *N-(3-[^{123}I]Iodoprop-2E-enyl)-2β-carbomethoxy-3β-(4-methylphenyl)nortropane ([^{123}I]PE2I) and N-(3-iodoprop-2E-enyl)-2β-[^{11}C]carbomethoxy-3β-(4-methylphenyl)nortropane ([^{11}C]PE2I)*

[^{123}I]PE2E was synthesized by iododestannylation of its tributyltin precursor using chloramine-T (hydrogen peroxide and Iodogen have also been reported) as oxidizing agent in 60–85% yield, with specific activity around 9 TBq µmol^{-1} and radiochemical purity greater than 98%.[180] [^{11}C]PE2I was prepared by alkylation of its carboxylic acid precursor with [^{11}C]methyl triflate in acetone in 49–74% yield,[181] with specific activity of 30–40 GBq mmol^{-1} and radiochemical purity 98%. Evaluation in monkey revealed fast brain uptake peaking at 4% ID at 5 min pi in all parts of the brain except the striatum, which reached maximum uptake of 6% ID at 10 min pi. Washout was relatively fast from all brain regions except the striatum, where the washout was very slow and reached equilibrium within 1 h. Displacement with carrier PE2I (1 mg kg^{-1} i.v.) produced a rapid and pronounced washout of striatal activity, whereas only a small and slow displacement was observed in the thalamus, and no effect on cerebellar uptake were detected. Similar results were observed by pretreatment with β-CIT (0.5 mg kg^{-1} i.v.) and cocaine (2 mg kg^{-1} i.v.). Pretreatment with cocaine or GRB12909 resulted in significant reduction of striatal uptake as well as an increase in washout rates in all parts of the brain; whereas citalopram (5 mg kg^{-1} i.v.) or maprotiline (5 mg kg^{-1} i.v.) did not produce significant changes.[181–183] Evaluation in MPTP treated monkey showed reduction in striatal uptake to cerebellar levels.[184] Whole body distribution in humans showed highest uptake in the liver, urinary bladder, kidney and stomach. For [^{11}C]PE2I, the highest absorbed dose estimate were found for the urinary bladder (0.018 mGy MBq^{-1}), kidneys (0.016 mGy MBq^{-1}), and stomach (0.014 mGy MBq^{-1}), with an EDE of 6.6 µSv MBq^{-1}.[185] For [^{123}I]PE2I, the highest radiation exposure was also to the urinary bladder (0.07 mGy MBq^{-1}) and the EDE was 0.022 mSv MBq^{-1}.[186] Kinetic modeling and quantification in human of [^{123}I]PE2I was evaluated after bolus injection using compartmental analysis (one and two tissue), area under the curve or Logan plot; the latest was preferred since it gives similar results than the other methods without blood sampling.[187,188] More precise quantification has been achieved using a bolus and constant infusion paradigm (B: I ratio of 2.7 h). Outcome measures were similar to those from bolus studies using a kinetic analysis (BP1 = 21.0 *vs.* 21.1, BP2 = 4.3 *vs.* 4.1 for bolus/infusion *vs.* bolus).[189] Using this bolus plus constant infusion protocol, good test-retest reproducibility was observed in striatal binding potential.[190] For [¹¹C]PE2I binding potential, kinetic compartment analysis (three compartment fitted better than two) with plasma input resulted in 30% higher values than a reference tissue approach (cerebellum as reference), presumably due to an over-correction of the plasma input or the presence of metabolites in the brain.[191] Recently, a thorough study evaluating different kinetic (1TC and 2TC iteratively or not, basis pursuit, iterative 2TC-constrained algorithm), graphical (Logan, likelihood estimation in graphical analysis (LEGA)), and non-invasive (simplified reference tissue model, blood less Logan and LEGA) model was published, and the best fit of the TAC was observed for the 2TC noniterative and the basis pursuit models.[192] Evaluation in PD patients revealed significant reduction of striatal uptake in the PD *versus* control group within 60 min; the reduction was more pronounced in the contralateral side of hemilateral patient, and the reduction was slightly higher in caudate than putamen (mean 48% and 44% for caudate and putamen).[188] In a group of young onset Parkinson disease (YOPD, with and without Parkin mutation) a significant reduction of binding potential was observed (56% caudate, 41% putamen) *versus* control, but no differences were detected between the two mutation groups.[193] In a pilot study, attention-deficit/hyperactivity disorder adolescent showed low but significant reduction of binding in the midbrain (−16%) as well as a non-significant increase in binding in both caudate (7%) and putamen (8%), which should be further investigated.[194]

3.4.1.4   (2-((2-(((3-(4-chlorophenyl)-8-methyl-8-azabicyclo-[3.2.1]oct-2-yl)methyl)(2-mercaptoethyl)amino)ethyl)-amino)-ethanethiolato-(3)-N2,N2',S2,S2')oxo-(1R-(exo-exo))-[⁹⁹mTc]technetium ([⁹⁹mTc]TRODAT-1)

[⁹⁹mTc]TRODAT-1 was obtained by *trans*-chelation with [⁹⁹mTc]Tc-gluconate generated by reduction of sodium [⁹⁹mTc]pertechnetate with stannous chloride with the soft N_2S_2 chelating tropane precursor, in 80% yield, with radiochemical purity 98%.[195,196] An improved purification method was introduced, leading to increased radiochemical purity with a simple setup.[197] Evaluation in human showed relatively slow brain uptake peaking at 25–30 min pi followed by a gradual washout from all brain regions (slowest rate for the DAT rich striatum).[198] The striatum delineation became apparent after background washout 60–110 min, and improved with time; the caudate:occipital and putamen: occipital ratios reached 2.85 and 2.15 respectively at 280 min pi.[199,200] In PD patients, a clear decrease in [⁹⁹mTc]TRODAT-1 was observed in the striatum, with no overlap in the posterior putamen compared to age-matched healthy controls (sensitivity 100%, specificity 96%). The uptake reduction was more pronounced in the contralateral side of the symptoms (contra putamen–81%, ipsi putamen–67%, contra caudate–46%, ipsi caudate–40%); and a good correlation of the uptake ratio was found with the stage of the disease (both H&Y and UPDRS).[201–204] Comparison between early-onset Parkinson's disease and late-onset with the same disease duration show a significant reduction (-34%) in the early-onset group.[205] Human biodistribution revealed the highest absorbed dose estimate for the liver (0.047 mGy MBq^{-1}), kidneys (0.035 mGy MBq^{-1}), and upper large intestine (0.028 mGy MBq^{-1}), with an effective dose of 0.053 rad mCi^{-1}.[206] Test-retest showed good reproducibility with a mean variability around 10% in PD patients.[207] Differentiation between PD and vascular Parkinsonism (VP) was accomplish with this tracer; the VP group show a slightly lower striatal uptake (not significant) than healthy control, and are clearly distinguishable from the PD group.[208] Similar results were observed between PD and essential tremor patients, with sensitivity and selectivity of 96% and 91%, respectively, between the two diagnostics.[209] Evaluation of untreated ADHD patients revealed increased [^{99m}Tc]TRODAT-1 striatal uptake when compared to healthy control, and upon methylphenidate treatment this uptake was shifted below control levels; good correlation between the reduction of tracer uptake during treatment and improvement of the symptoms was observed, as well as a good prediction of the treatment outcome from the initial tracer uptake unresponsive patients present lower initial DAT levels than the responsive group.[210–212] Tourette's syndrome patients were evaluated with this tracer; however, no differences were found when compared with aged matched control group.[213–215]

3.4.2 Norepinephrine Transporter (NET)

3.4.2.1 (R)-N-[^{11}C]Methyl-3-(2-methoxyphenoxy)-3-phenyl-1-propylamine/[N-^{11}C]nisoxetine, and (R)-N-Methyl-3-(2-[^{11}C]methoxyphenoxy)-3-phenyl-1-propylamine ([O-^{11}C]nisoxetine)

Both tracers were synthesized by alkylation of their nor precursors with [^{11}C]methyliodide in 63–72% and 23–29% yield ([N-^{11}C]nisoxetine and [O-^{11}C]nisoxetine respectively) with specific activity 1.7–3.7 and 2.1–2.5 Ci µmol^{-1} and radiochemical purity exceeding 99%. Evaluation of the racemic [N-^{11}C]nisoxetine in mouse showed fast brain uptake peaking at 2 min pi (0.71% ID g whole brain), followed by a gradual washout. Regional brain distribution revealed rank order of hypothalamus > cortex > thalamus $\approx$ striatum, reaching a maximum hypothalamus:striatum ratio of 1.96 and cortex:striatum ratio of 1.72 at 20 min pi. Pretreatment with nisoxetine (7 mg kg^{-1} i.v.) resulted in significant reduction of tracer uptake in cortex and hypothalamus. Evaluation in baboon showed higher distribution volume in the occipital and basal ganglia than hypothalamus, indicating high non-specific binding; pretreatment with nisoxetine only resulted in 10–20% decrease in hypothalamic DVR.[216–218]

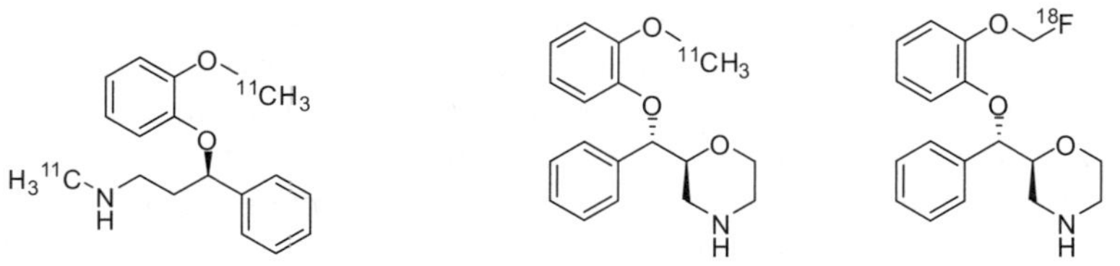

N or *O*[^{11}C]Nisoxetine   K$_d$ = 1.63 nM [a]   [^{11}C]MeNER K$_i$ = 2.5 nM [b]   [^{18}F]FMeNER K$_i$ = 2.5 nM [c]

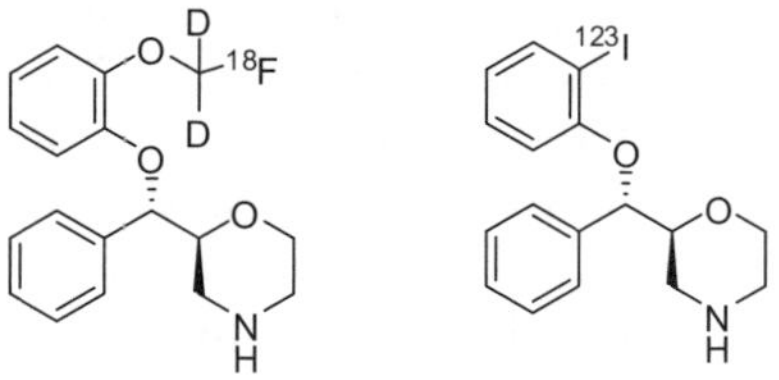

[^{18}F]FMeNER-D2 K$_d$ = 3.6 nM [c]   [^{123}I]INER K$_i$ = 2.5 nM [d]

Figure 3.5   Norepinephrine Transporter Ligands.
[a]Dissociation constant on rat vas deferens membrane homogenate. [b]Dissociation constant using [^{3}H]nisoxetine on HEK-293 cells expressing human NET. [c]Dissociation constant on human locus coeruleus slice. [d]Inhibition constant using [^{3}H]nisoxetine on rat forebrain homogenate.

3.4.2.2   *(S,S)-2-((2-[^{11}C]Methoxyphenoxy)phenylmethyl)-morpholine ([^{11}C]MeNER or [^{11}C]MRB)*

This tracer was synthesized either in one step by methylation of its desmethyl precursor by [^{11}C]methyliodide, or in two steps, first methylation of the desmethyl-*N*-Boc protected precursor with [^{11}C]methyltriflate, then deprotection of the *N*-Boc amide with trifluoroacetic acid. Both methods produced almost quantitative yield, with specific activity around 74 GBq μmol^{-1}, and radiochemical purity >99%. *In vitro* and *in vivo* evaluation in rats of each enetiomeric pair (*S,S* and *R,R*) clearly indicates the higher affinity of the *S,S* configuration (the *R,S* and *S,R* pair show even more reduced affinity).[219,220] Evaluation in baboons showed fast but moderate brain uptake, peaking around 0.02% ID ml^{-1} at 10 min pi and followed by a relatively fast washout from the low NET density regions, and relatively slow from NET rich brain region. The signal specificity was evaluated by pretreatment with nisoxetine (0.5 mg kg^{-1}, i.v.) and results in a significant uptake reduction in the thalamus and cerebellum, whereas no effects were observed in the cerebellum, nor by pretreatment with GRB12909 (selective DAT inhibitor). Metabolite analysis reveals only more polar metabolites and a relatively high level of unchanged tracer in plasma over time (98%, 59%, 41%, 30%, 26% @ 1, 5, 10, 30, 60 min respectively).[218,221,222] Similar results were observed in human with the highest uptake detected in the thalamus and midbrain, pretreatment with various dose of atomoxetine results in significant decrease of the tracer binding, but no significant differences in DVR between each doses were found (25, 50, 100 mg) reflecting a low signal-to-noise ratio for this tracer.[223] Moreover, this tracer was shown to not always reach its peak, resulting in non stable binding potential, during a two hour scan.[224] Evaluation of the effect of age and cocaine consumption in human show a significant reduction of NET density in the locus coeruleus, hypothalamus, and pulvinar with aging; comparison of BP_{ND} between cocain user and age match control reveals a significant NET upregulation in the cocaine user group, reaching 63% in the locus coeruleus and 55% in the pulvinar (aged normalized).[225] Metabolite analysis in human shows a good stability of the tracer with 88% and 82% of unchanged tracer at 4 and 40 min respectively; a transient more lipophilic metabolite was detected in human, accounting for 9% and 1% of the plasma activity at 4 and 40 min.[226]

3.4.2.3 *(S,S)-2-(α-(2-[^{18}F]Fluoromethoxyphenoxy)-benzyl)morpholine ([^{18}F]FMeNER) and (S,S)-2-(α-(2-[^{18}F]Fluoro-dideutero-methoxyphenoxy)benzyl)-morpholine ([^{18}F]FMeNER-D2)*

These tracers have been synthesized in one or two steps; the one step synthesis was achieved by direct alkylation of desmethyl reboxetine with [^{18}F]bromofluoromethan or its deutero analogue; while reaction of [^{18}F]fluoromethyltiflate or its deutero analogue require the protection of the morpholine nitrogen with Boc, and subsequent deprotection step using trifluoroacetic acid. Both methodology leads to the desired tracer in yield greater than 90%, with specific activity between 111–185 GBq µmol^{-1}, and radiochemical purity exceeding 95%. Evaluation of both tracers in monkey shows a better regional brain distribution of [^{18}F]FMeNER-D2 than [^{18}F]FMeNER as well as a lower bone absorption for the deuteron analogue, which was therefore selected for further evaluation. [^{18}F]FMeNER-D2 showed a relatively high brain uptake in monkey reaching 3.6% ID 12 min pi, and reaching the peak specific binding between 120–160 min and the regional brain distribution consistent with NET distribution; the highest ratio to striatum were observed at 160 min for the mesencephalon (1.6), lower brainstem (1.5), temporal cortex (1.5), and thalamus (1.3). The signal specificity was assessed by pretreatment with desipramine (5 mg kg^{-1}, i.v.) leading to a significant uptake reduction in all NET reach regions, whereas citalopram and GRB12909 (SERT and DAT selective inhibitors, 5 mg kg^{-1}, i.v.) did not result in any significant change. Plasma analysis reveals a good stability of the tracer with 85% and 76% of unchanged tracer at 45 and 90 min, with the free fraction representing 22%.[227] Receptor occupancy of atomoxetine was evaluated with this tracer in monkey using a constant infusion paradigm, and for the first time a dose dependent occupancy was observed in NET rich brain regions, with an estimate amount of atomoxetine required to occupy 50% of NET of 16 ng ml^{-1} (plasma concentration).[228,229] Human biodistribution and dosimetry shows a high initial uptake in the lungs, liver, brain, kidneys, urinary bladder and heart, all well identify within the first scan; the radiation absorbed estimate was the highest for the urinary bladder (0.039 mGy MBq^{-1}, based on a 2.4 h urine voiding intervals), lungs (0.031 mGy MBq^{-1}), osteogenic cells (0.023 mGy MBq^{-1}), and the red marrow (0.022 mGy MBq^{-1}), with an effective dose estimate of 0.017 mGy MBq^{-1}.[230] Using either a template method or anatomical ROIs to generate BP$_{ND}$ a clear mapping of NET in the human brain emerged *in vivo* with the highest binding found in the medial thalamus (lower levels in the anterior and pulvinar division), in the cerebral aqueduct of the midbrain and within the dorsal pons (comprising the locus coeruleus).[231] Kinetic modeling and quantification were evaluated, and a three compartment model can fit TAC of all region, whereas a two compartment only fitted the caudate TAC; binding potential calculated by the indirect kinetic, simplified reference-tissue model, multilinear reference-tissue model, or ratio method all provide similar estimates.[232] Dose occupancy of nortriptyline was evaluated in human, and the ED$_{50}$ was estimated at 77 mg (single dose p.o.) corresponding to a plasma concentration of 60 ng ml^{-1}.[233]

3.4.2.4   *(S,S)-2-(α-(2-[^{123}I]Iodophenoxy)benzyl)morpholine ([^{123}I]INER or [^{123}I]IPBM)*

This tracer was synthesized in two steps: first, iododestannylation of its tri-methyltin *N*-Boc protected precursor using peracetic acid as oxidizing agent, then deprotection by removal of the Boc with trifluoroacetic acid in 47% yield of [^{123}I]INER with radiochemical purity 98%.[234] This tracer was also obtained by halogen exchange from its bromine precursor in presence of ammonium sulfate–copper sulfate in 65% yield with radiochemical purity 98%.[235] Evaluation in mice showed relatively rapid brain uptake peaking at 0.6% ID 30 min pi, followed by a gradual washout. *Ex vivo* regional brain distribution revealed highest tracer concentration in the locus coeruleus and anteroventricular thalamic nucleus and low levels in the striatum. The specificity was assessed by co-injection of nisoxetine (10 mg kg^{-1}), resulting in significant reduction of activity in NET-rich regions of the brain, whereas fluoxetine (10 mg kg^{-1}) and GBR12909 (1 mg kg^{-1}) did not affect the tracer accumulation.[235] Evaluation in baboon showed relatively faster brain uptake reaching 1% ID after 10 min pi, followed by a relatively slow washout, with the highest uptake in the locus coeruleus and diencephalon. In a bolus plus constant infusion paradigm, steady levels of activity were obtained up to 6 h pi, with similar distribution as for the bolus experiment. Injection of reboxetine (1 mg kg^{-1}, i.v.) at 210 min produced significant displacement of the tracer from all brain regions but was more pronounced in NET-rich regions. In contrast, injection of citalopram (5 mg kg^{-1}, i.v.) or methylphenidate (0.5 mg kg^{-1}, i.v.) did not result in displacement of the tracer.[234] The ratio of diencephalon to cerebellum during equilibrium reached 2.0, and could be reduced in a dose-dependent fashion with reboxetine (33% at 0.5 mg kg^{-1} and 42% at 1.5 mg kg^{-1}).[236]

3.4.3 Serotonin Transporter (SERT or 5-HTT)

3.4.3.1 (+)-1,2,3,5,6b,10bb-Hexahydro-6a-((4-[^{11}C]methylthio)phenyl)pryrolo[2,1-a]isoquinoline ((+)[^{11}C]McN)

(+)[^{11}C]McN was synthesized by methylation with [^{11}C]methyliodide of the freshly prepared thiol precursor in 12% yield, with specific activity 4250 mCi mmol^{-1} and radiochemical purity 95%. Due to the instability of the thiol precursor, an improved synthesis was published using either the methyl, butyryl, or benzoate thioether as precursor, which generates *in situ* the thiol intermediate by reaction with tetrabutylammonium hydroxide or sodium hydroxide during the methylation step.[237–240] Evaluation in monkey before and after treatment with fenfluramine (5 mg kg^{-1} twice per day subcutaneously for 4 days) produced a drastic reduction in tracer uptake in SERT-rich parts of the brain (hypothalamus, midbrain, pons, thalamus, caudate and putamen). Similar results were observed after MDMA treatment.[241] Evaluation in human showed gradual uptake of the tracer in SERT-rich regions (thalamus, hypothalamus, midbrain, putamen, caudate) peaking around 70 min pi and followed by a relatively slow washout thereafter. In SERT-poor regions, the uptake was lower and faster, peaking 30–40 min pi and followed by a gradual washout.[242,243] The time activity curves were best fitted by either a one- or two-compartment model, or a non-compartmental model using white matter as reference region, all with plasma input. The mirror image ligand (−)[^{11}C]McN showed a fast uptake peaking at 20 min for all brain region, followed by a relatively fast washout, with no significant difference between brain region, assumed to represent the non-specific binding. First reports of this tracer estimated the binding potential (BP) of (+)[^{11}C]McN by the difference between the (+) and (−) V_T, but this method results in an overestimation of specific

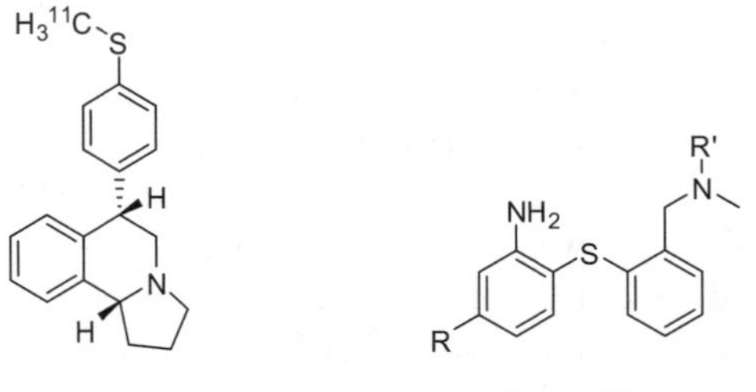

	R	R'	
[^{11}C]McN5652	$K_i = 0.4$ nM ^{a}		
[^{123}I]ADAM	^{123}I	CH$_3$	$K_i = 0.013$ nM
[^{11}C]DASB	CN	[^{11}C]CH$_3$	$K_d = 0.54$ nM

Figure 3.6 Serotonin Transporter Ligands.
^{a}Competition radioligand binding against [^{3}H]5-HT on rat cerebral cortex synaptosomes membrane. ^{b}Competition radioligand binding against [^{125}I]IDAM on LLC-PK1 cells expressing human SERT. ^{c}Dissociation constant on rat cortical membrane homogenates.

binding, especially in the low SERT density parts of the brain.[244–247] Effect of normal aging was investigated in healthy human (20–79 years of age) and a significant reduction of BP was observed in thalamus and midbrain ofaround 10% per decade.[248]

Evaluations of this tracer in depressed patients have shown variable results. Initial reports indicated an increase in BP in thalamus (22–23%) of depressed subjects with no significant differences in midbrain,[249] frontal cortex and cingulate.[250] All subsequent reports have observed a significant reduction of binding in depressed patients, especially in the amygdala and midbrain, less pronounced in the thalamus and putamen, and lower significant change in the hippocampus and anterior cingulate. This decrease was even more pronounced in depressed patients reporting childhood abuse.[251–253] In a one year follow-up study of depressed patients, relatively good prediction of remission was achieved since all non-remitters show lower binding potential (midbrain, amygdale, anterior cingulate) than control and remitter group at baseline.[254] Dose occupancy of paroxetine was evaluated with (+)-[^{11}C]McN in patient with social phobia receiving 20–40 mg day during 3–6 month period; tracer binding was significantly lower in all patients and in all regions (midbrain, hippocampus, amygdale, thalamus, cingulate, striatum) in a dose dependent manner (all subjectsshow clinical improvement at the time of paroxetine scan). The plasma concentration of paroxetine associated with 50% occupancy of SERT (calculated as V_3" difference between baseline and paroxetine scan) was estimated at 2.9 ng ml^{-1} and the average mean occupancy of all subject was 98% in midbrain, 94% amygdale, 92% hippocampus, 81% in cingulate and thalamus, and 75% in striatum.[255] A significant reduction of BP and V_3" in the anterior cingulate cortex, was also observed in patient with impulsive aggression disorder, the other parts of the brain show lower binding than control without reaching significance.[256] The effect of MDMA abuse was evaluated and all studies reveal a significant reduction in SERT density, correlated with the extend of MDMA exposure; this reduction was evident in all brain region but particularly marked in the striatum and thalamus.[257–260] A relatively small number of Parkinson's disease patients have been imaged with this tracer and show a significant reduction of distribution volume in both caudate (50%) and putamen (34%).[261] A multitracer (DAT, D2, SERT, 5-HT2A) study in Tourette syndrome patient showed significant reduction of [^{11}C]McN BP in midbrain, putamen and caudate.[262] Patients affected by chronic fatigue syndrome demonstrated a significant reduction of SERT density in the rostral subdivision of the anterior cingulate (26%) as well as good correlation between the reduction of SERT in the dorsal anterior cingulate and pain level.[263] In autism patients, SERT density was lower in all brain regions compared to age and IQ matched controls; good correlation between the reduction in anterior and posterior cingulate and the impairment of social cognition, as well as thalamic reduction and repetitive/obsessive disorder were observed.[264] The effect of SERT polymorphism has been studied in many of those reports, and the triallelic 5-HTTLPR genotype did not seem to affect SERT expression in the human brain using this tracer.[265]

3.4.3.2 3-Amino-4-(2-[^{11}C]dimethylaminomethylphenyl-sulfanyl)-benzonitrile ([^{11}C]DASB)

This tracer was obtained by methylation of its monomethyl precursor using [^{11}C]methyliodide in 66% yield, with specific activity around 86 GBq μmol^{-1} and radiochemical purity 98%.[266–268] Evaluation in monkey showed relatively slow brain uptake constantly increasing over the duration of the scan (90 min) in all SERT-rich parts of the brain (midbrain > thalamus > caudate > putamen), while the SERT-poor regions had faster peak uptake around 20–30 min (frontal cortex, cerebellum) followed by gradual washout. Pretreatment with paroxetine (10 mg kg^{-1}, i.v.) resulted in drastic reduction of tracer binding in SERT rich areas (76% hypothalamus, 72% midbrain, 70% pons, 69% thalamus, 62% caudate, 54% putamen). Metabolite analysis showed a relatively fast metabolic rate ($t_{1/2}$ = 30 min), with detection of highly polar metabolites as well as at least one lipophilic metabolite.[269] Similar curves were obtained in humans, with relatively slow uptake peaking first for the SERT-poor regions (frontal cortex, cerebellum around 30 min) followed by a gradual washout, whereas the SERT-rich regions peaked later (raphe, striatum, thalamus around 60 min) followed by a gradual washout. White matter (even though low in SERT) showed a low but constant increase in uptake peaking at around 75–90 min pi.[270,271] [^{11}C]DASB kinetic modeling was successfully achieved using either a 1-tissue compartment model, constrained 2-tissue compartment, data-driven estimation of parametric image based on compartmental theory (DEPICT), simplified reference tissue model (SRTM), or multilinear reference tissue model (MRTM). However, the most useful and commonly used method to date is MRTM2. Test-retest reproducibility provides BP estimates with variability 4–13%.[271–275] Biodistribution and dosimetry in human revealed high initial peak uptake in the lungs (53% ID) followed by the heart (5.3% ID), liver (4.3% ID), brain (4.0% ID), kidney (1.4% ID), and spleen (0.8% ID); the highest dose estimates was found in the lungs (0.0328 mGy MBq^{-1}), urinary bladder (0.012 mGy MBq^{-1}), kidneys (0.00928 mGy MBq^{-1}), gall bladder (0.00927 mGy MBq^{-1}), and liver (0.00641 mGy MBq^{-1}), with an EDE of 0.00698 mGy MBq^{-1}, within the limits of international guidelines.[276]

Seasonal variation in SERT expression was evaluated with this tracer in human, and a significant negative correlation was found between the number of daylight hours and the [^{11}C]DASB BP in the caudate, putamen and mesencephalon, which could in part explain seasonal affective disorder, especially in short 5-HTTLPR allele genotype carriers.[277,278] Evaluation of this tracer in patients suffering from major depression episodes (MDE) and healthy controls showed no significant differences between the two groups, although a significant correlation was observed in the MDE group between increase in tracer binding and their negativist dysfunctional attitude (DAS score and SERT BP in prefrontal cortex, anterior cingulate, putamen, thalamus).[279] More recently, an increase in BP was reported in MDE patients, reaching significance in the thalamus, ventral caudate, insular cortex, and periaqueductal gray matter.[280] However, significant reduction in thalamic BP was also recently reported, which correlated with patient anxiety levels (STAI score).[281] In bipolar depressed patients, a significant increase in BP was observed in thalamus, dorsal cingulate cortex, medial prefrontal cortex, and insula.[280,282] In a small group of depressed and non-depressed HIV patients, a reduction of BP was observed in both groups, but the reduction was less pronounced in HIV depressed than HIV non-depressed patients, and no correlation with the duration of the illness was found.[283] No differences in BP were found between healthy volunteers and one year recovered depressed patients.[284] The occupancy of different clinical antidepressant drugs has been evaluated with [11C]DASB. Paroxetine and citalopram (20 mg per day) showed a similar 80% occupancy of SERT after 4 weeks of treatment in all brain regions (caudate, putamen, thalamus, prefrontal cortex, insula, anterior cingulate).[285] Similar results were observed when comparing five antidepressant (citalopram, fluoxetine, sertaline paroxetine, extended-release venlafaxine) at different dosages, with a minimum therapeutic dose (80% striatal occupancy) of 30, 20, 50, 20, and 75 mg per day respectively; the daily dose to reach 50% transporter occupancy was estimated at 3.4, 2.7, 9.1, 5, and 5 mg, respectively.[286] Higher dose of venlafaxin (400 mg per day), sertaline (150–200 mg per day), and citalopram (60–80 mg per day) resulted in a higher striatal occupancy reaching around 85% occupancy for the three drugs, but the therapeutic significance of this 5% increase needs to be evaluated.[287] The occupancy of the appetite suppressant sibutramine was evaluated, and a relatively low but significant reduction in BP was observed in brainstem (33%), caudate (32%), putamen (29%), and thalamus (24%) when compared to placebo; the SERT occupancy only correlates with the plasma levels of di-desmethylsibutramine, which seems to be responsible for the SERT effect of sibutramine.[288] [11C]DASB has been proposed as a measurement of fluctuation of endogenous serotonin *in vivo*, with the goal of providing measurement of serotonin presynaptic activity. Rapid tryptophan depletion (RTD) has been proven to rapidly produce a profound decrease in plasma and CSF tryptophan levels as well as a decrease in CSF of serotonin metabolite; using this paradigm [11C]DASB binding were compared before and after RTD but a very small reduction of BP was observed (− 4.5%), suggesting that this tracer is not affected by endogenous serotonin.[289,290] The effect of MDMA consumption was studied and results in a significant reduction of tracer binding in multiple parts of the brain and negatively correlates with the duration of MDMA use in the hippocampal and thalamic region for a two-week abstinent user *versus* control. This decrease seems reversible since no differences were observed between a one-year abstinent user and drug naïve controls.[259,291,292] In Parkinson's disease patients, this tracer showed widespread reduction in binding, reaching significance in the caudate (30%), putamen (26%), midbrain (29%), and orbitofrontal cortex (22%) with no asymmetry detected; however, in depressed Parkinson's patients, a widespread increase of [11C]DASB binding was observed, reaching significance in the prefrontal (68%) and dorsolateral (37%) cortices, compared to healthy subjects.[293–295] In patients affected with obsessive-compulsive disorder, the tracer showed significant reduction in binding in the orbitofrontal (42%) and insular cortex (30%); however, this study did not agree with previously published significant reductions in midbrain (15%) and thalamus (15%), nor the correlation between BP reduction and disease severity assessed by Y-BOCS score.[296,297] Interestingly, a positive correlation between thalamic binding of [^{11}C]DASB and neuroticism personality (NEO PI-R score) was observed, while midbrain binding negatively correlated with openness trait.[298,299] In alcoholic and schizophrenic patients, no differences in binding could be detected *versus* healthy controls.[300,301]

3.4.3.3 2-((2-((Dimethylamino)methyl)phenyl)thio)-5-[^{123}I]-iodophenylamine ([^{123}I]ADAM)

This tracer was synthesized by iododestannylation of the trimethyl or tributyl tin precursor, using hydrogen peroxide as oxidizing agent, in 90% yield with specific activity greater than 12,000 Ci mmol^{-1}, and radiochemical purity exceeding 99%. Evaluation in monkey showed fast brain uptake peaking at 30–60 min (the peak was faster in SERT-poor part of the brain, whereas a higher and slower uptake was observed in SERT-rich regions), followed by fast washout from low SERT density parts of the brain (cerebellum, frontal and temporal cortices), a gradual washout from medium density regions (striatum and thalamus), and a slow washout from the high density midbrain; equilibrium was achieved in midbrain, thalamus and striatum after 210 min pi.[302–304] Evaluation in human shows similar TAC with an average brain uptake reaching 4.1% ID, however a relatively faster washout from midbrain was observed, reaching pseudoequilibrium 240 min pi, whereas thalamus, striatum, and mesial temporal regions stabilized at 180 min, with an excellent delineation of those brain regions. Pretreatment with citalopram results in significant reduction of tracer accumulation.[305–307] The tracer kinetic was modeled using tissue reference models (simplified reference tissue model, Logan reference tissue model, ratio method), as well as one and two tissue compartment models (four and three parameters were evaluated). The ratio method seems to overestimate the binding by 10% on average, whereas a slight underestimation was observed with the simplified reference tissue and Logan reference tissue models (3–5% for data between 200–240 min pi); the one and two tissue compartment (three and four parameters) as well as Logan model (for acquisition between 0–120 min, and fixed cerebellar clearance rates) leads to the best estimation of [^{123}I]ADAM binding potential.[308–311] Biodistribution and dosimetry in human showed high uptake in the lungs immediately after injection as well as high to moderate uptake in the brain, liver, gallbladder, and intestine (all peaking at different times between 30 min to 6 h). The radiation absorbed dose estimates were the highest for gall bladder (0.0835 mSv MBq^{-1}), upper and lower intestine wall (0.0668 and 0.0588 mSv MBq^{-1} respectively), and the urinary bladder (0.0514 mSv MBq^{-1}), and the estimate effective dose was of 0.0302 mSv MBq^{-1} (slight differences were observed between the different studies).[312–314] Test-retest shows an acceptable variability between 13–22% depending on the brain regions.[306] [^{123}I]ADAM was evaluated in patients with major depression and shows a reduction of tracer binding in midbrain (after age correction, 8% reduction per decade) *versus* healthy controls; and a good correlation between midbrain binding and the degree of depression symptoms was observed.[315,316] However, two other reports failed to detect any significant change in [^{123}I]ADAM binding between major depression and healthy control.[317,318] Similarly no differences were observed within subjects between summer and winter.[319] The brain occupancy of various SSRI was evaluated with this tracer; citalopram, escitalopram, and paroxetine all reduce the tracer binding in a similar way, with a maximum occupancy between 60–80% depending on the study and the brain region; interestingly 10 mg per day of escitalopram (S-enantiomer of citalopram) produced greater occupancy than 20 mg per day of citalopram (81% *vs.* 64%).[305,317,318,320–322] In a well designed study comparing this tracer in bulimic/healthy twins and women from a healthy twin pair, no differences could be found among the three groups, but a significant increase in uptake ratio midbrain to cerebellum was found in patients with night eating syndrome, confirming the implication of SERT in this disorder, since SSRI is to date the only effective treatment for this condition.[323] The uptake of this tracer was also significantly increased in the mesopontine brainstem of migraineurs (mesopontine: occipital ratio: 0.88 *vs.* 0.58).[324] Borderline personality patients compared with control patients have been reported to have a 43% and 12% increase in brainstem and hypothalamus uptake ratio, which correlates with the age and impulsivity of patients, but not with their depression status.[325]

3.5   Vesicular Monoamine Transporter Type 2 (VMAT2)

Vesicular monoamine transporters are responsible for the reaccumulation into the synaptic vesicle of cytosolic monoamine neurotransmitter (catecholamine, serotonin, histamine), and this active transport is driven by an electrochemical proton gradient across the vesicle membrane. Two closely related vesicular monoamine transporters VMAT1 and VMAT2 have been cloned and characterized. Both VMAT1 and VMAT2 are acidic glycoproteins with high sequence homology and seem to be transmembrane proteins with twelve transmembrane domains (crystal structures are not available) similar to the plasma membrane monoamine transporter. While both transporters are able to store dopamine, serotonin, and norepinephrine with similar pharmacological profile, only VMAT2 is able to use histamine as a substrate at low concentration and only VMAT2 is sensitive to the transport inhibitor tetrabenazine. VMAT1 and VMAT2 differ largely in their tissue distribution with VMAT1 being the predominant transporter in the peripheral nervous system and VMAT2 being preferentially expressed in the CNS. However, caution should be taken since species differences have been reported in the tissue distribution of both transporters. Although VMAT2 does not discriminate between any particular monoamine, its localization in the striatum is restricted to nigrostriatal dopaminergic neuron, and therefore imaging ligand of VMAT2

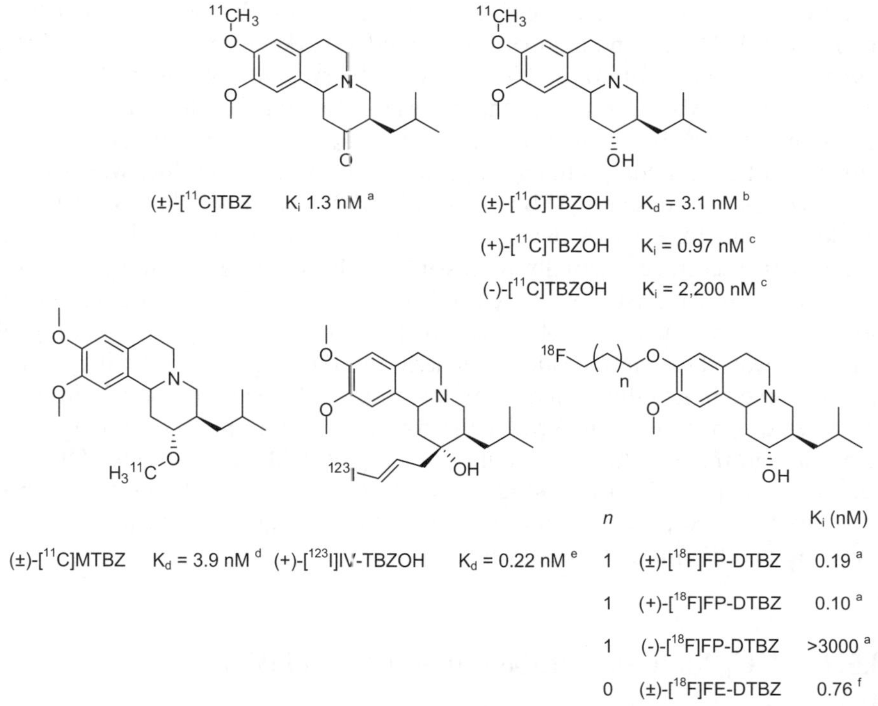

Figure 3.7   VMAT2 Ligands.
[a]Inhibition constant using (±)-[^{3}H]TBZ on rat striatal homogenate. [b]Dissociation constant on bovine chromaffin granule membrane. [c]Inhibition constant using (±)-[^{3}H]MTBZ on rat striatal slide. [d]Dissociation constant on rat striatal slide. [e]Dissociation constant on rat striatal homogenate. [f]Inhibition constant using (±)-[^{3}H]TBZ on rat striatal homogenates.

could play a crucial role in pathology affecting dopamine neurotransmission such as drug addiction, Parkinson's disease, Huntington's disease, and schizophrenia. For general reviews on the topic see bibliography.[326–332]

3.5.1 [^{11}C]-Tetrabenazine ([^{11}C]TBZ)

This tracer was synthesized by [^{11}C]methylation of 9-O-desmethylTBZ,[333] with specific activity >1000 Ci mmol^{-1} and radiochemical purity >95%. [^{11}C]TBZ shows a rapid brain uptake after i.v. injection in mice, reaching 3.2% of injected dose at 2 min pi, similarly the washout is very fast with only 0.21% of injected dose at 60 min pi.[334] However, the tracer shows a slower clearance from striatum and hypothalamus. The general order of distribution was consistent with autoradiography studies with highest accumulation of [^{11}C]TBZ at 10 min pi in the striatum > hypothalamus > hippocampus > cortex ≈ cerebellum (striatum:cerebellum = 2.93; hypothalamus:cerebellum = 1.65 @10min). The specific binding was assessed by competition studies using unlabeled TBZ and ketanserin injected at the same time as the tracer and by pre-treatment with reserpine, GRB12935 and haloperidol. A moderate (ketanserin) to sharp (TBZ, reserpine) reduction of tracer binding was observed for the drugs known to interact with VMAT2, but no change was observed for the D2 and DAT antagonists.[334,335] Blood analysis shows a fast metabolization rate of the tracer with 57% of parent compound after 15 min, and those metabolites were clearly identified as 9-O-desmethylTBZ, [^{11}C]α-TBZOH, and [^{11}C]β-TBZOH and would account for part of the signal detected in the brain, making quantification with this tracer virtually impossible.[335] In an elegantly designed study the same team demonstrates the feasibility of imaging VMAT2 in non human primate using a unilateral MPTP-lesioned monkey where specific binding of [^{11}C]TBZ was absent in the side of the lesion in the striatum.[336] Similar results were observed in healthy human, with fast brain uptake and fast washout in all brain regions but slower in VMAT2 rich regions (striatum, and thalamus).[337] Even though the qualitative feasibility of imaging VMAT2 with [^{11}C]TBZ has been demonstrated by those studies, the fast metabolism of this tracer into active and labeled parent compounds proscribes any quantitative analysis and therefore greatly limits its usefulness.

3.5.2   [^{11}C]-Methoxytetrabenazine ([^{11}C]MTBZ)

[^{11}C]MTBZ was synthesized by [^{11}C]methylation of the sodium salt of α-TBZOH with specific activity >900 Ci mmol^{-1} and radiochemical purity >97%. Similar to [^{11}C]TBZ, [^{11}C]MTBZ showed fast brain uptake following i.v. injection in rodent, monkey and human followed by a fast washout from all brain regions except for those known to have high VMAT2 concentration, in which the washout was significantly slower.[338,339] The regional brain distribution was consistent in all three species with highest concentration found in striatum > hypothalamus > hippocampus > cerebral cortex ≈ cerebellum. In human, the initial intake was around 9% of the injected dose @ 10 min pi and the maximum uptake in the striatum was observed after 45 min reaching a maximum striatum to cerebellum ratio of 2.7. Human dosimetry,[338] was assessed in healthy subjects and shows the highest uptake in the testis (0.04 mGy MBq^{-1} dose limiting organ) and organs involved in the excretion of the tracer (liver > small intestine > kidney 0.014–0.017 mGy MBq^{-1}), with an EDE of 0.03 mGy MBq^{-1}. The specificity of the signal was assessed by acute pretreatment or co-injection of various compounds, in all cases treatment with VMAT2 ligand,[339] (reserpine, tetrabenazine, dihydrotetrabenazine) resulted in a sharp decrease of [^{11}C]MTBZ binding especially in the VMAT2 rich regions; conversely, treatment with haloperidol or pargyline did not affect tracer binding. The same team examined the influence of chronic treatment in rats with tetrabenazine (5 mg kg^{-1} i.p. twice per day, 3 days), pargyline (80 mg kg^{-1} per day s.c. 14 days), deprenyl (10 mg kg^{-1} per day infusion, 5 days), and L-DOPA (100 mg kg^{-1} per day infusion, 5 days) showing no significant differences in the binding and distribution of [^{11}C]MTBZ,[339] proving the absence of upregulation of VMAT2 by those drugs and the potential use of this tracer for imaging VMAT2 in medicated patients. A study,[340] using Tottering mice (mutant mice characterized by an increase in noradrenergic innervations, and increased levels of norepinephrine in striatum, hippocampus, cortex, anterior hypothalamus) show a 150–195% increase of [11C]MTBZ in those regions compared to normal mice, in accordance with the physiological specification of this transgenic strain. Metabolite analysis shows a marked species difference between rat and monkey.[338] In rats both polar and non polar labeled metabolites were detected in plasma and account for 53% of the activity and only 18% in the brain 15 min pi, by contrast only highly polar metabolites were detected in monkey plasma, accounting for 38% of the activity 15 min pi.[338]

3.5.3 [125I]-Iodovinyltetrabenazine ([123I]IV-TBZOH)

[125I]Iodovinyltetrabenazine is the only SPECT tracer developed to image VMAT2; it was synthesized by iododestannylation of its tributyltin precursor using hydrogen peroxide as oxidant with specific activity of 2,200 Ci mmol^{-1} and radiochemical purity >98%.[341] The racemic mixture of this tracer was resolved, and the two resulting enantiomers (named [125I]IV-TBZOH-I and [125I]IV-TBZOH-II from their elution order) were evaluated *in vitro* and *in vivo*, and in all cases, [125I]IV-TBZOH-I was more potent and selective than its counterpart.[341,342] This tracer showed a good *in vitro* profile with a low Kd, and accumulation in VMAT2 rich region using rat *ex vivo* autoradiography. After i.v. injection the tracer shows a fast brain uptake, followed by a relatively fast washout (1.12%ID @ 5 min, 0.58% ID @ 20 min, 0.33% ID @ 60 min, 0.18% ID @ 120 min) in rat brain. However the regional brain distribution shows a low degree of specificity with a maximum signal ratio striatum: cerebellum of 1.31 @ 20 min, presumably due to the high lipophilicity of this ligand.[341]

3.5.4 [11C]Dihydrotetrabenazine ([11C]TBZOH)

This compound was by far the most studied, and was introduced to overcome the metabolic issue of [11C]TBZ (as we mention earlier [11C]TBZ is rapidly metabolized into active [11C]TBZOH impairing any quantification). This tracer was synthesized by [11C]methylation of 9-*O*-desmethylTBZOH with specific activity >1600 Ci mmol^{-1} and radiochemical purity >95%.[343] The resolution of the racemic mixture and subsequent *in vitro* assay of each enantiomer revealed a highly stereospecific binding of this compound to VMAT2 since only the (+)-enantiomer is active.[344] *In vivo* injection of the (−)-enantiomer resulted in uniform brain distribution in mice whereas the (+)-enantiomer showed good accumulation in striatum and hypothalamus, and intermediate values were obtained for the racemic mixture. This tracer was first evaluated in Parkinson's disease; initial uptake following i.v. injection is fast and homogeneous throughout the brain, followed by a fast washout and slowest clearance in the caudate and putamen (those structures can be easily identify after 3–7 min).

The tracer is rapidly metabolized in the plasma and liver to a more polar species but only authentic [^{11}C]TBZOH was identified in the brain enabling quantification of VMAT2 in the brain.[345,346] A study by the Parkinson's group showed a drastic reduction of distribution volume in putamen (−61%) and caudate (−43%) compared to age matched controls, with an asymmetric binding in striatal and midbrain region and greater reduction contralateral to the clinically most affected side. Interestingly in this study a good correlation was observed between reductions of binding in the striatum and Schwab and England Activities of Daily Living Scale as well as with the duration of the disease, but not with the Unified Parkinson Disease Rating Scale or the modified Hoehn and Yahr Scale.[345] The reduction of striatal binding in healthy elderly compared to young subjects was around 0.5–0.7% per year; in accordance with results obtained from postmortem analysis of dopamine neurons in the substantia nigra pars compacta.[346] This tracer was also evaluated in multiple system atrophy (MSA) and olivopontocerebellar atrophy (OPCA), MSA is characterized by degeneration in the cerebellum, brainstem, basal ganglia, and spinal cord whereas OPCA shows loss of inferior olivary neurons and their cerebellar projection, decrease in the population of pontine neurons and their projection to cerebellum, and reduction of Purkinje and granule cells. Both groups show a significant reduction in distribution volume in caudate when compared to healthy subjects, and in putamen only for the MSA group. While the OPCA group shows a large range of binding in the striatum, the MSA group shows consistent decrease in specific binding, well correlated with the severity of extrapyramidal symptoms in this group.[347] [^{11}C]TBZOH was also evaluated in DOPA-responsive dystonia (DRD) characterized by a dopamine deficiency which is not associated with neuronal loss as opposed to Parkinson's disease. In this study an increase in striatal binding potential was observed, and demonstrates the absence of coregulation between the dopamine synthesis system and VMAT2 expression. The higher binding potential in DRD was attributed to the decreased competition for [^{11}C]TBZOH binding by endogenous dopamine.[348] [^{11}C]TBZOH was used to evaluate the hypothesis of potential increase of striatal dopamine innervation in schizophrenic patients; however no difference was observed in the binding potential of twelve schizophrenic subjects and the same number of age and sex match control.[349] Despite excellent results obtained with this tracer, recent studies clearly point out certain limitations; firstly, rat studies have shown variation of striatal binding following variation of endogenous dopamine levels. Dopamine depletion results in increased binding of [^{11}C]TBZOH (+14% with α-methyl-*p*-tyrosine; +12% with d-amphetamine) whereas dopamine elevation results in a decrease of binding (−16% with γ-hydroxybutyrate; −20% with levodopa),[350] and those results clearly illustrate the future complication in quantifying VMAT2 in patients under treatment. Secondly, the human radiation dosimetry estimates from baboon whole body distribution shows high accumulation of the tracer in the liver and lungs, followed by intestines, brain, and kidneys. The highest absorbed radiation dose was in the stomach wall at a level that might exceed regulatory guidelines in multiple injection protocols.[351]

3.5.5 Fluoroalkyl dihydrotetrabenazine ([^{18}F]FE-DTBZ and [^{18}F]FP-DTBZ)

Two ^{18}F analogues of tetrabenazine have been reported to date; both were prepared by nucleophilic substitution of the corresponding mesylate precursor by [^{18}F]fluoride, with specific activity around 1,500–2,000 Ci mmol^{-1} and radiochemical purity exceeding 99%.[352] Both compounds displayed excellent binding affinity in rat striatal homogenates, and show accumulation in VMAT2 rich region (caudate, putamen) from autoradiography in mouse which can be blocked by pretreatment with tetrabenazine. *In vivo* evaluation in mice revealed a fast and high brain uptake for both tracers (4.66% ID and 7.08% ID @ 2 min for [^{18}F]FE-DTBZ and [^{18}F]FP-DTBZ respectively), however [^{18}F]FP-DTBZ displayed a faster washout than [^{18}F]FE-DTBZ leading to a better target to background ratio (striatum:cerebellum of 3.0 and 1.7 @ 30 min for [^{18}F]FE-DTBZ and [^{18}F]FP-DTBZ respectively). The specificity of the signal was evaluated by blockade experiment using dihydrotetrabenazine leading to unit ratio in all regions examined, whereas raclopride leaves the ratio unchanged.[352] In unilateral 6-hydroxydopamine-lesioned mice [^{18}F]FE-DTBZ shows a selective reduction of binding on the lesioned side, which correlates (r = 0.95) with results obtained with [^{125}I]IPT (selective marker of dopamine transporter) when both tracers are injected at the same time.[353] Metabolite analyses at 30 min pi in mouse brain indicate that only 5% of the radioactivity was due to parent compounds. (±)-[^{18}F]FE-DTBZ, (±)-[^{18}F]FP-DTBZ, and (+)-[^{11}C]TBZOH were compared *in vivo* in the same monkey, while (±)-[^{18}F]FP-DTBZ and (+)-[^{11}C]TBZOH show similar brain uptake and regional distribution (max striatum: cerebellum of 5.25 and 5.15 respectively), (±)-[^{18}F]FE-DTBZ was significantly lower (max striatum : cerebellum of 2.55). Those results encouraged the resolution of the two enantiomers of (±)-[^{18}F]FP-DTBZ, not surprisingly only the (+) enantiomer was active *in vitro* and shows the highest distribution volume ratio of 6.2 ever obtained for VMAT2 tracers *in vivo* in monkey.[354] All those results clearly indicate the great potential of (+)-[^{18}F]FP-DTBZ as an imaging agent for VMAT2.

3.6 Post-Synaptic Dopamine Receptor D3 (D3r)

Dopaminergic neurotransmission is mediated through five receptor subtypes (D1-D5) that belong to two receptor subfamilies. D1-like receptors include D1 and D5 subtypes and are positively linked to adenylate cyclase through a G_s/G_{olf} protein. D2-like receptors comprise D2, D3, and D4 subtypes and they are all inhibitors of adenylate cyclase *via* coupling to a $G_{i/o}$ protein. D2-like receptors overall share little sequence homology (30–50%) when looking at the full length receptors, but a dramatic increase is revealed when focusing on the transmembrane domain implicated in ligand recognition (70–90%). D2 receptors are localized in the dorsal striatum and nucleus accumbens as well as in the prefrontal and temporal areas of the neocortex, where they co-localized with D4 receptors. D3 receptors are mostly expressed in the accumbens and ventral side of the putamen and substantia nigra. Almost all the antipsychotic and antiparkinsonian drugs act through the D2-like family and strong evidence of the involvement of the D3 receptor in those pathologies have been accumulated in recent years. D3 receptors have been implicated in Parkinson's disease, since the observation that in the MPTP model of parkinsonism, the dopamine D3 receptor is dramatically upregulated (66–77% increase), and D3 selective agonists have shown neuroprotective activity in MPTP treated animals. None of the dopamine antagonists used in the treatment of schizophrenia are selective for D2 receptors, and they all show high affinity for the D3 subtype as well and clozapine (atypical antipsychotic) seems to mediate its effect principally through neurons expressing the D3 subtype. The development of specific D3 imaging agents has been hindered by the lack of selectivity over D2 receptors, however the last five years have seen the emergence of D3 selective ligands with potential imaging application in Parkinson's and schizophrenic patients. For general reviews on the topic, see bibliography.[355–361]

3.6.1   [^{11}C](+)-4-Propyl-3,4,4a,5,6,10b-hexahydro-2H-naphto-[1,2-b][1,4]oxazin-9-ol ([^{11}C](+)-PHNO) (Figure 3.8)

This tracer was first developed as a D2 agonist imaging agent,[362,363] but later experiments revealed a certain degree of preference for the D3 subtype *versus* D2 *in vivo*; in this paragraph we only report the works focusing on D3 imaging with [^{11}C](+)-PHNO. It was synthesized in two steps: first, reaction of the (+)-hexahydro-naphthoxazine with [^{11}C]propionyl chloride leads to an amide intermediate which was reacted with lithium aluminum hydride to provide [^{11}C](+)-PHNO in 10% yield, with specific activity 900–1800 mCi mmol^{-1} and radiochemical purity >99%.[362] Evaluation in monkey shows a good and fast brain uptake peaking at 500 nCi mCi^{-1} ID at 10 min followed by gradual washout. The signal specificity was first demonstrated to only D2/D3 receptors by pretreatment with raclopride, which decreased the specific to non specific equilibrium partition coefficient V_3", (using a simplified reference tissue model) by the same value in ventral striatum, dorsal striatum, and globus pallidus than pretreatment with raclopride and imaging with [^{11}C]raclopride. In contrast, pretreatment with the selective D3 partial agonist BP897 (0.25 mg kg^{-1}) leads to significantly more decrease of V_3" for [^{11}C](+)-PHNO than for

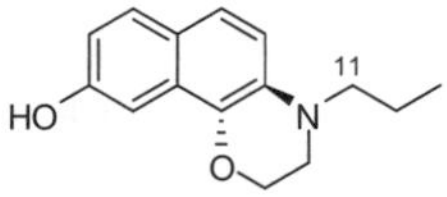

	K$_i$ D3 (nM)	K$_i$ D2 (nM)
[^{11}C]-(+)-PHNO	0.16 [a]	8.5 [a]

Figure 3.8   Dopamine D3 Ligands.
[a]Using [^{125}I]odosulpiride on hD2 and hD3 receptor expressed in CHO cells.

[^{11}C]raclopride especially in the D3 rich globus pallidus (57% *vs.* 29%) and ventral striatum (30% *vs.* 19%) with no significant differences in other brain regions. Those results demonstrate the selectivity of [^{11}C](+)-PHNO for D3 receptors over D2 in those brain regions and the selectivity was estimated to be between 4–12 fold, with the majority of the binding in the globus pallidus (60–90%) specific for D3 receptors.[364] Based on the hypothesis that in early Parkinson's disease, D3, but not D2 populations are decreased in the striatum, a small group of non-depressed, non-demented, dopaminergic drug-naïve subjects with early PD were imaged with both [^{11}C](+)-PHNO (D3 > D2) and [^{11}C]raclopride (D2 = D3). Despite a large overlapping pattern of both tracers significant differences were observed, while [^{11}C]raclopride shows high signal in the dorsal caudate (DC) and dorsal putamen (DP), [^{11}C](+)-PHNO uptake was higher in the ventral striatum (VS) and globus pallidus (GP); in the substantia nigra the signal was quantifiable for [^{11}C](+)-PHNO but not for [^{11}C]raclopride. Comparison of the binding potential (using the Mann-Whitney non-parametric U-test) of both tracers in control and PD groups show that for both tracers the BP was increased by 25% in the DP of Parkinson's group with a more pronounced effect in the contralateral side of parkinson's symptoms. [^{11}C](+)-PHNO but not [^{11}C]raclopride was associated with a decrease in BP in GP and VS (− 42% and − 11% respectively). Comparison of BP ratio (BP [^{11}C](+)-PHNO: BP [^{11}C]raclopride) in the D2 compartment (DP, DC) and D3 compartment (GP and VS) show no difference between control and PD group for the D2 compartment, whereas a significant 23% decrease was observed for the D3 compartment; moreover this decrease in ratio correlates well with the motor deficit of PD patients (Purdue pegboard task, symptomatic hand).[365] The same team examined D3 (GP and VS) as well as D2 (caudate, putamen, thalamus, cerebellum) binding potential in drug free schizophrenic patients using [^{11}C](+)-PHNO and did not find any difference *when compared with the* age- and sex-matched control group.[366] Comparison of [^{11}C](+)-PHNO and [^{11}C]raclopride in long term treated schizophrenic patients (clozapine, resperidone, olanzapine) show a significant decrease in BP for all drugs and both tracers in caudate and putamen (caudate −71%, and −53%; putamen −69% and −41% for [^{11}C]raclopride and [^{11}C](+)-PHNO respectively); in the D3 rich region (GP), [^{11}C]raclopride shows a 59% reduction of BP for all three treatments, but surprisingly [^{11}C](+)-PHNO shows a significant increase of 70% in GP binding potential (clozapine 100%, olanzapine 68%, risperidone 38%); in the ventral striatum a consistent (all three drugs) 72% reduction of BP was observed for [^{11}C]raclopride, whereas only 17% reduction was detected for [^{11}C](+)-PHNO. The resperidone treated group was also imaged 2h after pretreatment with the D3 receptor-preferring agonist pramipaxole and shows very little change in BP in the caudate, putamen and ventral striatum, but a significant decrease of 45% in the globus pallidus was observed.[367] Further elucidation of the specificity of [^{11}C](+)-PHNO for D3 *versus* D2 receptors was published in an elegant study coupling autoradiography on wild, D2 knock-out, D3 knock-out mice and monkey imaging both with pretreatment with selective D3 (SB-277011) and D2 (SV-156) ligands. This study clearly demonstrates the presence of two distinct binding sites (D2 and D3) for [^{11}C](+)-PHNO and possibly a third one attributed to D4 (marginal residual activity of ^{3}H-(+)-PHNO in the ventral pallidum and ventral striatum of D2KO-SB-277011 treated mice); in both species very similar results were observed and conclude to a D3 specific signal of [^{11}C](+)-PHNO in extra-striatal region especially subatantia nigra, ventral tegmental area (95–98% monkey), thalamus (88–92% monkey), globus pallidus (72–77% monkey), intermediate contribution in ventral striatum (49–57% monkey) and low contribution in caudate (23–33% monkey) and ptamen (8–21% monkey) providing for the first time a map of D3 receptor in monkey using *in vivo* imaging.[368] Despite the obvious non selectivity of [^{11}C](+)-PHNO for D3 *vs.* D2 this tracer is an acceptable tool to quantify D3 receptors *in vivo*.

3.7   Post-Synaptic Serotonin Receptor Targets

The physiological effect of serotonin (5-hydroxy tryptamine, 5HT) are medi-ated by seven receptor subtypes 5-HT1–7, all members of the G protein coupled receptor superfamily with the exception of 5-HT3 which is a ligand gate channel. 5-HT1 and 5-HT5 subtypes are negatively coupled to adenylate cyclase, 5-HT2 subtypes are positively associated with phospholipase C, while 5-HT4,6,7 are positively coupled to adenylate cyclase. The 5-HT4 gene is localized on chromosome 5 and nine splice variants have been identified (eight of those variants only differ on the length of the C terminal domain, after the residue L).[358] In the brain 5-HT4 are expressed in two distinct structures, the extrapyramidal system comprising the striatum, globus pallidus and substantia nigra, and the mesolimbic system comprising the nucleus accumbens, hippo-campus, amygdale, and olfactory tubercle. Subcellular localization has placed 5-HT4 receptors on GABAergic and cholinergic neurons, and many studies have demonstrated the ability of 5-HT4 to modulate the release of various neurotransmitters (dopamine, serotonin, acetylcholine). 5-HT4 has long been postulated as a target for various neurological disorders including Parkinson's, Alzheimer's and Huntington's disease where drastic reduction of 5-HT4 receptor binding sites have been shown in post mortem studies. However to date all 5-HT4 modulators on the market only target the peripheral receptors in the GI tract. The 5HT6 gene was located on chromosome 1, with no functional splice variant. Within the brain high levels of 5HT6 receptors have been found in the striatum, olfactory tubercule, nucleus accumbens, and hippocampus. Since lesions induced by 5,7-dihydroxytrypatamine on serotoninergic neurons failed to reduce 5HT6 mRNA in the hippocampus and striatum, 5HT6 seems to be positioned post synaptically to serotoninergic neurons, and immunohis-tochemistry have placed it mostly on GABAergic neurons. One of the first putative applications of 5HT6 modulator was schizophrenia but mixed results have been reported, and recent research seems to indicate a major role of 5HT6 in learning and memory deficit, as well as feeding behavior. For general reviews on the topic, see bibliography.[369–377]

3.7.1 Serotonin Receptor Subtype 4 (5-HT4)

3.7.1.1 (1-[^{11}C]Methylpiperidin-4-yl)methyl 8-amino-7-chloro-2,3-dihydrobenzo[b][1,4]dioxine-5-carboxylate ([^{11}C]SB207145)

This tracer was synthesized by methylation with [^{11}C]methyliodide of the corresponding desmethyl precursor in 60% yield with specific activity around 107 GBq µmol^{-1} and radiochemical purity exceeding 99%. [^{11}C]SB207145 was first evaluated in pig (Yorkshire and Danish Landrace crossbreed) and shows a fast brain uptake peaking at 10–20 min pi followed by a gradual washout. The regional brain distribution matches the 5HT4 distribution with the highest receptor concentration found in the striatum > thalamus > cortical regions > cerebellum, and was reduced to cortical levels by pretreatment with the selective 5HT4 antagonist SB207040 (0.5 mg kg^{-1}).[378] Quantification was also accomplished in pig (Gottingen minipig) and comparison between one and two-tissue compartment models, and a Logan plot using three different reference tissue models (multilinear reference tissue model (MRTM), simplified reference tissue model (SRTM), Logan noninvasive model). Both one and two-tissue compartment models lead to highly similar results, with no mean difference in the V_T values, and the Logan plot has a fairly good correlation but with an underestimation of V_T for the high receptor concentration regions. Both MRTM and SRTM were more sensitive to noise than the plasma input model, however those results were ameliorated by fixing k_2 to a global mean value and the best fit was obtained for SRTM with a fixed k_2. The highest distribution volume was found in the striatum (23.2), thalamus (14.8), diencephalon (14.0), mesencephalon (13.1), hippocampus (12.9), cortical regions (12.3–10.7) and the lowest in cerebellum (8.6).[379] Evaluation in humans shows a good brain uptake peaking at around 20 kBq ml^{-} at 10 min except for the striatum which displays slower kinetics reaching a maximum uptake of 27 kBq ml^{-1} at 20 min pi. Similarly the washout kinetics were faster for all brain regions other than striatum (the cerebellum devoid of the 5HT4 receptor shows the fastest

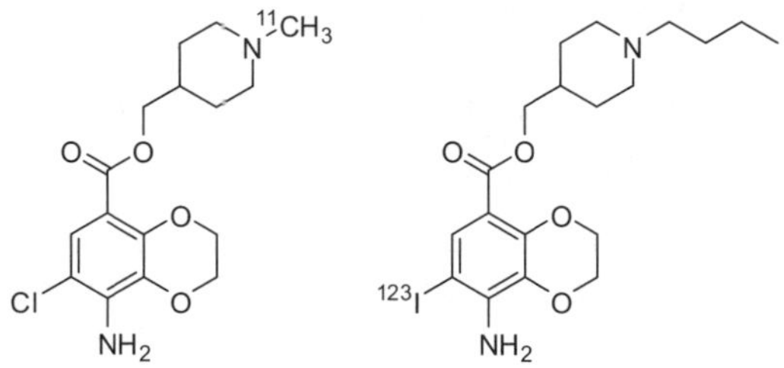

[^{11}C]SB207145 K_d = 0.39 nM [a]

[^{123}I]SB207710 K_d = 0.2 nM [b]

Figure 3.9   Serotonin Receptor Subtype 4 (5-HT$_4$) Ligands.
[a]Dissociation constant in striatal pig brain homogenate. [b]Dissociation constant in human atrium homogenate.

washout rates). Pretreatment with 150 mg of piboserod (5HT4 inverse agonist, oral, 4h before injection) results in significant reduction of activity in all brain regions to baseline cerebellar levels. Quantification using a two-tissue compartment model gave the best fit of experimental data with binding potential of 3.38, 0.82, 0.3, and 0.23 for striatum, hippocampus, parietal and superior frontal cortex respectively. The tracer was quickly metabolized with only 15% unchanged parent compound in plasma at 30 min and fit a biexponential function, and the free fraction was around 25%. Test-retest was good and no significant differences on the area under the curve were observed.[380] In a relatively small group (16 subjects) no sex differences were observed but a significant (p = 0.046) decreas in binding with age was observed and the inter- and intra-subject variability was low. Increase of endogenous serotonin levels following citalopram treatment failed to produce any significant change on [^{11}C]SB207145 binding *in vivo*.

3.7.1.2 *(1-Butyl-4-piperidinylmethyl)-8-amino-7-[^{123}I]iodo-1,4-benzodioxan-5-carboxylate ([^{123}I]SB207710)*

[^{123}I]SB207010 was obtained by classical iodo destannylation of its tributyltin precursor using chloramine-T as oxidizing agent with a yield of 60–90% and radiochemical purity exceeding 99%. The brain uptake in rats was relatively low peaking at 0.25% ID g at 10 min pi, followed by a relatively fast washout especially from regions of the brain with low 5HT4 receptor concentration. The ratio to cerebellum was maximal at 60 min reaching 3.6 for striatum, 3.0 for hypothalamus and 2 for hippocampus. Evaluation in monkey shows a brain uptake peaking at 2.3% ID between 5 and 10 min followed by gradual washout with similar rate in all brain regions, and the highest activity was found in the striatum > frontal cortex > temporal cortex > cerebellum. The ratio striatum to cerebellum was around 3 at the beginning of the scan and slowly increased up to 4 at 170 min; signal specificity was assessed by pretreatment with SA204070 (0.5mg kg^{-1}) and leads to a significant reduction of uptake in all brain regions.[381] No further reports have been published.

3.7.2   Serotonin Receptor Subtype 6 (5-HT6)

3.7.2.1   *3-(3-Fluorophenylsulfonyl)-8-(4-[^{11}C]methylpiperazin-1-yl)quinoline ([^{11}C]GSK-215083)*

Only preliminary reports have been presented for this tracer and it should be noted that GSK-215083 is not specific to 5HT6 receptors and also binds to the 5HT2A receptor with a Ki = 0.79 nM. [^{11}C]GSK-215083 was synthesized by methylation of its desmethyl precursor using [^{11}C]methyl triflate. Evaluation in pig shows a peak uptake of around 12% ID L at 15 min pi followed by a slow washout; the highest activity was found in the striatum > cortical region > cerebellum in accordance with 5HT6 distribution. The signal specificity was assessed by pretreatment with the non specific 5HT6 antagonist clozapine

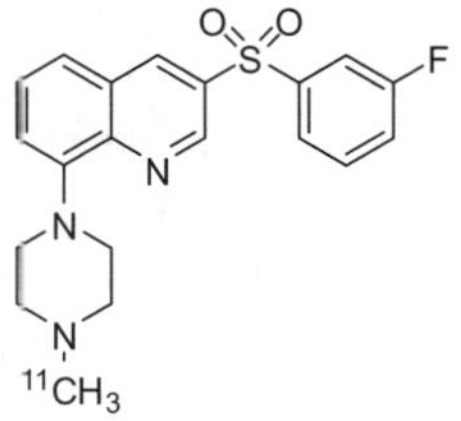

Figure 3.10   Serotonin Receptor Subtype 6 (5-HT$_6$) Ligands. [^{11}C]GSK-215083 pKi = 9.82 (no experimental details published).

(6.25 mg kg^{-1}), resulting in a significant reduction of striatal uptake, while pretreatment with 5HT2A antagonist ketanserin (0.3 mg kg^{-1}) did not affect striatal uptake but greatly (> 90%) reduced the cortical binding.[382] Evaluation in humans shows the highest binding potential in putamen (1.23) and caudate (1.08) followed by frontal cortex (0.29) and pretreatment with SB-742457 (selective 5HT6 antagonist, 175 mg kg^{-1} oral) leads to a decrease of BP in caudate and putamen of 85% and 60% respectively. Pretreatment with the 5HT2A antagonist ketanserin (0.1mg kg^{-1}, i.v.) did not affect the striatal uptake, demonstrating the specificity of the signal in the striatum.[383] Human dosimetry was examined following injection of 330–370 MBq^{-1} of [^{11}C]GSK-215083, the mean residence time was 0.103 h for the liver, 0.079 h for the lungs, 0.026 h for the brain and 0.005 h for the heart, and the effective dose was estimated between 5–8 mSv MBq^{-1}.[384]

3.8   Peripheral Benzodiazepine Receptor, PBR (Translocator Protein 18kD, TSPO)

The peripheral benzodiazepine receptor (PBR), also known as translocator protein 18 kDa (TSPO), was originally discovered as a binding site of diazepam in kidney cells and is now known to be widespread in peripheral tissues, including liver, heart, lungs, adrenal, blood cells, and glial cells in the brain. Its primary localization was reported to be mainly in the mitochondrial outer membranes in many tissues, however rat lung PBR are located on the inner membrane of the mitochondria and PBR were also found on plasma membranes which lack mitochondria, in heart, liver, adrenal, testis, and hematopoietic cells. PBR seems to be a heteromeric complex of at least three different subunits, including an isoquinoline binding subunit (18 kDa), a voltage-dependent anion channel (VDAC, 32 kDa), and an adenine nucleotide carrier (ANC, 30 kDa). It has been demonstrated, that a PBR specific ligand like PK11195 binds specifically to the 18 kDa subunit, whereas an unspecific PBR/CBR ligand like diazepam binds to a site consisting of both VDAC and 18 kDa subunits. The cDNA for the 850-nucleotide PBR mRNA as been cloned for various species, including humans and rodents. The genes for humans and rats have been partially cloned and characterized, and the human gene was found as a single copy located on chromosome 22 in the 22q13.31 band. The physiological functions of PBR have not been fully elucidated, and this receptor is thought to be associated with many biological functions, including cell growth and proliferation, steroidogenesis, bile acid synthesis, calcium flow, chemotaxia and cellular immunity, heme biosynthesis, and mitochondrial respiration. In the normal brain PBR are mainly identified in the olfactory bulb, periventricular regions, choroid plexus, and the prostrema and ependymal area. In the brain parenchyma, PBR are predominantly present in microglia and astrocytes and in low density under normal circumstances. However, during brain insult or inflammation, PBR expression is dramatically increased due to microglia activation. This response has been shown to be extremely rapid and robust after injury or inflammation (within minutes), regardless of brain cell type damaged. Saturation isotherms and Scatchard analysis of [^{3}H]PK11195 binding to PBR have demonstrated that this increase in binding was due to an increase in the maximal number of binding sites (B_{max}) without any change in the affinity of the ligand for the binding site. Numerous studies have demonstrated the presence of activated microglia in areas of neuronal degeneration in patients with a variety of neurodegenerative diseases including Parkinson's, Huntington's and Alzheimer's disease, amyotrophic lateral sclerosis, AIDS, multiple sclerosis, ischemia, stroke and epilepsy indicating that early, robust microglial activation may be a critical step in neurodegeneration. For general reviews on the topic, see bibliography.[385–392]

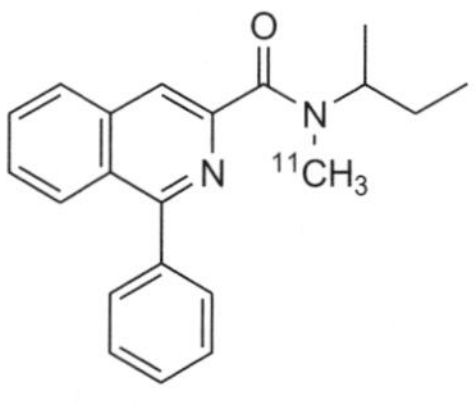

[^{11}C]PK11195   Ki = 8.9 nM [a]

Ki = 3.8 nM [b]

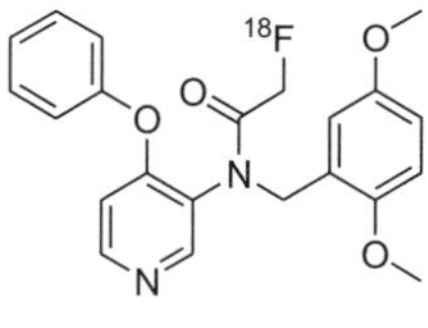

[^{18}F]PBR06   Ki = 0.30 nM [b]

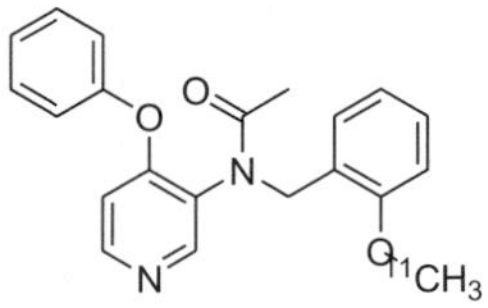

[^{11}C]PBR28 Ki = 0.2 nM [b]

Figure 3.11   Peripheral Benzodiazepine Receptor (TSPO) Ligands.
[a]Competition radioligand binding against [^{3}H]Ro5–4864 on rat aortic smooth muscle. [b]Competition radioligand binding against [^{3}H]PK11195 on rat brain mitochondrial membrane.

3.8.1 1-(2-Chlorophenyl)-N-[^{11}C]methyl-N-(1-methylpropyl)-3-isoquinoline carboxamide ([^{11}C]PK11195)

This tracer was synthesized by methylation with [^{11}C]methyliodide of its des-methyl precursor, using tetrabutylammonium hydroxide as base, in up to 80% yield with specific activity between 20 and 96 GBq µmol^{-1} and radiochemical purity 99%. Evaluation in baboon showed fast brain uptake, peaking within 5 min, followed by slow washout after the distribution phase; pretreatment or displacement with nonradioactive PK11195 did not affect the TAC, indicating a high non specific binding for this tracer.[393] Similar results were observed in healthy humans, with a relatively high and fast initial brain uptake, peaking at 1–3 min pi, followed by a fast distribution phase (ending around 10 min pi) and a slow washout thereafter. Plasma analysis revealed a relatively slow metabolic rate of this tracer, with unchanged tracer accounting for 96%, 70%, and 57% of the total activity at 5, 30, and 50 min pi respectively, with however a relatively high inter-subject variability, indicating the need of plasma sampling to accurately determine the input function.[394] The highest initial uptake was observed in the lungs, followed by the liver and urinary bladder at a later time, indicating both urinary and hepatobiliary excretion; the highest equivalent organ dose was observed in the kidney (14.0 µSv MBq^{-1}), spleen (12.4 µSv MBq^{-1}), and small intestine (12.2 µSv MBq^{-1}) with an effective dose of 4.8 µSv MBq^{-1} and 5.1 µSv MBq^{-1} (IPCR60 and IPCR 103 respectively).[395] In children, the effective dose was higher 11.6 µSv MBq^{-1} (aged 4 to 7) and 7.7 µSv MBq^{-1} (aged 8 to 12), with the highest dose received by the gallbladder wall (21.5 µSv MBq^{-1}).[396] Different kinetic models have been investigated. A two-tissue reversible compartment model with plasma input and fixed K_1/k_2 leads to the best fit when compared with a one-tissue, two-tissue irreversible model with or without fixing K_1/k_2 ratio or blood volume to cortex values.[397] A simplified reference tissue model, basis pursuit, and wavelet-based Logan plot gave the best intra-class correlation coefficient, whereas reference to target ratio and Logan graphical method were less reliable and sensitive.[398–400] Pathological states complicate the quantification since PBR are expressed in the brain vasculature, which is affected, for example, in Alzheimer's disease.[401] [^{11}C]PK11195 was evaluated in Parkinson's disease patients, and showed a significant increase in midbrain binding potential in the contralateral side of the symptoms when compared with healthy subjects, correlating with the severity of the symptoms in the early stage of the disease; moreover, a significant inverse correlation was found between the midbrain binding potential of this tracer and the binding potential of [^{11}C]CFT (dopamine transporter imaging agent) in the putamen of the same PD subjects.[402,403] However, in a two-year longitudinal study, no significant changes were observed in tracer uptake of the PD group (despite a significant increase in tracer binding in pons basal ganglia, frontal and temporal cortices of the PD group *versus* control), suggesting that activation of microglia is an early event in PD.[404] In a pilot study assessing the effect of two-month treatment with the COX-2 inhibitor celecoxib, no significant change in tracer uptake could be detected.[405] In Huntington's disease, significant increase in striatal binding potential has been observed in patients, correlating with the severity of the disease using the Huntington's Disease Rating Scale, patients' CAG index, and striatal [¹¹C]raclopride BP (dopamine D2 receptor imaging agent). [¹¹C]PK11195 binding was also significantly increased in cortical regions of HD when compared with the control group.[406] An elegant study evaluated this tracer in presymptomatic Huntington's disease gene carriers, showed a significant increase in striatal and cortical tracer binding *versus* control as well as an inverse correlation between [¹¹C]PK11195 and [¹¹C]raclopride in the striatum, indicating that the microglial activation in HD is an early event.[407,408] Glial activity was assessed with this tracer in demented and non demented HIV patients with contradictory outcome. In one study,[409] both demented and non demented HIV group showed a significant increase in binding in the thalamus, putamen, frontal, temporal and occipital cortices when compared to controls; with only a slight, non-significant, increase in BP observed between the demented group and non demented. However another study,[410] did not show any differences between the HIV group (demented or non demented) and the control group. In Alzheimer's patients, significant increase of [¹¹C]PK11195 binding was observed in the entorhinal, temporoparietal, and cingulate cortices, albeit with significant overlapping of the two groups.[411] A previous study using the racemic mixture of this tracer did not detect any differences between AD and healthy control.[412] A case study on a patient with glioblastoma showed a clear increase (two fold) of the tracer uptake at the tumor site; and part of the signal (30%) could be displaced by carrier levels of PK11195.[413] An increase of tracer uptake was also observed in other pathologies, including herpes encephalitis,[414] multiple sclerosis,[415] amyotrophic lateral sclerosis (ALS),[416] frontotemporal dementia,[417] corticobasal degeneration,[418] and schizophrenia.[419] A major limitation on the use of this tracer is its low signal-to-noise ratio and high non-specific binding, leading to its characterization as "an unsuitable tracer for accurate or reliable quantification of neuroinflammation."[405]

3.8.2   N-[¹⁸F]Fluoroacetyl-N-(2,5-dimethoxybenzyl)-2-phenoxyaniline ([¹⁸F]PBR06) and N-acetyl-N-(2-[¹¹C]-methoxybenzyl)-2-phenoxy-5-pyridinamine ([¹¹C]PBR28)

[¹⁸F]PBR06 was synthesized by nucleophilic displacement of its bromo precursor with [¹⁸F]fluoride in 69–79% yield with specific activity between 3.4 and 9 Ci μmol⁻¹ and radiochemical purity >99%. In nonhuman primates, this tracer showed fast and high brain uptake, peaking around 300% SUV one hour pi, followed by gradual washout thereafter. Pretreatment with DAA1106 (3 mg kg⁻¹ i.v.) led to higher and faster initial brain uptake, peaking around 600% SUV 5 min pi, followed by fast washout, leading to an estimated specific binding around 80–90% of the total activity. The distribution of [¹⁸F]PBR06 was relatively uniform in all brain regions, with higher uptake in the gray than white matter and the highest activity in the fourth ventricle of the choroid plexus.[420,421] Similar time-activity curves were observed in healthy humans, with a fast peak uptake 3 min pi (170% SUV), followed by a gradual washout (50% and 30% of initial peak were reached at 60 and 240 min pi, respectively), without significant bone uptake. Plasma analysis showed the presence of two polar metabolites, accounting for 47% and 16% of the total activity 30 min pi. The best fit of the data was obtained using an unconstrained two compartment model, and the optimal scanning time was estimated at 120 min, as later times might reflect possible accumulation of radiometabolites in the brain.[422–424] Whole body imaging in humans showed a high initial uptake in the lungs, quickly distributing and presumably representing the blood flow. Radiation dosimetry estimates placed the highest absorbed dose for the gallbladder wall (367μSv MBq^{-1}, potentially limiting organ), spleen (64 μSv MBq^{-1}), and liver (46 μSv MBq^{-1}), indicating hepatobiliary excretion of the tracer. The estimated effective dose equivalent (EDE) was 18.5 μSv MBq^{-1}.[425] To date, no evaluation of this tracer has been reported in pathological states.

3.8.3 N-Acetyl-N-(2-[^{11}C]methoxybenzyl)-2-phenoxy-5-pyridinamide ([^{11}C]PBR06)

This tracer was synthesized by methylation of its phenol precursor with [^{11}C]methyl triflate, (using sodium hydride as base) in 70–80% yield, with specific activity 5–15 Ci μmol^{-1} and radiochemical purity exceeding 93%.[426] Evaluation in monkey showed relatively slow brain uptake peaking at 300% SUV around 40 min pi followed by a slow washout. Pretreatment with DAA1106 (3 mg kg^{-1} i.v.) results in a fast brain uptake peaking at 500% SUV 4 min pi followed by fast washout from the brain, leading to an estimated 76% of specific signal.[427,428] Biodistribution in humans showed high initial uptake in the lungs, reaching 35% ID at 2 min, all other organs peaked within 10 min pi and reached a maximum of 15% ID in the liver, 14% kidneys, 5% brain, 4.5% spleen, and 3.7% in the heart. The estimated radiation exposure was highest in the kidney (52.6 μSv MBq^{-1}), spleen (25.9 μSv MBq^{-1}), lungs (22.0 μSv MBq^{-1}), and urinary bladder wall (13.3 μSv MBq^{-1}), with an effective dose equivalent (EDE) of 6.6 μSv MBq^{-1}.[429] The only report published to date of [^{11}C]PBR06 in a pathological state is the detection of subacute brain infarction as an incidental finding during a research study using this tracer.[430]

The promise of both PBR06 and PBR28 for broad application has been muted by the discovery of a subgroup of patients showing no binding of [^{11}C]PBR28 and representing around 10% of the population.[429,431] A thorough *in vitro* study demonstrates the existence of three different binding populations toward [^{3}H]PBR28, the high affinity group representing 46% of the 22 brain donors studied, low affinity 23%, and two-site binders 31% (high and low affinity), whereas all groups showed similar high affinity towards [^{3}H]PK11195.[432,433]

3.9 Phosphodiesterase Inhibitors

Cyclic nucleotide phosphodiesterase (PDEs) are enzymes that specifically degrade cyclic adenosine monophosphate (cAMP) and/or cyclic guanosine monophosphate (cGMP) produced during the signal transduction of a G-protein coupled receptor, and therefore regulate the intracellular levels of these ubiquitous second messengers. The concentration, localization and duration of action of cyclic nucleotide within the cell are extremely important in the transduction signal and will influence the activation of many enzyme (protein kinase A and C) and transcription factors. Any dysregulation of the tightly controlled cyclic nucleotide parameters results in aberrant cellular response. To date eleven phosphodiesterase (PDE1–11) families have been characterized, and most of these families are expressed by more than one gene, and have multiple splice variants leading to a total of more than 100 human PDE. All PDE families have been found in the brain with the exception of PDE6 and PDE11, with specific expression of different isoforms in distinct areas of the brain. The PDE families can be divided into three groups, the cAMP specific including PDE4, 7, and 8, the cGMP specific including PDE5 and 9, and the dual acting PDE1, 2, and 10. To date only a few families have selective inhibitors, that often show no selectivity for the different family isoforms (PDE2, 4, 5, 9, and 10), but the field is rapidly growing and isoform selective inhibitors will be available in the future. PDE4 has been one of the most studied phosphodiesterases, and is encoded by four distinct genes expressing PDE4A-C, all expressed in the human brain with the exception of PDE4C. PDE4 has been implicated in the signaling cascade mediating memory and learning both in healthy and pathological models of cognitive impairment. The selective PDE4 inhibitor rolipram, reverses the amyloid beta induced deficit of long term potentiation *in vitro*, and alle*via*tes cognitive deficit in double mutant mouse model of Alzheimer's disease. Antipsychotic drugs are, in their vast majority, antagonists of the D2 receptors and therefore increase the intracellular levels of cAMP since D2 receptors are negatively coupled to adenylate cyclase. It was postulated that inhibitors of PDE4 could accomplish similar effects without the intrinsic motor

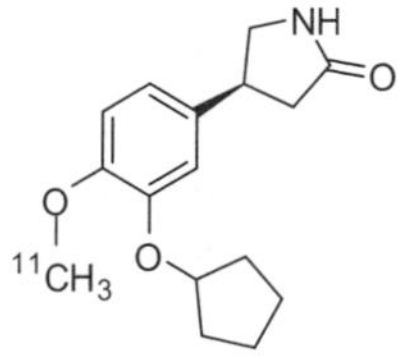

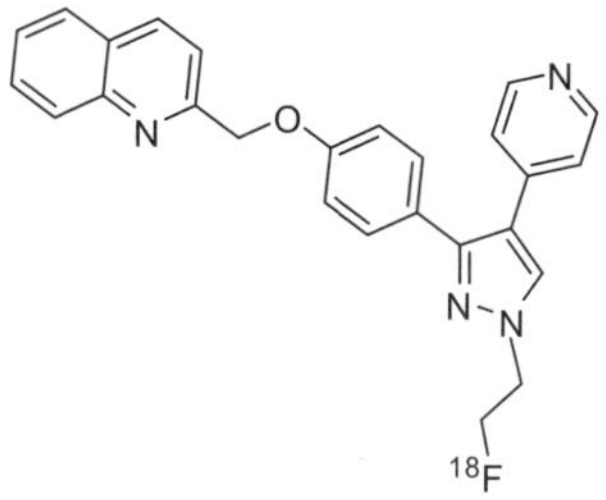

[^{11}C]-*R*-Rolipram	K_d = 0.93 nM [a]	[^{18}F]JNJ41510417 [c]
[^{11}C]-*S*-Rolipram	K_d = 4.3 nM [a]	
[^{11}C]-*RS*-Rolipram	K_d = 1.2-2.4 nM [b]	

Figure 3.12   Phosphodiesterase Inhibitor Ligands.
[a]Dissociation constant in rat forebrain membrane fractions. [b]Dissociation constant in rat forebrain membrane and cytosolic fraction, respectively. [c]No pharmacological data published.

side effect of all D2 antagonists, and recent studies have shown the efficacy of rolipram in pre-pulse inhibition models of schizophrenia. PDE10A is the only member of the PDE10 family with at least two splice variants (PDE10A1 and PDE10A2), and regulates both cAMP and cGMP. PDE10A is found in the brain predominantly in the caudate and putamen on the membrane of GABAergic medium spiny neurons, and this localization suggests an implication of PDE10A in several neurodegenerative and psychiatric disorders including Huntington's, Parkinson's disease, schizophrenia, and obsessive-compulsive disorders. Recent reports have shown that selective PDE10A inhibitors (papaverine, MP-10, TP-10) show promising results in animal models of schizophrenia. For general reviews on the topic, see bibliography.[434–443]

3.9.1 PDE4

To date, rolipram is the only imaging agent for PDE4, it binds to all four human isoforms (PDE4A-C) with comparable affinity, the R enantiomer is the more active (Kd = 1–5 nM),[444,445] and its image shows a decrease affinity of 10–30-fold.[445,446] A striking difference exists between the catalytic inhibition activity of PDE4 by rolipram and its binding affinity to this enzyme; its functional potency for inhibiting cAMP hydrolysis is around 1000 nM and its binding affinity for PDE4 is 1000 times more potent.[447,448] While this discrepancy needs to be addressed in the development of PDE4 inhibitors, it does not influence the development of a PDE4 imaging agent, where the crucial parameter is the affinity of the ligand to the enzyme and not the functional outcome.

[^{11}C]Rolipram is synthesized by [^{11}C]methylation of desmethyl-rolipram with specific activity > 400 Ci μmol^{-1} and radiochemical purity > 95%. This tracer was first evaluated in rats, where a good brain uptake was observed after i.v. injection followed by a gradual washout. The regional brain distribution was in accordance with the distribution of [^{3}H]rolopram *ex vivo* in rat (frontal cortex, remainder of the cortex, olfactory bulb), and specificity was demonstrated by pretreatment with specific PDE4 inhibitors (cold rolipram and Ro-20–1724 decreasing tracer uptake) and by PDE1,2,3,5 selective inhibitor (vinpocetine, Bay 60–7550, cilostazol, and zaprinast) and noradrenalin re-uptake inhibitor (desipramine) did not affect the signal.[449,450] Comparison between the different enantiomers R-, S- of rolipram in rat and pig, not surprisingly shows that the R-enantiomer is the active imaging agent of the racemic mixture, since S-enantiomer shows a uniform distribution in all brain regions and a faster washout than R-enantiomer (0.1% ID *vs.* 0.7% ID @ 45 min in the brain for S- and R-enantiomers respectively),[451] and [^{11}C]S-rolipram was proposed as a measurement of the non specific binding component of [^{11}C]R-rolipram.[452] [^{11}C]R-rolipram was extensively metabolized in the blood with only 20% of unchanged ligand after 30 min, but the metabolites did not appear to cross the blood-brain barrier (presumably due to their polar nature) since only authentic [^{11}C]R-rolipram was detected in the brain. *In vivo* quantification in rat was successfully achieved using [^{11}C]R-rolipram by applying an unconstrained two-tissue compartment model, using arterial input (since no reference tissue is available for PDE4), showing that up to 86% of the V'$_T$ was specific binding.[453] Consistent with rat results, [^{11}C]R-rolipram was successfully used to image PDE4 in human. The tracer shows a high and fast brain uptake picking at 10 min pi (with the highest uptake in the thalamus reaching 5%ID @ 10 min), followed by a gradual washout. The high thalamic uptake decreased to the level similar of other PDE4 rich regions (cortical and striatal areas) after 40 min, and the lowest binding was observed in the cerebellum. The tracer was metabolized up to 50% at 90 min into more polar metabolites, and the plasma analysis shows a high level of protein binding for the tracer around 96%. Kinetic modeling using a two-tissue compartment and arterial plasma input was successful in fitting the time activity curve.[454] Since PDE4 activity is closely related to the concentration of cAMP in its surroundings, it was postulated that an agent capable of increasing cAMP levels should produce an increase in [^{11}C]R-rolipram binding *in vivo*, and an elegant study in rats has proven this hypothesis. [^{11}C]R-rolipram biodistribution (brain, lung, heart) was evaluated after acute treatment that selectively increases the synaptic levels of noradrenaline (tranylcypromine (MOAI), desipramine, maprotiline), serotonin (tranylcypromine, fluoxetine, sertaline), histamine (thioperamide (H3 antago)), and dopamine (tranylcypromine, GBR 12909 (DAT inhib.), cocaine, SKF81297 (D1 agonists)). In all cases except for dopamine, a dose dependent increase of [^{11}C]R-rolipram binding in the brain was observed, proving the feasibility of imaging change in PDE4 activity in the brain using *in vivo* imaging.[455] Whole-body dosimetry was evaluated in both human and monkey and the radiation dose estimate was the highest in human for the gallbladder wall (23 µGy MBq^{-1}), kidneys (19.7 µGy MBq^{-1}), urinary bladder wall (19.3 µGy MBq^{-1}), liver (15.2 µGy MBq^{-1}), heart wall (11.8 µGy MBq^{-1}), and lungs (9.6 µGy MBq^{-1}), indicating an hepatobiliary and urinary excretion, with EDE 4.8 µGy MBq^{-1}. The distribution of radioactivity was different in monkey and human, leading to a 40% overestimate of the human effective dose when estimated from monkey distribution.[456] In a recent study, the effect of isoflurane anesthesia on Kd and B$_{max}$ was evaluated in rats and shows an increase of both parameters in conscious rats (59% and 29% respectively), as well as 3–7 fold higher Kd values obtained from *in vitro* binding compared with the conscious *in vivo* value (B$_{max}$ was not affected in this case.).[457]

3.9.2   PDE10

Only very recently PDE10 has became a target for *in vivo* imaging, and to our knowledge only two tracers have been reported at the time of this manuscript preparation,[458,459] with only the structure of [^{18}F]JNJ41510417 disclosed. [^{18}F]JNJ41510417, was prepared by nucleophilic substitution of the corresponding mesylate precursor by [^{18}F]fluoride, with specific activity around 8600 Ci mmol^{-1} and radiochemical purity exceeding 99%. This tracer was evaluated in rats and shows accumulation in the striatum after i.v. injection peaking at 100 min, with only residual signal in the cerebellum (reference region) and cortex. The maximum ratio striatum to cerebellum was around 5 and can be reduced to less than two by pre-treatment with specific PDE10 inhibitors; similarly displacement of the tracer rapidly brings the ratio from 3 to less than 1.5 after injection of a selective PDE10 inhibitor. *In vivo* imaging of PDE knock-out mice show no specific uptake when compared to wild type, and *in vitro* autoradiography on striatal slide shows a 30-fold lower binding for knock-out.[458]

In vivo evaluation in baboon of a [^{11}C]tracer for PDE10 was also recently reported; this tracer inhibits PDE10 with an Kd = 1 nM (human striatal slide binding) with a selectivity > 700-fold for the other PDE subtypes. Evaluation in rats shows an excellent striatal uptake with a ratio striatum: thalamus greater than 9 at 5 min pi. The tracer shows specific accumulation in baboon striatum which can be reduced by pre-treatment with a cold PDE10 inhibitor correlated with plasma levels of the inhibitor in a dose dependent manner. It should be noted that this tracer has a short residence time with the striatal signal reaching cerebellar and thalamic levels in less than an hour.[459] These preliminary results will encourage further evaluation of those tracers.

3.10 Adenosine Receptor A1 and A2A

Extracellular adenosine (released either by ecto-nucleotidase-mediated hydrolysis of extracellular adenosine nucleotides, or through an equilibrative transport) can bind to cell surface adenosine receptors (ARs) to produce a broad variety of physiological responses. ARs are members of the G-protein-coupled receptors (GPCRs) superfamily and are currently divided into four subtypes, A1, A2A, A2B, and A3 based on their sequence (all four ARs have been cloned), tissue distribution, and effector coupling. Activation of A1 receptor inhibits adenylate cyclase through a G_i protein, resulting in increased activity of the phospholipase C (PLC). Upon activation, adenosine A2A receptors increase adenylate cyclase through a G_s protein in the peripheral system and a closely relate G_{olf} protein in the striatum. The A2B receptors are positively coupled to adenylate cyclase and PLC, and interact with the arachidonic acid pathway upon activation. The last member of the family, adenosine A3 inhibits adenylate cyclase and stimulates calcium mobilization and PLC. All four ARs are wildly expressed throughout the human body, A1 was found in the brain, heart, aorta, liver, eye and bladder. A2A receptors were found in the spleen, thymus, leukocytes, heart, and blood platelets. In the brain, A2A is highly expressed in striatopallidal GABAergic neurons (caudate-putamen, nucleus accumbens) as well as in the olfactory bulb. A2B receptors are ubiquitous and express in virtually every tissue, with the highest mRNA expression in the cecum, colon, bladder, blood cells, lung, and mast cells. A3 adenosine receptors are found mainly in lung, and liver, and at a lower level in the brain and heart.

Brain imaging of adenosine receptors is limited to A1 and A2A subtypes, principally due to the lack of tissue specificity of A2B and low levels of A3 receptors in the brain, as well as the absence of selective tracers for those subtypes. A1 imaging ligands have mainly been developed as markers of ischemia in the brain, since its density and expression sharply decrease during cerebral ischemia and stroke. The tissue specific expression of A2A in the basal

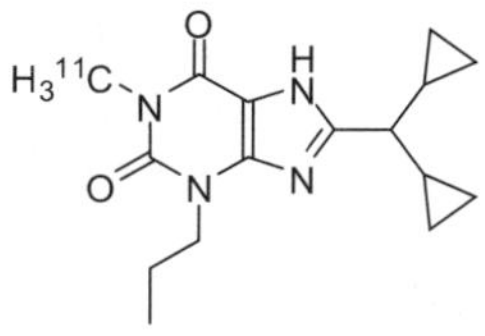

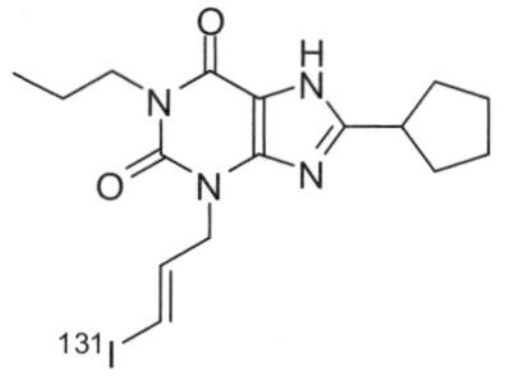

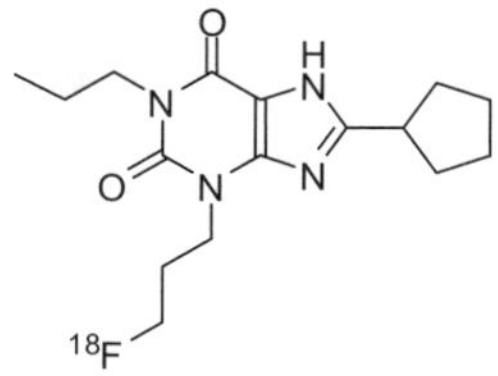

[^{11}C]MPDX K$_i$ = 4.2 nM ^{a} [^{131}I]CPIPX K$_d$ = 5.8 nM ^{b} [^{18}F]CPFPX K$_d$= 1.26 nM ^{c}

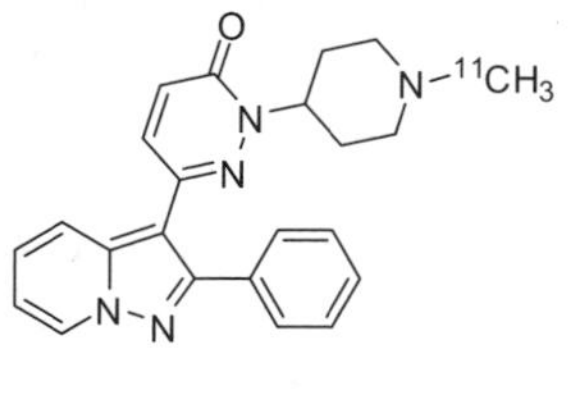

[^{11}C]FR194921 K$_i$ = 2.9 nM ^{d}

Figure 3.13   Adenosine Receptor A1 Ligands.
^{a}Inhibition constant using [^{3}H]N6-cyclohexyladenosine on rat forebrain membrane. ^{b}Dissociation constant in HEK cells expressing human A1 receptors. ^{c}Dissociation constant using CHO cells expressing human A1 receptors. ^{d}Inhibition constant using [^{3}H]DPCPX on CHO cells expressing human A1 receptors.

ganglia and the co-localization of this receptor with dopamine D2 receptors (95% of A2AmRNA expressing neurons express D2mRNA) in the striatum, have prompted the development of A2A imaging agents in the field of movement disorder (Parkinson's and Huntington's disease). For general reviews on the topic, see bibliography.[460–471]

3.10.1   8-Dicyclopropylmethyl-1-[^{11}C]methyl-3-propylxantine ([^{11}C]MPDX)

This ligand was originally selected from a set of three compounds ([^{11}C]MPDX, [^{11}C]EPDX, [^{11}C]KF15372 – the methyl, ethyl and propyl analogs respectively) due to its simplified synthesis leading to high radiochemical yield.[472] Initially [^{11}C]MPDX was synthesized by [^{11}C]methylation of 8-dicyclopropylmethyl-3-propylxantine using [^{11}C]methyl iodine,[472] and was later improved by using [^{11}C]methyl triflate,[473] (higher radiochemical yield, less precursor needed, shorter reaction time) leading to radiochemical yield of 34%, with specific activity 18–108 TBq mmol^{-1} and radiochemical purity > 95%. The tracer evaluated in rat, shows a fast brain uptake (2.5 %ID g tissue @ 5min) followed by a gradual clearance; the specificity of the signal was determined by pre-treatment with A1 and A2A antagonist (KF15372 and KF17837) leading to a decreased and unchanged tracer uptake respectively.[472] Regional brain distribution was evaluated in cat (comparison of distribution volume (DV) using a two-tissue compartment four parameter model) revealed a high DV in cerebral cortex, striatum, and cerebellum followed by thalamic and midbrain regions, in accordance with the *in vitro* distribution in cats. Displacement with DPCPX (A1 antagonist) rapidly displaces the activity from cerebral cortex as well as all the other brain regions, similarly pretreatment with DPCPX lead to a fast decrease of the activity after initial uptake for all brain regions. The tracer shows a fast metabolic rate in plasma, and only 19% of unchanged [^{11}C]MPDX was detected at 15 min pi.[474] Acute toxicity was assessed in rat at a dose 41 000 times higher than a routine tracer dose, and animals show no clinical differences from control over a 15 day period. Dosimetry in mice shows the highest initial uptake in the liver, kidneys, lungs, heart, pancreas, small intestine, and brain followed by a gradual washout from all regions except for the small intestine where accumulation increased for 60 min. From those data, the human radiation absorbed dose was extrapolated and revealed the highest dose for the upper and lower large intestine wall (6.13 and 6.59 μGy MBq^{-1} respectively). Evaluation of [^{11}C]MPDX in monkey shows a fast brain uptake peaking at 5 min pi, widely distributed in all brain regions, followed by relatively uniformed washout; while the distribution is mostly uniform, ranking can be distinguished in the cortex occipital > temoral > parietal > frontal and thalamus > cerebellum > brain stem. As in other species the tracer is relatively quickly metabolized, with 54% of unchanged tracer at 30 min pi.[475] In human, the tracer shows a similar initial brain uptake to other species followed by a faster washout than in rat or monkey. the metabolic rate was also different in human, where the tracer is more stable and accounts for 70% of the activity in plasma after 60 min (25% in rat, 6% in cat, 41% in monkey).[476] Quantitative analysis of the tracer in human shows a good correlation between the estimated binding potential (using a two-tissue three-compartment model) and the estimate made by a Logan plot. Using the cerebellum as reference region (the distribution volume was significantly lower), binding potential was the highest in striatum (1.04), thalamus (1.01), occipital cortex (0.94), parietal cortex (0.89), and temporal cortex (0.85).[477] Simplification of the quantification by omitting the arterial input was successfully achieved using independent component analysis.[478] [^{11}C]MPDX was finally evaluated in Alzheimer's patients where the binding potential was calculated for the following region of interest: frontal, medial frontal, temporal, medial temporal, parietal, and occipital cortices, striatum, thalamus, cerebellum, and pons. Alzheimer's patients show a significant reduction of binding potential in temporal, medial temporal cortices and thalamus when compared to aged match healthy subjects, and this reduction was greatest for the medial temporal cortex where the BP was reduced to the levels of the cerebellum in accordance with post mortem studies.[479]

3.10.2   8-cyclopentyl-3-[(E)-3-iodoprop-2-en-1-yl]-1-propylxanthine ([^{131}I]CPIPX)

[^{131}I]CPIPX was synthesized by classical iododestannylation of its tributyltin precursor, using chloramine-T as oxidant, with specific activity of 33 GBq μmol^{-1} radiochemical purity >98%, and a lipophilicity logP$_{oct}$ = 2.96.

Saturation experiments demonstrated its selectivity toward hA1 receptors *versus* hA2A and rA3 receptors (Kd of 5.8, 131, and 85 nM, respectively). Autoradiography studies revealed high non-specific binding for this tracer of 50–75%. *In vivo* evaluation showed good brain uptake at 5 min pi (1.6% ID g in mice and 1.3% ID g in rats) followed by a relatively fast washout; however, blocking with CPDPX showed that only 14% of the brain activity was specific at 5 min.[480] Those results clearly make this ligand unsuitable for SPECT imaging of A1 receptors, and no further evaluation was reported.

3.10.3   8-Cyclopentyl-3-(3-[18F]fluoropropyl)-1-propylxanthine ([18F]CPFPX)

Nucleophilic substitution of the corresponding tosylate precursor by [18F]fluoride led to [18F]CPFPX with radiochemical yield around 45%, specific activity > 7500 Ci mmol^{-1} and radiochemical purity $> 98\%$.[481] *In vitro* autoradiography in rat shows a high specific binding in the cerebellum, neocortex and thalamus, with only 7% of non specific binding, and low levels were found in midbrain and brain stem. *In vivo* the tracer shows a fast brain uptake peaking at 2.4% ID at 2 min pi and remaining constant for 40 min and then slowly washing out of the brain in mice;[481] similar results were observed in rats, with a high uptake peaking before 5 min and a relatively slow washout (67% reduction @ 60 min). Displacement with DPCPX leads to a rapid and significant decrease of binding (62% reduction with 1 mg kg^{-1} @ 6 min), but no significant changes were observed by pretreatment or displacement using the A2A antagonist DMPX.[482] Evaluation of this tracer in a rat model of glioma reveals an increased uptake in the zone surrounding the tumor (136–146%) attributed to activated astrocytes upregulating A1 receptors; consistent with those results from the imaging of a patient with recurrent glioblastoma showing high accumulation of [18F]CPFPX around the tumor accompanied by a lower accumulation in the center of the tumor co-localized by MRI.[483] Quantification of A1 receptors in healthy volunteers employing either a bolus injection,[484] (two-tissue compartment model, and Logan plot) or a bolus-to-infusion paradigm,[485] (equilibrium was attained within 60 min) all show good correlation and were further simplified by using venous instead of arterial blood sampling.[486] The highest receptor density (estimated from equilibrium total distribution volume) were found in the striatum (0.89), occipital (0.86), temporal (0.86), orbitofrontal (0.84) cortices, thalamus (0.83), and the lowest was found in cerebellum (0.45). Infusion of the unlabeled tracer results in a dose dependent reduction of tracer binding except for the cerebellum where floor effect seems to occur. Estimate of specific binding was around 66% for the cortical region and only 33% of the activity in the cerebellum.[487] Metabolite analysis in humans, rats, and mouse shows a fast degradation of the tracer in plasma with only 20% of unchanged tracer after 20 min. At least seven main metabolites were detected with variation in their relative amounts with each species, the estimated half-life of the tracer was 11 min in human. All metabolites were more polar than [18F]CPFPX and no defluorination or cleavage of the fluoropropyl moiety was observed.[488] Test-retest experiment shows a low (5.9% variation on BP2) to moderate (13.2% variation DV_t') variability of the tracer binding for eight brain regions (low to high density), in drug-free, non-smoking, and caffeine-refraining subjects.[489] Whole-body distribution and dosimetry in human shows a high liver uptake, peaking at 35% ID at 40 min. The highest radiation was received by the gallbladder (136 μSv MBq^{-1}), liver (84 μSv MBq^{-1}) and the urinary bladder (78 μSv MBq^{-1}), which limits the injected dose under US regulation to a comfortable 9.95 mCi.[490]

3.10.4 $[^{11}C]$2-(1-Methyl-4-piperidinyl)-6-(2-phenylpyrazolo[1,5-a]pyridine-3-yl)-3(2H)-pyridazinone ($[^{11}C]$FR194921)

This tracer is the only non-xanthine ligand developed for imaging A1 receptors to date. It was synthesized by $[^{11}C]$methylation of its desmethyl precursor, using $[^{11}C]$methyl iodine with radiochemical yield of 38%, with specific activity 25 GBq μmol^{-1} and radiochemical purity $>99\%$.[491] $[^{11}C]$FR194921 shows a good binding for human A1 receptors, and good selectivity toward A2A and A3 subtypes (no binding was detected up to 1 μM).[492] In rat experiments the tracer shows a relatively low brain uptake when compared to the other A1 tracers previously discussed, gradually increasing to reach a peak of 0.03% ID 30 min pi, followed by a gradual washout. The highest level was found in the hippocampus, striatum, and cerebellum in accordance with A1 distribution. Pretreatment with DPCPX (3.2 mg kg^{-1}) only reduce the tracer accumulation by 50%, which might

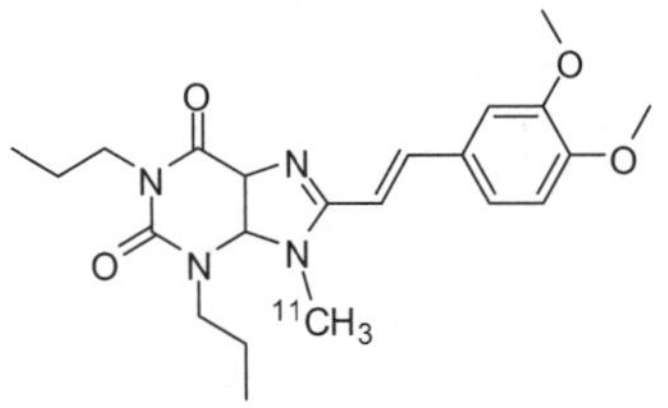

$[^{11}C]$KF17837 K_i = 1.0 nM [a] 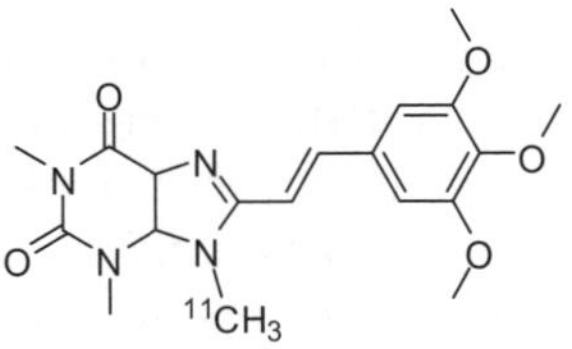 $[^{11}C]$KF18446/$[^{11}C]$TMSX K_i = 5.9 nM [a]

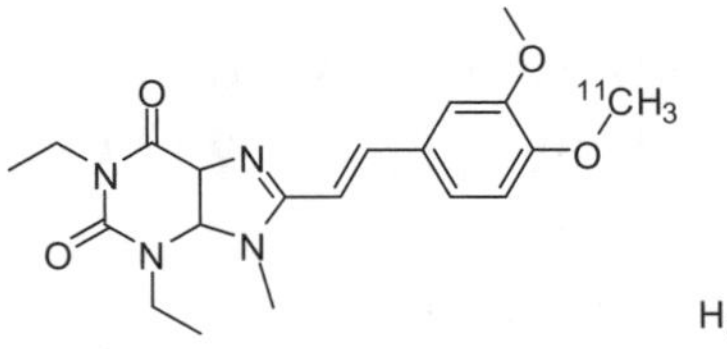

$[^{11}C]$KW6002 K_i = 2.2 nM [ε] $[^{11}C]$SCH442416 K_i = 0.048 nM [b]

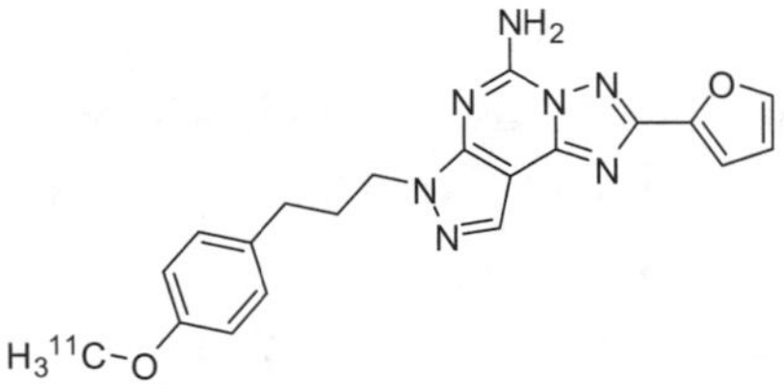

Figure 3.14 Adenosine Receptor A2A Ligands.
[a]Inhibition constant in rat striatal membrane using $[^3H]$CGS 21680.
[b]Inhibition constant in HEK-293 cells expressing human A2A using $[^3H]$SCH58261.

indicate a high non-specific binding. Imaging *in vivo* in conscious monkey shows a similar profile, with continuous accumulation of [^{11}C]FR194921 during the first 40 min, and a subsequent equilibrium up to 90 min (end point of the scan). All brain regions show relatively similar accumulation, with higher levels in the occipital and temporal cortices. Whole-body distribution in rat demonstrate accumulation in the liver after injection and subsequent displacement to the small intestine, suggesting an hepatobillaric excretion.[491] No further evaluation of this tracer was reported since the publication of those results.

3.10.5   (E)-8-(3,4-Dimethoxystyryl)-1,3-dipropyl-7-[^{11}C]-methylxanthine ([^{11}C]KF17837)

This tracer was synthesized by methylation of its nor-precursor using [^{11}C]methyliodide with radiochemical yield of 50% (sodium hydride) and 80% (potassium carbonate); the radiochemical purity exceeded 99% and the specific activity was 17–100GBq µmol^{-1} at the end of synthesis.[493] Evaluation in mice shows a slow brain uptake peaking at 1.74% ID after 30 min followed by gradual washout. No significant difference in regional brain uptake were observed during the first 30 min (uptake period), but washout was faster in cerebellum and cortex regions than in striatum after that time, leading to an increased ratio striatum/cerebellum with time reaching 2 at 60 min.[494] The specificity was confirmed by co-injection of cold KF17837, which led to increased tracer uptake in the first 5 min pi followed by a fast washout of the activity to levels below the untreated group in all brain regions including the striatum. The first evaluation of [^{11}C]KF17837 in cynomolgus monkeys was disappointing with low brain uptake, similar retention in all brain regions, and marginal decrease of binding when co-injected with cold compounds, and the authors conclude that this tracer was unsuitable for *in vivo* imaging.[493] However, another team reported slightly better results in rhesus monkey at higher tracer dose (28.5 MBq kg^{-1} *vs.* 4 MBq kg^{-1}), the highest uptake was found in the striatum followed by cerebellum, thalamus and cortex, with a ratio striatum: cerebellum of 1.32 at 30 min pi. Since the cerebellum is essentially devoid of A2A receptors, the cerebellar uptake can be assimilated to the non-specific binding leading to an estimated striatal specific binding of this tracer of only 28%.[495]

3.10.6   [7-Methyl-^{11}C]-(E)-8-(3,4,5-trimethoxystyryl)-1,3,7-trimethylxanthine ([^{11}C]KF18446 or [^{11}C]TMSX)

[^{11}C]KF18446 was prepared by methylation of its desmethyl precursor with [^{11}C]methyliodide or [^{11}C]methyltriflate (as we report earlier for [^{11}C]MPDX, the use of methyltriflate increased the yield) with radiochemical yield of 25–55%, a specific activity 10–72 GBq mmol^{-1}, and radiochemical purity >99%.[473,496] In mouse, the tracer shows a fast brain uptake peaking at 2.3% ID at 5 min pi, followed by a slow washout, with a distribution at all time points consistent with A2A distribution (high striatal uptake, equally low cerebellar and cortical uptake with a constant ratio striatum: cerebellum around 2 over 60 min). In mice, two polar radiometabolites were detected in plasma, with 80% of unchanged tracer at 30 min, but the tracer accounted for 98% of the activity in the brain at the same time. Displacement studies in mice, using cold tracer show a dose dependent reduction of [^{11}C]KF18446 in the striatum, without significant change in the other brain regions (cerebellum, cortex).[496] Further evaluation of the specificity of the signal was accomplished using various A2A and A1 ligands resulting in a significant reduction of tracer binding for all A2A ligands, as well as for the A1 antagonist KF15372.[497] Evaluation in rhesus monkey shows a fast uptake also, followed by a gradual washout of all brain regions, however the uptake ratio of striatum to other regions was significantly lower than in rat or mice, increasing with time up to 1.5 at 60 min pi. Metabolic rate was relatively higher in monkey with only 41% of unchanged [^{11}C]FK18446 at 30 min in plasma.[496] This tracer was evaluated in a rat model of Huntington's disease (unilateral quinolinic acid lesion of the striatum) where the binding potential was significantly lower in the lesion side (0.59 and 0.43 respectively).[498] Radiation dosimetry in mice from whole-body scans show the highest uptake in the kidneys and heart, and human extrapolation show radiation absorbed dose low enough for clinical use; acute toxicity and mutagenicity of [^{11}C]KF18446 were not found at doses 37 000 times higher than the imaging dose.[499] Evaluation in healthy humans showed a higher distribution volume in the head of the caudate (1.72), putamen (1.72), and thalamus (1.64) and lower in cerebellum (1.52) and cortical region (1.26).[500] Quantification in human was achieved using a two-compartment model for the centrum semiovale (use as reference region) and the main cortices, whereas a three-compartment model was used for the other brain regions. The estimated BP using a Logan plot was in accordance with the compartment analysis, and metabolite input only increased the BP by 5% (90% of unchanged tracer after 60 min)[501], and the ranking of BPs were: putamen (1.26), caudate (1.09), thalamus (1.06), cerebellum (0.85), brainstem (0.77), posterior cingulated gyrus (0.72), cortex (0.56–0.46).[501,502]

3.10.7 (E)-1,3-Diethyl-8-(3,4[^{11}C]-dimethoxystyryl)-7-methyl-3,7-dihydro-1H-purine-2,6-dione ([^{11}C]KW6002

This tracer was synthesized by methylation of its 4-desmethyl precursor using [^{11}C]methyliodide with radiochemical purity >95%.[503,504] Evaluation in rats shows a relatively good brain uptake with homogeneous distribution throughout the brain for 20 min followed by gradual washout from low A2A density brain region (cortex), and persisting in the A2A riche (striatum > cerebellum, olfactory tubercle), with a maximum ratio striatum: cortex around 2 from 60–120 min pi. The specificity was assessed by pretreatment with unlabeled KW6002 leading to a decrease of activity in all brain regions; while pretreatment with A1 antagonist KF15372 did not affect the tracer uptake and distribution. It should be noted that in this report, pretreatment with the selective A2A non-xanthine antagonist ZM241385 failed to show any effect on

[^{11}C]KW6002 distribution. Metabolite analysis revealed a relatively slow metabolic rate for this tracer with 66% of unchanged [^{11}C]KW6002 at 45 min pi in plasma; 97–87% of the activity in the brain was due to unchanged tracer at 30–90 min.[503] Evaluation in human with quantification and dose occupancy of the cold compound was studied using a two-tissue compartment model. The regional volume of distribution of this tracer was in accordance with A2A distribution (putamen > nucleus caudatus > nucleus accumbens) with lower but substantial activity in brain regions devoid of A2A receptors (thalamus, cerebellum). Daily chronic treatment with various doses of cold compound for 15 days before imaging, in healthy volunteers, show a uniform high occupancy >90% with doses greater than 5 mg, and the ED_{50} estimate from the total volume of distribution was 0.62 mg.[505]

3.10.8   5-Amino-7-(3-(4-[^{11}C]methoxy)phenylpropyl)-2-(2-furyl)pyrazolo[4,3-e]-1,2,4-triazolo[1,5-c]pyrimidine ([^{11}C]SCH442416)

[^{11}C]SCH442416 is the only non-xanthine imaging agent developed to date, and was synthesized by methylation of its desmethyl precursor using [^{11}C]methyliodide in 29% yield, with specific activity of 1500 mCi μmol^{-1}, with radiochemical purity greater than 95%.[506] Evaluation in rats shows high accumulation of the tracer in the liver (0.94% ID g), adrenal gland (0.85% ID g), kidney (0.69% ID g), lungs (0.38% ID g), and lower exposure to the brain (<0.2% ID g), all reaching a maximum uptake between 5 and 15 min, followed by gradual washout. Regional brain uptake was in accordance with A2A distribution with the highest uptake in the striatum reaching a ratio striatum: cerebellum of 4.6 at 15 min pi. The metabolic rate of [^{11}C]SCH442416 was relatively slow with 40% unchanged tracer at 60 min, the remaining activity was carried by two more-polar metabolites unlikely to cross the blood-brain barrier.[506] Signal selectivity was studied by pretreatment with xanthine (KW6002, caffeine) and non-xanthine (SCH58261) A2A antagonists, A1 antagonist (DPCPX), and dopamine D1 and D2 receptor antagonists (SCH23390, FESP respectively), leading to a reduction of striatal uptake for all A2A antagonist, whereas no apparent changes were observed for the A1 and dopamine antagonists. In quinolinic acid lesioned rats, a 50% reduction in striatum: cerebellum ratio was observed (QA 1.99 ratio, PBS 4.13 ratio). Evaluation in Macaca nemestrina shows a relatively low brain uptake reaching 0.035% ID g at 4 min; the tracer mainly accumulates in the striatum reaching a ratio striatum: cerebellum of 2.2 at 15 min and fairly stable for the following 30 min (1.5 @ 60 min). Similarly to the results observed in rats, the metabolic rate of [^{11}C]SCH442416 was slow in monkey, with the same two polar metabolites detected in rats. Quantification using a single exponential function for the striatum and a two-exponential function for the remainder of the brain gives an estimate of binding potential of 0.74 for the striatum, 0.16 cortex, and 0.13 for the cerebellum.[507]

3.11 Cannabinoid Receptors

Cannabis seems to be one of the oldest psychoactive substances, used (in traditional Chinese medicine) and abused (as recreational euphorigenic) for more than 4000 years. However, our understanding of the endocannabinoid system only took place in the last two decades. Two types of cannabinoid receptors have been cloned to date (CB1 and CB2 receptors) sharing an overall low homology (44% conserved sequence). Both receptors belong to the G-protein-coupled receptor super family and inhibit adenylate cyclase through a G_i/G_0 protein leading to a reduction of cAMP and an activation of mitogen-activated protein kinase (MAPK). In addition CB1 but not CB2, inhibits, upon activation, the presynaptic N and P/Q type calcium channels and activate G-protein-coupled inwardly rectifying potassium channels. The transduction pathway of cannabinoid receptors is not fully elucidated, and recent studies have shown that different CB1 agonists can stabilize certain pathways causing preferential transduction signals, and even the same agonist can elicit different transduction signals depending on its concentration. It was initially believed that CB1 receptors were expressed exclusively in the brain and testis, whereas CB2 receptors were limited to the periphery, but results in the last 10 years have indicated a broader range. CB1 receptors are widespread within the brain, with high expression in the hippocampus, cortical regions, cerebellum, and basal ganglia where they are predominantly expressed in a presynaptic position, and reduce the release of glutamate, norepinephrine, dopamine, and serotonin. In the periphery, CB1 has been found in heart, bladder, prostate, tonsils, and bone marrow. CB2 receptors are expressed mainly in immune cells (monocytes, B-cells, T-cells) and are almost absent in the healthy brain. However, upon activation, microglia (the resident macrophage of the brain) will express CB2 receptors. A growing body of evidence indicates that CB1 receptors can form homo and heterodimers (D2, A2A, opiate, GABA) leading to new physiological function of the dimer. The distribution pattern of CB1 receptor, mainly expressed in the extrapyramidal motor system (basal ganglia, cerebellum) makes it an attractive target for movement disorder such as Huntington's disease, Parkinson's disease, Tourette's syndrome, and multiple sclerosis. CB2 imaging agents are scarce, but the upregulation of this receptor in pathology such as Alzheimer's disease, multiple sclerosis, HIV-encephalitis, and astrocytic tumor, will surely encourage the development of CB2 selective imaging agents. For general reviews on the topic, see bibliography.[508-521]

3.11.1 CB1

3.11.1.1 N-(Piperidin-1-yl)-5-(4-[^{123}I]iodophenyl)-1-(2,4-dichlorophenyl)-4-methyl-1H-pyrazole-carboxamide ([^{123}I]AM251)

This tracer was synthesized by iododestannylation of its tributyltin precursor, using Chloramine-T as oxidizing agent in 50–60% yield, with specific

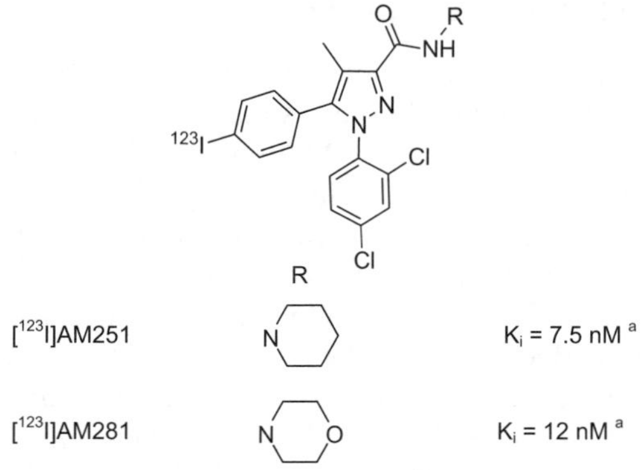

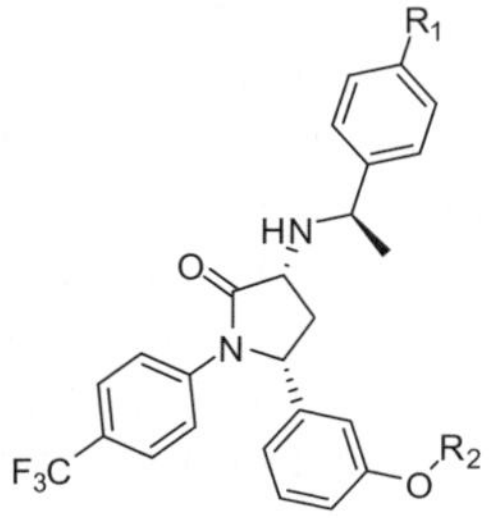

[¹²³I]AM251 K_i = 7.5 nM [a]

[¹²³I]AM281 K_i = 12 nM [a]

	R_1	R_2	K_i (nM) [b]
[¹¹C]MePPEP	H	[¹¹C]CH₃	0.47
[¹¹C]FMePPEP	F	[¹¹C]CH₃	0.21
[¹⁸F]FEPEP	H	[¹⁸F]FCH₂CH₂	0.42
[¹⁸F]FMPEP	H	[¹⁸F]FCH₂	0.18
[¹⁸F]FMPEP-D2	H	[¹⁸F]FCD₂	0.18

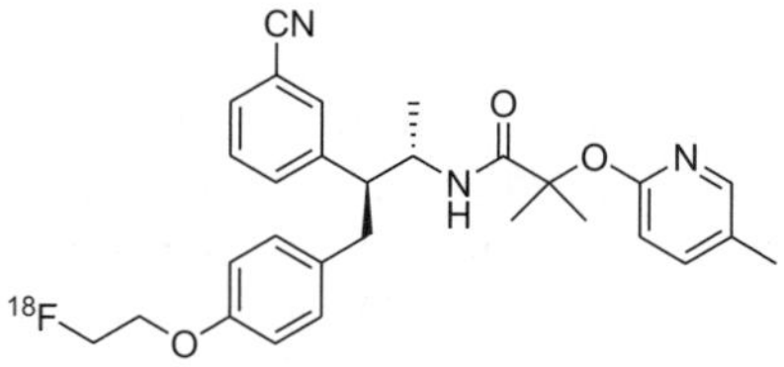

[¹⁸F]MK9470   IC₅₀ 0.7 nM

Figure 3.15 Cannabinoid CB1 Ligands.
[a]Inhibition constant using [³H]CP-55,940 on rat forebrain membrane homogenate. [b]Functional inhibition using GTPγ³⁵S in the presence of methanandamide in CHO cells expressing human CB1. [c]Half maximal inhibitory concentration using [³H]CP-55,940 on CHO cells expressing human CB1.

activity exceeding 5 Ci μmol⁻¹ and radiochemical purity >95%. Evaluation in mice show a relatively slow brain uptake, peaking at 1% ID at 60 min pi, followed by a slow gradual washout (half-clearance time of 8 h). The signal was specifically reduced by co-injection of the selective CB1 ligand SR141716A in a dose dependent manner in the brain (estimate ED_{50} = 0.1mg kg^{-1}) without any significant effect on the peripheral uptake (liver, kidneys, heart, lungs). The regional brain distribution was in accordance with CB1 distribution in mice, with high uptake in the cerebellum, striatum, and hippocampus. [^{123}I]AM251 is metabolized in the periphery into more polar metabolites accounting for 70–95% of the activity 2 h pi; the same metabolites were detected in the brain with only 70% of the activity due to unchanged tracer.[522] Despite those encouraging preliminary results, this tracer failed to show any accumulation in the baboon brain after i.v. injection.[523]

3.11.1.2   N-(Morpholin-4-yl)-5-(4-[^{123}I]iodophenyl)-1-(2,4-dichlorophenyl)-4-methyl-1H-pyrazole-carboxamide ([^{123}I]AM281)

This tracer, a close analog of [^{123}I]AM251, was synthesized by iododestannylation of the tributyltin precursor leading to similar results as for AM251 (yield > 50%, specific activity > 5 Ci μmol^{-1}, radiochemical purity > 95%). Evaluation in mice shows a fast brain uptake reaching a maximum of 3.3% ID at 30 min, followed by a gradual washout. The highest uptake was observed for the cerebellum and lower levels in the brainstem; pretreatment with SR141716A significantly reduced the uptake with an identical time activity curve for cerebellum and brainstem. In baboon [^{123}I]AM281 successfully accumulates in CB1-rich parts of the brain (cerebellum, cortex, striatum), with a peak uptake at 20 min and a clearance half-time of 90 min. Pretreatment and displacement (30 min pi) with SR141716A significantly reduce the cerebellar uptake.[523] This tracer was evaluated in a pilot study with patients suffering from Tourette's syndrome before and after Δ^9-THC therapy (15 days increasing dose from 2.5 mg day to 10 mg day). The maximum uptake was around 2% ID at 40 min and the tracer kinetic was well correlated using either a one-tissue compartment model with plasma input or a Logan plot; the highest distribution volume was observed for the lentiform nucleus (3.8), white matter (3.1), and occipital cortex (2.8) and Δ9-THC therapy modestly reduced the distribution volume in all regions (2.3, 1.8, 1.8 respectively). It should be noted that the authors of this study chose the lentiform nucleus as opposed to the cerebellum in order to avoid potential partial volume effect, and from post mortem results showing a greater density of CB1 in the lentiform nucleus than in cerebellum in human in contrast to non-human primates. Human dosimetry was estimated from whole-body imaging and revealed the highest target organ dose for the upper large intestine (9.2 mSv per 200MBq), spleen (8.3 mSv per 200 MBq), and red marrow (5.4 mSv per 200 MBq).[524] Despite these results, no further evaluations of this tracer have been reported, presumably because of its low signal-to-noise ratio.

3.11.1.3   ((3R, 5R)-5-(3-[^{11}C]Methoxyphenyl)-3-(R-1-phenylethylamino)-1-(4-trifluoromethylphenyl)-pyrrolidin-2-one) ([^{11}C]MePPEP)

This tracer was obtained by methylation of its desmethyl precursor using [^{11}C]methyliodide with a yield of 2.5%, a specific activity of 81 GBq mmol^{-1} and radiochemical purity greater than 99%. Evaluation in rodent show a maximum brain uptake of 190% SUV (rats) and 260% SUV (mouse) around 20 min, followed by relatively slow washout. Experiments in P-gp and CB1 knockout mice showed that [^{11}C]MePPEP is not a P-gp substrate, and that around 65% of the signal was specific. Displacement study using direct or indirect agonists anandamide, methanandamide, and URB597 (anandamide reuptake inhibitor) did not result in any significant change in brain uptake, whereas displacement with the inverse agonist rimonabant (3mg kg^{-1}) leads to a fast and pronounced reduction of tracer binding. Metabolism was fast in plasma with only 16% of unchanged tracer at 60 min; all metabolites were more polar than [^{11}C]MePPEP but were detected in brain at a constant level of around 13% of the total brain activity.[525] In monkey, the tracer showed high and rapid brain uptake, peaking at almost 600% SUV in the cerebellum at 10–20 min; the regional distribution was in accordance with the CB1 distribution with the highest activity in the cerebellum and striatum and the lowest in the thalamus and pons. The signal selectivity was demonstrated by pretreatment and displacement using rimonabant and ISPB show significant reduction of tracer accumulation in all brain regions. [^{11}C]MePPEP was quickly metabolized, with parent tracer accounting for 83%, 53%, 21%, and 14% of the activity at 5, 10, 30, and 60 min, respectively. Quantification was achieved by using a two-tissue compartment model with arterial input, and comparison of total distribution volume at baseline, pretreatment with rimonabant and with ISPB reveals an estimate of 89% of total brain uptake associated with CB1 binding in the CB1 rich region of the brain (pons and thalamus were excluded).[526] Evaluation in healthy human subjects shows a good brain uptake (greater than 3 SUV), and a distribution consistent with CB1 in human brain, followed by a slow washout; the highest uptake was found in the putamen (3.59 SUV), occipital cortex (2.96 SUV), prefrontal cortex (2.89 SUV), cerebellum (2.55), and thalamus (2.42 SUV), and the lowest were found in the pons (1.94 SUV) and white matter (1.19). The tracer was extensively metabolized (12% of unchanged tracer at 60 min) with at least one less-polar metabolite and five more-polar; the free fraction in plasma was low averaging at 0.05% in twelve subjects. An unconstrained two-tissue compartment model fitted the time activity curve of the tracer well, and the estimate BP$_{ND}$ reach 19 in healthy human. In test-retest and intersubject variability the simple brain uptake was better than the distribution volume (8% *vs.* 15% test-retest and 16% *vs.* 52% intersubject) presumably due to the slow washout of the tracer in all brain region and to the low concentration of the tracer in plasma samples.[527]

3.11.1.4   (3R,5R)-3-((R)-1-(4-Fluorophenyl)ethylamino)-5-
(3-[^{11}C]methoxyphenyl)-1-(4-(trifluoromethyl)-
phenyl)pyrrolidin-2-one ([^{11}C]FMePPEP); (3R,5R)-5-
(3-(2-[^{18}F]fluoroethoxy)phenyl)-3-((R)-1-phenylethy-
lamino)-1-(4-(trifluoromethyl)phenyl)pyrrolidin-2-one
([^{18}F]FEPEP); (3R,5R)-5-(3-([^{18}F]fluoromethoxy)-
phenyl)-3-((R)-1-phenylethylamino)-1-(4-(trifluoro-
methyl)phenyl)pyrrolidin-2-one ([^{18}F]FMPEP);
(3R,5R)-5-(3-([^{18}F]fluorodeuteromethoxy)phenyl)-3-
((R)-1-phenylethylamino)-1-(4-(trifluoromethyl)-
phenyl)pyrrolidin-2-one ([^{18}F]FMPEP-d2)

All four compounds were evaluated and compared simultaneously; they were prepared from their phenolic precursor by alkylation with either [^{11}C]methyliodide (16.5% yield), [^{18}F]fluoroethyl bromide (7.9 % yield), [^{18}F]fluoromethyl bromide (5.9% yield), or [^{18}F]fluorodeuteromethyl bromide (7.9% yield) with radiochemical purity >99% and specific activity of 284 GBq µmol^{-1} ([^{11}C]FMePPEP); 150 GBq µmol^{-1} ([^{18}F]FEPEP); 140 GBq µmol^{-1} ([^{18}F]FMPEP); and 127 GBq µmol^{-1} ([^{18}F]FMPEP-d2).[528] Evaluation of brain uptake in rats was performed with the nonradioactive compounds using the relatively new application of LC-MS to quantify brain extracts at low concentration without the need of radiolabeling. Injection of 30 µg kg^{-1} in rat and subsequent analysis of cortical regions revealed good brain uptake for all three tracers (FMPEP-d2 was not evaluated) peaking at 48 ng g (FMePPEP @ 15 min); 22 ng g (FEPEP @ 15 min); and 39 ng g (FMPEP @ 30 min) gradually decreasing to less than 10 ng g over 8 h. Pretreatment with rimonabant at a dose of 10 mg kg^{-1} resulted in significant reduction of brain uptake of 65% (FMePPEP); 70% (FEPEP) and 75% (FMPEP) at 15 min pi, and only FMePPEP could be quantified after 4 h, suggesting specificity of these tracers for CB1.[528] All four compounds were evaluated in monkey: [^{11}C]FMePPEP shows a peak uptake in the brain of 3.3 SUV at 30 min, followed by a slow washout; quantification using a two-tissue compartment model on base line and rimonabant pretreatment leads to an estimated specific binding of 73% (15.9 *vs.* 4.3 total distribution volume respectively). However the ligand shows a distribution moderately consistent with CB1 distribution in the brain and a lower brain uptake than [^{11}C]MePPEP, and was therefore not further evaluated. [^{18}F]FEPEP shows a relatively low brain uptake 2–3 SUV at 15 min, followed by a gradual washout, but only 60% specific binding (total distribution volume at baseline and rimonabant pretreatment). Despite a very low bone uptake (> 0.5 SUV mandible) which represents a good estimate of *in vivo* desfluorination of the tracer, this compound was not evaluated further because of its relatively low specificity. [^{18}F]FMPEP shows the greatest brain uptake of this series reaching 5–6.5 SUV at 20 min, followed by gradual washout, the specificity of the signal was estimated at 90% (total distribution volume at baseline and rimonabant pretreatment) with increasing V_T in the cerebellum and pons throughout the length of the scan. Significant mandibular uptake was observed and increased with time (1.4 SUV @ 10 min; 3.1 SUV @ 180 min). [18F]FMPEP-d2 also shows a good uptake peaking at 4.5–6.5 SUV at 20 min, followed by gradual washout and the specific signal was estimated at 80–90% (total distribution volume at baseline and rimonabant pretreatment). However this tracer shows one third less bone uptake than its non-deuterated analogues and was chosen for further evaluation in human. After injection in healthy volunteers the tracer readily entered the brain peaking over 4 SUV in CB1 rich areas of the brain, followed by a slow washout (85% of the peak @ 2 h and 60% @ 5 h). The regional brain distribution was in accordance with CB1 distribution with the highest concentration in putamen (V_T = 24.3), prefrontal cortex (V_T = 22.9), cerebellum (V_T = 14), and low levels in pons (V_T = 6.0) and white matter (V_T = 6.5); significant skull uptake was detected in the clivus (> 4 SUV @ 300 min), occipital (2 SUV @ 300 min), and parietal bone (1 SUV @ 300 min). The tracer is extensively metabolized in blood with only 11% of unchanged tracer at 60 min and declined thereafter, and the free fraction in plasma averages 0.63% in 9 subjects. A two-tissue compartment model was significantly better than a one-tissue compartment model to fit the measured data, and an unconstrained two-tissue model was better than constraining non-displaceable uptake. The minimum scanning time was estimated at 120 min since V_T was identified as stable between 60 and 120 min; V_T still increased after 120 min presumably due to accumulation of radiometabolite in the brain. Using this quantification model the estimated BP_{ND} was 7.3 in healthy human. Test-retest and intersubject variability were examined and found to be moderate with 14% (brain uptake) to 16% (V_T) in test-retest and 14% (brain uptake) to 26% (V_T) in inter-subject variability; the higher V_T variability seems to be in part the result of a relatively high variation in the plasma free fraction of the tracer, reaching 50% both in test-retest and inter-subject variation.[529]

3.11.1.5   N-((2S,3S)-3-(3-Cyanophenyl)-4-(4-(2-[18F]fluoroethoxy)phenyl)butan-2-yl)-2-methyl-2-(5-methylpyridin-2-yloxy)propanamide ([18F]MK-9470)

[18F]MK-9470 was synthesized by alkylation of its hydroxyl precursor using [18F]fluoroethyl bromide with radiochemical purity > 95% and a specific activity higher than 200 GBq μmol^{-1}.[530] This tracer shows a relatively slow brain uptake peaking at almost 2.5 SUV at 120 min, followed by a steady plateau up to 360 min. The regional brain distribution was comparable to CB1 distribution and correlates well with autoradiography experiment using this tracer (cerebellum > putamen > occipital cortex ≫ white matter). Steady-state plasma levels of approximately 1 μM of MK-0364 results in a significant reduction of signal in all brain region reaching the levels of white matter baseline; similarly displacement with a bolus plus constant infusion of MK-0364 (0.8 mg kg^{-1} then 0.64 mg kg^{-1} h) results in a rapid displacement of the tracer from all brain region. From those studies the specific binding of [^{18}F]MK-9470 was estimated around 80%. In human the tracer behaves similarly with a relatively slow uptake peaking at 1.2 SUV at 120 min and remaining constant over the next 4 h; the highest uptake was found in the striatum, frontal cortex and posterior cingulated, intermediate levels in cerebellum and lower levels in thalamus and hippocampus. The concentration of tracer declined with time in the blood (77% @ 10 min; 33% @ 60 min; 18% @ 120 min; 13% @ 180 min; 5% @ 360 min), however, the brain activity remained constant from 120 to 360 min suggesting that the tracer remained bound to the CB1 receptor during this time, and supported the use of the area under the curve analysis approach to estimate the receptor availability.[531] Test retest variability was low and around 7% in all brain regions, and inter-subject variation did not exceed 16%.[532] Dosimetry using whole-body distribution in human show the highest organ dose for the gallbladder wall (0.159 mGy MBq^{-1}), the upper large intestine (0.086 mGy MBq^{-1}), small intestine (0.087 mGy MBq^{-1}), and liver (0.086 mGy MBq^{-1}) suggesting an hepatobiliary excretion of [^{18}F]MK-9470. The mean effective dose was 22.8 µSv MBq^{-1} (16.4 to 31.3 in 8 subjects) and the brain uptake was around 4%, which allowed good brain imaging at doses within the international guidelines.[533] Gender and age variation of CB1 density in the brain were investigated, and show an increase in CB1 receptor binding with age in the basal ganglia, lateral temporal cortex, and thalamus, but only in women; whereas men show higher binding than woman, independently of age, in clusters of the limbic system and cortico-striato-thalamic-cortical circuit. Those variations are in accordance with studies suggesting interaction between the endocannabinoid system and sexual hormone. Marijuana smokers show higher infertility rates; for a recent review on the topic, see reference.[534,535]

3.11.2 CB2

Only a handful of compounds have been evaluated as imaging agents for CB2; and published reports are limited to preliminary evaluation in rodent models.

3.11.2.1 *N-[(1S)-1-[4-[[4-methoxy-2-[(4-[^{11}C]-methoxyphenyl)sulfonyl]-phenyl]sulfonyl]phenyl]ethyl] methansulfonamide ([^{11}C]SCH225336)*

This tracer was synthesized by methylation of its phenol precursor using [^{11}C]methyliodide with a yield of 30%, radiochemical purity >99%, and a specific activity of 88.8 GBq µmol^{-1}. Evaluation in mice shows negligible brain uptake limiting its usefulness as a central imaging agent; in the periphery high uptake were found in the liver and intestine but no uptake in the spleen (a peripheral CB2 rich organ) was observed.[536]

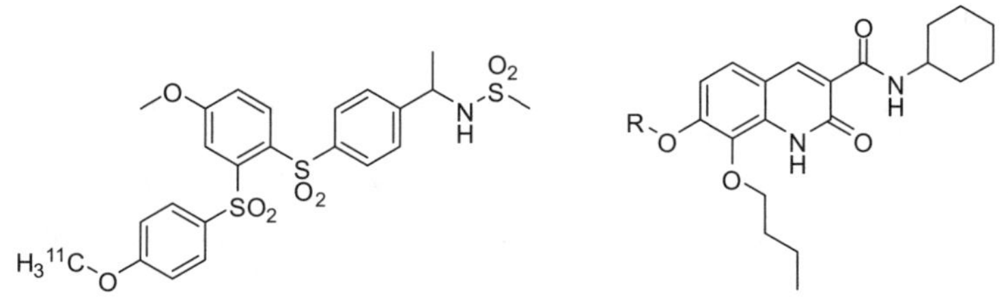

[¹¹C]SCH225336 K_i = 4.5 nM [a]

[¹¹C]**1**: R = ¹¹CH₃ K_i = 9.6 nM [a]

[¹⁸F]**2**: R = CH₂CH₂¹⁸F K_i = 35.8 nM [a]

Figure 3.16   Cannabinoid CB2 Ligands.
[a]Inhibition constant using [³H]CP55,940 in CHO cells expressing hCB2. *Abbreviations*: 1TC, one-tissue compartment (model); 2TC, two-tissue compartment (model); AD, Alzheimer's disease; ADHD, attention-deficit/hyperactivity disorder; ALS, amyotrophic lateral sclerosis; DV, distribution volume, DVR, distribution volume ratio; EDE, effective dose equivalent; [¹⁸F]FDG, [¹⁸F]fluorodeoxyglucose; FTLD, frontotemporal lobar degeneration; GABA, γ-aminobutyric acid; ID, injected dose; MCI, mild cognitive impairment; MOM, methoxymethyl-; MPTP, 1-methyl-4-phenyl-1,2,3,6-tetrahydropyridine; PBR, peripheral benzodiazepine receptor (same as TSPO); PD, Parkinson's disease; PDD, Parkinson's disease dementia; PET, positron emission tomography; pi, post-injection; PPA, primary progressive aphasia; ROI, region of interest; SPET, single-photon emission tomography; SPECT, single-photon emission computed tomography; SUV, standardized uptake value; TAC, time-activity curve; TSPO, translocator protein-18kDa, *see* PBR; VP, vascular Parkinsonism.

3.11.2.2   *2-Oxo-7-([¹¹C]methoxy and [¹⁸F]fluoroethyl)-8-butyloxy-1,2-dihydroquinoline-3-carboxylic acid cyclohexylamine ([¹¹C]1 and [¹⁸F] 2, Figure 3.16)*

Both tracers were prepared from their common hydroxyl precursor by alkylation with either [¹¹C]methyliodide or [¹⁸F]fluoroethyl bromide in 32% (specific activity 37 000 GBq mmol⁻¹) and 8.7% (specific activity 10 000 GBq mmol⁻¹) respectively, and with radiochemical purity >99%. Both compounds showed similar distribution in mice, with high uptake in the liver, kidney, and intestine, suggesting hepatobiliary excretion. The uptake in the spleen was high and could be selectively reduced by pretreatment with selective CB1 antagonists (no significant changes in other peripheral organs were observed). Both tracers were extensively metabolized, with only 20% of unchanged tracer at 20 min pi, and although all detected plasma metabolites appeared to be more polar than the parent compound, a relatively high level of metabolite was found in the brain (around 20% at 10 min pi), explained as either blood-brain-barrier permeable metabolites or apparent regional activity from the large fraction of metabolites in the vascular compartment of the brain.[537]

References

1. R. B. Innis, M. S. Al-Tikriti, S. S. Zoghbi, R. M. Baldwin, E. H. Sybirska, M. A. Laruelle, R. T. Malison, J. P. Seibyl, R. C. Zimmermann, E. W. Johnson, E. O. Smith, D. S. Charney, G. R. Heninger, S. W. Woods and P. B. Hoffer, *J. Nucl. Med.*, 1991, **32**(9), 1754.
2. R. B. Innis, J. P. Seibyl, E. Scanley, M. Laruelle, A. Abi-Dargham, E. Wallace, R. M. Baldwin, Y. Zea-Ponce, S. Zoghbi, S. Wang, Y. Gao, J. L. Neumeyer, D. S. Charney, P. B. Hoffer and K. Marek, *Proc. Natl. Acad. Sci. U. S. A.*, 1993, **90**(24), 11965.
3. H. W. Querfurth and F. M. LaFerla, *N. Engl. J. Med.*, 2010, **362**, 329.
4. M. Sakono and T. Zako, *FASEB J.*, 2010, **277**, 1348.
5. R. A. Armstrong, *Folia Neuropathol.a*, 2009, **47**, 289.
6. S. Krishnaswamy, G. Verdile, D. Groth, L. Kanyenda and R. N. Martins, *Crit. Rev. Clin Lab. Sci.*, 2009, **46**, 282.
7. K. M. Rodrigue, K. M. Kennedy and D. C. Park, *Neuropsychol. Rev.*, 2009, **19**, 436.
8. A. Koudinov, E. Kezlya, N. Koudinova and T. Berezov, *J. Alzheimers Dis.*, 2009, **18**, 381.
9. W. E. Klunk, H. Engler, A. Nordberg, Y. Wang, G. Blomqvist, D. P. Holt, M. Bergström, I. Savitcheva, G.-F. Huang, S. Estrada, B. Ausén, M. L. Debnath, J. Barletta, J. C. Price, J. Sandell, B. J. Lopresti, A. Wall, P. Koiviso, G. Antoni, C. A. Mathis and B. Långström, *Ann. Neurol.*, 2004, **55**(3), 306.
10. C. A. Mathis, D. P. Holt, Y. Wang, G.-F. Huang, M. Debnath, L. Shao and W. E. Klunk, *Neurobiol. Aging*, 2004, **25**(Suppl. 2), S277.
11. H. Toyama, D. Ye, M. Ichise, J.-S. Liow, L. Cai, D. Jacobowitz, J. L. Musachio, J. Hong, M. Crescenzo, D. Tipre, J.-Q. Lu, S. Zoghbi, D. C. Vines, J. Seidel, K. Katada, M. V. Green, V. W. Pike, R. M. Cohen and R. B. Innes, *Eur. J. Nucl. Med. Mol. Imaging*, 2005, **32**, 593.
12. W. E. Klunk, B. J. Lopresti, M. D. Ikonomovic, I. M. Lefterov, R. P. Koldamova, E. E. Abrahamson, M. L. Debnath, D. P. Holt, G.-F. Huang, L. Shao, S. T. DeKosky, J. C. Price and C. A. Mathis, *J. Neurosci.*, 2005, **16**, 10598.
13. B. J. Bacskai, M. P. Frosch, S. H. Freeman, S. B. Raymond, J. C. Augustinack, K. A. Johnson, M. C. Irizarry, W. E. Klunk, C. A. Mathis, S. T. DeKosky, S. M. Greenberg and B. T. Hyman, *Arch. Neurol.*, 2007, **64**, 431.
14. M. D. Ikonomovic, W. E. Klunk, E. E. Abrahamson, C. A. Mathis, J. C. Price, N. D. Tsopelas, B J. Lopresti, S. Ziolko, W. Bi, W. R. Paljug, M. L. Debnath, C. E. Hope, B. A. Isanski, R. L. Hamilton and S. T. DeKosky, *Brain*, 2008, **131**, 1630.
15. R. F. Rosen, B. J. Ciliax, T. S. Wingo, M. Gearing, J. Dooyema, J. J. Lah, J. A. Ghiso, H. Le Vine and L. C. Walker, *Acta Neuropathol.*, 2010, **119**, 221.
16. J. C. Price, W. E. Klunk, B. J. Lopresti, X. Lu, J. A. Hoge, S. K. Ziolko, D. P. Holt, C. C. Meltzer, S. T. DeKosky and C. A. Mathis, *J. Cereb. Blood Flow Metab.*, 2005, **25**, 1528.

17. Y. Zhou, S. M. Resnick, W. Ye, H. Fan, D. P. Holt, W. E. Klunk, C. A. Mathis, R. F. Dannals and D. F. Wong, *Neuroimage*, 2007, **36**, 298.

18. M. Yaqub, N. Tolboom, R. Boellaard, B. N. M. Van Berckel, E. W. Van Tilburg, G. Luurtsema, P. Scheltens and A. A. Lammertsma, *Neuroimage*, 2008, **42**, 76.

19. J. Fripp, P. Bourgeat, O. Acosta, P. Raniga, M. Modat, K. E. Pike, G. Jones, G. O'Keefe, C. L. Masters, D. Ames, K. A. Ellis, P. Maruff, J. Currie, V. L. Villemagne, C. C. Rowe, O. Salvodo and S. Ourselin, *Neuroimage*, 2008, **43**, 430.

20. P. Raniga, P. Bourgeat, J. Fripp, O. Acosta, V. L. Villemagne, C. Rowe, C. L. Masters, G. Jones, G. O'Keefe, O. Salvado and S. Ourselin, *Acad. Radiol.*, 2008, **15**, 1376.

21. P. Raniga, P. Bourgeat, J. Fripp, O. Acosta, S. Ourselin, C. Rowe, V. L. Villemagne and O. Salvado, *Proc. SPIE*, 2009, **7262**, 72621O.

22. H. Engler, A. Forsberg, O. Almkivst, G. Blomquist, E. Larsson, I. Savitcheva, A. Wall, A. Ringheim, B. Lagstrom and A. Nordberg, *Brain*, 2006, **129**, 2856.

23. N. M. Scheinin, S. Aalto, J. Koikkalainen, J. Lotjonen, M. Karrasch, N. Kemppainen, M. Viitanen, K. Nagren, S. Helin, M. Scheinin and J. O. Rinne, *Neurology*, 2009, **73**, 1186.

24. A. Okello, J. Koivunen, P. Edison, H. A. Archer, F. E. Turkheimer, K. Nagren, R. Bullock, Z. Walker, A. Kennedy, N. C. Fox, M. N. Rossor, J. O. Rinne and D. J. Brooks, *Neurology*, 2009, **73**, 754.

25. D. A. Wolk, J. C. Price, J. A. Saxton, B. E. Snits, J. A. James, O. L. Lopez, H. J. Aizenstein, A. D. Cohen, L. A. Weissfeld, C. A. Mathis, W. E. Klunk and S. T. DeKoskym, *Ann. Neurol.*, 2009, **65**, 557.

26. J. C. Morris, C. M. Roe, E. A. Grant, D. Head, M. Storandt, A. M. Goate, A. M. Fagan, D. M. Holtzman and M. A. Mintun, *Arch. Neurol.*, 2009, **66**, 1469.

27. S. M. Resnick, J. Sojkova, Y. Zhou, Y. An, W. Ye, D. P. Holt, R. F. Dannals, C. A. Mathis, W. E. Klunk, L. Ferrucci, M. A. Kraut and D. F. Wong, *Neurology*, 2010, **74**, 807.

28. A. Forsberg, O. Almkvist, H. Engler, A. Wall, B. Langstrom and A. Nordberg, *Curr. Alzheimer Res.*, 2010, **7**, 56.

29. G. D. Rabinovic, A. J. Furst, A. Alkalay, A. C. Racine, J. P. O'Neill, M. Janabi, S. L. Baker, N. Agarwal, S. Bonasera, E. C. Mormino, M. W. Weiner, M. L. Gorno-Tempini, H. J. Rosen, B. L. Miller and W. J. Juqust, *Brain*, 2010, **133**, 512.

30. C. R. Jack, V. J. Lowe, S. D. Weigand, H. J. Wiste, M. L. Senjem, D. S. Knopman, M. M. Shiung, J. L. Gunter, B. F. Boeve, B. J. Kemp, M. W. Weiner and R. C. Petersen, *Brain*, 2009, **132**, 1355.

31. Y. Li, J. O. Rinne, L. Mosconi, E. Pirraglia, H. Rusinek, S. DeSanti, N. Kemppainen, K. Nagren, B.-C. Kim, W. Tsui and M. J. De Leon, *Eur. J. Nucl. Med. Mol. Imaging*, 2008, **35**, 2169.

32. J. O. Rinne, D. J. Brooks, M. N. Rossor, N. C. Fox, R. Bullock, W. E. Klunk, C. A. Mathis, K. Blennow, J. Barakos, A. Okello, S. Rodriguez

Martinez de Llano, E. Liu, M. Koller, K. M. Gregg, D. Schenk, R. Black and M. Grundman, *Lancet Neurol.*, 2010, **9**, 363.

33. C. A. Raji, J. T. Becker, N. D. Tsopelas, J. C. Price, C. A. Mathis, J. A. Saxton, B. J. Lopresti, J. A. Hoge, S. K. Ziolko, S. T. DeKosky and W. E. Klunk, *J. Neurosci. Methods*, 2008, **172**, 277.

34. S. Y. Ng, V. L. Villemagne, C. L. Masters and C. C. Rowe, *Archives of Neurology*, 2007, **64**, 1140.

35. G. D. Rabinovici, A. J. Furst, J. P. O'Neill, C. A. Racine, E. C. Mormino, S. L. Baker, S. Chetty, P. Patel, T. A. Pagliaro, W. E. Klunk, C. A. Mathis, H. J. Rosen, B. L. Miller and W. J. Juqust, *Neurology*, 2007, **68**, 1205.

36. J. Koivunen, A. Verkkoniemi, S. Aalto, A. Paetau, J.-P. Ahonen, K. Viitanen, J. Rokka, M. Haaparanta, H. Kalimo and J. O. Rinne, *Brain*, 2008, **131**, 1845.

37. W. Maetzler, M. Reimold, I. Liepelt, C. Solbach, T. Leyhe, K. Schweitzer, G. W. Eschweiler, M. Mittelbronn, A. Gaenslen, M. Uebele, G. Reischl and T. Gasser, *Neuroimage*, 2008, **39**, 1027.

38. J. V. Ly, G. A. Donnam, V. L. Villemagne, J. A. Zavala, H. Ma, G. O'Keefe, S. J. Gong, R. M. Gunawan, T. Saunder, U. Ackerman, H. Tochon-Danguy, L. Churilov, T. G. Phan and C. C. Rowe, *Neurology*, 2010, **74**, 487.

39. E. D. Agdeppa, V. Kepe, J. Liu, S. Flores-Tores, N. Satyamurthy, A. Petric, G. M. Cole, G. W. Small, S.-C. Huang and J. R. Barrio, *J. Neurosci.*, 2001, **21**, RC189.

40. R. P. Klok, J. P. Klein. B. N. M. Van Berckel, N. Tolboom, A. A. Lammertsma and A. D. Windhorst, *Appl. Radiat. Isot.*, 2008, **66**, 203.

41. J. Liu, V. Kepe, A. Zabjek, A. Petric, H. C. Padgett, N. Satyamurthy and J. R. Barrio, *Mol. Imaging Biol.*, 2007, **9**, 6.

42. J. Vercouillie, C. Prenant, S. Maia, P. Emond, S. Guillouet, J. B. Deloye, L. Barre, and D. Guilloteau, *J. Labelled Compd. Radiopharm.*, 2010, **53**, 208–212.

43. E. D. Agdeppa, V. Kepe, J. Liu, S. Flores-Tores, N. Satyamurthy, A. Petric, G. M. Cole, G. W. Small, S. C. Huang and J. R. Barrio, *J. Neurosci.*, 2001, **21**(24), RC189.

44. J. Mukherjee, B. Easwaramoorthy, I. Chen, D. Collins, R. Pichika, C. Wang, V. Nguyen and E. Head, *J. Nucl. Med.*, 2006, **47**(Suppl. 1), 134P.

45. R. Kaghazala, I. Chen, C. Wang, B. Easwaramoorthy, M. Nistor, E. Head and J. Mukherjee, *J. Nucl. Med.*, 2007, **48**(Suppl. 2), 115P.

46. V. Kepe, G. M. Cole, J. Liu, D. G. Flood, S. P. Trusko, N. Satyamurthy, S.-C. Huang, G. W. Small, M. E. Phelps and J. R. Barrio, *Alzheimers Dement.*, 2005, **1**(Suppl. 1), S45.

47. K. Shoghi-Jadid, G. W. Small, E. D. Agdeppa, V. Kepe, L. M. Ercoli, P. Siddarth, S. Read, N. Satyamurthy, A. Petric, S.-C. Huang and J. R. Barrio, *Am. J. Geriatr. Psychiatry*, 2002, **10**, 24.

48. E. D. Agdeppa, V. Kepe, J. Liu, G. W. Small, S.-C. Huang, A. Petric, N. Satyamurthy and J. R. Barrio, *Mol. Imaging Biol.*, 2003, **5**(6), 404.

49. G. W. Small, V. Kepe, L. M. Ercoli, P. Siddarth, S. Y. Bookheimer, K. J. Miller, H. Lavretsky, A. C. Burggren, G. M. Cole, H. V. Vinters, P. M. Thompson, S.-C. Huang, N. Satyamurthy, M. E. Phelps and J. R. Barrio, *N. Engl. J. Med.*, 2006, **355**, 2652.

50. M. N. Braskie, A. D. Klunder, K. M. Hayashi, H. Protas, V. Kepe, K. J. Miller, S.-C. Huang, J. R. Barrio, L. M. Ercoli, P. Siddarth, N. Satya-murthy, J. Liu, A. W. Toga, S. Y. Bookheimer, G. W. Small and P. M. Thompson, *Neurobiol. Aging*, 2010, **31**, 1669–78.

51. H. D. Protas, S.-C. Huang, V. Kepe, K. Hayashi, A. Klunder, M. N. Braskie, L. Ercoli, S. Bookheimer, P. M. Thompson, G. W. Small and J. R. Barrio, *Neuroimage*, 2010, **49**, 240.

52. G. W. Small, P. Siddarth, A. C. Burggren, V. Kepe, L. M. Ercoli, K. J. Miller, H. Lavretsky, P. M. Thompson, G. M. Cole, S.-C. Huang, M. E. Phelps, S. Y. Bookheimer and J. R. Barrio, *Arch. Gen. Psychiatry*, 2009, **66**, 81.

53. M. Yaqub, R. Boellaard, B. N. M. Van Berckel, N. Tolboom, G. Luurtsema, A. A. Dijkstra, M. Lubberink, A. D. Windhorst, P. Scheltens and A. A. Lammertsma, *Mol. Imaging Biol.*, 2009, **11**, 322.

54. M. Yaqub, N. Tolboom, B. N. M. Van Berckel, P. Scheltens, A. A. Lammertsma and R. Boellaard, *Neuroimage*, 2010, **49**, 433.

55. J. Shine, S.-Y. Lee, S. H. Kim, Y.-B. Kim and S.-J. Cho, *Neuroimage*, 2008, **43**, 236.

56. P. M. Thompson, L. Ye, J. L. Morgenstern, L. Sue, T. G. Beach, D. J. Judd, N. J. Shipley, V. Libri and A. Lockhart, *J. Neurochem.*, 2009, **109**, 623.

57. G. Luurtsema, R. C. Schuit, K. Takkenkamp, M. Lubberink, H. H. Hendrikse, A. D. Windhorst, C. F. M. Molthoff, N. Tolboom, B. N. M. Van Berckel and A. A. Lammertsma, *Nucl. Med. Biol.*, 2008, **35**, 869.

58. M.-P. Kung, C. Hou, Z.-P. Zhuang, B. Zhang, D. Skovronsky, J. Q. Trojanowski, V. M.-Y. Lee and H. F. Kung, *Brain Res.*, 2002, **956**(2), 202.

59. Z. P. Zhuang, M. P. Kung, A. Wilson, C. W. Lee, K. Plossl, C. Hou, D. M. Holtzman and H. F. Kung, *J. Med. Chem.*, 2003, **46**(2), 237.

60. M. P. Kung, Z.-P. Zhuang, C. Hou, L. W. Jin and H. F. Kung, *J. Mol. Neurosci.*, 2003, **20**(3), 249.

61. H. F. Kung, M.-P. Kung, Z. P. Zhuang, C. Hou, C.-W. Lee, K. Plossl, B. Zhuang, D. M. Skovronsky, V. M.-Y. Lee and J. Q. Trojanowski, *Mol. Imaging Biol.*, 2003, **5**(6), 418.

62. K. W. Chang, C. C. Chen, S. Y. Lee, L. H. Shen and H. E. Wang, *Appl. Radiat. Isot.*, 2009, **67**, 1397.

63. M.-P. Kung, C. Hou, Z.-P. Zhuang, A. J. Cross, D. L. Maier and H. F. Kung, *Eur. J. Nucl. Med. Mol. Imaging*, 2004, **31**(8), 1136.

64. M.-P. Kung, Z.-P. Zhuang, C. Hou and H. F. Kung, *J. Mol. Neurosci.*, 2004, **24**(1), 49.

65. M.-P. Kung, C. Hou, Z.-P. Zhuang, D. Skovronsky and H. F. Kung, *Brain Res.*, 2004, **1025**(1–2), 98.

66. A. B. Newberg, N. A. Wintering, K. Plossl, J. Hochold, M. G. Stabin, M. Watson, D. Skovronsky, C. M. Clark, M. P. Kung and H. F. Kung, *J. Nucl. Med.*, 2005, **47**, 748.

67. Y. Kudo, N. Okamura, S. Furumoto, M. Tashiro, K. Furukawa, M. Maruyama, M. Itoh, R. Iwata, K. Yanai and H. Arai, *J. Nucl. Med.*, 2007, **48**, 553.

68. N. Okamura, S. Furumoto, Y. Funaki, T. Suemoto, M. Kato, Y. Ishikawa, S. Ito, H. Akatsu, T. Yamamoto, T. Sawada, H. Arai, Y. Kudo and K. Yanai, *Geriatr. Gerontol. Int.*, 2007, **7**, 393.

69. M. T. Fodero-Tavoletti, R. S. Mulligan, N. Okamura, S. Furumoto, C. Rowe, Y. Kudo, C. L. Masters, R. Cappai, K. Yanai and V. L. Villemagne, *Eur. J. Pharmacol.*, 2009, **617**, 54.

70. Y. Kudo, *Minimal. Invasive Ther.*, 2006, **15**, 209.

71. K. Furukawa, N. Okamura, M. Tashiro, M. Waragai, S. Furumoto, R. Iwata, K. Yanai, Y. Kudo, and H. Arai, *J. Neurol.*, 2009, DOI:10.1007/s00415–009–5396–8.

72. M. Waragai, N. Okamura, K. Furukawa, M. Tashiro, S. Furumoto, Y. Funaki, H. Kato, R. Iwata, K. Yanai, Y. Kudo and H. Arai, *J. Neurol. Sci.*, 2009, **285**, 100.

73. W. Zhang, S. Oya, M.-P. Kung, C. Hou, D. L. Maier and H. F. Kung, *Nucl. Med. Biol.*, 2005, **32**(8), 799.

74. C. C. Rowe, U. Ackerman, W. Browne, R. Mulligan, K. E. Pike, G. O'Keefe, H. Tochon-Danguy, G. Chan, S. U. Berlangieri, G. Jones, K. L. Dickinson-Rowe, H. P. Kung, J. Zhang, M. P. Kung, D. Skovronsky, T. Dyrks, G. Holl, S. Krause, M. Friebe, L. Lehman, S. Lindemann, L. M. Dinkelborg, C. L. Masters and V. L. Villemagne, *Lancet Neurol.*, 2008, **7**, 129.

75. C. Rowe, U. Ackerman, R. Mulligan, G. O'Keefe, H. Kung, D. Skovronsky, T. Dyrks, L. Dinkelborg, C. Masters and V. L. Villemagne, *J. Nucl. Med.*, 2008, **49**(Suppl. 1), 35P.

76. C. Rowe, U. Ackerman, R. S. Mulligan, T. Saunder, G. O'Keefe, H. F. Kung, D. Skovronsky, T. Dyrks, G. Holl, S. Krause, S. Lindemann and L. M. Dinkelborg, *et al., Alzheimer's Dementia*, 2008, **4**(Suppl.), T89.

77. G. J. O'Keefe, T. S. Saunder, S. Ng, U. Ackerman, H. J. Tochon-Danguy, G. Chan, S. Gong, T. Dyrks, S. Lindemann, G. Holl, L. Dinkelborg, V. L. Villemagne and C.C. Rowe, *J. Nucl. Med.*, 2009, **50**, 309.

78. M. Patt, A. Schildan, H. Barthel, G. Becker, M. H. Schultze-Mosgau, B. Rohde, C. Reininger and O. Sabri, *J. Radioanal. Nucl. Chem.*, 2010, DOI:10.1007/s10967–010–0514–8.

79. C. M. Niswender and P. J. Conn, *Annu. Rev. Pharmacol. Toxicol.*, 2010, **50**, 295.

80. P. J. Conn and J. P. Pin, *Annu. Rev. Pharmacol. Toxicol.*, 1997, **37**, 205.

81. F. Ferraguti, L. Crepaldi and F. Nicoletti, *Pharmacol. Rev.*, 2008, **60**, 536.

82. P. J. Conn, *Ann. N. Y. Acad. Sci.*, 2003, **1003**(1), 12.

83. P. J. Conn, G. Battaglia, M. J. Marino and F. Nicoletti, *Nat. Rev. Neurosci.*, 2005, **6**, 787.

84. D. D. Schoepp, *J. Pharmacol. Exper. Ther.*, 2001, **299**, 12.
85. C. J. Swanson, M. Bures, M. P. Johnson, A.-M. Linden, J. A. Monn and D. D. Schoepp, *Nat. Rev. Drug Discovery*, 2005, **4**, 131.
86. M. V. Catania, S. D'Antoni, C. M. Bonaccorso, E. Aronica, M. F. Bear and F. Nicoletti, *Mol. Neurobiol.*, 2007, **35**, 298.
87. H. Homayouna and B. Moghaddam, *Eur. J. Pharmacol.*, 2010, In Press (online doi:10.1016/j.ejphar.2009.12.042).
88. W. Eng, H. Kawamoto, S. O'Malley, S. Krause, C. Ryan, Z. Zeng, K. Riffel, T. Kimura, S. Ito, G. Suzuki, J. J. Cook, S. Ozaki,, *World Molecular Image Congress*, 2009, Montreal, Canada., September 23.
89. M. Ohgami, T. Haradahira, N. Takai, M.-R. Zhang, K. Kawamura, T. Yamasaki and K. Yanagimoto, *Annual Congress of the European Association of Nuclear Medicine*, 2009, Barcelona, Spain., October 10.
90. S. M. Ametamey, L. J. Kessler, M. Honer, M. T. Wyss, A. Buck, S. Hintermann, Y. P. Auberson, F. Gasparini and P. A. Schubiger, *J. Nucl. Med*, 2006, **47**, 698.
91. S. M. Ametamey, V. Treyer, J. Streffer, M. T. Wyss, M. Schmidt, M. Blagoev, S. Hintermann, Y. Auberson, F. Gasparini, U. C. Fischer and A. Buck, *J. Nucl. Med.*, 2007, **48**, 247.
92. V. Treyer, J. Streffer, M. T. Wyss, A. Bettio, S. M. Ametamey, U. Fischer, M. Schmidt, F. Gasparini, C. Hock and A. Buck, *J. Nucl. Med.*, 2007, **48**, 1207.
93. V. Treyer, J. Streffer, S. M. Ametamey, A. Bettio, P. Blauenstein, M. Schmidt, F. Gasparini, U. Fischer, C. Hock and A. Buck, *Eur. J. Nucl. Med. Mol. Imaging*, 2008, **35**, 766.
94. M. Honer, A. Stoffel, L. J. Kessler, P. A. Schubiger and S. M. Ametamey, *Nucl. Med. Biol.*, 2007, **34**, 973.
95. C. Lucatelli, M. Honer, J.-F. Salazar, T. L. Ross, A. P. Schubiger and S. M. Ametamey, *Nucl. Med. Biol.*, 2009, **36**, 613.
96. T. G. Hamill, S. Krause, C. Ryan, C. Bonnefous, S. Govek, T. J. Seiders, N. D. P. Cosford, J. Roppe, T. Kamenecka, S. Patel, R. E. Gibson, S. Sanabria, K. Riffel, W. Eng, C. King, X. Yang, M. D. Green, S. S. O'Malley, R. Hargreaves and H. D. Burns, *Synapse*, 2005, **56**(4), 205.
97. J.-Q. Wang, W. Tueckmantel, A. Zhu, D. Pellegrino and A. L. Brownell, *Synapse*, 2007, **61**, 951.
98. M. V. Patel, T. G. Hamill, B. Connolly, E. Jagoda, W. Li and R. E. Gibson, *Nucl. Med. Biol.*, 2007, **34**, 1009.
99. M. J. Belanger, S. M. Krause, C. Ryan, S. Sanabria-Bohorquez, W. Li, T. G. Hamill and H. D. Burns, *Nucl. Med. Commun*, 2008, **29**, 915.
100. S. Patel, O. Ndubizu, T. Hamill, A. Chaudhary, H. D. Burns, R. Hargreaves and R. E. Gibson, *Mol. Imaging Biol.*, 2005, **7**, 314.
101. F. G. Simeon, A. K. Brown, S. S. Zoghbi, V. M. Patterson, R. B. Innis and V. W. Pike, *J. Med. Chem.*, 2007, **50**, 3256.
102. H. U. Shetty, S. S. Zoghbi, F. G. Simeon, J.-S. Liow, A. K. Brown, P. Kannan, R. B. Innis and V. W. Pike, *J. Pharmacol. Exper. Ther.*, 2008, **327**, 727.

103. A. K. Brown, Y. Kimura, S. S. Zoghbi, F. G. Simeon, J.-S. Liow, W. C. Kreisl, A. Taku, M. Fujita, V. W. Pike and R. B. Innis, *J. Nucl. Med.*, 2008, **49**, 2042.

104. G. E. Torres, R. R. Gainetdinov and M. G. Caron, *Nat. Rev. Neurosci.*, 2003, **4**, 13.

105. R. R. Gainetdinov and M. C. Caron, *Annu. Rev. Pharmacol. Toxicol.*, 2003, **43**, 261.

106. C. E. Glatt and V. I. Reus, *Pharmacogenomics*, 2003, **4**, 583.

107. L. Norregaard and U. Gethe, *Curr. Opin. Drug Discovery Dev.*, 2001, **4**, 591.

108. R. D. Blakely, L. J. Defelice and A. Galli, *Physiol. Behav*, 2005, **20**, 225.

109. M. Jaber, S. Jones, B. Giros and M. C. Caron, *Movement Disord.*, 2004, **12**, 629.

110. *Dopamine Transporters: Chemistry, Biology, and Pharmacology*, ed. M. L. Trudell, S. Izenwasser and B. Wang, Wiley Series in Drug Discovery and Development, Wiley, 1st edition, 2008.

111. G. R. Uhl, *Movement Disord.*, 2003, **18**(Suppl. 7), S71.

112. B. K. Madras, G. M. Miller and A. J. Fischman, *Biol. Psychiatry*, 2005, **57**, 1397.

113. J. D. Foster, M. A. Cervinski, B. K. Gorentla and R. A. Vaughan, *Handb Exp Pharmacol.*, 2006, **175**, 197.

114. K.-P. Lesch, *Neuroscientist*, 1998, **4**, 25.

115. R. D. Blakely, L. J. De Felice and H. C. Hartzell, *J. Exp. Biol.*, 1994, **196**, 263.

116. A. Serretti, R. Calati, L. Mandelli and D. De Ronchi, *Curr. Drug Targets*, 2006, **7**, 1659.

117. D. J. Nutt, *J. Clin. Psychiatry*, 2006, **67**(Suppl. 6), 3.

118. R. Mössner, A. Schmitt, Y. Syagailo, M. Gerlach, P. Riederer and K. P. Lesch, *J. Neural. Transm. Suppl.*, 2000, **60**, 345.

119. B. Olivier, W. Soudijn and I. Van Wijngaarden, *Prog. Drug Res.*, 2000, **54**, 59.

120. K. J. Ressler and C. B. Nemeroff, *Biol. Psychiatry*, 1999, **46**, 1219.

121. *Brain Norepinephrine Neurobiology and Therapeutics*, ed. G. A. Ordway, M. A. Schwartz and A. Frazer, Cambridge University Press, 2007.

122. J. Zhou, *Drugs Future*, 2004, **29**, 1235.

123. M. Sieber-Blum and Z. Ren, *Mol. Cell. Biochem.*, 2000, **212**, 61.

124. J. L. Neumeyer, S. Wang, Y. Gao, R. A. Milius, N. S. Kula, A. Campbell, R. J. Baldessarini, Y. Zea-Ponce, R. M. Baldwin and R. B. Innis, *J. Med. Chem.*, 1994, **37**(11), 1558.

125. R. M. Baldwin, Y. Zea-Ponce, M. S. Al-Tikriti, S. S. Zoghbi, J. P. Seibyl, D. S. Charney, P. B. Hoffer, S. Wang, R. A. Milius, J. L. Neumeyer and R. B. Innis, *Nucl. Med. Biol.*, 1995, **22**, 211.

126. J. Booij, G. Andringa, L. J. M. Rijks, R. J. Vermeulen, K. Debruin, G. J. Boer, A. G. M. Janssen and E. A. Vanroyen, *Synapse (N. Y.)*, 1997, **27**(3), 183.

127. A. Abi-Dargham, M. S. Gandelman, G. A. DeErausquin, Y. Zea-Ponce, S. Zoghbi, R. M. Baldwin, M. Laruelle, D. S. Charney, P. B. Hoffer, J. L. Neumeyer and R. B. Innis, *J. Nucl. Med.*, 1996, **37**, 1129.

128. G. Tissingh, J. Booij, P. Bergmans, A. Winogrodzka, A. G. M. Janssen, E. A. Vanroyen, J. C. Stoof and E. C. Wolters, *J. Nucl. Med.*, 1997, **39**, 1143.

129. J. Booij, G. Tissingh, G. J. Boer, J. D. Speelman, J. C. Stoof, A. G. M. Janssen, E. C. Wolters and E. A. Vanroyen, *J. Neurol., Neurosurg. Psychiatry*, 1997, **62**(2), 133.

130. J. P. Seibyl, K. L. Marek, K. Sheff, S. S. Zoghbi, R. M. Baldwin, D. S. Charney, C. H. van Dyck and R. B. Innis, *J. Nucl. Med.*, 1998, **39**(9), 1500.

131. H. T. S. Benamer, J. Patterson, D. J. Wyper, D. M. Hadley, G. J. A. Macphee and D. G. Grosset, *Movement Disord.*, 2000, **15**, 692.

132. J. Booij, J. B. A. Habraken, P. Bergmans, G. Tissingh, A. Winogrodzka, E. C. Wolters, A. G. M. Janssen, J. C. Stoof and E. A. Van Royen, *J. Nucl. Med.*, 1998, **39**, 1879.

133. J. Booij, E. B. Sokole, M. G. Stabin, A. G. M. Janssen, K. Debruin and E. A. Vanroyen, *Eur. J. Nucl. Med.*, 1998, **25**(1), 24.

134. W. Robeson, V. Dhawan, A. Belakhlef, Y. Ma, V. Pillai, T. Chaly, C. Margouleff, D. Bjelke and D. Eidelberg, *J. Nucl. Med.*, 2003, **44**(6), 961.

135. S. J. Colloby, J. T. O'Brien, J. D. Fenwick, M. J. Firebank, D. J. Burn, I. G. McKeith and E. D. Williams, *Neuroimage*, 2004, **23**, 956.

136. S. Takada, M. Yoshimura, H. Shindo, K. Saito, K. Koizumi, H. Utsumi and K. Abe, *Ann. Nucl. Med.*, 2006, **20**, 477.

137. L. Tossici-Bolt, S. M. A. Hoffmann, P. M. Kemp, R. L. Metha and J. S. Fleming, *Eur. J. Nucl. Med. Mol. Imaging*, 2006, **33**, 1491.

138. A. Kas, P. Payoux, M.-O. Habert, Z. Malek, Y. Cointepas, G. El Fakhri, P. Chaumet-Riffaud, E. Itti and P. Remy, *J. Nucl. Med.*, 2007, **48**, 1459.

139. C. Crespo, J. Gallego, A. Cot, C. Falcon, S. Bullich, D. Pareto, P. Aguira, J. Sempau, F. Lomena, F. Calvino, J. Pavia and D. Ros, *Eur. J. Nucl. Med. Mol. Imaging*, 2008, **35**, 1334.

140. J. C. Dickson, L. Tossici-Bolt, T. Sera, K. Erlandsson, A. Varrone, K. Tatsch and B. F. Hutton, *Eur. J. Nucl. Med. Mol. Imaging*, 2010, **37**, 23.

141. W. Koch, M. Mustafa, C. Zach and K. Tatsch, *Nucl. Med. Commun.*, 2007, **28**, 603.

142. J. T. O'Brien, S. Colloby, J. Fenwick, E. D. Williams, M. Firbank, D. Burn, D. Aarsland and I. G. McKeith, *Arch. Neurol.*, 2004, **61**, 919.

143. I. McKeith, J. O'Brien, Z. Walker, K. Tatsch, J. Booij, J. Darcourt, A. Padovani, R. Giubbini, U. Bonuccelli, D. Volterrani, C. Holmes, P. Kemp, N. Tabet, I. Meyer, C. Reininger and DLB Study Group, *Lancet Neurol.*, 2007, **6**, 305.

144. Z. Walker, E. Jaros, R. W. H. Walker, L. Lee, D. C. Costa, G. Livingston, P. G. Ince, R. Perry, I. McKeith and C. L. E. Katona, *J. Neurol., Neurosurg., Psychiatry*, 2007, **78**, 1176.

145. M. Lorberboym, R. Djaldetti, E. Melamed, M. Sadeh and Y. Lampl, *J. Nucl. Med.*, 2004, **45**, 1688.

146. J. Booij, J. D. Speelman, M. W. I. M. Horstink and E. C. Wolters, *Eur. J. Nucl. Med. Mol. Imaging*, 2001, **28**, 265.

147. J. J. Diaz-Corrales, S. Sanz-Viedma, D. Garcia-Solis, T. Escobar-Delgado and P. Mir, *Eur. J. Nucl. Med. Mol. Imaging*, 2010, **37**, 556.
148. G. Kaig, K. P. Bhatia and E. Tolosa, *J. Neurol., Neurosurg., Psychiatry*, 2010, **81**, 5.
149. J. Booij, J. De Jonk, K. De Bruin, R. Knol, M. M. L. De Win and B. L. F. Van Eck-Smith, *J. Nucl. Med.*, 2006, **48**, 359.
150. S. Nikolaus, C. Antke, K. Kley, M. Beu, A. Wirrwar and H.-W. Muller, *Journal of Nucl. Med.*, 2009, **50**, 1147.
151. J. Booij and P. Kemp, *Eur. J. Nucl. Med. Mol. Imaging*, 2008, **35**, 424.
152. R. M. Baldwin, Y. Zea-Ponce, S. S. Zoghbi, M. Laruelle, M. S. Al-Tikriti, E. H. Sybirska, R. T. Malison, J. L. Neumeyer, R. A. Milius, S. Wang, M. Stabin and E. O. Smith, *Nucl. Med. Biol.*, 1993, **20**(5), 597.
153. J. L. Neumeyer, S. Wang, R. A. Milius, R. M. Baldwin, Y. Zea-Ponce, P. B. Hoffer, E. H. Sybirska, M. S. Al-Tikriti, M. Laruelle and R. B. Innis, *J. Med. Chem.*, 1991, **34**(10), 3144.
154. R. B. Innis, R. Baldwin, E. Sybirska, Y. Zea, M. Laruelle, M. Al-Tikriti, D. Charney, S. Zoghbi, E. Smith, G. Wisniewski, P. Hoffer and S. Wang, *Eur. J. Pharmacol.*, 1991, **200**, 369.
155. J. P. Seibyl, E. Wallace, P. B. Hoffer, S. S. Zoghbi, G. Zubal, Y. Gao, J. L. Neumeyer, E. O. Smith, R. M. Baldwin and R. B. Innis, *J. Nucl. Med.*, 1994, **35**(5), 764.
156. R. B. Innis, J. P. Seibyl, B. E. Scanley, M. Laruelle, A. Abi-Dargham, E. Wallace, R. Baldwin, Y. Zea-Ponce, S. Zoghbi, S. Wang, Y. Gao, J. L. Neumeyer, D. S. Charney, P. B. Hoffer and K. L. Marek, *Proc. Natl. Acad. Sci. U. S. A.*, 1993, **90**, 11965.
157. S. Asenbaum, T. Brucke, W. Pirker, I. Podreka, P. Angelberger, S. Wenger, C. Wober, C. Muller and L. Deecke, *J. Nucl. Med.*, 1997, **38**, 1.
158. K. L. Marek, J. P. Seibyl, S. S. Zoghbi, Y. Zea-Ponce, R. M. Baldwin, B. Fussell, D. S. Charney, C. H. van Dyck, P. B. Hoffer and R. B. Innis, *Neurology*, 1996, **46**(1), 231.
159. J. P. Seibyl, M. Laruelle, C. H. van Dyck, E. Wallace, R. M. Baldwin, S. S. Zoghbi, Y. Zea-Ponce, J. L. Neumeyer, D. S. Charney, P. B. Hoffer and R. B. Innis, *J. Nucl. Med.*, 1996, **37**(2), 222.
160. J. P. Seibyl, K. L. Marek, K. Sheff, R. M. Baldwin, S. S. Zoghbi, Y. Zea-Ponce, D. S. Charney, C. H. van Dyck, P. B. Hoffer and R. B. Innis, *J. Nucl. Med.*, 1997, **38**(9), 1453.
161. G. Tissingh, P. Bergmans, J. Booij, A. Winogrodzka, E. A. Van Royen, J. C. Stoof and E. C. Wolters, *J. Neurol.*, 1998, **245**, 14.
162. T. Muller, J. Farahati, W. Kuhn, E. G. Eising, H. Przuntek, C. Reiners and H. H. Coenen, *Eur. Neurol.*, 1998, **39**, 44.
163. W. Pirker, *Movement Disord.*, 2003, **18**(Suppl. 7), (S43-S51).
164. J. E. Ahlskog, R. J. Uitti, M. K. O'Connor, D. M. Maraganore, J. Y. Matsumoto, K. F. Stark, M. F. Turk and O. L. Burnett, *Movement Disord.*, 1999, **14**, 940.
165. R. B. Innis, K. L. Marek, K. Sheff, S. S. Zoghbi, J. Castronuovo, A. Feigin and J. P. Seibyl, *Movement Disord.*, 1999, **14**(3), 436.

166. K. L. Marek, R. B. Innis, C. H. van Dyck, B. Fussell, M. Early, S. Eberly, D. Oakes and J. P. Seibyl, *Neurology*, 2001, **57**(11), 2089.
167. W. Pirker, I. Holler, W. Gerschlager, S. Asenbaum, G. Zettinig and T. Brucke, *Movement Disord.*, 2003, **18**, 1266.
168. A. Winogrodzka, P. Bergmans, J. Booij, E. A. Van Royen, J. C. Stoof and E. C. Wolters, *J. Neurol., Neurosurg. and Psychiatry*, 2003, **74**, 294.
169. Parkinson Study Group, *J. Am. Med. Assoc.*, 2002, **287**(13), 1653.
170. K. Marek, D. Jennings and J. P. Seibyl, *Eur. J. Neurol.*, 2002, **9**(Suppl. 3), 15.
171. S. Fahn, D. Oakes, I. Shoulson, K. Kieburtz, A. Rudolph, A. Lang, C. W. Olanow, C. Tanner and K. Marek; Parkinson Study Group, *New Engl. J. Med.*, 2004, **351**, 2498.
172. S. Fahn and P. S. Group, *J. Neurol.*, 2005, **252**(Suppl. 4), IV37.
173. R. T. Malison, C. J. McDougle, C. H. van Dyck, L. Scahill, R. M. Baldwin, J. P. Seibyl, L. H. Price, J. F. Leckman and R. B. Innis, *Am. J. Psychiatry*, 1995, **152**(9), 1359.
174. B. Jeon, J. M. Kim, J. M. Jeong, K. M. Kim, Y. S. Chang, D. S. Lee and M. C. Lee, *J. Neurol., Neurosurg. Psychiatry*, 1998, **65**, 60.
175. W. Pirker, S. Asenbaum, G. Bencsits, D. Prayer, W. Gerschlager, L. Deeke and T. Brucke, *Movement Disord.*, 2000, **15**, 1158.
176. A. Varrone, K. L. Marek, D. Jennings, R. B. Innis and J. P. Seibyl, *Movement Disord.*, 2001, **16**(6), 1023.
177. C. Scherfler, K. Seppi, E. Donnemiller, G. Goebel, C. Brenneis, I. Virgolini, G. K. Wenning and W. Poewe, *Brain*, 2005, **128**, 1605.
178. K. Seppi, C. Scherfler, E. Donnemiller, I. Virgolini, M. F. H. Schocke, G. Goebel, K. J. Mair, S. Boesch, C. Brenneis, G. K. Wenning and W. Poewe, *Arch. Neurol.*, 2006, **63**, 1154.
179. W. Gerschlager, G. Bencsits, W. Pirker, B. R. Bloem, S. Asenbaum, D. Prayer, J. C. M. Zijlmans, M. Hoffmann and T. Brucke, *Movement Disord.*, 2002, **17**, 518.
180. E. Berthommier, C. Loc'h, S. Chalon, C. Olivier, P. Emond, H. Dao Boulanger, M. A. Lelait and L. Mauclaire, *J. Labelled Compd. Radiopharm.*, 2002, **45**, 1019.
181. F. Dolle, M. Bottlaender, S. Demphel, P. Emond, C. Fuseau, C. Coulon, M. Ottaviani, H. Valette, C. Loc'h, C. Halldin, L. Mauclaire and D. Guilloteau, *et al.*, *J. Labelled Compd. Radiopharm.*, 2000, **43**, 997.
182. D. Guilloteau, P. Emond, J. L. Baulieu, L. Garreau, Y. Frangin, L. Pourcelot, L. Mauclaire, J. C. Besnard and S. Chalon, *Nucl. Med. Biol.*, 1998, **25**(4), 331.
183. C. Halldin, N. Erixon-Lindroth, S. Pauli, Y.-H. Chou, Y. Okubo, P. Karlsson, C. Lundkvist, H. Olsson, D. Guilloteau, P. Emond and L. Farde, *Eur. J. Nucl. Med. Mol. Imaging*, 2003, **30**(9), 1220.
184. T. Poyot, F. Conde, M. C. Gregoire, V. Frouin, C. Coulon, C. Fuseau, F. Hinnen, F. Dolle, P. Hantraye and M. Bottlaender, *J. Cereb. Blood Flow Met.*, 2001, **21**, 782.

185. M.-J. Ribeiro, M. Ricard, M.-A. Lievre, S. Bourgeois, P. Emond, P. Gervais, F. Dolle and A. Syrota, *Nuc. Med. Biol.*, 2007, **34**, 465.

186. J. T. Kuikka, J. L. Baulieu, J. Hiltunen, C. Halldin, K. A. Bergstrom, L. Farde, P. Emond, S. Chalon, M. X. Yu, T. Nikula, T. Laitinen, J. Karhu, E. Tupala, T. Hallikainen, V. Kolehmainen, L. Mauclaire, B. Maziere, J. Tiihonen and D. Guilloteau, *Eur. J. Nucl. Med.*, 1998, **25**(5), 531.

187. L. H. Pinborg, C. Videbæk, C. Svarer, S. Yndgaard, O. B. Paulson and G. M. Knudsen, *Eur. J. Nucl. Med.*, 2002, **29**(5), 623.

188. C. Prunier, P. Payoux, D. Guilloteau, S. Chalon, B. Giraudeau, C. Majorel, M. Tafani, E. Bezard, J.-P. Esquerre and J.-L. Baulieu, *J. Nucl. Med.*, 2003, **44**(5), 663.

189. L. H. Pinborg, M. Ziebell, V. G. Frokjaer, R. de Nijs, C. Svarer, S. Haugbol, S. Yndgaard and G. M. Knudsen, *J. Nucl. Med.*, 2005, **46**(7), 1119.

190. M. Ziebell, G. Thomsen, G. M. Knudsen, R. De Nijs, C. Svarer, A. Wagner and L. H. Pinborg, *Eur. J. Nucl. Med. Mol. Imaging*, 2007, **34**, 101.

191. A. Jucaite, I. Odano, H. Olsson, S. Pauli, C. Halldin and L. Farde, *Eur. J. Nucl. Med. Mol. Imaging*, 2006, **33**, 657.

192. C. DeLorenzo, J. S. Dileep Kumar, F. Zanderigo, J. J. Mann and R. V. Parsey, *J. Cereb. Blood Flow Met.*, 2009, **29**, 1332.

193. M.-J. Ribeiro, S. Thobois, E. Lohmann, S. Tezenas du Montcel, S. Lesage, A. Pelissolo, B. Dubois, L. Mallet, P. Pollak, Y. Agid, E. Broussolle and A. Brice, P. Remy and French Parkinson's Disease Genetics Study Group, *J. Nucl. Med.*, 2009, **50**, 1244.

194. A. Jucaite, E. Fernell, C. Halldin, H. Forssberg and L. Farde, *Biol. Psychiatry*, 2004, **57**, 229.

195. S. Meegalla, K. Plossl, M.-P. Kung, D. A. Stevenson, L. M. Liable-Sands, A. L. Rheingold and H. F. Kung, *J. Am. Chem. Soc.*, 1995, **117**(44), 11037.

196. S. K. Meegalla, K. Plossl, M. P. Kung, S. Chumpradit, D. A. Stevenson, S. A. Kushner, W. T. McElgin, P. D. Mozley and H. F. Kung, *J. Med. Chem.*, 1997, **40**(1), 9.

197. G. Toth, Z. Szakonyi, B. Kanyo, F. Fulop, G. Jancso and L. Pavics, *J. Labelled Compd. Radiopharm.*, 2003, **46**, 1067.

198. P. D. Acton, S. R. Choi, K. Plossl and H. F. Kung, *Eur. J. Nucl. Med.*, 2002, **29**(5), 691.

199. P.-F. Kao, K.-Y. Tzen, T.-C. Yen, C.-S. Lu, Y.-H. Weng, S.-P. Wey and G. Ting, *Nucl. Med. Commun.*, 2001, **22**, 151.

200. H. F. Kung, H. J. Kim, M. P. Kung, S. K. Meegalla, K. Plossl and H. K. Lee, *Eur. J. Nucl. Med.*, 1996, **23**(11), 1527.

201. P. D. Mozley, J. S. Schneider, P. D. Acton, K. Plossl, M. B. Stern, A. Siderowf, N. A. Leopold, P. Y. Li, A. Alavi and H. F. Kung, *J. Nucl. Med.*, 2000, **41**(4), 584.

202. Y.-H. Weng, T.-C. Yen, M.-C. Chen, P.-F. Kao, K.-Y. Tzen, R.-S. Chen, S.-P. Wey, G. Ting and C.-S. Lu, *J. Nucl. Med.*, 2004, **45**, 393.

203. W.-S. Huang, S.-Z. Lin, J.-C. Lin, S.-P. Wey, G. Ting and R.-S. Liu, *J. Nucl. Med.*, 2001, **42**, 1303.

204. Y. Geng, G.-H. Shi, Y. Jiang, L.-X. Xu, X.-Y. Hu and Y.-Q. Shao, *J. Zhejiang Univ., Sci. B*, 2005, **6**(1), 22–27.

205. M. C. shih, L. A. Franco de Andrade, E. Amaro, A. Carvalho Felicio, H. Ballalai Ferraz, J. Wagner, M. Queiroz Hoexter, L. F. Lin, Y. K. Fu, J. Jesus Mari, S. Tufik and R. Affonseca Bressan, *Movement Disord.*, 2007, **22**, 863.

206. P. D. Mozley, J. B. Stubbs, K. Plossl, S. H. Dresel, E. D. Barraclough, A. Alavi, L. I. Araujo and H. F. Kung, *J. Nucl. Med.*, 1998, **39**(12), 2069.

207. W. J. Hwang, W. J. Yao, S. P. Wey and G. Ting, *J. Nucl. Med.*, 2004, **45**(2), 207.

208. K.-Y. Tzen, C.-S. Lu, T.-C. Yen, S.-P. Wey and G. Ting, *J. Nucl. Med.*, 2001, **42**(3), 408.

209. J. Wang, Y.-P. Jiang, X.-D. Liu, Z.-P. Chen, L.-Q. Yang, C.-J. Liu, J.-D. Xiang and H.-L. Su, *Acta Neurol. Scand.*, 2005, **112**, 380.

210. S. Dresel, J. Krause, K. H. Krause, C. LaFougere, K. Brinkbaumer, H. F. Kung, K. Hahn and K. Tatsch, *Eur. J. Nucl. Med.*, 2000, **27**(10), 1518.

211. J. Krause, C. La Fougere, K.-H. Krause, M. Ackenheil and S. H. Dresel, *Eur. Arch. Psy. Clin. N.*, 2005, **255**, 428.

212. C. La Fougere, J. Krause, K.-H. Krause, F. J. Gildehaus, M. Hacker, W. Koch, K. Hahn, K. Tatsc and S. Dresel, *Nucl. Med. Commun.*, 2006, **27**, 733.

213. C.-B. Yhe, C.-H. Lee, Y.-H. Chou, C.-J. Chang, K.-H. Ma and W.-S. Huang, *Nucl. Med. Commun.*, 2006, **27**, 779.

214. C.-B. Yeh, C.-S. Lee, K.-H. Ma, M.-S. Lee, C.-J. Chang and W.-S. Huang, *Psychiatry Res., Neuroimaging*, 2007, **156**, 75.

215. W.-J. Hwang, W.-J. Yao, Y.-K. Fu and A.-S. Yang, *Psychiatry Res., Neuroimaging*, 2008, **162**, 159.

216. M. S. Haka and M. R. Kilbourn, *Nucl. Med. Biol.*, 1989, **16**, 771.

217. K.-S. Lin and Y.-S. Ding, *Bioorg. Med. Chem.*, 2005, **13**(15), 4658.

218. J. Logan, Y.-S. Ding, K.-S. Lin, D. Pareto, J. S. Fowler and A. Biegon, *Nucl. Med. Biol.*, 2005, **32**(5), 531.

219. A. A. Wilson, D. Patrick Johnson, D. Mozley, D. Hussey, N. Ginovart, J. Nobrega, A. Garcia, J. Meyer and S. Houle, *Nucl. Med. Biol.*, 2003, **30**(2), 85.

220. K.-S. Lin and Y.-S. Ding, *Chirality*, 2004, **16**, 475.

221. M. Schou, C. Halldin, J. Sovago, V. W. Pike, B. Gulyas, P. D. Mozley, D. P. Johnson, H. Hall, R. B. Innis and L. Farde, *Nucl. Med. Biol.*, 2003, **30**(7), 707.

222. Y.-S. Ding, K.-S. Lin, V. Garza, P. Carter, D. Alexoff, J. Logan, C. Shea, Y. Xu and P. King, *Synapse*, 2003, **50**, 345.

223. J. Logan, G.-J. Wang, F. Telang, J. S. Fowler, D. Alexoff, J. Zabroski, M. Jayne, B. Hubbard, P. King, P. Carter, C. Shea, Y. Xu, L. Muench, D. Schlyer, S. Learned-Coughlin, V. Cosson, N. D. Volkow and Y. S. Ding YS, *Nucl. Med. Biol.*, 2007, **34**, 667.

224. A. Takano, B. Gulya, A. Varrone and C. Halldin, *Eur. J. Nucl. Med. Mol. Imaging*, 2009, **36**, 1885.
225. Y.-S. Ding, T. Singhal, B. Planeta-Willson, J.-D. Gallezot, N. Nabulsi, D. Labaree, J. Ropchan, S. Henry, W. Williams, R. E. Carson, A. Neumeister and R. T. Malison, *Synapse*, 2010, **64**, 30.
226. M. Schou, S. S. Zoghbi, H. U. Shetty, E. Shchukin, J.-S. Liow, J. Hong, B. A. Andree, B. Gulyas, L. Farde, R. B. Innis, K. E. Pike and C. Halldin, *Mol. Imaging Biol.*, 2008, **11**, 23.
227. M. Schou, C. Halldin, J. Sóvágó, V. W. Pike, H. Hall, B. Gulyás, P. D. Mozley, D. Dobson, E. Shchukin, R. B. Innis and L. Farde, *Synapse*, 2004, **53**(2), 57.
228. N. M. Seneca, B. Gulyas, A. Varrone, M. Schou, A. Airaksinen, J. Tauscher, F. Vandenhende, W. Kielbasa, L. Farde, R. B. Innis and C. Halldin, *Psychopharmacology (Berlin)*, 2006, **188**, 119.
229. A. Takano, B. Gulyas, A. Varrone, R. P. Maguire and C. Halldin, *Eur. J. Nucl. Med. Mol. Imaging*, 2009, **36**, 1308.
230. A. Takano, C. Halldin, A. Varrone, P. Karlsson, N. Sjoholm, J. B. Stubbs, M. Schou, A. J. Airaksinen, J. Tauscher and B. Gulyas, *Eur. J. Nucl. Med. Mol. Imaging*, 2007, **35**, 630.
231. A. Takano, A. Varrone, B. Gulyas, P. Karlsson, J. Tauscher and C. Halldin, *Neuroimage*, 2008, **42**, 474.
232. R. Arakawa, M. Okumura, H. Ito, C. Seki, H. Takahashi, H. Takano, R. Nakao, K. Suzuki, Y. Okubo, C. Halldin and T. Suhara, *J. Nucl. Med.*, 2008, **49**, 1270.
233. M. Sekine, R. Arakawa, H. Ito, M. Okumura, T. Sasaki, H. Takahashi, H. Takano, Y. Okubo, C. Halldin, T. Suhara, *Psychopharmacology (Berlin)*, 2010, in press, (online DOI: 10.1007/s00213–010–1824–9).
234. G. D. Tamagnan, E. Brenner, D. Alagille, J. K. Staley, C. Haile, A. Koren, M. Early, R. M. Baldwin, F. I. Tarazi, R. J. Baldessarini, N. Jarkas, M. M. Goodman and J. P. Seibyl, *Bioorg. Med. Chem. Lett.*, 2007, **17**(2), 533.
235. N. Kanegawa, Y. Kiyono, H. Kimura, T. Sugita, S. Kajiyama, H. Kawashima, M. Ueda, Y. Kuge and H. Saji, *Eur. J. Nucl. Med. Mol. Imaging*, 2006, **33**, 639.
236. G. Tamagnan, A. O. Koren, J. K. Staley, K. P. Cosgrove, C. Megyola, K. Marek and J. P. Seibyl, *Neuroimage*, 2006, **31**, T137.
237. M. Suehiro, H. T. Ravert, R. F. Dannals, U. A. Scheffel and H. N. Wagner, Jr., *J. Labelled Compd. Radiopharm.*, 1992, **31**(10), 841.
238. M. Suehiro, U. A. Scheffel, R. F. Dannals, H. T. Ravert, G. A. Ricaurte, H. N. Wagner and Jr. , *J. Nucl. Med.*, 1993, **34**(1), 120.
239. M. Suehiro, J. L. Musachio, R. F. Dannals, W. B. Mathews, H. T. Ravert, U. Scheffel and H. N. Wagner, *Nucl. Med. Biol.*, 1994, **22**, 543.
240. J. Zessin, P. Gucker, S. M. Ametamey, J. Steinbach, P. Brust, F. X. Vollenweider, B. Johannsen and P. A. Schubiger, *J. Labelled Compd. Radiopharm.*, 1999, **42**(13), 1301.

241. U. Scheffel, Z. Szabo, W. B. Mathews, P. A. Finley, R. F. Dannals, H. T. Ravert, K. Szabo, J. Yuan and G. A. Ricaurte, *Synapse*, 1998, **29**, 183.

242. Z. Szabo, P. F. Kao, U. A. Scheffel, M. Suehiro, W. B. Mathews, H. T. Ravert, J. L. Musachio, S. Marenco, S. E. Kim, G. A. Ricaurte, D. F. Wong, H. N. Wagner Jr., and R. F. Dannals, *Synapse*, 1995, **20**(1), 37.

243. Z. Szabo, P.-F. Kao, W. B. Mathews, H. T. Ravert, J. L. Musachio, U. Scheffel and R. F. Dannals, *Behav. Brain Res.*, 1996, **73**, 221.

244. Z. Szabo, U. Scheffel, W. B. Mathews, H. T. Ravert, K. Szabo, M. Kraut, S. Palmon, G. A. Ricaurte and R. F. Dannals, *J. Cerebr. Blood F. Met.*, 1999, **19**(9), 967.

245. R. Parsey, L. S. Kegeles, D.-R. Hwang, N. Simpson, A. Abi-Dargham, O. Mawlawi, M. Slifstein, R. L. Van Heertum, J. J. Mann and M. Laruelle, *J. Nucl. Med.*, 2000, **41**, 1465.

246. A. Buck, P. M. Gucker, R. D. Schonbachler, M. Arigoni, S. Kneifel, F. X. Vollenweider, S. M. Ametamey and C. Burger, *J. Cerebr. Blood F. Met.*, 2000, **20**(2), 253.

247. Y. Ikoma, T. Suhara, H. Toyama, T. Ichimiya, A. Takano, Y. Sudo, M. Inoue, F. Yasuno and K. Suzuki, *J. Cerebr. Blood F. Met.*, 2002, **22**, 490.

248. M. Yamamoto, T. Suhara, Y. Okubo, T. Ichimiya, Y. Sudo, M. Inoue, A. Takano, F. Yasuno, K. Yoshikawa and S. Tanada, *Life Sci.*, 2002, **71**, 751.

249. T. Ichimiya, T. Suhara, Y. Sudo, Y. Okubo, K. Nakayama, M. Nankai, M. Inoue, F. Yasuno, A. Takano, J. Maeda and H. Shibuya, *Biol. Psychiatry*, 2002, **51**, 715.

250. M. Reivich, J. D. Amsterdam, D. J. Brunswick and C. Y. Shiue, *J. Affective Disord.*, 2004, **82**, 321.

251. R. V. Parsey, R. S. Hastings, M. A. Oquendo, Y.-Y. Huang, N. Simpson, J. Arcement, Y. Huang, R. T. Ogden, R. L. Van Heertum, V. Arango and J. J. Aann, *Am. J. Psychiatry*, 2006, **163**, 52.

252. M. A. Oquendo, R. S. Hastings, Y.-Y. Huang, N. Simpson, R. T. Ogden, X.-Z. Hu, D. Goldman, V. Arango, R. L. Van Heertum, J. J. Mann and R. V. Parsey, *Arch. Gen. Psychiatry*, 2007, **64**, 201.

253. J. M. Miller, E. L. Kinnally, R. T. Ogden, M. A. Oquendo, J. J. Mann and R. V. Parsey, *Synapse*, 2009, **67**, 565.

254. J. M. Miller, M. A. Oquendo, R. T. Ogden, J. J. Mann and R. V. Parsey, *J. Psychiatr. Res.*, 2008, **42**, 1137.

255. J. M. Kent, J. D. Coplan, I. Lombardo, D.-R. Hwang, Y. Huang, O. Mawlawi, R. L. Van Heertum, M. Slifstein, A. Abi-Dargham, J. M. Gorman and M. Laruelle, *Psychopharmacology (Berlin)*, 2002, **164**, 341.

256. W. G. Frankle, I. Lombardo, A. S. New, M. Goodman, P. S. Talbot, Y. Huang, D.-R. Hwang, M. Slifstein, S. Curry, A. Abi-Dargham, M. Laruelle and L. J. Siever, *Am. J. Psychiatry*, 2005, **162**, 915.

257. U. D. McCann, Z. Szabo, U. Scheffel, R. F. Dannals and G. A. Ricaurte, *Lancet*, 1998, **352**, 1433.

258. R. Buchert, R. Thomasius, F. Wilke, K. Petersen, B. Nebeling, J. Obrocki, O. Schulze, U. Schmidt and M. Clausen, *Am. J. Psychiatry*, 2004, **161**, 1181.

259. U. D. McCann, Z. Szabo, E. Sekin, P. Rosenblatt, W. B. Mathews, H. T. Ravert, R. F. Dannals and G. A. Ricaurte, *Neuropsychopharmacology*, 2005, **30**, 1741.

260. R. Buchert, F. Thiele, R. Thomasius, F. Wilke, K. Petersen, W. Brenner, J. Mester, L. Spies and M. Clausen, *J. Psychopharmacol.*, 2007, **21**, 628.

261. L. Kerenyi, G. A. Ricaurte, D. J. Schretlen, U. McCann, J. Varga, W. B. Mathews, H. T. Ravert, R. F. Dannals, J. Hilton, D. F. Wong and Z. Szabo, *Arch. Neurol.*, 2003, **60**(9), 1223.

262. D. F. Wong, J. R. Brasic, H. S. Singer, D. J. Schretlen, H. Kuwabara, Y. Zhou, A. Nandi, A. A. Maris, M. Alexander, W. Ye, O. Rousset, A. Kumar, Z. Szabo, A. Gjedde and A. A. Grace, *Neuropsychopharmacology*, 2008, **33**, 1239.

263. S. Yamamoto, Y. Ouchi, H. Onoe, E. Yoshikawa, H. Tsukada, H. Takahashi, M. Iwase, K. Yamaguti, H. Kuratsune and Y. Watanabe, *NeuroReport*, 2004, **15**, 2571.

264. K. Nakamura, Y. Sekine, Y. Ouchi, M. Tsujii, E. Yoshikawa, M. Futatsubashi, K. J. Tsuchiya, G. Sugihara, Y. Iwata, K. Suzuki, H. Matsuzaki, S. Suda, T. Sugiyama, N. Takei and N. Mori, *Arch. Gen. Psychiatry*, 2010, **67**, 59.

265. R. V. Parsey, R. S. Hastings, M. A. Oquendo, X. Hu, D. Goldman, Y.-Y. Huang, N. Simpson, J. Arcement and Y. Huang, *Am. J. Psychiatry*, 2006, **163**, 48.

266. A. A. Wilson, N. Ginovart, M. Schmidt, J. H. Meyer, P. G. Threlkeld and S. Houle, *J. Med. Chem.*, 2000, **43**(16), 3103.

267. C. Solbach, G. Reischl and H.-J. Machulla, *Radiochim. Acta*, 2004, **92**, 341.

268. D. Haeusler, L.-K. Mien, L. Nics, J. Ungersboeck, C. Philippe, R. R. Lanzenberger, K. Kletter, R. Dudczak, M. Mitterhauser and W. Wadsak, *Appl. Radiat. Isot.*, 2009, **67**, 1654.

269. Z. Szabo, U. D. McCann, A. A. Wilson, U. Scheffel, T. Owonikoko, W. B. Mathews, H. T. Ravert, J. Hilton, R. F. Dannals and G. A. Ricaurte, *J. Nucl. Med.*, 2002, **43**(5), 678.

270. S. Houle, N. Ginovart, D Hussey, J. H. Meyer and A. A. Wilson, *Eur. J. Nucl. Med.*, 2000, **27**(11), 1719.

271. M. Ichise, J. S. Liow, J. Q. Lu, T. Takano, K. Model, H. Toyama, T. Suhara, T. Suzuki, R. B. Innis and T. E. Carson, *J. Cerebr. Blood F. Met.*, 2003, **23**(9), 1096.

272. W. G. Frankle, M. Slifstein, R. N. Gunn, Y. Huang, D.-R. Hwang, E. A. Darr, R. Narendran, A. Abi-Dargham and M. Laruelle, *J. Nucl. Med.*, 2006, **47**, 815.

273. T. Morimoto, H. Ito, A. Takano, Y. Ikoma, C. Seki, T. Okauchi, K. Tanimoto, A. Ando, T. Shiraishi, T. Yamaya and T. Suhara, *Ann. Nucl. Med.*, 2006, **20**, 237.

274. J. S. Kim, M. Ichise, J. Sangare and R. B. Innis, *J. Nucl. Med.*, 2006, **47**, 208.

275. R. V. Parsey, A. Ojha, R. T. Ogden, K. Erlandsson, D. Kumar, M. Landgrebe, R. Van Heertum and J. J. Mann, *J. Nucl. Med.*, 2006, **47**, 1796.

276. J.-Q. Lu, M. Ichise, J.-S. Liow, S. Ghose, D. Vines and R. B. Innis, *J. Nucl. Med.*, 2004, **45**(9), 1555.

277. J. Kalbitzer, D. Erritzoe, K. K. Holst, F. A. Nielsen, L. Marner, S. Lehel, T. Arentzen, T. L. Jernigan and G. M. Knudsen, *Biol. Psychiatry*, 2010, in press (online DOI: 10.1016/j.biopsych.2009.11.027).

278. N. Praschak-Rieder, M. Willeit, A. A. Wilson, S. Houle and J. H. Meyer, *Arch. Gen. Psychiatry*, 2008, **65**, 1072.

279. J. H. Meyer, S. Houle, S. Sagrati, A. Carella, D. F. Hussey, N. Ginovart, V. Goulding J. Kennedy and A. A. Wilson, *Arch. Gen. Psychiatry*, 2004, **61**, 1271.

280. D. M. Cannon, M. Ichise, D. Rollis, J. M. Klaver, D. K. Gandhi, D. S. Charney, H. K. Manji and W. C. Drevets, *Biol. Psychiatry*, 2007, **62**, 870.

281. M. Reimold, A. Batra, A. Knobel, M. N. Smalka, A. Zimmer, K. Mann, C. Solbach, G. Reischl, F. Schwarzler, G. Grunder, H.-J. Machulla, R. Bares and A. Heinz, *Mol. Psychiatry*, 2008, **13**, 606.

282. D. M. Cannon, M. Ichise, S. J. Fromm, A. C. Nugent, D. Rollis, S. K. Ganghi, J. M. Klaver, D. S. Charney, H. K. Manji and W. C. Drevets, *Biol. Psychiatry*, 2006, **60**, 207.

283. D. A. Hammoud, C. J. Endres, E. Hammond, O. Uzuner, A. Brown, A. Nath, A. I. Kaplin and M. G. Pomper, *Neuroimage*, 2010, **49**, 2588.

284. Z. Bhagwagar, N. Murthy, S. Selvaraj, R. Hinz, M. Taylor, S. Fancy, P. Grasby and P. Cowen, *Am. J. Psychiatry*, 2007, **164**, 1858.

285. J. H. Meyer, A. A. Wilson, N. Ginovart, V. Goulding, D. Hussey, K. Hood and S. Houle, *Am. J. Psychiatry*, 2001, **158**(11), 1843.

286. J. H. Meyer, A. A. Wilson, S. Sagrati, D. Hussey, A. Carella, W. Z. Potter, N. Ginovart, E. P. Spencer, A. Cheok and S. Houle, *Am. J. Psychiatry*, 2004, **161**, 826.

287. A. N. Voineskos, A. A. Wilson, A. Boovariwala, S. Sagrati, S. Houle, P. Rusjan, S. Sokolov, E. P. Spencer, N. Ginovart and J. H. Meyer, *Psychopharmacology (Berlin)*, 2007, **193**, 539.

288. P. S. Talbot, S. Bradley, C. P. Clarke, K. O. Babalola, A. W. Philipp, G. Brown, A. W. McMahon and J. C. Matthews, *Neuropsychopharmacology*, 2010, **35**, 741.

289. P. S. Talbot, W. G. Frankle, D.-R. Hwang, Y. Huang, R. F. Suckow, M. Slifstein, A. Abi-Dargham and M. Laruelle, *Synapse*, 2005, **55**, 164.

290. N. Praschak-Rieder, A. A. Wilson, D. Hussey, A. Carella, C. Wei, N. Ginovart, M. J. Schwarz, J. Zach, S. Houle and J. H. Meyer, *Biol. Psychiatry*, 2005, **58**, 825.

291. U. D. McCann, Z. Szabo, M. Vranesic, M. Palermo, W. B. Mathews, H. T. Ravert, R. F. Dannals and G. A. Ricaurte, *Psychopharmacology (Berlin)*, 2008, **200**, 439.

292. S. Selvaraj, R. Hoshi, Z. Bhagwagar, N. Venkatesha, R. Hinz, P. Cowen, H. V. Curran and P. Grasby, *Br. J. Pharmacol.*, 2009, **194**, 355.

293. M. Guttman, I. Boileau, J. Warsh, J. A. Saint-Cyr, N. Ginovart, T. McCluskey, S. Houle, A. Wilson, E. Mundo, P. Rusjan, J. Meyer and S. J. Kish, *Eur. J. Neurol.*, 2007, **14**, 523.

294. R. L. Albin, R. A. Koeppe, N. I. Bohnen, K. Wernette, M. A. Kilbourn and K. Frey, *J. Cerebr. Blood F. Met.*, 2008, **28**, 441.

295. I. Boileau, J. J. Warsh, M. Guttman, J. A. Saint-Cyr, T. McCluskey, P. Rusjan, S. Houle, A. A. Wilson, J. H. Meyer and S. J. Kish, *Movement Disord.*, 2008, **12**, 1776.

296. M. Reimold, M. N. Smolka, A. Zimmer, A. Batra, A. Knobel, C. Solbach, A. Mundt, H. U. Smoltczyk, D. Goldman, K. Mann, G. Reischl, H.-J. Machulla, R. Bares and A. Heinz, *J. Neural Transm.*, 2007, **114**, 1603.

297. R. Matsumoto, M. Ichise, H. Ito, T. Ando, H. Takahashi, Y. Ikoma, J. Kosaka, R. Arakawa, Y. Fujimura, M. Ota, A. Takano, K. Fukui, K. Nakayama and T. Suhara, *Neuroimage*, 2010, **49**, 121.

298. A. Takano, R. Arakawa, M. Hayashi, H. Takahashi, H. Ito and T. Suhara, *Biol. Psychiatry*, 2007, **62**, 588.

299. J. Kalbitzer, V. G. Frokjaer, D. Erritzoe, C. Svarer, P. Cumming, F. A. Nielsen, S. H. Hashemi, W. F. C. Baare, J. Madsen, S. G. Hasselbalch, M. L. Kringelbach, E. L. Mortensen and G. M. Knudsen, *Neuroimage*, 2009, **45**, 280.

300. A. K. Brown, D. T. George, M. Fujita, J.-S. Liow, M. Ichise, J. Hilbbeln, S. Ghose, J. Sangare, D. Hommer and R. B. Innis, *Alcohol: Clin. Exp. Res.*, 2007, **31**, 28.

301. W. G. Frankle, R. Narendran, Y. Huang, D.-R. Hwang, I. Lombardo, C. Cangiano, R. Gil, M. Laruelle and A. Abi-Dargham, *Biol. Psychiatry*, 2005, **57**, 1510.

302. S. Oya, S. R. Choi, C. Hou, M. Mu, M. P. Kung, P. D. Acton, M. Siciliano and H. F. Kung, *Nucl. Med. Biol.*, 2000, **27**(3), 249.

303. W. S. Huang, K. H. Ma, C. Y. Cheng, C. Y. Chen, Y. K. Fu, Y. H. Choue, S. P. Wey and J. C. Liu, *Nucl. Med. Commun.*, 2004, **25**(5), 515.

304. P. D. Acton, S.-R. Choi, C. Hou, K. Plossl and H. F. Kung, *J. Nucl. Med.*, 2001, **42**(10), 1556.

305. K. Erlandsson, T. Sivananthan, D. Lui, A. Spezzi, C. E. Townsend, S. Mu, R. Lucas, S. Warrington and P. J. Ell, *Eur. J. Nucl. Med. Mol. Imaging*, 2005, **32**, 1329.

306. A. M. Catafau, V. Perez, M. M. Penengo, S. Bullich, M. Danus, D. Puigdemont, J. C. Pascual, I. Corripio, J. Llop, J. Perich and E. Alvarez, *J. Nucl. Med.*, 2005, **46**, 1301.

307. J. Sacher, S. Asenbaum, N. Klein, T. Geiss-Granadia, N. Mosseheb, C. Poetzi, T. Attarbaschi, R. Lanzenberger, C. Spindelegger, A. Rabas, G. Heinze, R. Dudczak, S. Kasper and J. Tauscher, *Int. J. Neuropsychopharmacology*, 2007, **10**, 211.

308. T. Kauppinen, A. Koskela, M. Diemling, A. Keski-Rahkonen, E. Sihvola and A. Ahonen, *Nuklearmedizin*, 2005, **44**, 205.

309. B.-H. Yang, S.-J. Wang, Y.-H. Chou, T.-P. Su, S.-P. Chen, J.-S. Lee and J.-C. Chen, *Comput. Meth. Prog. Bio.*, 2008, **92**, 294.
310. V. G. Frokjaer, L. H. Pinborg, J. Madsen, R. De Nijs, C. Svarer, A. Wagner and G. M. Knudsen, *J. Nucl. Med*, 2008, **49**, 247.
311. Y.-H. Chou, B.-H. Yang, M.-Y. Chung, S.-P. Chen, T.-P. Su, C.-C. Chen and S.-J. Wang, *Psychiatry Res., Neuroimaging*, 2009, **172**, 38.
312. T. A. Kauppinen, K. A. Bergstrom, P. Heikman, J. Hiltunen and A. K. Ahonen, *Eur. J. Nucl. Med.*, 2002, **30**(1), 132.
313. A. B. Newberg, K. Plossl, P. D. Mozley, J. B. Stubbs, N. Wintering, M. Udeshi, A. Alavi, T. Kauppinen and H. F. Kung, *J. Nucl. Med.*, 2004, **45**(5), 834.
314. K.-J. Lin, C.-Y. Liu, S.-P. Wey, I.-T. Hsiao, J. Wu, Y.-K. Fu and T.-C. Yen, *Nucl. Med. Biol*, 2006, **33**, 193.
315. A. B. Newberg, J. D. Amsterdam, N. Wintering, K. Ploessl, R. L. Swanson, J. Shults and A. Alavi, *J. Nucl. Med.*, 2005, **46**, 973.
316. N. Wintering, J. Amsterdam, E. Breslow and A. Newberg, *J. Nucl. Med.*, 2009, **50**(Suppl.2), 1296.
317. H. Herold, K. Uebelhack, L. Franke, H. Amthauer, L. Luedemann, H. Bruhn, R. Felix, R. Uebelhack and M. Plotkin, *J. Neural Transm.*, 2006, **113**, 659.
318. A. M. Catafau, V. Perez, P. Plaza, J.-C. Pascual, S. Bullich, M. Suarez, M. M. Penengo, I. Corripio, D. Puigdemont, M. Danus, J. Perich and E. Alvarez, *Psychopharmacology (Berlin)*, 2006, **189**, 145.
319. A. Koskela, T. Kauppinen, A. Keski-Rahkonen, E. Sihvola, J. Kaprio, A. Rissanen and A. Ahonen, *Chronobiol. Int.*, 2008, **25**, 657.
320. N. Klein, J. Sacher, T. Geiss-Granadia, T. Attarbaschi, N. Mossaheb, R. Lanzenberger, C. Potzi, A. Holik, C. Spindelegger, S. Asenbaum, R. Dudczak, J. Tauscher and S. Kasper, *Psychopharmacology (Berlin)*, 2006, **188**, 263.
321. N. Klein, J. Sacher, T. Geiss-Granadia, N. Mossaheb, T. Attarbaschi, R. Lanzenberger, C. Spindelegger, A. Holik, S. Asenbaum, R. Dudczak, J. Tauscher and S. Kasper, *Psychopharmacology (Berlin)*, 2007, **191**, 333.
322. E. Van de Giessen and J. Booij, *Eur. J. Nucl. Med. Mol. Imaging*, 2010, in press (DOI 10.1007/s00259–010–1424–2).
323. J. D. Lundgren, A. B. Newberg, K. C. Allison, N. A. Wintering, K. Ploessl and A. J. Stunkard, *Psychiatry Res.*, 2008, **162**, 214.
324. S. Schuh-Hofer, M. Richter, L. Geworski, A. Villringer, H. Israel, R. Wenzel, D. L. Munz and G. Arnold, *J. Neurol.*, 2007, **254**, 789.
325. W. Koch, N. Schaaff, G. Popperl, C. Mulert, G. Juckel, M. Reicherzer, C. Ehmer-Von, H.-J. Moller, U. Hegerl, K. Tatsch and O. Pogarell, *J Psychiatr. Neurosci.*, 2007, **32**, 234.
326. G. Zheng, L. P. Dwoskin and P. A. Crooks, *AAPS J.*, 2006, **8**, E682.
327. J. P. Henry, D. Botton, C. Sagne, M. F. Isambert, C. Desnos, V. Blanchard, R. Raisman-Vozari, E. Krejci, J. Massoulie and B. Gasnier, *J. Exp. Biol.*, 1994, **196**, 251.

328. R. Yelin and S. Schuldiner, in *Neurotransmitter Transporters: Structure, Function, and Regulation*, ed. M. E. Reith, Humana Press, Totowa, New Jersey, 2nd edn, 2002, p. 313.

329. J. Yao and L. B. Hersh, *J. Neurochem.*, 2007, **100**, 1387.

330. L. E. Eiden, *FASEB J.*, 2000, **14**, 2396.

331. K. Wimalasena, *Med. Res. Rev.*, 2000, in press (Online DOI 10.1002/med.20187).

332. B. Gasnier, *Biochimie*, 2000, **82**, 327.

333. J. N. DaSilva, M. R. Kilbourn, T. J. Mangner and S. A. Toorongian, *J. Labelled Compd Radiopharm*, 1993, **32**, 257.

334. J. N. DaSilva and M. R. Kilbourn, *Life Sci.*, 1992, **51**, 593.

335. J. N. Dasilva, J. E. Carey, P. S. Sherman, T. J. Pisani and M. R. Kilbourn, *Nucl. Med. Biol.*, 1994, **21**(2), 151.

336. J. N. DaSilva, M. R. Kilbourn and E. F. Domino, *Synapse*, 1993, **14**, 128.

337. M. R. Kilbourn, J. N. Dasilva, K. A. Frey, R. A. Koeppe and D. E. Kuhl, *J. Neurochem.*, 1993, **60**(6), 2315.

338. T. M. Vander Borght, M. R. Kilbourn, R. A. Koeppe, J. N. DaSilva, D. D. Carey, D. E. Kuhl and K. A. Frey, *J. Nucl. Med.*, 1995, **36**, 2252.

339. M. R. Kilbourn, K. A. Frey, T. Vanderborght and P. S. Sherman, *Nucl. Med. Biol.*, 1996, **23**(4), 467.

340. M. R. Kilbourn, P. S. Sherman and L. C. Abbott, *Nucl. Med. Biol.*, 1995, **22**, 565.

341. D. J. Canney, Y. Z. Guo, M. P. Kung and H. F. Kung, *J. Labelled Compd. Radiopharm.*, 1993, **33**(4), 355.

342. M. P. Kung, D. J. Canney, D. Frederick, Z. Zhuang, J. J. Billings and H. F. Kung, *Synapse*, 1994, **18**(3), 225.

343. D. M. Jewett, M. R. Kilbourn and L. C. Lee, *Nucl. Med. Biol.*, 1997, **24**, 197.

344. M. R. Kilbourn, L. Lee, T. Vander Borght, D. Jewett and K. Frey, *Eur. J. Pharmacol.*, 1995, **278**, 249.

345. N. I. Bohnen, R. L. Albin, R. A. Koeppe, K. A. Wernette, M. R. Kilbourn, S. Minoshima and K. A. Frey, *J. Cerebr. Blood F. Met.*, 2006, **26**, 1198.

346. K. A. Frey, R. A. Koeppe, M. R. Kilbourn, T. M. Vander Borght, R. L. Albin, S. Gilman and D. E. Kuhl, *Ann. Neurol.*, 1996, **40**, 873.

347. S. Gilman, K. A. Frey, R. A. Koeppe, L. Junck, R. Little, T. M. Vander Borght, M. Lohman, S. Martorello, L. C. Lee, D. M. Jewett and M. R. Kilbourn, *Ann. Neurol.*, 1996, **40**, 885.

348. R. De La Fuente-Fernandez, S. Furtado, M. Guttman, Y. Furukawa, C. S. Lee, D. B. Calne, T. J. Ruth and A. J. Stoessl, *Synapse*, 2003, **49**, 20.

349. S. F. Taylor, R. A. Koeppe, R. Tandon, J.-K. Zubieta and K. A. Frey, *Neuropsychopharmacology*, 2000, **23**, 667.

350. J. Tong, A. A. Wilson, I. Boileau, S. Houle and S. Kish, *Synapse*, 2008, **62**, 873.

351. R. Murthy, P. Harris, N. Simpson, R. L. Van Heertum, R. Leibel, J. J. Mann and R. Parsey, *Eur. J. Nucl. Med. Mol. Imaging*, 2008, **35**, 790.

352. R. Goswami, D. E. Ponde, M.-P. Kung, C. Hou, M. R. Kilbourn and H. F. Kung, *Nucl. Med. Biol.*, 2006, **33**(6), 685.
353. M. P. Kung, C. Hou, R. Goswami, D. E. Ponde, M. R. Kilbourn and H. F. Kung, *Nucl. Med. Biol.*, 2007, **34**, 239.
354. M. R. Kilbourn, B. Hockley, L. Lee, C. Hou, R. Goswami, D. E. Ponde, M. P. Kung and H. F. Kung, *Nucl. Med. Biol.*, 2007, **34**, 233.
355. P. Seeman and H. H. Van Tol, *Curr Opin Neurol Neu.*, 1993, **6**, 602.
356. R. B. Mailman and X. Huang, *Handb Clin Neurol.*, 2007, **83**, 77.
357. P. Sokoloff, M. Andrieux, R. Besançon, C. Pilon, M.-P. Martres, B. Giros and J.-C. Schwartz, *Eur. J. Pharmacol., Mol. Pharmacol. Sect.*, 1992, **225**, 331.
358. J.-C. Schwartza, J. Diazb, C. Pilona and P. Sokoloffa, *Brain Res. Rev.*, 2000, **31**, 277.
359. D. R. Sibley, K. Neve, S. J. Enna and B. B. David, *xPharm: The Comprehensive Pharmacology Reference*, Elsevier, New York, 2007.
360. F. Boeckler and P. Gmeiner, *Pharmacol. Therapeut.*, 2006, **112**, 281.
361. N. M. Richtand, S. C. Woods, S. P. Berger and S. M. Strakowski, *Neurosci. Biobehav. R.*, 2001, **25**, 427.
362. A. A. Wilson, P. McCormick, S. Kapur, M. Willeit, A. Garcia, D. Hussey, S. Houle, P. Seeman and N. Ginovart, *J. Med. Chem.*, 2005, **48**, 4153.
363. N. Ginovart, L. Galineau, M. Willeit, R. Mizrahi, P. M. Bloomfield, P. Seeman, S. Houle, S. Kapur and A. A. Wilson, *J. Neurochem.*, 2006, **97**, 1089.
364. R. Narendran, M. Slifstein, O. Guillin, Y. Hwang, D.-R. Hwang, E. Scher, S. Reeder, E. Rabiner and M. Laruelle, *Synapse*, 2006, **60**, 485.
365. B. I., M. Guttman, P. Rusjan, J. R. Adams, S. Houle, J. Tong, O. Hornykiewicz, Y. Furukawa, A. A. Wilson, S. Kapur and S. Kish, *Brain*, 2009, **132**, 1366.
366. A. Graff-Guerrero, R. Mizrahi, O. Agid, H. Marcon, P. Barsoum, P. Rusjan, A. A. Wilson, R. Zipursky and S. Kapur, *Neuropsychopharmacology*, 2009, **34**, 1078.
367. A. Graff-Guerrero, D. Mamo, C. M. Shammi, R. Mizrahi, H. Marcon, P. Barsoum, P. Rusjan, S. Houle, A. A. Wilson and S. Kapur, *Arch. Gen. Psychiatry*, 2009, **66**, 606.
368. E. A. Rabiner, M. Slifstein, J. Nobrega, C. Plisson, M. Huiban, R. Raymond, M. Diwan, A. A. Wilson, P. McCormick, G. Gentile, R. N. Gunn and M. Laruelle, *Synapse*, 2009, **63**, 782.
369. N. M. Barnes and T. Sharp, *Neuropharmacology*, 1999, **38**, 1083.
370. J. Hannon and D. Hoyer, *Behav. Brain Res.*, 2008, **195**, 198.
371. J. Bockaert, S. Claeysen, V. Compan and A. Dumuis, *Neuropharmacology*, 2008, **55**, 922.
372. M. V. King, C. A. Marsden and K. C. F. Fone, *Trends Pharmacol. Sci.*, 2008, **29**, 482.
373. A. Dumuis, H. Ansanay, C. Waeber, M. Sebben, L. Fagni and J. Bockaert, *Pharmacochem. Libr.*, 1997, **27**, 261.
374. E. S. Mitchell and J. F. Neumaier, *Pharmacol. Therapeut.*, 2005, **108**, 320.

375. T. A. Branchek and T. P. Blackburn, *Annu. Rev. Pharmacol. Toxicol.*, 2000, **40**, 319.
376. K. C. F. Fone, *Neuropharmacology*, 2008, **55**, 1015.
377. J. Holenz, P. J. Pauwels, J. L. Diaz, R. Merce, X. Codony and H. Buschmann, *Drug Discovery Today*, 2006, **11**, 283.
378. A. D. Gee, L. Martarello, J. Passchier, M. Wishart, C. M. Parker, J. , R. Comley, R. Hopper and R. Gunn, *Curr. Radiopharm.*, 2008, **1**, 110.
379. B. R. Kornum, N. M. Lind, N. Gilling, L. Marner, F. Andersen and G. M. Knudsen, *J. Cerebr. Blood F. Met.*, 2009, **29**, 186.
380. L. Marner, N. Gillings, R. A. Comley, W. F. C. Baare, E. A. Rabiner, A. A. Wilson, S. Houle, S. G. Hasselbalch, C. Svarer, R. N. Gunn, M. Laruelle and G. M. Knudsen, *J. Nucl. Med.*, 2009, **50**, 900.
381. V. W. Pike, C. Halldin, K. Nobuhara, J. Hiltunen, R. S. Mulligan, C. G. Swahn, P. Karlsson, H. Olsson, S. P. Hume, E. Hirani, J. Whalley, L. S. Pilowsky, S. Larsson, P. O. Schnell, P. J. Ell and L. Farde., *Eur. J. Nucl. Med. Mol. Imaging*, 2003, **30**, 1520.
382. L. Martarello, M. Ahmed, A. T. Chuang, V. J. Cunningham, S. Jakobsen, C. N. Johnson, J. C. Matthews, A. Medhurst, S. F. Moss, E. A. Rabiner, A. Ray, D. Rivers, G. Stemp and A. D. Gee, *J. Labelled Compd. Radiopharm.*, 2005, **48**, S7.
383. C. A. Parker, V. J. Cunningham, L. Martarello, E. A. Rabiner, G. E. Searle, A. D. Gee, M. Davy, C. N. Johnson, M. Ahmed, R. N. Gunn and M. Laruelle, *Neuroimage*, 2008, **41**, T20.
384. R. Comley, R. Mizrahi, C. Salinas, I. Vitcu, A. Ng, W. Hallett, N. Keat, E. Rabiner, M. Laruelle and S. Houle, *J. Nucl. Med.*, 2009, **50**, 1850.
385. M. K. Chen and T. R. Guilarte, *Pharmacol Therapeut.*, 2008, **118**, 1.
386. V. Papadopoulos, M. Baraldi, T. R. Guilarte, T. B. Kunudsen, J. J. Lacapere, P. Lindemann, M. D. Norenberg, D. Nutt, A. Weizman, M. R. Zhang and M. Gavish, *Trends Pharmacol. Sci.*, 2006, **27**, 402.
387. A. M. Scarf, L. M. Ittner and M. Kassiou, *J. Med. Chem.*, 2009, **52**, 581.
388. S. Taliani, F. Da Settimo, E. Da Pozzo, B. Chelli and C. Martini, *Curr. Med. Chem.*, 2009, **16**, 3359.
389. V. Papadopoulos and L. Lecanu, *Exp. Neurol.*, 2009, **219**, 53.
390. M. Cosenza-Nashat, M. L. Zhao, H. S. Suh, J. Morgan, R. Natividad, S. Morgello and S. C. Lee, *Neuropath. Appl. Neuro.*, 2009, **35**, 306.
391. H. Akiyamaa, S. Bargera, S. Barnuma, B. Bradta, J. Bauera, G. M. Colea, N. R. Coopera, P. Eikelenbooma, M. Emmerlinga, B. L. Fiebicha, C. E. Fincha, S. Frautschya, W. S. Griffin, H. Hampel, M. Hull, G. Landreth, L. Lue, R. Mrak, I. R. Mackenzie, P. L. McGeer, M. K. O'Banion, J. Pachter, G. Pasinetti, C. Plata-Salaman, J. Rogers, R. Rydel, Y. Shen, W. Streit, R. Strohmeyer, I. Tooyoma, F. L. Van Muiswinkel, R. Veerhuis, D. Walker, S. Webster, B. Wegrzyniak, G. Wenk G and T. Wyss-Coray, *Neurobiol. Aging*, 2000, **21**, 383.
392. B. Cameron and G. E. Landreth, *Neurobiol. Dis.*, 2010, **37**, 503.
393. M. C. Petit-Taboue, J. C. Baron, F. Gourand, N. Ravenel, L. Barre, J. M. Travere and E. T. MacKenzie, *Euro. J. Pharmacol.*, 1991, **200**, 347.

394. A. Roivainen, K. Nagren, J. Hirvonen, V. Oikonen, P. Virsu, T. Tolvanen and J. O. Rinne, *Eur. J. Nucl. Med. Mol. Imaging*, 2009, **36**, 671.

395. J. Hirvonen, A. Roivainen, J. Virta, S. Helin, K. Nagren and J. O. Rinne, *Eur. J. Nucl. Med. Mol. Imaging*, 2010, **37**, 606.

396. A. Kumar, O. Muzik, D. Chugani, P. Chakraborty and H. T. Chugani, *J. Nucl. Med.*, 2010, **51**, 139.

397. M. A. Kropholler, R. Boellaard, A. Schuitemaker, B. N. M. Van Berckel, G. Luurtsema, A. D. Windhorst and A. A. Lammertsma, *J. Cerebr. Blood F. Met.*, 2005, **25**, 842.

398. A. N. Anderson, N. Pavese, P. Edison, Y. F. Tai, A. Hammers, A. Gerhard, D. J. Brook and F. E. Turkheimer, *Neuroimage*, 2007, **36**, 28.

399. F. E. Turkheimer, P. Edison, N. Pavese, F. Roncaroli, A. N. Anderson, A. Hammers, A. Gerhard, R. Hinz, Y. F. Tai and D. J. Brook, *J. Nucl. Med.*, 2007, **48**, 158.

400. A. Schuitemaker, B. N. M. Van Berckel, M. A. Kropholler, R. W. Kloet, C. Jonker, P. Scheltens, A. A. Lammertsma and R. Boellaard, *J. Cerebr. Blood F. Met.*, 2007, **27**, 1603.

401. G. Tomasi, P. Edison, A. Bertoldo, F. Roncaroli, P. Singh, A. Gerhard, C. Cobelli, D. J. Brooks and F. E. Turkheimer, *J. Nucl. Med.*, 2008, **49**, 1249.

402. Y. Ouchi, E. Yashikawa, Y. Sekine, M. Futatsubashi, T. Kanno, T. Ogusu and T. Torizuka, *Ann. Neurol.*, 2005, **57**, 168.

403. Y. Ouchi, S. Yagi, M. Yokokura and M. Sakamoto, *Parkinsonism Relat. D.*, 2009, **15S3**, S200.

404. A. Gerhard, N. Pavese, G. Hotton, F. Turkheimer, M. Es, A. Hammers, K. Eggert, W. Oertel, R. B. Banati and D. J. Brooks, *Neurobiol. Dis.*, 2006, **21**, 404.

405. A. L. Bartels, A. T. M. Willensen, J. Doorduin, E. F. J. De Vries, R. A. Dierckx and K. L. Leenders, *Parkinsonism Relat. D.*, 2010, **16**, 57.

406. N. Pavese, A. Gerhard, Y. F. Tai, A. K. Ho, F. Turkheimer, R. A. Barker, D. J. Brooks and P. Piccini, *Neurology*, 2006, **66**, 1638.

407. Y. F. Tai, N. Pavese, A. Gerhard, S. J. Tabrizi, R. A. Baraker, D. J. Brooks and P. Piccini, *Brain*, 2007, **130**, 1759.

408. Y. F. Tai, N. Pavese, A. Gerhard, S. J. Tabrizi, R. A. Barker, D. J. Brooks and P. Piccini, *Brain Res. Bull.*, 2007, **72**, 148.

409. D. A. Hammoud, C. J. Endres, E. R. Chander, T. R. Guilarte, D. F. Wong, N. C. Sacktor, J. C. McArthur and M. G. Pomper, *J. NeuroVirol.*, 2005, **11**, 346.

410. C. A. Wiley, B. J. Lopresti, J. T. Becker, F. Boada, O. L. Lopez, J. Mellors, C. C. Meltzer, S. R. Wisniewski and C. A. Mathis, *J. Neuro-Virol.*, 2006, **12**, 262.

411. A. Cagnin, D. J. Brooks, A. M. Kennedy, R. N. Gunn, R. Myers, F. E. Turkheimer, T. Jones and B. R. Banati, *Lancet*, 2001, **358**, 461.

412. G. N. Groom L. Junck, N. L. Foster, K. A. Frey and D. E. Kuhl, *J. Nucl. Med.*, 1995, **36**, 2207.

413. S. Pappata, P. Cornu, Y. Samson, C. Prenant, J. Benavides, B. Scatton, C. Crouzel, J. Hauw and A. Syrota, *J. Nucl. Med.*, 1991, **32**, 1608–1610.

414. A. Cagnin, R. Myers, R. N. Gunn, A. D. Lawrence, T. Stevens, G. W. Kreutzberg, T. Jones and R. B. Banati, *Brain*, 2001, **124**, 2014.

415. J. C. Debruyne, J. Versijpt, K. J. Van Laere, F. De Vos, J. Keppens, K. Strijckmans, E. Achten, G. Slegers, R. A. Dierckx, J. Korf and J. L. De Reuck, *Eur. J. Neurol.*, 2003, **10**, 257.

416. M. R. Turner, A. Cagnin, F. E. Turkheimer, C. C. J. Miller, C. E. Shaw, D. J. Brooks, P. N. Leigh and R. B. Banati, *Neurobiol. Dis.*, 2004, **15**, 601.

417. A. Cagnin, M. Rossor, E. Sampson, T. MacKinnon and R. B. Banati, *Ann. Neurol.*, 2004, **56**, 894.

418. A. Gerhard, J. Watts, I. Trender-Gerhard, F. Turkheimer, R. B. Banati, K. Bhatia and D. J. Brooks, *Movement Disord.*, 2004, **19**, 1221.

419. B. N. M. Van Berckel, M. G. Bossong, R. Boellaard, R. Kloet, A. Schuitemaker, E. Caspers, G. Luurtsema, A. D. Windhorst, W. Cahn, A. A. Lammertsma and R. S. Kahn, *Biol. Psychiatry*, 2008, **64**, 820.

420. M. Imaizumi, E. Briard, S. S. Zoghbi, J. P. Gourley, J. Hong, J. L. Musachio, R. Gladding, V. W. Pike, R. B. Innis and M. Fujita, *Synapse*, 2007, **61**, 595.

421. E. Briard, S. S. Zoghbi, F. G. Simeon, M. Imaizumi, J. P. Gourley, H. U. Shetty, S. Lu, M. Fujita, R. B. Innis and K. E. Pike, *J. Med. Chem.*, 2009, **52**, 688.

422. Y. Fujimura, S. S. Zoghbi, F. G. Simeon, A. Taku, K. E. Pike, R. B. Innis and M. Fujita, *J. Nucl. Med.*, 2009, **50**, 1047.

423. Y. Fujimura, S. S. Zoghbi, R. Gladding, F. G. Simeon, A. Taku, K. E. Pike, R. B. Innis and M. Fujita, *Neuroimage*, 2008, **41**, T111.

424. Y. Fujimura, S. Zoghbi, F. G. Simeon, A. Taku, K. E. Pike, R. B. Innis and M. Fujita, *J. Cerebr. Blood F. Met.*, 2009, **29**, S360.

425. Y. Fujimura, Y. Kimura, F. G. Simeon, L. P. Dickstein, K. E. Pike, R. B. Innis and M. Fujita, *J. Nucl. Med.*, 2010, **51**, 145.

426. M. Wang, K. K. Yoder, M. Gao, B. H. Mock, X.-M. Xu, A. J. Saykin, G. D. Hutchins and Q.-H. Zheng, *Bioorg. Med. Chem. Lett.*, 2009, **19**, 5636.

427. E. Briard, J. Shah, J. L. Musachio, S. S. Zoghbi, M. Fujita, M. Imaizumi, V. Cropley, R. B. Innis and K. E. Pike, *J. Labelled Compd Radiopharm.*, 2005, **48**(Suppl. 1), S4.

428. M. Imaizumi, E. Briard, S. S. Zoghbi, J. P. Gourley, J. Hong, Y. Fujimura, V. W. Pike, R. B. Innis and M. Fujita, *Neuroimage*, 2008, **39**, 1289.

429. A. K. Brown, M. Fujita, Y. Fujimura, J.-S. Liow, M. Stabin, Y. H. Ryu, M. Imaizumi, J. Hong, K. E. Pike and R. B. Innis, *J. Nucl. Med.*, 2007, **48**, 2072.

430. W. C. Kreisl, G. Mbeo, M. Fujita, S. Zoghbi, K. E. Pike, R. B. Innis and J. C. McArthur, *Arch. Neurol.*, 2009, **66**, 1288.

431. M. Fujita, M. Imaizumi, S. S. Zoghbi, Y. Fujimura, A. G. Farris, T. Suhara, J. Hong, K. E. Pike and R. B. Innis, *Neuroimage*, 2008, **40**, 43.

432. W. C. Kreisl, M. Fujita, Y. Fujimura, N. Kimura, K. L. Jenko, P. Kannan, J. Hong, C. L. Morse, S. S. Zoghbi, R. L. Gladding, S. Jacobson, U. Oh, V. W. Pike and R. B. Innis, *Neuroimage*, 2010, **49**, 2924.

433. D. R. Owen, O. W. Howell, S.-P. Tang, L. A. Wells, I. Bennacef, M. Bergstrom, R. N. Gunn, E. A. Rabiner, M. R. Wilkins, R. Reynolds, P. M. Matthews and C. A. Parker, *J. Cerebr. Blood F. Met.*, 2010, in press (Online doi:10.1038/jcbfm.2010.63).

434. T. H. Jeon, Y.-S. Heo, C. M. Kim, Y.-L. Hyun, T. G. Lee, S. Ro and J. M. Cho, *Cell. Mol. Life Sci.*, 2005, **62**, 1198.

435. M. P. Kelly and N. J. Brandon, *Prog. Brain Res.*, 2009, **179**, 67.

436. C. Lugnier, *Pharmacol. Therapeut.*, 2006, **109**, 366.

437. T. B. Halene and S. J. Siegel, *Drug Discovery Today*, 2007, **12**, 870.

438. F. S. Menniti, W. S. Faraci and C. J. Schmidt, *Nat. Rev. Drug Discovery*, 2006, **5**, 660.

439. O. A. Reneerkens, K. Rutten, H. W. M. Steinbusch, A. Blokland and J. Prickaerts, *Psychopharmacology (Berlin)*, 2009, **202**, 419.

440. M. D. Houslay, P. Schafer and K. Y. J. Zhang, *Drug Discovery Today*, 2005, **10**, 1503.

441. L. Pages, A. Gavalda and M. D. Lehner, *Expert Opin. Ther. Targets*, 2009, **19**, 1501.

442. J. A. Siuciak and C. A. Strick, *Drug Discovery Today: Ther. Strategies*, 2006, **3**, 527.

443. F. S. Menniti, T. A. Chappie, J. M. Humphrey and C. J. Schmidt, *Curr. Opin. Invest. Drugs*, 2007, **8**, 54.

444. M. D. Houslay, M. Sullivan and G. B. Bolger, *Adv Pharmacol (San Diego)*, 1998, **44**, 225.

445. F. Laliberte, Y. Han, A. Govindarajan, A. Giroux, S. Lui, B. Bobechko, P. Lario, A. Barlett, E. Gorseth, M. Gresser and Z. Huang, *Biochemistry*, 2000, **39**, 6449.

446. H. Schneider, H. R. Schmiechen, A. A. Wilson and J. N. DaSilva, *Eur. J. Pharmacol.*, 1986, **127**, 105.

447. M. S. Barnette, J. O'Leary Bartus, M. Burman, S. B. Christensen, L. B. Cieslinski, K. M. Esser, U. S. Prabhakar, J. A. Rush and T. J. Torphy, *Biochem. Pharmacol.*, 1996, **51**, 949.

448. T. J. Torphy, J. M. Stadel, M. Burman, L. B. Cieslinski, M. M. McLaughlin, J. R. White and G. P. Livi, *J. Biol. Chem.*, 1992, **267**, 1798.

449. M. Kenk, M. Greene, J. Thackeray, R. A. deKemp, M. Lortie, S. Thorn, R. S. Beanlands and J. N. DaSilva, *Nucl. Med. Biol.*, 2007, **34**, 71.

450. C. M. Lourenco, J. N. DaSilva, J. J. Warsh, A. A. Wilson and S. Houle, *Synapse*, 1999, **31**(1), 41.

451. C. M. Lourenco, S. Houle, A. A. Wilson and J. N. DaSilva, *Nucl. Med. Biol.*, 2001, **28**, 347.

452. C. A. Parker, J. C. Matthews, R. N. Gunn, L. Martarello, V. J. Cunningham, D. Dommett, S. T. Knibb, D. Bender, S. Jakobsen, J. Brown and A. D. Gee, *Synapse*, 2005, **55**, 270.

453. M. Fujita, S. S. Zoghbi, M. S. Crescenzo, J. Hong, J. L. Musachio, J.-Q. Lu, J.-S. Liow, N. Seneca, D. N. Tipre and V. L. Cropley, *Neuroimage*, 2005, **26**(4), 1201.

454. J. N. DaSilva, C. M. Lourenco, J. H. Meyer, D. Hussey, W. Z. Potter and S. Houle, *Eur. J. Nucl. Med*, 2002, **29**, 1680.

455. C. M. Lourenco, M. Kerk, R. S. Beanlands and J. N. DaSilva, *Life Sci.*, 2006, **79**, 356.

456. D. R. Sprague, M. Fujita, Y. H. Ryu, J.-S. Liow, V. W. Pike and R. B. Innis, *Nucl. Med. Biol.*, 2008, **35**, 493.

457. T. Itoh, K. Abe, S. S. Zoghbi, O. Inoue, J. Hong, M. Imaizumi, V. W. Pike, R. B. Innis and M. Fujita, *J. Nucl. Med.*, 2009, **50**, 749.

458. S. Celen, M. De Angelis, S. K. Chitneni, J. Alcazar, M. LKoole, S. Dedeurwaerdere, T. Steckler, M. Schmidt, K. Van Laere, X. Langlois, J. I. Andres and G. Bormans, presented at, *EANM'09*, Barcelona, Spain, 2009.

459. H. Chen, E. Hu, M. DSlifstein, D.-R. Hwang, C. Biorn, D. Lester-Zeiner, J. Ma, R. Cho, J. Shi, J. Wong, S. Miller, G. Hill Della Puppa, *presented at the, SFN*, 2009, Chicago, U.S.A.

460. A. M. Sebastiao and J. A. Ribeiro, *Handb. Exp. Pharmacol.*, 2009, **193**, 471.

461. T. W. Stone, S. Ceruti and M. P. Abbracchio, *Handb. Exp. Pharmacol.*, 2009, **193**, 535.

462. K. A. Jacobson and Z.-G. Gao, *Nat. Rev. Drug Discovery*, 2006, **5**, 247.

463. E. C. Klaasse, A. P. IJzerman, W. J. De Grip and M. W. Beukers, *Purinergic Signalling*, 2008, **4**, 21.

464. S. Moro, Z.-G. Gao, K. A. Jacobson and G. Spalluto, *Med. Res. Rev.*, 2006, **26**, 131.

465. E. Elzein and J. Zablocki, *Expert Opin. Invest. Drugs*, 2008, **17**, 1901.

466. D. R. Lara, O. P. Dall'Igna, E. S. Ghisolfi and M. G. Brunstein, *Prog. Neuro-Psychoph.*, 2006, **30**, 617.

467. R. A. Cunha, S. Ferre, J.-M. Vaugeois and J.-F. Chen, *Curr. Pharm. Des.*, 2008, **14**, 1512.

468. M. Morelli, A. R. Carta and P. Jenner, *Handb. Exp. Pharmacol.*, 2009, **193**, 589.

469. N. Simola, M. Morelli and A. Pinna, *Curr. Pharm. Des.*, 2008, **14**, 1475.

470. S. Ferre, C. Quiroz, A. S. Woods, R. Cunha, P. Popoli, F. Ciruela, C. Lluis, R. Franco, K. Azdad and S. N. Schiffmann, *Curr. Pharm. Des.*, 2008, **14**, 1468.

471. S. N. Schiffmann, G. Fisone, R. Moresco, R. A. Cunha and S. Ferre, *Prog. Neurobiol.*, 2007, **83**, 277.

472. J. Noguchi, K. Ishiwata, R. Furata, J.-I. Simada, M. KIyosawa, S.-I. Ishii, K. Endo, F. Suzuki and M. Sendo, *Nucl. Med. Biol.*, 1997, **24**, 53.

473. K. Kawamura and K. Ishiwata, *Ann. Nucl. Med.*, 2004, **18**, 165.

474. Y. Shimada, K. Ishiwata, M. Kiyosawa, T. Nariai, K. Oda, H. Toyama, F. Suzuki, K. S. Ono and M. , *Nucl. Med. Biol.*, 2002, **29**, 29.

475. K. Ishiwata, T. Nariai, Y. Kimura, K. Oda, K. Kawamura, K. Ishii, M. Senda, S. Wakabayashi and J. Shimada, *Ann. Nucl. Med.*, 2002, **16**, 377.

476. N. Fukumitsu, K. Ishii, Y. Kimura, K. Oda, T. Sasaki, Y. Mori and K. Ishiwata, *Ann. Nucl. Med.*, 2003, **17**, 511.

477. Y. Kimura, K. Ishii, N. Fukumitsu, K. S. Oda, T. , K. Kawamura and K. Ishiwata, *Nucl. Med. Biol.*, 2004, **31**, 975.

478. M. Naganawa, Y. Kimura, T. Nariai, K. Ishii, K. Oda, Y. Manabe, K. Chihara and K. Ishiwata, *Neuroimage*, 2005, **26**, 885.

479. N. Fukumitsu, K. Ishii, Y. Kimura, K. Oda, M. Hashimoto, M. Suzuki and K. Ishiwata, *Ann. Nucl. Med.*, 2008, **22**, 841.

480. W. Sihver, M. H. Holschbach, D. Bier, W. Wutz, A. Schulze, R. A. Olsson and H. H. Coenen, *Nucl. Med. Biol.*, 2003, **30**, 661.

481. M. H. Holschbach, R. A. Olsson, D. Bier, W. Wutz, W. Sihver, M. Schuller, B. Palm and H. H. Coenen, *J. Med. Chem.*, 2002, **45**, 5150.

482. A. Bauer, M. H. Holschbach, M. Cremer, S. Weber, C. Boy, N. J. Shah, R. A. Olsson, H. Halling, H. H. Coenen and K. Zilles, *J. Nucl. Med.*, 2003, **44**, 1682.

483. A. Bauer, K.-J. Langen, H. Bidmon, M. H. Holschbach, S. Weber, R. A. Olsson, H. H. Coenen and K. Zilles, *J. Nucl. Med.*, 2005, **46**, 450.

484. P. T. Meyer, D. Bier, M. H. Holschbach, C. Boy, R. A. Olsson, H. H. Coenen, K. Zilles and A. Bauer, *J. Cerebr. Blood F. Met.*, 2004, **24**, 323.

485. J. H. Meyer, D. Elmenhorst, D. Bier, M. Holschbach, A. Matusch, H. H. Coenen, K. Zilles and A. Bauer, *Neuroimage*, 2005, **24**, 1192.

486. P. T. Meyer, D. Elmenhorst, K. Zilles and A. Bauer, *Synapse*, 2005, **55**, 212.

487. P. T. Meyer, D. Elmenhorst, A. Matusch, O. Winz, K. Zilles and A. Bauer, *Neuroimage*, 2006, **32**, 1100.

488. D. Bier, M. H. Holschbach, W. Wutz, R. A. Olsson and H. H. Coenen, *Drug Metab. Dispos.*, 2006, **34**, 570.

489. D. Elmenhorst, P. T. Meyer, A. Matusch, O. H. Winz, K. Zilles and A. Bauer, *Eur. J. Nucl. Med. Mol Imaging*, 2007, **34**, 1061.

490. H. Herzog, D. Elmenhorst, O. Winz and A. Bauer, *Eur. J. Nucl. Med. Mol. Imaging*, 2008, **35**, 1499.

491. T. Matsuya, H. Takamatsu, Y. Murakami, A. Noda, R. Ichise, Y. Awaga and S. Nishimura, *Nucl. Med. Biol.*, 2005, **32**, 837.

492. T. Maemoto, M. Tada, T. Mihara, N. Ueyama, H. Matsuoka, K. Harada, T. S. Yamaji, K. , S. Kuroda, A. Akahane, A. Iwashita, N. Matsuoka and S. Mutho, *J. Pharm. Sci.*, 2004, **96**, 42.

493. S. Stone-Elander, J. O. Thorell, E. L., B. B. Fredholm and M. Ingvar, *Nucl. Med. Biol.*, 1997, **24**, 187.

494. K. Ishiwata, J. Noguchi, H. Toyama, Y. Sakiyama, N. Koike, S.-I. Ishii, K. Oda, K. Endo, F. Suzuki and M. Senda, *Appl. Radiat. Isot.*, 1996, **47**, 507.

495. J. Noguchi, K. Ishiwata, S.-I. Wakabayashi, T. Nariai, S. Shumiya, S.-I. Ishii, H. Toyama, K. Endo, F. Suzuki and M. Senda, *J. Nucl. Med.*, 1998, **39**, 498.

496. K. Ishiwata, J. Noguchi, S.-I. Wakabayashi, J. Shimada, N. Ogi, T. Nariai, A. Tanaka, K. Endo, F. Suzuki and M. Senda, *J. Nucl. Med.*, 2000, **41**, 345.

497. K. Ishiwata, N. Ogi, J. Shimada, H. Nonaka, A. Tanaka, F. Suzuki and M. Senda, *Ann. Nucl. Med.*, 2000, **14**, 81.

498. K. Ishiwata, N. Ogi, N. Hayakawa, K. Oda, T. Nagaoka, H. Toyama, F. Suzuki, K. Endo, A. Tanaka and M. Senda, *Ann. Nucl. Med.*, 2002, **16**, 467.

499. K. Ishiwata, W.-F. Wang, Y. Kimura, K. Kawamura and K. Ishii, *Ann. Nucl. Med.*, 2003, **17**, 205.

500. K. Ishiwata, M. Mishina, Y. Kimura, K. Oda, T. Sasaki and K. Ishii, *Synapse*, 2005, **55**, 133–136.

501. M. Mishina, K. Ishiwata, Y. Kimura, M. Naganawa, K. Oda, S. Kobayashi, Y. Katayama and K. Ishii, *Synapse*, 2007, **61**, 778.

502. M. Naganawa, Y. Kimura, M. Mishina, Y. Manabe, K. Chihara, K. Oda, K. Ishii and K. Ishiwata, *Eur. J. Nucl. Med. Mol. Imaging*, 2007, **34**(5), 679–687.

503. E. Hirani, J. Gillies, A. Karasawa, J. Shimada, H. Kase, J. Opacka-Juffry, S. Osman, S. K. Luthra, S. P. Hume and D. J. Brooks, *Synapse*, 2001, **42**, 164.

504. J. Gillies, S. K. Luthra, F. Brady, A. Karasawa, J. Shimada and H. Kase, *J. Labelled Compd. Radiopharm.*, 1999, **42**(Suppl. 1), S456.

505. D. J. Brooks, M. Dooder, S. Osman, S. K. Luthra, E. Hirani, S. Hume, H. Kase, J. Kilborn, S. Martindill and A. Mori, *Synapse*, 2008, **62**, 671.

506. S. Todde, R. M. Moresco, P. Simonelli, P. G. Baraldi, B. Cacciari, G. Spalluto, K. Varani, A. Monopoli, M. Matarrese, A. Carpinelli, F. Magni, M. G. Kienle and F. Fazio, *J. Med. Chem.*, 2000, **43**, 4359.

507. R. M. Moresco, S. Todde, S. Belloli, P. Simonelli, A. Panzacchi, M. Rigamonti, M. Galli-Kienle and F. Fazio, *Eur. J. Nucl. Med. Mol. Imaging*, 2005, **32**, 405.

508. G. D. Dalton, C. E. Bass, C. G. Van Horn and A. C. Howlett, *CNS Neurol. Disord.: Drug Targets*, 2009, **8**, 422.

509. N. Wegener and M. Koch, *Pharmacopsychiatry*, 2009, **42**(Suppl. 1), S79.

510. E. S. Graham, J. C. Ashton and M. Glass, *Front. Biosci.*, 2009, **14**, 944.

511. E. S. Onaivi, *Int. Rev. Neurobiol.*, 2009, **88**, 335.

512. D. Cota, *Diabetes/Metab. Res. Rev.*, 2007, **23**, 507.

513. E. Fernandez-Espejo, M. P. Viveros, L. Núñez, B. A. Ellenbroek and F. Rodriguez de Fonseca, *Psychopharmacology (Berlin)*, 2009, **206**, 531.

514. J. Fernandez-Ruiz, *Br. J. Pharmacol.*, 2009, **156**, 1029.

515. S. Ferre, S. R. Goldberg, C. Lluis and R. Franco, *Neuropharmacology*, 2009, **56**(Suppl. 1), 226.

516. V. Micale, C. Mazzola and F. Drago, *Pharmacol. Res.*, 2007, **56**, 382.

517. J. M. Orgado, J. Fernandez-Ruiz and J. Romero, *Int. Rev. Psychiatr.*, 2009, **21**, 172.

518. J. Fernandez-Ruiz, M. R. Pazos, M. García-Arencibia, O. Sagredo and J. A. Ramos, *Mol. Cell. Endocrinol.*, 2008, **286**(Suppl. 1), S91.

519. G. A. Cabral, E. S. Raborn, L. Griffin, J. Dennis and F. Marciano-Cabral, *Br. J. Pharmacol.*, 2008, **153**, 240.

520. C. Benito, R. M. Tolon, M. R. Pazos, E. Nunez, A. I. Castillo and J. Romero, *Br. J. Pharmacol.*, 2008, **153**, 277.

521. P. W. Brownjohn and J. C. Ashton, *Curr. Anaesth. Crit. Care*, 2009, **20**, 189.

522. S. J. Gatley, A. N. Gifford, N. D. Volkow, R. X. Lan and A. Makriyannis, *Eur. J. Pharmacol.*, 1996, **307**(3), 331.

523. S. J. Gatley, R. Lan, N. D. Volkow, N. Pappas, P. King, C. T. Wong, A. N. Gifford, B. Pyatt, S. L. Dewey and A. Makriyannis, *J. Neurochem.*, 1998, **70**(1), 417.

524. G. Berding, K. Muller-Vahl, U. Schneider, P. Gielow, J. Fitschen, M. Stuhrmann, H. Harke, R. Buchert, F. Donnerstag and M. Hofmann, *Biol. Psychiatry*, 2004, **55**(9), 904.

525. G. Terry, J.-S. Liow, E. Chernet, S. S. Zoghbi, S. Phebus, C. C. Felder, J. Tauscher, J. M. Schaus, V. W. Pike, C. Halldin and R. B. Innis, *Neuroimage*, 2008, **41**, 690.

526. F. Yasuno, A. K. Brown, S. S. Zoghbi, J. H. Krushinski, E. Chernet, J. Tauscher, J. M. Schaus, L. A. Phebus, A. K. Chesterfield, C. C. Felder, R. L. Gladding, J. Hong, C. Halldin, V. W. Pike and R. B. Innis, *Neuropsychopharmacology*, 2008, **33**, 259.

527. G. Terry, J.-S. Liow, S. Zoghbi, J. Hirvonen, A. G. Farris, A. Lerner, J. T. Tauscher, J. M. Schaus, L. F. Phebus, C. C., C. L. Morse, J. S. Hong, V. W. Pike, C. Halldin and R. B. Innis, *Neuroimage*, 2009, **48**, 362.

528. S. R. Donohue, J. H. Krushinski, V. W. Pike, E. Chernet, L. Phebus, A. K. Chesterfield, C. C. Felder, C. Halldin and J. M. Schaus, *J. Med. Chem.*, 2008, **51**, 5833.

529. G. E. Terry, J. Hirvonen, J.-S. Liow, S. S. Zoghbi, R. Gladding, J. T. Tauscher, J. M. Schaus, L. Phebus, C. C. Felder, C. L. Morse, S. R. Donohue, V. W. Pike, C. Halldin and R. B. Innis, *J. Nucl. Med.*, 2010, **51**, 112.

530. P. Liu, L. S. Lin, T. G. Hamill, J. P. Jewell, T. J. Lanza, R. E. Gibson, S. M. Krause, C. Ryan, W. Eng, S. Sanabria, X. Tong, J. Wang, D. A. Levorse, K. A. Owens, T. M. Fong, C. P. Shen, J. Lao, S. Kumar, W. Yin, J. F. Payack, S. A. Springfield, R. J. Hargreaves, H. D. Burns, M. T. Goulet and W. K. Hagmann, *J. Med. Chem.*, 2007, **50**, 3427.

531. S. M. Sanabria-Bohorquez, T. G. Hamill, K. Goffin, I. De Lepeleire, G. Bormans, H. D. Burns and K. Van Laere, *Eur. J. Nucl. Med. Mol. Imaging*, 2010, in press, DOI: 10.1007/s00259.

532. H. D. Burns, K. Van Laere, S. Sanabria-Bohorquez, T. G. Hamill, G. Bormans, W. Eng, R. Gibson, C. Ryan, B. Connolly, S. Patel, S. Krause, A. Vanko, A. Van Hecken, P. Dupont, I. De Lepeleire, P. Rothenberg, S. A. Stoch, J. Cote, W. K. Hagmann, J. P. Jewell, L. S. Lin, P. Liu, M. T. Goulet, K. Gottesdiener, J. A. Wagner, J. de Hoon, L. Mortelmans, T. M. Fong and R. J. Hargreaves, *Proc. Natl. Acad. Sci. U. S. A.*, 2007, **104**, 9800.

533. K. Van Laere, M. Koole, S. M. Sanabria-Bohorquez, K. Goffin, I. Guenther, M. J. belanger, J. Cote, P. Rothenberg, I. De Lepeleire, I. D. Grachev, R. J. Hargreaves, G. Bormans and H. D. Burns, *J. Nucl. Med.*, 2008, **49**, 439.
534. N. Battista, N. Pasquariello, M. Di Tommaso and M. Maccarrone, *J. Neuroendocrinol.*, 2008, **20**, 82.
535. K. Van Laere, K. Goffin, C. Casteels, P. Dupont, L. Mortelmans, J. D. Hoon and G. Bormans, *Neuroimage*, 2008, **39**, 1533.
536. N. Evens, B. Bosier, B. J. Lavey, J. A. Kozlowski, P. Vermaelen, L. Baudemprez, R. Busson, D. M. Lambert, K. Van Laere, A. M. Verbruggen and G. M. Bormans, *Nucl. Med. Biol.*, 2008, **35**, 793.
537. N. Evens, G. G. Muccioli, N. Houbrechts, D. M. Lambert, A. M. Verbruggen, K. Van Laere and G. M. Bormans, *Nucl. Med. Biol.*, 2009, **36**, 455.

CHAPTER 4

Design and Synthesis of Radiopharmaceuticals for SPECT Imaging

DAVID HUBERS, R.Ph, BCNP AND
PETER J. H. SCOTT, Ph.D

Department of Radiology, University of Michigan Medical School,
Ann Arbor, MI 48109, USA

4.1 Introduction

Single-photon emission computed tomography (SPECT) imaging is an exciting and expanding area of research in which bioactive molecules are labeled with short-lived radionuclides.[1,2] The resulting "radiopharmaceuticals" find widespread application in the noninvasive examination of biochemical processes in living human subjects.[3] The patient receives an injection of the radiopharmaceutical followed by a SPECT scan. After data processing, the resulting image from the SPECT scanner provides physicians with a 3D spatial distribution of the single-photon emitting radionuclide (and the bioactive molecule to which it is attached) within the human body.

Radiopharmaceuticals used in SPECT imaging are typically radiolabeled with γ-ray emitting radionuclides that have energies between 30 and 300 keV (Table 4.1.). SPECT radionuclides tend to have relatively short half-lives (hours to a few days), although they are somewhat longer than the half-lives (minutes to hours) of the higher energy radionuclides employed in corresponding PET scans. The radionuclides in Table 4.1 are typically either produced directly using nuclear reactors, or they are the daughter products of reactor prepared

RSC Drug Discovery Series No. 15
Biomedical Imaging: The Chemistry of Labels, Probes and Contrast Agents
Edited by Martin Braddock
© Royal Society of Chemistry 2012
Published by the Royal Society of Chemistry, www.rsc.org

Table 4.1 Radionuclides commonly employed in SPECT Imaging.

Radionuclide	γ [or β⁻] Energy	Half-life
^{67}Ga	93, 185, 300 keV	3.26 d
^{123}I	159 keV	13.2 h
^{125}I	35.5 keV	59.9 d
^{131}I	364 keV	8.02 d
^{111}In	171, 245 keV	2.8 d
^{177}Lu	208 keV	6.71 d
^{188}Re	155 keV	17.02 h
^{153}Sm	97, 103 keV	1.95 d
^{99m}Tc	140 keV	6.02 h
^{201}Tl	135, 167 keV	3.04 d
^{133}Xe	31, 35, 81 keV	5.25 d
^{90}Y	$\beta^- = 2180$ keV	2.67 d

parent radionuclides, and many are now commercially available in convenient generator or solution form.

Radiopharmaceuticals can range from the small and simple (*e.g.* small molecules, sugars, peptides, amino acids) to the large and complex (*e.g.* large peptides, stem cells, proteins, antibodies) and can be custom-synthesized for applications in, for example, neurology,[4–6] oncology,[7,8] and cardiology.[9,10] Table 4.2 showcases common SPECT radiopharmaceuticals employed at the University of Michigan on a routine basis. Radiochemistry is the process by which such molecules are tagged with a radionuclide and, owing to the high levels of radiation involved, it is frequently not a trivial process. In the last two or three decades, a broad range of ingenious radiochemical reactions and automated hardware solutions have been developed to simplify the process of SPECT radiopharmaceutical synthesis.[11,12] This chapter outlines strategies for SPECT radionuclide production, incorporation of radionuclides into bioactive molecules, and quality control of the resulting radiopharmaceuticals. Proof-of-concept is shown by highlighting examples of common SPECT radio-pharmaceuticals currently in clinical use.

4.2 Radiopharmaceuticals Labeled with Technetium-99m

4.2.1 Production of Technetium-99m

Technetium-99m (^{99m}Tc), the daughter of molybdenum-99 (^{99}Mo), is by far the most commonly used radionuclide in SPECT imaging procedures due to its favorable physical properties (Table 4.1) and ready availability. It is estimated that over 80% of radiopharmaceuticals used clinically are labeled with tech-netium-99m, and, reflecting this, the current state-of-the-art in [^{99m}Tc]labeled radiopharmaceuticals has been extensively reviewed.[12–16] Molybdenum-99 is typically produced in nuclear reactors according to the equation

Table 4.2 A Selection of the SPECT and Therapeutic Radiopharmaceuticals in use at the University of Michigan.

Radiopharmaceutical	Application
Technetium-99m	
[^{99m}Tc]Sodium pertechnetate	thyroid imaging and radiolabeling of red blood cells
[^{99m}Tc]Sestimibi (Cardiolite®)	imaging of the myocardium (cardiac rest/ stress test)
[^{99m}Tc] Methylene diphosphate (MDP)	bone imaging
[^{99m}Tc]Bicisate (Neurolite®)	brain imaging
[^{99m}Tc]Sulfur colloid	agent for imaging areas of functioning retriculoendothelial cells in the liver, spleen and bone marrow; and for lymphoscintigraphy
[^{99m}Tc]Tetrofosmin (Myoview™)	scintigraphic imaging of the myocardium (Cardiac rest / stress test)
[^{99m}Tc]Choletech (Mebrofen®)	hepatobiliary imaging
[^{99m}Tc] diethylenetriamine pentaacetic acid (DTPA) Aerosol	lung imaging
[^{99m}Tc] Macroaggregated albumin (MAA)	V/Q perfusion studies
[^{99m}Tc] Mercaptoacetyltriglycine (*MAG3*)	renal imaging
[^{99m}Tc]*Exametazime* (Ceretec™)	agent used for detecting altered regional cerebral perfusion in stroke and for labeling leukocytes
Iodine-123 / Iodine-131	
[$^{123/131}$I]Sodium Iodident]	Agent used for imaging thyroid, thyroid cancer and hyperthyroidism
[^{131}I]Tositumomab (Bexxar®)	radiotherapeutic treatment of follicular lymphoma
[^{123}I]*m*-Iodobenzylguanidine (MIBG)	tumor and myocardial imaging
Gallium-67	
[^{67}Ga]Citrate (Neoscan®)	tumor and infection/inflammation imaging
Yttrium-90	
[^{90}Y]Ibritumomab (Zevalin®)	radiotherapeutic treatment of non-Hodgkin's lymphoma
Indium-111	
[^{111}In]Pentetreotide (Octreoscan™)	tumor imaging
[^{111}In] Capromab Pendetide (ProstaScint®)	prostate cancer imaging
Samarium-153	
[^{153}Sm]Lexidronam (Quadramet®)	Palliation of osteoblastic metastatic bone lesions

^{235}U(n,fission)^{99}Mo. However, this process attracted a lot of unwanted attention in 2009 when an unscheduled shutdown of the Chalk River Nuclear Reactor in Canada led to a worldwide shortage of molybdenum-99 (and technetium-99m).[17,18] The result was that many patients did not receive the SPECT scans they required, leading many researchers to consider alternative methods for production of molybdenum-99. The outcome was that a number of high energy cyclotron accelerators have been equipped with targets that enable accelerator production of molybdenum-99 from molybdenum-100 *via*, for

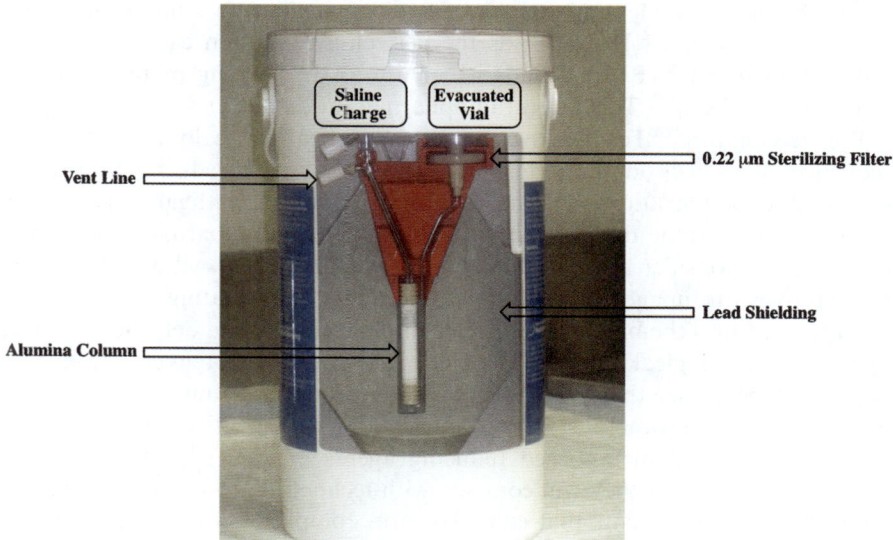

Figure 4.1 The ^{99}Mo/^{99m}Tc Generator.

example, the nuclear reaction ^{100}Mo(p,pn)^{99}Mo.[19] Investigators are also responding to the shortage by investigating direct production of technetium-99m using cyclotron accelerators.[20]

Molybdenum-99 is then adsorbed onto aluminum oxide columns, which are then packed into commercial ^{99}Mo/^{99m}Tc generators, and shipped out to radiopharmacies and radiochemistry laboratories all over the world.[21] Molybdenum-99 ($t_{1/2} = 67$ hours) decays to technetium-99m ($t_{1/2} = 6$ hours) according to a transient equilibrium, and equilibrium occurs within four half-lives. A typical generator (Figure 4.1), is calibrated with $0.5 - 19$ Ci ($18.5 - 703$ GBq) of activity and they expire in 2 weeks. The generator is equipped with an alumina column, from which technetium-99m is eluted with sterile saline (0.9% NaCl) as sodium pertechnetate ([^{99m}Tc]NaTcO$_4$). The more highly charged [^{99}Mo]MoO$_4^{2-}$ is retained on the alumina column, where it continues to undergo radioactive decay. The [^{99m}Tc]sodium pertechnetate solution is passed through a 0.22 μm sterile filter to ensure sterility of the eluate. It can then be used directly in certain imaging studies,[11] or it can be incorporated into radiopharmaceuticals using the strategies outlined in Section 4.2.2.

4.2.2 Radiolabeling Strategies using Technetium-99m

The large number of oxidation states that technetium can exist in make it an attractive and versatile transition metal with which to prepare radio-pharmaceuticals. As outlined above, technetium-99m is obtained from ^{99}Mo/^{99m}Tc generators as a sterile solution of [^{99m}Tc]sodium pertechnetate. Due to the similarities to iodine in its charge and size, pertechnetate gets taken up by the thyroid and, consequently, it finds application in scintigraphic imaging of the thyroid.[22] However, as the pertechnetate anion cannot cross the blood brain barrier (**BBB**) it finds wider use as the starting material for the synthesis of other [^{99m}Tc]labeled radiopharmaceuticals.

Preparation of [^{99m}Tc]labeled radiopharmaceuticals typically requires initial reduction of [^{99m}Tc]sodium pertechnetate using a suitable reducing agent, and subsequent complexation with an appropriate coordinating ligand. Therefore, when considering the design of new [^{99m}Tc]labeled radiopharmaceuticals it is important to consider not only the bioactive molecule that will target the site that you wish to image, but also how a pertinent coordinating ligand can be incorporated into the bioactive molecule of choice without a detrimental effect upon pharmacological activity and pharmacokinetics. Moreover, a reducing agent must be chosen that is capable of reducing the technetium but which does not cause any unwanted side reactions with the bioactive molecule. For example, transition metal-based reducing agents (Ti^{III}, Cr^{II}, Cu^{I}, Fe^{II} *etc.*) should be avoided, as they will compete with technetium to form complexes.[13] Similarly, reducing agents that can also form complexes with technetium are undesirable (oxalates, hydrazines, formates, hydroxylamines *etc.*).[13] Typically Sn^{II} is employed as the reducing agent of choice when preparing radiopharmaceuticals labeled with technetium-99m.[23] With a reduced form of technetium (frequently in the + V oxidation state) in hand, it is then mixed with the bioactive molecule bearing the coordinating appendage (Figure 4.2.). For example, if pertechnetate is reduced to [Tc=O]$^{3+}$, then it can be chelated with a range of tetradentate pure (*e.g.* N, **1**) or mixed (*e.g.* NS, **2**) ligands. Radiochemistry utilizing [Tc=O]$^{3+}$ has been reviewed.[16,24–29] Alternatively, treatment of pertechnetate with, for example, succinic dihydrazide generates a nitrido core ([Tc≡N]$^{2+}$) which can be complexed by PXP-type ligands such as **3** (where X = N, O or S). This chemistry has also been recently reviewed, and has found application in labeling peptides, oligonucleotides and small molecules.[25,30,31] The simplicity allows widespread use, but the phosphine ligands (like many phosphines) can be toxic and air sensitive. A related approach is the hydrazinonicotinamide (hynic) labeling strategy that finds widespread application in the context of peptide and protein radiolabeling. Typically hynic is attached to the peptide initially, and then technetium is introduced along with a co-ligand (*e.g.* tricine) to provide complexes such as **4**. A number of varying protocols have been developed for hynic labeling, particularly of peptides, and have been highlighted in numerous recent review articles.[25,27,32–35]

A number of labeling strategies have been developed which utilize technetium in the + III oxidation state.[25] For example, pertechnetate is treated with Sn^{II} and EDTA to obtain [Tc(EDTA)]$^-$. If the technetium is then stabilized with a tetradentate "umbrella" type (*e.g.* NS$_3$) then a biomolecule with a monodentate ligand can also be complexed (**5**).[36] Isocyanides are useful monodentate ligands, although thioethers and phosphines are also appropriate. As isocyanides can be volatile, it is common to stabilize them with a metal center (*e.g.* Cu^{I}), and this approach has been successfully utilized in the preparation of Cardiolite (discussed further in Section 4.2.3.3.). Such chemistry has been reviewed,[25] and finds widespread use in radiolabeling fatty acids or CNS ligands.

Finally, labeling of bioactive molecules has also been explored using the $[Tc(CO)_3]^+$ core ($+I$ oxidation state) which is attractive from a labeling perspective because large numbers of ligands can bind Tc^I. The intermediate core $[^{99m}Tc(OH_2)_3(CO)_3]^+$ is prepared by treating pertechnetate with, for example, CO/BH_4^-. Tridentate ligands such as **6** are the most effective for complexing Tc^I and the many possibilities have been reviewed.[25,28,29,37,38]

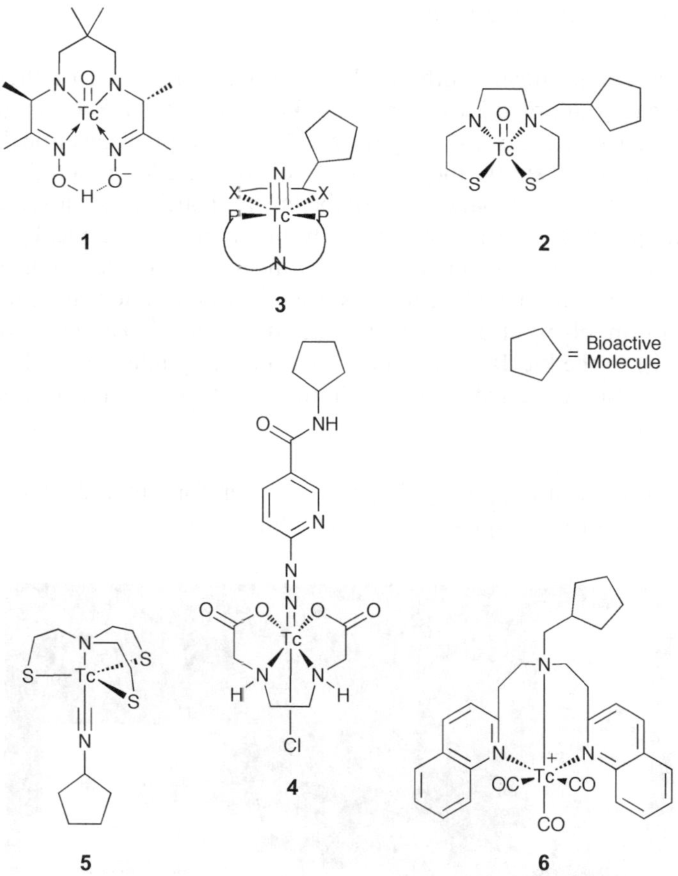

Figure 4.2 Common ^{99m}Tc complexing ligands.

When considering strategies for preparing [^{99m}Tc]radiopharmaceuticals that will ultimately be translated into clinical use, the following guidelines have been suggested by Alberto:[25]

- The labeling yield must be <95%;
- Labeling chemistry should not require purification processes;

- The radiolabeling must be highly efficient, even at low biomolecule concentrations;
- Toxic or unstable ingredients should be avoided whenever possible;
- Bio-incompatible organic solvents should be avoided;
- The complex label must be well defined by both high performance liquid chromatography (HPLC) and classical chemical analysis;
- Nonspecific labeling of the biomolecule (or other side reactions) should not occur;
- Compatibility of the process with both technetium and rhenium is desirable (to enable translation to radiotherapeutic applications using [188]Re);
- *In vivo* stability is required.

A radiopharmaceutical synthesis that complies with each of these criteria should be readily adapted into a single use kit (Figure 4.3.). Such single use kits are an attractive strategy for preparing [99mTc]radiopharmaceuticals because they can be prepared sterile and pyrogen free, and have long shelf lives.[39] These kits enable "shake and bake" radiolabeling and are how most (approved) [99mTc]radiopharmaceuticals are currently prepared. Kits typically contain a reducing agent and the bioactive molecule (including complexing ligand) to be labeled, and then radiolabeling itself is done just before administration to the patient and involves straightforward addition of [99mTc] generator eluent (Na[99mTc]O$_4$) to the kit. Building upon the synthesis guidelines outlined above, Alberto has also suggested 'rules of thumb' for transitioning radiochemistry into kit form:[25]

- Single use kits are typically lyophilized. Therefore, ingredients should not be volatile or decomposed during lyophilization;

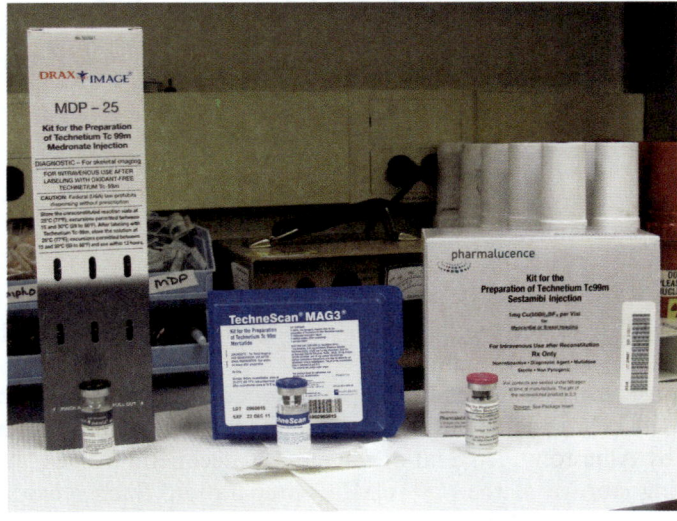

Figure 4.3   Examples of common kits used for [99mTc] radiolabeling.

- After labeling, the radiopharmaceutical should be ready to administer after straightforward quality control (QC) testing (see discussion of QC below);
- Very precise labeling conditions (*e.g.* 12.56 min at 94.3 °C) are undesirable.

Following preparation of the [^{99m}Tc]-labeled radiopharmaceutical, it is then necessary to conduct quality control (QC) testing to ensure that the product is suitable for clinical use. QC testing is completed in accordance with the local pharmacopeia (*e.g.* the U. S. Pharmacopeia, the European Pharmacopeia *etc.*) prior to administration of the dose, and therefore must be fast and efficient. If using an approved and appropriately validated radiopharmaceutical compounding kit, then limited QC testing will be required. QC testing has been reviewed,[40] and typically involves testing of both the generator eluent and the radiopharmaceutical dose. The generator eluent is tested for radionuclidic purity to ensure no breakthrough of ^{99}Mo (eluent must contain <0.15 µCi ^{99}Mo to 1 mCi ^{99m}Tc <u>at the time of administration</u>), and alumina content to confirm no leaching of the column material (eluent must contain <10 µg alumina mL^{-1}). The radiopharmaceutical is tested for radiochemical purity (using Thin Layer Chromtography (TLC) or Sep-Pak methods) to ensure there is no free technetium, hydrolyzed reduced technetium, or byproducts resulting from nonspecific radiolabeling (radiochemical purity typically must be $>90\%$). However, if preparing a research dose for use in a clinical study (*i.e.* a drug in clinical trials for which there is no approved kit), more extensive QC testing will be required to confirm suitability of the dose for human use. This testing includes determination of radiochemical purity (TLC or HPLC), dose pH, sterility, bacterial endotoxin levels, radionuclidic purity, absence of residual solvents, *etc.*, prior to administration of the dose to a patient.

4.2.3 Examples of Technetium-99m based Radiopharmaceuticals

4.2.3.1 [^{99m}Tc]MDP

[^{99m}Tc]Methylene diphosphate ([^{99m}Tc]MDP or [^{99m}Tc]medronic acid, (**8**) is a widely used bone imaging agent that selectively accumulates in bone by chemical adsorption onto (and into) the crystalline structure of hydroxyapatite.[41,42] [^{99m}Tc]MDP has been approved by the U.S. FDA for investigation of osteogenesis,[43] and it has also been used to localize bone metastases.[44–47] A kit for production of [^{99m}Tc]MDP is commercially available and contains the stannous chloride for reduction of pertechnetate, medronic acid (**7**) as a complexing ligand, and *p*-aminobenzoic acid as an antioxidant (Scheme 4.1.). *p*-Aminobenzoic acid is not involved in the complex formation, but is added to prevent reoxidation of reduced technetium back to pertechnetate during storage of the [^{99m}Tc]MDP preparation. Such reoxidation can result from small amounts of oxygen in the eluate or saline solution used for the reconstitution, or due to the introduction of air when dispensing patients' doses

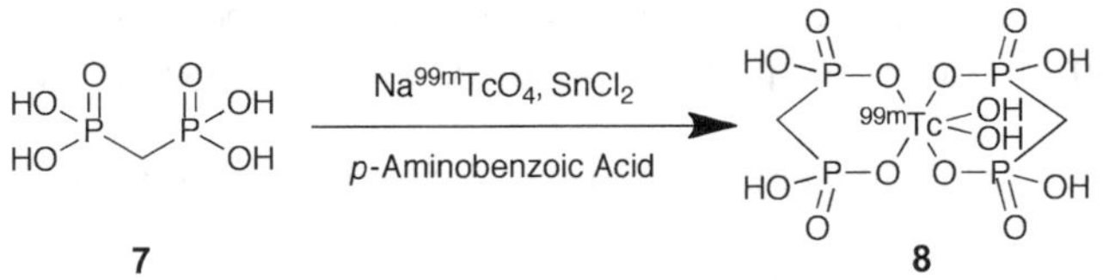

Scheme 4.1   Synthesis of [^{99m}Tc]MDP.

from the *vial*. Such reoxidation should be considered [and accounted for] when designing a SPECT radiopharmaceutical synthesis.

4.2.3.2 *[^{99m}Tc]Exametazime (Ceretec, HMPAO)*

[^{99m}Tc]Exametazime ([^{99m}Tc]Ceretec or [^{99m}Tc]HMPAO, **10**) is a radiopharmaceutical used to detect altered regional cerebral perfusion in stroke victims [and other related cerebrovascular diseases],[48] and for the labeling of leucocytes to diagnose inflammatory bowel disease,[49] or localize infections.[50] [^{99m}Tc]Exametazime is prepared using a commercially available kit and according to Scheme 4.2.[11] The kit contains exametazime (**9**), stannous chloride and sodium chloride, and [^{99m}Tc]sodium pertechnetate is mixed with the contents to generate [^{99m}Tc]exametazime. Kits also contain extra vials of methylene blue and phosphate buffer which, when mixed together, act as a stabilizer and allow [^{99m}Tc]exametazime to be formulated with or without stabilizer. [^{99m}Tc]Exametazime is prepared with or without stabilizer for brain imaging, and the stabilizer confers product stability for up to 4 h. When using [^{99m}Tc]exametazime to label leukocytes, it is prepared without stabilizer. This limits product life to 30 min and so it should be prepared fresh as needed.

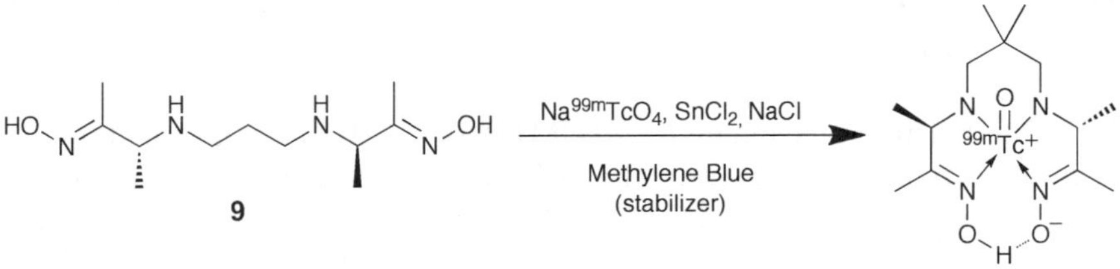

Scheme 4.2   Synthesis of [^{99m}Tc]exametazime.

4.2.3.3 *[^{99m}Tc]Sestamibi (Cardiolite)*

[^{99m}Tc]Sestamibi (**12**) is a coordination complex of technetium-99m and the most widely used [successful] SPECT radiopharmaceutical developed to date.[51] Typically it is employed in myocardial perfusion-imaging (MPI),[52] as well as oncology applications,[53,54] and is commercially available as "Cardiolite" or "Miraluma" for either application respectively. In 2008 [^{99m}Tc]sestamibi went off patent, and the generic drug is now also available. [^{99m}Tc]Sestamibi is a lipophilic complex, which when injected into a patient distributes in myocardium at a rate that is proportional to myocardial blood flow. In a typical MPI regimen, two sets of SPECT images are acquired. Myocardial imaging is conducted initially when the patient is at rest, and then again when the patient is stressed either pharmacologically or after exercising on a treadmill (rest/ stress test).[55] The SPECT images from the two studies can then be used to identify ischemic from infarcted areas of the myocardium. Beyond MPI, [^{99m}Tc]sestamibi imaging has also been used for parathyroid imaging and evaluation of breast nodules.

[^{99m}Tc]Sestamibi is prepared from a commercially available kit and the coordinating ligand employed in the synthesis is methoxyisobutylnitrile (MIBI). The kit contains a lyophilized mixture of tetrakis (2-methoxyisobutylnitrile) copper[I] tetrafluoroborate (**11**), sodium citrate dihydrate, L-cysteine HCL monohydrate, mannitol and stannous chloride. [^{99m}Tc]Sodium pertechnetate is then added to the kit and it is placed in boiling water for 10 min. Initially the [^{99m}Tc]pertechnetate is converted to [^{99m}Tc]citrate, which then displaces copper[I] to generate [^{99m}Tc]sestamibi (**12**) as shown in Scheme 4.3.[11]

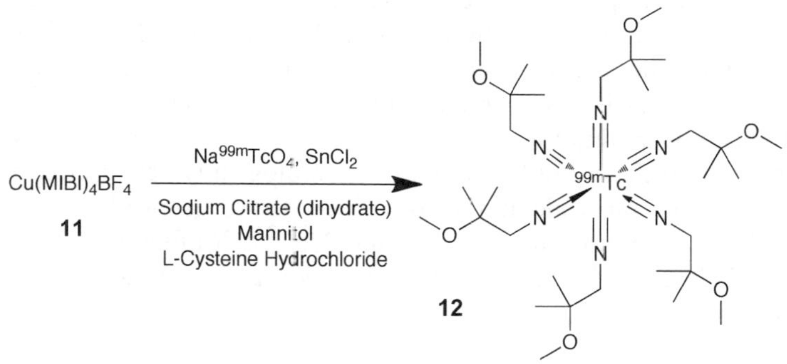

Scheme 4.3 Synthesis of [^{99m}Tc]sestamibi.

4.3 Radiopharmaceuticals Labeled with Radioactive Iodine

There are numerous radioactive isotopes of iodine that find application in molecular imaging and radiotherapy (Table 4.3).[56,57] For example, ^{123}I and ^{131}I have physical characteristics that make them attractive candidates for SPECT imaging, whilst ^{122}I and ^{124}I are suitable for PET imaging.[58] ^{125}I can also be used for SPECT imaging, but its long half-life (59 days) certainly favors use in radiotherapy or *in vitro* work. The chemical equivalence of all of these radionuclides allows investigators to easily produce PET and SPECT radiopharmaceuticals, or radiotherapeutics, from a common scaffold, simply by varying the isotope of

Table 4.3 Common isotopes of iodine used in nuclear medicine.

Isotope	γ Energy	Half-life	Application
^{122}I	511, 564 keV	3.63 min	PET
^{123}I	159 keV	13.2 h	SPECT / Therapy
^{124}I	511, 602, 722 keV	4.18 d	PET
^{125}I	35.5 keV	59.9 d	SPECT / Therapy
^{131}I	364 keV	8.02 d	SPECT / Therapy

iodine. Thus, one might select an ^{123}I-labeled bioactive molecule for initial diagnostic purposes due to the short half-life of ^{123}I, but then prepare the same bioactive molecule labeled with longer-lived ^{131}I for radiotherapeutic applications. This section will concentrate upon radiopharmaceuticals labeled with ^{123}I and ^{131}I and their use from a SPECT imaging perspective.

4.3.1 Production of Iodine-123 and Iodine-131

Iodine-123 is now the most commonly used isotope of iodine for SPECT imaging. Its half-life (13.2 h) and relatively low energy (159 keV) makes it safer for the patient in terms of radiation dose received during a SPECT scan, than the corresponding imaging procedures performed using iodine-131 ($t_{1/2} = 8.02$ days, $\gamma = 364$ keV, β- $= 600$ keV). Iodine-123 is typically produced using a cyclotron according to the equation: ^{124}Xe(p,2n)^{123}Cs → ^{123}X → ^{123}I although, depending upon the target material and proton energy, other approaches are possible. Iodine-123 is commercially available as a solution of [^{123}I]sodium iodide and can be used in the iodination reactions outlined in Section 4.3.2 that are typically used during radioiodination of biologically active molecules.

Iodine-131, like molybdenum-99 described above, is produced in nuclear reactors. Also a fission product of uranium-235, ^{235}U(n,fission)^{131}Te → ^{131}I, it is commercially available as [^{131}I]sodium iodide and finds widespread use in both diagnostic molecular imaging and radiation therapy applications.

4.3.2 Radiolabeling Strategies with Iodine-123 and Iodine-131

The relatively long half-lives of the isotopes of iodine described above allow distribution of radiopharmaceuticals with regulatory approval from a central manufacturing site. Therefore, unlike the technetium-based radiopharmaceuticals described above, local production of established iodine-based radiopharmaceuticals is rare, and iodine radiochemistry tends to be reserved for research and development purposes. Reflecting this, many strategies for radioiodination have been developed in the last few decades. A detailed review of all of these is beyond the scope of this chapter, but such radiochemistry has been the subject of books,[59] and many excellent review articles.[2,56,57,60–62] This article will highlight the most common strategies used to date. When designing iodine-labeled radiopharmaceuticals it should be taken into account that it is preferential to generate aryl- or vinyl iodides because the carbon–iodine bond strength is high and the compounds are typically stable. In contrast, generation of aliphatic iodides should be avoided because the carbon–iodide bond is weak and deiodination can occur either during storage, or *in vivo*, by nucleophilic substitution or β–elimination.

By way of introduction, radiolabeling of aromatic species with radioiodine can be achieved using no-carrier-added radioiodide anions, although the process is often slow.[59] Such reactions follow the same directing effects as standard nucleophilic aromatic substitution reactions and can be enhanced by putting electron withdrawing groups (EWG) in the *ortho-* or *para-* positions on the aromatic ring. The leaving group (X) can be iodine (isotopic exchange) although this limits achievable specific activity. Alternatively, if X = Br or Cl, then higher specific activities can be achieved, assuming that the iodinated product is separable from the bromo- or chloro- precursor (Scheme 4.4a.). Given these relatively sluggish nucleophilic aromatic substitution reactions, investigations into improving the reactivity discovered that copper[I]-mediated halogen exchange reactions were a more effective labeling strategy.[59,63,64] Once again, the reactions can proceed *via* isotopic (X = I) or halogen (X = Cl, Br) exchange processes (Scheme 4.4b), and involve preformation of a radioiodo–cuprate species **13**. Alternatively, treatment of aromatic diazonium salts such as **14** with radioiodide leads to radioiodo-dediazonisation (Wallach-type reaction) and has also been employed as a nucleophilic radiolabeling strategy. One pertinent example of such an approach is in the preparation of 3-iodoquinu-clidinyl benzilate (IQNB, **15**), a heart and brain-imaging agent used to visualize muscarinic cholinergic receptors (Scheme 4.5.).[65,66]

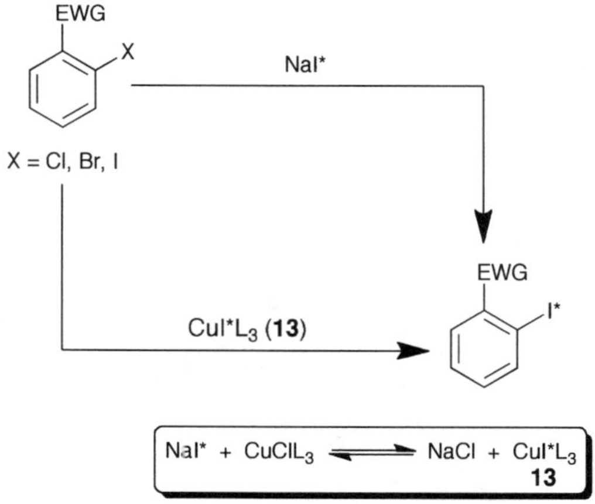

Scheme 4.4 Nucleophilic aromatic substitution with radioiodine.

Alternatively, mild and efficient electrophilic iodination reactions can also be used for radiolabeling, but such reactions do require preformation of an electrophilic source of radioiodine (*I⁺). Elemental iodine (I_2) is not typically used because of low reactivity and handling difficulties associated with volatility.[59] Moreover, it is by definition "carrier added" and so undesirable when it is

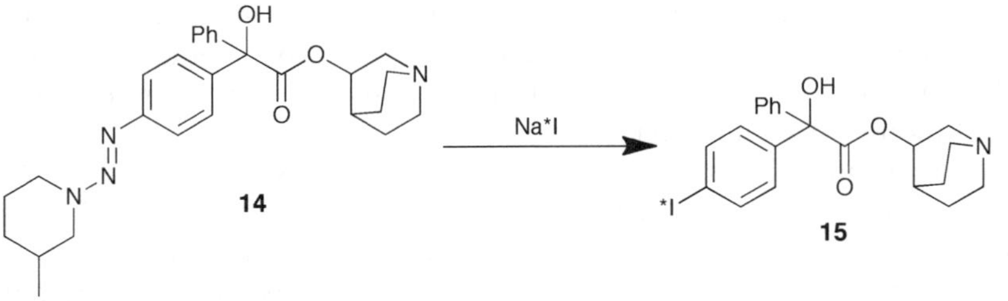

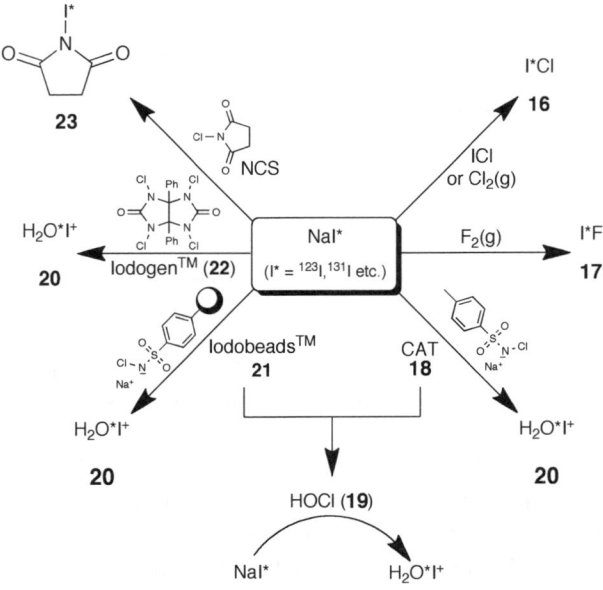

Scheme 4.5   Synthesis of 3-iodoquinuclidinyl benzilate (IQNB).

necessary to prepare high specific activity radiopharmaceuticals. Therefore it is more common to use radioiodide to preform other active electrophilic species (*e.g.* ICl, IF, IOH, H_2OI), typically *via* treatment with an oxidizing agent, which is then utilized in electrophilic substitution-type radiolabeling reactions (Scheme 4.6.). A range of oxidizing agents has been employed in such reactions.[57,59] For example, treatment of radioiodide with ICl results in an isotopic exchange reaction to give radioactive iodine monochloride (**16**). Alternatively, reaction of radioiodide with elemental halogens directly (Cl_2 or F_2) can also be used to generate iodine monohalides (**16** and **17**, Scheme 4.6.).[56,57,59,67]

Other commonly employed oxidizing agents include *N*-chloroamides (*e.g.* *N*-chloro-*p*-toluenesulfonic acid, chloramine-T (CAT, **18**), Scheme 4.6.).[56,57,59,67–69] These compounds are thought to release HOCl (**19**), which oxidizes iodide to H_2OI (**20**), an iodonium ion that can be used in electrophilic substitution

Scheme 4.6   Production of electrophilic radioiodine.

reactions. It should be noted that agents such as chloramine-T are strong oxidizing agents and so competing chlorination and oxidation reactions can generate unwanted oxidative byproducts.[59] To overcome this problem, polymer-supported analogs of *N*-chloroamides such as chloramine-T (Iodobeads™, **21**) have been developed.[68,70] These immobilized analogs reduce concentration of the oxidizing species in solution and also allow for straightforward filtration and removal following reaction to reduce contact time. Alternatively, milder oxidizing agents can be selected which include 1,3,4,6-tetrachloro-3α,6α-diphenylglycouril (Iodogen™, **22**),[69,71] and *N*-halosuccinimides such as *N*-chlorosuccinimide (NCS). In the latter case, radioactive *N*-iodosuccinimide (**23**) is thought to be the active species.[59] However, if chlorination side reactions are a problem then non-chlorine based oxidants such as peracids can be utilized, although radiochemical yields tend to be lower than those obtained using *N*-halooxidants.[59]

With a source of electrophilic iodine in hand, radiolabeling can then be achieved using electrophilic substitution reactions to prepare vinyl and aryl iodides (Scheme 4.7). All of the oxidizing agents described above generate sources of electrophilic radioiodine ($*I^+$) that can be employed in such reactions. The simplest form of electrophilic substitution is direct electrophilic radioiodination (Y=H), although this method is generally restricted to activated arenes.[59] For example, Mennicke and colleagues explored radioiodination of a range of substituted aromatic compounds using different combinations of solvents and oxidizing agents (Table 4.4).[72] These reactions follow the same activating / deactivating and directing rules governing electrophilic substitution reactions on aryl rings, which means that the formation

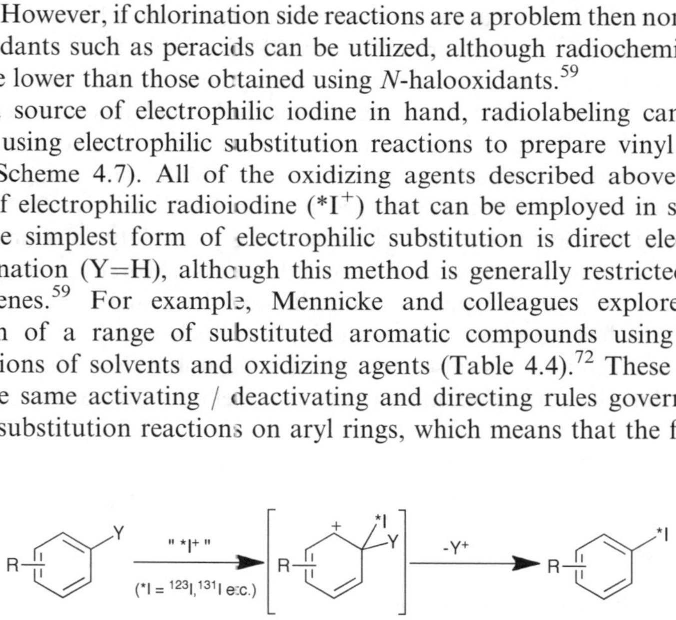

Scheme 4.7 Electrophilic radioiodination.

Table 4.4 Direct electrophilic radioiodination of arenes.

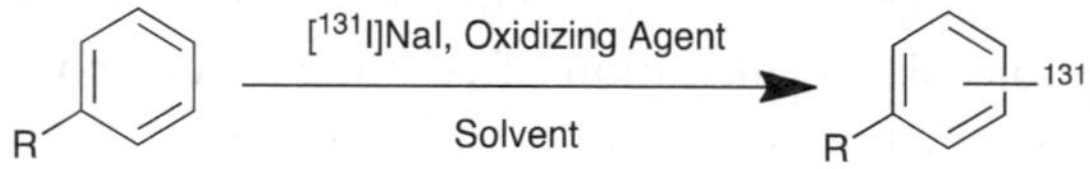

Solvent	TFAA	TFAA	TFAA	TFA	Triflic Acid
Oxidizing Agent	NCTFS	NCS	CAT	TTFA	NCS
Reaction Time	4 h	4 h	10 min	15 min	5 – 60 min
R					
– OMe	69 ± 5	72 ± 4	75 ± 8	83 ± 5	39 ± 5
– Me	47 ± 8	30 ± 6	49 ± 8	89 ± 7	84 ± 7
– H	~4	~1	~3	34 ± 2	81 ± 3
– Cl	ND	ND	ND	0	76 ± 5
– NO_2	ND	ND	ND	0	38 ± 1

of multiple regioisomers is possible and can lower the yield of the desired regioisomer. Purification of the desired radiopharmaceutical from mixtures of structurally related regioisomers is frequently not a trivial undertaking and this should be considered when designing the radiosynthesis.

One possibility for overcoming this non-regioselective radioiodination of arenes is the use of organometallic precursors to direct the radiolabeling (Scheme 4.8).[56,59,73] Introducing a suitable organometallic leaving group at the Y position (Scheme 4.7) will enhance electrophilic substitution reactions by reducing preparation of unwanted regioisomers, whilst concomitantly maximizing the radiochemical yield (RCY) of the desired regioisomer. When employing such demetallation strategies, the organometallics of choice are trialkylstannyl (**24**)[56,59,73] or organoboron (**25**) derivatives,[56,59,73–75] (although substitution with silicon, germanium, mercury and thallium has also been reported).[56,59,73] In each case, the carbon–metal bond has a lower binding energy than the corresponding carbon–hydrogen bond, and so demetallation reactions are more favored than the direct electrophilic substitution reactions described above, and typically proceed in high RCYs to give **26**. The main advantage of the demetallation technique is the possibility to regioselectively radiolabel both activated and deactivated arenes. Moreover, the reaction conditions are mild and, typically, the radioiodo-demetallation reaction is the final step, which is attractive when designing a radiochemical synthesis. Beyond arenes, this technique is also suitable for radiolabeling vinyl substrates assuming that the corresponding vinyl organometallics are easily accessible.

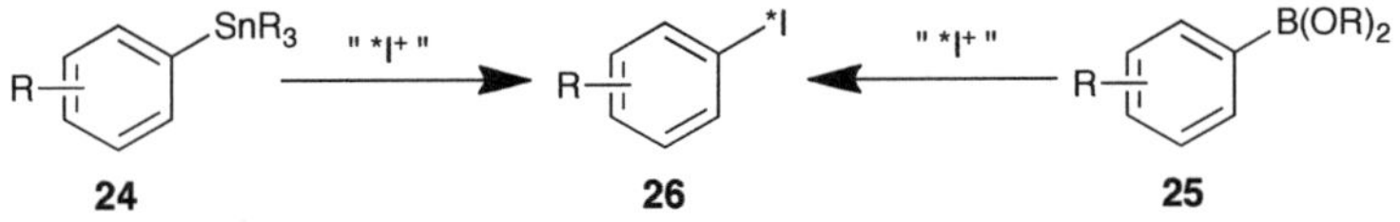

24   **26**   **25**

Scheme 4.8   Radioiodo-demetallation.

Beyond the direct radioiodination strategies reviewed above, a final radiolabeling approach that finds widespread use in radioiodination of macromolecules (*e.g.* proteins, antibodies, oligonucleotides) is initial formation of a smaller iodine-labeled prosthetic group, which is then subsequently attached to the macromolecule by alkylation of common nucleophilic groups (*e.g.* $-NH_2$, $-SH$).[59] This method is the approach of choice for radiolabeling macromolecules [and smaller molecules] that have no site suitable for direct radioiodination, or are not tolerant of, for example, the harsh oxidative conditions typically employed in electrophilic iodination.

A large number of prosthetic groups have been utilized in radioiodination, and have been reviewed.[59,62,76,77] Noticeably, in their book *Radioiodination Reactions for Radiopharmaceuticals,* Coenen and colleagues discuss the range of prosthetic groups that have been used to radioiodinate macromolecules to date.[59] For example, Bolton-Hunter-type activated esters (**27**), have been employed as prosthetic groups for alkylation of amino functionalities (Scheme 4.9.).[78–80] Related imidate esters (*e.g.* the Wood reagent **28**) also react with protein amino groups to form amidine bonds.[81] Other approaches include generation of activated halides (*e.g.* α-carbonyl halides such as (radio-iodophenyl)-α-bromoacetamide **29**), which have been used to alkylate both free amino groups and thiols;[82,83] aldehydes (*e.g.* radioiodinated (4-hydroxyphenyl)acetaldehyde **30**) which can form imines with free amino groups;[84] isothiocyanates (*e.g.* 3-iodophenyl isothiocyanate **31**) which Ram reported form thiourea bonds with protein amino groups;[85,86] and maleimides (*e.g.* N-(m-[125I]iodophenyl) maleimide **32**) which selectively alkylate thiol groups to generate thioethers (Scheme 4.9.).[87]

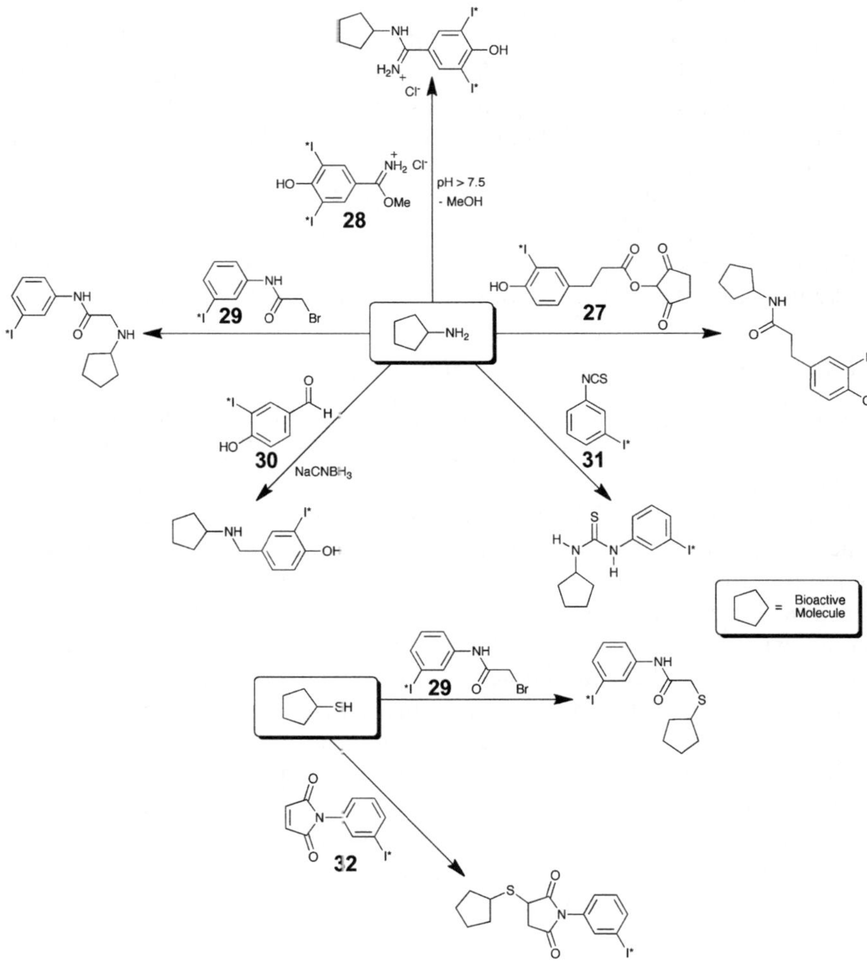

Scheme 4.9 Prosthetic groups for radioiodination.

Commercially available iodinated radiopharmaceuticals will have quality control testing completed by the manufacturing site prior to distribution to customers for clinical studies. However, as in the case of technetium-labeled radiopharmaceuticals, if preparing a research dose for a clinical study, more extensive QC testing will be required to confirm suitability of the dose for human use. This testing includes determination of radiochemical purity (TLC or HPLC), dose pH, sterility, bacterial endotoxin levels, radionuclidic purity, absence of residual solvents, *etc.* prior to administration of the dose to a patient.

4.3.3 Examples of Iodine-123 and Iodine-131 Based Radiopharmaceuticals

4.3.3.1 Sodium Iodide

The short half-life of $[^{123}I]$sodium iodide (when compared to $[^{131}I]$sodium iodide) makes it ideal for imaging the thyroid and aiding both radiologists and nuclear medicine physicians in the diagnosis and evaluation of thyroid function and/or morphology.[88–90] $[^{123}I]$Sodium iodide is trapped and organically bound to the thyroid, enabling measurement of radioactive iodine uptake and quantitative evaluation of thyroid function. $[^{123}I]$Sodium iodide for this type of diagnostic application is provided in capsule form and administered orally. Reflecting the half-life difference and β- particle emission, $[^{131}I]$sodium iodide can be used for treatment of thyroid cancer with radiation therapy.[91]

4.3.3.2 Meta-Iodobenzylguanidine (MIBG)

Meta-iodobenzylguanidine (MIBG, **36**), a norepinephrine analog, was developed at the University of Michigan by Weiland and colleagues. MIBG is now commercially available and, since it localizes to adrenergic tissue, can be used in both the diagnosis (when labeled with ^{123}I) and therapy (when labeled with ^{131}I) of neuroblastoma[92] and pheochromocytoma.[93] MIBG also finds widespread use in myocardial imaging of noradrenergic innervation.[94,95]

In 1986, Weiland prepared $[^{131}I]$MIBG using the radioiodide exchange reaction with non-radioactive MIBG (**33**) (Scheme 4.10.).[96] MIBG can also be labeled with iodine-123 and iodine-124 using analogous chemistry. However, specific activity of MIBG prepared using this method is typically low because of the "carrier added" nature of the synthesis. Since its introduction, many methods have been described for the synthesis and purification of MIBG to the extent that the topic has been reviewed.[97] For example, specific activity can be improved by preparing MIBG from the corresponding bromo-MIBG precursor (**34**) in a copper^I-mediated halogen exchange reaction. Alternatively, iododestannylation of **35** is also possible (Scheme 4.10.).[98]

4.3.3.3 $[^{123}I]$Ioflupane (DaTSCAN)

$[^{123}I]$Ioflupane (**38**), a radiolabeled analog of cocaine, is a phenyltropane that is radiolabeled with iodine-123 (Scheme 4.11.). It is approved for clinical use in Europe and the United States, and marketed as DaTSCAN. Nuclear medicine physicians use $[^{123}I]$ioflupane to diagnose Parkinson's disease (PD),[99] and to

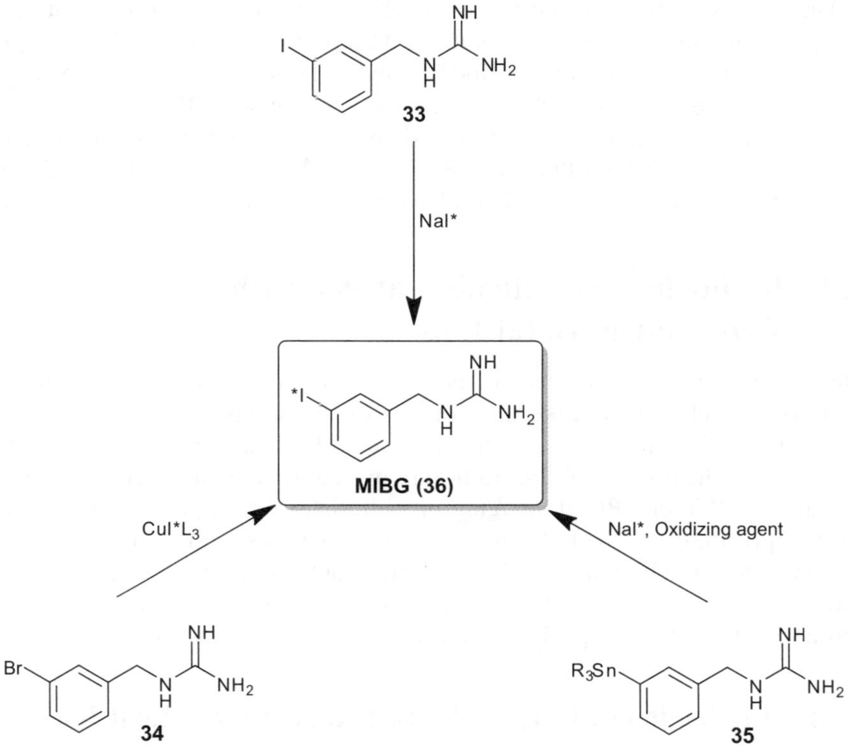

Scheme 4.10 Synthesis of meta-iodobenzylguanidine (MIBG).

differentiate PD from other related neurological disorders that present with similar clinical symptoms (*e.g.* dementia with Lewy bodies).[100] [^{123}I]Ioflupane has a high binding affinity for presynaptic dopamine transporters (DAT). Thus, a SPECT scan conducted using [^{123}I]ioflupane provides physicians with a quantitative measure and the spatial distribution of DAT in the brain. A marked reduction in DAT in the striatal region of the brain is indicative of PD, allowing physicians to diagnose or differentiate a patient's neurological condition with improved diagnostic confidence when compared to diagnosing from clinical symptoms alone.

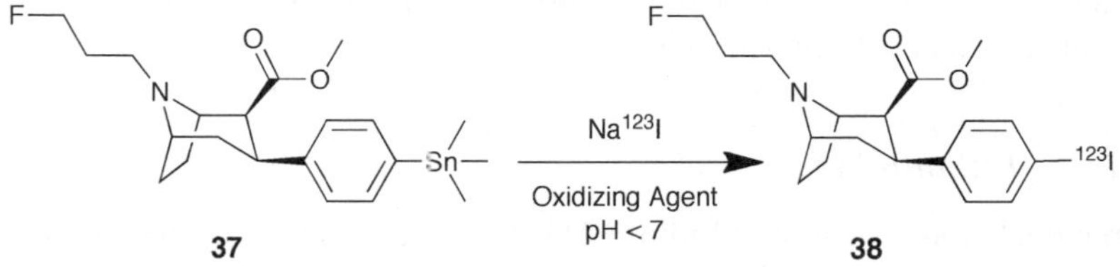

Scheme 4.11 Synthesis of [^{123}I]ioflupane.

The relatively long half life of iodine-123 allows commercial distribution of [^{123}I]ioflupane as an injectable solution. The synthesis of [^{123}I]ioflupane proceeds *via* the oxidative radioiododestannylation mechanism described and illustrated in Scheme 4.11.[101] The stannane precursor (**37**) is treated with commercially available [^{123}I]sodium iodide, in the presence of an oxidizing agent, to provide [^{123}I]ioflupane (**38**). The pH was also found to influence radiochemical purity and an acidic pH was discovered to be most favorable.

4.4 Radiopharmaceuticals Labeled with Radioactive Metal Ions

Many radioactive metal ions can be used to radiolabel bioactive molecules using related chemistry. Like the isotopes of iodine discussed above, having interchangeable radioactive metal ions available with a range of properties allows radiochemists to tailor radiopharmaceuticals to a given application (diagnostic PET or SPECT imaging or radiotherapy) simply by selecting the most appropriate metal. Labeling bioactive molecules with radioactive metal ions involves attaching a suitable chelating group, followed by similar "shake and bake" chelation chemistry to that described above in the context of technetium-99m radiolabeling. Representative examples are highlighted below.

4.4.1 Production of Commonly used Radioactive Metal Ions

The main radioactive metal ions employed in SPECT imaging applications are gallium-67,[11,102] indium-111[11,102] and yttrium-90,[11] although others are available. Due to the long half-lives (Table 4.1), all three of these radionuclides are commercially available as solutions that are typically mixed with single use kits to prepare radiopharmaceuticals.

4.4.1.1 Gallium-67

Gallium-67 ($t_{1/2}$ = 3.26 days) is typically prepared in a cyclotron using a solid target and according to the following nuclear reaction: $^{68}Zn(p,2n)^{67}Ga$.[102] Gallium-67 is commercially available as [^{67}Ga]GaCl$_3$, and is typically provided in a mild solution of HCl. Gallium chloride can be reacted with sodium citrate to produce gallium citrate which finds application in imaging various inflammatory conditions and tumors such as Hodgkin's disease and lymphoma.[102] Alternatively, [^{67}Ga]GaCl$_3$ can be used as a precursor to other radiopharmaceuticals as discussed below.

4.4.1.2 Indium-111

The relatively long half-life of Indium-111 ($t_{1/2}$ = 2.80 days) means it finds use in specialized diagnostic applications, where longer half-lives are required (*e.g.* antibody labeling or for radiolabeling blood cell components), or radiotherapeutic strategies. Indium-111 is produced in a cyclotron using a cadmium enriched target and according to the following nuclear reaction: $^{112}Cd(p,2n)^{111}In$.[102] Like gallium, indium-111 is also commercially available as the chloride salt and $[^{111}In]InCl_3$ is also typically provided in a mild solution of HCl.

4.4.1.3   Yttrium-90

Yttrium-90 represents an ideal isotope for nuclear medicine applications because it has a short half-life ($t_{1/2} = 64$ hours) and is a pure beta-emitter with high average beta energy (934 keV). Yttrium-90 can be used to label radio-pharmaceuticals for diagnostic purposes, but also finds use in therapy where it can be attached to monoclonal antibodies for targeting cancer cells, or incorporated into glass microspheres (TheraSphere™) used for the treatment of liver tumors.[103] Yttrium-90 is the daughter product of strontium-90 decay, which, in turn, is a fission product of uranium obtained from nuclear reactors. Yttrium-90 is therefore formed according to the following nuclear reaction: $^{90}Sr(28.74$ years, Beta-$)^{90}Y$. Yttrium-90 can be obtained as $[^{90}Y]YCl_3$,[104] typically provided in a mild solution of HCl or, more recently, in $^{90}Sr/^{90}Y$ generator form.[105]

4.4.2   Radiolabeling Strategies with Radioactive Metal Ions

All of the radioactive metals described above are available as the MCl_3 salt. Other pertinent but less common radioactive metal ions are also available in this convenient form (*e.g.* $^{153}SmCl_3$, $^{177}LuCl_3$). In certain cases, metal ions can be administered as the salt form for molecular imaging (*e.g.* ^{67}Ga citrate, ^{89}Sr chloride). However, it is much more common to radiolabel a bioactive molecule with the radioactive metal ion. Labeling of bioactive molecules with any of these radioactive metal ions involves attaching a chelating group, *via* a suitable linker, followed by "shake and bake" chelation chemistry as shown in Scheme 4.12. As the radioactive metals are all in the M^{3+} state, a few common chelating groups can be used to radiolabel molecules with any of them.[106] As discussed above, this provides scope for use of a given bioactive molecular scaffold in both diagnostic and therapeutic applications simply by switching the radioactive metal ion component.

Chelating groups employed in this capacity are often referred to as bifunctional chelants (BFCs) and their use to prepare radiopharmaceuticals has been reviewed.[2,102,106] This reflects the functionality on the chelating group that serves as a point of attachment to bioactive molecules, as well as its ability to

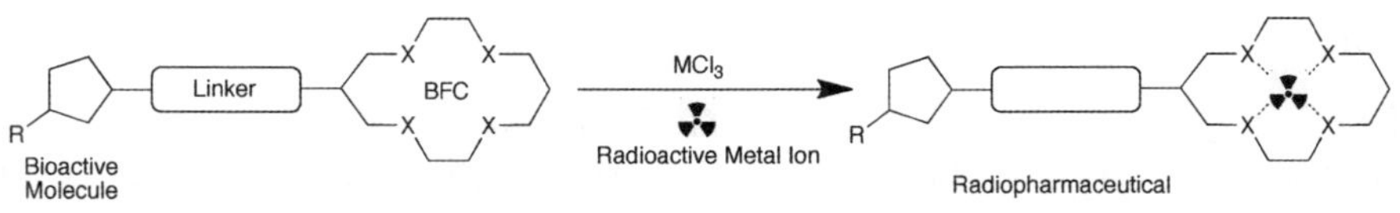

Scheme 4.12   Chelation of radioactive metal ions.

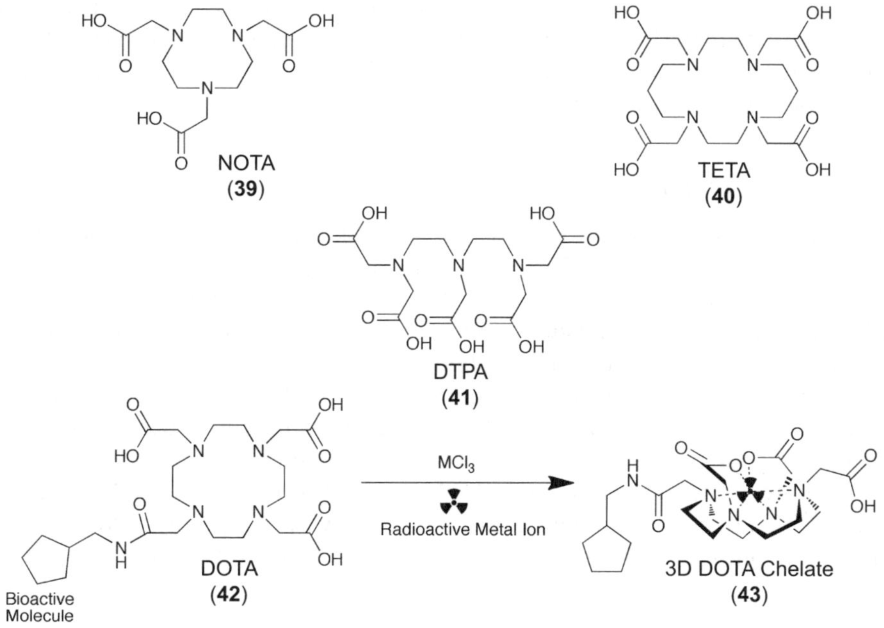

Figure 4.4 Common bifunctional chelants (BFCs) used to complex radioactive metal ions.

complex radioactive metal ions. Common BFCs are illustrated in Figure 4.4, and groups such as 1,4,7,-tri*aza*cyclododecane-*N, N″, N‴, N⁗*-tetraacetic acid (**39**, NOTA), 1,4,8,11-tetra*aza*cyclotetrado-decane-*N, N″, N‴, N⁗*-tetraacetic acid (**40**, TETA), acyclic diethylenetriaminepentaacetic acid (**41**, DTPA) and, most commonly, 1,4,7,10-tetra*aza*cyclodocane-*N, N″, N‴, N⁗*-tetraacetic acid (**42**, DOTA), have been conjugated with, for example, small molecules, peptides, monoclonal antibodies, proteins and nanoparticles to prepare novel radiopharmaceuticals for SPECT imaging. For simplicity, the chelating groups in Figure 4.4 are shown in 2 dimensions. However, in reality, the carboxylic acids are also involved in chelation to provide stable 3-dimensional chelates such as **43**. Beyond application in the radiopharmaceutical arena, these BFCs can also be used to coordinate non-radioactive metal ions for other applications (*e.g.* Gd^{3+} for MRI).[106,107]

4.4.3 Examples of Radiopharmaceuticals Labeled with Radioactive Metal Ions

4.4.3.1 *DOTATOC*

DOTA-(Tyr^3)-octreotide acetate (DOTATOC, **44**) is a large cyclic peptide bearing a DOTA appendage that is a substrate for the somatostatin receptors (SSTRs). Neuroendocrine tumors like gastrinomas, insulinomas and carcinoids overexpress $SSTR_{1-5}$, with $SSTR_2$ and $SSTR_5$ being the most abundant.

Therefore, DOTATOC labeled with diagnostic radioactive metal ions can be used to visualize such tumors, whilst labeling with therapeutic radionuclides allows use of peptide receptor mediated radionuclide therapy (PRRT) to treat such cancers. For example, SPECT and PET imaging of somatostatin receptors (SSTRs) using gallium-67 (**45a**) and gallium-68 (**45b**) radiolabeled DOTATOC, respectively, have been employed as sensitive and specific techniques for imaging neuroendocrine tumors.[108,109] Similarly SPECT imaging and/or radiation therapy can be conducted using yttrium-90 (**45c**), indium-111 (**45d**) or lutetium-177 (**45e**) DOTATOC (Scheme 4.13.).[110–112] Preparation of the radiopharmaceutical is the same in each case and simply involves combining DOTATOC with a solution of the radioactive MCl$_3$ salt, typically in dilute hydrochloric acid. The reaction is heated ($\sim 95\,^{\circ}$C) and subsequent purification

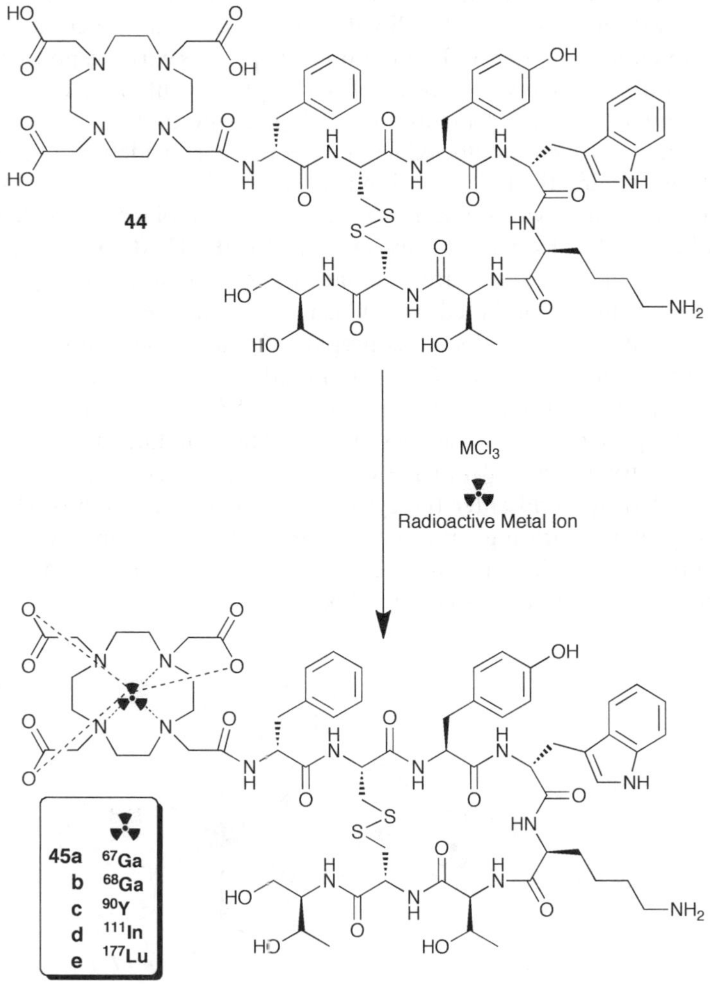

Scheme 4.13 Synthesis of radiolabeled DOTATOC.

and reformulation is achieved using solid-phase extraction (SPE) techniques. If using an approved and appropriately validated radiopharmaceutical compounding kit, then limited QC testing will be required to determine radiochemical purity. However, if preparing a research dose for a clinical study, more extensive QC testing will be required to confirm suitability of the dose for human use. This testing includes determination of radiochemical purity (TLC or HPLC), dose pH, sterility, bacterial endotoxin levels, radionuclidic purity, absence of residual solvents, *etc.* prior to administration of the dose to a patient.

4.4.3.2 [¹¹¹In]Labeled Leukocytes (White Blood Cells)

The range of possible targets of biological significance that can be radiolabeled, and exploited in SPECT imaging [and/or gamma scintigraphy], is perhaps best highlighted with a discussion of radiolabeled leukocytes and their use to image infection and inflammation. Inflammation is the body's early response to injury [and infection] and involves delivery of leukocytes to the infected area where they can clear infectious agents and degrade necrotic tissues. Therefore, scintigraphy using *ex vivo* radiolabeled leukocytes has been widely used as the gold standard for imaging the inflammation process.[113–116] Blood is drawn from the patient and the leukocyte fraction is isolated from the plasma. The leukocytes can then be radiolabeled with, for example, [⁹⁹ᵐTc]HMPAO or [¹¹¹In]oxine. In the latter case, the leukocytes are radiolabeled by incubating with [¹¹¹In]oxine (**47**) for 15–20 min, recombined with plasma and then centrifuged to remove unbound indium-111. [¹¹¹In]oxine is prepared by using 8-quinolinol (**46**) as a chelating group (Scheme 4.14.). Green and Huffman elucidated the structure of [¹¹¹In]oxine [recrystallized from ethanol] in 1988,[117] and showed that the indium is complexed with 3 molecules of 8-quinolinol making it quite lipophilic. This lipophilicity allows [¹¹¹In]oxine to cross the cell membrane [*via* an intercalation mechanism] and enter the cytoplasm. There, the indium-111 is translocated to proteins with higher affinity for the metal ion than oxine. The free oxine can pass out of the cell, whilst the indium-111 remains bound to proteins and trapped in the leukocytes (Figure 4.5.).

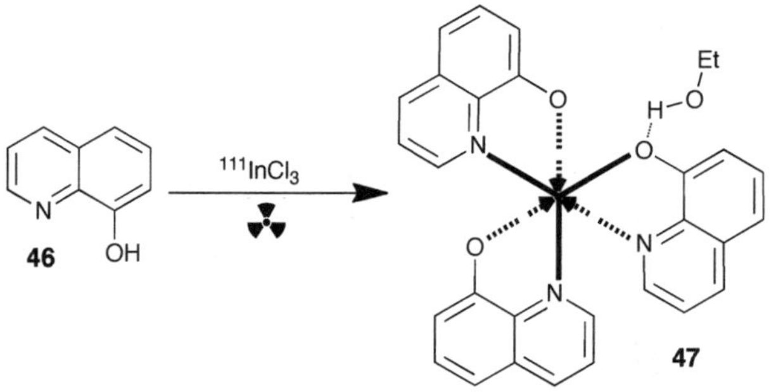

Scheme 4.14   Synthesis of [¹¹¹In]oxine.

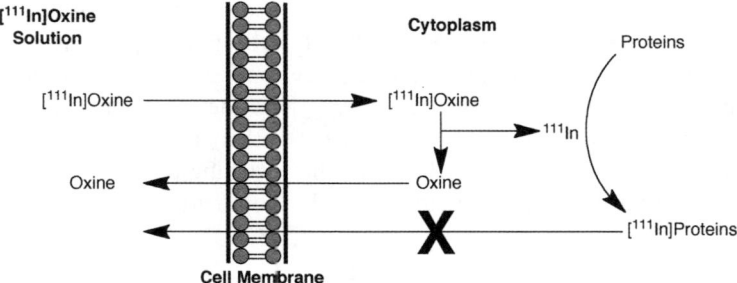

Figure 4.5 Radiolabeling of leukocytes with [^{111}In]oxine.

Following quality control testing, the labeled leukocytes are then taken up in additional plasma and reinjected into the patient where they become distributed in the intravascular space. Patients then typically receive their scan 4 hours post injection. This lapse gives time for blood pool activity to decrease and improves image quality. An area of increased activity is indicative of inflammation and infection. This has made indium-111-labeled leukocyte scintigraphy the method of choice for diagnosing osteomyelitis (infection of the bone). However, using radiolabeled leukocytes in this capacity also has associated challenges, because there is always the question of whether the radiolabeled cells remain viable and there is also the loss of specificity associated with imaging a system as complicated as the leukocyte response to inflammation. Moreover, *ex vivo* labeling entails cumbersome isolation or reinjection of the cells, and this also involves significant radiation exposure and increased risk of infection to both the patient and the staff.[113,114,118] Therefore, alternative strategies have been developed, including the use of radiolabeled nanocolloids,[21] and introduction of radiolabeled antigranulocyte monoclonal antibodies (MAbs),[119] with the aim of addressing some of these issues.

4.5 Summary

SPECT imaging is a valuable technique for functional molecular imaging that is used routinely in the global radiological community. Crucial to the success of SPECT imaging has been creating global access, and logistical support for, a broad range of sophisticated radiopharmaceuticals for a variety of indications in diagnostic medicine, a concept showcased in this chapter by highlighting representative examples. This process has been greatly facilitated by ready availability of radionuclides, and the development of single use, sterile, pyrogen free kits for on-site radiopharmaceutical compounding by trained personnel such as radiochemists, nuclear pharmacists and nuclear medicine technologists. However, many diseases still cannot be reliably diagnosed and patient response to therapy in such cases remains uncertain. Both SPECT and PET imaging have potential to impact all such disease states and improve diagnostic confidence for radiologists and nuclear medicine physicians. In order to do so however, imaging must be supported by concomitant programs aimed at developing novel radiopharmaceuticals, a process very much still in the hands of highly trained synthetic organic and radiopharmaceutical chemists. Reflecting this, radiopharmaceutical discovery is an aggressive area of research and the last few decades have seen impressive development of both new radiopharmaceuticals, and novel techniques for their preparation. This article introduces the tools available to those chemists wishing to explore the arena of SPECT radiopharmaceutical research and development.

References

1. R. J. Jaszczak and R. E. Coleman, *Invest. Radiol.*, 1985, **20**, 897–910.
2. S. Vallabhajosula *Molecular Imaging: Radiopharmaceuticals for PET and SPECT*, Springer, New York, 2009.
3. E. L. Kramer and J. J. Sanger, *Clinical SPECT Imaging*, Raven Press, New York, 1995.
4. R. L. Van Heertum, R. S. Tikofsky and M. Ichise, *Functional Cerebral SPECT and PET Imaging*, Lippincott Williams & Wilkins, Philadelphia PA, 2009.
5. S. L. Pimlott and K. P. Ebmeier, *Br. J. Radiol.*, 2007, **80**, S153–S159.
6. E. E. Camargo, *J. Nucl. Med.*, 2001, **42**, 611–623.
7. R. C. Mease, in *Molecular Imaging in Oncology*, ed. M. G. Pomper and J. G. Gelovani, Informa Healthcare, USA, New York, 2008.
8. M. E. Van Dort, A. Rehemtulla and B. D. Ross, *Curr. Comput.-Aided Drug Des.*, 2008, **4**, 46–53.
9. E. G. De Puey, E. V. Garcia and D. S. Berman, *Cardiac SPECT Imaging*, Lippincott Williams & Wilkins, Philadelphia PA, 2001.
10. D. R. Hwang and S. R. Bergmann, in *Handbook of Radiopharmaceuticals: Radiochemistry and Applications*, ed. M. J. Welch and C. S. Redvanly, Wiley, Chichester, 2003.
11. R. J. Kowalsky and S. W. Falen, *Radiopharmaceuticals in Nuclear Pharmacy and Nuclear Medicine*, American Pharmacists Association, Washington DC, 3rd edn., 2011.
12. I. Zolle, *Technetium-99m Pharmaceuticals*, Springer, Berlin-Heidelberg, 2007.
13. K. Schwochau, *Angew. Chem. Int. Ed.*, 1994, **33**, 2258–2267.
14. A. Mahmood and A. G. Jones, in *Handbook of Radiopharmaceuticals: Radiochemistry and Applications*, ed. M. J. Welch and C. S. Redvanly, Wiley, Chichester, 2003.
15. *Technetium-99m Radiopharmaceuticals: Status and Trends*, International Atomic Energy Agency, Vienna, 2009.
16. K. Schwochau, *Technetium: Chemistry and Radiopharmaceutical Applications*, Wiley-VCH, Weinheim, 2000.
17. P. Gould, *Nature*, 2009, **460**, 312–313.
18. M. Voith, *Chem. Eng. News*, 2009, **87**, 8.
19. K. Bertsche, *Proc. IPAC'10, Kyoto, Japan*, 2010, **MOPEA025**, 121.

20. P. Schaffer, S. Zeisler, S. McQuarrie, M. Kovacs, T. Ruth and F. Bernard, *J. Nucl. Med.*, 2010, **51**(Suppl. 2), 1468.
21. I. Zolle, in *Technetium-99m Pharmaceuticals*, ed. I. Zolle, Springer, Berlin-Heidelberg, 2007.
22. W. J. Dodds and M. R. Powell, *Radiology*, 1968, **91**, 27–31.
23. H. Spies and H.-J. Pietzsch, in *Technetium-99m Pharmaceuticals*, ed. I. Zolle, Springer, Berlin-Heidelberg, 2007.
24. R. E. Weiner and M. L. Thakur, *Biodrugs*, 2005, **19**, 145–163.
25. R. Alberto, in *Technetium-99m Radiopharmaceuticals: Status and Trends*, International Atomic Energy Agency, Vienna, 2009.
26. H. R. Maecke and M. Eisenhut, *Bioinorg. Chem.*, 1995, **2**.
27. R. C. Mease and C. Lambert, *Semin. Nucl. Med.*, 2001, **31**, 278–285.
28. S. R. Banerjee, K. P. Maresca, L. Francesconi, J. Valliant, J. W. Babich and J. Zubieta, *Nucl. Med. Biol.*, 2005, **32**, 1–20.
29. P. Blower, *Dalton Trans.*, 2006, 1705–1711.
30. R. Pasqualini, A. Duatti, E. Bellande, V. Comazzi, V. Brucato, D. Hoffschir, D. Fagret and M. Comet, *J. Nucl. Med.*, 1994, **35**, 334–341.
31. C. Bolzati, E. Benini, E. Cazzola, C. Jung, F. Tisato, F. Refosco, H.-J. Pietzsch, H. Spies, L. Uccelli and A. Duatti, *Bioconjug. Chem.*, 2004, **15**, 628–637.
32. H. R. Maecke, *Eur. J. Nucl. Med. Mol. Imaging*, 2004, **31**, S296.
33. S. Liu and D. S. Edwards, *Chem. Rev.*, 1999, **99**, 2235–2268.
34. J. Fichna and A. Janecka, *Bioconjug. Chem.*, 2003, **14**, 3–17.
35. S. M. Okarvi, *Med. Res. Rev.*, 2004, **24**, 357–397.
36. H. Spies and M. Glaser, *Inorg. Chim. Acta.*, 1995, **240**, 465–478.
37. S. Liu, *Chem. Soc. Rev.*, 2004, **33**, 445–461.
38. K. H. Thompson and C. Orvig, *Dalton Trans.*, 2006, 761–764.
39. *Technetium-99m Radiopharmaceuticals: Manufacture of Kits*, International Atomic Energy Agency, Vienna, 2008.
40. J. Hung, in *Radiopharmaceuticals in Nuclear Pharmacy and Nuclear Medicine*, ed. R. J. Kowalsky and S. W. Falen, American Pharmacists Association, Washington D. C., 2011.
41. A. Chopra, in *Molecular Imaging and Contrast Agent Database (MICAD) [database online]*, National Library of Medicine (US), NCBI., Bethesda (MD), 2004.
42. D. Kanishi, *Oral Surg., Oral Med., Oral Pathol.*, 1993, **75**, 239–246.
43. J. K. Lee and S. S. Sun, *Clin. Nucl. Med.*, 1998, **23**, 619.
44. O. Karacalioglu, S. Ilgan, O. Kuzhan, O. Emer and M. Ozguven, *Ann. Nucl. Med.*, 2006, **20**, 437–440.
45. C. W. Choi, D. S. Lee, J. K. Chung, M. C. Lee, N. K. Kim, K. W. Choi and C. S. Koh, *Clin. Nucl. Med.*, 1995, **20**, 310–314.
46. T. Uematsu, S. Yuen, S. Yukisawa, T. Aramaki, N. Morimoto, M. Endo, H. Furukawa, Y. Uchida and J. Watanabe, *Am. J. Roentgenol.*, 2005, **184**, 1266–1273.
47. E. Even-Sapir, U. Metser, E. Mishani, G. Lievshitz, H. Lerman and I. Leibovitch, *J. Nucl. Med.*, 2006, **47**, 287–297.

48. B. Infeld and S. M. Davis, in *Stroke: Pathophysiology, Diagnosis, and Management*, ed. J. P. Mohr, D. W. Choi, J. C. Grotta, B. Weir and P. A. Wolf, Churchill Livingstone, Philadelphia, PA, 3rd edn., 1998.
49. J. C. Mansfield, M. H. Giaffer, W. B. Tindale and C. D. Holdsworth, *Gut*, 1995, **37**, 679–683.
50. A. M. Peters, *Semin. Nucl. Med.*, 1994, **24**, 110–127.
51. J. Bucerius, H. Ahmadzadehfar and H.-J. Biersack, *99mTc-Sestamibi*, Springer, Dordrecht, 2011.
52. J. A. Leppo, E. G. De Puey and L. L. Johnson, *J. Nucl. Med.*, 1991, **32**, 2012–2022.
53. O. Schillaci, A. Spanu and G. Madeddu, *Q. J. Nucl. Med. Mol. Imaging*, 2005, **49**, 133–144.
54. L. Maffioli, J. Steens, E. Pauwels and E. Bombardieri, *Tumori*, 1996, **82**, 12–21.
55. Y. Sheikine, D. S. Berman and M. F. D. Carli, *Clin. Cardiol.*, 2008, **33**, E39-E45.
56. R. Finn, in *Handbook of Radiopharmaceuticals: Radiochemistry and Applications*, ed. M. J. Welch and C. S. Redvanly, Wiley, Chichester, 2003.
57. M. Bourdoiseau, *Int. J. Radiat. Appl. Intrum. Part B: Nucl. Med. Biol.*, 1986, **13**, 83–88.
58. L. Koehler, K. Gagnon, McQuarrie and F. Wuest, *Molecules*, 2010, **15**, 2686–2718.
59. H. Coenen, J. Mertens and B. Mazieère, *Radioiodination Reactions for Radiopharmaceuticals*, Springer, Dordrecht, 2006.
60. M. H. Bourguignon, E. K. J. Pauwels, C. Loc'h and B. Mazière, *Eur. J. Nucl. Med.*, 1997, **24**, 331–344.
61. V. Ardisson and N. Lepareur, in *Comprehensive Handbook of Iodine*, ed. V. Preedy, G. Burrow and R. Watson, Elsevier, London, 2009.
62. J. Grassi and P. Pradelles, in *Handbook of Experimental Pharmacology*, ed. C. Patrono and P. A. Peskar, Springer-Verlag, Berlin-Heidelberg-New York, 1987.
63. S. Moerlein, *Radiochim. Acta*, 1990, **50**, 55–61.
64. S. Moerlein, D. R. Hwang and M. J. Welch, *Int. J. Radiat. Appl. Intrum. Part A: Appl. Radiat. Isot.*, 1988, **39**, 369–372.
65. B. F. Francis, W. J. Rzeszotarski, W. C. Eckelman and R. C. Reba, *J. Labelled Compd. Radiopharm.*, 1982, **19**, 1499–1500.
66. W. J. Rzeszotarski, W. C. Eckelman, B. F. Francis, D. A. Simms, R. E. Gibson, E. M. Jagoda, M. P. Grissom, R. R. Eng, J. J. Conklin and R. C. Reba, *J. Med. Chem.*, 1984, **27**, 156–160.
67. D. M. Doran and I. L. Spar, *J. Immunol. Methods*, 1980, **39**, 155–163.
68. A. A. Hussain, J. A. Jona, A. Yamada and L. W. Dittert, *Anal. Biochem.*, 1995, **224**, 221–226.
69. G. S. Bailey, in *The Protein Protocols Handbook*, ed. J. M. Walker, Humana Press Inc., Totowa, NJ, 1996.
70. M. A. K. Markwell, *Anal. Biochem.*, 1982, **125**, 427–432.

71. G. W. Visser, R. P. Klok, G. W. Gebbink, T. Ter Linden, G. A. Van Dongen and C. F. Molthoff, *J. Nucl. Med.*, 2001, **42**, 509–519.
72. E. Mennicke, M. Holschbach and H. H. Coenen, *J. Labelled Compd. Radiopharm.*, 2000, **43**, 721–737.
73. M. M. Goodman, G. W. Kabalka, X. Meng, R. N. Waterhouse, F. F. Knapp Jr. and C. P. D. Longford, in *Synthesis and Application of Isotopically Labeled Comp. 1991*, ed. E. Buncel and G. W. Kabalka, Elsevier Science Publishers, Amsterdam, 1992.
74. G. W. Kabalka and E. E. Gooch, *J. Org. Chem.*, 1981, **46**, 2582–2584.
75. G. W. Kabalka, E. E. Gooch, H. C. Hsu, L. C. Washburn, T. T. Sun and R. L. Hayes, *Appl. Nucl. Radiochem.*, 1982, **1982**, 197–203.
76. D. S. Wilbur, *Bioconjug. Chem.*, 1992, **3**, 433–470.
77. T. M. Behr, M. Gotthardt, W. Becker and M. Behe, *Nuklearmedizin*, 2002, **41**, 71–79.
78. A. E. Bolton and W. M. Hunter, *Biochem. J.*, 1973, **133**, 529–533.
79. J. J. Langone, *Methods Enzymol.*, 1981, **73**, 113–127.
80. J. Rudinger and U. Ruegg, *Biochem. J.*, 1973, **133**, 538–539.
81. F. T. Wood, M. M. Wu and J. J. Gerhart, *Anal. Biochem.*, 1975, **69**, 339–349.
82. L. A. Khawli, F.-M. Chen, M. M. Alaudin and A. L. Stein, *Antibody Immunoconjugates Radiopharm.*, 1991, **4**, 163–182.
83. P. Wyeth and S. G. Douglas, in *Proceedings of the 5th International Symposium on Radiopharmaceutical Chemistry*, Tokyo, Japan, 1984.
84. J. R. Panuska and C. W. Parker, *Anal. Biochem.*, 1987, **10**, 192–201.
85. S. Ram, E. Fleming and D. J. Buchsbaum, *J. Nucl. Med.*, 1992, **33**, 1029.
86. S. Ram and D. J. Buchsbaum, *Int. J. Radiat. Appl. Intrum. Part A: Appl. Radiat. Isot.*, 1992, **43**, 1337–1391.
87. L. A. Khawli, A. D. Van Der Abeele and A. I. Kassis, *Nucl. Med. Biol.*, 1992, **19**, 289–295.
88. H.-M. Park, *J. Nucl. Med.*, 2002, **43**, 77–78.
89. H. N. Wellman and R. T. Anger Jr, *Semin. Nucl. Med.*, 1971, **1**, 356–378.
90. S. J. Mandel, L. K. Shankar, B. F. Yamamoto and A. Alavi, *Clin. Nucl. Med.*, 2001, **26**, 6–9.
91. J. D. Pineda, T. Lee, K. Ain, J. C. Reynolds and J. Robbins, *J. Clin. Endocrinol. Metab.*, 1995, **80**, 1488–1492.
92. R. Howman-Giles, P. J. Shaw, R. F. Uren and D. K. V. Ching, *Semin. Nucl. Med.*, 2007, **37**, 286–302.
93. P. D. Mozley, C. K. Kim, J. Mohsin, A. Jatlow, E. Gosfield III and A. Alavi, *J. Nucl. Med.*, 1994, **35**, 1138–1144.
94. D. M. Raffel and D. M. Wieland, *J. Am. Coll. Cardiol. Img.*, 2010, **3**, 111–116.
95. H. J. Verberne, L. M. Brewster, G. A. Somsen and B. L. F. Van Eck-Smit, *Eur. Heart J.*, 2008, **29**, 1147–1159.
96. *United States Pat.*, 4584187, 1986.
97. A. R. Wafelman, M. C. Konings, C. A. Hoefnagel, R. A. Maed and J. H. Beijnen, *Appl. Radiat. Isot.*, 1994, **45**, 997–1007.
98. S. Samnick, J. B. Bader, M. Müller, C. Chapot, S. Richter, A. Schaefer, B. Sax and C.-M. Kirsch, *Nucl. Med. Commun.*, 1999, **20**, 537–545.

99. R. A. Hauser and D. G. Grosset, *J. Neuroimaging*, 2011, **21**, DOI: 10.1111/j.1552-6569.2011.00583.x.
100. A. Antonini, *Neuropsychiatr. Dis. Treat.*, 2007, **3**, 287–292.
101. J. L. Neumeyer, S. Wang, Y. Gao, R. A. Milius, N. S. Kula, A. Campbell, R. J. Baldessarini, Y. Zea-Ponce, R. M. Baldwin and R. B. Innis, *J. Med. Chem.*, 1994, **37**, 1558–1561.
102. R. E. Weiner and M. L. Thakur, in *Handbook of Radiopharmaceuticals: Radiochemistry and Applications*, ed. M. J. Welch and C. S. Redvanly, Wiley, Chichester, 2003.
103. R. Murthy, R. Nunez, J. Szklaruk, W. Erwin, D. C. Madoff, S. Gupta, K. Ahrar, M. J. Wallace, A. Cohen, D. G. Coldwell, A. S. Kennedy and M. E. Hicks, *Radiographics*, 2005, **25**, S41–S55.
104. J. Šrank, F. Mellchar, A. T. Filyanin, M. Tomeš and M. Beran, *Appl. Radiat. Isot.*, 2010, **68**, 2163–2168.
105. D. Jin, M. Chinol, A. Savonen and J. Hiltunen, *Radiochim. Acta*, 2005, **93**, 111–113.
106. R. K. Frank, P. S. Athey, G. Gulyas, G. E. Kiefer, K. McMillan and J. Simón, in *ACS Symposium Series: Polymeric Drug Delivery I*, ed. S. Svenson, American Chemical Society, Washington DC, 2006.
107. P. Caravan, J. J. Ellison, T. J. McMurry and R. B. Lauffer, *Chem. Rev.*, 1999, **99**, 2293–2352.
108. H. Zhang, M. A. Moroz, I. Serganova, T. Ku, P. Smith-Jones and R. Blasberg, *J. Nucl. Med.*, 2009, **50**(Suppl. 2), 323.
109. A. Versari, L. Camellini, G. Carlinfante, A. Frasoldati, F. Nicoli, E. Grassi, C. Gallo, F. P. Giunta, A. Fraternali, D. Salvo, M. Asti, F. Azzolini, V. Iori and R. Sassatelli, *Clin. Nucl. Med.*, 2010, **35**, 321–328.
110. A. Otte, R. Herrmann, A. Heppeler, M. Behe, E. Jermann, P. Powell, H. R. Maecke and J. Muller, *Eur. J. Nucl. Med.*, 1999, **26**, 1439–1447.
111. A. Frilling, F. Weber, F. Saner, A. Bockisch, M. Hofmann, J. Mueller-Brand and C. E. Broelsch, *Surgery*, 2006, **140**, 968–976.
112. O. Akinlolu, K. Ottolino-Perry, J. A. McCart and R. M. Reilly, *Cancer Biother. Radiopharm.*, 2010, **25**, 325–333.
113. B. Sammak, M. A. E. Bagi, M. A. Shahed, D. Hamilton, J. Al Nabulsi, B. Youssef and M. Al Thagafi, *Eur. J. Radiol.*, 1999, **9**, 894–900.
114. D. Marshall and D. O. Haskard, in *Methods in Molecular Biology. Volume 225: Inflammation Processes*, ed. P. G. Winyard and D. A. Willoughby, Humana Press, Totowa, NJ, 2003.
115. C. J. Palestro, *J. Nucl. Med.*, 2007, **48**, 332–334.
116. N. Prandini, E. Lazzeri, B. Rossi, P. Erba, M. G. Parisella and A. Signore, *Nucl. Med. Commun.*, 2006, **27**, 633–644.
117. M. A. Green and J. C. Huffman, *J. Nucl. Med.*, 1988, **29**, 417–420.
118. W. S. Richter and B. I. Humplik, *Newsletter of the International Society of Radiolabeled Blood Elements*, 2004, **10**, 5–9.
119. A. Blazeski, K. M. Kozloff and P. J. H. Scott, *Rep. Med. Imag.*, 2010, **3**, 17–27.

CHAPTER 5.1

MRI Contrast Agents: Synthesis, Applications and Perspectives

PIER LUCIO ANELLI,* LUCIANO LATTUADA AND
MASSIMO VISIGALLI

Bracco Imaging SpA – Centro Ricerche Bracco, Via Ribes 5,
10010 Colleretto Giacosa (TO), Italy

5.1.1 Introduction

Magnetic Resonance Imaging (MRI), with well above 50 million procedures per year on a worldwide basis, plays a relevant role among *in vivo* diagnostic techniques. Although with noticeable geographical differences, in some countries as many as one third of the procedures are so-called contrast enhanced procedures, meaning that they are performed in association with the administration of a contrast agent.

Images of the human body with the MRI technique originate from the almost ubiquitous presence of water molecules in the body. Indeed, signal intensities which allow reconstruction of the images depend on differences in concentration of water molecules in different tissues as well as on differences in the relaxation times (*i.e.* longitudinal, T_1, and transversal, T_2, relaxation times) of water protons located in different environments (*e.g.* fat *vs.* muscular tissue).[1]

It was discovered in the late 1970s that paramagnetic species can be administered to animals and human beings to increase the relaxation rate of water protons of those body regions in which the species diffuse.[2] Initially, paramagnetic metal ions such as Mn^{2+} and Fe^{3+} were investigated. Paramagnetic

RSC Drug Discovery Series No. 15
Biomedical Imaging: The Chemistry of Labels, Probes and Contrast Agents
Edited by Martin Braddock
© Royal Society of Chemistry 2012
Published by the Royal Society of Chemistry, www.rsc.org

aquo-ions, with several water molecules in the first coordination sphere, are ideal contrast agents provided that the coordinated water molecules are in fast enough exchange with bulk water molecules. However, it was soon realized that from a safety standpoint, owing to their low tolerability, simple salts of para-magnetic metal ions could not be directly used as contrast agents. From earlier work on the use of chelates of heavy metals and lanthanides as possible X-ray contrast agents it was known that chelation of metal ions with poly-aminopolycarboxylic ligands such as ethylenediaminetetraacetic acid (EDTA) and diethylenetriaminepentaacetic acid (DTPA) could be an effective way to increase tolerability of the metal ions.[3] In addition, thermodynamic stability constants for these chelates were known to be very good.[4]

Despite Mn^{2+} and Fe^{3+} being endogenous metal ions, the exogenous Gd^{3+} ion, owing to its suitable magnetic properties (*i.e.* magnetic moment and electron spin relaxation time), quickly became the most studied paramagnetic ion for encapsulation in highly stable complexes. Gadolinium aquo-ion is coordinated to eight/nine water molecules. On the basis of the experience of several research groups it was shown that gadolinium complexes of stability suitable for *in vivo* applications require hepta- or octa-dentate polyamino-polycarboxylic ligands (*vide infra*). These complexes accommodate at least one water molecule in the first coordination sphere of the metal ion.[5]

The use of paramagnetic species as contrast agents relies on properties which deserve some discussion. The observed water proton relaxation rate $(1/Ti)_{obs}$ in the presence of a paramagnetic species stems from the sum of a diamagnetic $(1/Ti)_d$ and a paramagnetic $(1/Ti)_p$ contribution. The latter linearly depends on the concentration of the paramagnetic species, [ps], through a coefficient r_i which is called relaxivity (eqn (5.1.1)).

$$(1/T_i)_{obs} = (1/T_i)_d + (1/T_i)_p = (1/T_i)_d + r_i[ps] \quad i = 1, 2 \qquad (5.1.1)$$

Relaxivity, which is magnetic field and temperature dependent, is usually measured in $mM^{-1} s^{-1}$ units. Large relaxivity values characterize paramagnetic species which are good candidates as contrast agents. Several parameters affect relaxivity of metal ion complexes. For a thorough treatment of these aspects reference to some relevant reviews is suggested.[6] However, in order to under-stand how contrast agents were and are designed, some of the parameters influencing relaxivity need to be mentioned, namely: (i) number of water molecules directly in contact with the metal ion (q, inner coordination sphere) as well as water molecules in the second coordination sphere and those diffusing nearby the complex (outer coordination sphere); (ii) distance between the protons of the coordinated water molecules and the paramagnetic centre; (iii) residence time on the complex of the water molecules in the first coordination sphere (τ_M); (iv) rotational time of the complex that is linked to its tumbling (τ_r); (v) electron spin relaxation time of the complex.

5.1.2 Currently Available Contrast Agents

5.1.2.1 Complexes of Paramagnetic Metal Ions

In all gadolinium complexes approved for clinical applications, the metal ion is chelated by an octadentate acyclic or cyclic polyaminopolycarboxylic ligand. All such ligands, as sketched in Figure 5.1.1, are variously functionalized derivatives of either diethylenetriamine or 1,4,7,10-tetraazacyclododecane (cyclen). Besides the nitrogen atoms, the additional four or five coordinating sites (*i.e.* drawn in Figure 5.1.1 as empty circles) are mainly carboxylic acids, but also hydroxy and carboxamide groups have been used.

The gadolinium complexes that are currently available for use in clinical applications are listed in Table 5.1.1 Their chemical structures are reported in Figures 5.1.2 and 5.1.3.

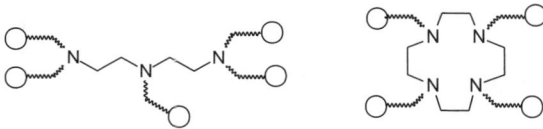

Figure 5.1.1 Cartoons of ligands in gadolinium complexes approved as MRI contrast agents.

Table 5.1.1 Metal complexes currently available as MRI contrast agents.

Compd #	Acronym	Trademark	CAS #	INN name	Ligand Structure
1	Gd-DTPA	Magnevist®	[86050-77-3]	Gadopentetate dimeglumine	acyclic
2	Gd-DTPA-BMA	Omniscan®	[131410-48-5]	Gadodiamide	acyclic
3	Gd-DTPA-BMEA	OptiMARK®	[131069-91-5]	Gadoversetamide	acyclic
4	Gd-DOTA	Dotarem®	[92943-93-6]	Gadoterate meglumine	cyclic
5	Gd-HP-DO3A	ProHance®	[120066-54-8]	Gadoteridol	cyclic
6	Gd-BT-DO3A	Gadovist®	[138071-82-6]	Gadobutrol	cyclic
7	Gd-BOPTA	MultiHance®	[127000-20-8]	Gadobenate dimeglumine	acyclic
8	Gd-EOB-DTPA	Primovist®	[135326-11-3][a]	Gadoxetate disodium	acyclic
9	MS-325	Vasovist®	[201688-00-8][a]	Gadofosveset trisodium	acyclic
10	Mn-DPDP	Teslascan®	[155319-91-8][a]	Mangafodipir trisodium	acyclic

[a]For the acid form.

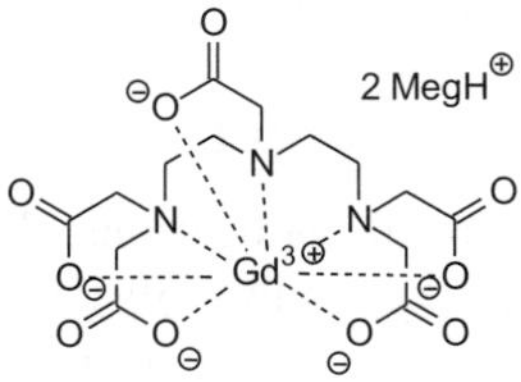

1

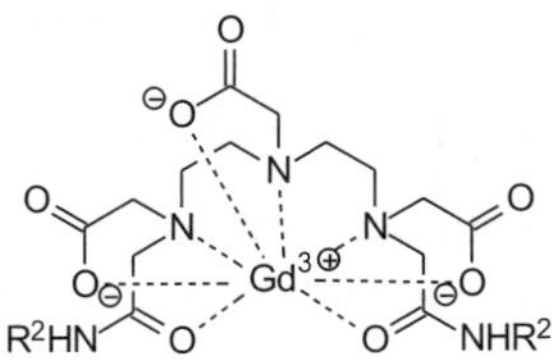

2 R^2 = Me
3 R^2 = CH_2CH_2OMe

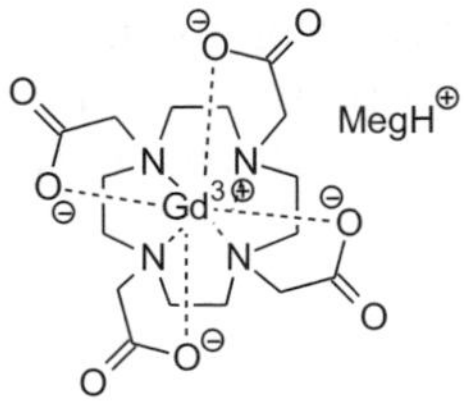

4

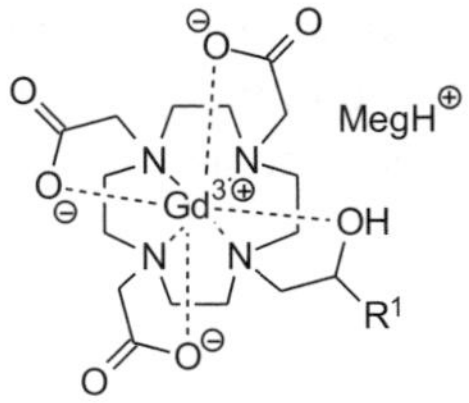

5 R^1 = Me
6 R^1 = $CHOHCH_2OH$

Figure 5.1.2 Approved MRI contrast agents: gadolinium complexes – part a.

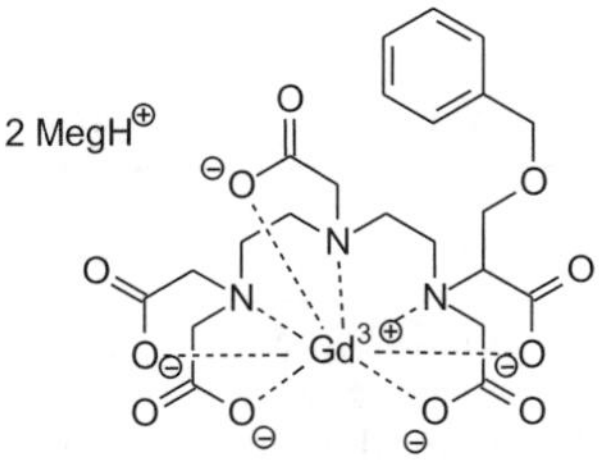

7

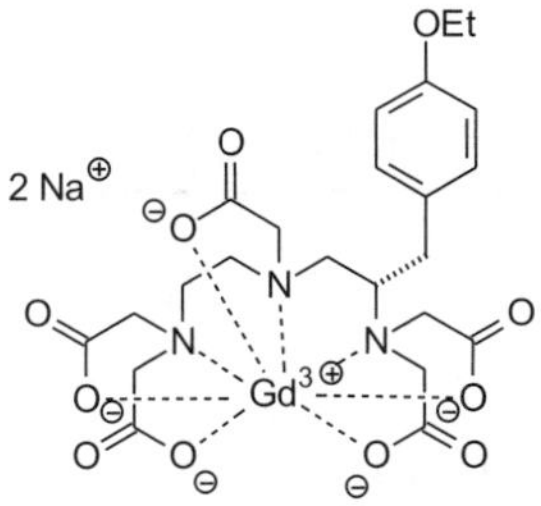

8

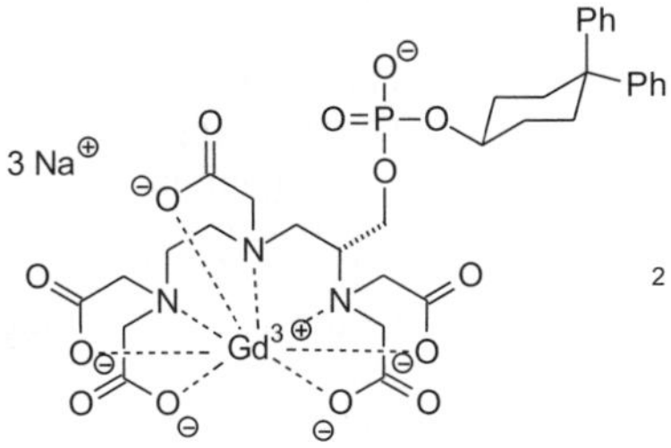

9

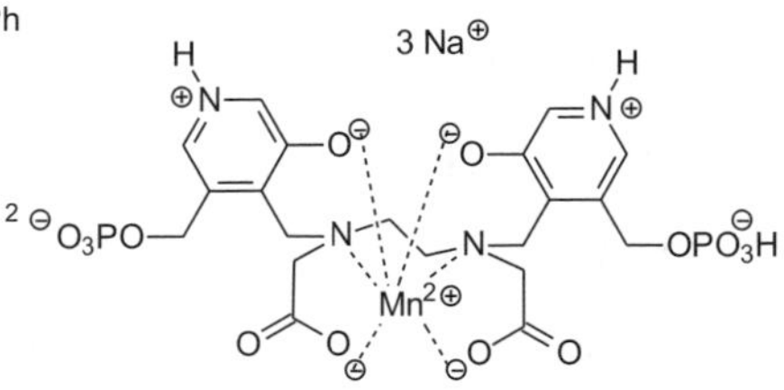

10

Figure 5.1.3 Approved MRI contrast agents: gadolinium complexes – part b.

5.1.2.1.1   Synthesis

The first complex to be approved in the late 1980s was gadopentetate dimeglumine, **1**.[7] Complex **1** is easily obtained by complexation of DTPA, a largely available chelating agent, with Gd_2O_3. Conversely, the routes used to synthesize complexes **2** to **9** deserve some comment. For the sake of simplicity the complexation step is omitted and discussion is limited to the synthesis of the corresponding ligands. Indeed, complexation of these ligands is usually carried out in water using either Gd_2O_3 or a water soluble salt like $GdCl_3$. In the former case the complexation step directly affords a solution which, after a suitable sterilization process, is ready for the intended use. On the other hand, use for example of $GdCl_3$ originates three equivalents of HCl which need to be neutralized with a suitable base. Accordingly, the solution thus generated requires a desalting step. Reaction of the appropriate amine and DTPA bisanhydride **11**,[8,9] that is easily obtained by dehydration of DTPA with acetic anhydride (Scheme 5.1.1), generates ligands **12**,[10] and **13**,[11] *i.e.* the precursors of **2** and **3**, respectively.

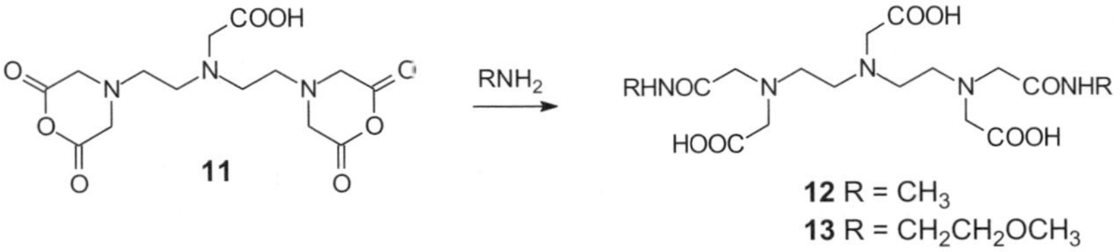

12 R = CH_3
13 R = $CH_2CH_2OCH_3$

Scheme 5.1.1

The synthesis of complex **4** requires the availability of cyclen **14**, a key building block in this field. For many years the unique access to **14** has been through the so-called Richman and Atkins route (Scheme 5.1.2, path a).[12] According to this synthetic approach the bis sodium salt of diethylenetriamine tris-*p*-toluenesulfonamide is cyclised by reaction with the tris-*p*-toluenesulfonyl derivative of diethanolamine in *N,N*-dimethylformamide to give the tetra-*p*-toluenesulfonamide **15** in over 90% yield. Compound **15** is then deprotected in conc. H_2SO_4 at high temperature to afford **14**. Although not appealing in terms of atom economy, the Richman and Atkins route is still the main industrial process to afford **14**. However, to overcome the above drawbacks some alternative routes have been developed.[13–15] For example, triethylenetetramine reacts with glyoxal (Scheme 5.1.2, path b) to give a mixture of tricyclic products **16** that by alkylation with dichloroethane (or dibromoethane) afford tetracycle **17**. Simple deprotection, for example with ethylenediamine, yields **14**.[16]

Ligand **18** is easily obtained from **14** by carboxymethylation in water, *e.g.* with chloroacetic acid at pH 9–10 by continuous addition of NaOH (Scheme 5.1.3).[17]

The synthesis of ligands **19** and **20**, *i.e.* the precursors of complexes **5** and **6**, respectively, requires the selective functionalization of one of the nitrogen atoms of cyclen (Scheme 5.1.4). By reaction with *N,N*-dimethylformamide

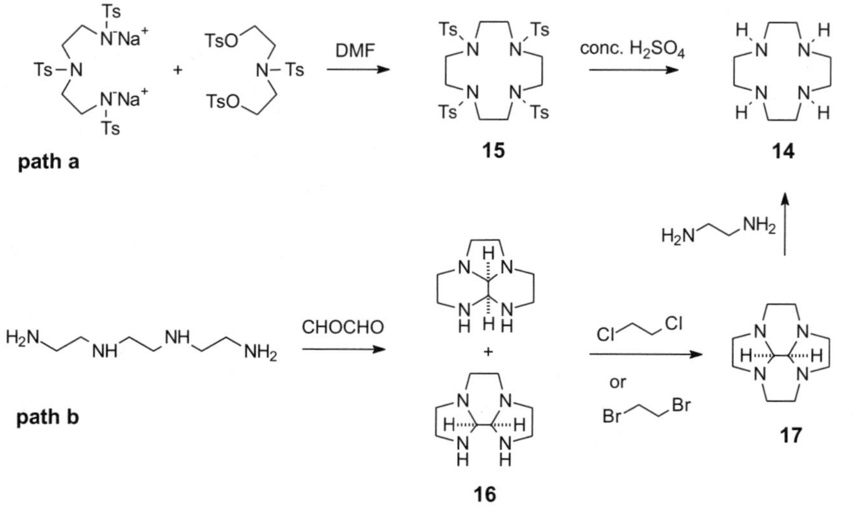

path a

path b

Scheme 5.1.2

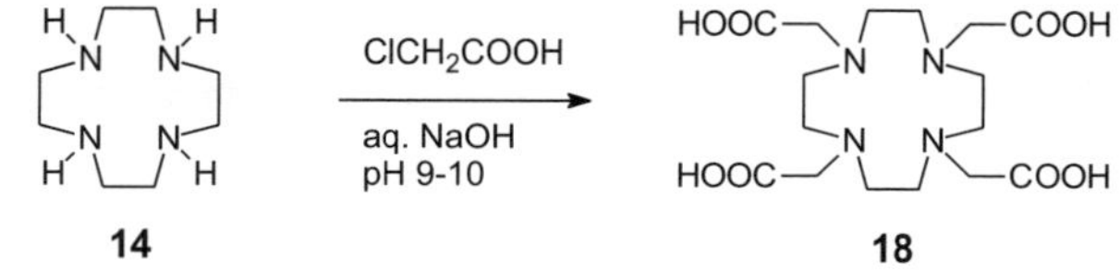

Scheme 5.1.3

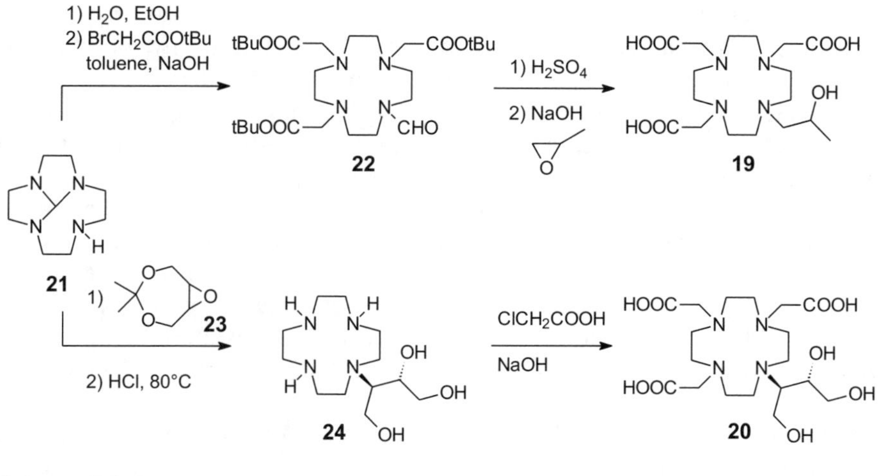

Scheme 5.1.4

dimethylacetal, cyclen is converted into the tricyclic **21** which is then hydro-lyzed to the formyl derivative and alkylated with *t*-butyl bromoacetate. Extensive deprotection of **22** and reaction with propylene oxide gives **19**.[18] Similarly, reaction of **21** with the epoxide **23** followed by hydrolysis leads to **24** which by carboxymethylation affords ligand **20**.[19]

The ligand **25**, that is required to produce complex **7**, is prepared by reaction of the α-chloroacid **26** with diethylenetriamine, followed by exhaustive alky-lation of **27** with bromoacetic acid at controlled pH (Scheme 5.1.5).[20]

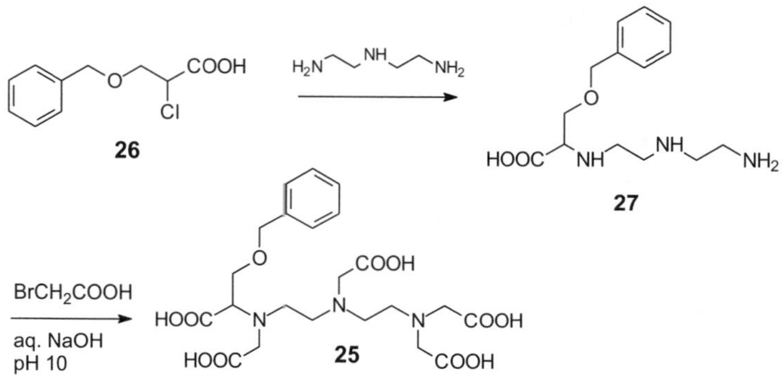

Scheme 5.1.5

The ligand precursor of complex **8** is prepared from the *N*-Z-protected methyl ester of *O*-ethyl tyrosine, **28**, which is reacted with ethylenediamine to give the intermediate **29**. Subsequent hydrogenolysis of the benzyloxycarbonyl protecting group and reduction of the carboxamide moiety affords triamine **30**. Peralkylation with *t*-butyl bromoacetate gives pentaester **31** which by depro-tection of the ester groups yields ligand **32** (Scheme 5.1.6).[21]

Scheme 5.1.7 describes the synthetic route to produce the ligand precursor of complex **9**. Penta *t*-butyl ester **33**,[22] which is prepared in analogy to **31**, is coupled with diphenylcyclohexanol through a phosphate moiety taking advantage of phosphoramidite chemistry. Subsequent treatment of inter-mediate **34** with NH_3 and deprotection of the *t*-butyl esters yields ligand **35**.[23]

5.1.2.1.2   General Properties

Solid state structures determined by means of single crystal X-ray diffraction studies were reported for several of the gadolinium complexes listed in Table 5.1.1. In complexes formed by acyclic ligands like DTPA or **25** the gadolinium ion is nine coordinated and the coordination geometry is best approximated by a distorted tricapped trigonal prism.[20,24] Eight coordination sites are provided by the ligand, *i.e.* three nitrogen atoms and five oxygen atoms of the carbo-xylate groups. The oxygen atom of a water molecule occupies the ninth

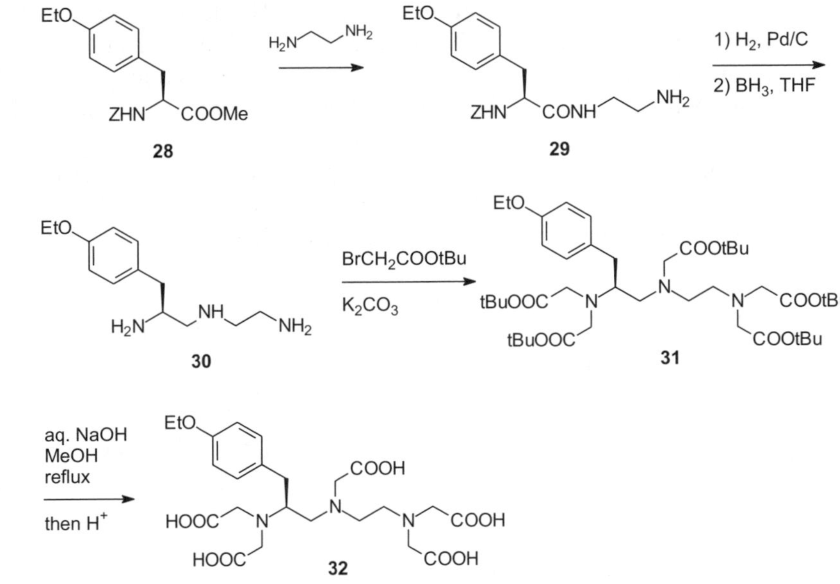

Scheme 5.1.6

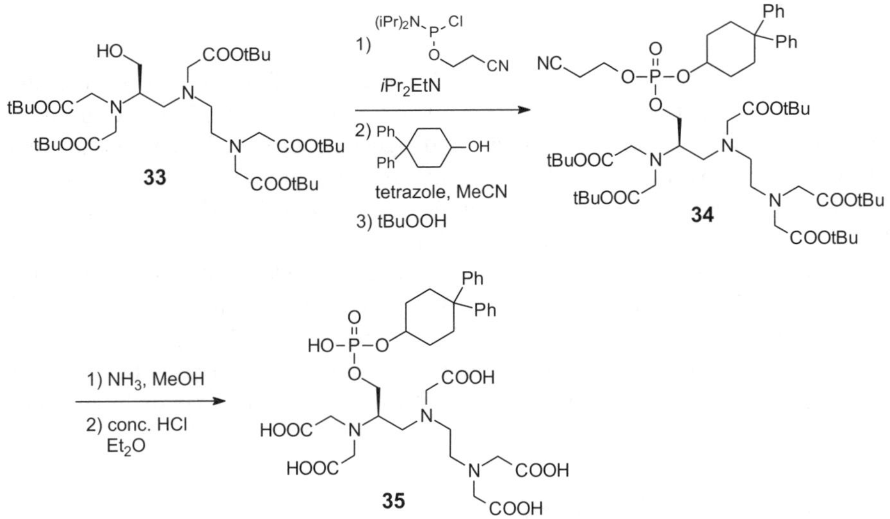

Scheme 5.1.7

coordination site. The presence of this molecule, which brings the two protons to the closest possible distance from the paramagnetic centre, plays a key role in the application of such complexes as MRI contrast agents (*vide supra*). Also gadolinium complexes of acyclic ligands containing two carboxamide residues like **12** show similar coordination geometry. Noticeably, for this complex it is found that the two carboxamide oxygen atoms participate in the coordination of the metal ion.[25]

Solid state structures of complexes formed by cyclic ligands like **19** show that again the gadolinium ion is coordinated by the eight binding sites of the ligand (*e.g.* for complex **5**, the four nitrogen atoms, three carboxylates and the oxygen of the hydroxypropyl residue).[26] Also in this structure a water molecule is present in the first coordination sphere. However, for this complex the coordination geometry is best described by a capped distorted square antiprism.

As already anticipated, stability is a key property for complexes used as MRI contrast agents. Indeed, stability in biological systems is a very complicated phenomenon if one takes into account that a variety of other, possibly competing, endogenous species are encountered by the complex after administration. In the early times, the thermodynamic stability constants for the formation of the complexes in water were considered (see eqn (5.1.2) and (5.1.3), where L is the chelating ligand, M is the metal ion and electric charges of the species are omitted for clarity).

$$L + M \rightleftharpoons ML \qquad (5.1.2)$$

$$K_{ML} = [ML]/[M][L] \qquad (5.1.3)$$

Log K_{ML} values for gadolinium complexes **1** to **8** are in the range of 16.8 to 25.3.[27] Complexes **2** and **3**, derived from the two diamides of DTPA, feature the lowest stability constants whereas the highest value belongs to Gd-DOTA, **4**. However, several research groups have stressed that "global conditional stability constants" that are calculated at the physiological pH (*i.e.* 7.4) and take into account the degree of protonation of the ligand, are more appropriate to assess the relative stabilities of different contrast agents.[28] In addition, although extremely stable from a thermodynamic standpoint, in accordance with eqn (5.1.3), the complexes are in equilibrium with tiny amounts of ligand and metal ion. Therefore, when *in vivo*, more complicated equilibria involving endogenous species have to be taken into account. This means for example that the ligand can chelate other endogenous metal ions (*e.g.* Ca^{2+}, Zn^{2+}, Cu^{2+}) or that the gadolinium ion can be bound by endogenous ligands which have a high affinity for it. In addition, Gd^{3+} forms very insoluble salts with some ubiquitous anions (*e.g.* phosphate). These speculations have led to in depth exploration of the kinetic stability of gadolinium complexes. Often, in order to see differences, experimental conditions stressed in comparison with those likely experienced by the complexes *in vivo* have been used. It was found that gadolinium complexes in which the ligand has a cyclic structure are less prone to metal and proton catalyzed dissociation than complexes in which the ligand has an acyclic structure.[29]

For all gadolinium complexes reported in Table 5.1.1, during preclinical development, it was shown that only negligible quantities of gadolinium are released and accumulate in target organs like liver, spleen and bones. However, *in vivo* stability of MRI contrast agents has recently become a hot issue. Indeed, it was found that the onset of a highly disabling and sometimes lethal disorder called nephrogenic systemic fibrosis (NSF) was unusually frequent after contrast-enhanced MRI procedures in renally impaired patients. Since the half life of permanence in blood of the contrast agent (*vide infra*) is much longer in these patients, it was postulated that *in vivo* dissociation of the complexes and the consequent gadolinium ion release could somehow trigger the onset of the disorder.[30] This hypothesis is supported by the evidence that NSF cases are more common after the administration of the complexes characterized by lower stability values. However, to date no sound evidence of a real dependence of the pathology from the *in vivo* release of free gadolinium has been found.[30]

For complexes **1–6** relaxivity, r_1, values in water (25 or 40 °C, 20 MHz)[31] are in the range of 3.5–4.5 mM^{-1} s^{-1} and such values do not change much once measurements are performed in human serum. Differently, complexes **7** and **8** feature r_1 values in human serum which are almost twice in comparison with those measured in water.[32] This is the result of a weak binding of **7** and **8** with human serum albumin (HSA) which reflects in an increase of the rotational correlation time, τ_r, of the complexes. This effect is much more remarkable for complex **9** for which r_1 value in 4.5% HSA (37 °C, 20 MHz) is 42.0 mM^{-1} s^{-1} (*vide infra*).[23]

5.1.2.1.3   Applications

Gadolinium complexes reported in Table 5.1.1 are formulated as concentrated (0.25 to 1 M) aqueous solutions, in some cases with the addition of excipients (*e.g.* the free ligand or the corresponding calcium complex). These solutions are administered by intravenous injection at dose levels typically in the range of 0.05–0.1 mmol kg^{-1} of body weight, although some of the complexes have been approved for up 0.3 mmol kg^{-1} of body weight for specific diagnostic procedures. After administration, complexes **1–6** equilibrate between plasma and the interstitial space. They diffuse across the normal endothelium but they do not enter cells. For this reason they are called "extravascular extracellular contrast agents". They are not metabolized and they are almost exclusively excreted through the renal route by glomerular filtration. In healthy human volunteers the plasma elimination is a quick process with a half life of about 1.5–2 h.[33] Compounds **1–6** were initially approved for the detection of central nervous system pathologies. Subsequently, other indications, such as angiography for example in different body regions, have been approved for some of the agents.

Complexes **7** and **8** immediately after administration behave like the extravascular extracellular contrast agents. However, by virtue of the presence of a lipophilic residue in their structure, they are, to a different degree, excreted by the hepatic route. Indeed, it was shown that on their way to the bile and faecal excretion they are transported through the hepatocytes.[32] In this respect they cannot be strictly considered "extracellular" contrast agents. The partial hepatic elimination makes **7** and **8** suitable for liver imaging. However, complex **7** is also approved for use as classical extravascular extracellular agents, with an additional advantage over complexes **1–6** since it features higher relaxivity values in blood as the result of the above mentioned weak binding to HSA.

Also the Mn^{II} complex **10** (see Figure 5.1.3), *i.e.* the only non-gadolinium complex approved as an MRI contrast agent, is specifically used as a hepato-biliary agent for liver imaging. Unlike the gadolinium complexes seen above, this manganese complex is not stable *in vivo* and its mechanism of action relies on this feature. Indeed, the lability of the complex gives rise to a slow release in blood of Mn^{2+} ions that are subsequently taken up by the liver thus allowing for contrast enhancement of this organ.

Complex **9**, which features a phosphate and a diphenylcyclohexyl moiety, is characterized by a very strong binding to HSA, the most abundant plasma protein. This binding almost entirely prevents the extravasation of the complex and makes it a true vascular MRI contrast agent (*i.e.* so-called "blood pool agent").[23] This agent is useful for angiographic procedures aimed at imaging tiny and tortuous vessels such as coronary arteries for example.

Other two complexes (see Figure 5.1.4) have reached an advanced stage of development as candidates for imaging of the vascular system. Complex **36** contains a bile acid residue which allows for a strong binding to human serum albumin.[34] Differently, the mechanism of action of complex **37** (*vide infra*) relies on the high molecular volume of the complex that leads to a peculiar phar-macokinetic profile resulting in a limited diffusion across the normal endothelium.[35,36]

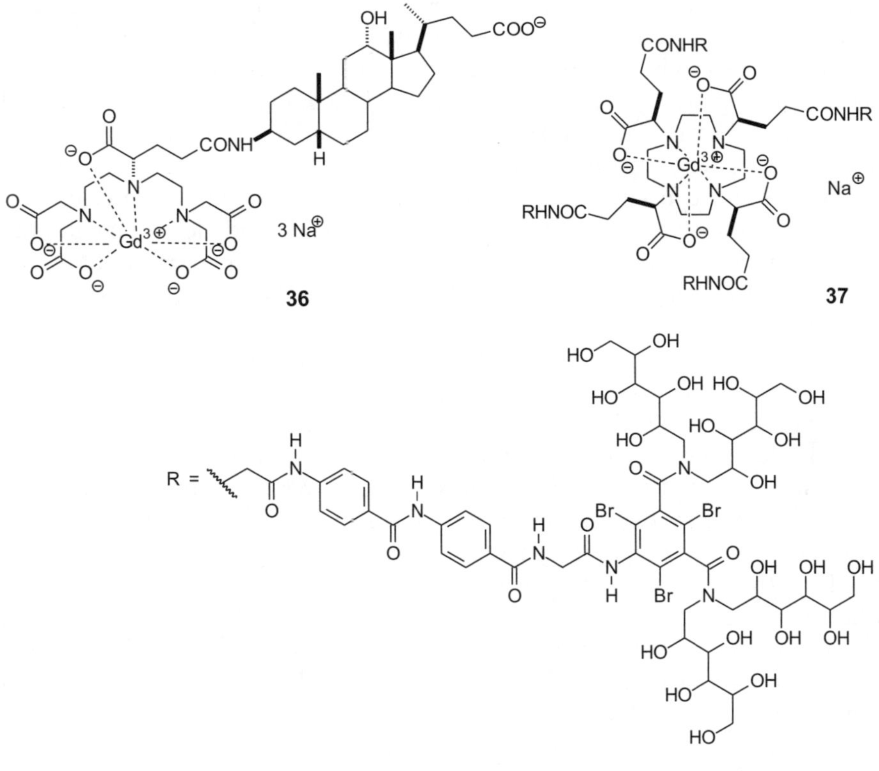

Figure 5.1.4   Examples of gadolinium complexes developed as blood pool agents.

5.1.2.2   Superparamagnetic Particles

In addition to complexes of paramagnetic metal ions, other contrast agents that are currently approved for use in clinical routine are based on super-paramagnetic particles. In these compounds the magnetically active species are nanocrystals, also called "grains", of ferrites that are characterized by having a size smaller (in the range of 4–10 nm) than the domains of ferromagnetic materials. Ferrite nanocrystals are constituted by mixtures of Fe^{II}/Fe^{III} oxides (*e.g.* magnetite, Fe_3O_4, and maghemite, γ-Fe_2O_3) that are obtained under carefully controlled crystallization conditions. In addition, as a part of the crystallization step or as a subsequent production step, the nanocrystals are embedded in a biocompatible coating which is required to stabilize the colloidal suspensions and to prevent the formation of aggregates.[37] Several organic and inorganic coatings have been and still are investigated in depth.[38] Nowadays, dextrane and carboxydextrane are successfully used as coatings for commercially available agents (*vide infra*). As a result of the coating process the nanoparticles are characterized by hydrodynamic diameters which range from few hundred nanometres (the so-called superparamagnetic iron oxide, SPIO, particles) down to 15–30 nm (the so-called ultrasmall superparamagnetic iron oxide, USPIO, particles). According to the production process the particles can contain either a single nanocrystal or several of them. Production of super-paramagnetic particles is a complex process which must reliably deliver a monodisperse population of iron oxide nanocrystals. In this respect several production methods have been investigated and to date the chemical co-pre-cipitation of mixtures of iron salts has proved the most successful one. Indeed, it has been shown that nanocrystal size (and shape) results from the combination of a variety of parameters such as: anion of the iron salts as well as Fe^{II}/Fe^{III} ratio and concentration, pH and temperature. Besides co-precipitation methodologies, especially for formulations which are under development, other synthetic procedures like hydrothermal reactions which are carried out at high temperature and high pressure as well as reactions in confined environments have been thoroughly investigated.[38]

Owing to the size of the embedded nanocrystals, superparamagnetic particles, *per se*, do not have magnetic properties. However, in the presence of an external magnetic field they feature strong magnetic moments, which are responsible for field distortions in the close surroundings of the particle. Water molecules, which diffuse nearby the particle, experience the magnetic field distortions and, as a consequence, their transversal relaxation time, T_2, is largely reduced.[39] Superparamagnetic particles are mostly used as negative T_2 MRI contrast agents. The term negative for a contrast agent indicates that, when using an appropriate instrumental sequence, the presence of the agent in a particular body region appears on the MR image as a black spot due to signal suppression. Conversely, positive contrast agents, like gadolinium complexes, generate white spots due to signal enhancement.

Dose levels of superparamagnetic particles required for performing MRI procedures account for administration of 50 to 200 mg of iron. This amount is relatively small if compared to the overall pool of iron stored in the human body (*i.e.* about 3.5 g). After intravenous administration superparamagnetic particles, depending on their structural features (*e.g.* size, coating material,...) have blood half lives that can be as short as 1 h but also in the 24–36 h range. However, a characteristic which is common to different nanoparticles is that their blood half life is dose dependent.[37] Like other particulate materials, they are taken up from blood by macrophages which are located mainly in the liver, spleen and bone marrow. Since this process can be slowed down using very small particles and suitable coatings, superparamagnetic particles have also been thoroughly investigated as blood pool agents.

Once inside the macrophages iron oxide nanoparticles are degraded. In the case of ferumoxtran-10, using doubly labelled (*i.e.* with ^{14}C and ^{59}Fe) nanoparticles, it was possible to assess that: (i) ^{14}C labelled dextrane is slowly degraded and eliminated through the kidneys (89% in the urine after 56 days); (ii) ^{59}Fe coming from the nanoparticles enters the iron pool of the human body. Accordingly, its elimination is very slow (around 20% is found in the faeces after 84 days) as for endogenous iron.

Various imaging applications are possible with currently available products (see Table 5.1.2): (i) liver imaging with SPIO characterized by fast uptake by macrophages in the liver (*i.e.* Kupffer cells); (ii) imaging of lymph nodes and of degenerative diseases involving an alteration of macrophage population in tissues as well as imaging of the vascular system taking advantage of the blood pool effect with USPIO characterized by a prolonged half life in blood; (iii) imaging of the gastrointestinal tract after oral administration of relatively large particles.

In addition to the products reported in Table 5.1.2 several other formulations containing iron oxide nanoparticles are under scrutiny at advanced preclinical and clinical level. Furthermore, great efforts are being devoted to research activities focused on the use of targeted superparamagnetic particles as molecular imaging agents.[38]

Table 5.1.2 Superparamagnetic particles currently available as MRI contrast agents.

Name	Trademark	Coating agent	Size^a (nm)	Administration route
Ferumoxides	Endorem®/ Feridex®	Dextran	120–180	intravenous
Ferumoxtran-10^b	Sinerem®/ Combidex®	Dextran	15–30	intravenous
Ferucarbotran	Resovist®	Carboxydextran	60	intravenous
Ferumoxsil	Lumirem®/ Gastromark®	Silicon	300	oral

^{a}Hydrodynamic size by laser light scattering, from ref 37.
^{b}Withdrawn from the European market in December 2007.

5.1.3   Future Trends

5.1.3.1   New Ligands

A selection of the most promising and recently published ligands for gadolinium complexation is given.

All current commercial gadolinium contrast agents, which are based on polyaminopolycarboxylic ligands, have only one coordinated water molecule. An alternative class of purely oxygen donor ligands, such as tris-hydroxypyridinone, **38** (HOPO, Scheme 5.1.8), which is able to coordinate two water molecules, has been reported.[40] The original synthesis,[41] (Scheme 5.1.8) involves carboxylation of 1-methyl-3-hydroxy-2(1*H*)-pyridinone, **39** followed by selective benzylation of the hydroxy group. The carboxylic group of **41** is activated to intermediate **42** that is then coupled to tris(2-aminoethyl)amine (TREN) to yield the protected ligand **43**. Catalytic hydrogenation of **43** gives the ligand **38**. The corresponding gadolinium complex can be easily achieved by treatment of **38** with $GdCl_3$ or $Gd(NO_3)_3$. The gadolinium complex of **38** shows an enhanced relaxivity (r_1, $= 10.5 \, \text{mM}^{-1} \, \text{s}^{-1}$, 37 °C, 20 MHz), which is more than twice that of commercial contrast agents, and a high stability (log $K_{ML} = 20.3$). One of the disadvantages of the first Gd-HOPO complexes was their poor solubility in water. In order to overcome the solubility issue, several modifications of the structure of HOPO ligands have been reported.[40] As an example, the replacement of TREN with 1,4,7-triazacyclononane as the ligand cap gives a new product with a 1000-fold increase in water solubility and an even better relaxivity, $r_1 = 13.1 \, \text{mM}^{-1} \, \text{s}^{-1}$ (25 °C, 20 MHz), due to the coordination of three water molecules ($q = 3$).[42]

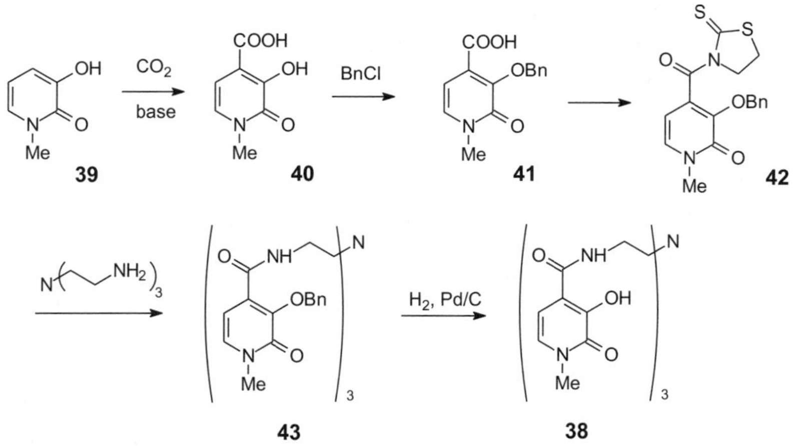

Scheme 5.1.8

Heptadentate polyaminopolycarboxylic ligands, such as diethylene-triaminetetracetic acid (DTTA) and 1,4,7,10-tetraazacyclododecane-1,4,7-triacetic acid (DO3A), give rise to gadolinium complexes with higher relaxivity because of two coordinated water molecules in solution. In spite of this behaviour, they are not the best choice for the synthesis of MRI contrast agents based on gadolinium. In fact, the decrease of the thermodynamic stability of the complex may lead to *in vivo* release of toxic Gd^{3+} ion. Moreover, the *in vivo* replacement of the two coordinated water molecules by endogenous bidentate anions, such as lactate or phosphate, largely quenches their relaxivity.

The innovative heptadentate ligand 6-amino-6-methylperhydro-1,4-diazepi-netetraacetic acid, **44** (AAZTA, Scheme 5.1.9),[43] overcomes both these problems. The complex Gd-AAZTA shows a good relaxivity ($r_1 = 7.1$ mM^{-1} s^{-1}, 25 °C, 20 MHz), a high thermodynamic stability in aqueous solution (log $K_{ML} = 20.2$),[44] and a nearly complete inertness towards the influence of bidentate anions. Moreover, the synthesis of AAZTA is very straightforward (Scheme 5.1.9). The key step is the formation of the seven-membered ring, **45**, through a nitro-Mannich condensation of nitroethane, formaldehyde and *N,N'*-dibenzylethylenediamine. The nitro compound **45** is then reduced and debenzylated by catalytic hydrogenation to give triamine **46**, which is peralkylated with *t*-butyl bromoacetate. Final deprotection of all the *t*-butyl esters, by treatment with trifluoroacetic acid, gives AAZTA ligand. The easy synthesis, coupled with the properties of the gadolinium complex, make AAZTA one of the most promising candidates for the development of a new class of MRI contrast agents.

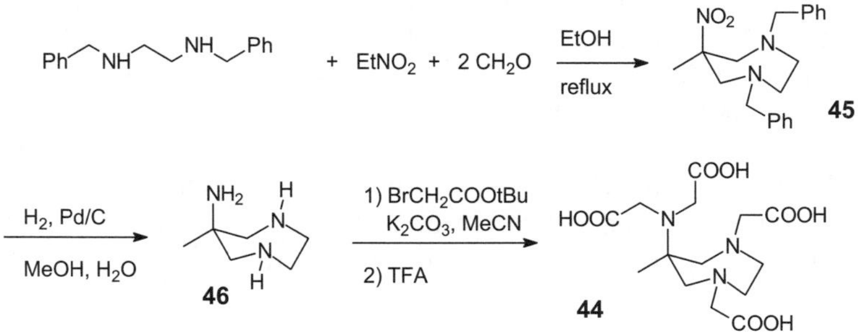

Scheme 5.1.9

Another interesting heptacoordinating ligand is based on a macrocyclic structure containing a pyridine ring. The first synthesis of Pyridine-Containing TriAza macrocycle triacetate ligand **47** (PCTA) dates back to 1981 and is based on a Richman and Atkins condensation between the bis sodium salt of diethylene-triamine tris-*p*-toluenesulfonamide **48** and 2,6-bis(bromomethyl)pyridine **49** (Scheme 5.1.10).[45] The intermediate **50** so obtained is deprotected with refluxing 48% hydrobromic acid and then alkylated with chloroacetic acid at pH 10 to give the ligand **47**. Improved syntheses of PCTA employing trinosylderivative of diethylenetriamine and less harsh conditions have recently appeared in literature.[46,47]

PCTA gives a stable gadolinium complex (log $K_{ML} = 20.8$) which coordinates two water molecules ($q = 2$) and for this reason shows a high relaxivity ($r_1 = 6.9$ mM^{-1}s^{-1}, 25 °C, 20 MHz).[48]

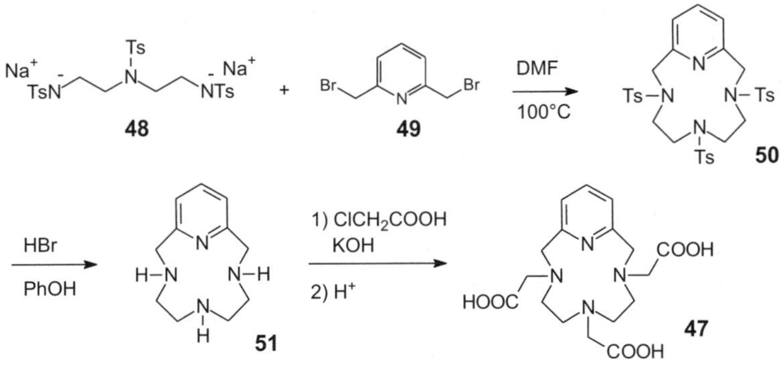

Scheme 5.1.10

An evolution of PCTA is ligand **52** (PCP2A), with a phosphonic group replacing the central carboxylic moiety. The synthesis of PCP2A,[49] (Scheme 5.1.11) involves the bisalkylation of diethyl aminomethylphosphonate **53** with *N*-tosylaziridine **54** to give the key intermediate **55**, which is then reacted with 2,6-bis(chloromethyl)pyridine, **56**, to give the cyclic intermediate **57**. Deprotection of tosylamides and phosphonic ester with sulfuric acid and alkylation with chloroacetic acid affords ligand **52**. The gadolinium complex of PCP2A shows: (i) increased relaxivity ($r_1 = 8.3$ mM^{-1} s^{-1}, 25 °C, 20 MHz); (ii) fast exchange of the coordinated water molecules ($\tau_M = 60$ ns); (iii) very high stability constant (log $K_{ML} = 23.4$).[49] Its unusually high relaxivity is the result of two water molecules in the inner coordination sphere ($q = 2$) and a significant contribution of water molecules hydrogen bonded to the phosphonate group. Moreover, Gd-PCP2A displays no change in relaxivity in a 3 mM phosphate buffer solution, which means that there is no transmetallation and no displacement of the two coordinated water molecules.

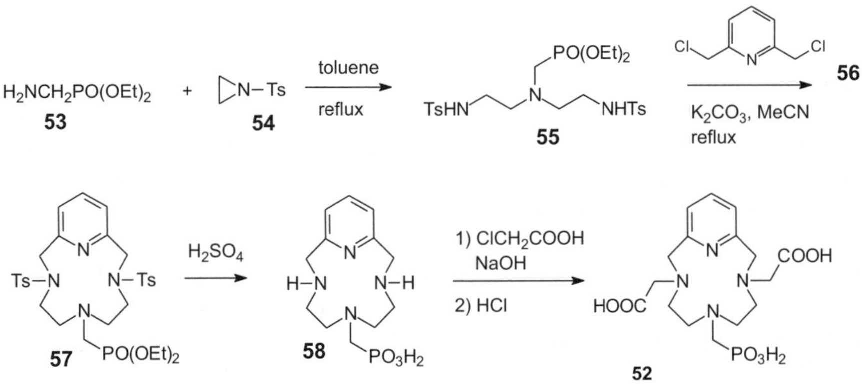

Scheme 5.1.11

The "bimodal" ligand **59** (NETA) has been designed to merge the advantages of the macrocyclic DOTA, **18**, (high thermodynamic stability) and the acyclic DTPA (favourable formation kinetics) for the complexation of metal ions. Gd-NETA complex displays enhanced relaxivity compared to Gd-DOTA, **4**, and the complex is exceptionally stable in serum and after administration to mice.[50]

An efficient and short preparation of NETA has recently appeared in literature,[51] (Scheme 5.1.12), replacing the initial multistep synthesis.[52] Partial alkylation of 1,4,7-triazacyclononane with *t*-butyl bromoacetate gives compound **60** that is reacted with bromide **61**. Both benzyl and *t*-butyl protection groups are removed by refluxing **62** in 6 M HCl thus affording NETA, **59**. A bifunctional chelating agent based on NETA has been recently synthesised and exploited in radioimmunotherapy and targeted MRI.[53]

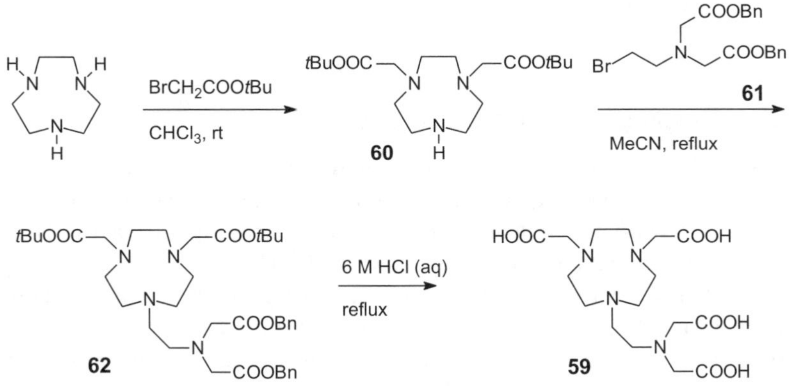

Scheme 5.1.12

Texaphyrins,[54] are large planar porphyrin-like macrocycles that coordinate trivalent lanthanide metal cations, including Gd^{3+}. As an example, texaphyrin **63**, designed to give water soluble complexes, has been obtained by an acid-catalyzed Schiff-base condensation between diamine **64** and tripyrrane dialdehyde **65** in quantitative yield (Scheme 5.1.13). The five nitrogen atoms of macrocycle **63** are able to coordinate Gd^{3+} ion in an almost coplanar arrangement. On the basis of this structure, several water molecules, between four and five, can interact with the metal centre, accounting for an unusually high relaxivity. Gadolinium complex of **63** displays in water a relaxivity, r_1, of 19.0 mM^{-1} s^{-1} (25 °C, 20 MHz). However, this favourable property cannot be exploited *in vivo*, since phosphate ions displace some of the coordinated water molecules, lowering r_1 to 5.3 mM^{-1} s^{-1} (25 °C, 20 MHz, phosphate buffer). Although showing good stability and high relaxivity, gadolinium complexes of texaphyrins have not been developed as pure MRI contrast agents. In fact, Gd-**66** (motexafin gadolinium, Xcytrin®), which is obtained from **65** and **67**, is MRI-detectable and has completed clinical trials as an agent for treating brain metastases in conjunction with X-ray radiation therapy.[55]

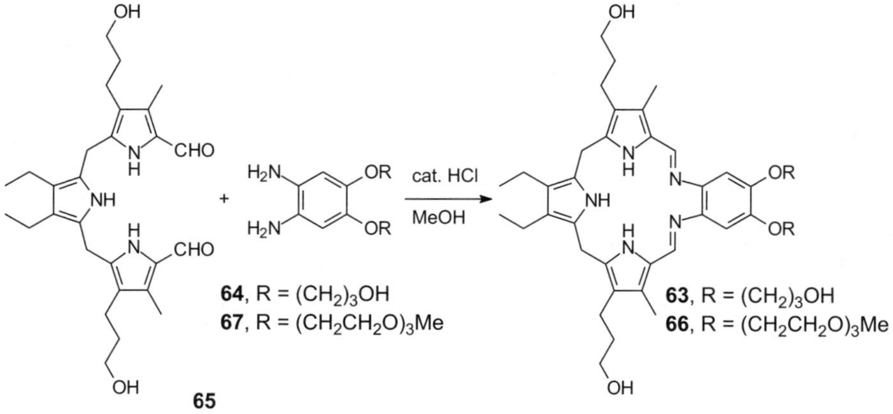

Scheme 5.1.13

5.1.3.2   Responsive Contrast Agents

In the past few years, the term "molecular imaging", that can be broadly defined as the *in vivo* characterization and measurement of biologic processes at the cellular and molecular level,[56] has become very popular in the MRI literature. Unfortunately, the poor sensitivity of MRI and the low number of target molecules located in suitable concentration on the cellular membrane prevent the visualisation of target cells (*e.g.* tumour cells) by MRI imaging. Accordingly, researchers have devoted many efforts in defining more efficient gadolinium complexes as well as delivery systems in order to achieve this goal.[57] Nevertheless, biologic phenomena, such as changes in concentration or activity of proteins, enzymes, metabolites, oxygen and metal ions, as well as changes in pH and temperature, can be "visualised" by responsive gadolinium contrast agents.[58,59]

5.1.3.2.1   Enzyme Responsive Agents

One of the pioneering examples of a responsive MRI contrast agent is the gadolinium complex of ligand **68** (Scheme 5.1.14).[60] This compound contains a galactopyranose residue that blocks the access of water to the first coordination sphere of gadolinium ion. Exposure of Gd-**68** to β-galactosidase causes a three-fold increase in relaxivity. This finding has been exploited *in vivo* in *Xenopus Laevis* embryos to obtain MR images of the regions expressing the β-galactosidase marker enzyme. The synthesis of ligand **68** involves the reaction of 2,3,4,6-tetraacetyl-1-α-bromogalactose, **69**, with 1-bromo-2-propanol, **70**, in the presence of silver triflate and collidine. After flash chromatography, the purified β anomer **71** is reacted with cyclen to give **72**, which after hydrolysis of the acetyl groups and carboxymethylation with bromoacetic acid yields ligand **68**.

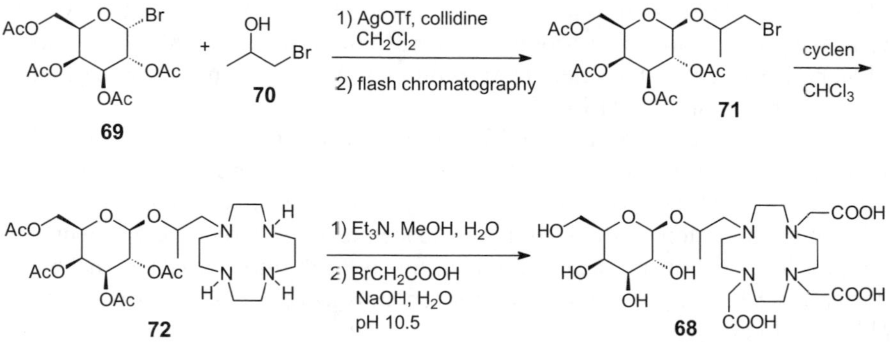

Scheme 5.1.14

Since sulfonamides are known to be specific inhibitors of carbonic anhydrase, a contrast agent designed for the selective targeting of this enzyme has been achieved coupling DTPA tetra-*t*-butyl ester **73**,[61] to *p*-aminoethylbenzenesulfonamide **74** (Scheme 5.1.15).[62] Deprotection of *t*-butyl esters of intermediate **75** affords ligand **76**. The gadolinium complex of **76** shows a high binding constant toward carbonic anhydrase and the tests performed in blood revealed that it interacts with the small amount of carbonic anhydrase present on the surface of the red cells, while the interaction with serum proteins is negligible.

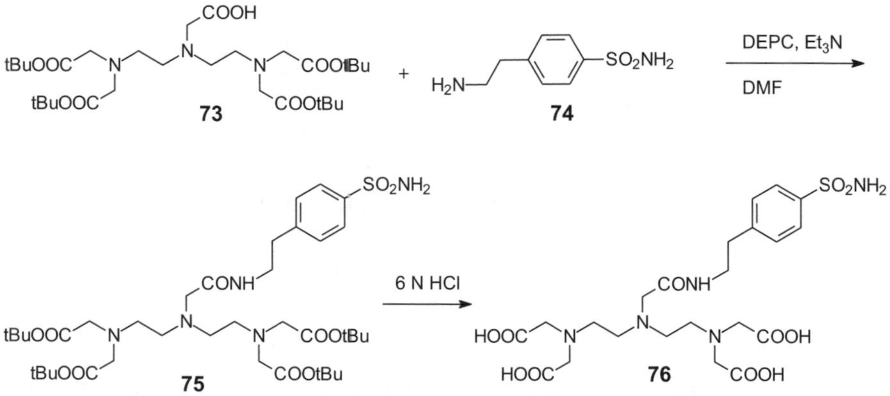

Scheme 5.1.15

Another interesting example of a responsive agent is the gadolinium complex of **77** (Scheme 5.1.16), which is activated by the alkaline phosphatase.[63] As already mentioned, one approach to increase the relaxivity of a contrast agent is taking advantage of binding to HSA. Indeed, the hydrophilic phosphate in **77** lowers the affinity for HSA but, in the presence of alkaline phosphatase and

HSA, a 70% increase in relaxivity is observed. This is the result of enzymatic cleavage of the phosphate group to generate the more hydrophobic *p*-hydroxybiaryl residue which confers to the complex a strong affinity for the hydrophobic binding sites of HSA.

The starting key intermediate **78**,[22] after protection of the triamine groups with Boc anhydride, is coupled to *p*-bromophenol **79** under Mitsunobu conditions to give **80**. Deprotection of *t*-butoxycarbonyl groups with trifluoroacetic acid and alkylation with *t*-butyl bromoacetate give derivative **81**, which is coupled to the boronic acid **82** under Suzuki conditions to give the biaryl derivative **83**. Phosphorylation of the phenol group of **83** with phosphoryl chloride, hydrolysis and deprotection with trifluoroacetic acid give ligand **77**.

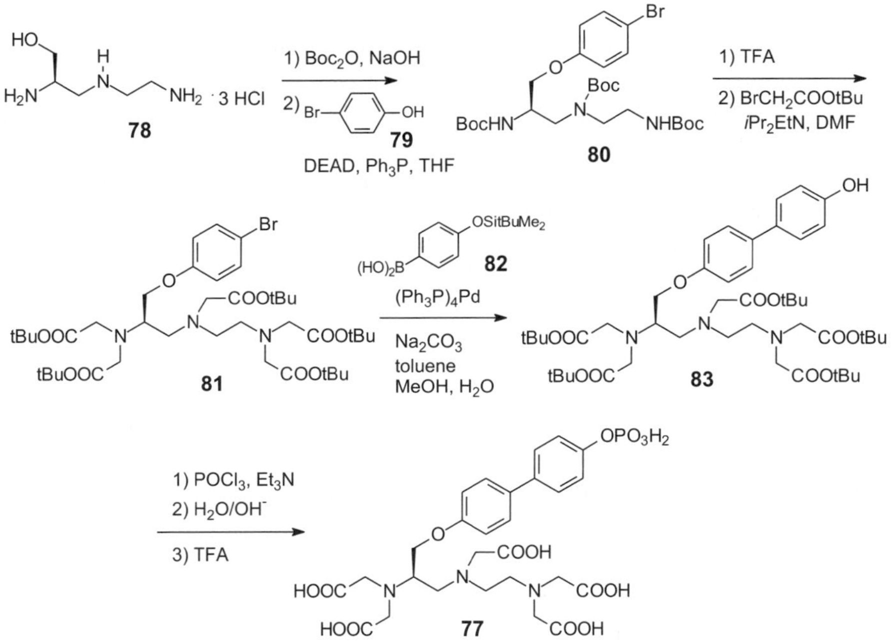

Scheme 5.1.16

A different example of enzyme-mediated relaxivity enhancement was achieved with the gadolinium complex of **84** (Scheme 5.1.17), which gives a relaxivity increase as the consequence of polymerization in the presence of peroxidases.[64] The increase in tumbling time caused by the polymerization is sufficient to give a three-fold increase in relaxivity (*i.e.* for Gd-**84**, r_1 from 4.6 to 15.9 mM^{-1} s^{-1}, 40 °C, 20 MHz) in the presence of horseradish peroxidase. The synthesis of ligand **84** (Scheme 5.1.17) is very straightforward as all the syntheses of DTPA-bisamides reported in literature, where DTPA-bisanhydride **11** is reacted with the amine of choice in a solvent.

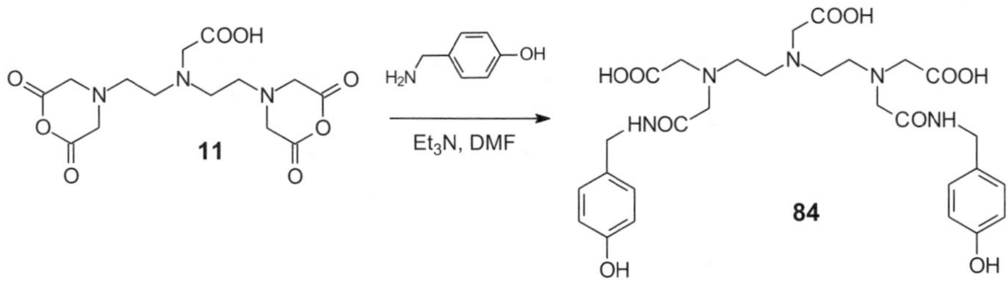

Scheme 5.1.17

5.1.3.2.2 Metal Ion Responsive Agents

One of the first examples of metal ion responsive MRI contrast agents is the gadolinium complex of ligand **85** (Scheme 5.1.18), which is selective for calcium.[65] Intracellular Ca^{2+} plays an important role in signal transduction and the determination of its concentration in the cytosol may be helpful to study cell metabolism and signalling by MRI. In fact, relaxivity of Gd-**85** is selectively modulated by Ca^{2+} concentration. When $[Ca^{2+}]$ is low (less than 0.1 μM), the carboxylates of the iminodiacetic moieties shield the Gd^{3+} ion to bulk water giving an r_1 value of 3.26 mM^{-1} s^{-1}. At higher levels of $[Ca^{2+}]$ (above 10 μM) the same carboxylates are engaged in Ca^{2+} complexation, allowing more water molecules to interact with the Gd^{3+} ion. As a consequence, the relaxivity rises to a maximum of 5.76 mM^{-1} s^{-1}. Importantly, Gd-**85** is selective for binding Ca^{2+} ions *versus* Mg^{2+} and H^+ and results revealed a 1:1 binding stoichiometry for Ca^{2+}. The synthesis of ligand **85** (Scheme 5.1.18) starts with monoalkylation of 2-nitroresorcinol, **86**, with 3-bromopropanol. Phenol **87** is then reacted with 1,2-dibromoethane and subsequent catalytic hydrogenation gives aniline **88**, which is fully alkylated with ethyl bromoacetate. Compound **89** is then brominated with carbon tetrabromide in the presence of triphenylphosphine and the resulting dibromoderivative **90** is reacted with cyclen. Final alkylation with bromoacetic acid at controlled pH provides ligand **85**.

Another essential metal ion is Zn^{2+} that is a component of many enzymes and regulates synaptic transmission and cell death. The gadolinium complex of ligand **91** has been designed to selectively sense Zn^{2+} by MRI imaging. The principle of action is similar to the one discussed for Gd-**85** and involves a modulation of the accessibility of water molecules to the chelated Gd^{3+} ion. In this case, opposite to the previous one, the presence of one equivalent of Zn^{2+} decreases the relaxivity (approximately 33%). Indeed, the formation of the pyridines-Zn^{2+} complex above the Gd^{3+} ion prevents direct coordination of water molecules to the latter. The synthesis exploits the well known reaction of DTPA-bisanhydride **11** with an amine (Scheme 5.1.19).[66] In this case the required amine **92** is easily obtained by alkylation of mono protected ethylenediamine **93** with 2-(chloromethyl)pyridine, **94**, followed by deprotection of the *t*-butoxycarbonyl group with trifluoroacetic acid.

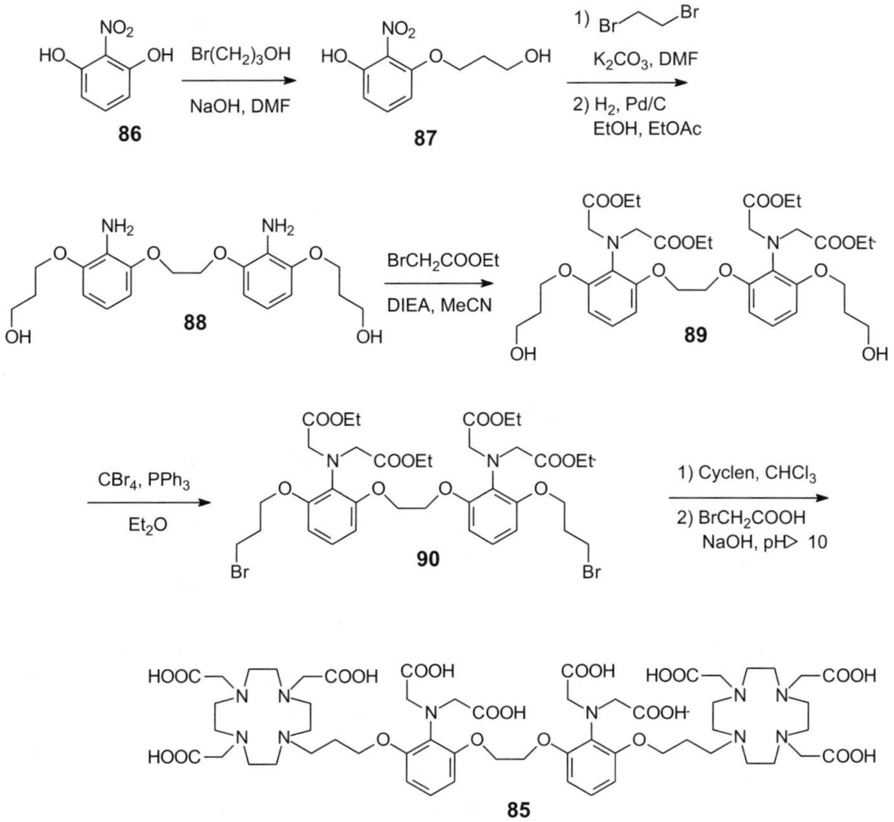

Scheme 5.1.18

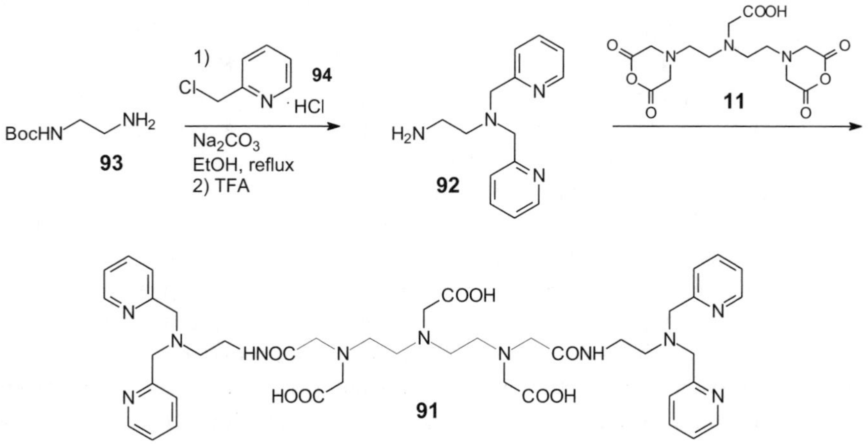

Scheme 5.1.19

More examples of gadolinium complexes selective for Fe^{II},[67] Cu^{II},[68] and other metal ions have been reported.[69]

5.1.3.2.3 pH Responsive Agents

The knowledge that the physiological pH in tumours is somewhat more acidic than that of healthy tissue (typically 6.8 *vs.* 7.4) has prompted researchers to investigate several gadolinium complexes which are able to evaluate pH variations over small tissue volumes due to the high spatial resolution of MRI imaging.[58]

The gadolinium complex of DOTA tetramide **95** represents one of the first examples of a contrast agent responsive to pH.[70] Although showing an unusual pH dependence, the relaxivity of Gd-**95** is monotonic between pH 5.75 (4.5 $mM^{-1} s^{-1}$) and pH 8.0 (3.2 $mM^{-1} s^{-1}$) (25 °C, 20 MHz) and this evidence has been exploited to measure the pH of kidneys in mice by MRI.[71] The synthesis of the ligand is very straightforward and involves alkylation of cyclen with bromoderivative **96**, easily prepared from bromoacetyl bromide and diethyl aminomethylphosphonate **53**. Deprotection of **97** with 30% HBr gives ligand **95** (Scheme 5.1.20).

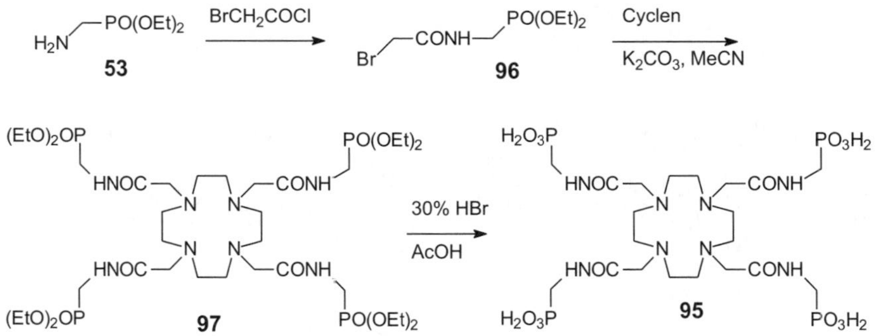

Scheme 5.1.20

5.1.3.3 Nanosized Contrast Agents

5.1.3.3.1 Macromolecular Systems

As already mentioned, most commercially available contrast agents show a pharmacokinetic behaviour characterized by rapid extravasation in the interstitial space and this feature limits their potential application in MR angiography. One of the avenues to limit or to prevent diffusion across vascular walls is to increase the molecular volume of the complex which is correlated to both molecular weight and three-dimensional shape (*i.e.* globular *vs.* monodirectional).

Compound **37** is characterized by a Gd–DOTA like core in which the four acetic arms are modified by insertion of four large and highly hydrophilic moieties.[35,36] Recently, a new synthesis of the ligand **98** (Scheme 5.1.21) has been reported.[72] Benzyl ester **99** is converted into bromo diester **100** which is used for peralkylation of cyclen. Subsequent hydrogenation gives tetra-car-boxy intermediate **101**. The latter is coupled with hydrophilic amine **102**,[73] after activation with HOBt/EDCI of **101**. Final deprotection of the *t*-butyl esters with trifluoroacetic acid gives ligand **98**. The gadolinium complex **37** shows a pretty high molecular weight (6.47 kDa) and a very interesting relaxivity $r_1 = 39$ mM^{-1} s^{-1} (37 °C, 20 MHz), which is more than ten times higher than that of Gd-DOTA, **4**. This reflects the increase in the τ_r of the complex due to slowed tumbling induced by the high molecular weight. The limitation in the diffusion through the endothelium relies on the diameter of the molecule (*i.e.* 50.5 Å).[74]

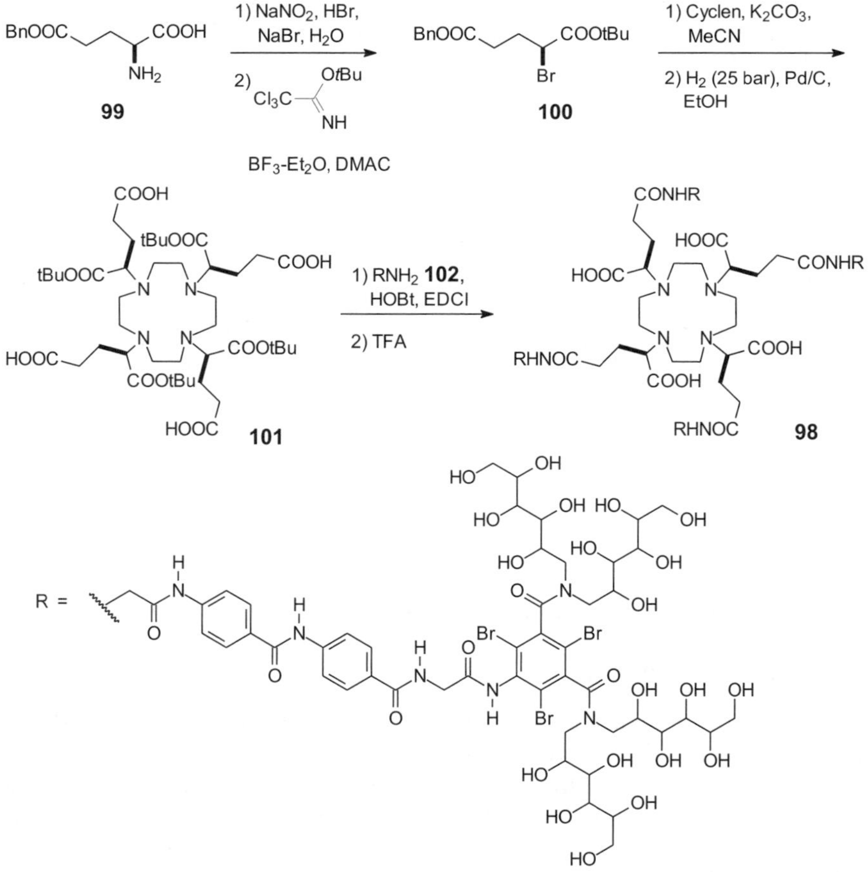

Scheme 5.1.21

In the last couple of decades dendrimers have assumed growing importance as a new class of non conventional polymers and their main features are: (i) the normally defined molecular weight, compared to Gaussian dispersion for conventional polymers; (ii) the almost spherical shape of the molecules at least in case of high degree of replication. The possibility to functionalize the "outer" surface of the molecules, for instance with paramagnetic moieties, allows to exploit the peculiarity of dendrimers in the development of new contrast agents. Gadomer-17 is a dendritic gadolinium complex based on a trimesic acid tria-mide central core to which two generations of lysine (six in the 1st and twelve in the 2nd generation for a total of eighteen Lys moieties) are linked and the external amino groups bind, through a suitable linker, twenty four Gd-DOTA-monoamide units. The synthesis of ligand **103** starts with the building of the 24-amine dendrimer **104** (Scheme 5.1.22)[75] obtained by coupling between trimesic acid triamide **105** and the fully *N*-protected tri-lysine **106** followed by *Z*-deprotection. The ligand **103** is achieved by acylation of peripheral amino groups of **104** with the macrocyclic derivative **107** followed by deprotection of *t*-butyl esters. More recently, a new synthesis of the ligand **103**, based on a slightly different strategy for the synthesis of dendrimer **104**, was reported.[76] The gadolinium complex of **103** (Gadomer-17) is characterized by a very high molecular weight (around 17.5 kDa) and a good relaxivity $r_1 = 17.3 \, \text{mM}^{-1} \, \text{s}^{-1}$ (40 °C, 20 MHz) per Gd^{3+} ion,[77] that corresponds to an overall relaxivity of 415 $\text{mM}^{-1} \, \text{s}^{-1}$. Pharmacokinetic,[77] and MRI studies,[78] demonstrate that Gadomer-17 features promising characteristics as a blood pool agent.

Dendritic contrast agents based on different frameworks like poly-amidoamine (PAMAM) or polypropyleneimine diaminobutane (DAB) have also been proposed as blood pool agents,[79] and the pharmacokinetic behaviour depends on both hydrophilicity and size of the molecule. In particular, only when the diameter ranges between 7 and 12 nm, the contrast agent is efficiently retained in the vascular compartment.

Different approaches to macromolecular contrast agents consist in: (i) the conjugation of suitably functionalized gadolinium complexes to linear poly-mers derived from polyamino acids like polylysine,[80] or polyornithine;[81] (ii) the incorporation in linear copolymers,[82] of gadolinium complexes. As the main drawback, these contrast agents show relaxivities much lower than those expected on the basis of the increased molecular weight. This is due to the extremely high flexibility of the molecules.

5.1.3.3.2 Supramolecular Assemblies

An alternative approach to large size high relaxivity contrast agents exploits the capability shown by amphiphilic molecules of self organization in supramole-cular structures like micelles or liposomes in water. Although commonly used gadolinium chelates are generally very hydrophilic, once suitably modified with the introduction of hydrophobic moieties, they can assume the amphiphilic characteristics required for self-assembling. Amphiphilic complexes showing

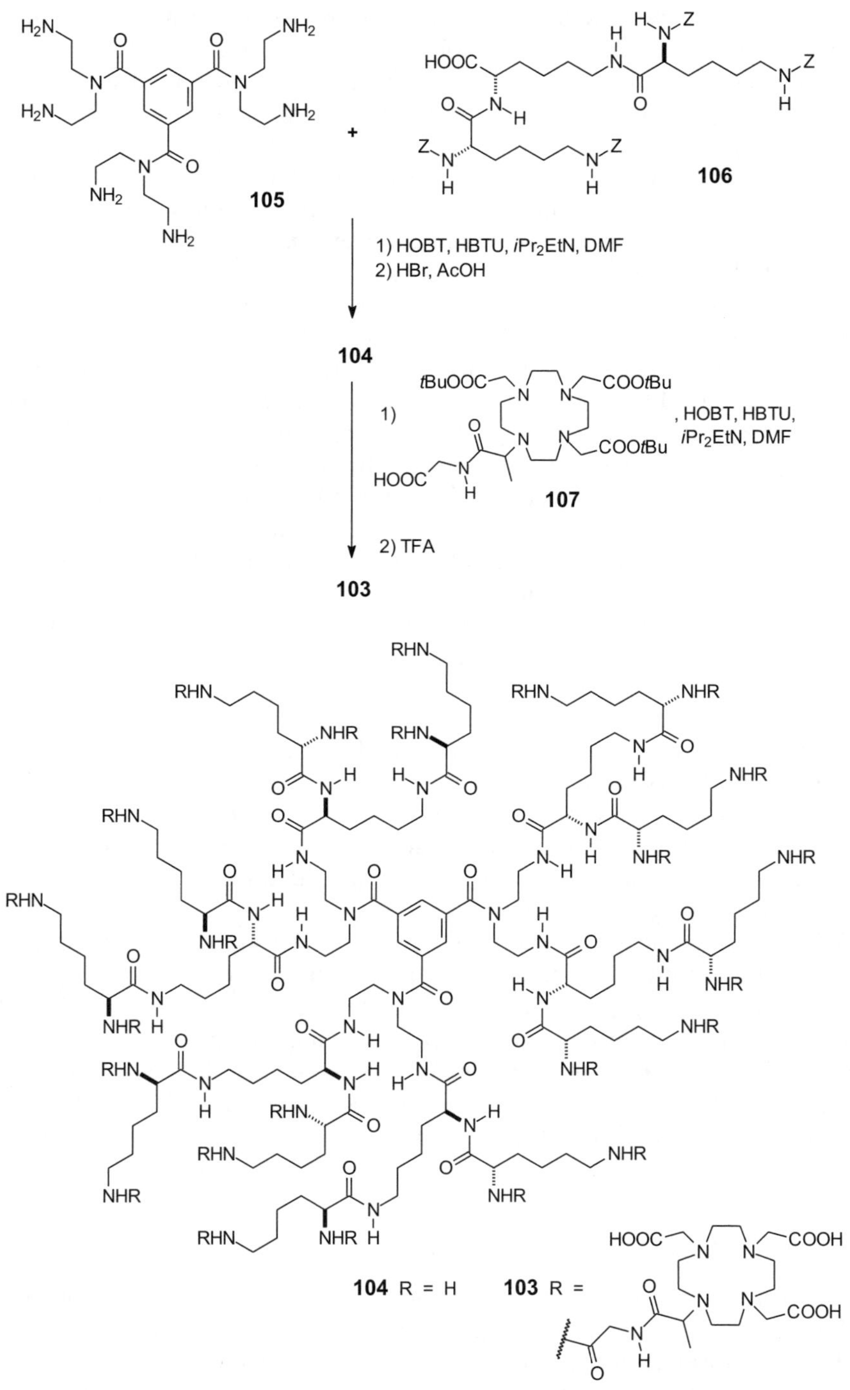

1) HOBT, HBTU, *i*Pr₂EtN, DMF
2) HBr, AcOH

104

1)   , HOBT, HBTU,
                                         *i*Pr₂EtN, DMF

107

2) TFA

103

104 R = H   **103** R =

Scheme 5.1.22

simple coordination cages like DTPA,[83,84] DOTA, **18**,[85,86] PCTA, **47**,[87] linked
to hydrophobic side chains derived from long alkyl chains (C_{12}–C_{18}), fatty acids
(C_{16}–C_{18}) or phospholipids (C_{18}) have been reported. Most of them are able to
self-assemble in single component micelles but more commonly they are formulated with suitable surfactants to form mixed micelles. As the consequence
of both τ_r increase of the supramolecular assembly and restricted mobility of
the single gadolinium complex, micellar contrast agents show very high relaxivities compared to the parent gadolinium chelates with $r_1 = 12$–25 mM^{-1} s^{-1}
(25–$39\,°C$, 20–$60\,MHz$) for $q = 1$ DTPA and DOTA complexes and
$r_1 = 28$–33 mM^{-1} s^{-1} (25–$37\,°C$, $20\,MHz$) for $q = 2$ PCTA complexes. Preliminary MRI and pharmacokinetic studies led to investigate micellar contrast
agents as candidate blood pool agents. However, some concern arises from
clearance studies that revealed the elimination from the body is not complete even
after several days. For a more extensive overview on the use of contrast agents
based on supramolecular assemblies including micelles, liposomes and other
nanoparticles reference is made to specific reviews.[88,89]

5.1.3.3.3 Encapsulated Systems

A relatively new class of contrast agents consists of gadolinium-containing
metallofullerenes or gadofullerenes which feature a Gd^{3+} ion encapsulated in
the almost spherical all-carbon cage of fullerenes.[90] Unlike the previously
discussed gadolinium based agents that are coordination compounds, in
gadofullerenes the metal ion is entrapped in the interior space of the fullerene
cage whose stability prevents any *in vivo* gadolinium release. Gadofullerenes are
conventionally indicated as $Gd@C_{2n}$ where C_{2n} represents the empty fullerenes
of different size, the most important of which is C_{60}. Gadofullerenes are usually
generated by standard direct current arc discharge of Gd_2O_3-impregnated
graphite rods under reduced pressure followed by sublimation to give a mixture
of different $Gd@C_{2n}$ and empty C_{2n} fullerenes. The most abundant insoluble
$Gd@C_{60}$ is purified from soluble empty C_{2n} and higher $Gd@C_{2n}$ by extraction
with solvents. As gadofullerenes are essentially insoluble in water, the deriva-
tisation of the C_{2n} cage with solubilising functional groups is required. Two
different derivatisations have been reported (Scheme 5.1.23). The cycloaddition
of ethyl bromomalonate to C–C double bonds of $Gd@C_{60}$ followed by
hydrolysis of ethyl esters gives water soluble deca-added derivative
$Gd@C_{60}[C(CO_2H)_2]_{10}$.[91] Alternatively, the treatment of $Gd@C_{82}$ with concentrated aq. NaOH in the presence of tetrabutylammonium hydroxide and 15-
crown-5 as catalyst followed by dialysis gives gadofullerenol $Gd@C_{82}(OH)_n$.[92]
A similar process using K_2O in the presence of 18-crown-6 is employed to
produce $Gd@C_{60}(OH)_n$.[93]

Gadofullerenes show pretty high relaxivities, in particular
$Gd@C_{60}[C(CO_2H)_2]_{10}$ and $Gd@C_{60}(OH)_n$ have $r_1 = 24$ and 83 mM^{-1} s^{-1}
($37\,°C$, $60\,MHz$, pH 7.4), respectively, while $Gd@C_{82}(OH)_n$ has $r_1 = 73$ mM^{-1}
s^{-1} ($19\,°C$, $20\,MHz$, pH 7). These values strongly depend on pH and on the

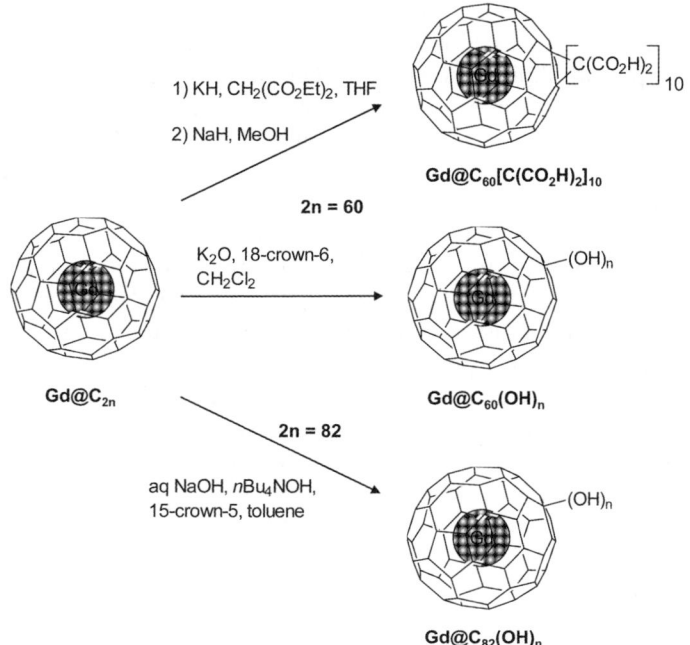

1) KH, CH₂(CO₂Et)₂, THF

2) NaH, MeOH

2n = 60

Gd@C₆₀[C(CO₂H)₂]₁₀

K₂O, 18-crown-6, CH₂Cl₂

Gd@C₆₀(OH)ₙ

Gd@C₂ₙ

2n = 82

aq NaOH, *n*Bu₄NOH, 15-crown-5, toluene

Gd@C₈₂(OH)ₙ

Scheme 5.1.23

concentration of salts in solution. Noticeably, r_1 values are smaller moving from acidic to basic pH values as well as in the presence of electrolytes, especially in case of salts (*e.g.* phosphate) that are able to form hydrogen bonds. In gadofullerenes the Gd^{3+} ion is completely isolated from the external molecules. Therefore, no direct interaction with water molecules is allowed in solution and the classical inner-sphere relaxation mechanism is prevented. The effective relaxation mechanism is still under study but the above mentioned behaviour suggests that different phenomena contribute to relaxivity: (i) proton exchange between bulk water molecules and the external hydroxy or carboxylic functions which are able to link a large number of water molecules through hydrogen bonds; (ii) formation of large clusters in solution which leads to increased rotational times. Preliminary MRI and pharmacokinetic studies demonstrate that gadofullerenes behave like clinically used gadolinium chelates.[91]

Another class of all-carbon three-dimensional materials able to encapsulate Gd^{3+} ions is the so called "single-walled carbon nanotubes" and among them the ultra-short nanotubes which feature a length ranging from 20 to 100 nm.[94] Ultra-short nanotubes are produced as empty structures with a process similar to the one used for fullerenes and loaded with Gd^{3+} ions by sonication of the nanotubes suspended in an aqueous solution of $GdCl_3$ to give the so-called gadonanotubes. Gd^{3+} aquo-ions are not uniformly distributed inside the nanotubes but are present as clusters located nearby the side-wall or end openings through which the loading takes place. Gadonanotubes show a very high relaxivity with r_1 in the range between 160 and 180 mM^{-1} s^{-1} (40 °C, 60 MHz) per Gd^{3+} ion depending on the surfactant employed to solubilise these almost water-insoluble materials. The high relaxivity is compatible with Gd^{3+} ions well confined inside the nanotubes but in the same time easily accessible by bulk water molecules. Although controversial evidences are reported in the literature, empty nanotubes are known to be taken up by cells with low cytotoxicity,[95] and gadonanotubes could represent promising contrast agent candidates for cellular imaging.

5.1.3.4 Agents for innovative MRI approaches

In the last decade, growing interest has been devoted to the development of innovative MRI approaches which are alternative to the conventional proton relaxation based MRI techniques. Examples of such approaches are: (i) relaxation of different nuclei (*e.g.* ^{19}F);[96] (ii) hyperpolarization;[97] (iii) Chemical Exchange Saturation Transfer (CEST).[98,99] For the first two techniques reference is made to fundamental works whereas for the last one some general features are discussed here. The chemical exchange of a nucleus, namely a proton, between two chemically different systems having different environments is at the origin of the CEST phenomenon. The CEST approach relies on the effects caused by pre-saturation of a proton pool A, which is in chemical exchange with a proton pool B. When the exchange rate is not greater than the difference between the chemical shifts of the two proton pools, saturation in proton pool A (proton on hypothetical CEST contrast agents) is transferred to proton pool B (proton of bulk H$_2$O molecules) with consequent decrease of signal intensity of the latter. From a practical point of view this effect may be used to diminish the signal intensity of tissue water, producing contrast, by saturating the proton signal of endogenous diamagnetic molecules like sugars, amino acids, and nucleosides. However, generally either the chemical shift or exchange rate prerequisites are not fulfilled, or the local concentration is too low to ensure the minimum detectable 5% effect on water signal intensity.[98] Some advantages come from the use of paramagnetic lanthanide chelates, the so called PARACEST agents, as the paramagnetic ion induces large chemical shifts in the exchangeable protons surrounding it such as: (i) protons present in the ligand structure like in Dy-DOTAM, **112**;[99] (ii) protons of coordinated H$_2$O molecules like in Eu or Tb-DOTAMGly, **113a,b**;[100] (iii) protons present in molecules interacting with the paramagnetic chelate, like in the supramolecular adduct formed by cationic polyarginine and Tm-DOTP, **114**[101] (Figure 5.1.5). Nevertheless, in all these cases the sensitivity is still below that of the conventional gadolinium based contrast agents.

A great improvement in sensitivity is obtained by encapsulating the lanthanide shift agents inside liposomes. The exchange of H$_2$O molecules through the lipid bilayer is generally slow, even though tuneable to some extent varying the bilayer composition. In such systems, called LIPOCEST agents, the resonance of the H$_2$O protons inside the liposome is shifted from that of external bulk

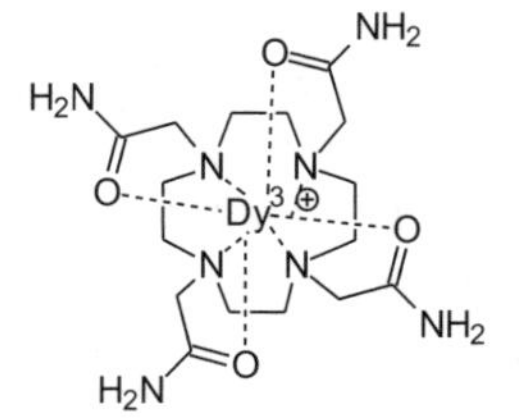

Figure 5.1.5   Lanthanide complexes under study as CEST contrast agents.

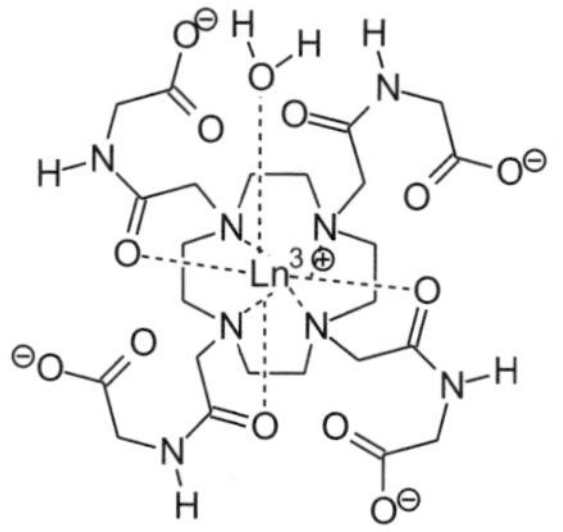

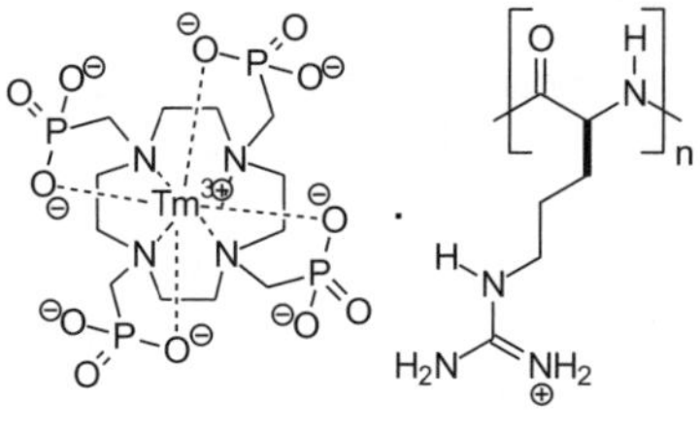

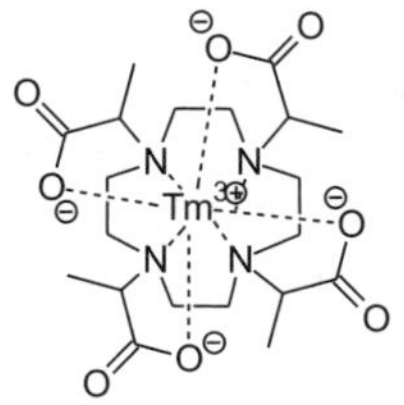

H_2O due to the presence of the entrapped shift reagent. As a consequence two different NMR signals are recorded in the water suspension of such agents thus generating the essential feature for the CEST phenomenon. Suitable candidates as shift reagents are lanthanide chelates, which feature at least one highly shifted and fast exchanging coordinated H_2O molecule like Tm-DOTMA, **115**, (Figure 5.1.5).[102] A dramatic increase in sensitivity is achieved with liposomes having diameters around 270 nm and entrapping 0.1 M Tm-DOTMA, **115**. Using these liposomes the above mentioned minimum detectable 5% effect on water signal intensity is obtained at a concentration as low as 90 pM. LIPO-CEST agents can be further optimized producing unsymmetrical liposomes by osmotic shrinkage or by incorporating amphiphilic lanthanide complexes in the liposomal bilayer wall.[103,104] Recent developments make CEST contrast agents promising candidates for innovative MRI applications.

References

1. R. B. Lauffer, *Chem. Rev.*, 1987, **87**, 901; and references therein.
2. P. C. Lauterbur, M. H. Mendonça Dias, and A. M. Rudin, in *Frontiers of Biological Energetics*, ed. P. L. Dutton, L. S. Leigh and A. Scarpa, Academic Press, New York, 1978, vol. 1.
3. M. Rubin and G. Di Chiro, *Ann. N. Y. Acad. Sci.*, 1959, **78**, 764.
4. A. E. Martell, and R. M. Smith, *Critical Stability Constants, Volume 1: Amino Acids*, Plenum Press, New York, 1974.

5. P. Caravan, J. J. Ellison, T. J. McMurry and R. B. Lauffer, *Chem. Rev.*, 1999, **99**, 2293.
6. S. Aime, M. Botta, and E. Terreno, *Adv. Inorg. Chem.*, 2005, **57**, 173; and references therein.
7. H.-J. Weinmann, R. C. Brasch, W.-R. Press and G. E. Wesbey, *Am. J. Roentgenol.*, 1984, **142**, 619.
8. W. C. Eckelman, S. M. Karesh and R. C. Reba, *J. Pharm. Sci.*, 1975, **64**, 704.
9. C. F. G. C. Geraldes, A. M. Urbano, M. C. Alpoim, A. D. Sherry, K.-T. Kuan, R. Rajagopalan, F. Maton and R. N. Muller, *Magn. Reson. Imaging*, 1995, **13**, 401.
10. C. A. Chang, *Invest. Radiol.*, 1993, **28**, Suppl. 1, S21.
11. M. Periasamy, D. White, L. deLearie, D. Moore, R. Wallace, W. Lin, J. Dunn, W. Hirth, W. Cacheris, G. Pilcher, K. Galen, M. Hynes, M. Bosworth, H. Lin and M. Adams, *Invest. Radiol.*, 1991, **26**, Suppl. 1, S217.
12. J. E. Richman and T. J. Atkins, *J. Am. Chem. Soc.*, 1974, **96**, 2268.
13. V. Jacques, J.-F. Desreux in *The Chemistry of Contrast Agents in Medical Magnetic Resonance Imaging*, ed. A. E. Merbach and É. Tóth, John Wiley and Sons Ltd, Chichester, 2001, pp 157–191.
14. P. S. Athey and G. E. Kiefer, *J. Org. Chem.*, 2002, **67**, 4081.
15. D. P. Reed and G. R. Weisman, *Org. Synth.*, 2002, **78**, 73.
16. M. Argese, G. Ripa, A Scala, and V. Valle, Int. Pat. WO 98/45296, 1998.
17. J. F. Desreux, *Inorg. Chem.*, 1980, **19**, 1319.
18. D. D. Dischino, E. J. Delaney, J. E. Emswiler, G. T. Gaughan, J. S. Prasad, S. K. Srivastava and M. F. Tweedle, *Inorg. Chem.*, 1991, **30**, 1265.
19. J. Platzek, P. Blaszkiewicz, H. Gries, P. Luger, G. Michl, A. Muller-Fahrnow, B. Radüchel and D. Sülzle, *Inorg. Chem.*, 1997, **36**, 6086.
20. F. Uggeri, S. Aime, P. L. Anelli, M. Botta, M. Brocchetta, C. de Haën, G. Ermondi, M. Grandi and P. Paoli, *Inorg. Chem.*, 1995, **34**, 633.
21. H. Schmitt-Willich, M. Brehm, C. L. J. Ewers, G. Michl, A. Müller-Fahrnow, O. Petrov, J. Platzek, B. Radüchel and D. Sülzle, *Inorg. Chem.*, 1999, **38**, 1134.
22. J. C. Amedio Jr., P. J. Bernard, M. Fountain and G. Van Wegenen Jr., *Synth. Commun.*, 1999, **29**, 2377.
23. T. J. McMurry, D. J. Parmelee, H. Sajiki, D. M. Scott, H. S. Ouellet, R. C. Walovitch, Z. Tyeklár, S. Dumas, P. Bernard, S. Nadler, K. Midelfort, M. Greenfield, J. Troughton and R. B. Lauffer, *J. Med. Chem.*, 2002, **45**, 3465.
24. H. Gries and H. Miklautz, *Physiol. Chem. Phys. Med. NMR*, 1984, **16**, 105.
25. M. S. Konings, W. C. Dow, D. B. Love, K. N. Raymond, S. C. Quay and S. M. Rocklage, *Inorg. Chem.*, 1990, **29**, 1488.
26. K. Kumar, C. A. Chang, L. C. Francesconi, D. D. Dischino, M. F. Malley, J. Z. Gougoutas and M. F. Tweedle, *Inorg. Chem.*, 1994, **33**, 3567.

27. P. Hermann, J. Kotec, V. Kubíček, and I. Lukeš, *J. Chem. Soc. Dalton*, 2008, 3027 and references therein.
28. C. de Haën and L. Gozzini, *J. Magn. Reson. Imaging*, 1993, **3**, 179.
29. E. Brücher, *Top. Curr. Chem.*, 2002, **221**, 103.
30. M. Port, J.-M. Idée, C. Medina, C. Robic, M. Sabatou and C. Corot, *Biometals*, 2008, **21**, 469.
31. Z. Zhang, S. A. Nair and T. J. McMurry, *Curr. Med. Chem.*, 2005, **12**, 751.
32. V. Lorusso, L. Pascolo, C. Fernetti, P. L. Anelli, F. Uggeri and C. Tiribelli, *Curr. Pharm. Design*, 2005, **11**, 4079.
33. M.-F. Bellin and A. J. Van Der Molen, *Eur. J. Radiol.*, 2008, **66**, 160.
34. P. L. Anelli, M. Brocchetta, L. Lattuada, G. Manfredi, P. Morosini, M. Murru, D. Palano, M. Sipioni and M. Visigalli, *Org. Process Res. Dev.*, 2009, **13**, 739.
35. M. Port, C. Corot, O. Rousseaux, I. Raynal, L. Devoldere, J.-M. Idée, A. Dencausse, S. Le Greneur, C. Simont and D. Meyer, *Magn. Reson. Mater. Phys., Biol. Med.*, 2001, **12**, 121.
36. J.-C. Pierrard, J. Rimbault, M. Aplincourt, S. Le Greneur and M. Port, *Contrast Media Mol. Imaging*, 2008, **3**, 243.
37. C. Corot, P. Robert, J.-M. Idée and M. Port, *Adv. Drug Delivery Rev.*, 2006, **58**, 1471.
38. S. Laurent, D. Forge, M. Port, A. Roch, C. Robic, L. Vander Elst and R. N. Muller, *Chem. Rev.*, 2008, **108**, 2064.
39. R. N. Muller, L. Vander Elst, A. Roch, J. A. Peters, E. Csajbok, P. Gillis and Y. Gossuin, *Adv. Inorg. Chem.*, 2005, **57**, 239.
40. K. N. Raymond, and V. C. Pierre, *Bioconjugate Chem.*, 2005, **16**, 3 and references therein.
41. J. Xu, S. J. Franklin, D. W. Whisenhunt and K. N. Raymond, *J. Am. Chem. Soc.*, 1995, **117**, 7245.
42. E. J. Werner, S. Avedano, M. Botta, B. P. Hay, E. G. Moore, S. Aime and K. N. Raymond, *J. Am. Chem. Soc.*, 2007, **129**, 1870.
43. S. Aime, L. Calabi, C. Cavallotti, E. Gianolio, G. B. Giovenzana, P. Losi, A. Maiocchi, G. Palmisano and M. Sisti, *Inorg. Chem.*, 2004, **43**, 7588.
44. Z. Baranyai, F. Uggeri, G. B. Giovenzana, A. Bényei, E. Brücher and S. Aime, *Chem. Eur. J.*, 2009, **15**, 1696.
45. H. Stetter, W. Frank and R. Mertens, *Tetrahedron*, 1981, **37**, 767.
46. S. Aime, M. Botta, L. Frullano, S. Geninatti Crich, G. B. Giovenzana, R. Pagliarin, G. Palmisano and M. Sisti, *Chem. Eur. J.*, 1999, **5**, 1253.
47. J.-M. Siaugue, F. Segat-Dioury, A. Favre-Reguillon, C. Madic, J. Foos and A. Guy, *Tetrahedron Lett.*, 2000, **41**, 7443.
48. S. Aime, M. Botta, S. Geninatti Crich, G. B. Giovenzana, G. Jommi, R. Pagliarin and M. Sisti, *Inorg. Chem.*, 1997, **36**, 2992.
49. S. Aime, M. Botta, L. Frullano, S. Geninatti Crich, G. Giovenzana, R. Pagliarin, G. Palmisano, F. Riccardi Sirtori and M. Sisti, *J. Med. Chem.*, 2000, **43**, 4017.

50. H.-S. Chong, K. Garmestani, L. H. Bryant Jr., D. E. Milenic, T. Overstreet, N. Birch, T. Le, E. D. Brady and M. W. Brechbiel, *J. Med. Chem.*, 2006, **49**, 2055.
51. H.-S. Chong, H. A. Song, N. Birch, T. Le, S. Lim and X. Ma, *Bioorg. Med. Chem. Lett.*, 2008, **18**, 3436.
52. H.-S. Chong, K. Garmestani, D. Ma, D. E. Milenic, T. Overstreet and M. W. Brechbiel, *J. Med. Chem.*, 2002, **45**, 3458.
53. H.-S. Chong, H. A. Song, X. Ma, D. E. Milenic, E. D. Brady, S. Lim, H. Lee, K. Baidoo, D. Cheng and M. W. Brechbiel, *Bioconjugate Chem.*, 2008, **19**, 1439.
54. J. L. Sessler, G. Hemmi, T. D. Mody, T. Murai, A. Burrell and S. W. Young, *Acc. Chem. Res.*, 1994, **27**, 43.
55. M. P. Metha, P. Rodrigus, C. H. J. Terhaard, A. Rao, J. Suh, W. Roa, L. Souhami, A. Bezjak, M. Leibenhaut, R. Komaki, C. Schultz, R. Timmerman, W. Curran, J. Smith, S.-C. Phan, R. A. Miller and M. F. Renschler, *J. Clin. Oncol.*, 2003, **21**, 2529.
56. R. Weissleder and U. Mahmood, *Radiology*, 2001, **219**, 316.
57. S. Aime, C. Cabella, S. Colombatto, S. Geninatti-Crich, E. Gianolio and F. Maggioni, *J. Magn. Reson. Imaging*, 2002, **16**, 394.
58. B. Yoo and M. D. Pagel, *Front. Biosci.*, 2008, **13**, 1733.
59. M. P. Lowe, *Curr. Pharm. Biotech.*, 2004, **5**, 519.
60. A. Y. Louie, M. M. Huber, E. T. Ahrens, U. Rothbacher, R. Moats, R. E. Jacobs, S. E. Fraser and T. J. Meade, *Nat. Biotechnol.*, 2000, **18**, 321.
61. P. L. Anelli, F. Fedeli, O. Gazzotti, L. Lattuada, G. Lux and F. Rebasti, *Bioconjugate Chem.*, 1999, **10**, 137.
62. P. L. Anelli, I. Bertini, M. Fragai, L. Lattuada, C. Luchinat and G. Parigi, *Eur. J. Inorg. Chem.*, 2000, 625.
63. R. B. Lauffer, T. J. McMurry, S. O. Dunham, D. M. Scott, D. J. Parmelee, and S. Dumas, PCT Int. Appl. WO 97/36619, 1997.
64. M. Querol, J. W. Chen, R. Weissleder and A. Bogdanov Jr., *Org. Lett.*, 2005, **7**, 1719.
65. W.-H. Li, G. Parigi, M. Fragai, C. Luchinat and T. J. Meade, *Inorg. Chem.*, 2002, **41**, 4018.
66. K. Hanaoka, K. Kikuchi, Y. Urano and T. Nagano, *J. Chem. Soc. Perkin Trans. 2*, 2001, 1840.
67. V. Comblin, D. Gilsoul, M. Hermann, V. Humblet, V. Jacques, M. Mesbahi, C. Sauvage and J. F. Desreux, *Coord. Chem. Rev.*, 1999, **185–186**, 451.
68. E. L. Que and C. J. Chang, *J. Am. Chem. Soc.*, 2006, **128**, 15942.
69. E. L. Que and C. J. Chang, *Chem. Soc. Rev.*, 2010, **39**, 51.
70. S. Zhang, K. Wu and A. D. Sherry, *Angew. Chem. Int. Ed.*, 1999, **38**, 3192.
71. N. Raghunand, S. Zhang, A. D. Sherry and R. J. Gillies, *Academic Radiology*, 2002, **9** Suppl 2, S481.
72. J. C. Pierrard, J. Rimbault, M. Aplincourt, S. Le Greneur and M. Port, *Contrast Media Mol. Imaging*, 2008, **3**, 243.

73. O. Rousseaux and C. Simonot, USP 6827927, 2004.
74. M. Port, C. Corot, I. Raynal, J.-M. Idée, A. Dencausse, E. Lancelot, D. Meyer, B. Bonnemain and J. Lautrou, *Invest. Radiol.*, 2001, **36**, 445.
75. H. Schmitt-Willich, J. Platzek, B. Radüchel, A. Mühler and T. Frenzel USP 5820849, 1998.
76. G. M. Nicolle, É. Tóth, H. Schmitt-Willich, B. Radüchel and A. E. Merbach, *Chem. Eur. J.*, 2002, **8**, 1040.
77. B. Misselwitz, H. Schmitt-Willich, W. Ebert, T. Frenzel and H. J. Weinmann, *Magn. Reson. Mater. Phys., Biol. Med.*, 2001, **12**, 128.
78. S. E. Stiriba, H. Frey and R. Haag, *Angew. Chem. Int. Ed.*, 2002, **41**, 1329.
79. H. Kobayashi and M. W. Brechbiel, *Adv. Drug Delivery Rev.*, 2005, **57**, 2271.
80. P. F. Sieving, A. D. Watson and S. M. Rocklage, *Bioconjugate Chem.*, 1990, **1**, 65.
81. S. Aime, M. Botta, S. Geninatti Crich, G. Giovenzana, G. Palmisano and M. Sisti, *Bioconjugate Chem.*, 1999, **10**, 192.
82. É. Tóth, I. van Uffelen, L. Helm, A. E. Merbach, D. Ladd, K. Briley-Saebo and K. E. Kellar, *Magn. Reson. Chem.*, 1998, **36**, S125.
83. H. Tournier, R. Hyacinthe and M. Schneider, *Academic Radiology*, 2002, **9**, S20.
84. S. Torres, M. I. M. Prata, A. C. Santos, J. P. André, J. A. Martins, L. Helm, É. Tóth, M. L. Garcia-Martin, T. G. Rodrigues, P. Lopez-Larrubia, S. Cérdan and C. F. G. C. Geraldes, *NMR Biomed.*, 2008, **21**, 322.
85. J. P. André, É. Tóth, H. Fischer, A. Seeling, H. R. Mäcke and A. E. Merbach, *Chem. Eur. J.*, 1999, **5**, 2977.
86. P. L. Anelli, L. Lattuada, V. Lorusso, M. Schneider, H. Tournier and F. Uggeri, *Magn. Reson. Mater. Phys., Biol. Med.*, 2001, **12**, 114.
87. C. Ferroud, H. Borderies, E. Lasri, A. Guy and M. Port, *Tetrahedron Lett.*, 2008, **49**, 5972.
88. W. J. M. Mulder, G. J. Strijkers, G. A. F. van Tilborg, A. W. Griffioen and K. Nicolay, *NMR Biomed.*, 2006, **19**, 142.
89. W. J. M. Mulder, A. W. Griffioen, G. J. Strijkers, D. P. Cormode, K. Nicolay and Z. A. Fayad, *Nanomedicine (London U.K.)*, 2007, **2**, 307.
90. R. D. Bolskar, *Nanomedicine (London U.K.)*, 2008, **3**, 201.
91. R. D. Bolskar, A. F. Benedetto, L. O. Husebo, R. E. Price, E. F. Jackson, S. Fallace, L. J. Wilson and J. M. Alford, *J. Am. Chem. Soc.*, 2003, **125**, 5471.
92. H. Kato, Y. Kanazawa, M. Okumura, A. Taninaka, T. Yokawa and H. Shinohara, *J. Am. Chem. Soc.*, 2003, **125**, 4391.
93. É. Tóth, R. D. Bolskar, A. Borel, G. Gonzalez, L. Helm, A. E. Merbach, B. Sitharaman and L. J. Wilson, *J. Am. Chem. Soc.*, 2005, **127**, 799.
94. B. Sitharaman and L. J. Wilson, *Int. J. Nanomed.*, 2006, **1**, 291.
95. N. Wong Shi Kam, T. C. Jessop, P. A. Wender and H. Dai, *J. Am. Chem. Soc.*, 2004, **126**, 6850.
96. J.-X. Yu, V. D. Kodibagkar, W. Cui and R. Mason, *Curr. Med. Chem.*, 2005, **12**, 819.

97. K. Golman, L. E. Olsson, O. Axelsson, S. Mansson, M. Karlsson and J. S. Petersson, *Br. J. Radiol.*, 2003, **76**, S118.
98. K. M. Ward, A. H. Aletras and R. S. Balaban, *J. Magn. Reson.*, 2000, **143**, 79.
99. M. Woods, D. E. Woessner and A. D. Sherry, *Chem. Soc. Rev.*, 2006, **35**, 500.
100. S. Aime, C. Carrera, D. Delli Castelli, S. Geninatti Crich and E. Terreno, *Angew. Chem. Int. Ed.*, 2005, **44**, 1813.
101. S. Aime, D. Delli Castelli and E. Terreno, *Angew. Chem. Int. Ed.*, 2003, **42**, 4257.
102. S. Aime, D. Delli Castelli and E. Terreno, *Angew. Chem. Int. Ed.*, 2005, **44**, 5513.
103. E. Terreno, C. Cabella, C. Carrera, D. Delli Castelli, R. Mazzon, S. Rollet, J. Stancanello, M. Visigalli and S. Aime, *Angew. Chem. Int. Ed.*, 2007, **46**, 966.
104. E. Terreno, D. Delli Castelli, E. Violante, H. M. H. F. Sanders, N. A. J. M. Sommerdijk and S. Aime, *Chem. Eur. J.*, 2009, **15**, 1440.

CHAPTER 5.2

The Future of Biomedical Imaging: Synthesis and Chemical Properties of the DTPA and DOTA Derivative Ligands and Their Complexes

E. BRÜCHER,* ZS. BARANYAI AND GY. TIRCSÓ

University of Debrecen, Department of Inorganic and Analytical Chemistry, Egyetm tér 1, P. O. Box 21, Debrecen, 4010, Hungary

5.2.1 Introduction

The open-chain DTPA, the macrocyclic DOTA, and their derivatives are universal chelating agents, which form high stability complexes with a variety of metal ions. Over the past two decades a number of metal chelates, formed with these ligands, have become highly important in different fields of medical diagnosis and therapy. The metal ions of interest include first of all the lanthanide[III], Sc[III], Y[III], Ga[III] and In[III] ions. The biomedical use of these metal chelates is based on the special magnetic, optical or nuclear properties of the metal ions and the most important applications include magnetic resonance imaging (MRI) contrast agents, diagnostic and therapeutic radiopharmaceuticals, and optical imaging probes. Metal ions can be used in biological systems only in the form of stable complexes that practically do not

RSC Drug Discovery Series No. 15
Biomedical Imaging: The Chemistry of Labels, Probes and Contrast Agents
Edited by Martin Braddock
© Royal Society of Chemistry 2012
Published by the Royal Society of Chemistry, www.rsc.org

dissociate in the body because both the free metal ions and ligands are toxic and the M^{3+}(aq) ions readily hydrolyze at physiological pH. The M(DTPA) or M(DOTA) chelates, which keep at least partly the special properties of the M^{3+} ions, must meet stringent requirements, of which high thermodynamic stability and kinetic inertness are the most important for all fields of application.

In MRI investigations the most frequently used contrast agents are the Gd^{3+} complexes (Scheme 5.2.1).[1,2] The intensity of the MRI signal depends on the proton density in the body (mainly water protons), and on the longitudinal and transverse relaxation rates of protons ($1/T_1$ and $1/T_2$) which may differ in healthy and diseased tissues, leading to differences in the image contrast. This difference can be increased with the use of Gd^{3+} complexes as contrast enhancing agents (CA), which are administered into the body by i.v. injection. The seven unpaired electrons of the Gd^{3+} have a relatively long electronic relaxation time and their fluctuating magnetic fields increase the nuclear relaxation rates of the surrounding protons. In the Gd^{3+} chelates used as CA, all the donor atoms of the octadentate ligands are coordinated. The ninth coordination site of Gd^{3+} is occupied by a water molecule. The protons of this water molecule relax rapidly, and *via* the fast exchange of this inner sphere water with the bulk the relaxation effect of Gd^{3+} is transferred to the surrounding molecules. For expressing the relaxation effect of CAs, the term relaxivity is used. The relaxivity ($1/T_{1,2}$, mM^{-1} s^{-1}) is the increase in the proton relaxation rate, when the concentration of the CA increases by 1 mM. All small molecular weight CAs distribute in the extracellular and intravascular space of the body in about 20–30 min after injection and the hydrophilic complexes are rapidly excreted through the kidneys ($t_{1/2} \approx 1.5$ h). The lipophilic chelates are eliminated partly through the liver and bile. Recent research of CAs is focused on the development of tissue specific agents and the so-called "smart" or responsive CAs, which can be used for the *in vivo* determination of metabolites, endogenous ligands and metal ions, pH, enzyme activity, *etc.*[1–3]

The ligands DTPA, DOTA, and their derivatives are widely used in nuclear medicine for complexation of radiometals. Although the most important diagnostic radioisotope is the ^{99m}Tc, more recently chelates of several other metals, such as ^{47}Sc, ^{67}Cu, ^{67}Ga and ^{111}In are also used in SPECT (Single-Photon Emission Computed Tomogaraphy).[4] The other important diagnostic modality is PET (Positron Emission Tomography), where mainly ^{11}C and ^{18}F isotopes are used, but due to the availability of the ^{68}Ge/^{68}Ga and ^{44}Ti/^{44}Sc generators, there is an increasing interest in the use of ^{68}Ga and ^{44}Sc.[5,6] Several β^--emitting radiometals (^{47}Sc, ^{90}Y, ^{153}Sm, ^{166}Ho, ^{177}Lu) are used in radiotherapy for the local treatment of cancer.[7–10] The metal ions are generally complexed with DTPA and DOTA derivative bifunctional ligands (BFCs), attached to monoclonal antibodies or smaller peptides, which deliver the radiometal to the tumor cells. Similar targeted radiopharmaceuticals, prepared with γ- or positron emitting isotopes (*e.g.* ^{67}Ga, ^{68}Ga, ^{67}Cu, ^{68}Cu, ^{88}Y, ^{99m}Tc, ^{111}In), are used for tumor imaging.[7–9]

The use of Ln^{3+} ions for optical imaging is based on their unique luminescence properties, characterized by the well-separated emission and absorption

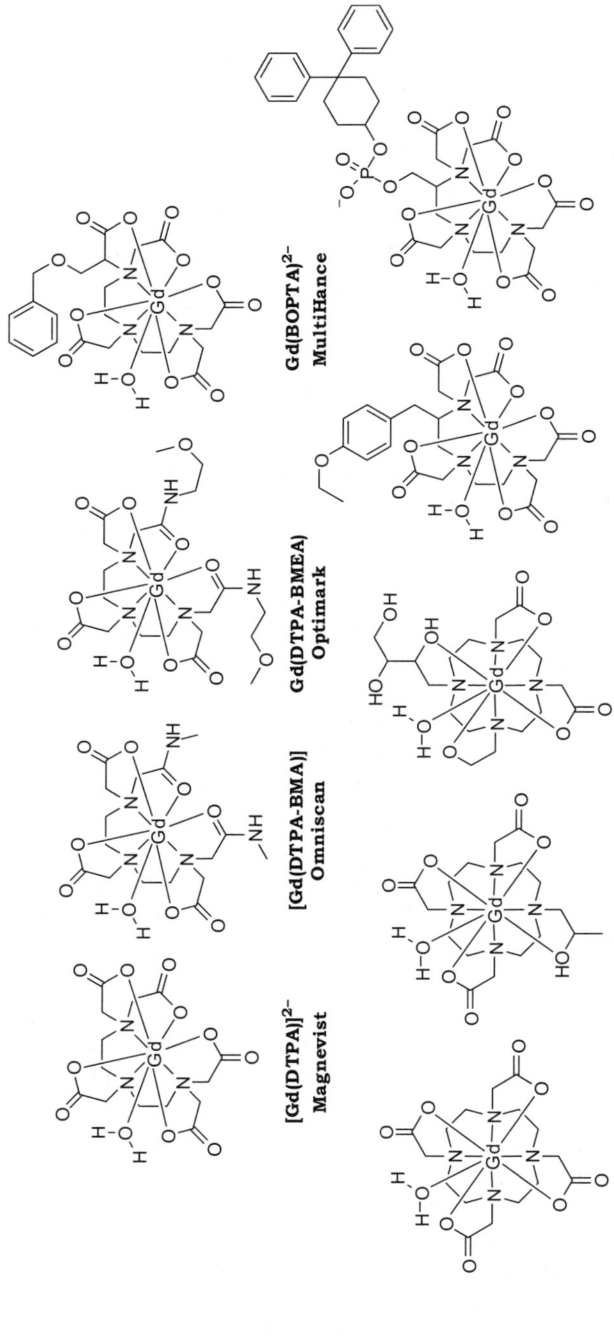

Scheme 5.2.1 Clinically used Gd^{3+} containing contrast agents.

bands, the long excited-state life-times (a few milliseconds for the Eu^{3+} and Tb^{3+}), which allow the use of time-resolved spectroscopy and microscopy. The application of time-delay prior to detection of the emitted light eliminates the interference from the tissue fluorescence. This technique is used for immuno assays, where the detection limit is $10^{-12} - 10^{-15}$ M. For the excitation of Ln^{3+} ions sensitizing chromophores are used (pyridine, bipyridine, terpyridine, triphenylene, quinoline, substituted phenyl and naphthyl, *etc.*) which are capable of transferring their excited state energies to the Ln^{3+} ions. Since the interaction between the chromophores and Ln^{3+} ions is generally weak, the Ln^{3+} ions are bound to the chromophores in the form of chelate complexes. The chelating agents are generally DOTA derivatives, which form highly stable, inert complexes with the Ln^{3+} ions. The optical probes, consisting of Ln^{3+} complexes and sensitizing chromophore, have been attached to cell penetrating peptides for live cell imaging. The structure of the chelating agents has also been modified to generate sensors for bioactive ions.[11–14]

Since in all fields of application listed above, the chelating agents used for metal complexation are DTPA and DOTA derivatives, in this chapter we shall focus on the synthesis of these ligands and the equilibrium and kinetic properties of their lanthanide complexes.

5.2.2 Synthesis of the DTPA and DOTA Derivative Ligands and their Complexes

5.2.2.1 Synthesis of Substituted DTPA and DOTA Derivatives

The advent of Gd^{III} complexes as MRI contrast agents has stimulated worldwide research to develop Gd^{III}-based agents with improved efficacy. The development of new generations of CAs has been achieved through the modification of the basic ligand structures of the two most common contrast agents, $Gd(DTPA)^{2-}$ and $Gd(DOTA)^{-}$. DTPA was first synthesized by *Frost* in 1956 by reacting diethylenetriamine (dien) with formaldehyde and sodium cyanide under alkaline conditions.[15] Since then the cyanomethylation of various polyamines (both cyclic and acyclic) and subsequent hydrolysis of the nitriles has become one of the most important synthetic routes to high purity chelating agents like EDTA or DTPA while DOTA is prepared by reacting cyclen with haloacetic acid sodium salt and NaOH in water.[16–18]

Mannich-type reaction of amines (both dien and 1,4,7,10-tetra-*aza*cyclododecane (cyclen)) with formaldehyde and phosphorous acid in acidic medium seems to be the most convenient route to derivatives with methylenephosphonate pendant arms.[19,20] Dialkyl phosphonates or alkyl phosphinates can be prepared by a similar approach using trialkyl phosphite or alkyl-phosphinic acids (or diethoxymethylphosphine) instead of phosphorous acid. Monoalkyl phosphonates can be obtained by partially hydrolyzing the dialyl phosphates with sodium or potassium hydroxide.[21–23]

Preparation of DTPA-*penta*amides or DOTA-*tetra*amides requires alkylation of the amines, dien and cyclen with the 2-bromo-acetamide or its derivatives.[24–28a] However, the amide bond can also be prepared by converting the DTPA or DOTA to methyl (or ethyl) esters and reacting these esters with the appropriate amines or ammonia (aminolysis). DOTA *tetra*amides were also prepared recently by activating the DOTA acetate arms with various peptide coupling agents (BOP (benzotriazol-1-yloxytris(dimethyloamino)phosphonium hexafluorophosphate) and HBTU (*N*-[(lH-benzotriazol-l-yl) (dimethylamino)-methylene]-*N*-methylmethanaminium hexafluorophosphate *N*-oxide) *etc.*).[28b]

Preparation of functionalized ligands with mixed side arms (*N*-functionalization) has opened new directions in research areas ranging from fundamental coordination chemistry to diagnostic and therapeutic nuclear medicine.[1,2,4,10,29–36] The synthesis of mixed side arm derivatives requires the selective protection of the *N*-atoms in the polyamine. The synthetic methods that have been developed for the selective *N*-functionalization of cyclen have recently been reviewed in detail.[37] Here we will mention only the synthesis of the most important intermediates.

5.2.2.2 Synthesis of the Most Important DTPA Based Intermediates and CAs

The large gap between the protonation constants of the amine nitrogen atoms of certain polyamines (cyclen, piperazine) provides a very convenient access to selectively functionalized derivatives. In these examples, the proton itself acts as a "protecting group" since the protonated nitrogens of the polyamine will not react. The protonation sequence of dien and cyclen differ considerably, which in turn are responsible for the large differences in their functionalization chemistry. The protonation constants of cyclen were found to be: log $K_1^H = 10.65$, log $K_2^H = 9.64$, log $K_3^H = 1.4$, and log K_4^H probably being even lower.[38] The first and the second protonation constants do not differ considerably and this may be responsible for the difficulties observed during mono alkylation (the 1,7-diprotected products are also formed). However the difference between the second and third protonation constants is reasonably high, which can be exploited to selectively obtain 1,7-diprotected cyclen derivatives.[39–41] The protonation constants determined for dien are as follows: log $K_1^H = 9.84$, log $K_2^H = 9.02$, and log $K_3^H = 4.25$. The first proton protonates the central nitrogen and when the second proton protonates one of the terminal nitrogen atoms, the proton from the central nitrogen moves to the other terminal nitrogen in order to achieve better charge separation. The protonation constants are very close and because of the delocalization of the protons pH-controlled protection of the dien is not feasible. However, monoalkylation of terminal nitrogens can be accomplished with a large dien excess. This allows the preparation of mono- and tetra-substituted DTPA based ligands through regioselective alkylation with α-haloacids.[42] The preparation of the ligand BOPTA is a good example of such a protection procedure (Scheme 5.2.2).[43]

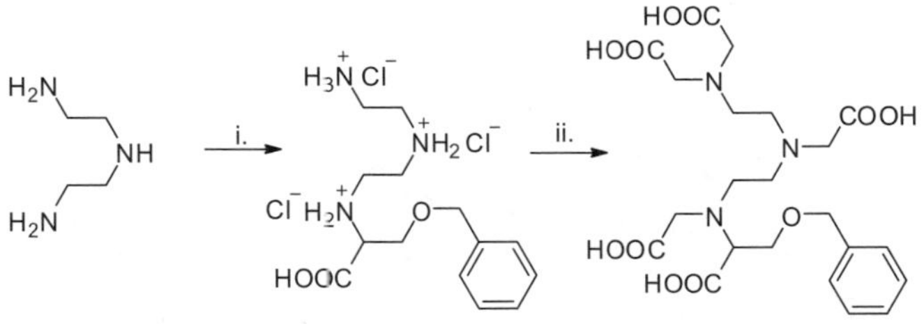

Scheme 5.2.2 Preparation of the BOPTA ligand (i) 2-chloro-3-(phenylmethoxy)-propanoic acid in H_2O followed by ion exchange on Amberlite IRA 400; (ii) bromoacetic acid, NaOH (pH = 10) followed by ion exchange on Amberlite IR 120.

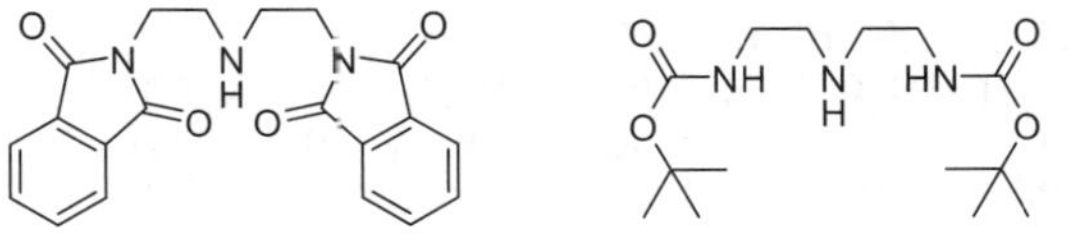

Figure 5.2.1 Structures of Bis(phthaloyl)- and *N,N'*-Boc- protected diethylenetriamine.

DTPA-*N*-monoamides were prepared according to a similar scheme while bis-amides including two ligands used for the preparation of commercial CAs, DTPA-BMA and DTPA-BMEA are obtained from DTPA-bis(anhydride), which is commercially available or can be prepared from DTPA by reacting it with acetic anhydride.[44] For instance, the cyclic 15-DTPA-EAM ligand was prepared using this method.[45] Selective functionalizaton of dien at the central *N*-atom has been achieved through the protection of the two terminal nitrogens with Boc or phthaloyl groups (Figure 5.2.1).[46,47] A large number of DTPA-*N'*-substituted derivatives including DTPA-*N'*-monoamides and some bifunctional ligands were prepared from dien in this way.[48]

The ligand EOB-DTPA, a precursor to the hepatospecific MRI contrast agent Gd(EOB-DTPA)$^{2-}$, has been prepared following two synthetic routes, which differ in the preparation of the chiral triamine 1-*N*-(2-amino-ethyl)-3-(4-ethoxy-phenyl)-propane-1,2-diamine. (Scheme 5.2.3).[49] First, aminolysis of the benzyloxycarbonyl-protected amino acid methyl ester, prepared from tyrosine with an excess of ethylendiamine, was performed. Addition of HCl to the reaction mixture afforded the [1-(2-amino-ethylcarbamoyl)-2-(4-ethoxy-phenyl)-ethyl]-carbamic acid benzyl ester hydrochloride in good yield. Catalytic hydrogenation and reduction of the amide bond resulted in the *C*-derivatized chiral triamine. The second scenario was developed in order to circumvent the use of diborane. The 2-benzyloxycarbonylamino-3-(4-ethoxy-phenyl)-propionic acid methyl ester was reduced to the corresponding alcohol which in turn was transformed to the mesylate. The reaction of the latter compound with an excess of ethylenediamine and subsequent catalytic hydrogenation of the product afforded the key triamine in the form of the crystalline dihydrochloride. Alkylation of the triamine with *t*-butyl bromoacetate in THF/H_2O in the

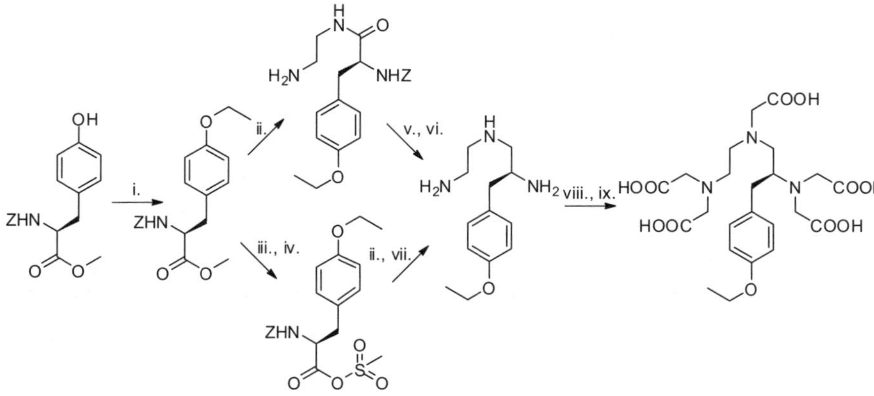

Scheme 5.2.3   Synthesis of the EOB-DTPA ligand (i) EtI, K$_2$CO$_3$, DMF; (ii)
NH$_2$CH$_2$CH$_2$NH$_2$; (iii) NaBH$_4$, THF; (iv) CH$_3$SO$_3$Cl, Et$_3$N, THF; (v)
H$_2$, Pd/C, toluene–aq. KOH; (vi) BH$_3$·THF, THF; (vii) H$_2$, Pd/C,
MeOH; (viii) *tert*-butyl bromoacetate, THF/H$_2$O, K$_2$CO$_3$; (ix) NaOH,
H$_2$O/MeOH reflux, Amberlite IR 120 (H$^+$ form).

presence of K$_2$CO$_3$ afforded the pentaester, which was deprotected in the
presence of NaOH and purified by ion-exchange chromatography.

The backbone of the MS-325 ligand (Figure 5.2.1), which shows strong binding
ability to human serum albumin, was synthesized by a similar procedure.[50]

5.2.2.3   N-functionalization of DOTA Derivative Macrocyclic Ligands

Synthesis of DOTA-based ligands is of great importance in the field of CA
research, because of the high thermodynamic stability and kinetic inertness of
the lanthanide complexes formed with this type of ligand. *N*-derivatization has
been accomplished by either direct derivatization or a protection-derivatiza-
tion-deprotection sequence. Several 1-mono-, 1,7-disubstituted, and 1,4,7-tri-
substituted key intermediates (Figure 5.2.2) were synthesized for the convenient
modification of the basic cyclen skeleton. In some cases more than one procedure has been developed to introduce the same type of functionality/pendant
arm. Hence, it may be difficult to choose the right procedure to solve an actual
synthetic problem.

The two most important monosubstituted cyclen derivatives are *N*-formyl and
N-benzyl cyclen. The synthesis of the former compound is surprisingly easy and
straightforward. The reaction between *N,N*-dimethylformamide dimethylacetal
and cyclen results in a triprotected (tricyclic) compound which can be hydrolyzed
in a mixture of water and alcohol at low temperature, to form quantitatively the
monoprotected *N*-formyl cyclen.[51] Benzyl protected cyclen can be prepared
either by introducing the benzyl protection before the cyclization step (with the
use of 4-benzyl-1,7-ditosyl-1,4,7-diethylenetriamine) or by alkylating cyclen

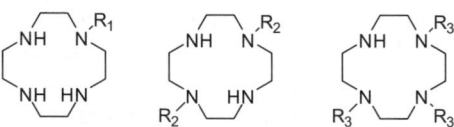

Figure 5.2.2 Structure of 1-, 1,7- and 1,4,7-substitiuted protected cyclens and inter-mediates (R_1 = CHO, $CH_2C_6H_5$ or $COOCH_2C_6H_5$, R_2 = $COOCH_2C_6H_5$ or CH_2SO_3H while R_3 = CHO, CH_2COO^tBu, COO^tBu or in some cases CH_2COOH.

with benzyl–bromide. *Dischino et al.*,[51] have tried several synthetic approaches to this product according to the first scenario, but the desired compound could only be obtained in low yields. Monoalkylation of the cyclen with benzyl–bromide can be achieved in the presence of a large excess of cyclen.[52–54] However, cyclen is relatively expensive and using it in large excess is not economical. Although some of the latest attempts to prepare mono alkylated products by direct alkylation of cyclen have been very promising, the tris-Boc or tris-formyl intermediates are still frequently used to prepare monoalkylated cyclens.[55–58]

In the case of diprotection, the formation of two regioisomers, the 1,4- and 1,7-substituted cyclen derivatives, is expected.[59,60] The 1,7-dialkylated cyclen derivatives are more useful in the synthesis of CA's and two intermediates in particular were found to be of great importance. Mannich-type reaction of cyclen with the bisulfite adduct of formaldehyde affords 1,7-disulfomethylated cyclen. This intermediate can be used to synthesize di- (or even mono-) sub-stituted cyclen derivatives (acetates, phosphonates or phosphinates).[61] On the other hand, 1,7-difunctionalization can be performed through the 1,7-Boc or Cbz protected derivatives. Originally, the 1,7-diprotection of cyclen was achieved by reacting cyclen with various chloroformates under acidic condi-tions (pH ≈ 2 − 3).[62–64] Later, in improved procedures, almost quantitative yields of diprotected (Boc and Cbz) products were obtained when the chlor-oformates were replaced by benzyl- or *tert*-butyl-(oxycarbonyl) succinimides.[39]

Triprotection of the cyclen is frequently performed by using a carbamate-type protecting group. The triprotected intermediates are subject to further func-tionalization followed by removal of the protecting groups. This step can be performed under mild conditions, which makes these carbamate protecting groups very convenient in various synthetic procedures. The tri-*N*-benzylox-ycarbonyl (tri-*N*-Cbz-cyclen) can be obtained by reacting the cyclen with 3.2 equiv. of benzyl chloroformate in the presence of Et_3N (6.4 equiv.) in dichlor-omethane. The synthesis of tri-Boc-cyclen was accomplished according to a similar procedure. Switching from CH_2Cl_2 to $CHCl_3$ as solvent and addition of 3.0 equiv. Et_3N base resulted in almost quantitative yield of the tri-Boc-cyclen.[55,65–67] The tris-formyl protected cyclen was prepared initially in excellent yields by reacting the cyclen with the excess chloral hydrate in hot EtOH.[68] Difficulties associated with the removal of the formyl protecting group (alkaline hydrolysis, oxidative or reductive deprotection) may limit its use, although acid hydrolysis was found to be an efficient route for deprotection.[57,69–71]

The most important tri-alkylated cyclen derivatives are 1,4,7,10-tetra-*aza*cyclododecane-1,4,7-triacetic acid and its tri-*tert*-butyl and tribenzyl esters. Their synthesis involves one-step alkylation of cyclen with an appropriate alky-lating agent.[51,72–74] Some trifunctionalized cyclen derivatives could also be pre-pared without using any kind of protection. The earlier syntheses suffered from low yields and time-consuming column chromatography to separate the di-, tri- and tetra-substituted cyclen derivatives formed under these conditions.[75–77] However, the synthetic procedure was improved recently by using slightly larger excess of the alkylating agents and considerably larger amount of the base (triethylamine).[73]

These important building blocks have been used in CA research (MRI, optical, X-ray and PET) and some of them are now available commercially. Starting from these intermediates and protected cyclen derivatives (Figure 5.2.2) a large number of DOTA derivatives have been synthesized. For instance, derivatiza-tion of DO3A was performed by reacting it with 2-Methyl-oxirane (alkylating agent) in aqueous NaOH which resulted in a derivative known as HP-DO3A, whose Gd[III]-complex is currently used as a CA in MRI (Scheme 5.2.4).[51]

The ligand BT-DO3A, used for preparation of the nonionic contrast agent Gd(BT-DO3A), was synthesized starting from a triprotected tricyclic inter-mediate as a nucleophile in a ring opening reaction with 4,4-dimethyl-3,5,8-trioxabicyclo[5,1,0]octane. The resulting product was hydrolyzed in aqueous MeOH and the formyl protection was removed subsequently in aqueous KOH. The triacid was obtained by reacting the compound obtained in the previous step, with chloroacetic acid in water at pH $= 9 - 10$, followed by hydrolysis to give BT-DO3A (Scheme 5.2.5).[78,79]

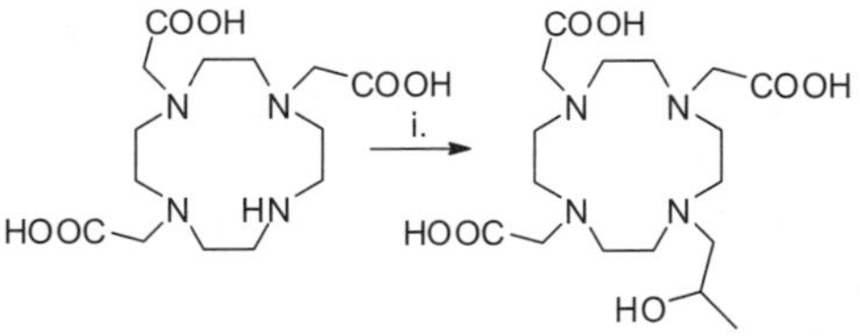

Scheme 5.2.4 Synthesis of the HP-DO3A (i) 2-Methyl-oxirane, NaOH/H$_2$O than anion-exchange chromatography.

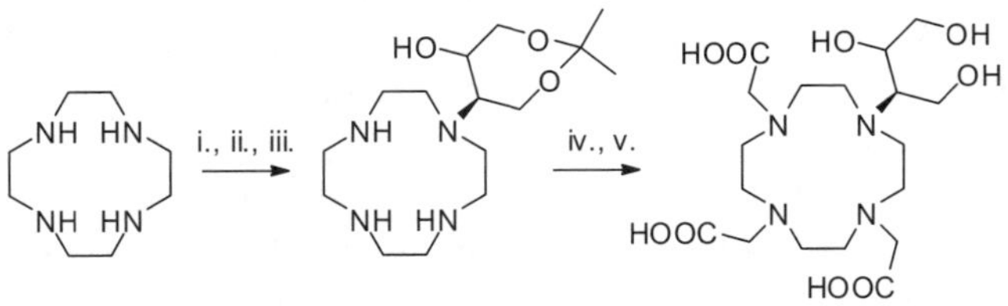

Scheme 5.2.5 Synthesis of the BT-DO3A ligand (i) *N,N*-dimethylformamide dimethyl acetal and 4,4-dimethyl-3,5,8-trioxabicyclo[5,1,0]octane, 120 °C; (ii) H$_2$O/MeOH, 20 °C; (iii) KOH, H$_2$O/MeOH, 80 °C; (iv) ClCH$_2$COOH, NaOH, H$_2$O 60 °C, pH $= 9 - 10$; (v) HCl and ion-exchange.

Ligands representing stepwise replacement of the acetates by phosphonate pendant arms have recently been synthesized and studied. The synthesis was performed either from DO3A by reacting it with diethyl phosphite in the presence of paraformaldehyde (DO3AP)[80] or by reacting the 1,7-DO2A-*tert*-butyl ester with triethyl phosphite in the presence of paraformaldehyde (DO2A2P).[81] The DO3P and DOA3P ligands were prepared with the use of similar procedures and conditions and studied recently.[82,83]

Pyclen is an interesting cyclen derivative in which a pyridine ring is fused to the tetra*aza*cyclododecane macrocycle. Pyclen was first isolated by *Stetter*,[84] and its triacetate derivative, PCTA, was later synthesized by *Aime et al.*,[85,86] by alkylating the macrocycle with chloroacetic acid in the presence of sodium carbonate. The free acid form of PCTA was obtained after acidifying the reaction mixture. Ln complexes of PCTA have reasonably high thermodynamic stability and satisfactory kinetic inertness for biomedical applications.[85,87] These chelates contain two inner sphere water molecules, which cannot be replaced by bioligands. In addition, the pyridine chromophore can act as an antenna for Ln sensitization in the UV-vis (Nd Eu, Tb and Yb). Several other pyclen based ligands, including the monophosphonate and a bifunctional ligand derived from the PCTA, have also been prepared and studied.[88,89] The synthesis of PCTA derivatives has recently been reviewed in detail by Guy *et al.*[90,91]

5.2.2.4 Structure and Synthesis of Bifunctional Ligands Derived from DTPA and DOTA

The term "bifunctional chelating agent" (BFC) has been used to denote ligands that can perform two different functions: they can sequester a therapeutic or diagnostic metal ion in a thermodynamically stable and kinetically inert complex, and besides metal binding they also possess a reactive functional group (anchor) that can be used for the covalent attachment of the complex to a targeting vector (peptides, proteins, dendrimers, nano-particles, or even another complex). The targeting vector is attached to the chelate through a spacer that is often an aliphatic or aromatic moiety (Figure 5.2.3).

The targeting molecules may contain different groups available for conjugation, hence the bifunctional ligands synthesized and investigated to date can be very different as far as their reactive groups are concerned. BFCs may be

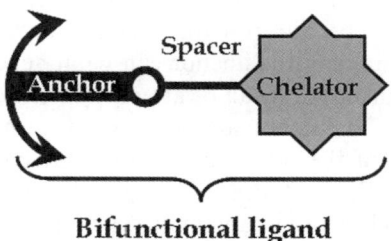

Bifunctional ligand

Figure 5.2.3   Schematic representation of a bifunctional ligand.

classified by the functional groups of the targeting vector they will react with. The most frequently used functionalities of biomolecules for bioconjugation include amines (*e.g.* lysine or other reactive amines), carboxylic acids (glutamic acid, aspartic acid or other carboxylic acids), thiols (cysteine or other reactive thiols) or alcohols.

Vectors containing carboxylate groups may be derivatized by converting them into amides (any amide bond forming reaction) or through an active ester such as hydroxysuccinimide active ester (NHS-ester) or reactive carbonyl intermediates (with the use of carbodiimide coupling agents such as EDC (1-ethyl-3-(3-dimethylaminopropyl)carbodiimide hydrochloride), DCC (dicyclohexyl carbodiimide) or DIC (diisopropyl carbodiimide). In these reactions the carboxylate group acts as an acylating agent toward the group to be modified. The activated intermediates can be reacted with alcohols (the product of the reaction will be an ester), thiols (the product will be a thioester), amines (an amide bond will be formed) or even hydrazide derivatives.[92]

For bioconjugation purposes the modification of alcohols is less important than the modification of any other functional groups discussed earlier, but at the same time not impossible, as these derivatives can be alkylated or acylated quite easily. In addition, the aromatic alcohols (phenols) may be targeted by Mannich condensation reactions or electrophilic reagents such as iodine or diazonium ions.

The maleimide group reacts specifically and efficiently with the sulfhydryl group (–SH moiety) of the biomolecule in the Michael-type addition reaction and has been widely used to form protein conjugates through thioether linkage. Conjugation reactions involving sulfhydryl groups may also involve simple alkylation or acylation reactions to form stable thioethers and relatively unstable thioesters, respectively.

Amines can be derivatized quite easily using alkylation and acylation reactions. It is worth noting that BFCs with an aryl isothiocyanate group can also be used to achieve conjugation to primary and secondary amines. Although several functional groups (carboxylic acids, OH, SH, amines) available for bioconjugation are discussed above, it should be emphasized that DOTA and DTPA complexes are almost exclusively conjugated to peptides and proteins through an amino group and the most commonly used bifunctional ligands are the isothiocyanato derivatives.

The site where the spacer and the anchor is attached to the bifunctional ligand seems to be an important issue, as it may slightly affect the complexation and other physico-chemical properties of the chelates formed by the bifunctional ligand. The spacer containing the anchor can be attached to the backbone of the ligand (backbone linked) or it can be attached to one of the side arms on the ligand (usually to the alpha carbon atom of the side arm – side arm linked). In the third scenario one of the side arms used for the complexation acts as an anchor (Figure 5.2.4). For instance, bioconjugation through one of the carboxylate groups often gives rise to an amide functionality, which is still capable of coordinating to metal ions. Among these options the last two are more cost effective than the first one and therefore, they are more frequently used.

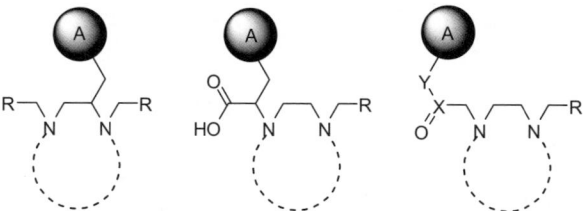

Figure 5.2.4 Places available for the attachment of the "anchor" to the chelator (A denotes anchor while R is a group with a donor atom capable of coordination, X can be a C atom (Y = NH) or P-OH moiety (Y = CH$_2$)).

In the past three decades there has been an increasing demand for bifunctional ligands, particularly for nuclear medicine applications. Several bifunctional versions of well-known chelating scaffolds, such as EDTA, DTPA and DOTA, have been synthesized and studied in detail. These ligands offer polyaminocarboxylate type binding for a variety of metal ions with donor atoms ranging from 6 (EDTA) to 8 (DTPA and DOTA) and they also provide acyclic (EDTA and DTPA) or macrocyclic (DOTA) encapsulation for the selected metal ion.

BFCs based on EDTA ligands were first synthesized by *Meares* and co-workers,[93–95] however the EDTA based BFCs do not offer enough donor atoms for complexation of metal ions with coordination number (CN) larger than six (Y^{III}, In^{III} and lanthanide^{III} ions). As a result, DTPA based BFCs were designed and synthesized because they provide a better match between the number of donor binding groups and the CN for larger ions such as Y^{III}, In^{III} and Ln^{III} ions. The structure of the most important DTPA based BFCs is shown in Figure 5.2.5.

DTPA-bis anhydride is often considered as the first DTPA based bifunctional ligand, and it was frequently used for bioconjugation in the early days of MRI contrast agent research. Depending upon the molar ratio of the amines selected for the reaction, one or two amide bonds will be present in the product as a result of the conjugation. This can be a disadvantage, since the cross-linking may considerably affect the biological activity of the conjugate. In addition, the thermodynamic stability (and probably the kinetic inertness) of the resultant DTPA-mono- and bis- amide complexes will be significantly lower than those of the corresponding DTPA chelates.[96–101]

Considerable improvement in the stability of the chelates was achieved by *Brechbiel et al.*,[102] who have reported the synthesis of 1-(*p*-isothiocyanatobenzyl)-DTPA (1B-DTPA), a bifunctional ligand that can be conjugated to the biomolecule through a thiourea linkage. Inclusion of the *p*-isothiocyanatobenzyl group not only provides a reactive functionality for bioconjugation but the backbone substitution increases the stability of the resulting complexes. An isomeric form of 1B-DTPA in which the protein-linking functional group is attached to one of the terminal carboxylate arms was synthesized somewhat later by *Westberg* and *Keana*.[103–104] The synthetic

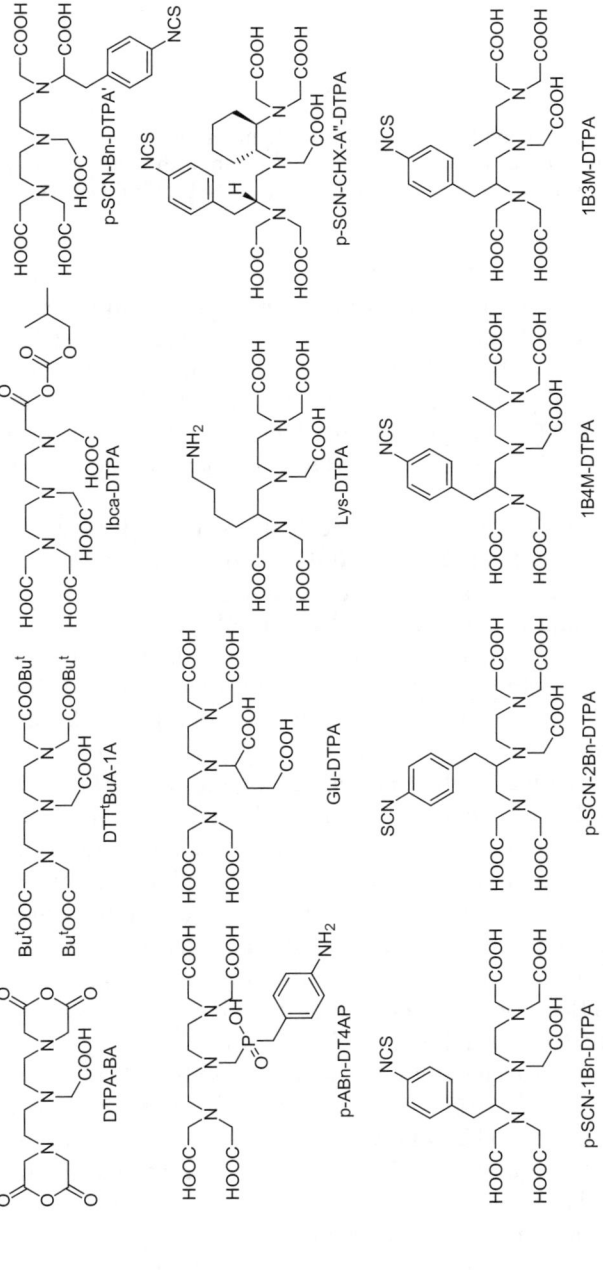

Figure 5.2.5 Selection of **DTPA** based **BFC** ligands.

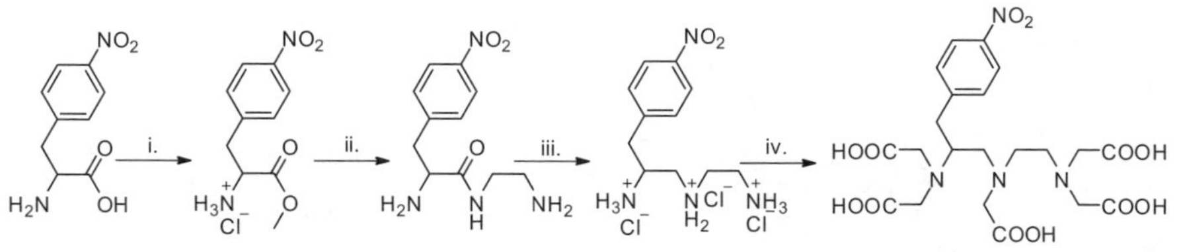

Scheme 5.2.6 Synthesis of the *p*-NO$_2$-Bn-DTPA ligand (i) HCl/MeOH; (ii) Et$_3$N, MeOH, NH$_2$CH$_2$CH$_2$NH$_2$; (iii) BH$_3$·THF in THF than HCl in EtOH; (iv) BrCH$_2$COOH/C$_6$H$_5$CH$_3$, KOH/H$_2$O than ion-exchange on AG50W-X8 followed by HPLC.

procedure developed by *Brechbiel et al.*,[102] is shown in Scheme 5.2.6 and it is used very often when backbone modified (*C*-functionalized) DTPA derivatives are synthesized. Using this procedure, *p*-nitrophenylalanine was esterified to give methyl ester hydrochloride. The free base was liberated with a triethylamine (TEA) and reacted with ethylene diamine in methanol to generate *N*-(2-aminoethyl)-p-nitrophenylalanine amide. Reduction of the amide to (*p*-nitrobenzyl)diethylenetriamine (*p*-NO$_2$-Bn-dien) was accomplished using excess borane–THF complex. The product was isolated as the hydrochloride salt after decomposing the borane adduct in refluxing ethanol saturated with HCl gas. Alkylation of the amine under standard conditions with bromo-acetic acid and aqueous KOH resulted in 1-(*p*-nitrobenzyl)diethylene-triaminepentaacetic acid (*p*-NO$_2$-Bn-DTPA), which was purified by ion exchange and HPLC chromatographic techniques (note: the *p*-NO$_2$-Bn-DTPA is the precursor of *p*-NCS-Bn-DTPA.). Reduction of the nitro group with Pd/C under basic conditions afforded the amino derivative *p*-NH$_2$-Bn-DTPA, which was converted to isothiocyanate (BFC) with thiophosgene (these steps are not indicated in the scheme).

Considerable improvements in the thermodynamic and *in vivo* stability of the complexes were achieved by replacing one of the flexible ethane-1,2-diamine units of DTPA with the more rigid propane-1,2-diamine (1B4M-DTPA) or cyclohexane-1,2-diamine (CHX-A or CHX-B ligands) units.[105,106] Synthesis of the 1B4M-DTPA was performed following a route that is very similar to the one reported for *p*-NO$_2$-Bn-DTPA (Scheme 5.2.6), but propane-1,2-diamine was used instead of ethylene diamine.[107] The first step in synthesis of *p*-NO$_2$-Bn-CHX-A-DTPA (Scheme 5.2.7) involved the reaction carbamate-protected (−)-*p*-nitrophenylalanine with *N*-benzyloxycarbonyl-(*R*,*R*)-*trans*-cyclohexane-1,2-diamine using a carbodiimide coupling agent. The carbamate protecting groups were removed with HBr in acetic acid and the resulting amide was reduced with borane–THF complex. Alkylation of the triamine (*p*-NO$_2$-Bn-CHX-A-triene) was achieved with *tert*-butyl bromoacetate. Treatment of the penta-ester with trifluoroacetic acid resulted in the BFC precursor. HPLC purification of the crude product afforded an analytically pure sample. This synthetic procedure was reported to give better yields and was used to prepare diastereoisomers of 1B4M-DTPA.[108]

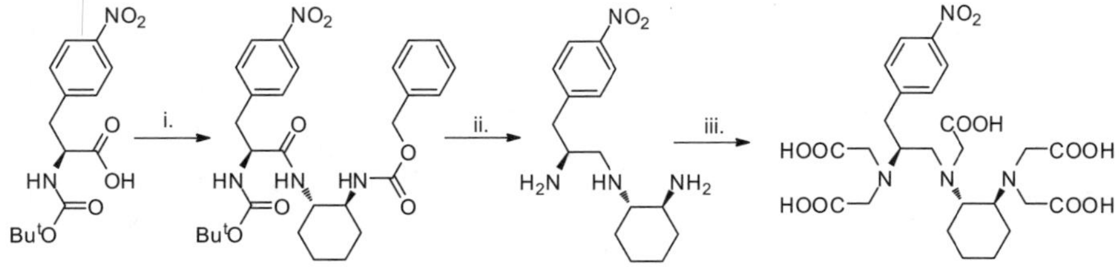

Scheme 5.2.7 Synthesis of the *p*-NO₂-Bn-CHX-A''-DTPA bifunctional ligand pre-
cursor (i) *N*-benzyloxycarbonyl-cyclohexane-1,2-diamine, HOBT, EDC,
EtOAc, DMF, 25 °C; (ii) HBr, AcOH; (iii) BH₃ · THF, THF, 50°C than
HCl in dioxane; (iv) *tert*-butyl bromoacetate, Na₂CO₃, DMF; 80°C; (v)
TFA, r.t. than HPLC and ion-exchange chromatography.

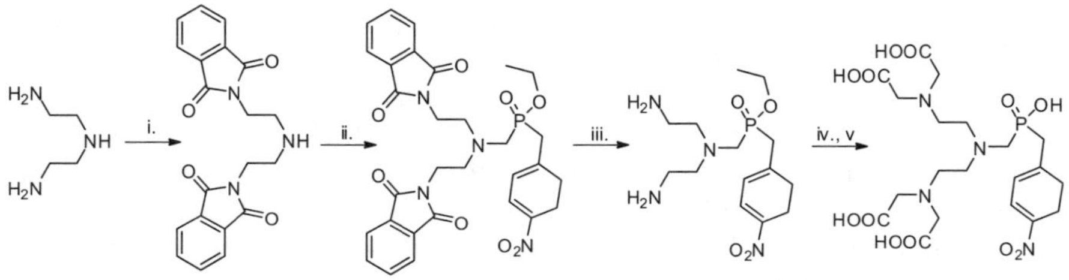

Scheme 5.2.8 Synthesis of the *p*-NO₂-Bn-DT4AP BFC precursor (i) phtalic anhydride, toluene, reflux; (ii) ethyl *p*-nitrobenzylphosphinate, (CH₂O)*n*,
toluene, reflux; (iii) N₂H₄ · H₂O, EtOH, reflux; (iv) ethyl bromoacetate,
K₂CO₃, DMF, rt; (v) aq. HCl, reflux).

From a synthetic chemical point of view, modification through the central
nitrogen of the 1-*N*-(2-amino-ethyl)-ethane-1,2-diamine backbone is also
interesting, since this approach provides access to a large number of side arm
linked BFC's. The synthesis requires simultaneous blocking of both of the
terminal nitrogens by protecting them with phthalic anhydride or Boc protecting group.[109,110] For example, the phthaloyl-protected diethylenetriamine
was reacted with *p*-nitrobenzyl-phosphinic acid ethyl ester in a Mannich-type
reaction to give the ethyl *N,N''*-bis(phthaloyl)diethylenetriamine-*N'*-methyle-
ne(*p*-nitrobenzyl)phosphinate intermediate. Removal of the phthaloyl protective groups with hydrazine hydrate followed by alkylation of the resultant
diamine with ethyl bromoacetate gave an ester intermediate that was converted
in one hydrolysis step to the BFC precursor, *p*-NO₂-Bn-DT4AP (Scheme 5.2.8).

Despite the successes achieved with acyclic DTPA based BFC's, the macrocyclic ligands have been shown to form more stable complexes. The insufficient kinetic inertness of chelates formed with radionuclides could be
potentially harmful to the body. To overcome this disadvantage, a large
number of macrocyclic BFCs were designed and synthesized. A selection of
DOTA based BFCs to complex radionuclides such as [86/90]Y, ^{111}In, the radiolanthanides, [212]Pb, [213]Bi, and [225]Ac is shown in Figure 5.2.6.

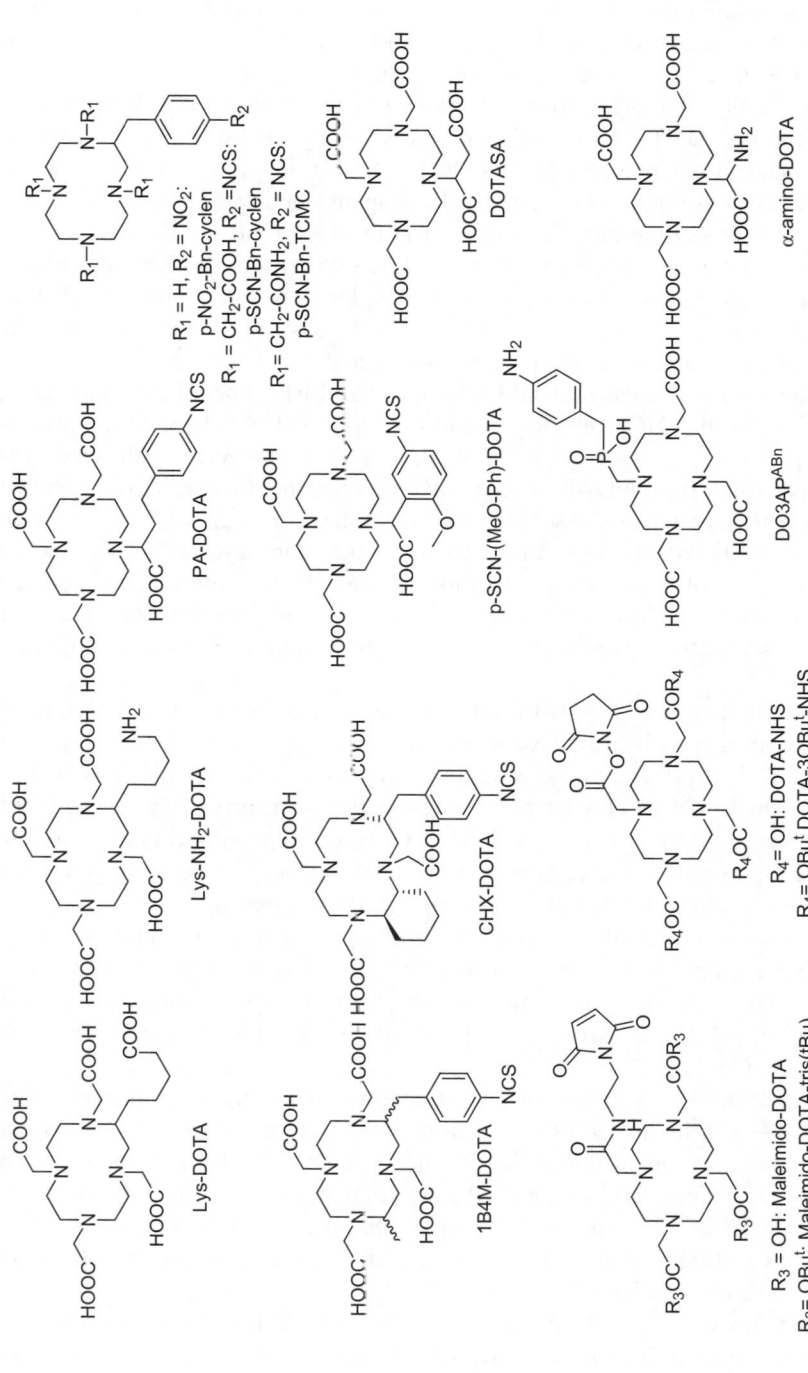

Figure 5.2.6 DOTA based BFC ligands (1B4M-DOTA201, CHX-DOTA201 and α-amino-DOTA202 ligands are not discussed in detail in the text).

The side arm linked BFCs can be prepared either by alkylating DO3A (or DO3A-tris-*tert*-butyl ester) or cyclen with the reactive functional group precursor containing the appropriate side arm (usually an alpha-halo carboxylic acid derivative). As discussed earlier, mono substitution of cyclen can be achieved using a large excess of the tetraamine).

In the family of DOTA BFCs the Lys- and Lys-NH_2-DOTA ligands were among the first DOTA based BFCs prepared, but these are rarely used nowadays.[111] The commercially available DOTA-NHS ester is increasingly used for side arm conjugation, however, it should be mentioned that the product of the conjugation involving this BFC will be a DOTA-mono amide type ligand. Substitution of an amide for one of the acetate side arms in DOTA resulted in a slight decrease of the stability constant of complexes. In spite of the successively lower thermodynamic stability of DOTA-mono-, bis- and tetra- amides, the kinetic inertness of their complexes is not affected negatively.[112–115]

The key step in the synthesis of backbone linked BFC chelating agents based on DOTA (both p-NO_2-Bn-DOTA and p-NO_2-Bn-TCMC ligands) is the formation of their precursor cyclic tetraamine, p-NO_2-Bn-cyclen. Published synthetic procedures for p-NO_2-Bn-cyclen can be divided into three groups: 1) peptide based synthesis, which often uses high dilution cyclization methods;[116–119] 2) use of a Richman-Atkins type cyclization to form the macrocycle;[116,120,121] and 3) a combination of the first two methods, in which the intermediate for the Richman-Atkins cyclization is obtained from a peptide precursor. The first published synthesis of p-NO_2-Bn-cyclen is a good example of this combination approach.[122]

The synthesis of the p-NO_2-Bn-DOTA is accomplished by the alkylation of p-NO_2-Bn-cyclen with bromoacetic acid or its *tert*-butyl ester followed by the subsequent hydrolysis of the resultant tetra-ester. The ligand p-NO_2-Bn-TCMC, which is a DOTA-tetraamide based BFC precursor, was synthesized by reacting p-NO_2-Bn-cyclen with an excess of 2-bromo-acetamide in the presence of excess triethylamine base in acetonitrile.[123] The isothiocyanato functionality was obtained by reducing the corresponding nitro derivatives with hydrogen over Pd/C catalyst and reacting the resulting aniline (p-NH_2-Bn-TCMC) with thiophosgene. The BFC ligand derived from p-NO_2-Bn-TCMC was suggested to sequester $^{203/212}$Pb isotopes, since the parent DOTAM ligand forms extremely stable complexes with heavy metal ions such as Pb^{2+} or Cd^{2+}.[27,124]

Owing to the slow complex formation observed under mild conditions (25 °C and pH $= 4 - 5$) the use of DOTA and its derivatives in radioimmuno-therapy or -imaging may be problematic. In the past couple of years, new ligand scaffolds were designed, synthesized and studied to overcome this problem (Figure 5.2.7).[87,125–127] This is a rapidly developing research area, as the BFC versions of virtually all ligand structures with documented fast complexation kinetics have already been synthesized.[128–131]

The synthesis of a PCTA based BFC precursor with improved complexation kinetics was accomplished by a Richman-Atkins type cyclization as shown in Scheme 5.2.9. The p-toluenesulfonyl protected tritosyl-nitrobenzyl-pyclen was

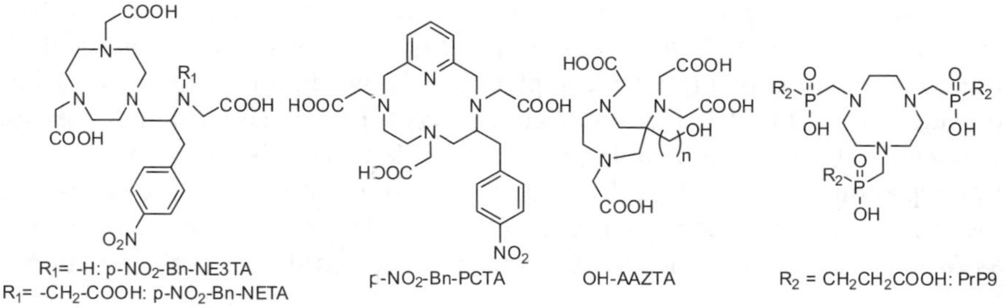

Figure 5.2.7 BFC ligands with advanced complex formation kinetics (OH-AAZTA ligand structures with n = 1,[203,204] and n = 2,[205] were prepared recently).

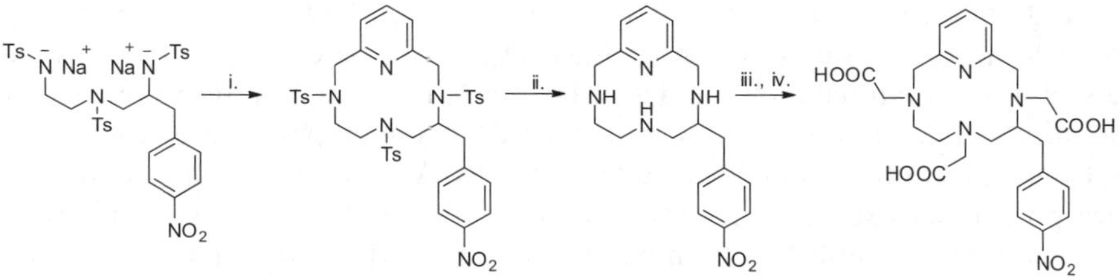

Scheme 5.2.9 Synthesis of *p*-NO$_2$-Bn-PCTA, BFC precursor with accelerated complexation kinetics (i) 2,6-bis(chloromethyl)pyridine, DMF, 55 °C; (ii) H$_2$SO$_4$, 180 °C; (iii) *tert*-butyl bromoacetate, anhydrous K$_2$CO$_3$, MeCN; (iv) HCl. H$_2$O than HPLC.

obtained by reacting the disodium salt of *N,N',N''*-tritosyl-(*S*)-2-(*p*-nitrobenzyl)-diethylenetriamine with 2,6-bis(chloromethyl)pyridine in DMF. The tosyl groups of the product were removed in concentrated sulfuric acid at elevated temperatures and the free amine was alkylated with *tert*-butyl bromoacetate. The resulting triester was hydrolyzed in aqueous hydrochloric acid to afford the final product as an HCl salt. An analytically pure sample of *p*-NO$_2$-Bn-PCTA was obtained by HPLC purification.[131]

5.2.2.5 Synthesis of the Complexes

The general method of preparing the complexes depends on the form of the ligand that is accessible for the complexation (in acidic (protonated H$_x$L) or basic (deprotonated L) form). It should be noted that the kinetics of complexation of Ln^{III} ions with open chain ligands (DTPA and DTPA based ligands) is nearly instantaneous in solution while it can be exceedingly slow for DOTA derivatives. Hence, complexations with DOTA derivatives are often carried out at elevated temperatures and the base used to neutralize the protons is added slowly, over a period of several hours.

Preparation of the complexes is often performed in a neutralization reaction. The complexation occurs between two reactants – the protonated ligand and a base such as metal hydroxides ($Ln(OH)_3$), carbonates ($Ln_2(CO_3)_3$) or base anhydrides such as lanthanide[III]-oxides (Ln_2O_3). These reactions are particularly preferred because besides the complexes, only water, or water and carbon dioxide, are produced. Generally, slight excess of the oxide is used and its excess is filtered off when the complexation is complete. The diprotonated complex (monoprotonated for DOTA derivatives) formed in this reaction is converted to the final complex with strong bases (NaOH or KOH) or weak organic bases such as *N*-methyl-D-glucosamine (NMG), which is often used for CA formulation.

$$LN_2O_3 + 2H_5DTPA = 2H_2\{Ln(DTPA)\} + 3H_2O \qquad (5.2.1)$$

$$\{H_2(LnDTPA)\} + 2NMG = \{H(NMG)^+\}_2 \cdot \{Ln(DTPA)\}^{2-} \qquad (5.2.2)$$

Hydrated chlorides ($LnCl_3 \cdot 6H_2O$), nitrates ($Ln(NO_3)_3 \cdot 6H_2O$) or rarely, perchlorates (aqueous solution of $Ln(ClO_4)_3$), can be easily prepared by the dissolution of lanthanide[III]-oxides in HCl, HNO_3 or $HClO_4$, respectively, and solutions obtained this way are widely used for the preparation of complexes. Extreme precautions should be taken when working with perchlorates as they are potentially explosive materials. Preparation of the complexes with these salts can be accomplished by mixing solutions of the lanthanide salt and the ligand in the acidic form followed by raising the pH to about 6.0 ($pH = 9 - 10$ for phosphonate ligands such as DOTP) to scavenge the protons released during the complex formation.[132] As a result, the solution of the desired complex also contains by-products (salts such as $NaNO_3$, NaCl *etc.*), which can be removed with desalting columns or HPLC purification of the complexes if salt-free compounds are required.

There are some special cases when either the ligand or the complex formed is not soluble in water, the complex hydrolyses in aqueous media due to its low stability, or the complexes are simply not formed below the pH where the hydrolysis of the metal ions starts.[133–135] Nevertheless, it is still possible to prepare the complexes in anhydrous organic solvents, such as pyridine, alcohols (*e.g.* methanol, ethanol), acetonitrile or even a mixture of solvents (*e.g.* $CHCl_3$/ $EtOH/H_2O$).[136,137] For this purpose Ln[III] triflates (trifluoromethanesulfonates, $Ln(CF_3SO_3)_3$) are the best choice. In this procedure the Ln[III] complexes are prepared by refluxing $Ln(CF_3SO_3)_3$ and the ligand (usually a macrocyclic chelator) in dry organic solvent for several hours. The complexes are often obtained by crystallization or by precipitation after adding another solvent.

Preparation of Ga^{3+} and In^{3+} complexes requires slightly different conditions than those applied for lanthanide complex preparations, because these metal ions hydrolyze at a much lower pH.[138] A recent paper suggests that Ga^{3+} complexes should be prepared at $pH = 3.5$ in the presence of HEPES buffer, which is also suitable for human use.[139] Radiolabeling of ligands conjugated to biological vectors is often performed in buffered solutions, most commonly in acetate-based buffer systems such as sodium acetate–acetic acid or ammonium acetate–acetic acid. In some reports, sodium or ammonium citrate buffer was used for preparations involving radioisotopes. It should be emphasized, however, that the citrate ligand forms stable complexes with a number of cations (including Ga^{3+}, In^{3+} and Ln^{3+} ions)[38] and may compete with the macrocyclic ligand for the metal ion under labeling conditions and therefore, its use should be avoided.

5.2.3 Equilibrium Properties of the DTPA and DOTA Derivative Complexes

The formation equilibria of the metal complexes are characterized by their thermodynamic stability constants as defined by the Equation 5.2.3:

$$K_{ML} = \frac{[ML]}{[M][L]} \tag{5.2.3}$$

where [ML], [M] and [L] are the equilibrium concentration of the complex, the metal ion and deprotonated ligand, respectively (the charges of the species are omitted for simplicity). Because of the multidentate nature of ligands, one or more donor atoms can be protonated at low pH values, when protonated complexes are formed. By considering the protonation equilibria, the protonation constants of complexes are defined as follows:

$$K_{MH_iL} = \frac{[MH_iL]}{[MH_{i-1}L][H^+]} \tag{5.2.4}$$

where $i = 1, 2, \ldots, n$, and where $[H^+]$, $[MH_{i-1}L]$ and $[MH_iL]$ are the equilibrium concentration of H^+, $MH_{i-1}L$ and MH_iL species, respectively.

To calculate the stability constants, the protonation constants of the ligands must be known, which are defined by Equation 5.2.5.

$$K_i^H = \frac{[H_iL]}{[H_{i-1}L][H^+]} \tag{5.2.5}$$

The conditional stability constants (K^c_{ML}), which reflect the competition between the metal ion and protons for the ligand, can be used to compare the behaviour of metal complexes at physiological pH:

$$K^c_{ML} = \frac{[ML]}{[M][L]_t} = \frac{[ML]}{[M][L]\alpha_H} \tag{5.2.6}$$

where $[L]_t = [L] + [HL] + [H_2L] + \ldots + [H_nL]$ $\alpha_H = 1 + K_1[H^+] + K_1K_2[H^+]^2 + \ldots + K_1K_2\ldots K_n[H^+]^n$ and so $K^c_{ML} = K_{ML}/\alpha_H$.

In biological systems, the free ligand L can be protonated or it can form complexes with the endogenous metal ions (Ca^{2+}, Zn^{2+} and Cu^{2+}), while the M^{z+} metal ion can interact with endogenous ligands A, B, ... (citrate, phosphate, carbonate, etc.). In addition, the metal complex can be protonated and it can also form ternary complexes with the endogenous ligands. By taking into account all the possible side-reactions, a more general conditional stability constant (K^*) can be defined.[140]

$$K^* = \frac{[ML]_t}{[M]_t[L]_t} = \frac{[ML]}{[M][L]}\frac{\alpha_{ML}}{\alpha_M\alpha_L} = K_{ML}\frac{\alpha_{ML}}{\alpha_M\alpha_L} \qquad (5.2.7)$$

where:

$[M]_t = [M] + [MA] + [MB] + \ldots$

$[L]_t = [L] + [HL] + [H_2L] + \ldots + [H_nL] + [CaL] + [CaHL] + \ldots + [ZnL]$
$\qquad + [ZnHL] + \ldots$

$[ML]_t = [ML] + [MHL] + [MLA] + [MLB] + \ldots$

$\alpha_M = 1 + K_{MA}[A] + K_{MB}[B] + \ldots$

$\alpha_L = 1 + K_1[H^+] + K_1K_2[H^+]^2 + \ldots + K_1K_2\ldots K_n[H^+]^n + K_{CaL}[Ca^{2+}] +$
$\qquad K_{CaHL}[H^+] + \ldots$

$\alpha_{ML} = 1 + K_{MHL}[H^+] + K_{MLA}[A] + K_{MLB}[B] + \ldots$

The stability and protonation constants of the complexes, the conditional stability constant (K^*), and the concentration of the free metal ion ($[M^{z+}]$) can be calculated using the protonation constants of the ligand. It is generally assumed that the toxicity of metal complexes is related to the concentration of the "free" metal ion ($[M^{z+}]$), which can be expressed by the pM value ($-\log[M^{z+}] = pM$). The pM values are calculated for the special condition as proposed by *Raymond et al.*:[141] $[M]_t = 10^{-6}$ M, $[L]_t = 10^{-5}$ M, physiological concentration of the $[Ca^{2+}]$, $[Zn^{2+}]$ and $[Cu^{2+}]$, pH $= 7.4$.

5.2.3.1   Experimental Methods and Computer Programs used for the Characterization of Complexation Equilibria

a. Methods and conditions:

The stability constants of the complexes and the protonation constants of the ligands are defined by the concentration of the species formed in the equilibrium (Equations 5.2.3 – 5.2.5). In order to keep the activity coefficients at a constant value during the measurements, an inert electrolyte is used in relatively high concentration, which maintains a constant ionic strength. The inert electrolyte is generally KCl, KNO_3, Me_4NCl, Me_4NNO_3, NaCl or $NaClO_4$. Their concentration is most often 0.1 M or 1.0 M, depending on the concentration of the ligand L and metal ion M^{z+} being investigated. The cations of the salts are prone to form weak complexes with the multidentate ligands, the stability of which decreases in the following order: $Na^+ > K^+ > Me_4N^+$.

In the complex formation reactions, there is generally a competition between the metal and H^+ ions for the donor atoms of ligands, which results in a change in the $[H^+]$ of the solutions. To study the complexation, the most widely used method is the pH-potentiometric titration of the ligand in the presence and absence of the metal ion. The instrumental set for the pH-potentiometric titration consists of a glass and a reference electrode (or a combined electrode), a pH-meter, and an autoburette. Before the pH-potentiometric titration is performed, the electrode system must be calibrated with two or three standard buffer solutions. In the practice of pH-potentiometric titrations, the pH meters are frequently used to measure the electromotive force (E), but measuring the pH values (pH_r) is also quite common. Both of these methods give H^+ ion activities from the titration data; however, for the calculation of the protonation or stability constants, H^+ concentrations ($[H^+]$) are needed. The simplest procedure to convert the measured pH_r values to $[H^+]$ using the relationship between pH_r and the activity coefficient (f) of the background electrolyte solution: $p[H^+] = pH_r + \log f$.[142]

When the electromotive force is measured, the E values are related to the H^+ concentration as:[143]

$$E = E'_0 + Q \log[H^+] + j_H[H^+] + j_{OH} \frac{K_W}{[H^+]} \qquad (5.2.8)$$

where E'_0 contains the standard potential, the activity coefficients and liquid-junction potential of the inert electrolyte, Q is the *Nernstian* slope, K_w is the stoichiometric water ionic product, and $j_H[H^+]$ and $j_{OH}[OH^-]$ express the contribution of the H^+ and OH^- ions to the liquid-junction potential. The H^+ concentrations are calculated with the Equation (5.2.8), but the E'_0, Q, K_w, j_H and j_{OH} values must be determined by titrating a strong acid with the same strong base that is used in the titration experiments.

The other frequently used method to obtain $[H^+]$ from the measured pH_r values was proposed by *Irving et al.*[144] A strong acid (HCl or HNO_3) of known concentration is titrated with the base used for the titration experiments and the differences between the measured and calculated pH values (A) are determined in acidic solutions. The A values are used to calculate the $[H^+]$ from the measured pH_r values ($pH_r = pH + A$). Using the term A, the K_w is calculated from the titration data of the strong acid obtained in basic solutions.[144] A prerequisite of pH-potentiometric titration techniques is that the equilibrium must be attained rapidly (in a few minutes) after the addition of the titrant. The metal: ligand concentration ratios are generally 1:1 but for some metal complexes, titrations are performed at other metal: ligand ratios. The volume of samples is generally 2 – 20 mL and the temperature of the samples must be kept constant (25 °C). The solutions are stirred and N_2 or Ar is bubbled through to prevent absorption of the CO_2 during the titration.

The complex formation reactions of the macrocyclic ligands are generally slow at low pH values and so for determining the stability constants the "out-of-cell" (or batch) method is used.[145] Several samples are prepared in the pH range where complexation equilibria exist and the closed samples are kept until the equilibria are reached. Reliable protonation and stability constants can be obtained by pH-potentiometry in the pH range of about 1.7 – 12. To determine the stability constants, competition reactions can also be used if the complexation equilibria exist outside of the well-measurable pH range (*e.g.* at pH 2). For the competition, another metal ion (or another ligand) can be used if the stability constants of their complexes are known and in their presence the equilibria are shifted into the well-measurable pH range.

The protonation of ligands and the formation of complexes can be studied by spectrophotometry if the metal ions or ligands have absorption band/s in the UV-vis range.[146] The absorbance values at each wavelength can be expressed by Equation 5.2.9:

$$A = \sum_1^n \varepsilon_i x_i l \qquad (5.2.9)$$

where l is the path length of the cell, and ε_i and x_i are the molar absorptivity and molar fraction of the species i, respectively. Equation 5.2.9 can be used to calculate the protonation constants of the ligand or the stability constants of complexes if the molar fractions are expressed with protonation or the stability constants and spectrophotometric measurements are performed at different pH values.[146] Spectrophotometric studies can be carried out at higher H^+ or OH^- concentrations, where pH-potentiometry cannot be used.

Multinuclear NMR spectroscopy is a convenient method to study the protonation of ligands and the formation of complexes. To obtain information about the protonation sites and protonation constants of the free ligand, the chemical shifts of the non-labile NMR active nuclei (^1H, ^{13}C, ^{31}P, *etc.*) are followed as a function of H^+ or OH^- concentration. The NMR "titration curve" displays sharp changes at different pH values, which are related to the protonation/deprotonation of the ligand. The protonation sites and protonation sequence of the ligand can be identified by the assignments of the signals. Since the protonation/deprotonation of the ligand is generally fast on the NMR time scale, the chemical shifts of the observed signals represent a weighted average of the shifts of the different H_iL species involved in a specific protonation step (Equation 5.2.10).[147]

$$\delta_{obs} = \sum_1^n x_{H_iL} \delta^{H_iL} \qquad (5.2.10)$$

where $i = 0, 1, 2, \ldots, n$, δ_{obs} is the observed chemical shift of a given signal, and x_{H_iL} and δ^{H_iL} are the molar fraction and the chemical shift of the involved species, respectively. After expressing the molar fraction with the $[L]_t$ and the protonation constants, the obtained equation is suitable for the calculation of protonation constants and δ^{H_iL} values of the ligand.[147]

b. Calculation of the equilibrium constants:

To calculate the equilibrium constants (protonation and stability constants), several computer programs (MINIQUAD, PSEQUAD, SUPERQUAD, HYPERQUAD OPIUM, *etc.*) are available, which generally operate on the basis of nonlinear least squares principle.[148–150]

In the calculations a chemical model is presumed, which involves the chemical reactions for the formation of species ($M_pH_qL_r$) from the components

(M, L, H) characterized by the cumulative formation constants β_{pqr}. The programs calculate a series of datapoints (*e.g.* titration volume, chemical shift, *etc.*) from the analytical concentrations of the components and series of measured data (*e.g.* pH values) using the model. The differences between the measured and calculated data (the residuals) are used for further calculations, when the minimum of the square sum of residuals is obtained by the variation of the β_{pqr} values.

$$\beta_{pqr} = \frac{[M_pH_qL_r]}{[M]^p[H^+][L]^r} \tag{5.2.11}$$

The programs calculate the statistical data characterizing the goodness of fitting (χ^2 or the standard deviation of fitted data) and the reliability of the estimated β_{pqr} constants. The calculations can be repeated with different models and the model that has the best statistical parameters is accepted. The individual stability constants are calculated from the log β_{pqr} values with the related protonation and equilibrium constants. The standard deviation values calculated for the stability constants originate from random experimental error and reflect only a part of the total uncertainty. The true errors of the calculated equilibrium constants can be evaluated by comparing the constants obtained with different methods, or obtained in independent laboratories.

5.2.3.2 Protonation Sequence and Protonation Constants of the DTPA and DOTA Derived Ligands

The protonation constants of the DTPA and DOTA based ligands, defined by Equation 5.2.5, are listed in Table 5.2.1 and 5.2.2.

The basicity of the donor atoms of DTPA and DOTA derived ligands are affected by several factors, such as the strength of H-bonds between the non-bonding electron pairs of the donor atoms and the protonated donor atoms, the electrostatic repulsion between the protonated donor atoms, and the electron donating or withdrawing effects of the neighbouring groups. The contributions of the different effects determine the protonation constant values.

The data in Table 5.2.1 show that the first three protonation constants of the DTPA derivatives are markedly influenced by the nature of the substituent/s (for the structures of the ligands please refer to Schemes 5.2.10 and 5.2.11). The processes related to the variation of the protonation constants were identified by pH dependent NMR studies of the ligands.[147] Starting with the deprotonated ligand, the protonation sequence of the DTPA derivatives generally follows this trend: the first protonation takes place at the central nitrogen atom. The second protonation occurs at one of the terminal nitrogens with the concurrent shift of proton from the central nitrogen to the other terminal nitrogen due to the electrostatic repulsion between the two protons. The third protonation occurs at the central nitrogen (with the partial protonation of the carboxylate attached to the central nitrogen). Further protonations take place at

Table 5.2.1 Protonation constants of the DTPA derivate ligands at 25 °C.

Ligands	Electrolyte	$\log K_1^H$	$\log K_2^H$	$\log K_3^H$	$\log K_4^H$	$\log K_5^H$	Method	Ref.
DTPA	0.1 M KCl	10.41	8.37	4.09	2.51	2.04	pH	206
BOPTA	0.1 M KCl	10.71	8.27	4.35	2.83	2.07	pot.(20 °C)	43,207
	–	10.51	8.17	4.19	1.94	1.5	NMR	
EOB-DTPA	0.1 M KCl	10.95	8.62	4.23	2.80	2.06	pH	208
MS-325	0.1M Me$_4$NCl	11.15	8.62	4.51	2.96	2.37	pH	209
p-NO$_2$-Bn-DTPA	0.1M Me$_4$NCl	11.16	8.30	4.44	2.75	2.53	pH	105
CHX-DTPA	0.1M Me$_4$NCl	12.3	9.24	5.23	3.32	2.18	pH	105
p-NO$_2$-Bn-CHX-DTPA	0.1M Me$_4$NCl	12.3	8.99	4.99	2.84	2.35	pH	105
DTPA-N-MA	0.1 M KCl	10.18	6.19	3.55	2.0	–	pH	48
DTPA-N'-MA	0.1 M KCl	10.04	8.41	2.73	1.94	–	pH	48
DTPA-BMA	0.1 M NaCl	9.37	4.38	3.31	1.43	–	pot.	140b
DTPA-BMEA	0.1 M NaClO$_4$	9.26	4.54	3.40	2.09	–	pH	210,99
	0.1 M NaClO$_4$	9.33	4.36	–	–	–	NMR	
DTPA-B(BbuA)	0.1 M KCl	9.77	6.72	4.08	–	–	pH	158
DTPA-BBzA	0.1 M KCl	9.39	4.57	3.54	–	–	pH	101
DTPA-BAMA	0.1 M KCl	9.35	4.85	3.73	–	–	pot.	211
DTPA-TrA	0.1 M KCl	8.50	6.53	2.82	–	–	pH	158
15-DTPA-EAM	0.1M KCl	9.45	4.21	3.39	1.96	–	pot.	45, 212
EPTPA	0.1 M Me$_4$NNO$_3$	10.60	8.92	5.12	2.80	–	pot.	159
p-NO$_2$-Bn-EPTPA	0.1 M Me$_4$NCl	10.86	8.91	4.70	3.25	2.51	pH	160

the carboxylate groups of the terminal nitrogen atoms. The protonation sequence is modified by the replacement of the central carboxylate with an amide group. For DTPA-*N'*-MA both the first and second protons protonate the terminal nitrogens. The third proton presumably protonates the central nitrogen.[48] The lower basicity of the central nitrogen of the DTPA-*N'*-MA and the terminal nitrogen of the DTPA-*N'*-MA can be interpreted by the electron withdrawing effect of the amide group and the formation of weak H-bonds between the amine nitrogen and the amide hydrogens, respectively. The replacement of two carboxylates with two amides at the terminal nitrogen atoms (*e.g.* DTPA-BMA, DTPA-BMEA, DTPA-BBzA, 15-DTPA-EAM) results in further lowering of the basicity of the nitrogens. The substitution of both amide hydrogens with alkyl (or aryl) groups increases the basicity of the

Table 5.2.2 Protonation constants of the DOTA derivative ligands at 25 °C.

Ligands	Electrolyte	log K_1^H	log K_2^H	log K_3^H	log K_4^H	log K_5^H	log K_6^H	log K_7^H	Method	Ref.
DOTA	0.1M Me$_4$NCl	12.6	9.70	4.50	4.14	2.32	–	–	[b]pH	[182]
p-NO$_2$-Bn-DOTA	1.0M Me$_4$NCl	10.93	9.14	4.44	4.19	2.33	1.4	–	[b]pH	[198]
DO3A	0.1M Me$_4$NCl	11.59	9.24	4.43	3.48	–	–	–	[b]pH	[197, 213]
HP-DO3A	0.1M Me$_4$NCl	11.96	9.43	4.30	3.26	–	–	–	[b]pH	[197, 213]
BT-DO3A	0.1M Me$_4$NCl	11.75	9.23	4.13	2.97	–	–	–	[b]pH	[79]
P-730	0.1 M Me$_4$NCl	12.22	9.18	6.29	5.69	5.04	4.80	4.17	[a]pot.	[214]
DO2A	0.1M Me$_4$NCl	10.94	9.55	3.85	2.55	–	–	–	[b]pH	[215]
DO3AP	0.1M Me$_4$NCl	13.83*	10.35	6.54	4.34	3.09	1.63	1.07	[a]pot.	[155]
DO2A2P	1.0 M KCl	12.6	11.43	5.95	6.15	2.88	2.77	–	[b]pH	[82]
	1.0 M KCl	12.8	11.7	5.93	6.15	–	–	–	[c]NMR	
DOA3P	1.0 M KCl	13.6*	11.42	7.69	6.33	5.13	2.73	1.62	[b]pH	[83]
DOTP	0.1 M Me$_4$NNO$_3$	14.65*	12.4*	9.28	8.09	6.12	5.22	–	[b]pH	[152,153]
DO3AP^{ABn}	0.1 M Me$_4$NCl	12.55	9.60	5.11	4.11	2.71	1.54	–	[a]pot.	[154]
DOTEP	0.1 M KNO$_3$	10.94	8.24	3.71	–	–	–	–	[b]pH	[216]
	1.0 M KCl	10.87	8.21	–	–	–	–	–	[c]NMR	
PCTA	1.0 M KCl	11.36	7.35	3.83	2.12	1.29	–	–	[b]pH	[87]
p-NO$_2$-Bn PCTA	1.0 M KCl	11.29	6.70	3.96	2.08	1.82	–	–	[b]pH	[131]
DOTAM	1.0 M KCl	9.08	6.44	–	–	–	–	–	[b]pH	[113]
DTMA	1.0 M KCl	9.56	5.95	–	–	–	–	–	[b]pH	[113]
TRITA	0.1 M Me$_4$NCl	11.81	9.21	4.03	2.62	–	–	–	[a]pot.	[145,217]

[a]potentiometry.
[b]pH-potentiometry.
[c]NMR titration.

terminal nitrogen atoms because of the formation of fewer hydrogen bonds (DTPA-B(BBuA) and DTPA-BMA).[99] The effect of the substituents attached to the ligand backbone is mainly controlled by the position of the substitution and the properties of the substituent. In the case of BOPTA, the presence of the benzyl-oxy-methyl group on the pendant arm has practically no effect on the basicity of the nitrogen atoms. However, the replacement of a hydrogen atom of the ligand backbone by the etoxy-benzyl- (EOB-DTPA) and diphenyl-cyclohexyl-phosphine groups (MS-325) results in a slight increase in the basicity of the central nitrogen atom which is probably caused by the electron donating behaviour of the substituents. The replacement of an ethylene group of the DTPA with cyclohexyl moiety results in an increase of the basicity of the central and the terminal nitrogen atoms because of the electron donating behaviour of the alkyl ring and the favourable position of the nitrogen atoms for hydrogen bond formation.[105]

DOTA based ligands most commonly contain amino nitrogen, carboxylate oxygen, phosphonate or phosphinate oxygen, and amide oxygen (and nitrogen) donor atoms. The NMR titration of DOTA indicates that the first and second

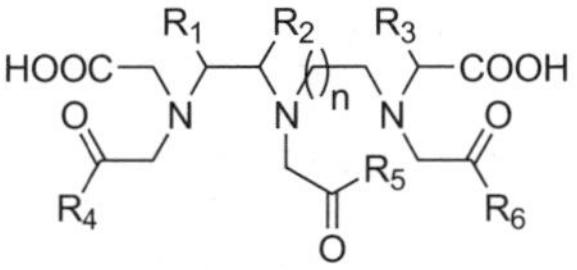

Ligands	n	R$_1$	R$_2$	R$_3$	R$_4$	R$_5$	R$_6$
DTPA	1	-H	-H	-H	-OH	-OH	-OH
DTPA-N-MA	1	-H	-H	-H	-OH	-OH	-NH-Me
DTPA-N'-MA	1	-H	-H	-H	-OH	-NH-Me	-OH
DTPA-BMA	1	-H	-H	-H	-NH-Me	-OH	-NH-Me
DTPA-BMEA	1	-H	-H	-H	-NH-Et-O-Me	-OH	-NH-Et-O-Me
DTPA-B(BbuA)	1	-H	-H	-H	-N(n-Bu)$_2$	-OH	-N(n-Bu)$_2$
DTPA-BBzA	1	-H	-H	-H	-NH-Bn	-OH	-NH-Bn
DTPA-BAMA	1	-H	-H	-H	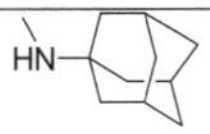 HN—[adamantyl]	-OH	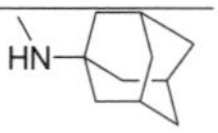 HN—[adamantyl]
DTPA-TrA	1	-H	-H	-H	-N(n-Bu)$_2$	-NH-Me	-N(n-Bu)$_2$
BOPTA	1	-H	-H	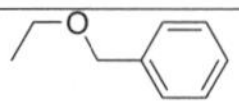 [CH$_2$-O-CH$_2$-C$_6$H$_5$]	-OH	-OH	-OH
EOB-DTPA	1	-H	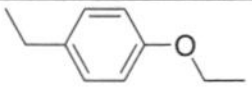 [CH$_2$-C$_6$H$_4$-O-Et]	-H	-OH	-OH	-OH
MS-325	1	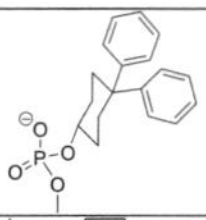 [phosphate-diphenylcyclohexyl group]	-H	-H	-OH	-OH	-OH
p-NO$_2$-Bn-DTPA	1	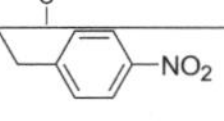 [CH$_2$-C$_6$H$_4$-NO$_2$]	-H	-H	-OH	-OH	-OH
EPTPA	2	-H	-H	-H	-OH	-OH	-OH
p-NO$_2$-Bn-EPTPA	2	 [CH$_2$-C$_6$H$_4$-NO$_2$]	-H	-H	-OH	-OH	-OH

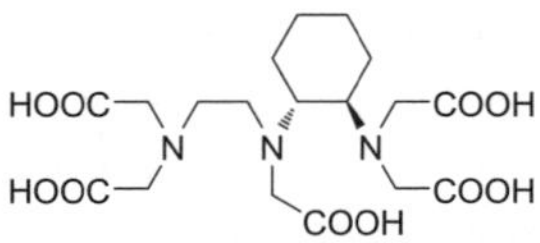

CHX-DTPA 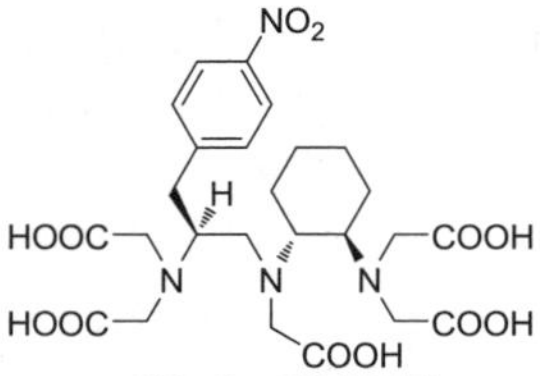   p-NO$_2$-Bn-CHX-DTPA 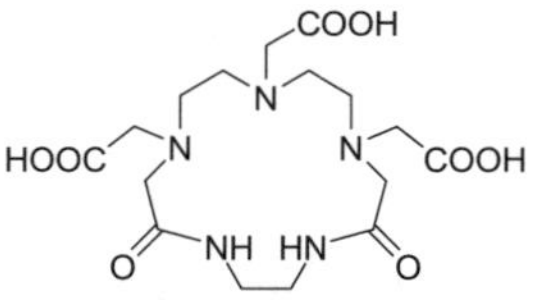   15-DTPA-EAM

Scheme 5.2.10   Structure of the DTPA derivative ligands discussed in the current Chapter.

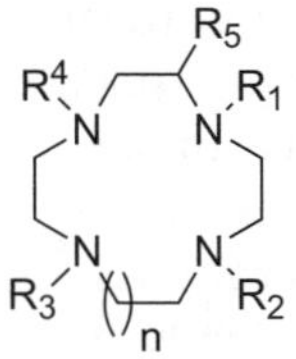

Ligands	n	R_1	R_2	R_3	R_4	R_5
DOTA	1	-CH₂-COOH	-CH₂-COOH	-CH₂-COOH	-CH₂-COOH	-H
***p*-NO₂-Bz-DOTA**	1	-CH₂-COOH	-CH₂-COOH	-CH₂-COOH	-CH₂-COOH	
P730	1	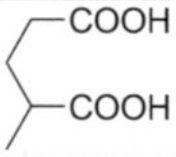	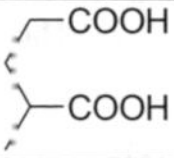	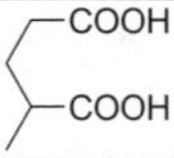	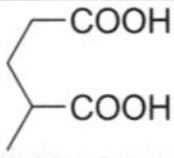	-H
HP-DO3A	1		-CH₂-COOH	-CH₂-COOH	-CH₂-COOH	-H
BT-DO3A	1		-CH₂-COOH	-CH₂-COOH	-CH₂-COOH	-H
DO3A	1	-H	-CH₂-COOH	-CH₂-COOH	-CH₂-COOH	-H
DO2A	1	-H	-CH₂-COOH	-H	-CH₂-COOH	-H
DO3AP	1	-CH₂-PO₃H₂	-CH₂-COOH	-CH₂-COOH	-CH₂-COOH	-H
DO3AP^ABn	1		-CH₂-COOH	-CH₂-COOH	-CH₂-COOH	-H
DO2A2P	1	-CH₂-PO₃H₂	-CH₂-COOH	-CH₂-PO₃H₂	-CH₂-COOH	-H
DOA3P	1	-CH₂-PO₃H₂	-CH₂-PO₃H₂	-CH₂-PO₃H₂	-CH₂-COOH	-H
DOTP	1	-CH₂-PO₃H₂	-CH₂-PO₃H₂	-CH₂-PO₃H₂	-CH₂-PO₃H₂	-H
DOTEP	1	-CH₂-PO₂H-CH₂-CH₃	-CH₂-PO₂H-CH₂-CH₃	-CH₂-PO₂H-CH₂-CH₃	-CH₂-PO₂H-CH₂-CH₃	-H
DOTAM	1	-CH₂-CO-NH₂	-CH₂-CO-NH₂	-CH₂-CO-NH₂	-CH₂-CO-NH₂	-H
DTMA	1	-CH₂-CO-NH-CH₃	-CH₂-CO-NH-CH₃	-CH₂-CO-NH-CH₃	-CH₂-CO-NH-CH₃	-H
TRITA	2	-CH₂-COOH	-CH₂-COOH	-CH₂-COOH	-CH₂-COOH	-H

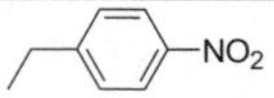

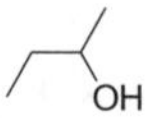

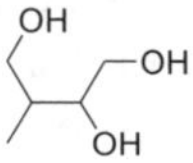

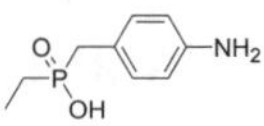

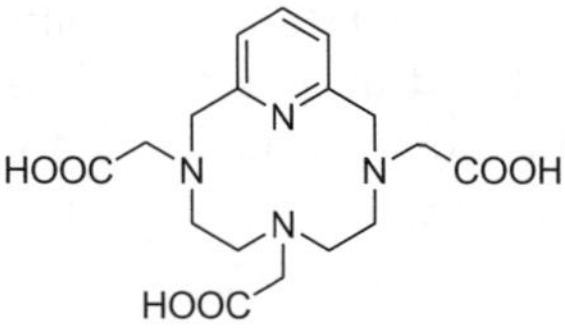

PCTA

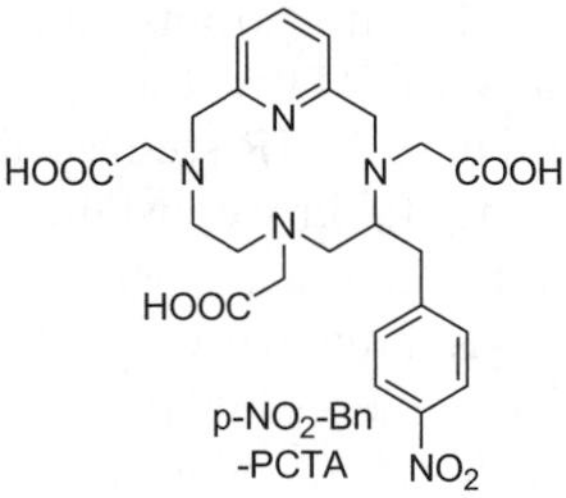

p-NO₂-Bn -PCTA

Scheme 5.2.11 Structure of the macrocyclic ligands discussed in the current Chapter.

protonations occur at two diagonal nitrogen atoms of the ring followed by the protonation of carboxylate groups attached to the non-protonated nitrogens.[151] The protonation sequence of DO3A, HP-DO3A and BT-DO3A is similar to that of DOTA. By the stepwise replacement of the acetate arms with phosphonate groups, the total basicity ($\Sigma \log K_i^H$), the $\log K_1^H$ and $\log K_2^H$ values of the ligands increase gradually because of the presence of the more basic phosphonate/s and the formation of a strong hydrogen bond between the protonated nitrogen atoms and phosphonate group/s. The protonation scheme of the phosphonate derivatives is somewhat different from that of DOTA.[81,83,132,152–155]

The presence of amide groups also decreases the basicity of the amine nitrogens of DOTA derivatives.[99,156] The first two protonation constants of DOTAM and DTMA are indeed significantly lower than those of DOTA. The incorporation of a pyridine ring into the macrocycle increases the rigidity of the ligand and decreases the basicity of the donor atoms due to the electron withdrawing effect of the aromatic ring. The protonation schemes of the PCTA and DOTA are slightly different. The first protonation of PCTA occurs at the nitrogen atom opposite to the pyridine ring. The addition of a second proton results in the protonation of the tertiary nitrogen atoms, positioned *trans-* to each other. The third protonation occurs at the carboxylate pendant of the non-protonated nitrogen atom. Further protonations of PCTA occur at the non-protonated carboxylate groups.[85]

5.2.3.3 Complexation Equilibria of the DTPA and DOTA Based Ligands

The ligands DTPA, DOTA, and their derivatives form high stability ML complexes with trivalent metal ions in which generally all the donor atoms of the hepta- or octadentate ligands are coordinated. The stability constants ($\log K_{ML}$) of the lanthanide[III] and some divalent metal complexes formed with DTPA and DOTA derivative ligands are listed in Tables 5.2.3 and 5.2.4.

The stability constants of complexes of the ligands DTPA, EOB-DTPA, BOPTA and MS-325, which contain three amine nitrogen and five carboxylate oxygen donor atoms, are very similar, showing that the side chain only slightly affects the basicity of donor atoms and the coordination environment of the metal ions. Similarly to DTPA, the ligands listed in Table 5.2.3 are all octadentate and according to NMR studies, all the donor atoms are coordinated to the Ln^{3+} ions.[157] The stability constants of complexes strongly depend on the charge of the ligand. The replacement of one or two carboxylate group/s with non-ionic –CO–NHR amide group/s leads to the decrease of the stability constants of Gd^{3+} complexes by about 2 – 3 and 5 – 6 log K units, respectively. The decrease in the log K_{ML} values is smaller when the amide group does not contain amide hydrogen (*e.g.* DTPA-B(BBuA)) since in this case the H-bond cannot form and thus the coordinated amide oxygen is more basic. (The electron withdrawing effect of the amide groups increases in the following

Table 5.2.3 Stability constants of the Ca^{2+}, Eu^{3+}, Gd^{3+}, Yb^{3+}, Zn^{2+} and Cu^{2+} complexes formed with DTPA and their derivates at 25 °C.

Ligands	Electrolyte	$logK_{CaL}$	$logK_{EuL}$	$logK_{GdL}$	$logK_{YbL}$	$logK_{ZnL}$	$logK_{CuL}$	Ref.
DTPA	0.1 M KCl	10.75	22.39	22.46	22.62	18.6	21.5	206
BOPTA	0.1 M KCl	–	–	22.59	–	17.04	21.94	43,218
EOB-DTPA	0.1 M KCl	11.82	23.1	23.6	23.0	18.78	20.2	208
MS-325	0.1M NaClO$_4$	10.45	22.21	22.06	–	17.82	21.3	209
DTPA-N-MA	0.1 M KCl	–	18.70	19.40	19.5	16.00	18.71	48
DTPA-N'-MA	0.1 M KCl	–	19.90	19.0	20.4	16.82	18.50	48
DTPA-BMA	0.1 M NaCl	7.17	–	16.85	–	12.04	13.05	140b
DTPA-BMEA	0.1 M NaClO$_4$	–	–	16.84	–	–	–	99
DTPA-B(BbuA)	0.1 M KCl	–	–	19.0	–	16.38	17.85	158
DTPA-BBzA	0.15 M NaCl	7.13	–	16.48	–	11.98	12.28	101
DTPA-BAMA	0.1 M KCl	7.49	–	16.85	–	11.9	12.86	211
DTPA-TrA	0.1 M KCl	–	–	17.93	–	13.36	13.75	158
15-DTPA-EAM	0.1M KCl	5.65	11.7	11.4	10.6	12.08	15.1	45, 212
EPTPA	0.1 M Me$_4$NNO$_3$	14.45	–	22.77	–	18.59	19.31	159
p-NO$_2$-Bn-EPTPA	0.1 M Me$_4$NCl	9.38	–	19.20	–	16.01	18.47	160

Some other log K_{ML} values of M(DTPA) complexes: ScL: 23.9, InL:29.5, GaL: 23.32 (0.1 M KCl and KNO$_3$; 25 °C)[219,220]; Y(p-NO$_2$-Bn-DTPA): 21.5, Y(CHX-DTPA): 24.2, Y(p-NO$_2$-Bn-CHX-DTPA): 24.4 (0.1 M NaClO$_4$; 25 °C)[105].

order: $-CO-NR_2 < -CO-NHR < -CO-NH_2$.) The coordination number of Cu^{2+} and Zn^{2+} is lower than the denticity of DTPA, so the free donor atoms can coordinate to another Cu^{2+} or Zn^{2+} ion, giving rise to dinuclear complexes. The formation of similar dinuclear complexes was not detected for the DTPA–bis-amide derivatives.[140] However, more recently the formation of binuclear Cu_2L and Zn_2L has been reported with DTPA–bis(butyl) amide and DTPA–bis(bis-butyl) amide.[158]

The log K_{ML} values of the EPTPA complexes are higher than expected, because the presence of a six-membered chelate ring should normally result in the decrease of the stability.[159] However, the complexes of p-NO$_2$-Bn-EPTPA are less stable than those of DTPA, showing the expected loss in the stability.[160]

The stability constants of the complexes of DOTA derivatives strongly depend on the number of pendant arms attached to the cyclen ring and on the charge of the ligand. The log K_{ML} values decrease in the order DOTA > DO3A > DO2A. The stability constants of the Ln(DOTA) complexes increase

Table 5.2.4 Stability constants of the Ca^{2+}, Eu^{3+}, Gd^{3+}, Yb^{3+}, Zn^{2+} and Cu^{2+} complexes formed with DOTA and their derivates at 25 °C.

Ligands	Electrolyte	log K_{CaL}	log K_{EuL}	log K_{GdL}	log K_{YbL}	log K_{ZnL}	log K_{CuL}	Ref.
DOTA	0.1M NaCl	16.37	23.5	24.7	25.0	18.7	22.72	[145,217,221]
p-NO₂-Bn-DOTA	1.0M Me₄NCl	–	–	24.2	–	–	–	[198]
DO3A	0.1M Me₄NCl	–	20.69	21.0	–	–	–	[197, 213]
HP-DO3A	0.1M Me₄NCl	–	–	23.8	–	17.32[218]	20.55[218]	[197, 213]
BT-DO3A	0.1M Me₄NCl	14.3	21.2	20.8	–	19.0	21.1	[79]
P-730	0.1M Me₄NCl	–	24.01	24.03	–	–	–	[214]
DO2A	0.1M Me₄NCl	7.16	12.99	13.06	13.26	–	–	[215]
DO3AP	0.1M Me₄NCl	–	27.8	27.5	–	–	–	[155]
DO2A2P	1.0 M KCl	15.1	25.6	25.7	–	22.5	24.9	[81]
DOA3P	1.0 M KCl	14.5	27.5	27.3	–	22.9	27.3	[83]
DOTP	0.1 M Me₄NNO₃	11.12	28.1	28.8	29.5	24.8	25.4	[132,152,153]
DO3AP^{ABn}	0.1M Me₄NCl	–	25.0	24.04	–	–	–	[154]
DOTEP	0.1 M KNO₃	9.39	–	16.5	–	15.8	19.57	[216]
PCTA	1.0 M KCl	12.72	20.26	20.39	20.63	20.48	18.79	[87]
p-NO₂-Bn PCTA	1.0 M KCl	12.72	19.02	19.42	–	21.36	19.11	[131]
DOTAM	1.0 M KCl	10.32	13.80	13.12	–	13.77	14.50	[113]
DTMA	1.0 M KCl	10.11	13.67	13.54	–	13.66	14.61	[113]
TRITA	0.1 M KCl	11.99	–	19.17	–	18.04	22.49	[145,217]

Some other log K_{ML} values of M(DOTA) complexes: ScL: 24.2, InL:23.9, GaL:21.33 (0.1 M KCl, 25 °C).[217,222]

from the La^{3+} to the Sm^{3+} – Eu^{3+}, indicating that the best fit of the Ln^{3+} ions in the coordination cage of DOTA is realised for the ions Sm^{3+} – Eu^{3+}. The replacement of one carboxylate with an alcoholic OH results in the decrease of the log K_{ML} values by 1 – 3 log K unit (HP-DO3A and BT-DO3A).

The substitution of phosphonate groups for the acetate arms results in the increase of the stability constants of complexes in the following order: DO2A2P < DOA3P < DOTP (the log K_{ML} values of the DOA3P complexes are larger than expected in comparison with those of the DOTP complexes). The replacement of the carboxylate groups of DOTA with amide groups results in a significant drop in the log K_{ML} values. The amide groups decrease the basicity of the ring nitrogens and the stability constants of the complexes are very low. However, the trend of the stability constants of the $Ln(DOTAM)^{3+}$ complexes is similar to that of the $Ln(DOTA)^{-}$ complexes. The stability constants increase from La^{3+} to Sm^{3+} – Eu^{3+} and for the heavier elements the log K_{ML} values are similar.[113,161] In these $Ln(DOTAM)^{3+}$ complexes the Ln^{3+} ion is in the coordination cage defined by the 4 macrocyclic N atoms and the 4 O atoms of the amide groups. The ninth coordination site is occupied by an inner sphere water molecule. Generally, the water exchange rate of Ln-DOTA tetraamide complexes is significantly slower than that of the corresponding DOTA complexes and in particular, the Eu-complexes have the lowest water exchange rate. The favorable paramagnetic properties of Eu^{3+} (relatively large Ln induced shift and negligible relaxation enhancement effects) combined with the extremely slow water exchange rate makes Eu-DOTA tetraamides ideal candidates for paramagnetic chemical exchange (PARACEST) imaging agents.[162–164]

5.2.3.4 Equilibria of the Transmetallation Reactions of the DTPA and DOTA Derivative Complexes

The metal complexes of DTPA and DOTA derivatives used in biomedicine may react *in vivo* with endogenous metals and ligands. The body fluids, where these reactions occur, are very complicated systems, which contain several metal ions and a great number of complex forming ligands. The new species formed in these reactions cannot be exactly predicted, but some assumptions can be made on the basis of our experiences and knowledge in chemistry. The concentration of various complexes, the "species distribution", can be calculated using the known stability constants. Such calculations give some information about the *in vivo* fate of the complexes used as imaging agents. These species distribution calculations are particularly interesting for Gd^{3+}-based MRI contrast agents, because the amount of Gd^{3+} administered into the body is high (0.1 – 0.3 mmol Gdkg^{-1} body weight), so decomplexation as low as 1–2% of the Gd^{3+} complex would result in a relatively significant amount of residual Gd^{3+} in the body.

The study of the complexation equilibria in the blood plasma has attracted renewed interest lately due to a new disease, Nephrogenic Systemic Fibrosis (NSF), which is assumed to be related to the release of Gd^{3+} from CAs.[165–167] The elimination of the CAs from the body is generally fast (the half-time of excretion, $t_{1/2} = 1.5$ h) and so the extent of de-chelation in the body due to the slow dissociation of Gd^{3+} complexes is negligible. However, in patients with severe renal insufficiency, the half-time of elimination (with dialysis) is 30–40 h,[165–167] and consequently the Gd^{3+} complex spends longer in the body and the extent of dissociation can be observable. Practically all known cases of NSF are associated with the use of the Gd^{3+} complexes of DTPA derivatives, first of all Gd(DTPA-BMA),[165–168] which has the lowest stability constant among commercial CAs. The Gd^{3+} complexes formed with the macrocyclic DOTA derivatives are extremely inert, so their de-chelation in the body does not take place.

It is generally assumed that the *in vivo* dissociation of the Gd^{3+} complexes occurs through transmetallation reactions, first of all with Zn^{2+}.[169] However, the equilibrium of:

$$Gd(DTPA)^{2-} + Zn^{2+} \rightleftharpoons Zn(DTPA)^{3-} + Gd^{3+} \qquad (5.2.12)$$

is shifted to the left hand side, because $\log K_{GdL} \gg \log K_{ZnL}$. Transmetalla-tion reactions with Zn^{2+} occur only in the presence of a second ligand (*e.g.* citrate), which forms more stable complexes with Gd^{3+} than with Zn^{2+}:

$$Gd(DTPA)^{2-} + Zn(cit)^- \rightleftharpoons Zn(DTPA)^{3-} + Gd(cit) \qquad (5.2.13)$$

This is well illustrated by the following example: if $[Gd(DTPA)^{2-}] = 0.3$ mM and $[Zn^{2+}] = 0.01$ mM, pH $= 7.4$, then in the equilibrium (Equation 5.2.11) only 4% of the Zn^{2+} is in the form of $Zn(DTPA)^{3-}$. If citrate is also present at a concentration of $[Cit] = 0.11$ mM, then 75% of the Zn^{2+} is in the form of $Zn(DTPA)^{3-}$ and 2.5% of the Gd^{3+} forms Gd(cit). These data clearly show that the transmetallation reactions and the release of Gd^{3+} from CAs can take place in the blood plasma, but for the calculations of the species distribution, some plasma models should be used.

The blood plasma model, developed by *May et al.*,[170] takes into account seven metal ions and 40 endogenous ligands, which can form more than 5000 different complexes. The program ECCLES calculates the concentration of all species formed in the equilibrium, using the stability constants of complexes and the concentration of the components.[170] This blood plasma model was applied for the $Gd(DTPA)^{2-}$ as CA, but only the free Gd^{3+} concentration $(10^{-13}$ M, if $[Gd(DTPA)^{2-}] = 1$mM) was reported; the formation of other Gd^{3+} containing species was not mentioned.[171] A simplified blood plasma model does not predict significant de-chelation of Gd(DTPA-BMA).[140] However, the experimental data clearly contradict this result.[172] Another simplified model, where the more probable reactions were taken into account, indicated sig-nificant de-chelation of $Gd(DTPA)^{2-}$.[173] The results of the species distribution calculations performed with the stability constants are valid only for equili-brium systems. Since the de-chelation rates of Gd^{3+}-based CAs are much lower than the rates of their elimination from the body, the residual Gd^{3+} in the body due to de-chelation is generally very low. The situation is different for patients with severe renal impairment, when the elimination of the CAs from the body is 20–30 times slower. The longer residence time of the complexes formed with DTPA derivatives can lead to partial dissociation, which should be considered when these complexes are used as CAs.

5.2.4 Kinetic Properties of the Complexes

The trivalent lanthanides, Sc^{III}, Y^{III}, Ga^{III} and In^{III} have closed, symmetric outer electronic shells so the kinetic behaviour of their complexes is determined mainly by the nature of the ligands and only slightly influenced by the size of the metal ions. The formation reactions of the complexes are fast with flexible multidentate ligands, but the complexation with rigid ligands is relatively slow.[174] The rapid complex formation is particularly important for the synth-esis of radiopharmaceuticals, when the complex forming BFC ligand is bound to a temperature sensitive protein (*e.g.* to monoclonal antibody) and the radioisotope has a short half-life. In such cases the slow formation reaction and the need for high temperature processing is very unfavorable. The formation rates of the complexes of DTPA and its derivatives are high, but the kinetic inertness of these complexes is not always sufficient for nuclear medicine applications. The complexes formed with the macrocyclic DOTA derivatives are kinetically extremely inert, which makes them suitable for use in medicine. In order to delineate the problems of complexation, we shall discuss briefly the kinetics of formation of the DOTA based complexes. Since the kinetic inertness is a general requirement for imaging agents and in this respect the behaviour of DTPA and DOTA complexes differ considerably, the kinetics of their decomplexation will be discussed separately.

5.2.4.1 Formation Kinetics of Complexes of DOTA Derivatives

In the formation reactions of the complexes of DOTA, the rigid "spider like" structure of the ligand and the large difference between the first two and the remaining protonation constants (Table 5.2.2) play a crucial role. During the complex formation, the metal ion has to enter the coordination cage formed by the four nitrogen atoms of the 12-membered macrocycle and the four carboxylate oxygens of the four acetate groups. The formation of the DOTA complexes with Ln^{3+}, In^{3+}, Sc^{3+} In^{3+} and Ga^{3+} ions is usually slow in the pH range 3–6, where DOTA is present in the form of protonated species H_4DOTA, H_3DOTA and H_2DOTA. The kinetic data indicated the fast formation of a reaction intermediate which was detected by spectrophotometry,[175] ^{1}H-NMR[176] and luminescence spectroscopy,[177,178] and EXAFS.[179] Due to the slow rearrangement at pH < 4, pH-potentiometric studies could be used to show that the intermediate was a diprotonated complex, $Ln(H_2DOTA)^{+*}$ and its stability constants could be determined.[175,178] Similarly, the stability constants of the intermediates $Ln(H_2L)^*$, formed with several other DOTA derivatives, have been determined.[155,180,181] Regarding the structure of the intermediate, it has been assumed that the Ln^{3+} ion is positioned outside of the coordination cage and only the four carboxylates are coordinated to the Ln^{3+}, while the two protons are attached to two diagonal nitrogens.[175,178] With luminescence decay studies it was found that in the intermediate $Eu(H_2DOTA)^{+*}$ four or five water molecules remained coordinated to the Eu^{3+} ion beside the four carboxylate oxygens.[177,178] At higher pH (pH > 7) monoprotonated intermediates, $Ln(HDOTA)^*$ are also formed.[182]

The mechanism of the formation of the $Ln(DOTA)^-$ complexes through the deprotonation of the di- and mono-protonated intermediates has been disputed.[24,176,183] However, all the authors agree that the rate determining step of the formation of the $Ln(DOTA)^-$ complexes is the loss of the proton from the monoprotonated $Ln(HDOTA)^*$ intermediate, which is followed by the rearrangement of the deprotonated intermediate to the product. Similar assumptions were made for the formation of complexes with other DOTA based ligands.[155,177,179,181] The deprotonation presumably occurs *via* the transfer of proton from the nitrogen to a surrounding water molecule (or OH^- ion) in which a carboxylate oxygen may play an important role. For the formation of the $Ln(DOTA)^-$ complexes the general base catalysis is valid.[182]

The formation of complexes of DOTA and DOTA derivatives occurs in a first-order reaction because the rate determining step is the deprotonation of the monoprotonated intermediate. The experimentally measured first-order rate constants, k_{obs} values, are generally directly proportional to the OH^- concentration, so $k_{obs} = k_{OH} \cdot [OH^-]$, where the k_{OH} (M^{-1} s^{-1}) rate constant characterizes the formation rate. The k_{OH} rate constants have been determined for the formation of a number of DOTA derivatives and are presented in Table 5.2.5.

A few general correlations can be established between the ligand structures and formation rates. The data presented in Table 5.2.5 reveal that for the complexes of ligands containing carboxylate and alcoholic OH oxygen donor atoms, the formation rates increase with the decrease in the size of the Ln^{3+} ions. The replacement of the carboxylate group(s) of DOTA with alcoholic OH group(s) leads to a decrease in the formation rates of Ln^{3+} complexes. The formation rates decrease progressively with the gradual replacement of the carboxylate groups with phosphonates, probably because of an increase in the stability of the intermediates (DO3AP > DO2A2P > DOA3P > DOTP). The substitution of a propionate group for an acetate in DOTA results in the ligand DO3A-Nprop, which forms complexes faster than DOTA.[184] The ligand TRITA obtained by the enlargement of the 12-membered ring of DOTA to a 13-membered macrocycle also forms complexes somewhat faster than DOTA.[180] The results are very promising with the ligand PCTA, which forms Ln^{3+} complexes significantly faster than DOTA. The kinetic behaviour of the

Table 5.2.5   Formation rate constants k_{OH} (M^{-1} s^{-1}) of the DOTA derivative complexes.

Ligand	Ce^{3+}	Eu^{3+}	Gd^{3+}	Yb^{3+}
DOTA[175]	3.5×10^6	1.1×10^7 $7.2 \times 10^{6,178}$	$5.9 \times 10^{6,\,183}$	4.1×10^7 $9.3 \times 10^{7,183}$
DO3A[183]	–	–	2.1×10^7	
DO2A[181]	2.8×10^5	–	–	2.5×10^5
HP-DO3A[183]	–	–	1.2×10^7	–
BT-DO3A[181]	2.1×10^6	4.8×10^6	–	1.6×10^7
DO3A-Nprop[184]	1.7×10^7	–	2.9×10^7	3.9×10^7
PCTA[87]	9.7×10^7	1.7×10^8	–	1.1×10^9
p-NO$_2$-Bz-PCTA[131]	1.0×10^7	1.4×10^8	–	5.6×10^8
TRITA[180]	6.9×10^6		2.6×10^7	5.0×10^7
DO3AP[155]	9.6×10^6	$2.7 \times 10^{6,177}$	9.0×10^4	
DO2A2P[81]	1.7×10^5	–	6.6×10^4	–
DOA3P[83]	–	–	2.2×10^4	–
DOTP[223]	–	–	7.2×10^3	–
DOTPMB[223]	–	–	1.3×10^3	–
DOTAM[185]	7.7×10^3	2.7×10^4	–	6.6×10^3 (Lu)
DTMA[113]	3.0×10^4	4.8×10^4	–	6.5×10^4 (Lu)

bifunctional p-NO$_2$-Bz-PCTA is similar; that is, the attachment of the side arm to the PCTA has practically no effect on the reactivity of the ligand.[87,131]

The mechanism of formation of the complexes of the DOTA-*tetra*amide ligands differs considerably from that of the DOTA complexes. In the reactions of DOTAM and DTMA the formation of a diprotonated intermediate cannot be detected in solution. The reactions occur with the direct encounter of the Ln^{3+} ion and the fully deprotonated ligand (L), so the rate is directly proportional to [Ln^{3+}] and [L].[113,185] DOTAM and DTMA contain four amide groups and the basicity of the amide oxygens (which coordinate to the Ln^{3+} ions) is much lower than that of the carboxylate oxygens of DOTA. So the proton loss from the protonated intermediate with the proton transfer from a ring nitrogen to an amide oxygen is not possible. In this reaction the diprotonated intermediate is a "dead end" complex.[176] However, the diprotonated [Gd(H$_2$DOTAM)(H$_2$O$_4$)](ClO$_4$)$_5$ has been prepared in the solid state. The X-ray crystal structure has shown that the Gd^{3+} is coordinated by four amide oxygens and four H$_2$O molecules but the metal ion is located outside the coordination cage of the diprotonated H$_2$DOTAM^{2+}.[186] The structure of this complex is similar to that of the intermediates assumed in the formation reactions of the Ln(DOTA)$^-$ complexes.

5.2.4.2 Kinetics of Dissociation of Complexes

The metal chelates used in medical diagnosis and therapy must be very inert, which means they essentially should not dissociate into the free metal ion and ligand once administered into the body. The Gd^{3+} used for increasing the relaxation rate of protons in MRI must be complexed to ensure its elimination from the body. The complexes of radiometals bound to proteins or monoclonal antibodies, used for diagnosis or therapy, must be also very inert. The delivery of the radiopharmaceuticals to the target site is relatively slow and in the case of dissociation of the complex, the free radioactive isotope would damage the healthy organs and the efficiency of the treatment would also decrease.

The importance of the kinetic inertness of the Gd^{3+} complexes was recognized quite early in the research into MRI contrast agents.[187] Animal experiments performed with different Gd^{3+} containing CAs indicated that the elimination of Gd^{3+} from the body of mice was not complete. The long term (14 days) whole body deposition was about $0.01 - 1.0\%$ depending on the properties of the Gd^{3+} complex.[187] The amount of the residual Gd was always lower when Gd^{3+} complexes of macrocyclic ligands were used. For characterizing the rates of decomplexation, first order-rate constants (k_d) were determined in 0.1 M HCl, where the complexes are not thermodynamically stable. The half-time of dissociation ($t_{1/2} = 0.693/k_d$) for the complexes are given in parenthesis: Gd(DOTA)$^-$ (338 h); Gd(BT-DO3A) (43 h); Gd(HP-DO3A) (3.9 h); complexes of DTPA derivatives (<5 s).[169] These data clearly show that the complexes formed with the DOTA derivatives are extremely inert compared to the complexes of DTPA derivatives.

It is generally assumed that the de-chelation of complexes in biological fluids can take place *via* transmetallation reactions with Zn^{2+} and Cu^{2+}. Another possibility is a ligand exchange reaction, when an endogenous ligand displaces the chelating agent in the Gd^{3+} complex. Phosphate ions were found to compete with the open chain ligands, DTPA, and DTPA-BMA for the Gd^{3+}, because of the formation of insoluble $GdPO_4$ in the presence of $ZnCl_2$ or $CuCl_2$. Under similar conditions the formation of $GdPO_4$ was not observed from the Gd^{3+} complexes of the macrocyclic DOTA and HP-DO3A.[188] Based on the formation of $GdPO_4$ precipitate, a simple relaxometric method was proposed by *Laurent et al.*,[189] for the comparison of the rates of de-chelation of Gd^{3+} complexes. In phosphate buffer at pH $= 7$ the free Gd^{3+}, released from the Gd^{3+} chelate in the presence of Zn^{2+}, forms $GdPO_4$, when the relaxivity of the solution decreases with time. In the case of $Gd(DOTA)^-$ and $Gd(HP-DO3A)$ the dissociation of the complexes could not be detected.[189] This method is suitable for a fast comparison, but it cannot be used for kinetic studies, because $Zn_3(PO_4)_3$ precipitate is also formed in the reaction, and so the concentration of Zn^{2+} decreases with time. Besides, the large phosphate excess significantly increases the rate of dissociation of the Gd^{3+} complexes. The kinetic properties of the complexes formed with other lanthanide[III] and Y^{3+} ions are very similar to those observed for the Gd^{3+} complexes.[190]

5.2.4.3   Kinetics of Decomplexation of Complexes of DTPA Derivatives

The kinetics of metal exchange reactions of the aminopolycarboxyate complexes of transition metals, lanthanides and Y^{3+} have been studied in detail for 40–50 years.[174] The isotopic exchange reactions of $Ln(EDTA)^-$ and $Ln(DTPA)^{2-}$ occur predominantly through proton assisted dissociation. The rates of transmetallation reactions were found to be inversely proportional to the stability constants of the complexes.[191] The kinetics of metal exchange reactions between the Gd^{3+} complex of DTPA and DTPA derivatives and Cu^{2+}, Zn^{2+} or Eu^{3+} have been studied more recently.[158,173,192] These studies were carried out in the presence of Cu^{2+}, Zn^{2+} or Eu^{3+} excess. The rates of the exchange reactions can be expressed by Equation 5.2.14 (k_d is the pseudo-first-order rate constant and $[LnL]_t$ is the total concentration of the complex.):

$$-\frac{d[LnL]_t}{dt} = k_d[LnL]_t \qquad (5.2.14)$$

The rates of metal exchange reactions have been studied by varying the H^+, Zn^{2+}, Cu^{2+} or Eu^{3+} concentration and following the reaction by spectrophotometry (Cu^{2+}, Eu^{3+}) or relaxometry (Zn^{2+}). The k_d values increase with increasing H^+ and metal concentration, but at pH values higher than about 4.5–5 the Cu^{2+} and Zn^{2+} assisted reactions predominate. The increase of the k_d values with increasing $[H^+]$ indicates that the transmetallation can take place with the proton-assisted dissociation of the Ln^{3+} complexes, which is followed by a fast reaction between the free ligand L and the Zn^{2+} or Cu^{2+}. The effect of H^+ can be interpreted by the formation of a protonated complex, which dissociates faster than the non-protonated species because one or more functional group(s) are decoordinated by protonation:

$$LnL + H^+ \overset{K_{LnHL}}{\rightleftharpoons} LnHL \overset{k_{LnHL}}{\longrightarrow} Ln^3 + HL \tag{5.2.15}$$

At pH < about 4, the k_d values show a second order dependence on the $[H^+]$, which can be explained with the proton assisted dissociation of the mono-protonated complexes, but this pathway is not important at pH around 7. The increase in the k_d values with increasing Zn^{2+} or Cu^{2+} concentration demonstrates the contribution of the reactions taking place with the direct attack of the exchanging metal M^{2+} (Zn^{2+} or Cu^{2+}) on the complex:

$$LnL + M^{2-} \overset{K_{LnLm}}{\rightleftharpoons} LnLM \overset{k_{LnLM}}{\longrightarrow} Ln^3 + ML \tag{5.2.16}$$

The dinuclear complexes LnLM are formed in an equilibrium reaction but during the intramolecular rearrangement of the complex LnL, the functional groups of the ligand can be slowly transferred to the attacking M^{2+} metal ion, step-by-step. Considering all the reaction pathways, the rate of transmetallation can be expressed as:

$$-\frac{d[LnL]_t}{dt} = k_0[LnL] + k_{LnHL}[LnHL] + k_{LnLM}[LnLM] \tag{5.2.17}$$

where the term $k_0[LnL]$ is characteristic for the spontaneous dissociation of the complex. By comparing equations 5.2.14 and 5.2.17, considering the total concentration of LnL ($[LnL]_t = [LnL] + [LnHL] + [LnLM]$) and the equations which define the K_{LnHL} and K_{LnLM} stability constants, the k_d value can be expressed as follows:[173]

$$k_d = \frac{k_0 + k_1[H^+] + k_3^M[M^{2+}]}{1 + K_{LnHL}[H^+] + K_{LnLM}[M^{2+}]} \tag{5.2.18}$$

where $k_1 = k_{LnHL} \cdot K_{LnHL}$ and $k_3^M = k_{LnLM} \cdot K_{LnLM}$. The rate constants k_0, k_1, k_3^M and the stability constant K_{LnLM} can be calculated by fitting the k_d values to Equation 5.2.18 (K_{LnHL} is often known from equilibrium studies, but if K_{LnHL} is too low, the term $K_{LnHL} \cdot [H^+]$ in the denominator can be neglected at pH > 4). The rate constants k_1 and k_3^M (M^{2+} is Zn^{2+} or Cu^{2+}) determined for the different DTPA derivative complexes are presented in Table 5.2.6. The k_0 values are not shown because their calculated values are generally very low and often have a negative sign and high error limits.

Comparing the rate constants obtained for different Gd^{3+} complexes (Table 5.2.6) reveals that the attachment of a hydrophobic group to an acetate group

Table 5.2.6   Rate constants, characterizing the decomplexation of the DTPA
derivative complexes.

Complex	$k_1(M^{-1}\,s^{-1})$	$k_3^{Zn}(M^{-1}\,s^{-1})$	$k_3^{Cu}(M^{-1}\,s^{-1})$
Gd(DTPA)$^{2-,}$[173]	0.58	0.056	0.93
Gd(BOPTA)$^{2-,}$[224]	0.41	0.029	0.68
Gd(EOB-DTPA)$^{2-,}$[208]	0.16	[a]N	[a]N
Gd(DTTA-Nprop)$^{2-,}$[184]	48	0.64	[a]N
Gd(DTPA-N-MA)$^{-,}$[192]	1.5	0.032	1.9
Gd(DTPA-N'-MA)$^{-,}$[192]	1.6	0.08	0.62
Gd(DTPA-BMA)[192]	12.7	0.0078	0.63
Gd(DTPA-BMEA)[225]	8.6	[a]N	[a]N
Gd(DTPA-TrA)$^{+,}$[158]	0.40	0.0087	0.063
Gd(DTPA-N'-P)$^{3-,}$[193]	3380	[a]N	33
Gd(DTPA-N'-PhPi)$^{2-,}$[193]	1600	[a]N	[a]N
Gd(DTPA-EAM)[226]	0.12	[a]N	1.3

[a]not investigated.

or to the amine chain results in an increase in the kinetic inertness, which increases in the order Gd(DTPA)$^{2-}$ < Gd(BOPTA)$^{2-}$ < Gd(EOB-DTPA)$^{2-}$. However, replacing an acetate of DTPA with a propionate group (DTTA-NProp) leads to a significant increase in the lability of the complex.

The replacement of a carboxylate with a phosphonate (DTPA-*N'*-P) or a phenylphosphinate (DTPA-*N'*-PhPi), results in a very significant increase in the rates of decomplexation, and both the proton and the Cu^{2+} assisted dissociation of the Gd^{3+} complexes becomes very fast. This significant change is probably caused by the high basicity of the phosphonate group (the complex is in protonated form at pH around 7) and the large steric requirement of the phenylphosphinate moiety.[193]

The replacement of one, two or three carboxylates of DTPA with amide groups leads to the mono-, bis- and tris-amides of DTPA. Although the protonation constants of the DTPA-amide complexes are very low (their formation cannot be detected by pH-potentiometry), the proton assisted dissociation of their Gd^{3+} complexes is significantly faster than that of Gd(DTPA)$^{2-}$. However, the k_3^{Cu} and k_3^{Zn} rate constants for the DTPA-amide derivative complexes are lower than those of the Gd(DTPA)$^{2-}$, so the rates of de-chelation are very similar. This finding is surprising if we consider the results of the *in vivo* studies, when *e.g.* the residual Gd^{3+} in mice was the largest for Gd(DTPA-BMA) of all CAs.[187] This contradiction is, however, only apparent. The rate constants presented in Table 5.2.6 have been determined generally in KCl solutions (at constant ionic strength), when the conditions are far from the biological ones. In body fluids both the equilibrium and kinetic properties of the Gd^{3+}-chelates are influenced by the presence of endogenous ligands. The rates of transmetallation reactions between the complexes Gd(DTPA)$^{2-}$ and Gd(BOPTA)$^{2-}$ and Zn^{2+} or Cu^{2+} in the presence of citrate and histidinate are significantly lower than in the absence of these ligands. However, the transmetallation reactions of Gd(DTPA-BMA) with Zn^{2+} or Cu^{2+} in the presence of citrate are approximately two orders of magnitude faster than in the absence of citrate.[194] These results clearly show that the endogenous ligands significantly accelerate the dissociation of the DTPA-amide derivative complexes.

In order to increase the kinetic inertness of the DTPA complexes several new, backbone substituted DTPA derivatives were synthesised. By replacing an ethylene group of DTPA with a cyclohexane ring (like in DCTA), the more rigid, cyclohexyl-DTPA derivatives were prepared. The acid assisted dissociation of the Y^{3+} complexes of the more rigid ligands is slower than that of $Y(DTPA)^{2-}$.[2-105]

5.2.4.4 Kinetics of Decomplexation of DOTA Derivative Complexes

The complexes of tripositive metals formed with DOTA and DOTA derivative ligands are extremely inert, so the rates of their substitution reactions cannot be studied at pH around 7. The high inertness of the complexes originates from their rigid structure. The Gd^{3+} or other metal ion in the coordination cage is inaccessible for another multidentate ligand and in addition, the coordinated functional groups of DOTA are not flexible enough to be transferred to an attacking metal ion. In contrast to metal ions, protons can successfully compete for the coordinated DOTA, when protonated complexes Ln(HL) are formed in the first step. The stability constants characterizing the formation of Ln(HL) complexes are relatively low (K_{LnHL} values are about 10–400) because of the strong carboxylate oxygen–Ln^{3+} interaction. The protons can be attached to a non-coordinated carboxylate oxygen atom as has been shown by ^{1}H-NMR spectroscopy for Lu(HDOTA).[195] The protons from this oxygen can then be transferred to a ring nitrogen when the metal ion moves out of the coordination cage, and with the attachment of a second proton to a diagonally positioned nitrogen, a diprotonated intermediate is formed, in which only the four carboxylates are coordinated to the Ln^{3+}, which is now outside of the coordination cage. This diprotonated intermediate dissociates into Ln^{3+} and H_2DOTA^{2-}, or the complex $Ln(DOTA)^-$ can be re-formed as it was discussed in Section 5.2.4.1. Since the complexes of DOTA and its derivatives are thermodynamically unstable at about $[H^+] > 0.05$ M, the kinetics of dissociation can be studied in the H^+ concentration range of 0.05–1.0 M by spectrophotometry, fluorescence spectroscopy, and relaxometry, or by separating the Ln^{3+} ion from the complex by ion-exchange or HPLC.[175,176,189,196] In the presence of excess acid, first-order rate constants (k_d) are obtained which are directly proportional to the $[H^+]$ up to about $0.2 - 0.3$ M, and the k_d values can be given as follows:

$$k_d = k_0 + k_1[H^+] \qquad (5.2.19)$$

where k_0 and k_1 are the rate constants, characterizing the spontaneous and proton assisted dissociation of complexes. At higher H^+ concentration k_d shows a saturation type dependence on $[H^+]$, which is interpreted by an accumulation of the mono- and diprotonated complexes.[175,197] However, at pH values around 7 only the dissociation of the monoprotonated complexes can play a role, so for characterizing the kinetic inertness of complexes, we shall use only the k_1 rate constants. The k_1 values obtained in the studies of dissociation of different DOTA based complexes are presented in Table 5.2.7.

The rates of dissociation of complexes can also be studied at higher pH values, if the concentration of the Ln^{3+} complex is relatively high (Equation 5.2.14). The rates of dissociation of Gd(HP-DO3A) and Gd(BT-DO3A) have been studied in the pH range of 3.2–5.3. The metal exchange between the complexes and Eu^{3+} was followed by spectrophotometry at 0.1 M Gd(HP-DO3A) or Gd(BT-DO3A) and 0.01 M Eu^{3+} concentration. The first-order rate constants were found to be independent of the $[Eu^{3+}]$ while the k_d values linearly increased with increasing $[H^+]$.

There are several reported studies on the dissociation of Sc^{3+}, Y^{3+}, Ga^{3+} and In^{3+} complexes of DOTA derivatives. The rates of proton assisted de-chelation of Ga(DO3A) and In(DO3A) are even lower than those of the lanthanide complexes.[196] The kinetic behaviour of the Y^{3+} complexes formed with DOTA, DO3A, PCTA, and several phosphonate and phosphinate derivatives of DOTA are very similar to that of the corresponding Eu^{3+} and Gd^{3+} complexes, and these chelates have satisfactory kinetic inertness for nuclear medicine applications.[87,154,190,196]

The comparision of the k_1 data presented in Table 5.2.7 shows that the complexes of DOTA derivatives are particularly inert if all four ring nitrogens possess pendant functional groups. The k_1 values of the complexes of DO3A, containing only three acetates, are 2 – 3 orders of magnitude higher than those

Table 5.2.7 Rate constants k_1 (M^{-1} s^{-1}) characterizing the proton assisted dissociation of the DOTA derivative complexes.

Ligand	Ce^{3+}	Eu^{3+}	Gd^{3+}
DOTA	8×10^{-4},[227] 3.4×10^{-4},[229]	1.4×10^{-5},[177]	8.4×10^{-6},[176] 2.0×10^{-5} (37 °C)[175], 3.6×10^{-5} (37 °C)[228]
DO3A	1.1×10^{-1},[197]	–	1.6×10^{-3},[197] 1.2×10^{-2},[196]
HP-DO3A	2.0×10^{-3},[197]	–	6.4×10^{-4},[197] 2.6×10^{-4},[79]
BT-DO3A[79]	–	–	2.8×10^{-5}
DO3A-Nprop[184]	7.3×10^{-3}	–	–
PCTA[87]	9.6×10^{-4}	5.1×10^{-4}	–
p-NO$_2$-Bz-PCTA[131]	4.8×10^{-5}		1.7×10^{-4}
TRITA[180]	–	–	0.35
DO3AP	1.2×10^{-3},[155]	9.8×10^{-5},[177]	2.8×10^{-3},[155]
DO2A2P[81]	–	–	1.9×10^{-4}
DOA3P[83]	–	–	2.7×10^{-4}
DOTP	4.6×10^{-2},[155]	1.3×10^{-3},[177]	5.4×10^{-4},[223]
DOTPMB[223]	–	–	2.1×10^{-4}
DTMA[113]	2.6×10^{-5}	5.6×10^{-7}	–

of the DOTA complexes. The secondary nitrogen of DO3A can be more easily protonated directly, which explains the faster dissociation of the DO3A complexes.[196,197]

The coordination of an alcoholic OH group is weaker than that of a carboxylate oxygen, so the dissociation of the complexes of HP-DO3A and BT-DO3A is faster than the de-chelation of the DOTA complexes. The k_1 value for Gd(BT-DO3A) is approximately ten times lower than that of Gd(HP-DO3A), although the log K_{GdL} value of the latter is higher. Gd(BT-DO3A) is kinetically more inert than the Gd(HP-DO3A), probably because the size of the butrol group is larger than the size of the 2-hydroxy-propyl group, which makes Gd(BT-DO3A) more rigid.[79]

In DOTA complexes the coordination of the nitrogen and oxygen donor groups result in the formation of five membered chelate rings, which contribute to the rigidity of complexes. The replacement of one acetate with a propionate or the enlargement of the 12-membered ring to a 13-membered macrocycle results in the formation of 6-membered chelate rings which decrease the rigidity and the kinetic inertness of the complexes of DO3A-Nprop and TRITA.[180,184]

The influence of the bifunctional linker on the kinetic inertness of the resulting complexes is an important issue in the design of the bifunctional ligands. Unfortunately, there are only very few systematic studies in this respect. The data known so far show *e.g.* that k_1 values of the complexes of DOTA and p-NO$_2$-Bn-DOTA or PCTA and p-NO$_2$-Bn-DOTA do not differ considerably.[131,198] Similar results were obtained in the study of the phosphonate and phosphinate derivatives of DOTA.[154,199]

The most inert complexes are formed with the DOTA-*tetra*amide derivative ligands. The half-times of dissociation of Gd(DOTA)$^-$, Gd(DOTAM)$^{3+}$ and Gd(DTMA)$^{3+}$ in 2.5 M HNO$_3$ at 25 °C were found to be 4.5 h, 68 h, and 155 h, respectively.[200] The k_1 values known for the DTMA complexes are also very low.[113] The slow proton assisted dissociation and the slow formation of the DTMA or DOTAM complexes can be interpreted in terms of low basicity of the ligands. The basicity of the amide oxygens is extremely low, so the protonation of the complex and the transfer of a proton to a ring nitrogen occurs with very low probability, which leads to the high kinetic inertness of the complexes. The inertness increases in the following order of substituents: –CONH$_2$ < –CONHMe < –CONMe$_2$.[113,200]

5.2.5 Summary

In recent years a number of metal complexes and metal complex bioconjugates have been developed for diagnostic imaging modalities such as MRI, SPECT, PET and optical imaging. Many of these agents contain a lanthanide ion; for example, in MRI the paramagnetic Gd^{3+} is used as contrast agent, luminescent lanthanide complexes are involved in optical probes, and several radiolanthanides are applied in diagnosis and therapy. The chelating agents used for the complexation of these metal ions are generally the open-chain DTPA and the macrocyclic DOTA, and their derivatives. The metal ions released in the body by de-chelation are very harmful (Gd^{3+} is toxic, radiometals cause radiation damage), so high kinetic inertness of the complexes is very important for their safety. The de-chelation of Gd^{3+} complexes (which are administered in large amounts) must be much slower than their elimination from the body ($t_{1/2} = 1.5$ h). This requirement is realized with the use of complexes of DOTA derivatives, since their proton-assisted dissociation is extremely slow at physiological pH ($t_{1/2}$ is about $10^5 - 10^7$ h).

The de-chelation of Gd^{3+} complexes of DTPA derivatives pressumably occurs *via* transmetallation with Zn^{2+} or Cu^{2+}. The dechelation of $Gd(DTPA)^{2-}$ and $Gd(BOPTA)^{2-}$ in the presence of citrate is slower, while that of Gd(DTPA-BMA) is much faster (about two orders of magnitude), than in the absence of citrate. The effect of citrate explains the larger residual Gd^{3+} observed with the use of Gd(DTPA-BMA) in animal and human experiments.

The complexation equilibria in the blood plasma can be characterized by the species distribution calculations performed using the stability constants of all the complexes formed with the endogenous ligands and metal ions. However, the results of such calculations are not valid, because the real systems are far from the equilibrium, due to the fast elimination and slow de-chelation of contrast agents. For patients with severe renal impairment, the half-time of elimination is about $30 - 40$ h and the amount of Gd^{3+} released from complexes of DTPA derivatives, first of all from Gd(DTPA-BMA), is not negligible. It was assumed that in very few cases the residual Gd^{3+} may give rise to the newly observed disease, Nephrogenic Systemic Fibrosis (NSF).

The permutations of DTPA and DOTA derived ligand structures made in the past two or three decades have resulted in a large number of bifunctional chelators designed for labeling proteins, protein fragments, monoclonal antibodies, dendrimers or fabricating multifunctional constructs and ligands for multimodal imaging. We believe that open chain bifunctional ligands are going to be replaced by suitable macrocyclic or bimodal bifunctional chelators that possess fast complexation kinetics under mild conditions and advanced kinetic inertness. On the other hand, conceptually different labeling reactions ("click chemistry") or labeling at specific amino acid moiety for instance need newly-designed reactive groups and coupling reactions besides the ones that are currently being used to conjugate bifunctional ligands to biological vectors. Finally, less common but potentially useful radionuclides such as [44,47]Sc, [89]Zr, [212]Pb, [223]Ra, [225]Ac *etc.* require the design and synthesis of more efficient bifunctional ligands capable of sequestering these metals in the form of stable chelates.

Acknowledgements

The authors are greatful to Dr. Zoltán Kovács (Advanced Imaging Research Center, University of Texas Southwestern Medical Center) and Mr. László Zékány (Department of Inorganic and Analytical Chemistry, University of Debrecen) for helpful discussions. The support of the Hungarian National Research Foundation (OTKA K 84291) is gratefully acknowledged. This publication is supported by the TÁMOP 4.2.1/B-09/1/KONV-2010–0007 project. The project is co-financed by the European Union and the European Social Fund. This book chapter was supported by the János Bolyai Research Scholarship of the Hungarian Academy of Sciences.

References

1. *The Chemistry of Contrast Agents in Medical Magnetic Resonance Imaging*, ed. A. E. Merbach and E. Toth, Wiley and Sons, New York, 2001.
2. P. Caravan, J. J. Ellison, T. J. McMurry and R. B. Lauffer, *Chem. Rev.*, 1999, **99**, 2293.
3. M. J. Allen and T. J. Meade, in *Metal Ions in Biological Systems*, ed. A. Sigel and H. Sigel, Marcel Dekker Inc., New York, 2004, vol. 42, p. 1.
4. C. J. Anderson and M. J. Welch, *Chem. Rev.*, 1999, **99**, 2219.
5. T. K. Nayak and M. W. Brechbiel, *Bioconjugate Chem.*, 2009, **20**, 825.
6. K. Tanaka and K. Fukase, *Org. Biomol. Chem.*, 2008, **6**, 815.
7. M. Ginj and H. R. Maecke, in *Metal Ions in Biological Systems*, ed. A. Sigel and H. Sigel, Marcel Dekker Inc., New York, 2004, vol. 42, p. 109.
8. K. L. Kolsky, V. Joshi, L. F. Mausner and S. C. Srivastava, *Appl. Radiat. Isot.*, 1998, **49**, 1541.
9. S. Liu and D. S. Edwards, *Bioconjugate Chem.*, 2001, **12**, 7.
10. W. A. Volkert and T. J. Hoffman, *Chem. Rev.*, 1999, **99**, 2269.
11. J.-C. G. Bunzli and C. Piguet, *Chem. Soc. Rev.*, 2005, **34**, 1048.
12. C. P. Montgomery, B. S. Murray, E. J. New, R. Pal and D. Parker, *Acc. Chem. Res.*, 2009, **42**, 925.
13. S. Pandya, J. H. Yu and D. Parker, *Dalton Trans.*, 2006, 2757.
14. D. Parker and J. A. G. Williams, in *Metal Ions in Biological Systems,* ed. A. Sigel and H. Sigel, Marcel Dekker Inc., New York, 2003, vol. 40, p. 233.
15. A. E. Frost, *Nature*, 1956, **178**, 322.
16. H. Distler and K. L. Hock, EP 45386 A1 19820210, 1982.
17. R. C. Mease, L. F. Mausner, and S. C. Srivastava, US 5428156 A 19950627., 1995.
18. J. J. Singer and M. Weisberg, US 3061628, 1962.
19. C. F. G. C. Geraldes, A. D. Sherry and W. P. Cacheris, *Inorg. Chem.*, 1989, **28**, 3336.
20. I. Lazar, D. C. Hrncir, W. D. Kim, G. E. Kiefer and A. D. Sherry, *Inorg. Chem.*, 1992, **31**, 4422.
21. L. Burai, J. M. Ren, Z. Kovacs, E. Brucher and A. D. Sherry, *Inorg. Chem.*, 1998, **37**, 69.
22. J. Huskens, D. A. Torres, Z. Kovacs, J. P. Andre, C. F. G. C. Geraldes and A. D. Sherry, *Inorg. Chem.*, 1997, **36**, 1495.
23. W. D. Kim, G. E. Kiefer, J. Huskens and A. D. Sherry, *Inorg. Chem.*, 1997, **36**, 4128.
24. A. Bianchi, L. Calabi, C. Giorgi, P. Losi, P. Mariani, P. Paoli, P. Rossi, B. Valtancoli and M. Virtuani, *J. Chem. Soc., Dalton Trans.*, 2000, 697.

25. D. Burdinski, J. Lub, J. A. Pikkemaat, D. Moreno Jalon, S. Martial and C. Del Pozo Ochoa, *Dalton Trans.*, 2008, 4138.

26. S. Laurent, F. Botteman, L. Vander Elst and R. N. Muller, *Helv. Chim. Acta*, 2004, **87**, 1077.

27. H. Maumela, R. D. Hancock, L. Carlton, J. H. Reibenspies and K. P. Wainwright, *J. Am. Chem. Soc.*, 1995, **117**, 6698.

28. (a) G. Tircsó, E. Tircsóné Benyó, Z. Baranyai, A. K. Barker, A. D. Sherry, and E. Brücher in *Abstracts of Annual Workshop of COST Chemistry D38, Metal-Based Systems for Molecular Imaging Applications*, Warsaw, Poland, 2009, p. 49.; (b) L. M. De León-Rodríguez, S. Viswanathan and A. D. Sherry, Contrast Media Mol. Imaging, 2010, **5**(3), 121.

29. *Lanthanide Probes in Life, Chemical and Earth Sciences: Theory and Practice*, ed. J. C. G. Bunzli and G. R. Choppin, Elsevier, Amsterdam, 1989.

30. V. Alexander, *Chem. Rev.*, 1995, **95**, 273.

31. V. Comblin, D. Gilsoul, M. Hermann, V. Humblet, V. Jacques, M. Mesbahi, C. Sauvage and J. F. Desreux, *Coord. Chem. Rev.*, 1999, **186**, 451.

32. M. G. Duarte, M. I. M. Prata, M. H. M. Gil and C. F. G. C. Geraldes, *J. Alloys. Compd.*, 2002, **344**, 4.

33. R. B. Lauffer, *Chem. Rev.*, 1987, **87**, 901.

34. S. Liu and D. S. Edwards, *Chem. Rev.*, 1999, **99**, 2235.

35. D. Parker, *Chem. Soc. Rev.*, 1990, **19**, 271.

36. S. B. Yu and A. D. Watson, *Chem. Rev.*, 1999, **99**, 2353.

37. M. Suchy and R. H. E. Hudson, *Eur. J. Org. Chem.*, 2008, 4847.

38. A. E. Martell, R. M. Smith, and R. J. Motekaitis, *Critically Selected Stability Constants of Metal Complexes, Database Version 8.0*, 2004.

39. L. M. De Leon-Rodriguez, A. C. Esqueda-Oliva and A. D. Miranda-Vera, *Tetrahedron Lett.*, 2006, **47**, 6937.

40. Z. Kovacs and A. D. Sherry, *Synthesis*, 1997, **7**, 759.

41. Z. Kovacs and A. D. Sherry, *J. Chem. Soc., Chem. Commun.*, 1995, 185.

42. S. Aime, S. G. Crich, E. Gianolio, E. Terreno, A. Beltrami and F. Uggeri, *Eur. J. Inorg. Chem.*, 1998, 1283.

43. F. Uggeri, S. Aime, P. L. Anelli, M. Botta, M. Brocchetta, C. Dehaen, G. Ermondi, M. Grandi and P. Paoli, *Inorg. Chem.*, 1995, **34**, 633.

44. E. R. Andersen, L. T. Holmaas, and V. Olaisen, WO/2005/058846, 2005.

45. J. F. Carvalho, S. H. Kim and C. A. Chang, *Inorg. Chem.*, 1992, **31**, 4065.

46. T. H. Cheng, Y. M. Wang, W. T. Lee and G. C. Liu, *Polyhedron*, 2000, **19**, 2027.

47. E. Perez-Mayoral, E. Soriano, S. Cerdan and P. Ballesteros, *Molecules*, 2006, **11**, 345.

48. L. Sarka, I. Banyai, E. Brucher, R. Kiraly, J. Platzek, B. Raduchel and H. Schmitt-Willich, *J. Chem. Soc., Dalton Trans.*, 2000, 3699.

49. H. Schmitt-Willich, M. Brehm, C. L. J. Ewers, G. Michl, A. Muller-Fahrnow, O. Petrov, J. Platzek, B. Raduchel and D. Sulzle, *Inorg. Chem.*, 1999, **38**, 1134.

50. P. Caravan, N. J. Cloutier, M. T. Greenfield, S. A. McDermid, S. U. Dunham, J. W. M. Bulte, J. C. Amedio, R. J. Looby, R. M. Supkowski, W. D. Horrocks, T. J. McMurry and R. B. Lauffer, *J. Am. Chem. Soc.*, 2002, **124**, 3152.

51. D. D. Dischino, E. J. Delaney, J. E. Emswiler, G. T. Gaughan, J. S. Prasad, S. K. Srivastava and M. F. Tweedle, *Inorg. Chem.*, 1991, **30**, 1265.

52. P. L. Anelli, M. Murru, F. Uggeri and M. Virtuani, *J. Chem. Soc., Chem. Commun.*, 1991, 1317.

53. W. J. Kruper, P. R. Rudolf and C. A. Langhoff, *J. Org. Chem.*, 1993, **58**, 3869.

54. C. Li and W. T. Wong, *Tetrahedron Lett.*, 2002, **43**, 3217.

55. S. Aoki, H. Kawatani, T. Goto, E. Kimura and M. Shiro, *J. Am. Chem. Soc.*, 2001, **123**, 1123.

56. M. Woods, G. E. Kiefer, S. Bott, A. Castillo-Muzquiz, C. Eshelbrenner, L. Michaudet, K. McMillan, S. D. K. Mudigunda, D. Grin, G. Tircso, S. R. Zhang, P. Zhao and A. D. Sherry, *J. Am. Chem. Soc.*, 2004, **126**, 9248.

57. J. Yoo, D. E. Reichert and M. J. Welch, *J. Med. Chem.*, 2004, **47**, 6625.

58. J. Yoo, D. E. Reichert and M. J. Welch, *Chem. Commun.*, 2003, 766.

59. F. Bellouard, F. Chuburu, N. Kervarec, L. Toupet, S. Triki, Y. Le Mest and H. Handel, *J. Chem. Soc. Perkin Trans. 1*, 1999, 3499.

60. A. Dumont, V. Jacques, Q. X. Peng and J. F. Desreux, *Tetrahedron Lett.*, 1994, **35**, 3707.

61. J. Vanwestrenen and A. D. Sherry, *Bioconjugate Chem.*, 1992, **3**, 524.

62. Z. Kovacs and A. D. Sherry, *Synthesis*, 1997, 759.

63. Z. Kovacs and A. D. Sherry, *J. Chem. Soc., Chem. Commun.*, 1995, 185.

64. S. R. Zhang, X. Y. Jiang and A. D. Sherry, *Helv. Chim. Acta*, 2005, **88**, 923.

65. S. Aoki, Y. Honda and E. Kimura, *J. Am. Chem. Soc.*, 1998, **120**, 10018.

66. S. Brandes, C. Gros, F. Denat, P. Pullumbi and R. Guilard, *B. Soc. Chim. Fr.*, 1996, **133**, 65.

67. E. Kimura, S. Aoki, T. Koike and M. Shiro, *J. Am. Chem. Soc.*, 1997, **119**, 3068.

68. V. Boldrini, G. B. Giovenzana, R. Pagliarin, G. Palmisano and M. Sisti, *Tetrahedron Lett.*, 2000, **41**, 6527.

69. G. Losse and D. Nadolski, *J. Prakt. Chem./Chem.-Ztg.*, 1964, **24**, 118.

70. M. Suchy, A. X. Li, R. Bartha and R. H. E. Hudson, *Org. Biomol. Chem.*, 2008, **6**, 3588.

71. J. O. Thomas, *Tetrahedron Lett.*, 1967, **8**, 335–336.

72. A. Dadabhoy, S. Faulkner and P. G. Sammes, *J. Chem. Soc., Perkin Trans. 2*, 2002, 348.

73. C. Li and W. T. Wong, *Tetrahedron*, 2004, **60**, 5595.

74. S. J. Ratnakar and V. Alexander, *Eur. J. Inorg. Chem.*, 2005, 3918.

75. S. Aime, A. Barge, M. Botta, J. A. K. Howard, R. Kataky, M. P. Lowe, J. M. Moloney, D. Parker and A. S. de Sousa, *Chem. Commun.*, 1999, 1047.

76. J. I. Bruce, R. S. Dickins, L. J. Govenlock, T. Gunnlaugsson, S. Lopinski, M. P. Lowe, D. Parker, R. D. Peacock, J. J. B. Perry, S. Aime and M. Botta, *J. Am. Chem. Soc.*, 2000, **122**, 9674.

77. T. Gunnlaugsson, J. P. Leonard, S. Mulready and M. Nieuwenhuyzen, *Tetrahedron*, 2004, **60**, 105.
78. J. Platzek, P. Blaszkiewicz, H. Gries, P. Luger, G. Michl, A. Muller-Fahrnow, B. Raduchel and D. Sulzle, *Inorg. Chem.*, 1997, **36**, 6086.
79. E. Toth, R. Kiraly, J. Platzek, B. Raduchel and E. Brucher, *Inorg. Chim. Acta*, 1996, **249**, 191.
80. J. Rudovsky, P. Cigler, J. Kotek, P. Hermann, P. Vojtisek, I. Lukes, J. A. Peters, L. Vander Elst and R. N. Muller, *Chem. Eur. J.*, 2005, **11**, 2373.
81. F. K. Kalman, Z. Baranyai, I. Toth, I. Banyai, R. Kiraly, E. Brucher, S. Aime, X. K. Sun, A. D. Sherry and Z. Kovacs, *Inorg. Chem.*, 2008, **47**, 3851.
82. X. K. Sun, M. Wuest, Z. Kovacs, A. D. Sherry, R. Motekaitis, Z. Wang, A. E. Martell, M. J. Welch and C. J. Anderson, *J. Biol. Inorg. Chem.*, 2003, **8**, 217.
83. F. K. Kalman, *Thesis*, University of Debrecen, 2007.
84. H. Stetter, W. Frank and R. Mertens, *Tetrahedron*, 1981, **37**, 767.
85. S. Aime, M. Botta, S. G. Crich, G. B. Giovenzana, G. Jommi, R. Pagliarin and M. Sisti, *Inorg. Chem.*, 1997, **36**, 2992.
86. S. Aime, N. Botta, S. G. Crich, G. B. Giovenzana, G. Jommi, R. Pagliarin and M. Sisti, *J. Chem. Soc., Chem. Commun.*, 1995, 1885.
87. G. Tircso, Z. Kovacs and A. D. Sherry, *Inorg. Chem.*, 2006, **45**, 9269.
88. S. Aime, M. Botta, L. Frullano, S. G. Crich, G. Giovenzana, R. Pagliarin, G. Palmisano, F. R. Sirtori and M. Sisti, *J. Med. Chem.*, 2000, **43**, 4017.
89. S. Aime, E. Gianolio, D. Corpillo, C. Cavallotti, G. Palmisano, M. Sisti, G. B. Giovenzana and R. Pagliarin, *Helv. Chim. Acta*, 2003, **86**, 615.
90. J. M. Siaugue, F. Segat-Dioury, A. Favre-Reguillon, C. Madic, J. Foos and A. Guy, *Tetrahedron Lett.*, 2000, **41**, 7443.
91. J. M. Siaugue, F. Segat-Dioury, I. Sylvestre, A. Favre-Reguillon, J. Foos, C. Madic and A. Guy, *Tetrahedron*, 2001, **57**, 4713.
92. G. T. Hermanson, *Bioconjugate Techniques*, Elsevier, Amsterdam, 2008.
93. C. F. Meares, D. A. Goodwin, C. S. H. Leung, A. Y. Girgis, D. J. Silvester, A. D. Nunn and P. J. Lavender, *Proc. Natl. Acad. Sci. U. S. A.*, 1976, **73**, 3803.
94. M. W. Sundberg, C. F. Meares, D. A. Goodwin and C. I. Diamanti, *Nature*, 1974, **250**, 587.
95. M. W. Sundberg, C. F. Meares, D. A. Goodwin and C. I. Diamanti, *J. Med. Chem.*, 1974, **17**, 1304.
96. S. Aime, F. Benetollo, G. Bombieri, S. Colla, M. Fasano and S. Paoletti, *Inorg. Chim. Acta*, 1997, **254**, 63.
97. S. Aime, E. Gianolio, A. Barge, D. Kostakis, I. C. Plakatouras and N. Hadjiliadis, *Eur. J. Inorg. Chem.*, 2003, 2045.
98. J. M. Couchet, J. L. Azema, P. Tisnes and C. Picard, *Inorg. Chem. Commun.*, 2003, **6**, 978.
99. H. Imura, G. R. Choppin, W. P. Cacheris, L. A. deLearie, T. J. Dunn and D. H. White, *Inorg. Chim. Acta*, 1997, **258**, 227.
100. S. Laurent, L. V. Houze, N. Guerit and R. N. Muller, *Helv. Chim. Acta*, 2000, **83**, 394.

101. Y. M. Wang, T. H. Cheng, G. C. Liu and R. S. Sheu, *J. Chem. Soc., Dalton Trans.*, 1997, 833.

102. M. W. Brechbiel, O. A. Gansow, R. W. Atcher, J. Schlom, J. Esteban, D. E. Simpson and D. Colcher, *Inorg. Chem.*, 1986, **25**, 2772.

103. J. F. W. Keana and J. S. Mann, *J. Org. Chem.*, 1990, **55**, 2868.

104. D. A. Westerberg, P. L. Carney, P. E. Rogers, S. J. Kline and D. K. Johnson, *J. Med. Chem.*, 1989, **32**, 236.

105. T. J. McMurry, C. G. Pippin, C. C. Wu, K. A. Deal, M. W. Brechbiel, S. Mirzadeh and O. A. Gansow, *J. Med. Chem.*, 1998, **41**, 3546.

106. L. Camera, S. Kinuya, K. Garmestani, M. W. Brechbiel, C. C. Wu, L. H. Pai, T. J. Mcmurry, O. A. Gansow, I. Pastan, C. H. Paik and J. A. Carrasquillo, *Eur. J. Nucl. Med.*, 1994, **21**, 640.

107. C. H. Cummins, E. W. Rutter and W. A. Fordyce, *Bioconjugate Chem.*, 1991, **2**, 180.

108. M. W. Brechbiel and O. A. Gansow, *Bioconjugate Chem.*, 1991, **2**, 187.

109. P. Lebduskova, J. Kotek, P. Hermann, L. V. Elst, R. N. Muller, I. Lukes and J. A. Peters, *Bioconjugate Chem.*, 2004, **15**, 881.

110. C. Miranda, F. Escart:, L. Lamarque, M. J. R. Yunta, P. Navarro, E. Garcia-Espana and M. L. Jimeno, *J. Am. Chem. Soc.*, 2004, **126**, 823.

111. J. P. L. Cox, A. S. Craig, I. M. Helps, K. J. Jankowski, D. Parker, M. A. W. Eaton, A. T. Millican, K. Millar, N. R. A. Beeley and B. A. Boyce, *J. Chem. Soc., Perkin Trans. 1*, 1990, 2567.

112. D. A. Keire and M. Kobayashi, *Bioconjugate Chem.*, 1999, **10**, 454.

113. A. Pasha, G. Tircso, E. T. Benyo, E. Brucher and A. D. Sherry, *Eur. J. Inorg. Chem.*, 2007, 4340.

114. A. D. Sherry, R. D. Brown, C. F. G. Geraldes, S. H. Koenig, K. T. Kuan and M. Spiller, *Inorg. Chem.*, 1989, **28**, 620.

115. G. Tircso, E. Tircsone Benyo and A. D. Sherry, in *Abstracts of Papers, 232nd ACS National Meeting, San Francisco, CA, United States*, 2006.

116. T. J. McMurry, M. Brechbiel, K. Kumar and O. A. Gansow, *Bioconjugate Chem.*, 1992, **3**, 108.

117. A. K. Mishra, J. F. Gestin, E. Benoist, A. FaivreChauvet and J. F. Chatal, *New J. Chem.*, 1996, **20**, 585.

118. O. Renn and C. F. Meares, *Bioconjugate Chem.*, 1992, **3**, 563.

119. K. Takenouchi, M. Tabe, K. Watanabe, A. Hazato, Y. Kato, M. Shionoya, T. Koike and E. Kimura, *J. Org. Chem.*, 1993, **58**, 6895.

120. M. H. Ansari, M. Ahmad and K. A. Dicke, *Bioorg. Med. Chem. Lett.*, 1993, **3**, 1067.

121. M. L. Garrity, G. M. Brown, J. E. Elbert and R. A. Sachleben, *Tetrahedron Lett.*, 1993, **34**, 5531.

122. M. K. Moi, C. F. Meares and S. J. Denardo, *J. Am. Chem. Soc.*, 1988, **110**, 6266.

123. L. L. Chappell, E. Dadachova, D. E. Milenic, K. Garmestani, C. C. Wu and M. W. Brechbiel, *Nucl. Med. Biol.*, 2000, **27**, 93.

124. L. Carlton, R. D. Hancock, H. Maumela and K. P. Wainwright, *J. Chem. Soc. Chem. Commun.*, 1994, 1007.

125. S. Aime, L. Calabi, C. Cavallotti, E. Gianolio, G. B. Giovenzana, P. Losi, A. Maiocchi, G. Palmisano and M. Sisti, *Inorg. Chem.*, 2004, **43**, 7588.
126. Z. Baranyai, F. Uggeri, G. B. Giovenzana, A. Benyei, E. Brucher and S. Aime, *Chem. Eur. J.*, 2009, **15**, 1696.
127. H. S. Chong, K. Garmestani, D. S. Ma, D. E. Milenic, T. Overstreet and M. W. Brechbiel, *J. Med. Chem.*, 2002, **45**, 3458.
128. H. S. Chong, X. Ma, T. Le, B. Kwamena, D. E. Milenic, E. D. Brady, H. A. Song and M. W. Brechbiel, *J. Med. Chem.*, 2008, **51**, 118.
129. H. S. Chong, H. A. Song, X. Ma, D. E. Milenic, E. D. Brady, S. Lim, H. Lee, K. Baidoo, D. Cheng and M. W. Brechbiel, *Bioconjugate Chem.*, 2008, **19**, 1439.
130. J. Notni, P. Hermann, J. Havlícková, J. Kotek, V. Kubícek, J. Plutnar, N. Loktionova, P. J. Riss, F. Rösch and I. Lukeš, *Chem. Eur. J.*, 2010, **16**, 7174.
131. G. Tircso, E. T. Benyo, E. H. Suh, P. Jurek, G. E. Kiefer, A. D. Sherry and Z. Kovacs, *Bioconjugate Chem.*, 2009, **20**, 565.
132. A. D. Sherry, J. Ren, J. Huskens, E. Brucher, E. Toth, C. F. C. G. Geraldes, M. M. C. A. Castro and W. P. Cacheris, *Inorg. Chem.*, 1996, **35**, 4604.
133. S. Amin, C. Marks, L. M. Toomey, M. R. Churchill and J. R. Morrow, *Inorg. Chim. Acta*, 1996, **246**, 99.
134. L. Huang, L. L. Chappell, O. Iranzo, B. F. Baker and J. R. Morrow, *J. Biol. Inorg. Chem.*, 2000, **5**, 85.
135. J. R. Morrow, S. Amin, C. H. Lake and M. R. Churchill, *Inorg. Chem.*, 1993, **32**, 4566.
136. G. W. Kabalka, M. A. Davis, T. H. Moss, E. Buonocore, K. Hubner, E. Holmberg, K. Maruyama and L. Huang, *Magn. Reson. Med.*, 1991, **19**, 406.
137. R. W. Storrs, F. D. Tropper, H. Y. Li, C. K. Song, J. K. Kuniyoshi, D. A. Sipkins, K. C. P. Li and M. D. Bednarski, *J. Am. Chem. Soc.*, 1995, **117**, 7301.
138. C. F. Baes, Jr. and R. E. Mesmer, *The Hydrolysis of Cations*, John Wiley & Sons Inc., New York, 1976.
139. I. Velikyan, H. Maecke and B. Langstrom, *Bioconjugate Chem.*, 2008, **19**, 569.
140. (a) A. Ringbom in, *Complexation in analytical Chemistry: A Guide for the Critical Selection of Analytical Methods, Based on Complexation Reactions in Chemical Analysis, A Series of Monographs on Analytical Chemistry and its Applications*, ed. P. J. Elving, and I. M. Kolthoff, John Wiley and Sons, New York, London, 1963, Vol. XVI. (b) W. P. Cacheris, S. C. Quay, and S. M. Rocklage, *Magn. Reson. Imaging*, 1990, **8**, 467.
141. D. M. J. Doble, M. Botta, J. Wang, S. Aime, A. Barge and K. N. Raymond, *J. Am. Chem. Soc.*, 2001, **123**, 10758.
142. A. E. Martell and R. J. Motekaitis, *The Determination and Use of Stability Constants*, VCH, New York, 1988.
143. L. Pehrsson, F. Ingman and A. Johansson, *Talanta*, 1976, **23**, 769.

144. H. M. N. H. Irving, M. G. Miles and L. D. Pettit, *Anal. Chim. Acta*, 1967, **38**, 475.
145. E. T. Clarke and A. E. Martell, *Inorg. Chim. Acta*, 1991, **190**, 27.
146. M. Beck and I. Nagypal, *Chemistry of Complex Equlibria*, Akadémia Kiadó and Nostrand Reinhold Company Ltd., Budapest and London, 1990.
147. J. L. Sudmeier and C. N. Reilley, *Anal. Chem.*, 1964, **36**, 1698.
148. *Computational Methods for the Calculation of Stability Constants*, ed. D. Leggett, Plenum Press, New York, 1985.
149. P. Gans, A. Sabatini and A. Vacca, *Talanta*, 1996, **43**, 1739.
150. J. Rohovec, M. Kyvala, P. Vojtisek, P. Hermann and I. Lukes, *Eur. J. Inorg. Chem.*, 2000, 195.
151. J. F. Desreux, E. Merciny and M. F. Loncin, *Inorg. Chem.*, 1981, **20**, 987.
152. R. Delgado, J. Costa, K. P. Guerra and L. M. P. Lima, *Pure Appl. Chem.*, 2005, **77**, 569.
153. R. Delgado, L. C. Siegfried and T. A. Kaden, *Helv. Chim. Acta*, 1990, **73**, 140.
154. M. Forsterova, I. Svobodova, P. Lubal, P. Taborsky, J. Kotek, P. Hermann and I. Lukes, *Dalton Trans.*, 2007, 535.
155. P. Taborsky, P. Lubal, J. Havel, J. Kotek, P. Hermann and I. Lukes, *Collect. Czech. Chem. Commun.*, 2005, **70**, 1909.
156. C. F. G. C. Geraldes, A. M. Urbano, M. C. Alpoim, A. D. Sherry, K. T. Kuan, R. Rajagopalan, F. Maton and R. N. Muller, *Magn. Reson. Imaging*, 1995, **13**, 401.
157. C. F. G. C. Geraldes, A. M. Urbano, M. A. Hoefnagel and J. A. Peters, *Inorg. Chem.*, 1993, **32**, 2426.
158. Z. Jaszberenyi, E. Toth, T. Kalai, R. Kiraly, L. Burai, E. Brucher, A. E. Merbach and K. Hideg, *Dalton Trans.*, 2005, 694.
159. Y. M. Wang, C. H. Lee, G. C. Liu and R. S. Sheu, *J. Chem. Soc. Dalton Trans.*, 1998, 4113.
160. S. Laus, R. Ruloff, E. Toth and A. E. Merbach, *Chem. Eur. J.*, 2003, **9**, 3555.
161. D. A. Voss, E. R. Farquhar, W. D. Horrocks and J. R. Morrow, *Inorg. Chim. Acta*, 2004, **357**, 859.
162. S. Aime, A. Barge, D. D. Castelli, F. Fedeli, A. Mortillaro, . U. Nielsen and E. Terreno, *Magn. Reson. Med.*, 2002, **47**, 639.
163. S. R. Zhang, M. Merritt, D. E. Woessner, R. E. Lenkinski and A. D. Sherry, *Acc. Chem. Res.*, 2003, **36**, 783.
164. F. A. Dunand, S. Aime and A. E. Merbach, *J. Am. Chem. Soc.*, 2000, **122**, 1506.
165. I. Erguen, K. Keven, I. Uruc, Y. Ekmekci, B. Canbakan, I. Erden and O. Karatan, *Nephrol., Dial., Transplant.*, 2006, **21**, 697.
166. C. Thakral, J. Alhariri and J. L. Abraham, *Contrast Media Mol. Imaging*, 2007, **2**, 199.
167. H. S. Thomsen, S. K. Morcos and P. Dawson, *Clin. Radiol.*, 2006, **61**, 905.
168. J.-M. Idee, M. Port, I. Raynal, M. Schaefer, S. Le Greneur and C. Corot, *Fundam. Clin. Pharmacol.*, 2006, **20**, 563.

169. M. Port, J.-M. Idee, C. Medina, C. Robic, M. Sabatou and C. Corot, *BioMetals*, 2008, **21**, 469.

170. P. M. May, P. W. Linder and D. R. Williams, *J. Chem. Soc., Dalton Trans.*, 1977, 588.

171. G. E. Jackson, S. Wynchank and M. Woudenberg, *Magn. Reson. Med.*, 1990, **16**, 57.

172. N. R. Puttagunta, W. A. Gibby and V. L. Puttagunta, *Invest. Radiol.*, 1996, **31**, 619.

173. L. Sarka, L. Burai and E. Brucher, *Chem. Eur. J.*, 2000, **6**, 719.

174. D. W. Margerum, G. R. Caylay, D. C. Weatherburn, and G. K. Pagenkopf, in *Coordination Chemistry*, ed. A. E. Martell, American Chemical Society, Washington D. C., 1978.

175. E. Toth, E. Brucher, I. Lazar and I. Toth, *Inorg. Chem.*, 1994, **33**, 4070.

176. X. Y. Wang, T. Z. Jin, V. Comblin, A. Lopezmut, E. Merciny and J. F. Desreux, *Inorg. Chem.*, 1992, **31**, 1095.

177. P. Taborsky, I. Svobodova, P. Lubal, Z. Hnatejko, S. Lis and P. Hermann, *Polyhedron*, 2007, **26**, 4119.

178. S. L. Wu and W. D. Horrocks, *Inorg. Chem.*, 1995, **34**, 3724.

179. J. Moreau, E. Guillon, J. C. Pierrard, J. Rimbault, M. Port and M. Aplincourt, *Chem. Eur. J.*, 2004, **10**, 5218.

180. E. Balogh, R. Tripier, R. Ruloff and E. Toth, *Dalton Trans.*, 2005, 1058.

181. E. Szilagyi, E. Toth, Z. Kovacs, J. Platzek, B. Raduchel and E. Brucher, *Inorg. Chim. Acta*, 2000, **298**, 226.

182. L. Burai, I. Fabian, R. Kiraly, E. Szilagyi and E. Brucher, *J. Chem. Soc. Dalton Trans.*, 1998, 243.

183. K. Kumar and M. F. Tweedle, *Inorg. Chem.*, 1993, **32**, 4193.

184. E. Balogh, R. Tripier, P. Fouskova, F. Reviriego, H. Handel and E. Toth, *Dalton Trans.*, 2007, 3572.

185. Z. Baranyai, I. Banyai, E. Brucher, R. Kiraly and E. Terreno, *Eur. J. Inorg. Chem.*, 2007, 3639.

186. P. A. Stenson, A. L. Thompson and D. Parker, *Dalton Trans.*, 2006, 3291.

187. P. Wedeking, K. Kumar and M. F. Tweedle, *Magn. Reson. Imaging*, 1992, **10**, 641.

188. M. F. Tweedle, J. J. Hagan, K. Kumar, S. Mantha and C. A. Chang, *Magn. Reson. Imaging*, 1991, **9**, 409.

189. S. Laurent, L. V. Elst, F. Copoix and R. N. Muller, *Invest. Radiol.*, 2001, **36**, 115.

190. K. P. Pulukkody, T. J. Norman, D. Parker, L. Royle and C. J. Broan, *J. Chem. Soc. Perkin Trans. 2*, 1993, 605.

191. R. H. Betts, O. F. Dahlinger, and D. M. Munro, *Radioisotopes in Scientific Research*, ed. E. C. Exterman, Pergamon Press, London, 1958, p. 326.

192. L. Sarka, L. Burai, R. Kiraly, L. Zekany and E. Brucher, *J. Inorg. Biochem.*, 2002, **91**, 320.

193. J. Kotek, F. K. Kalman, P. Hermann, E. Brucher, K. Binnemans and I. Lukes, *Eur. J. Inorg. Chem.*, 2006, 1976.

194. Z. Baranyai, E. Brucher and Z. Palinkas, *Abstracts of Annual Workshop of COST Chemistry D38, Metal-Based Systems for Molecular Imaging Applications*, Warsaw, Poland, 2009, p. 34.

195. E. Szilagyi, E. Toth, E. Brucher and A. E. Merbach, *J. Chem. Soc., Dalton Trans.*, 1999, 2481.

196. H. Z. Cai and T. A. Kaden, *Helv. Chim. Acta*, 1994, **77**, 383.

197. K. Kumar, C. A. Chang and M. F. Tweedle, *Inorg. Chem.*, 1993, **32**, 587.

198. M. Woods, Z. Kovacs, R. Kiraly, E. Brucher, S. R. Zhang and A. D. Sherry, *Inorg. Chem.*, 2004, **43**, 2845.

199. T. J. Norman, D. Parker, L. Royle, A. Harrison, P. Antoniw and D. J. King, *J. Chem. Soc., Chem. Commun.*, 1995, 1877.

200. S. Aime, A. Barge, J. I. Bruce, M. Botta, J. A. K. Howard, J. M. Moloney, D. Parker, A. S. de Sousa and M. Woods, *J. Am. Chem. Soc.*, 1999, **121**, 5762.

201. L. L. Chappell, D. Ma, D. E. Milenic, K. Garmestani, V. Venditto, M. P. Beitzel and M. W. Brechbiel, *Nucl. Med. Biol.*, 2003, **30**, 581.

202. B. Yoo and M. D. Pagel, *Tetrahedron Lett.*, 2006, **47**, 7327.

203. G. Gugliotta, M. Botta, G. G. Giovenzana and L. Tei, *Abstracts of Annual Workshop of COST Chemistry D38, Metal-Based Systems for Molecular Imaging Applications*, Warsaw, Poland, 2009, p. 44.

204. L. Tei, G. Gugliotta, and M. Botta, *Abstracts of Annual Workshop of COST Chemistry D38, Metal-Based Systems for Molecular Imaging Applications*,Warsaw, Poland, 2009, p. 29.

205. R. S. Sengar, A. Nigam, S. J. Geib and E. C. Wiener, *Polyhedron*, 2009, **28**, 1525.

206. A. E. Martell and R. M Smith, *Critical Stability Constants*, Plenum Press, New York, 1974.

207. L. Alderighi, A. Bianchi, L. Biondi, L. Calabi, M. De Miranda, P. Gans, S. Ghelli, P. Losi, L. Paleari, A. Sabatini and A. Vacca, *J. Chem. Soc., Perkin Trans. 2*, 1999, 2741.

208. L. Burai, E. Brucher, R. Kiraly, P. Solymosi and T. Vig, *Acta Pharm. Hung.*, 2000, **70**, 89.

209. P. Caravan, C. Comuzzi, W. Crooks, T. J. McMurry, G. R. Choppin and S. R. Woulfe, *Inorg. Chem.*, 2001, **40**, 2170.

210. E. N. Rizkalla, G. R. Choppin and W. Cacheris, *Inorg. Chem.*, 1993, **32**, 582.

211. Y. M. Wang, S. T. Lin, Y. J. Wang and R. S. Sheu, *Polyhedron*, 1998, **17**, 2021.

212. S. T. Frey, C. A. Chang, J. F. Carvalho, A. Varadarajan, L. M. Schultze, K. L. Pounds and W. D. Horrocks, *Inorg. Chem.*, 1994, **33**, 2882.

213. K. Kumar, C. A. Chang, L. C. Francesconi, D. D. Dischino, M. F. Malley, J. Z. Gougoutas and M. F. Tweedle, *Inorg. Chem.*, 1994, **33**, 3567.

214. J. Moreau, E. Guillon, P. Aplincourt, J. C. Pierrard, J. Rimbault, M. Port and M. Aplincourt, *Eur. J. Inorg. Chem.*, 2003, 3007.

215. C. A. Chang, Y. H. Chen, H. Y. Chen and F. K. Shieh, *J. Chem. Soc., Dalton Trans.*, 1998, 3243.

216. I. Lazar, A. D. Sherry, R. Ramasamy, E. Brucher and R. Kiraly, *Inorg. Chem.*, 1991, **30**, 5016.
217. E. T. Clarke and A. E. Martell, *Inorg. Chim. Acta*, 1991, **190**, 37.
218. Z. Pálinkás, Zs. Baranyai, E. Brücher, and B. Rózsa, *Inorg. Chem.*, 2011, **50**(8), 3471.
219. R. Delgado, M. d. C. Figueira and S. Quintino, *Talanta*, 1997, **45**, 451.
220. W. T. Kurmina, K. V. Astakhov and S. A. Barkov, *Russ. J. Phys. Chem.*, 1969, **43**, 611.
221. W. P. Cacheris, S. K. Nickle and A. D. Sherry, *Inorg. Chem.*, 1987, **26**, 958.
222. N. Viola-Villegas and R. P. Doyle, *Coord. Chem. Rev.*, 2009, **253**, 1906–1925.
223. L. Burai, R. Kiraly, I. Lazar and E. Brucher, *Eur. J. Inorg. Chem.*, 2001, 813.
224. Zs. Baranyai, Z. Palinkas, F. Uggeri and E. Brucher, *Eur. J. Inorg. Chem.*, 2010, 1948.
225. L. Rothermel, E. N. Rizkalla and G. R. Choppin, *Inorg. Chim. Acta*, 1997, **262**, 133.
226. K. Y. Choi, K. S. Kim and J. C. Kim, *Polyhedron*, 1994, **13**, 567.
227. E. Brucher, G. Laurenczy and Z. Makra, *Inorg. Chim. Acta*, 1987, **139**, 141.
228. W. Schwizer, R. Fraser, H. Maecke, K. Siebold, R. Funck and M. Fried, *Magn. Reson. Med.*, 1994, **31**, 388.
229. C. A. Cheng and Y.-L. Liu, *J. Chin. Chem. Soc.*, 2000, **47**, 1001.

CHAPTER 5.3

MRI Contrast Agents Based on Metallofullerenes

CHUN-YING SHU AND CHUN-RU WANG

Key Laboratory of Molecular Nanostructure and Nanotechnology, Institute of Chemistry, Chinese Academy of Sciences, Beijing 100190, China

5.3.1 Introduction

Magnetic resonance imaging (MRI) has evolved into one of the most powerful techniques as a noninvasive diagnostic tool by providing high quality anatomical images of soft tissue;[1,2] and the rapid expansion of medical MRI has prompted the development of contrast agents (CAs). These agents can shorten the relaxation time of nearby water molecules, thereby enhancing the contrast between areas containing the contrast agent and the surrounding tissues, and so increasing diagnostic confidence.[2-6] Currently used MRI contrast agents include extracellular fluid (ECF) agents, intravascular blood pool agents and tissue-specific agents. The typical ECF agents are mainly gadolinium poly-(aminocarboxylate) chelates with lower molecular weight such as Magnevist (gadolinium-diethylenetriaminepentaacetic acid, Gd-DTPA), Prohance (Gadoteridol), and Omniscan (Gd-diethylenetriaminepentaacetate-bismethylamide), which have a relatively short residence time in the vascular system; the developed intravascular contrast agents include Gd-DTPA labeled albumin, Gd-DTPA labeled dextran, and chromium-labeled red blood cells, which have longer residence times and allow extended imaging procedures as a result of having a molecular weight of approximately 70 000 and above; as for tissue-specific agents, superparamagnetic iron oxides (SPIOs) have been used as

RSC Drug Discovery Series No. 15
Biomedical Imaging: The Chemistry of Labels, Probes and Contrast Agents
Edited by Martin Braddock
© Royal Society of Chemistry 2012
Published by the Royal Society of Chemistry, www.rsc.org

darkening contrast agents for liver imaging and for darkening the bowel due to the reticuloendothelial system (RES) uptake.

Annually, approximately 40–50% of MRI procedures use a contrast agent, most of which are based on gadolinium complexes.[2] These nonspecific agents accumulate passively throughout the body and are normally excreted intact through the kidneys. Recent reports show that the use of these agents increases the risk of the development of a serious medical condition, nephrogenic systemic fibrosis (NSF) in patients with acute or chronic severe renal insufficiency and patients with renal dysfunction.[7,8] Therefore, there is a continuing need for safer gadolinium-based contrast agents with higher sensitivity (relaxivity) and target specificity to increase the ability for the molecular/cellular diagnosis (9–12).[9–12] To address these issues, efforts have been directed mainly toward the development of alternative contrast materials, such as gadobenate dimeglumine,[13–16] and gadofosveset (MS-325).[17–19]

Since 1997, a promising new class of MRI contrast-enhancing agents based on gadolinium-containing metallofullerenes or gadofullerenes,[20–30] and gadonanotubes,[31,32] were reported, which exhibit much higher relaxivities than commercial Magnevist® and Ominiscan®. Most significantly, the fullerene cage is believed to hinder both chemical attack on the lanthanide ion and the escape of the lanthanide ion, which should effectively suppress the toxicity of naked Gd^{3+} ions even in the most extreme chemical environments. Moreover, the bio-distribution can be improved by modifying the cages with biologically active groups that exhibit high and specific affinity for a particular tissue, resulting in accumulation of targeting probes and thus local contrast enhancement.[33,34] Detailed studies of the water-proton relaxivity properties and intermolecular nanoclustering behavior of gadofullerene derivatives,[35–39] have revealed valuable information about their relaxivity mechanisms and given a deeper understanding of this new class of paramagnetic contrast agent. Herein, the latest findings on water-solubilized gadofullerene materials, such as $Gd@C_{82}$,[20–24,26,33] $Gd@C_{60}$,[25] $Sc_xGd_{3-x}@C_{80}$ (x = 0, 1, 2)[27,28,40,41] and how these findings relate to their future applications in MRI are reviewed and discussed.

5.3.2   MRI Contrast Agents Based on Gadofullerenes

In order to explore the potential of gadofullerenes as MRI contrast agents, the proton relaxivities,[2] r_1 and r_2 values (the paramagnetic longitudinal and transverse relaxation rate enhancement of water protons, respectively, referred to 1 mM concentration), usually were measured and compared with a commercially available MRI contrast agent. Generally, the relaxivity, r_i, can be defined by eqn (5.3.1):

$$\frac{1}{T_{i,obsd}} = \frac{1}{T_{i,d}} + \frac{1}{T_{i,para}} = \frac{1}{T_{i,d}} + r_i[M] \quad i = 1, 2 \tag{5.3.1}$$

in which $1/T_1$ and $1/T_2$ are the longitudinal and transverse relaxation rates of the solvent. $1/T_{i,\text{obsd}}$, $1/T_{i,\text{d}}$ and $1/T_{i,\text{para}}$ are the observed relaxation rates in the presence of the paramagnetic species, the values in the absence of the para-magnetic species and the contribution from paramagnetic compound, respectively. [M] is the concentration of the paramagnetic species.

5.3.2.1 MRI Contrast Agent Based on Gadofullerene Gd@C$_{82}$

Gd@C$_{82}$ as one of the abundant endohedral metallofullerenes has attracted great attention due to the encapsulation of paramagnetic Gd into a stable carbon cage, where the so-called intrafullerene electron transfer from the encaged metal atoms to fullerene cages occurred to form "superatoms" with $K_{dissoc} = 0$. Zhang *et al.*,[20] reported the first example of functionalized gadolinium encapsulated fullerene (Gd@C$_{82}$), gadolinium fullerenols, as a novel and potential contrast agent for magnetic resonance imaging. This new species exhibited excellent efficiency in enhancing water proton relaxation with a relaxivity of *ca.* 47.0 mM^{-1} s^{-1}. Subsequently, Wilson's group,[42] and Shinohara's group,[22] functionalized the same material *via* the same method to produce Gd@C$_{80}$(OH)$_n$ as Scheme 5.3.1 shows, and the relaxivities of r$_1$ are 20 mM^{-1} s^{-1} at 40 °C and 67 mM^{-1} s^{-1} at 25 °C at the same field strength (20 MHz, 0.47 T), respectively. The different relaxivities may be due to the different number of hydroxyl groups on the cage surface,[22] and/or the differences in the aqueous environments, such as pH and ionic strength.[36–38]

Figure 5.3.1 shows the *in vitro* T$_1$-weighted images of gadofullerenols compared with that of Gd-DTPA. Extremely strong signals of Gd@C$_{82}$(OH)$_{40}$ were demonstrated at the Gd concentration of 0.05 mM whereas Gd-DTPA at the same concentration showed only a slight enhancement of MRI signals as compared with pure water. This result is consistent with the reported

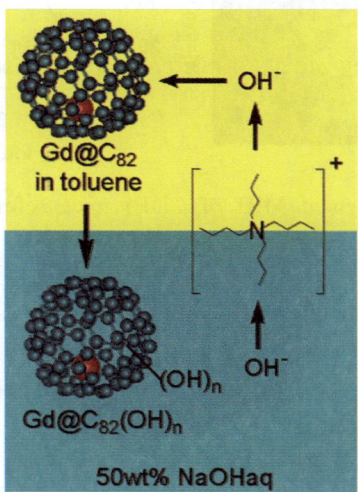

Scheme 5.3.1 Synthesis of the water-soluble polyhydroxylated Gd@C$_{82}$, Gd@C$_{82}$(OH)$_n$ (Gd-fullerenols) by the tetrabutylammonium hydroxide (40% in water, TBAH) phase transfer reaction. Reproduced with permission from Ref. 22.

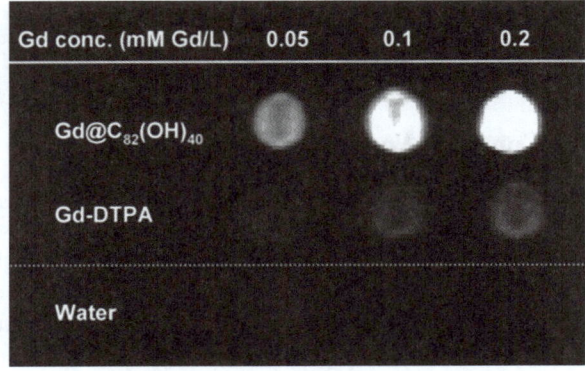

Figure 5.3.1 T_1-weighted MRI of $Gd@C_{82}(OH)_{40}$ and Gd-DTPA by 4.7 T Unity
INOVA (Varian, by conventional spin-echo with $T_R/T_E = 300$ ms/9
ms, FOV $= 10 \times 10$ cm^2, 256×64, points zero-filled to 1024×1024,
24 °C). Reproduced with permission from Ref. 22.

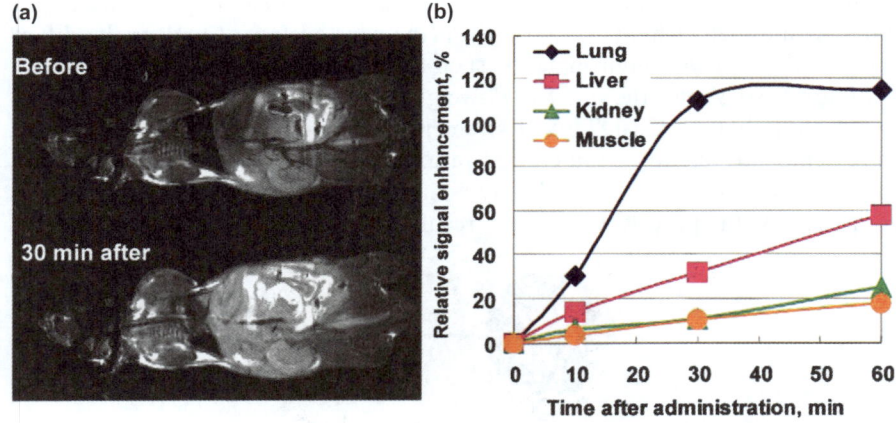

Figure 5.3.2   (a) T_1-weighted MRI of CDF1 mice before and 30 min after i.v.
administration of $Gd@C_{82}(OH)_{40}$ *via* tail vein as the dose of 5μmol Gd
kg^{-1} (n = 3), and (b) its time dependent signal intensity change in
various organs. MRI conditions: 4.7 T Unity INOVA (Varian), at $T_R/$
$T_E = 300$ ms/11 ms. Reproduced with permission from Ref. 22.

relaxivities. Further studies revealed that the gadofullerenols exhibited strong
contrast enhancement at lung, liver, spleen and kidney after i.v. administration
at a dosage of 5 μmol Gd kg^{-1}, which was 1/20 of a typical clinical dosage of
Gd-DTPA (100 μmol Gd kg^{-1}) (Figure 5.3.2). The biodistribution indicated
that the $Gd@C_{82}(OH)_{40}$ tends to be entrapped in the reticular-endothelial
system (RES) maybe due to the formation of large particles by aggregation or
the interaction with the plasma components.[22] For Gd-chelate complexes as the
RES agents, there is a risk of *in vivo* Gd^{3+} dissociation, whereas gadofullerenols

will avoid the release of toxic Gd^{3+} due to the solid carbon cage structure. Their preliminary results also revealed contrast-enhanced rat blood vessels within a few minutes after i.v. administration of gadofullerenols at the dosage of 10 μmol Gd kg^{-1}, indicating that the angiogram may be further improved by surface modification of carbon cage for the prolongation of the blood circulation period and the avoidance of RES uptake.

Subsequently, Shinohara's group investigated other lanthanoid endohedral metallofullerenols as potential MRI contrast agents and compared results with those of the corresponding lanthanoid-DTPA complexes and intact ions.[24] The results indicated an extremely strong signal enhancement for gadofullerenols but only a slight enhancement for other metallofullerenols such as Er, Ce, La and Dy compared with pure water (Figure 5.3.3). In contrast, the corresponding lanthanoid-DTPA complexes and naked ions have extremely low r_1. The strong relaxivities of the current metallofullerenols are mainly due to the dipole-dipole relaxation as well as a substantial decrease of the overall molecular rotational motion by forming aggregates *via* hydrogen bonding.

Zhang *et al.*,[43] synthesized a $Gd@C_{82}(OH)_x$ with a lower OH number, $Gd@C_{82}(OH)_{16}$, which exhibits longitudinal relaxivity of 19.3 mM^{-1} s^{-1} together with liver and kidney uptake behavior. They predicated that this water-soluble gadofullerenol with adequate hydrophilic groups could satisfy both conditions of good stability *in vivo* and high efficiency for MRI. In addition to being a high efficient MRI contrast agent, multihydroxylated metallofullerenol particles $[Gd@C_{82}(OH)_{22}]_n$ (22 nm in a saline solution) at a dose level as low as 10^{-7} mol kg^{-1} exhibit a very high antineoplastic efficiency ($\sim 60\%$),[44] and antioxidative function in tumor bearing mice *in vivo*,[45] or tumor cells *in vitro*.[46] More recently, $[Gd@C_{82}(OH)_{22}]_n$ nanoparticles were proved to overcome the acquired resistance to cisplatin, a commonly used

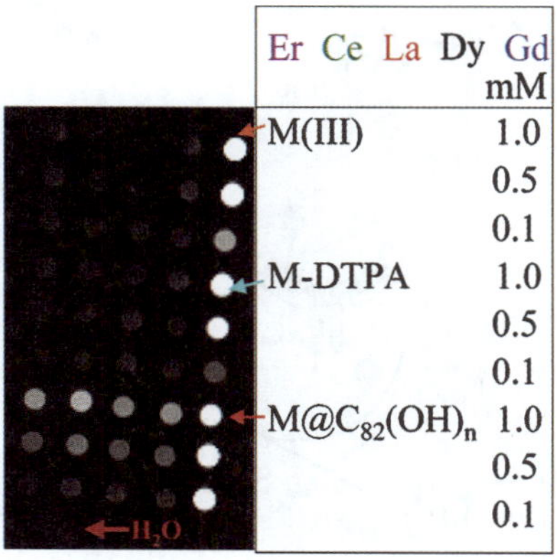

	Er Ce La Dy Gd	
		mM
M(III)		1.0
		0.5
		0.1
M-DTPA		1.0
		0.5
		0.1
M@C$_{82}$(OH)$_n$		1.0
		0.5
		0.1
←H$_2$O		

Figure 5.3.3 Phantom NMR images of various metallofullerenols (together with those of lanthanoid ions and lanthanoid-DTPA complexes) solutions. All MRI images were obtained by 1 T (MAGNETOM Harmony, Siemens), spin-echo with T_R/T_E = 200 ms/6 ms, FOV = 19 × 19 cm^2, slice thickness of 6 mm, 256 × 256 points, and at 20 °C. Reproduced with permission from Ref. 24.

chemotherapeutic drug by reactivating the impaired endocytosis of cisplatin-resistant human prostate cancer (CP-r) cells.[47]

Gu *et al.*,[48] functionalized $Gd@C_{82}$ to produce water soluble $Gd@C_{82}(OH)_6(NHCH_2CH_2SO_3H)_8$, which exhibited comparable relaxivity $(4.5\,mM^{-1}\,s^{-1}$ at 1.5 T and 37 °C) to that of commercially available Gd-complex MRI contrast agents. We also reported a new contrast agent derived from $Gd@C_{82}$ and amino acid, $Gd@C_{82}O_{\sim6}(OH)_{\sim16}(NHCH_2CH_2COOH)_{\sim8}$, which shows longitudinal relaxivities of $9.1\,mM^{-1}\,s^{-1}$ at 1.5 T,[26] and $16.0\,mM^{-1}\,s^{-1}$ at 0.35 T.[34] Though the relaxivities of these functionalized gadofullerenes are lower than those of gadofullerenols, the carboxyl functionalized gadofullerenes can work as a precursor of tissue-targeting MRI diagnostic agents (Scheme 5.3.2) and thus resulting in accumulation of targeting probes and local contrast enhancement at a low dosage of administration. In addition, the resulting targeted contrast agent itself can lead to a little higher contrast enhancement compared with the parent molecules due to the increase of overall molecular rotational motion.[34]

Another example of $Gd@C_{82}$ based contrast agent is $Gd@C_{82}O_2(OH)_{16}$ $(C(PO_3Et_2)_2)_{10}$,[33] revealing an r_1 of $37.0\,mM^{-1}\,s^{-1}$ at 0.35 T and potential bone-targeting due to the high affinity of the phosphonate to the surface of hydroxylapatite–bone tissue. This is a typical example of specific tissue-targeting by direct modification of carbon cage.

Molecular MRI offers the potential to image some pathological details at the cellular and subcellular level.[49,50] Anderson *et al.*,[10] used a transfection agent for labeling cells in culture with $Gd@C_{82}$ fullerenol and found that protamine sulfate

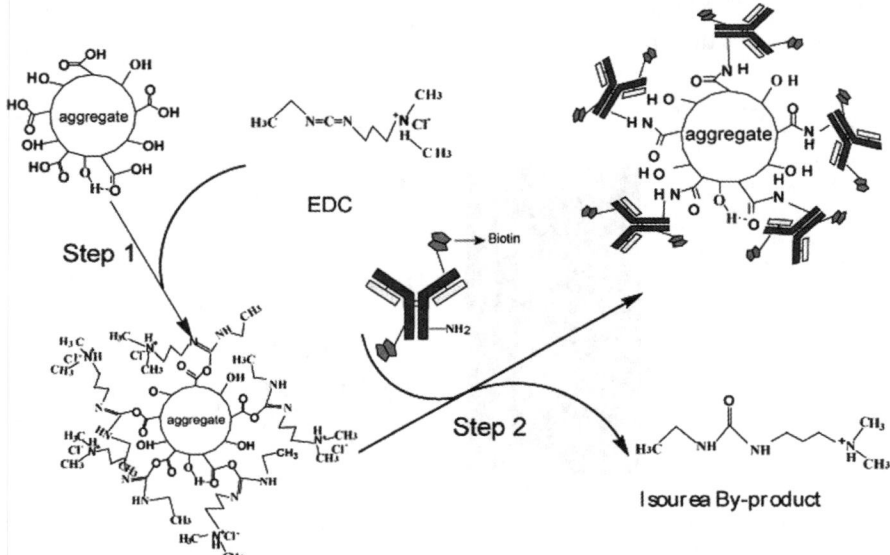

Scheme 5.3.2 Conjugation reaction of carboxylated gadofullerene aggregates with the antibodies to produce specific targeting diagnostic agent. Reproduced with permission from Ref. 34.

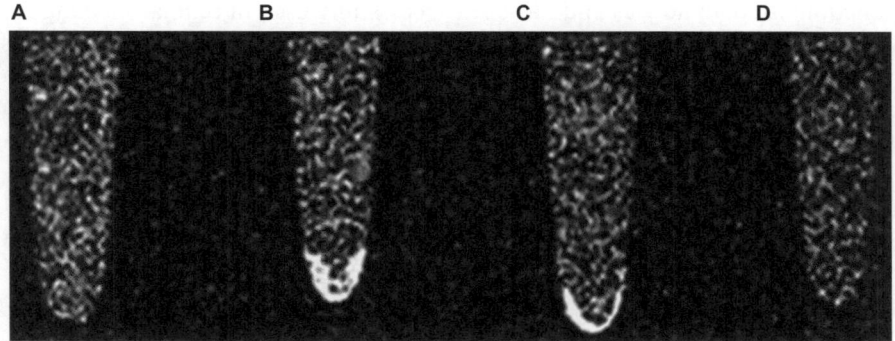

Figure 5.3.4   Spin echo image of HeLa cells. Control cells with no Gd-fullerenol (A); 35:6 ratio GdFPro (B); 35:2 ratio GdFPro (C); cells incubated with Gd fullerenol alone (D). Reproduced with permission from Ref. 10.

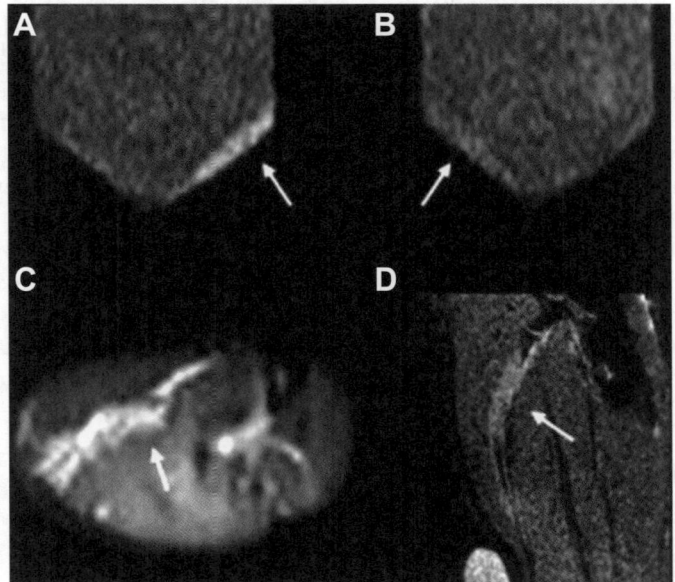

Figure 5.3.5   MRI of live mesenchymal stem cells in culture and injected in rat thigh. (A) 2×10^6 labeled and (B) 2×10^6 unlabeled cells. (C) 3×10^6 Gd-fullerenol-labeled cells injected in rat thigh muscle at 1.5 T; and (D) 3×10^6 Gd-fullerenol-labeled cells injected in rat thigh muscle at injected in rat thigh at 7.0 T. Reproduced with permission from Ref. 10.

transfection increased cell uptake of gadofullerenols and produced the T_1-enhanced cell on MRI (Figures 5.3.4 & 5.3.5). Their study suggested that gadofullerenols are suitable for tracking studies of viable cells in therapeutic applications.

In conclusion, the polyhydroxylated $Gd@C_{82}$ exhibits much higher relaxivities than those of carboxylated, sulfonated or organophosphonated functionalized analogues and has been successfully used in cellular labeling for MRI study at molecular level. This dramatic difference between relaxivities among these gadofullerenes are mainly due to the different tumbling–rotation time related to aggregation phenomena in aqueous solution, which will in turn contribute to the biodistribution of diagnostic agent in tissue due to the different blood circulation period. The carboxylated $Gd@C_{82}$ can further conjugate with other biomolecules to produce specific-targeted MRI contrast agent, which will reduce the dose of administration with high efficient contrast enhancement.

5.3.2.2   MRI Contrast Agent Based on Gadofullerene $Gd@C_{60}$

The more abundant $Gd@C_{60}$ has been ignored for a long time due to its insolubility and air sensitivity. Bolskar *et al.*,[25] first modified this previously unused and insoluble species to water-soluble $Gd@C_{60}[C(COOH)_2]_{10}$ and evaluated it as an MRI contrast agent. Figure 5.3.6 (left) shows the possible structure of this new contrast agent. Though $Gd@C_{60}[C(COOH)_2]_{10}$ exhibits a relaxivity of $4.6\,mM^{-1}\,s^{-1}$ at 20 MHz comparable to commercially available Gd^{III}-complex contrast agents and far lower than gadofullerenols, an *in vivo* MRI biodistribution study in a rodent model reveals $Gd@C_{60}[C(COOH)_2]_{10}$ to possess the first non-reticuloendothelial system (RES) localizing behavior for the water-soluble endohedral metallofullerene species (Figure 5.3.6, right.). They ascribed this phenomenon to lack of intermolecular aggregation in solution.[35,36,38]

Sitharaman *et al.*,[11] used anionic gadofullerene $Gd@C_{60}[C(COOH)_2]_{10}$ as an *in vitro* cellular MRI label. They found that the cellular uptake of this gadofullerene was nonspecific and 98–100% labeling efficiency was achieved without a transfecting agent. Figure 5.3.7 shows that the signal intensity of the T_{1w} images at 1.5 T was 250% greater in gadofullerenes labeled cells compared

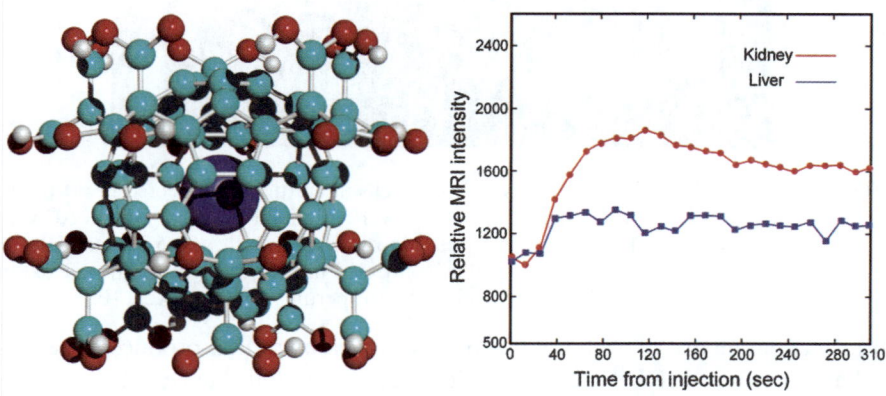

Figure 5.3.6   Ball-and-stick depiction of $Gd@C_{60}[C(COOH)_2]_{10}$ (left) and representative *in vivo* MRI intensity-derived biodistribution, revealing rapid renal uptake with a minimum of liver uptake (red circles, kidney; blue squares, liver) (right). Reproduced with permission from Ref. 25.

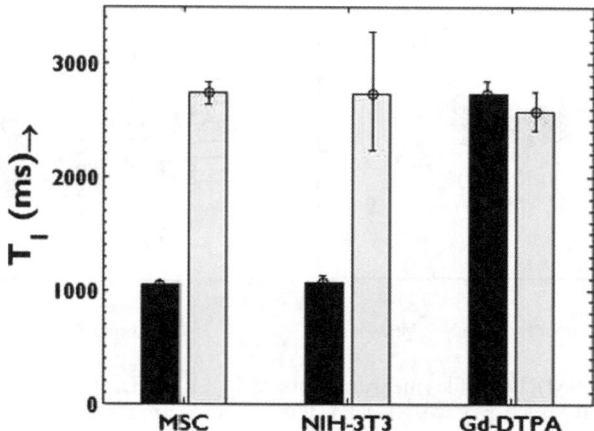

Figure 5.3.7 T_1 of MSCs and NIH-3T3 cells labeled with gadofullerenes (light gray bars) compared with controls at 25 °C (dark gray bars) and those with conventional Gd-DTPA and controls. Reproduced with permission from Ref. 11.

with controls, whereas no observable difference was seen between conventional Gd-DTPA labeled and unlabeled cells.

Detailed study of the water–proton relaxivity properties and intermolecular nanoclustering behavior of gadofullerene derivatives, $Gd@C_{60}(OH)_x$ and $Gd@C_{60}[C(COOH)_2]_{10}$, under different conditions were explored,[35–38] which provided us valuable information about their relaxivity mechanism. Sitharaman *et al.*,[35] investigated the aggregation of $Gd@C_{60}(OH)_x$ and $Gd@C_{60}[(COOH)_2]_{10}$ as a function of pH, temperature and concentration by dynamic and static light scattering (DLS and SLS) experiments. The results indicated that at pH = 9, only $Gd@C_{60}[C(COOH)_2]_{10}$ aggregates showed a slight concentration- and temperature-dependance, whereas $Gd@C_{60}(OH)_x$ aggregates are less affected by concentration and temperature. However, both $Gd@C_{60}(OH)_x$ and $Gd@C_{60}[C(COOH)_2]_{10}$ aggregates were highly pH-dependent, and at pH = 9, the aggregate sizes ranged from 30 to 90 nm. SLS and TEM results suggest that the intermolecular forces between $Gd@C_{60}(OH)_x$ molecules within aggregates are stronger than those between $Gd@C_{60}[C(COOH)_2]_{10}$ molecules within its aggregates, which correspondingly leads to larger aggregate sizes, slow tumbling and higher relaxivities.

Toth *et al.* studied the relaxivities of $Gd@C_{60}(OH)_x$ and $Gd@C_{60}[C(COOH)_2]_{10}$ as a function of magnetic field and temperature.[37] Both of them show relaxivity maxima at high magnetic fields (30–60 MHz) with a maximum relaxivity of 38.5 mM^{-1} s^{-1} for the former and 10.4 mM^{-1} s^{-1} for the latter at 299 K (Figure 5.3.8), which provides valuable information for separating the different interaction mechanism and dynamic processes influencing the relaxation behavior. For both $Gd@C_{60}(OH)_x$ and $Gd@C_{60}[C(COOH)_2]_{10}$ at all temperatures, the full-field nuclear magnetic relaxation dispersion (NMRD) profiles show high field maxima centered at around 40 MHz, typical of slowly

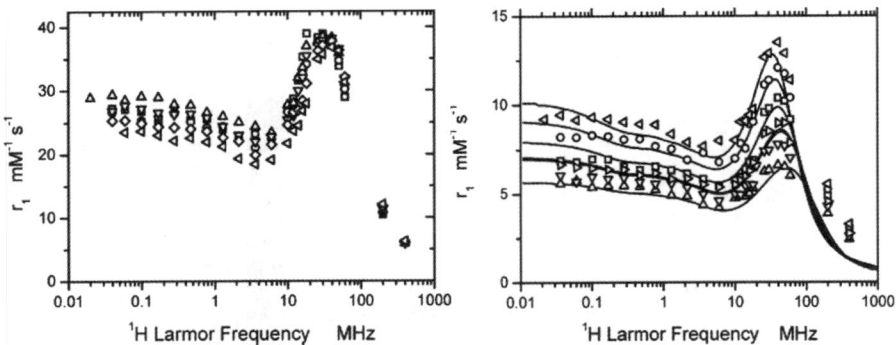

Figure 5.3.8   NMRD profiles measured for a Gd@C$_{60}$(OH)$_x$ (x ≈ 27) solution (c$_{Gd}$ =
0.5 mM; pH = 9.8) (left), temperatures are 278.0 K (square), 287.5 K
(circle), 299.3 K (up pointing triangle), 310.6 K (down pointing triangle), 322.0 K (diamond), and 335.3 K (left pointing triangle); and for a
Gd@C$_{60}$[C(COOH)$_2$]$_{10}$ solution (c$_{Gd}$ = 3.6 mM; pH = 9.3) (right),
temperatures are 278.0 K (left pointing triangle), 287.5 K (circle), 299.3 K
(square), 310.6 K (right pointing triangle), 322.0 K (down pointing triangle), and 335.3 K (up pointing triangle). Reproduced with permission
from Ref. 37.

rotating gadofullerenes and Gd-chelates with slow molecular motion.[1] For the
former, the high-field (> 10 MHz) relaxivities are temperature-independent and
somewhat temperature-dependent at lower frequencies. In contrast, the relaxivities of Gd@C$_{60}$[C(COOH)$_2$]$_{10}$ decrease with increasing temperature at all
fields. A detailed analysis of the Nuclear Magnetic Resonance Dispersion
(NMRD) profiles for Gd@C$_{60}$[C(COOH)$_2$]$_{10}$ resulted in a rotational correlation time of τ_R^{298} = 2.6 ns and a proton exchange rate of k$_{ex}^{298}$ = 1.4 × 10^7 s^{-1}.
The longer rotational correlation time (the order of nanoseconds) as a result of
aggregation in contrast with the conventional Gd-DTPA complex (τ_R 58 ps),[2]
contributes a lot to the higher relaxivity of this species. In addition, the proton
relaxivities of these gadofullerenes display a remarkable pH dependency
derived from the different aggregating sizes, increasing dramatically with
decreasing pH. The strong pH dependency of the proton relaxivities makes
these gadofullerene derivatives excellent candidates for pH-responsive MRI
contrast agent applications.

Laus *et al.*,[36] studied the salt effect on gadofullerene aggregates of
Gd@C$_{60}$(OH)$_x$ and Gd@C$_{60}$[C(COOH)$_2$]$_{10}$ and corresponding relaxivites.
They found that though NaCl can disrupt aggregation of gadofullerenes
(Figure 5.3.9) to some extent, the efficiency is far lower than phosphate. For
example, in 10 mM phosphate and in 150 mM sodium chloride, the longitudinal
relaxivity–hydrodynamic diameters of Gd@C$_{60}$(OH)$_x$ are 14.1 mM^{-1} s^{-1}/90.9 nm
and 31.6 mM^{-1} s^{-1}/120.8 nm, respectively. Their findings revealed an important
clue that the real biological fluids present a rather high salt concentration and
may influence the properties of gadofullerenes related to the aggregation
phenomenon.

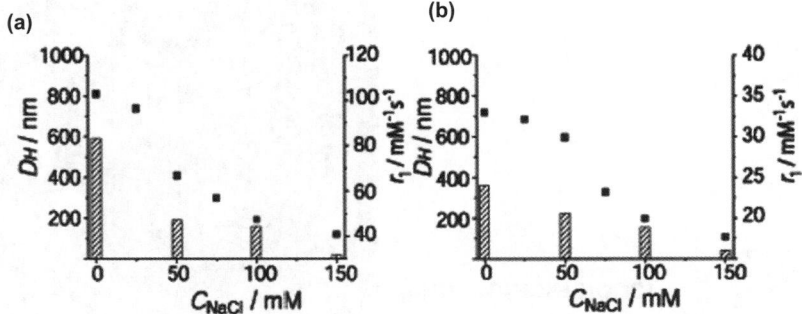

Figure 5.3.9 Hydrodynamic diameters (left y-axis, ■) and ^1H relaxivities (right y-axis, histograms) of $Gd@C_{60}(OH)_x$ (c_{Gd} = 0.5 mM) (a), and $Gd@C_{60}[C(COOH)_2]_{10}$ (c_{Gd} = 0.4 mM) (b), aqueous solutions at variable NaCl concentration; pH 7.4, 37 °C, 60 MHz. Reproduced with permission from Ref. 36.

In summary, the wide studies of $Gd@C_{60}$ based on MRI contrast agents, especially in the field of aggregation closely related to the relaxivities, will give us deep understanding on the novel relaxation mechanism since there is no direct contact of Gd to water.

5.3.2.3 MRI Contrast Agent Based on Gadofullerene Gd3N@C$_{80}$

$Gd_3N@C_{80}$ as one of the important members of trimetallic nitride templated endohedrals will exhibit much higher paramagnetic properties than $Gd@C_{60}$ and $Gd@C_{82}$ due to the presence of three Gd^{3+} ions within one cage, where the charge transfer between the Gd_3N cluster and the C_{80} cage leads to a very stable nanoparticle and the ferromagnetic coupling gives rise to a large magnetic moment, 21 μB.[51,52]

Fatouros *et al.*,[27] first reported pegylated-hydroxylated gadofullerene $Gd_3N@C_{80}$ (Gd_3N fMF) as a high efficient MRI contrast agent. Figure 5.3.10 shows the schematic structure of Gd_3N fMF and T_{1w}-images at different concentrations compared with gadodiamide. Gd_3N fMF exhibited significantly higher contrast enhancement at a very low concentration compared with gadodiamide, which is consistent with the measured longitudinal relaxivity (143 mM^{-1} s^{-1} vs. 4 mM^{-1} s^{-1} at 2.4 T). In order to understand the distribution of the new Gd_3N fMF upon direct infusion by convection-enhanced delivery (CED) into the extracellular space of the brain, agarose gel (0.6%) was used as a surrogate for *in vivo* brain tissue. With the CED method, a bilateral infusion of 0.0261 mM Gd_3N fMF at 0.5 μL min^{-1} and 1 mM gadodiamide at 0.2 μL min^{-1} (left side of each image) was administered into the agarose gel phantom. Figure 5.3.11 displays successive T_1 computed maps of Gd_3N fMF (right side of each image) and gadodiamide (left side of each image) in the agarose gel. The displayed times are in minutes after the start of infusion. Positive pressure infusion

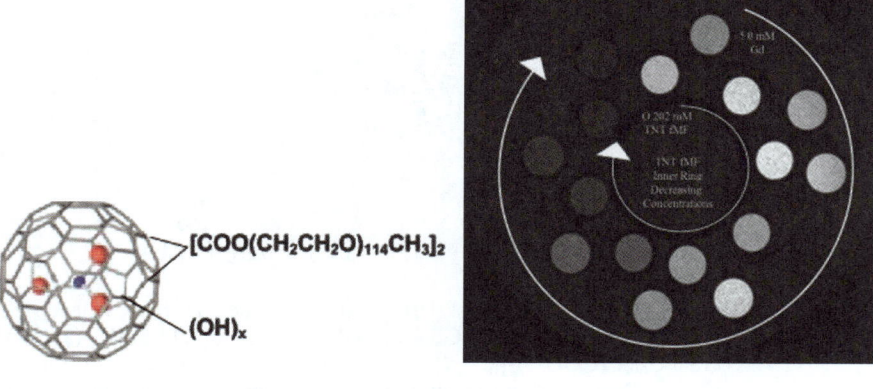

Figure 5.3.10   Pegylated-hydroxylated gadofullerene: $Gd_3N@C_{80}$ (left); T_1-weighted
MR image (700/10) of aqueous solutions of Gd_3N fMF (inner ring,
concentration decreasing in clockwise direction: 0.2020, 0.0101, 0.0505,
0.0252, 0.0126, 0.0063, 0.0032, and 0.0016 mMol L^{-1}) and gadodiamide (Gd) (outer ring, concentration decreasing in clockwise direction:
5.0, 3.0, 1.0, 0.70, 0.50, 0.30, 0.10, and 0.050 mMol L^{-1}). Reproduced
with permission from Ref. 27.

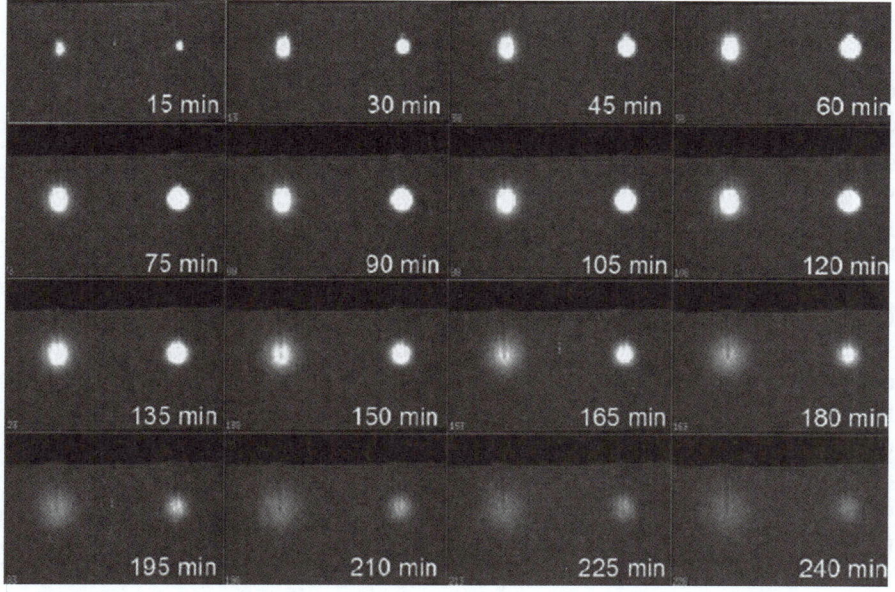

Figure 5.3.11 T_1-weighted MR images (700/10) of bilateral infusion into 0.6%
agarose gel of 0.0261 mM. Gd_3N fMF at 0.5 μL min^{-1} (right side of
each image) and 1 mM gadodiamide at 0.2 μL min^{-1} (left side of each
image). Displayed times are in minutes from the start of bilateral
infusion. Reproduced with permission from Ref. 27.

Table 5.3.1 Relaxivity Data of $Gd_3N@C_{80}[DiPEG(OH)_x]$ Series.

mol. weight of PEG	conc. range (μM)	r_1 (0.35 T) (mM^{-1} s^{-1})	r_2 (0.35 T) (mM^{-1} s^{-1})	r_1 (2.4 T) (mM^{-1} s^{-1})	r_2 (2.4 T) (mM^{-1} s^{-1})	r_2 (9.4 T) (mM^{-1} s^{-1})	r_2 (9.4 T) (mM^{-1} s^{-1})
5000	0.4–6.5	107±8	127±36	139±6	221±11	52,5±2.4	186±12
5000 (27)	1.6–12.6	102	144	143	222	32	137
2000	1.0–15.2	130±4	148±8	158±6	249±12	41.9±3,0	218±11
750	1.1–17.4	152±5	169±20	232±10	398±22	63.3±1,8	274±9
350	1.5–23.5	227±31	268±19	237±9	460±23	68.2±3.3	438±5

*a*The relaxivities were measured at room temperature in water.

was performed for 120 minutes, with an additional 120 minutes of imaging follow-up for studying diffusion. The result indicates the greater diffusion and faster disappearance rate of the conventional gadolinium compound compared with the fMF nanoparticle and suggests that the prolonged residence for the Gd_3N fMF within the tumor volume may work as a long term diagnostic agent.

Furthermore, toward optimized functionality, Zhang et al.,[53] synthesized pegylated and hydroxylated $Gd_3N@C_{80}$ with different PEG molecular weights (350–5000 Da) and measured the relaxivities of these derivatives at three different magnetic field strengths. As shown in Table 5.3.1, the 350–750 Da PEG derivatives have the highest relaxivities among the derivatives, $237/232\,mM^{-1}$ s^{-1} for r_1 and $460/398\,mM^{-1}$ s^{-1} for r_2 ($79/77\,mM^{-1}$ s^{-1} and $153/133\,mM^{-1}$ s^{-1} based on per Gd^{3+} ion), respectively, at a clinical-range magnetic field of 2.4 T. These represent some of the highest relaxivities reported for commercial or investigational MRI contrast agents.

MacFarland et al.,[28] reported a novel endohedral metallofullerene platform based on $Gd_3N@C_{80}$, $Gd_3N@C_{80}$-R_x (Hydrochalarones), where $R = -[N(OH)(CH_2CH_2O)_nCH_3]_x$, n = 1,3,6 and x is 10–22, for enhancing MRI contrast. The optimal member of this series, Hydrochalarone-6, has r_1 of $205\,mM^{-1}$ s^{-1} per molecule ($68\,mM^{-1}$ s^{-1} per Gd ion compared with $3.8\,mM^{-1}$ s^{-1} for Magnevist under the same conditions). Hydrochalarone-6 was administered intravenously to a BALB/c mouse at a dose of $0.01\,mmol\,kg^{-1}$, 1/10 times of a typical clinical dose of commercial Gd-complex MRI contrast agent. Figure 5.3.12 shows a coronal slice from a T_1-weighted whole-body image of a mouse before and after Hydrochalarone-6 administration. The persistent image enhancement of the aorta (arrow) indicates Hydrochalarone-6 remains in the vasculature, consistent with the observation that polyethylene glycol modifications increase the circulation time of proteins. Since the Hydrochalarones are stable and appear well-tolerated *in vivo* as well as provide excellent MRI enhancement, the development of Hydrochalarones as imaging agent platforms will improve clinical diagnosis and management of a number of diseases upon various attachment of targeting species.

For tissue-targeting purposes, amino and carboxyl functional groups are very important for conjugation with biomolecules. $Gd@C_{82}$,[26,34] and $Gd@C_{60}$,[25] have been successfully modified with carboxyl functional groups, but the contrast enhancement is far from optimal. One of the important

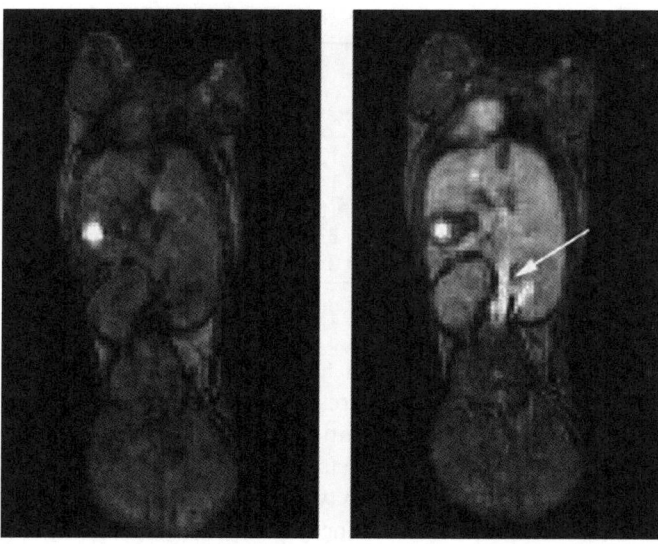

Figure 5.3.12   Coronal slice from a whole-body MRI of a mouse before (left) and 30
min after (right) administration of Hydrochalarone-6. The unidentified
bright spot in both the pre- and post-contrast images is likely an
artifact. Reproduced with permission from Ref. 28.

factors is the limited paramagnetic Gd ion in one cage. In 2009, carboxylated
$Gd_3N@C_{80}$,[41] was synthesized *via* the free radical reaction as Scheme 5.3.3
shows, and exhibited r_1 of 207 mM^{-1} s^{-1} per molecule (69 mM^{-1} s^{-1} per Gd)
which is 50 times larger than Gd-DTPA BMA (4.1 mM^{-1} s^{-1}). Figure 5.3.13
shows T_1 weighted images (700/10) and T_2 weighted images (6000/100) of
direct infusion of 0.0475 mM $Gd_3N@C_{80}(OH)_{26}(CH_2CH_2COOM)_{16}$ into T9
tumor bearing rat brain. The results indicate that the carboxylated
$Gd_3N@C_{80}$ provides contrast enhancement (red circles) at very low concentration (0.0475 mM, 36 µL), similar to but better than other reported
$Gd_3N@C_{80}$-based MRI contrast agents. To investigate the diffusion of the
carboxylated $Gd_3N@C_{80}$ quantitatively, agarose gel infusion experiments
were performed on the carboxylated $Gd_3N@C_{80}$ and the clinically used
Omniscan. The slower diffusion of the carboxylated $Gd_3N@C_{80}$ is consistent
with the dependence of the width of the diffusion profile on diffusion coef-
ficient D and time of diffusion t: full width at half-maximum (fwhm) $\propto (D \times t)^{1/2}$ assuming an approximate Gaussian profile. From the Stokes-Einstein
equation and the known particle radii, the diffusion time for the carboxylated
$Gd_3N@C_{80}$ is 40–50 times longer than Omniscan and slightly longer than
formerly reported $Gd3N@C_{80}[Peg5000(OH)_x]$.[23] The imaged bigger tumor
bolus for a long time post injection (even 7 days, Figure 5.3.13) indicated the
long term stay of this contrast agent around tumor, which is consistent with
its slower diffusion.

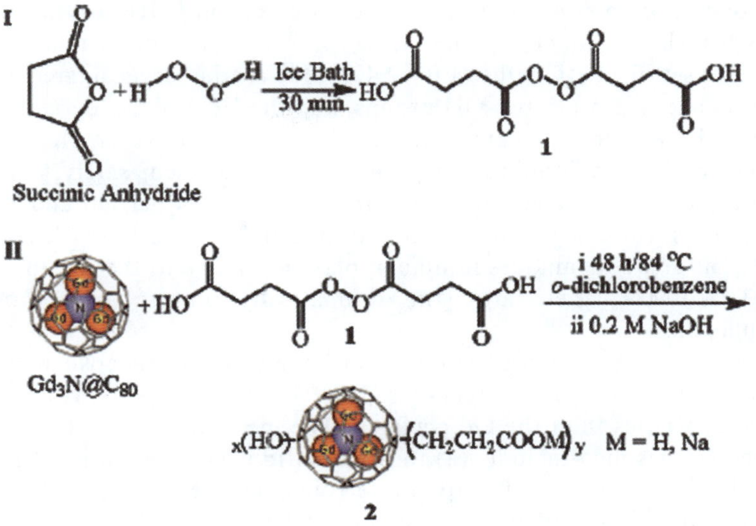

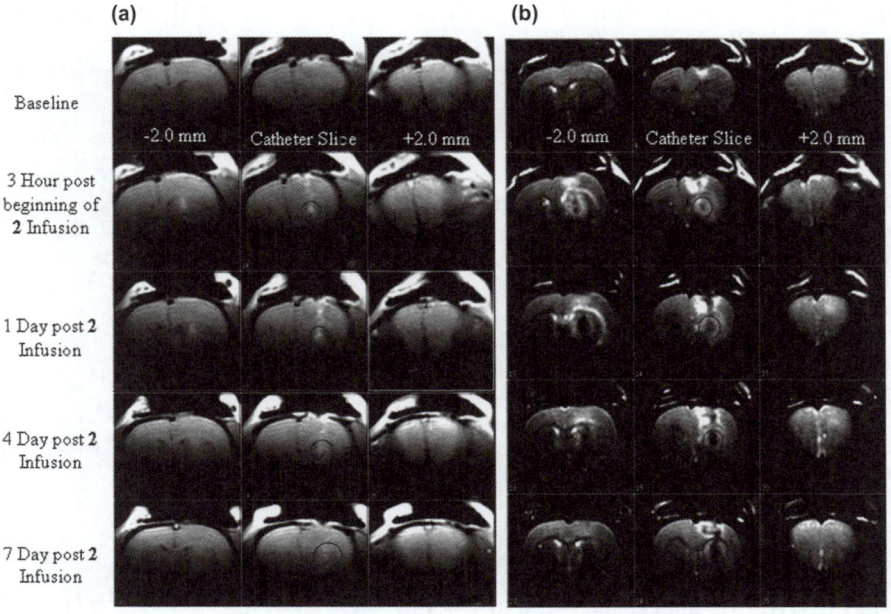

Scheme 5.3.3 Synthesis of Succinic Acid Acyl Peroxide, **1** and Gd$_3$N@C$_{80}$(OH)x(CH$_2$CH$_2$COOM)y (M = H, Na), **2**. Reproduced with permission from Ref. 41.

Figure 5.3.13 T_1 weighted images (700/10) (a), and T_2 weighted images (6000/100) (b), of direct infusion into T9 tumor bearing rat brain of 0.0475 mM Gd$_3$N@C$_{80}$(OH)$_{26}$(CH$_2$CH$_2$COOM)$_{16}$. Infusion was applied for 180 minutes at 0.2 µL min^{-1}. Reproduced with permission from Ref. 41.

In conclusion, among the gadofullerenes based on MRI contrast agents, polyhydroxylated $Gd@C_{82}$,[22] and $Gd@C_{60}$,[36] exhibit highest relaxivities at clinically used field strengths (1.0~2.4 T) due to the larger aggregates and fast exchange with water *via* OH groups, and the $Gd_3N@C_{80}$ derivatives are the second member of agents with high relaxivities. However, the C_{80} cage can entrap three gadolinium ions and concentrate them safely, thus providing dramatically higher relaxivities (r_1/r_2 from 69/94 per Gd and 207/282 per cage), as Table 5.3.2 shows. In contrast, though Gd-BOPTA and MS-325 exhibit great advantages in human plasma relative to the routinely used Gd-DTPA-BMA, they are still dramatically lower than those of gadofullerenes.

Water molecules are not directly coordinated to the encapsulated metal ion, why do these gadofullerene-based MRI contrast agents exhibit much higher relaxivities than the Gd chelates? The mechanisms ascribed to these high relaxivities may include: (a) the charge transfer between paramagnetic Gd and carbon cage leading to the ferromagnetic coupling of Gd^{3+} and $[C_{60,82}]^{3-}$ or Gd-Gd-Gd in C_{80} cage and a large magnetic moment; (b) a large number of hydroxyl groups attached to the carbon cage which facilitate exchange of protons with surrounding water protons; (c) a longer rotational correlation time (the order of nanoseconds) as a result of

Table 5.3.2 Comparison of Relaxivities ($mM^{-1} s^{-1}$ per mM of gadolinium ions).

| Compound | Relaxivity (r_1/r_2) $mM^{-1} s^{-1}$ | | |
	0.35 T 15 MHz	2.4 T 100 MHz	9.4 T 400 MHz
$Gd@C_{82}(OH)_x$,[20]	—	—	47/—
$Gd@C_{82}(OH)_x$,[42]	20/— (0.47 T)	—	—
$Gd@C_{82}(OH)_{40}$,[22]	67/79 (0.47 T)	81/108 (1.0 T)	31/131 (4.7 T)
$Gd@C_{82}(OH)_{16}$,[43]	—	—	19.3/44.9 (4.7 T)
$Gd@C_{82}O_2(OH)_{16}[C(PO_3Et_2)]_{10}$,[33]	37/42	39/68	20/74
$Gd@C_{82}O_6(OH)_{16}(NHC_2H_4 COOH)_8$,[26]	16/—	—	—
$Gd@C_{60}(OH)_x$,[52]	—	83.2/— (1.5 T)	—
$Gd@C_{60}[C(COOH)_2]_{10}$,[55]	4.6/— (0.47 T)	24.0/— (1.5 T)	—
$Sc_2GdN@C_{80}O_m(OH)_n$,[40]	—	—	20.7/— (14.1 T)
$ScGd_2N@C_{80}O_m(OH)_n$			17.6/— (14.1 T)
$Gd_3N@C_{80}(OH)_{\sim 26}(CH_2CH_2 COOM)_{\sim 16}$,[41]	51/68	69/94	25/77
$Gd_3N@C_{80}Peg5000(OH)_x$,[27]	34/48	48/74	11/49
$Gd_3N@C_{80}R_x$,[28] R=N(OH) $(CH_2CH_2O)_6CH_3$	68/79 (0.47 T)	—	—
Gd-BOPTA,[56] (in human blood plasma)	10.9/— (0.2 T) 4.39/5.56[13*]	7.9/— (1.5 T)	5.9/— (3T)
MS-325(in human plasma),[18]	53.5/— (0.47 T)	—	—
Gd-DTPA BMA	—	4.1/4.7	—

Notes: * is under the condition of 0.47 T, pH = 7.4, 39 °C.

Table 5.3.3 Characterization of the Gadonanotubes and Their Derivatives.

Compound	R group (Abbreviation)	Number[a] of groups nm^{-1}		Solubility[b] in water at RT, mg/ml	Relaxivity[c] r_1/Gd^{3+}, $mM^{-1} s^{-1}$
		x	y		
Gd@US-tube	n/a	0	0	0	64
I	-OEt	14–28	–	0	28
II	-OH	14–19	–	0	n/a
III	-OC$_6$F$_5$	14–19	–	reacts	60
IVa	-NHMe	12–19	–	0	> 30
IVb	-NHCH(CH$_2$OH)$_2$	8–13	–	1–2	21
IVc	-NHCH(CO$_2$HXCH$_2$)$_2$SMe (Met)	8–12	–	0	47
IVd	-NHCH(CO$_2$H)CH$_2$OH (Ser)	12–15	–	1–2	49
IVe	-NHCH(CO$_2$H)CH$_2$(Me)OH (Thr)	8–10	–	1–3	33
IVf	-NHCH(CO$_2$H)(CH$_2$)$_3$NHC(NH$_2$)=NH (Arg)	10–13	–	1–2	—[d]
IVg	-NH-CH-CO-NH-CH-CO-NH-CH-CO$_2$H (Met-Ala-Ser) (CH$_2$)$_2$ Me CH$_2$OH SMe	13–11	–	0.05–0.2	
Va	RGD	–	0.65–1	0	—[d]
Vb	RGD + Met	8–12	0.65–1	0	—[d]
vc	RGD + Ser	8–12	0.65–1	1–2	33

Note: [a]From combined XPS (for C, N, and S) and TGA data.
[b]Determined by UV-vis spectroscopy.
[c]T_1-weighted relaxivity per Gd^{3+} ion at 1.5 T and 40 °C. All compounds, except **IVd** and **IVf**, were suspended in 1% Pluronic F 108 solution (aq.); **IVd** and **IVf** were suspended in DI water.
[d]The sample was prepared using US-tubes (no Gd^{3+}).

aggregation and (d) the presence of a pool of water molecules that are trapped within the aggregates are relaxed by the gadofullerenes and exchange rapidly with the bulk water molecules; (e) the larger surface area of cage to relax the surrounding bulk water molecules.[55] This is in contrast with the conventional Gd-DTPA complex which has a single water molecule bound in the first coordination sphere of the metal ion and the entire complex tumbles very rapidly ($\tau_R \sim 58$ ps).[2]

5.3.3 MRI Contrast Agents Based on Confined Gadonanotubes and Silicon Nanoparticles

Another new class of Gd-based carbonaceous nanomaterial for MRI contrast agent is ultrashort (20–80 nm) gadonanotubes.[31,32] It was reported that ultrashort gadonanotubes (Gd@US-tubes) exhibit very high relaxivities ($r_1 = 180$ mM^{-1} s^{-1}, 1.5 T) at pH 6.5 and 37 °C.[31] The unprecedented relaxivity is not readily rationalized using classic Solomon-Bloembergen-Morgan (SBM) theory.[56] Wilson and coworkers suggested that gadonanotubes are likely to be an example of special properties (magnetic–relaxivity) arising from the nanoscalar confinement of Gd^{3+} ion clusters within their carbon capsule sheaths.[31] Recently, Mackeyev et al.,[32] catalytically synthesized amino acid and peptide derivatized gadonanotubes, as Schemes 5.3.4 and 5.3.5 show. The r_1 relaxivities for the underivatized Gd@US-tubes (64 mM^{-1} s^{-1}) and several of the amino acid and peptide derivatized Gd@US-tubes are shown in Table 5.3.2. The data in this Table demonstrate that the gadonanotubes retain high-performance T_1-weighted MRI contrast enhancement even when derivatized, although the r_1 relaxivities vary depending upon the size and the hydrophilicity of the R group.

On the other hand, single-walled carbon nanohorns (SWNHs) are new carbonaceous materials with a large surface area and cavity. Zhang et al. first reported the preparation of SWNHs encapsulating Lu$_3$N@C$_{80}$ or Gd$_3$N@C$_{80}$, and conjugated with CdSe/ZnS quantum dots externally. It provides a dual diagnostic platform for in vitro and in vivo biomedical applications of these new carbonaceous materials.[57]

Ananta et al.,[58] developed a general method for increasing relaxivity by confining contrast agents inside the nanoporous structure of silicon particles. Three different Gd-CAs: Magnevist (MAG), gadolinium fullerenols based on Gd@C$_{60}$ (GFs) and gadonanotube (GNTs) were confined within two SiMPs: quasi-hemispherical (H-SiMPs) and discoidal (D-SiMPs) particles, respectively. The longitudinal relaxivity of the six different nanostructures was measured; the results are shown in Figures 5.3.14 & 5.3.15. Compared to the Gd-CA alone, a significant increase in r_1 was observed for all nanoconstructs. For MAG, r_1 increased by a factor of about four with the H-SiMP and two with the D-SiMP. For GFs, r_1 increased by a factor of about three with the

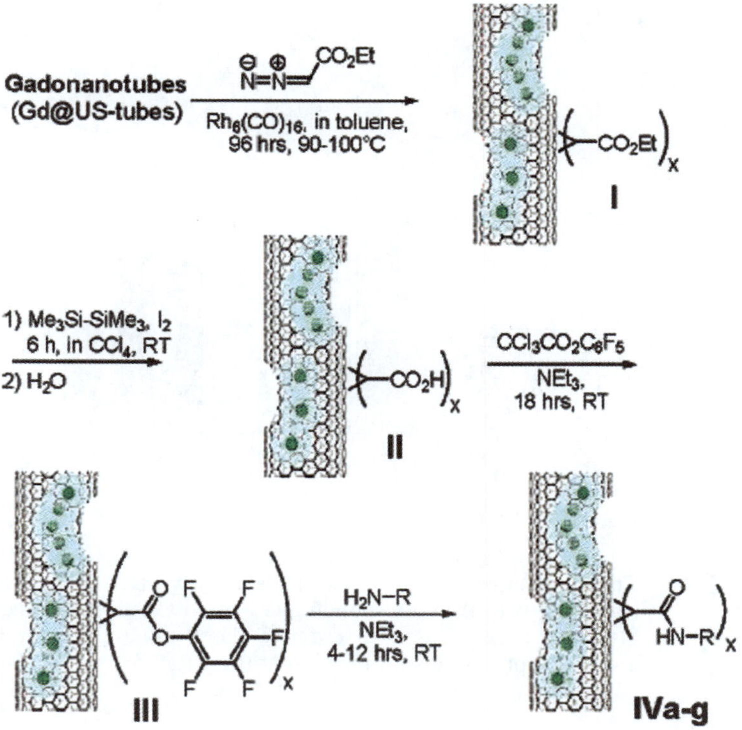

Scheme 5.3.4 Catalytic Functionalization of the Gadonanotubes. Reproduced with permission from Ref. 32.

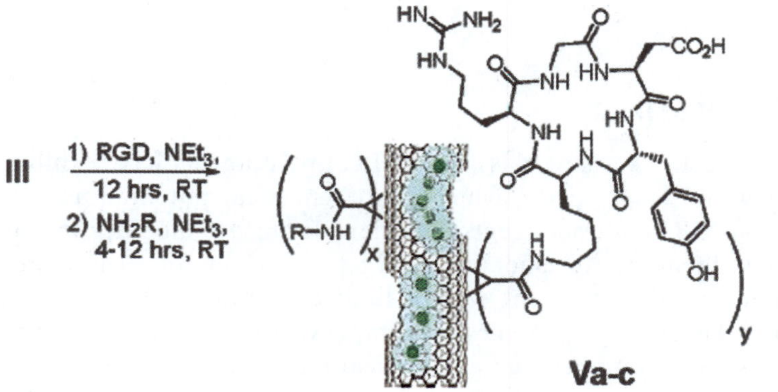

Scheme 5.3.5 Synthesis of the Water-Soluble RGD-Peptide Derivatives of the Gadonanotubes. Reproduced with permission from Ref. 32.

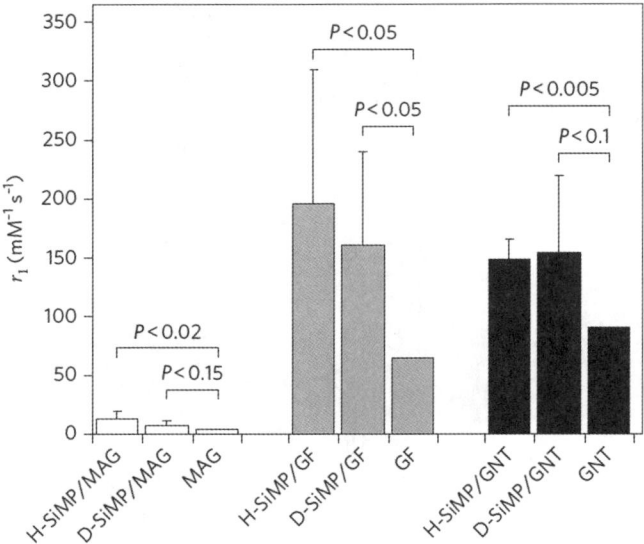

Figure 5.3.14   The longitudinal relaxivity, r_1, of the six new MRI nanoconstructs is compared with the corresponding Gd-based CAs (1.41 T and 37 °C). Student's t-test is used to estimate the P-values between the two groups. Reproduced with permission from Ref. 58.

H-SiMP and about 2.5 with the D-SiMP, and for GNTs, r_1 increased by a factor of approximately 1.5 for both SiMPs. The enhancement in contrast is attributed to the geometrical confinement of the agents, which influences the paramagnetic behavior of the Gd^{3+} ions, including q, τ_R, τ_m, r_{GdH} for the inner-sphere relaxivity r_1^{IS}, and τ_D for the outer-sphere relaxivity r_1^{OS}, and offers a new and general strategy for enhancing the contrast of gadolinium-based contrast agents.

5.3.4   Prospect

Gadofullerenes as a new class of MRI contrast agents have exhibited high-efficiency and low-toxicity, which give them great potential as a next generation of MRI contrast agents. The flexible surface modification of carbon cage will facilitate specific tissue-targeting, which in turn increases the local contrast enhancement with a limited administered dosage. The cellular uptake of gadofullerenes will improve the diagnosis and therapy of some diseases. For real clinic application, gadofullerene MRI contrast agents with accurate molecular structure should be confirmed and the *in vivo* metabolized process and long term toxicity should be further investigated.

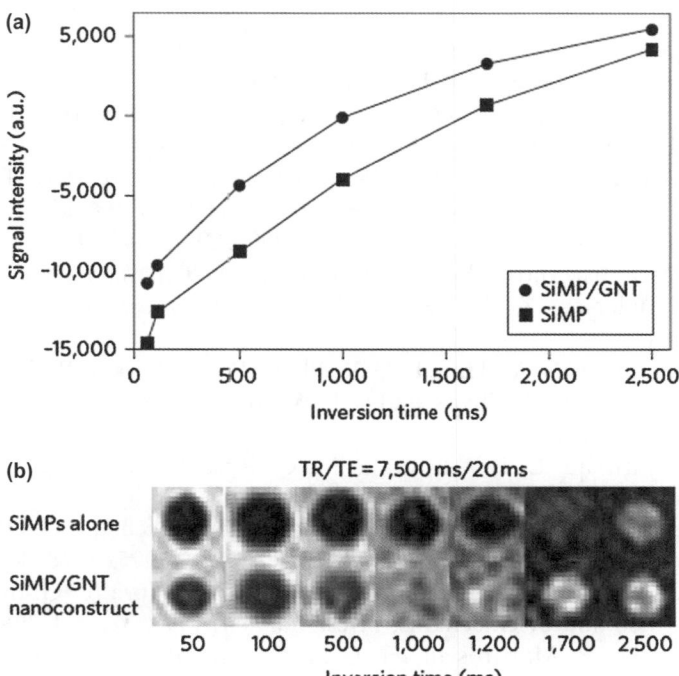

Figure 5.3.15 MRI characterization of the H-SiMP–GNT nanoconstruct in a clinical scanner; (a) Inversion recovery fits for SiMPs (black squares) and SIMP–GNT(black circles) nanoconstructs were acquired using an inversion recovery pulse sequence and plotted as a function of their inversion time T_{inv} (time at which the signal is completely suppressed). (b) Inversion recovery phantoms for SiMP and SiMP–GNT nanoconstruct, clearly showing faster recovery for the nanoconstruct. Data were obtained using a 1.5 T commercial clinical scanner with $T_R = 7$ 500 ms and $T_E = 20$ ms. Reproduced with permission from Ref. 58.

Acknowledgements

I'm grateful for support of this work by the National Science Foundation (Nos. 51072200, 20821003) and NSAF (No.11076027).

References

1. *The Chemistry of Contrast Agents in Medical Magnetic Resonance Imaging*, ed. E. Toth and A. E. Merbach, Wiley, Chichester 2001.
2. P. Caravan, J. J. Ellision, T. J. McMurry and R. B. Lauffer, *Chem. Rev.*, 1999, **99**, 2293–2352.
3. K. N. Raymond and V. C. Pierre, *Bioconjugate Chem.*, 2005, **16**, 3–8.
4. T. J. Meade, A. K. Taylor and S. R. Bull, *Curr. Opin.Neurobiol.*, 2003, **13**, 597–602.

5. S. Aime, C. Cabella, S. Colombatto, S. G. Crich, E. Gianolio and F. Maggioni, *J. Mag. Res. Imaging*, 2002, **16**, 394–406.
6. V. Comblin, D. Gilsoul, M. Hermann, V. Humblet, V. Jacques, M. Mesbahi, C. Sauvage and J. F. Desreux, *Coord. Chem. Rev.*, 1999, **185–6**, 451–470.
7. Gadolinium-Based Contrast Agents for Magnetic Resonance Imaging (MRI): marketed as Magnevist, MultiHance, Omniscan, OptiMARK, ProHance; http://www.fda.gov/medwatch/safety/2007/safety07.htm#Gadolinium.
8. T. Grobner, *Nephrol. Dial. Transplant*, 2006, **21**, 1104–1108.
9. D. E. Sosnovik and R. Weissleder, *Curr. Opin. Biotechnol.*, 2007, **18**, 4–10.
10. S. A. Anderson, K. K. Lee and J. A. Frank, *Invest. Radiol.*, 2006, **41**, 332–338.
11. B. Sitharaman, L. A. Tran, L. Q. P. Pham, R. D. Bolskar, R. Muthupillai, S. D. Flamm, A. G. Mikos, and L. J. Wilson, *Contrast Media Mol. Imaging* 2, 2007, 139–146.
12. K. B. Hartman, L. J. Wilson and M. G. Rosenblum, *Mol. Diagn. Ther.*, 2008, **12**, 1–14.
13. F. Uggeri, S. Aime, P. L. Anelli, M. Botta, M. Brocchetta, C. Dehaen, G. Ermondi, M. Grandi and P. Paoli, *Inorg. Chem.*, 1995, **34**, 633–642.
14. F. M. Cavagna, F. Maggioni, P. M. Castelli, M. Dapra, L. G. Imperatori, V. Lorusso and B. G. Jenkins, *Invest. Radiol.*, 1997, **32**, 780–796.
15. J. Pintaske, P. Martirosian, H. Graf, G. Erb, K. P. Lodemann, C. D. Claussen and F. Schick, *Invest. Radiol.*, 2006, **41**, 213–221.
16. F. L. Giesel, H. von Tengg-Kobligk, I. D. Wilkinson, P. Siegler, C. W. von der Lieth, M. Frank, K. P. Lodemann and M. Essig, *Invest. Radiol.*, 2006, **41**, 222–228.
17. D. J. Parmelee, R. C. Walovitch, H. S. Ouellet and R. B. Lauffer, *Invest. Radiol.*, 1997, **32**, 741–747.
18. R. B. Lauffer, D. J. Parmelee, S. U. Dunham, H. S. Ouellet, R. P. Dolan, S. Witte, T. J. McMurry and R. C. Walovitch, *Radiology*, 1998, **207**, 529–538.
19. P. Caravan, N. J. Cloutier, M. T. Greenfield, S. A. McDermid, S. U. Dunham, J. W. M. Bulte, J. C. Amedio, R. J. Looby, R. M. Supkowski, W. D. Horrocks, T. J. McMurry and R. B. Lauffer, *J. Am. Chem. Soc.*, 2002, **124**, 3152–3162.
20. S. R. Zhang, D. Y. Sun, X. Y. Li, F. K. Pei and S. Y. Liu, *Fullerene Sci. Technol.*, 1997, **5**, 1635–1643.
21. L. J. Wilson, *The Electrochem. Soc. Interface Winter*, 24–28.
22. M. Mikawa, H. Kato, M. Okumura, M. Narazaki, Y. Kanazawa, N. Miwa and H. Shinohara, *Bioconjugate Chem.*, 2001, **12**, 510–514.
23. M. Okumura, M. Mikawa, T. Yokawa, Y. Kanazawa, H. Kato and H. Shinohara, *Acad. Radiol.*, 2002, **9**, S495–S497.
24. H. Kato, Y. Kanazawa, M. Okumura, A. Taninaka, T. Yokawa and H. Shinohara, *J. Am. Chem. Soc.*, 2003, **125**, 4391–4397.
25. R. D. Bolskar, A. F. Benedetto, L. O. Husebo, R. E. Price, E. F. Jackson, S. Wallace, L. J. Wilson and J. M. Alford, *J. Am. Chem. Soc.*, 2003, **125**, 5471–5478.

26. C. Y. Shu, L. H. Gan, C. R. Wang, X. L. Pei and H. B. Han, *Carbon*, 2006, **44**, 496–500.

27. P. P. Fatouros, F. D. Corwin, Z. J. Chen, W. C. Broaddus, J. L. Tatum, B. Kettenmann, Z. Ge, H. W. Gibson, J. L. Russ, A. P. Leonard, J. C. Duchamp and H. C. Dorn, *Radiology*, 2006, **240**, 756–764.

28. D. K. MacFarland, K. L. Walker, R. P. Lenk, S. R. Wilson, K. Kumar, C. L. Kepley and J. R. Garbow, *J. Med. Chem.*, 2008, **51**, 3681–3683.

29. L. Dunsch and S. Yang, *Small*, 2007, **3**, 1298–1320.

30. R. D. Bolskar, *Nanomedicine*, 2008, **3**, 201–213.

31. K. B. Hartman, S. Laus, R. D. Bolskar, R. Muthupillai, L. Helm, E. Toth, A. E. Merbach and L. J. Wilson, *Nano Lett.*, 2008, **8**, 415–419.

32. Y. Mackeyev, K. B. Hartman, J. S. Ananta, A. V. Lee and L. J. Wilson, *J. Am. Chem. Soc.*, 2009, **131**, 8342–8343.

33. C. Y. Shu, C. R. Wang, J. F. Zhang, H. W. Gibson, H. C. Dorn, F. D. Corwin, P. P. Fatouros and T. J. S. Dennis, *Chem. Mat.*, 2008, **20**, 2106–2109.

34. C. Y. Shu, X. Y. Ma, J. F. Zhang, F. D. Corwin, J. H. Sim, E. Y. Zhang, H. C. Dorn, H. W. Gibson, P. P. Fatouros, C. R. Wang and X. H. Fang, *Bioconjugate Chem.*, 2008, **19**, 651–655.

35. B. Sitharaman, R. D. Bolskar, I. Rusakova and L. J. Wilson, *Nano Lett.*, 2004, **4**, 2373–2378.

36. S. Laus, B. Sitharaman, V. Toth, R. D. Bolskar, L. Helm, S. Asokan, M. S. Wong, L. J. Wilson and A. E. Merbach, *J. Am. Chem. Soc.*, 2005, **127**, 9368–9369.

37. E. Toth, R. D. Bolskar, A. Borel, G. Gonzalez, L. Helm, A. E. Merbach, B. Sitharaman and L. J. Wilson, *J. Am. Chem. Soc.*, 2005, **127**, 799–805.

38. S. Laus, B. Sitharaman, E. Toth, R. D. Bolskar, L. Helm, L. J. Wilson and A. E. Merbach, *J.Phys.Chem. C*, 2007, **111**, 5633–5639.

39. C. Y. Shu, E. Y. Zhang, J. F. Xiang, C. F. Zhu, C. R. Wang, X. L. Pei and H. B. Han, *J. Phys. Chem. B*, 2006, **110**, 15597–15601.

40. E. Y. Zhang, C. Y. Shu, L. Feng and C. R. Wang, *J. Phys. Chem. B*, 2007, **111**, 14223–14226.

41. C. Y. Shu, F. D. Corwin, J. F. Zhang, Z. J. Chen, J. E. Reid, M. H. Sun, W. Xu, J. H. Sim, C. R. Wang, P. P. Fatouros, A. R. Esker, H. W. Gibson and H. C. Dorn, *Bioconjugate Chem.*, 2009, **20**, 1186–1193.

42. L. J. Wilson, *The Electrochemical Society Interface Winter*, 24–28.

43. J. Zhang, K. Liu, G. Xing, T. Ren and S. Wang, *J. Radioanal. Nucl. Chem.*, 2007, **272**, 605–609.

44. C. Y. Chen, G. M. Xing, J. X. Wang, Y. L. Zhao, B. Li, J. Tang, G. Jia, T. C. Wang, J. Sun, L. Xing, H. Yuan, Y. X. Gao, H. Meng, Z. Chen, F. Zhao, Z. F. Chai and X. H. Fang, *Nano Lett.*, 2005, **5**, 2050–2057.

45. J. X. Wang, C. Y. Chen, B. Li, H. W. Yu, Y. L. Zhao, J. Sun, Y. F. Li, G. M. Xing, H. Yuan, J. Tang, Z. Chen, H. Meng, Y. X. Gao, C. Ye, Z. F. Chai, C. F. Zhu, B. C. Ma, X. H. Fang and L. J. Wan, *Biochem. Pharmacol.*, 2006, **71**, 872–881.

46. J. J. Yin, F. Lao, P. P. Fu, W. G. Wamer, Y. L. Zhao, P. C. Wang, Y. Qiu, B. Y. Sun, G. M. Xing, J. Q. Dong, X. J. Liang and C. Y. Chen, *Biomaterials*, 2009, **30**, 611–621.

47. X. J. Liang, H. Meng, Y. Z. Wang, H. Y. He, J. Meng, J. Lu, P. C. Wang, Y. L. Zhao, X. Y. Gao, B. Y. Sun, C. Y. Chen, G. M. Xing, D. W. Shen, M. M. Gottesman, Y. Wu, J. J. Yin and L. Jia, *Proc. Natl. Acad. Sci. U.S.A.*, 2010, **107**, 7449–7454.

48. X. Lu, J. X. Xu, Z. J. Shi, B. Y. Sun, Z. N. Gu, H. D. Liu and H. B. Han, *Chem. J. Chin. Univ.*, 2004, **25**, 697–700.

49. D. E. Sosnovik and R. Weissleder, *Curr. Opin.Biotechnol.*, 2007, **18**, 4–10.

50. R. Weissleder, *Radiology*, 1999, **212**, 609–614.

51. M. C. Qian, and S. N. Khanna, *J. Appl. Phys.*, 2007, **101**, 09E105 (1–3).

52. M. C. Qian, S. V. Ong, S. N. Khanna, and M. B. Knickelbein, *Phys. Rev. B 75*, 104424 (1–6).

53. J. F. Zhang, P. P. Fatouros, C. Y. Shu, J. Reid, L. S. Owens, T. Cai, H. W. Gibson, G. L. Long, F. D. Corwin, Z. J. Chen and H. C. Dorn, *Bioconjugate Chem.*, 2010, **21**, 610–615.

54. G. Schneider, K. Altmeyer, M. A. Kirchin, R. Seidel, L. Grazioli, G. Morana and S. Saini, *Invest. Radiol.*, 2007, **42**, 105–115.

55. S. Laus, B. Sitharaman, E. Toth, R. D. Bolskar, L. Helm, L. J. Wilson and A. E. Merbach, *J. Phys. Chem. C*, 2007, **111**, 5633–5639.

56. *The Chemistry of Contrast Agents in Medical Magnetic Resonance Imaging*, ed. E. Toth and A. E. Merbach, Wiley, Chichester, 2001.

57. J. F. Zhang, J. C. Ge, M. D. Shultz, E. Chung, G. Singh, C. Y. Shu, P. P. Fatouros, S. C. Henderson, F. D. Corwin, D. B. Geohegan, A. A. Puretzky, C. M. Rouleau, K. More, C. Rylander, M. N. Rylander, H. W. Gibson and H. Dorn, *Nano Lett.*, 2010, **10**, 2843–2848.

58. J. S. Ananta, B. Godin, R. Sethi, L. Moriggi, X. W. Liu, R. Krishnamuthy, R. Muthupillai, R. D. Bolskar, L. Helm, M. Ferrari, L. J. Wilson and P. Decuzzi, *Nat. Nanotechnol.*, 2011, **5**, 815–821.

CHAPTER 5.4

Application of Magnetic Resonance Imaging (MRI) to Radiotherapy

JENGHWA CHANG, Ph.D.,*[a] GABOR JOZSEF, Ph.D.,*[a] NICHOLAS SANFILIPPO, M.D.,[a] KERRY HAN, Ph.D.,[a] BACHIR TAOULI, M.D.,[b] ASHWATHA NARAYANA, M.D.[a] AND KEITH DEWYNGAERT, Ph.D.[a]

[a] Department of Radiation Oncology, and [b] Radiology, NYU School of Medicine, 566 First Avenue, TCH 114 HC, New York, NY 10016, USA

5.4.1 Introduction to Radiotherapy

Radiotherapy has been used as a cancer treatment modality for more than 100 years. The physical basis of radiotherapy is based on two discoveries made at the end of the 19th century: the discovery of X-rays in 1895 (W.C. Rontgen) and radioactivity a year later (H. Becquerel, P. and M. Curie). M. Curie was the first who suggested the use of radiation (radium) for treating superficial and gynecological tumors. By the 1930s there were well-established techniques for interstitial and intracavitary radium implants, and superficial applicators for radiotherapy for a large variety of diseases. Treatments with implanted sources (any kind) are called *"brachytherapy"* [brachy = close (Greek)].

In the following we will focus on *teletherapy* [tele = far (Greek)] which refers to the use of distant radiation sources, external to the body, and directing their beams to the tumors. Before World War II, there were only two methods to do

RSC Drug Discovery Series No. 15
Biomedical Imaging: The Chemistry of Labels, Probes and Contrast Agents
Edited by Martin Braddock
© Royal Society of Chemistry 2012
Published by the Royal Society of Chemistry, www.rsc.org

that: using X-ray machines of different voltage (30–300 kV or kilo voltages) and the use of a high activity radium source.

5.4.1.1 Treatment Equipments

Conventional X-ray tubes reached their technological limit at a voltage of about 500 kV. Unfortunately, the usual mass producible 200–250 kV X-ray units have a penetration depth (half value layer) of about 6 cm in tissue. Teletherapy cobalt units are still in use, although not very often in the developed world. In those machines the source of the photons is the high activity Co-60 isotope, produced in nuclear reactors. It emits gamma rays of 1.25 MeV (average) energies. A modern teletherapy application of this source is the Gamma Knife (see later).

Today the primary teletherapy units are linear accelerators (Figure 5.4.1). Electrons are accelerated along an electromagnetic wave guide (a specially formed linear metal tube with multiple cavities). High-power microwaves are fed into the tube and the electrons can draw energy from the wave as they travel along the tube. At the end of the tube the electrons are either used directly for irradiation or they are slowed down in a target (usually tungsten), and emit high energy X-rays. The so-called accelerating potential is between 4 and 20 megavolts (MV), and most machines have 2 or 3 photon (X-ray) and 5 or 6 electron energies.

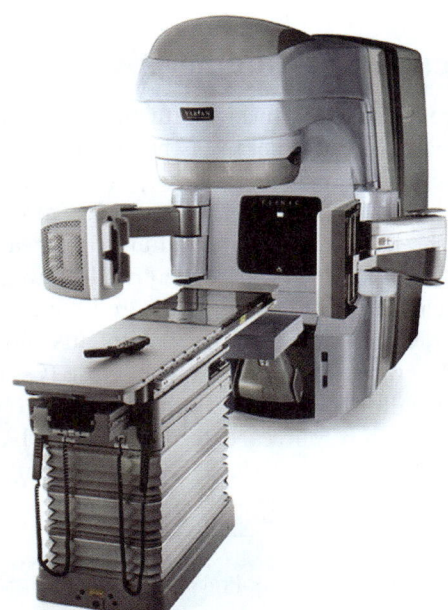

Figure 5.4.1   A modern medical linear accelerator.

New (and very costly, as yet) teletherapy units are using beams of very high energy protons. The protons are accelerated in a circular tube (cyclotron), where a strong magnetic field keeps them in a circular orbit, and in each cycle they pass twice through an accelerating potential gaining more and more energy. The resulting proton beams have a very advantageous dosimetric property: most of their energy is absorbed at a certain energy dependent depth (Bragg-peak) delivering relatively low dose proximally and low dose distally from that peak.

The source of radiation in these devices is mounted on a gantry rotating in one plane; therefore it is possible to have cross multiple beams from different directions. A combination of the gantry rotation with the rotation of the treatment couch results in 3D or non-coplanar beam directions. Rectangular beams are formed with two movable pairs of heavy metal blocks (jaws). Secondary beam shaping is usually achieved with manually cut blocks or with two series of narrow (0.3–1 cm projected width), individually movable blocks (leaves). The leaves can be moved while the beam is on, creating the desired dose distribution within the field.

5.4.1.2 Radiotherapy Process

The four classical steps of the radiotherapy process is target localization, treatment planning, treatment and monitoring–follow up.

"Target localization" is most frequently based on CT (computed tomography) scans. More and more often, however, other modalities such as magnetic resonance imaging (MRI), and position emission tomography (PET) are also employed. The images can be combined (registered) in one single coordinate system and overlaid to obtain more accurate location and extent of the tumor and the surrounding organs. The tumor must be surrounded by an appropriate margin to account for microscopic involvements, internal motion and setup inaccuracies. The volume encompassing the gross tumor volume and all added margin is called the planning target volume (PTV). The purpose of radiotherapy is therefore to deliver the desired radiation dose (or prescription dose) to the PTV. The unit for radiation dose is the Gray (Gy), which is defined as the amount of energy absorbed per unit mass of the irradiated tissue (J kg^{-1}).

"Treatment planning" is a process to select the most appropriate doses, beam angles and shapes and dynamic leaf patterns for the treatment. The goal is to give a tumoricidal dose to the PTV while keeping the doses to the normal structures as low as possible but at least below their radiation tolerance limit. There are various computer programs to aid that process, capable of just calculating the dose distribution from a certain field arrangement or perform some kind of automatic plan optimization. The planning process therefore heavily depends on the accuracy of the localization process.

"Treatment" is the execution of the treatment plan. The plan is transferred to a "record and verify system" from which the treatment machine can obtain the machine setting parameters as required by the plan. The radiation therapist positions the patient according to the localization CT and some external markers. The accuracy of the setup is aided by on-board imagers, capable of cone-beam CT (CBCT) reconstruction or taking regular X-rays. The prescribed dose is usually delivered in multiple fractions. In each fraction the setup must be accurately reproduced. Within fractions the motion of the patient needs to be minimal. Imaging can help in resolving both issues.

"Monitoring" includes the treatment accuracy checks, mentioned above, and checking and adapting to anatomical changes during the treatment process. Follow-ups include imaging studies, where MR imaging has a major role in following the efficacy of the treatment.

5.4.1.3   Radiobiology

Each step of the complex process from the initial ionization events to tumor killing and/or organ damage through chemical/cellular/physiological changes,[1] is under intense research and not yet fully understood. Radiation ejects electrons from their orbits, which in turn create some free radicals. These free radicals then interact with the macromolecules, most importantly causing single or double breaks on the DNA strands. The breaks are either repaired or lead to cell death. Incorrect repair in theory can lead to genetic modification, if it happens to a stem cell, it may cause cancer in the long term.

We mention here only two important findings for radiation therapy: (a) fast reproducing cells are more susceptible to radiation and (b) the dose tolerance for extended dose delivery (*e.g.* multiple fractions) is higher than for a single dose irradiation (due to the repair and repopulation between fractions). Since tumor cells usually divide faster, in theory they are more radiosensitive. By employing an optimal fractionation scheme one can maximize tumor killing and minimize normal tissue damage.[2]

There are a lot of empirical data of individual organ tolerance doses and cell death measurements with a number of different fractionation schemes.[3] Computational models based on those data are available to calculate probabilities of tumor control and normal tissue complication.[4-6] The reliability of those models and their most appropriate input parameters for different tissues and endpoints is still under investigation.

5.4.2   Radiotherapy Treatment Planning Process

Treatment planning is the process that transforms the radiation oncologist's cognitive formulation of the therapeutic plan for the patient into a deliverable treatment. Once it is understood the course of radiation therapy that would benefit the patient, the process of planning the treatment logistics can begin.

Before the radiotherapy treatment planning, a patient first undergoes the "simulation" process during which the patient is positioned on a treatment simulator in the treatment position to determine the treatment parameters and obtain a simulation image set. While there has been some development toward MRI simulators or teletherapy units with associated MRI imaging, most planning systems base their calculations on a CT image set taken using a CT simulator. An exception to this is the Gamma Knife which bases its dose calculations on a skull surface generated from measurements using a helmet which measures the depth to the isocenter along different points of the skull.

5.4.2.1 Steps of Radiotherapy Treatment Planning

Treatment planning is a composite of the physicians' prescription, patient data, beam placement and dose calculation, and can be temporally separated into six stages: (1) Import and registration of simulation CT and additional imaging modalities, (2) Target and organ delineation, (3) Beam placement, (4) Isodose planning, (5) Patient setup images created and (6) Plan completion and validation.

We need a target, a dose for the target and dose constraints for surrounding normal tissues. These can then be applied to the simulation CT image set acquired with the patient positioned in the treatment position. More specifically we need accurate information about the spatial extent of the target. MRI or PET-CT may be used to better identify and delineate the target volume as well as certain normal structures.

Treatment beams may be selected with the aid of digitally reconstructed radiographs (DRRs) or using an algorithm for optimizing beam positions. IMRT,[7] (intensity modulated radiation therapy) planning may use a class solution approach that applies a template containing a standard set of beams, as input for an inverse planning solution for delivering the dose. The recent introduction of volumetric intensity modulated arc therapy programs,[8] has reduced the problem to selecting beam arcs that are acceptable and not acceptable without the need for individual beam angle optimization.

The current generation of dose calculation algorithms requires geometry superimposed with electron densities to perform dose calculations. These range in sophistication from pencil beam algorithms,[9] with corrections for tissue density heterogeneities to convolution based algorithms, Monte Carlo algorithms,[10] and diffusion based algorithms.[11] The major differences between the qualities of the dose calculation occur in regions of density heterogeneities and at the interfaces between such regions of different electron densities.

The inherent value of MRI imaging for treatment planning is not increased precision in the dose calculation but in improved structure and target delineation. An understanding of the target and the more precise identification and localization of structures can take advantage of the power of IMRT to shape the dose distribution, or integrate areas of different dose into the inverse planning solution.

5.4.2.2 Target Definition

MRI and PET-CT provide different and perhaps complementary information to the CT acquired during simulation to improve target and structure delineation. For example MRI's ability to image in sagittal or coronal planes can help with defining some structures such as optic nerves and chiasm as well as improved target definition. The availability of different contrast agents may offer other information including perfusion, hypoxia or proliferation which can be used to better define the target or for designating differential dose regions within the target.

It is important to assess the source of the information and the uncertainties of the information they provide. If the structure is a suspect for respiratory or cardiac associated motion, the images must be evaluated with an understanding of potential artifacts resultant from this physiological motion. Gated acquisition of images or retrospective binning of images may be used to better understand the extent of the motion and remove geometric artifacts associated with organ and tumor motion. Both CT and MRI have been used to capture the variation of position with respiratory motion for tumors within the lung.

5.4.2.3   Image Fusion

Image fusion is a process of merging information from different image sets or modalities, each providing information that is complementary to the others. These may be the same imaging modality separated in time or different imaging modalities such as CT, CBCT, PET, Ultrasound and MRI.

The value added from additional imaging modalities is in more precise target or organ delineation, assessment of patient setup or in assessment of the response to therapy. These imaging sets are at a minimum separated in time, but often with the patient in different postures or supported by different tabletop designs. There may be deformations between image studies due to changes in internal structures or due to a different setup. These differences combined with the inherent variation in the information presented may make any registration of the image sets difficult. For example, the MRI images of interest may be in a coronal plane where the CT images are in an axial plane as well as the many different contrast and sequence possibilities between studies.

Image fusion starts with a registration between the different imaging studies. That is, a common coordinate system is established and then the associated information of each study may be shared as there is a relationship between the voxels of one image and the voxels of the other. These may be simple rigid transformations (affine transformation) or require a more complex deformable transformation. Rigid body transformations are the standard in radiation oncology although in nuclear medicine there is a longer history of clinical practice using deformable image registration.

These registrations for rigid body transformations may be feature based,[12] with the user selecting common anatomic locations on the two studies or mutual information based,[13] where the overlap of intensity values are used to define the best match. It may be possible to restrict the region of interest (ROI) to a small volume where an acceptable rigid registration between two sets of data may be achieved even though the two sets are quite different. Examples include segments of the vertebrae as well as the prostate.

After registration it is now possible to share or fuse the data information as there is now a relationship between the coordinates of one imaging study to the other. For many brain lesions MRI is used to define targets and some normal structures. Structures segmented on the MRI study may be directly overlaid with the CT simulation study. PET-CT may be used to assess nodal involvement and in general, the CT simulation study becomes the host study for other biological, functional and physiological information, enriching the simulation process and making ready for treatment planning.

There is an uncertainty implicit to the registration process. This will vary with body site and the presence of other factors such as motion and acquisition time for imaging. This can hopefully be limited to perhaps 1 mm for the brain and 2–3 mm for other sites, subject to the resolution of the studies. However, careful scrutiny of the registration is necessary to assess which data is well matched and acceptable for data mapping between studies. This uncertainty must be incorporated in the PTV when a structure delineated using MRI, for example, is copied to a CT set used for dose calculation and verification of patient setup. It should also be noted that different imaging studies may suggest different target volumes or volumes that do not completely overlap leaving the user to decide to generate a PTV that encompasses all the possible target volumes.

5.4.3 Chemoradiation

5.4.3.1 Chemotherapy

Chemotherapy involves the use of chemical agents to stop cancer cells from growing. As a form of systemic treatment (treatment using substances that travel through the bloodstream, reaching and affecting cells all over the body), chemotherapy can eliminate cancer cells at sites great distances from the original cancer.

The majority of chemotherapeutic drugs can be divided into alkylating agents, antimetabolites, anthracyclines, plant alkaloids, topoisomerase inhibitors, and other antitumor agents.[14] Although these agents differ in their mechanism and side-effects, most of them act by killing rapidly-dividing cells, one of the main properties of cancer cells, as well as some normal cells (*e.g.*, cells in the bone marrow, digestive tract and hair follicles). The most common side-effects of chemotherapy are therefore decreased production of blood cells (myelosuppression), inflammation of the lining of the digestive tract (mucositis) and hair loss (alopecia). More than half of all people diagnosed with cancer receive chemotherapy.

5.4.3.2 Combining Chemotherapy and Radiotherapy

Since radiotherapy is a local treatment, very frequently there is a need to combine it with a systemic therapy (*e.g.*, chemotherapy). Chemoradiotherapy, also called chemoradiation, is a cancer treatment modality that combines chemotherapy with radiation therapy, similar to treating a patient with a number of different drugs simultaneously. The biggest advantage is minimizing the chances of resistance developing to any one agent. It is also useful in killing any cancerous cells which have spread to other parts of the body.

Chemoradiation is effective for many locally advanced tumors that have a low cure rate with traditional surgery-, radiation- and chemotherapy-alone treatment. For example, chemoradiation is often the standard treatment for patients with stage III or nonmetastatic stage IV head and neck cancer, and has been proven to be superior to radiotherapy alone for patients with advanced nasopharyngeal cancers with respect to progression-free survival and overall survival.[15]

Because chemotherapy agents are usually highly cytotoxic, the need for normal tissue protection has arisen. For example, some success was achieved by the drug Amifostine to prevent serious damage to the mucosa during radiotherapy of head and neck tumors.[16]

5.4.3.3   Administration of Chemoradiation

Some of the general chemotherapy agents also have radiosensitizing effects. It is very important therefore to find optimal administration dosages and schedules to avoid harmful interactions between radiation and chemotherapy and maximize therapeutic benefit. To lower the toxicity, one could decrease the radiation fractionation size or total dose, reduce the chemotherapeutic dose per cycle or the number of cycles or temporally separate the radiotherapy.[17]

Timing of administration can also be optimized for chemoradiation. Neoadjuvant chemotherapy refers to chemotherapy prior to primary treatment therapy (surgery or radiotherapy). The initial chemotherapy is designed to shrink the primary tumor, so that surgery or radiotherapy will be less destructive or more effective. Adjuvant chemotherapy is a secondary treatment given after the primary treatment. The main goal is to kill any remaining cancerous cells and/or to prevent the cancer from recurrence. Simultaneous (or concurrent) chemoradiation delivers the radiation and chemotherapy agents simultaneously. Evidence has shown that temporizing radiotherapy and chemotherapy with the least temporal separation is most likely to achieve superior results.[17] Palliative chemotherapy without radiotherapy, on the other hand, is given without curative intent, but simply to decrease tumor load and increase life expectancy. For these regimens, a better toxicity profile is generally expected.

5.4.4   MRI for Radiotherapy

CT is the main imaging modality for radiotherapy treatment planning because it can provide accurate electron density information for radiation dose calculation, and has good visualization of bony anatomy for treatment setup.

However, CT is poor in soft tissue differentiation, where electron densities are not very different. MRI, on the other hand, is more sensitive to chemical differences, therefore it is better in producing anatomic and biological information for tumor staging, target or normal tissue identification, and monitoring treatment outcomes, which are essential to the success of radiotherapy.

5.4.4.1 Anatomic MRI

By proper adjustment of MR imaging parameters, tissue difference in magnetic characteristics can be enhanced to show anatomic details. Anatomic MRI routinely used for radiotherapy includes proton density images, T_1- and T_2-weighted MRI, and fluid-attenuated-inversion-recovery (FLAIR) images. For T_1-weighted MRI that employs a short TR and a short TE, most of the contrast between tissues is due to differences in the T_1 value. T_2-weighted MRI use long TR and long TE to enhance the T_2 contrast. Proton density weighted MRI, on the other hand, use a long TR and short TE to show the difference in proton density of different tissues. FLAIR MRI uses an inversion-recovery pulse sequence to suppress signal from fluids.

Figure 5.4.2 shows the T_1-, T_2- and FLAIR MRI of the same slice of a brain. T_1-weighting causes white matter (myelinated axons) to appear white, and the congregations of gray matter (neurons) to appear gray, while cerebrospinal fluid (CSF) appears dark. The contrast of white matter, gray matter and cerebrospinal fluid is reversed using T_2-weighted imaging. For FLAIR images, the contrast of white matter and gray matter is similar to T_2-weighted images but the CSF is dark (suppressed by the imaging sequence). Suppression of CSF can enhance the periventricular hyperintense lesions and edema that are difficult to identify using other imaging sequence.

A contrast agent may be administered if the image contrast is not adequate to show the anatomy or pathology of interest. When presented in tissues or body

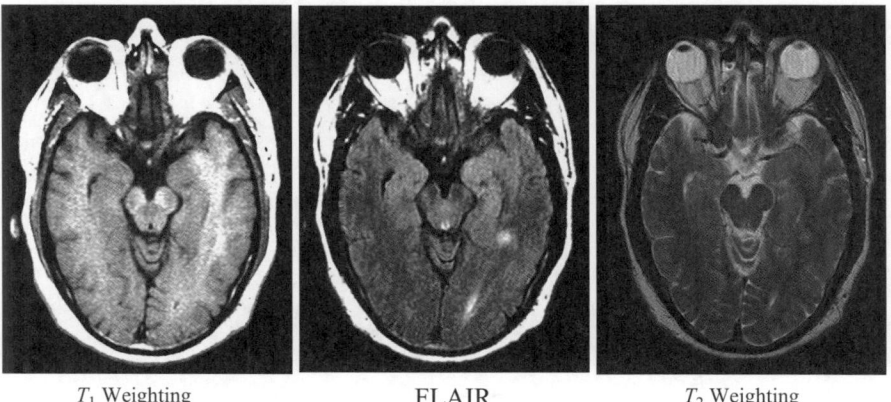

T_1 Weighting   FLAIR T_2 Weighting

Figure 5.4.2 T_1-, T_2- and FLAIR MRI of the same slice of a brain.

cavity, MRI contrast agents alter the relaxation times and lead to a higher or lower signal depending on the imaging parameters. Gadolinium chelates, *e.g.*, Gadopentetate dimeglumine or Gd-DTPA, are commonly used as MRI contrast agents because Gadolinium is paramagnetic, which reduces the T_1 relaxation time and enhances T_1-weighted signals. Once injected intravascularly, gadolinium-based contrast agents accumulate in abnormal tissues of the brain and body, providing a greater contrast between normal and abnormal tissues (*e.g.*, tumors). Other chemical compounds that have been used for MRI contrast include iron oxide and manganese chelates.

5.4.4.2   Functional MRI

In addition to anatomic information, MRI can also be used to obtain physiological and metabolic information. These imaging techniques, including magnetic resonance spectroscopic imaging (MRSI), perfusion weighted imaging (PWI), diffusion weighted imaging (DWI)–diffusion tensor imaging (DTI) and functional MRI (fMRI), can be used to identify tumor cells with abnormal metabolic or angiogenic activities for radiotherapy targeting, or locate functional active normal tissues that should be spared.

MRSI is an emerging diagnostic tool that can obtain magnetic resonance spectra of protons (^{1}H) from selected regions within the region of interest. Magnetic resonance spectra contain signatures of metabolites unique to normal or abnormal tissue and therefore can be used to diagnose certain metabolic disorders of cancer cells. For example, MRSI can provide information regarding tumor activity based upon the levels of cellular metabolites, including choline-containing compounds (Cho), total creatine (Cr), *N*-acetylaspartate (NAA). As shown in Figure 5.4.3, for a brain tumor, the Cho intensity is much higher than that of the Cr or NAA intensity.

PWI uses specially designed MR pulse sequences to measure blood flow at microscopic levels, the most common technique being dynamic susceptibility

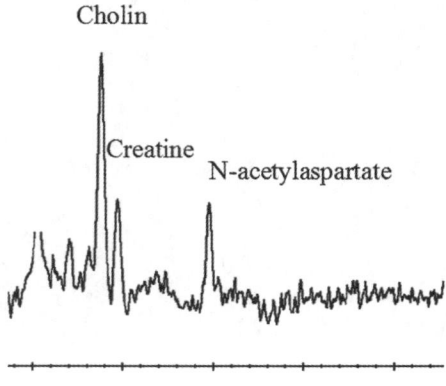

Figure 5.4.3   A typical proton MR spectrum for abnormal brain tumor area.

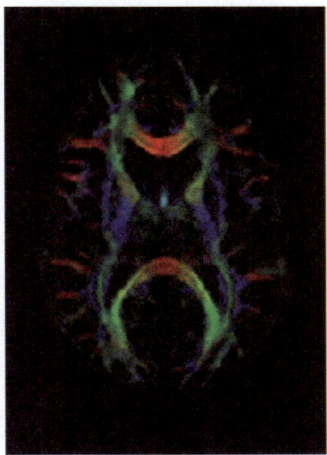

Figure 5.4.4 DTI images of white matter track.

contrast imaging (DSC MRI). Relative cerebral blood volume (rCBV) measurements from DSC MRI have been shown to be closely correlated with histological measurements of microvascular density in gliomas,[18] and can provide physiologic information about neovascularity and angiogenesis of the entire brain and have been shown to correlate well with glioma grade in humans.[19,20]

DWI and DTI are MRI methods that produce images weighted with water diffusion. In DWI, the image intensity of each voxel reflects the rate of microscopic water diffusion (apparent diffusion coefficient or ADC) at that location. In addition to the rate of diffusion, DTI also measures the preferred direction of water diffusion for each voxel and describes the measurements by a tensor. DTI enables the visualization of white matter fibers in the brain and can map subtle changes in the white matter associated with diseases (Figure 5.4.4). Diffusion metrics are clearly altered within the vasogenic edema surrounding both gliomas and metastatic tumors,[21] which enables the differentiation of solitary intraaxial metastatic brain tumors and meningiomas from gliomas.[22,23]

fMRI is an imaging technique that usually uses a gradient echo-echo planar imaging (GE-EPI) sequence to define the locations of functionally eloquent cortices, such as the motor cortex, Broca's area, Wernicke's area, and the Visual Cortex in the brain. Figure 5.4.5 shows an fMRI image of primary motor cortex and supplementary motor area overlaid on T_2-weighted MRI. fMRI is based on the blood oxygen level dependent (BOLD) mechanism during brain function activation,[24,25] which uses paramagnetic deoxyhemoglobin in venous blood as a naturally occurring contrast agent for MRI. Incorporation of fMRI information into the radiotherapy treatment process can possibly allow a radiation oncologist to properly plan and deliver an adequate radiation dose while avoiding damage to the adjacent functional cortices in the treatment of gliomas.[26–29]

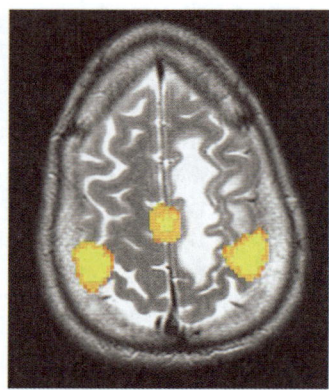

Figure 5.4.5   An fMRI image of primary motor cortex and supplementary motor area
overlaid on T_2-weighted MRI.

5.4.4.3   Applications of MRI in Radiotherapy

MRI is widely used to assist tumor staging, which takes into account the tumor
size, tumor penetration, tumor invasion of adjacent organs, lymph nodes
involvement, and tumor spread to distant organs. Proper tumor staging is the
most important prognostic factor and is critical to the determination of optimal
treatment for cancer patients. For radiotherapy, patients with low-stage tumors
usually receive radiotherapy only, while chemo-radiation is generally considered for patients with locally advanced tumors. Biopsy can determine the
pathology grade in the gross tumor region but is not suitable for detecting
tumor invasion to adjacent organs and lymph nodes, which is usually the key
for staging advanced tumors. Imaging study like CT-MRI is therefore usually
used to assist the determination of the high grade tumors that require
chemoradiation.

MRI is also routinely used for target or normal tissue definition in radiotherapy treatment planning, for tumors in brain, spinal cord, and head & neck
regions. MRI is also widely used to monitor the treatment responses during
routine follow-up to detect recurrences or normal tissue complication. A few
examples will be given in the following sections to illustrate the clinical use of
MRI for radiotherapy treatment planning and monitoring treatment responses.

5.4.5   Chemoradiation of Head & Neck Tumors

5.4.5.1   MRI Evaluation of Head and Neck Cancer

Radiologic evaluation is a critical component of head and neck cancer treatment with chemoradiation therapy. While CT and MRI have both been used in
staging, MRI provides more accurate definition in certain clinical situations. In
nasopharynx cancer, for example, MRI better demonstrates early invasion
beyond the nasopharynx, differentiation of retropharyngeal nodes from the

primary tumor, and assessment of the parapharyngeal space, skull base, paranasal sinuses, and cranial invasion.[30–33] For these reasons, most centers use MRI systematically in the pretreatment evaluation of nasopharyngeal cancer.

The adaptation of computerized treatment planning and IMRT further allows clinicians to refine treatment by concentrating the dose on the tumor and reducing the dose to normal structures. These techniques have improved some clinical outcomes. Patients treated with IMRT had superior global quality of life with fewer cases of fatigue, loss of taste, and dry mouth, compared to those treated with conformal RT.[34,35] Further research will determine if these advancements will translate into improved local tumor control or overall survival.

5.4.5.2   A Clinical Case

The patient was 41-year old Asian male with a right sided neck mass. Head and neck examination revealed a 3 cm × 3 cm × 3 cm irregular mucosal lesion in the nasopharynx which extended from midline to the right fossa of Rosenmuller and involved the posterior right nasal cavity. The patient underwent biopsy of the nasopharyngeal mass which was consistent with undifferentiated naso- pharyngeal carcinoma. The patient underwent diagnostic imaging with MRI of the neck, which revealed an enhancing mass in the nasopharynx with ipsilateral paranasopharyngeal extension and involvement of the posterior aspect of the nasal cavity and maxillary sinus (Figure 5.4.6). The study showed the right sided lymph node measuring 4 cm and demonstrated bilateral shotty adeno- pathy measuring less than 1 cm, which was not considered pathologic. The patient's stage was thus T3N1M0, Stage III.[36]

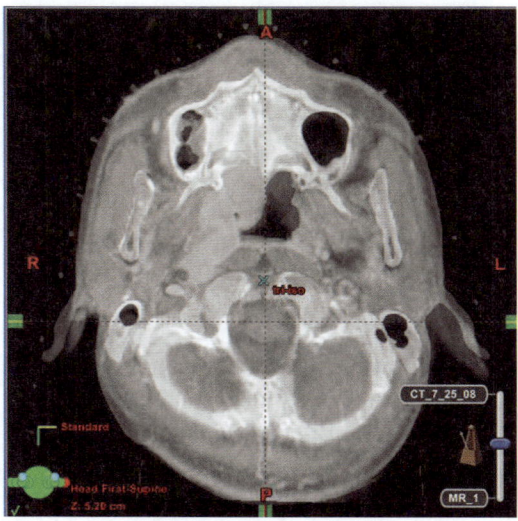

Figure 5.4.6   MRI Neck demonstrating nasopharyngeal carcinoma.

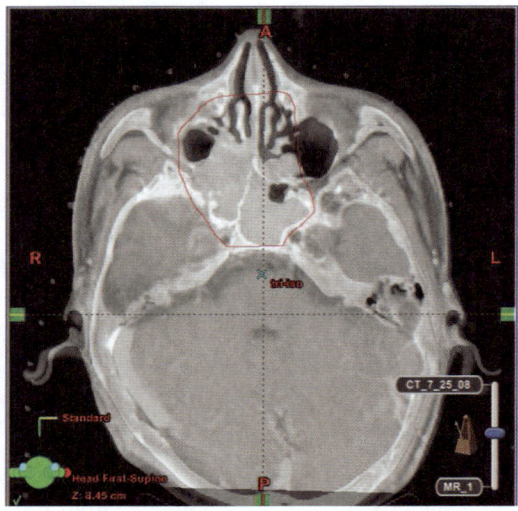

Figure 5.4.7   MRI-CT fusion with nasopharyngeal tumor contoured in red for radiation therapy planning.

The patient was referred to medical oncology evaluation and the treatment team agreed on a course of concurrent chemotherapy and radiotherapy followed by adjuvant systemic chemotherapy as is considered standard in the United States.[15]

The patient had a plastic mask made for immobilization and underwent CT scanning for simulation using 2.5 mm slice thickness. Images were transferred to the treatment planning system. To better delineate the tumor volume, the diagnostic MRI scan and treatment planning CT were fused so that the definition seen on MRI, particularly extension to the paranasal sinuses and soft tissue, could be outlined directly on the planning study. The fusion involves selecting fixed points on both image sets, such as the mastoid tip or orbital apex, and the system then fuses the images. An example of image fusion with contoured tumor volume is shown in Figure 5.4.7.

The patient received concurrent cisplatin and radiotherapy for 7 weeks. Cisplatin was delivered in 3 cycles and radiation therapy was delivered in fraction sizes of 2 Gy. IMRT was employed. Initially both sides of the neck and primary region were treated to 50 Gy. This was followed by a cone down to only gross tumor as shown by MRI to a total dose of 70 Gy. This was followed by an additional 3 cycles of 5-Fluorouracil and Cisplatin given over 3 months.

5.4.6   The Use of MRI for Gamma Knife Treatment Planning

5.4.6.1   Leksell Gamma Knife

The Leksell Gamma Knife instrument was first developed in the 1950s by Lars Leksell, a Swedish neurosurgeon. It is a teletherapy device that uses multiple

(~ 200 depending on the model) Cobalt-60 sources to deliver a combined high dose of gamma ray radiation to brain lesions. The sources are arranged in a spherical or conical geometry such that any single photon beam enters only a small portion of the brain but all sources intersect at a single point in three-dimensional space creating an ellipsoidal or spherical "cloud" of radiation dose. The lesion in the patient's brain is moved to this point for treatment. The size of the radiation dose cloud can be modified by using different diameters in the collimation system for the beams.

5.4.6.2 Imaging for Gamma Knife

To plan a Gamma Knife treatment, image data must be acquired and transferred to the treatment planning system. The majority of images are from MRI although CT images can be used when clinically necessary. In the case of an arteriovenous malformation (AVM), MR images and two orthogonal, diagnostic-quality angiographic X-ray images are used. In the following we will focus on the use of MR images for Gamma Knife planning.

To prepare a patient for scanning as well as for treatment, the Leksell Coordinate Head Frame is firmly attached to the patient's skull using titanium screws and rigid fixation posts attached to this rectangular base frame. This base frame serves two purposes: (1) It provides a base to attach the imaging indicator box that defines a 3-dimensional Cartesian coordinate system whereby a lesion in the brain can be uniquely located and (2) It immobilizes a patient during treatment, preventing even the smallest head movements from happening.

With the MRI indicator box attached to the coordinate base frame surrounding the head, the patient is escorted to the MRI scanner. The patient's head is placed into a head coil using an adapter. Standard anatomic (T_1-, T_2- or FLAIR) MRI with contrast is usually obtained. The volume data is re-formatted into 1 mm contiguous axial images and electronically delivered to the Gamma Knife Planning workstation for treatment planning. Before any planning is done, the MR images are validated to confirm that there are no large spatial distortions. Typically spatial discrepancies are less than 1mm.

5.4.6.3 Treatment Planning for Gamma Knife

The treatment planning of Gamma Knife treatment involves the neurosurgeon, radiation oncologist and medical physicist. The neurosurgeon usually outlines the boundary of the target (tumor) volume. Once the volume is defined either the neurosurgeon, radiation oncologist or physicist can plan the irradiation scheme using different size beam collimators or "shots" to determine the best radiation dose coverage of the defined treatment volume. Depending on the treatment protocol, the best plan is the one where a specified isodose line (*e.g.*, the 50% of the maximum dose) conformally covers the periphery of the defined target volume. The radiation oncologist then determines the absolute treatment dose. This value will depend on the type of tumor, the size and the location within the brain.

MRI can be used alone for Gamma Knife treatment planning. In this case, skull measurement is performed at a couple of points and a mesh is generated by interpolation to approximate the shape of the skull. Dose calculation is done assuming that the tissue inside the skull is water equivalent. Alternatively, CT scan can be imported and fused with the MRI scan. With CT images, the skull boundary can be automatically detected and precisely defined. In addition, more accurate dose calculation with heterogeneous correction can be performed by converting the CT number to electron density for each CT voxel.

Figure 5.4.8 shows that mesh (in red) generated from skull measurement overlaid on the axial, coronal, and sagittal slices of the MRI scan for a patient with multiple brain metastases. The prescription is to deliver 20 Gy radiation dose to encompass the tumor enhancement on the MRI scan with a maximum dose of 40 Gy inside the tumor. Figure 5.4.9 shows the dose distribution for the axial (left), coronal (upper right) and sagittal (lower right) slices of a tumor. The numbers in red circles are the number and location of different "shots," each shot involving all or part of the 200 sources converging at the center location of the shot specified by the planner. The yellow curve is the 20 Gy (or prescription) isodose line while the green curve is the 10 Gy line. It is observed from Figure 5.4.9 that, by choosing the shots properly, the prescription isodose line can be conformed to the tumor enhancement so that dose to the surrounding normal tissue can be minimized.

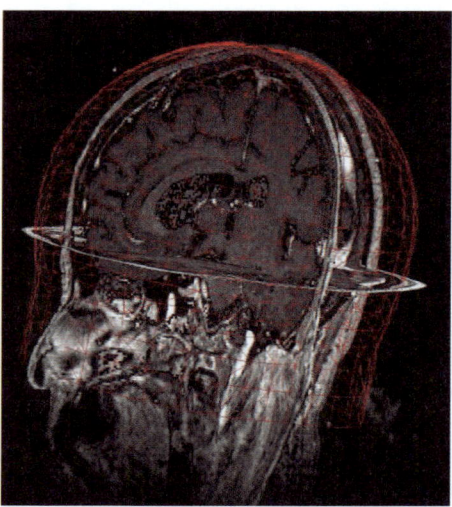

Figure 5.4.8   Mesh (in red) generated from skull measurement overlaid on the axial, coronal, and sagittal slices of the MRI scan.

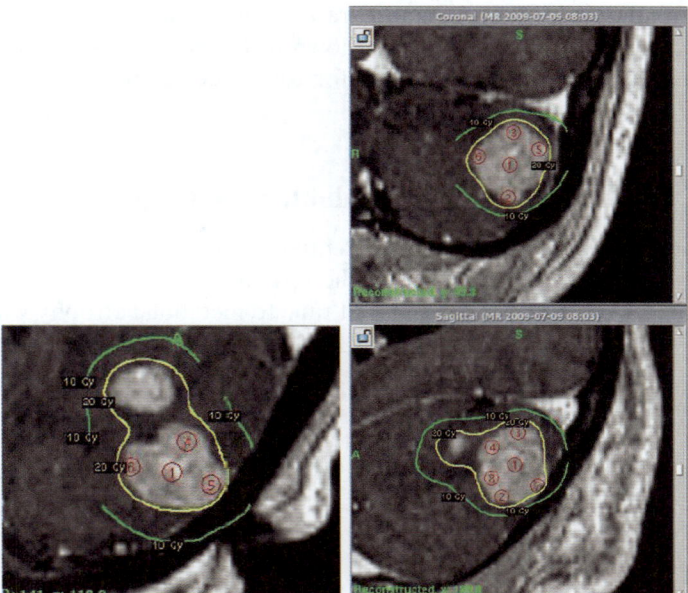

Figure 5.4.9   Dose distribution for the axial (left), coronal (upper right) and sagittal (lower right) slices.

5.4.7   MRI for Monitoring Radiation Therapy in Prostate Cancer

5.4.7.1   Radiotherapy of Prostate Cancer

Both external beam radiotherapy and brachytherapy are important treatment options for patients with prostate cancer. Patients with prostate cancer treated with radiotherapy are usually followed with serial prostatic-specific antigen (PSA), to assess for biochemical failure (rising serum PSA after a nadir level has been reached),[37,38] which may be due to local and/or systemic recurrence.[38] However, serum PSA is limited in accuracy, and thus, an alternative method to PSA would be preferable for post-radiation monitoring. In addition, the detection of locally recurrent prostate cancer after external beam radiotherapy is essential since other treatment options are available, including hormonal therapy, salvage prostatectomy and local ablation therapy.

MRI has emerged as a powerful tool for detection and staging of prostate cancer. There is limited data on the use of MRI for predicting or for following response to radiotherapy. However, because post-radiation changes including prostatic atrophy, and diffuse low T_2 signal changes (in relation with fibrosis) and loss of zonal anatomy, this limits the accuracy of conventional (mostly based on T_2-weighted imaging) MRI for the detection of recurrent tumor. Functional MR methods including MRSI, DWI and DCE (dynamic contrast-enhanced) MRI show considerable promise in assessment of patients with prostate cancer, and have been shown to be effective in the detection of prostate cancer; with better accuracy for tumor detection and staging when compared to T_2-weighted MRI.

5.4.7.2   MRSI for Assessing Radiotherapy Response

Few studies have demonstrated the potential role of MRSI in determining the response of prostate cancer to radiotherapy and for diagnosing recurrence earlier than PSA, by demonstrating that time to metabolic atrophy with MRSI is shorter than time to PSA nadir.[39,40] In addition, a recent study by Joseph *et al.*,[41] demonstrated that MRI and MRSI findings before external radiotherapy in patients with prostate cancer are more accurate independent predictors of outcome than clinical variables, with findings of seminal vesicle invasion and extensive tumor metabolism at MRSI shown to be independent predictors of worse prognosis.

Figure 5.4.10 illustrates the (A) Axial T_2-weighted image and (B) MRSI of an 80-year old man with prostate cancer treated with brachytherapy. Axial T_2-weighted image (left) shows atrophic prostate, with diffuse low T_2 signal and seed implants (arrows). Spectral map (right) obtained with MRSI shows diffuse metabolic atrophy, in relation with radiotherapy, without abnormal metabolism to suggest tumor recurrence.

5.4.7.3   DWI and DCE MRI for Assessing RT Response

There is limited data on the use of DWI to follow patients with prostate cancer treated with external beam radiotherapy. Recently, Kim *et al.*[42] showed better sensitivity of the combined T_2-weighted and diffusion-weighted approach

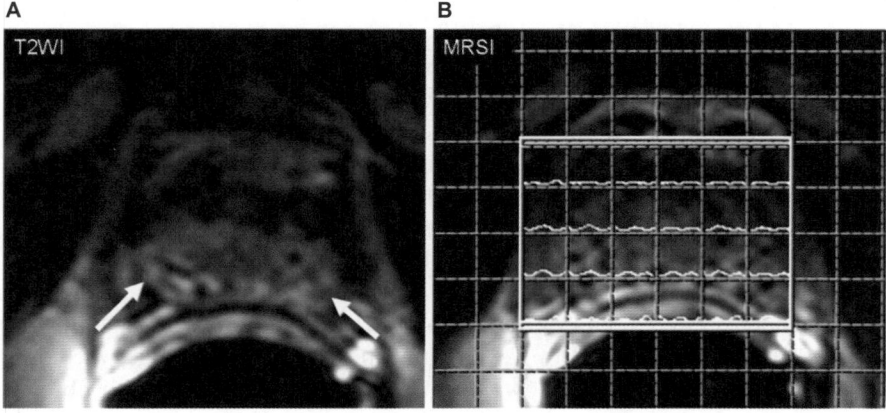

Figure 5.4.10   (A) Axial T_2-weighted image and (B) MRSI of a prostate case for 80-year old man with prostate cancer.

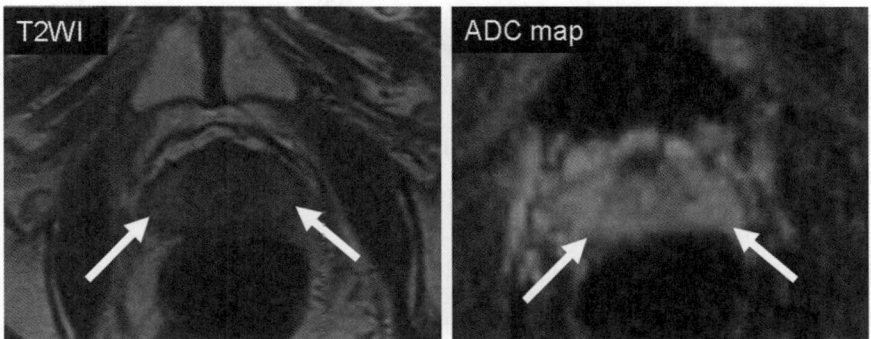

Figure 5.4.11 (A) Axial T_2-weighted image and (B) ADC (apparent diffusion coefficient) of a prostate case for 75-year old man with prostate cancer. The arrows surround the prostate edge.

compared to T_2-weighted images alone for predicting recurrent cancer after external beam radiotherapy. In addition, Haider *et al.*[43] showed better performance of DCE MRI compared to T_2-weighted MRI for detection and localization of prostate cancer in the peripheral zone after external beam radiotherapy.

Figure 5.4.11 demonstrates (A) Axial T_2-weighted image and (B) ADC (apparent diffusion coefficient) map of the prostate for a 75-year old man with prostate cancer treated with external beam radiotherapy. Axial T_2-weighted image (left) shows atrophic prostate, with diffuse low T_2 signal. ADC map (right) obtained with DWI shows normal ADC of the prostate, without focal lesion to suggest tumor recurrence.

5.4.8 MRI for Monitoring Chemoradiation of High-Grade Glioma

In addition to prostate cancers, anatomic and functional MRI has also been used to monitor the treatment response of brain tumors. In this section, we will report a clinical case on using MRI to follow up the treatment outcome of a patient with high-grade glioma.[44]

5.4.8.1 Chemoradiation of Glioma

The combination of surgery, radiation therapy, and temozolomide chemotherapy represents the standard approach to the treatment of patients with high-grade gliomas.[45] Because malignant gliomas are highly vascular and express vascular endothelial growth factor (VEGF), targeting the vascular endothelium offers an interesting option for these patients. Bevacizumab (Avastin; Genentech, San Francisco, CA), a humanized immunoglobulin G1 monoclonal antibody that inhibits VEGF, is the first Food and Drug

Administration–approved antiangiogenic agent that has shown its effectiveness in metastatic colorectal, breast, and lung cancers. Preclinical data showed that it regressed microvascular density, normalized existing mature vasculature, and inhibited vessel regrowth in a glioma model.[46] Marked improvement in radiologic response after bevacizumab therapy in patients with recurrent high-grade gliomas has been reported.[47,48]

5.4.8.2   A Clinical Case

The patient was a 58-year old man with right parietal glioblastoma multiforme treated with bevacizumab, radiation, and temozolomide. Radiation therapy was started within 1 month of surgery. The prescription dose was 59.4 Gy over 6.5 weeks, using 1.8 Gy/fraction, and five fractions per week. Treatment planning was done with five noncoplanar IMRT beams. The clinical target volume was defined as the gross disease observed in the FLAIR MRI series plus a 1.5 cm margin. The PTV was determined by adding an additional 0.5 cm

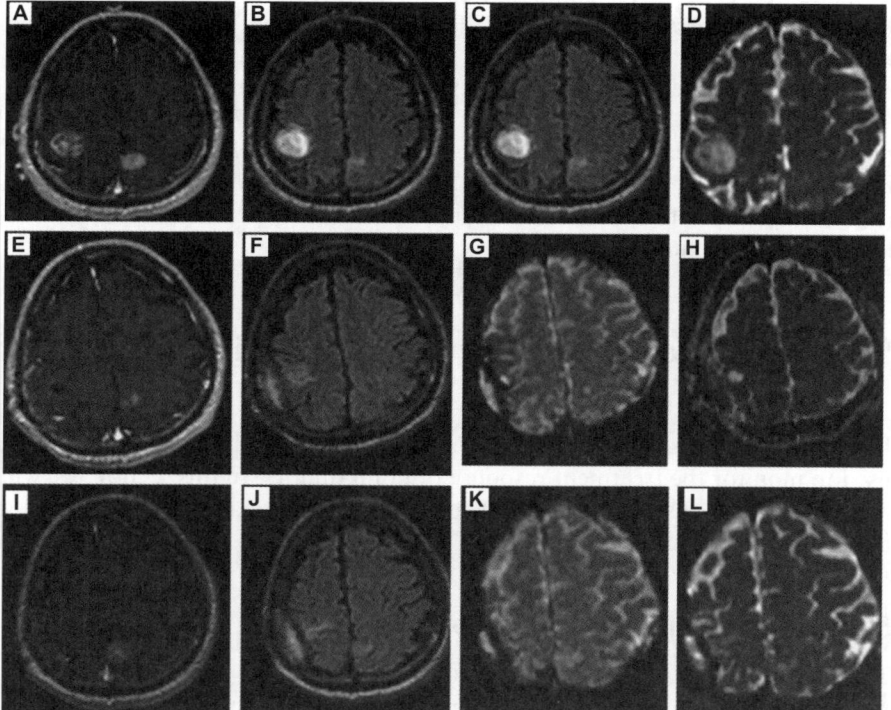

Figure 5.4.12   Radiologic response – T_1-weighted, FLAIR, rCBV and Ktrans – of the glioma patient treated with bevacizumab, radiation, and temozolomide. The patient has an incidental left parietal meningioma. A–D: before surgery; E–H: after completion of radiation; I–L: after completion of planned therapy.

margin to the clinical target volume to account for treatment uncertainties. Based on MRI, normal tissues delineated included the spinal cord, brain stem, optical structures, and "normal brain," defined as all brain tissue outside the planning target volume. The dose limits that defined an acceptable plan included a maximum point dose of 45 Gy to the spinal cord, 54 Gy to the optic nerves and chiasm, 45 Gy to the retina, and 55 Gy to the brain stem. Bevacizumab was administered after 28 days after surgery at 10 mg kg^{-1} as a continuous intravenous infusion on days 14 and 28 of radiation therapy.

Gadolinium-enhanced MRI was performed 1 month after the completion of radiotherapy and subsequently at 8-week intervals. DSC MRI studies were performed in addition to conventional MRI. Relative cerebral blood volume (rCBV) and vascular permeability (Ktrans) measurements were made from perfusion data and assessed by a neuroradiologist.[49]

Figure 5.4.12 shows the radiologic response (T_1-weighted, FLAIR, rCBV and Ktrans) of this patient. Images A–D were obtained before surgery, images E–H one month after completion of radiation, and images I–L after completion of planned therapy. It is observed from Figure 5.4.12 that the tumor responded to the planned treatment radiologically and the changes of rCBV and Ktrans correlated well with the decrease in T_1 contrast enhancement with conventional T_1-weighted and FLAIR MRI.

5.4.9   Conclusion

Anatomic and functional MRI has been widely used in modern radiotherapy for target and normal tissue delineation and monitoring treatment outcomes. With a computerized treatment planning system, it is possible to fuse MRI with simulation CT so that more accurate anatomic, physiological and pathological information can be incorporated into radiotherapy treatment planning. This capability, combined with IMRT, allows selective targeting of the tumor while avoiding surrounding critical organs. Functional MRI has shown promise in detecting early responses from radiotherapy and identifying non-responders early on so that additional treatment can be planned in advance.

References

1. W. C. Deqey, and J. S. Bedford in *Radiobiologic Principles: Textbook of Radiation Oncology*, ed. S. A. Leibel, and T. L. Philips, Saunders, Philadelphia, 2nd edn, 2004.
2. K. K. Ang in *Fractionation Effects in Clinical Practice: Textbook of Radiation Oncology*, ed. S. A. Leibel, and T. L. Philips, Saunders, Philadelphia, 2nd edn, 2004.
3. B. Emami, J. Lyman, A. Brown, L. Coia, M. Goitein, J. E. Munzenrider, B. Shank, L. J. Solin and M. Wesson, *Int. J.Radiat. Oncol. Biol. Phys.*, 1991, **21**, 109–122.

4. J. T. Lyman and A. B. Wolbarst, *Int. J. Radiat. Oncol. Biol. Phys.*, 1987, **13**, 103–109.
5. C. Burman, G. J. Kutcher, B. Emami and M. Goitein, *Int. J. Radiat. Oncol. Biol. Phys.*, 1991, **21**, 123–135.
6. R. Dale and A. Carabe-Fernandez, *Cancer Biother. Radiopharm.*, 2005, **20**, 47–51.
7. C. C. Ling, C. Burman, C. S. Chui, G. J. Kutcher, S. A. Leibel, T. LoSasso, R. Mohan, T. Bortfeld, L. Reinstein, S. Spirou, X. H. Wang, Q. Wu, M. Zelefsky and Z. Fuks, *Int. J. Radiat. Oncol. Biol. Phys.*, 1996, **35**, 721–730.
8. K. Otto, *Med. Phys.*, 2008, **35**, 310–317.
9. R. Mohan, C. Chui and L. Lidofsky, *Med. Phys.*, 1986, **13**, 64–73.
10. I. J. Chetty, B. Curran, J. E. Cygler, J. J. DeMarco, G. Ezzell, B. A. Faddegon, I. Kawrakow, P. J. Keall, H. Liu, C. M. C. Ma, D. W. O. Rogers, J. Seuntjens, D. Sheikh-Bagheri and J. V. Siebers, *Med. Phys.*, 2007, **34**, 4818–4853.
11. M. L. Williams, D. Ilas, E. Sajo, D. B. Jones and K. E. Watkins, *Med. Phys.*, 2003, **30**, 3183–3195.
12. D. L. Hill, P. G. Batchelor, M. Holden and D. J. Hawkes, *Phys. Med. Biol.*, 2001, **46**, R1–45.
13. W. M. Wells III, P. Viola, H. Atsumi, S. Nakajima and R. Kikinis, *Med. Image Anal.*, 1996, **1**, 35–51.
14. C. H. Takimoto and E. Calvo, in *Cancer Management: A Multidisciplinary Approach*, ed. R. Pazdur, L. Wagman and K. Camphausen, CMPMedia, 11th edn, 2008.
15. M. Al-Sarraf, M. LeBlanc, P. G. Giri, K. K. Fu, J. Cooper, T. Vuong, A. A. Forastiere, G. Adams, W. A. Sakr, D. E. Schuller and J. F. Ensley, *J.Clin. Oncol.*, 1998, **16**, 1310–1317.
16. L. G. Marcu, *Eur. J. Cancer Care (Engl.)*, 2009, **18**, 116–123.
17. M. J. John, in *Radiotherapy and Chemotherapy: Textbook of Radiation Oncology*, ed. S. A. Leibel, and T. L. Philips, Saunders, Philadelphia, 2nd edn, 2004.
18. S. Cha, E. A. Knopp, G. Johnson, S. G. Wetzel, A. W. Litt and D. Zagzag, *Radiology*, 2002, **223**, 11–29.
19. E. A. Knopp, S. Cha, G. Johnson, A. Mazumdar, J. G. Golfinos, D. Zagzag, D. C. Miller and P. J. Kelly, *Radiology*, 1999, **211**, 791–798.
20. M. V. Lev, Y. Ozsunar, J. W. Henson, A. A. Rasheed, G. D. Barest, G. R. I. V. Harsh, M. M. Fitzek, E. A. Chiocca, J. D. Rabinov, A. N. Csavoy, B. R. Rosen, F. H. Hochberg, P. W. Schaefer and R. G. Gonzalez, *Am. J. Neuroradiol.*, 2004, **25**, 214–221.
21. S. Lu, D. Ahn, G. Johnson and S. Cha, *Am. J. Neuroradiol.*, 2003, **24**, 937–941.
22. S. Lu, D. Ahn, G. Johnson, M. Law, D. Zagzag and R. I. Grossman, *Radiol.*, 2004, **232**, 221–228.
23. J. M. Provenzale, P. McGraw, P. Mhatre, A. C. Guo and D. Delong, *Radiology*, 2004, **232**, 451–460.

24. S. Ogawa, T. M. Lee, A. R. Kay and D. W. Tank, *Proc. Natl. Acad. Sci. U.S.A.*, 1990, **87**, 9868–9872.
25. D. G. Norris, *J. Magn. Reson. Imaging*, 2006, **23**, 794–807.
26. J. Chang, A. Kowalski, B. Hou and A. Narayana, *Med. Dosim.*, 2008A, **33**, 42–47.
27. A. Narayana, J. Chang, S. Thakur, W. Huang, S. Karimi, B. Hou, A. Kowalski, G. Perera, A. Holodny and P. H. Gutin, *Br. J. Radiol.*, 2007, **80**, 347–354.
28. M. M. Miften, S. K. Das, M. Su and L. B. Marks, *Phys. Med. Biol.*, 2004, **49**, 1711–1721.
29. F. Dhermain, D. Ducreux, F. Bidault, A. Bruna, F. Parker, T. Roujeau, A. Beaudre, J. P. Armand and C. Haie-Meder, *Bull. Cancer*, 2005, **92**, 333–342.
30. P. Olmi, C. Fallai, S. Colagrande and G. Giannardi, *Int. J. Radiat. Oncol. Biol. Phys.*, 1995, **32**, 795–800.
31. A. D. King, P. Teo, W. W. Lam, S. F. Leung and C. Metreweli, *Clin. Oncol. (R. Coll. Radiol.)*, 2000, **12**, 397–402.
32. V. F. H. Chong and Y. F. Fan, *Clin. Radiol.*, 1996, **51**, 625–631.
33. V. F. Chong, Y. F. Fan and J. B. Khoo, *J. Comput. Assist. Tomo.*, 1996, **20**, 563–569.
34. F.-M. Fang, C.-Y. Chien, W.-L. Tsai, H.-C. Chen, H.-C. Hsu, C.-C. Lui, T.-L. Huang and H.-Y. Huang, *Int. J. Radiat. Oncol. Biol. Phys.*, 2008, **72**, 356–364.
35. P. Graff, M. Lapeyre, E. Desandes, C. Ortholan, R.-J. Bensadoun, M. Alfonsi, P. Maingon, P. Giraud, J. Bourhis, V. Marchesi, A. Mège and D. Peiffert, *Int. J. Radiat. Oncol. Biol. Phys.*, 2007, **67**, 1309–1317.
36. J. Cooper, I. Flemming and D. Henson, in *AJCC Cancer Staging Manual*, ed. F. L. Greene, D. L. Page, I. D. Fleming, A. G. Fritz, C. M. Balch, D. G. Haller and M. Morrow, Springer-Verlag, New York, 2002, 6th Edn.
37. L. L. Kestin, F. A. Vicini, E. L. Ziaja, J. S. Stromberg, R. C. Frazier and A. A. Martinez, *Cancer*, 1999, **86**, 1557–1566.
38. J. W. Moul, *J. Urol.*, 2000, **163**, 1632–1642.
39. B. Pickett, J. Kurhanewicz, F. Coakley, K. Shinohara, B. Fein and M. Roach, *Int. J. Radiat. Oncol. Biol. Phys.*, 2004, **60**, 1047–1055.
40. B. Pickett, R. K. Ten Haken, J. Kurhanewicz, A. Qayyum, K. Shinohara, B. Fein and M. Roach, *Int. J. Radiat. Oncol. Biol. Phys.*, 2004, **59**, 665–673.
41. T. Joseph, D. A. McKenna, A. C. Westphalen, F. V. Coakley, S. Zhao, Y. Lu, I. C. Hsu, M. Roach III and J. Kurhanewicz, *Int. J. Radiat. Oncol. Biol. Phys.*, 2009, **73**, 665–671.
42. C. K. Kim, B. K. Park and H. M. Lee, *J. Magn. Reson. Imaging*, 2009, **29**, 391–397.
43. M. A. Haider, P. Chung, J. Sweet, A. Toi, K. Jhaveri, C. Ménard, P. Warde, J. Trachtenberg, G. Lockwood and M. Milosevic, *Int. J. Radiat. Oncol. Biol. Phys.*, 2008, **70**, 425–430.

44. A. Narayana, J. G. Golfinos, I. Fischer, S. Raza, P. Kelly, E. Parker, E. A. Knopp, P. Medabalmi, D. Zagzag, P. Eagan and M. L. Gruber, *Int. J. Radiat. Oncol. Biol. Phys.*, 2008, **72**, 383–389.
45. R. Stupp, W. P. Mason, M. J. van den Bent, M. Weller, B. Fisher, M. J. Taphoorn, K. Belanger, A. A. Brandes, C. Marosi, U. Bogdahn, J. Curschmann, R. C. Janzer, S. K. Ludwin, T. Gorlia, A. Allgeier, D. Lacombe, J. G. Cairncross, E. Eisenhauer and R. O. Mirimanoff, *N. Engl. J. Med.*, 2005, **352**, 987–996.
46. D. W. Siemann and W. Shi, *Semin. Radiat. Oncol.*, 2003, **13**, 53–61.
47. W. Chen, S. Delaloye, D. H. S. Silverman, C. Geist, J. Czernin, J. Sayre, N. Satyamurthy, W. Pope, A. Lai, M. E. Phelps and T. Cloughesy, *J. Clin. Oncol.*, 2007, **25**, 4714–4721.
48. W. B. Pope, A. Lai, P. Nghiemphu, P. Mischel and T. F. Cloughesy, *Neurology*, 2006, **66**, 1258–1260.
49. M. Law, R. J. Young, J. S. Babb, N. Peccerelli, S. Chheang, M. L. Gruber, D. C. Miller, J. G. Golfinos, D. Zagzag and G. Johnson, *Radiology*, 2008, **247**, 490–498.

CHAPTER 6

Autoradiography in Pharmaceutical Discovery and Development

ERIC G. SOLON, PhD., QPS, LLC

110 Executive Drive, Suite 7, Newark, Delaware, 19702, USA

6.1 Introduction

Autoradiography (ARG) is a powerful, high resolution, quantitative molecular imaging technique used to study the tissue distribution and pharmacokinetics of new radiolabeled chemical entities in biological models. Although ARG has applications in botany, entomology, microbiology, biology, and the material sciences to name a few, this chapter will focus on its use for the discovery and development of new pharmaceutical compounds, also known as xenobiotics. ARG is the first molecular imaging technique used for the localization of diffusible and bound xenobiotics in pharmaceutical research and had been used for decades before the more recent field of *in vivo* molecular imaging (*e.g.* Positron Emission Topography, or PET imaging). In its most basic form, ARG is defined as "self radioactivity writing" and has become a collection of techniques to visually localize radioactivity within solid samples. In the classic technique, the radioactivity in a sample produces an image on a detection media such as X-ray film, or another type of media that contains a photographic emulsion. In most cases, the specimen contains a compound that had been intentionally labeled with an isotope having low energy beta radioactivity and that is incorporated into the molecule without damaging its chemical characteristics and/or intended effect(s). Most often the isotope is ^{14}C, ^{3}H, or

RSC Drug Discovery Series No. 15
Biomedical Imaging: The Chemistry of Labels, Probes and Contrast Agents
Edited by Martin Braddock
© Royal Society of Chemistry 2012
Published by the Royal Society of Chemistry, www.rsc.org

^{125}I, however ^{35}S, ^{33}P, and other beta emitters can and have been used. In pharmaceutical research, the specimens can be single cells, individual organs, organ systems, and/or the whole-body of lab animals, such as rats, mice, rabbits, dogs, and monkeys.

ARG can be separated into two basic categories: macro-autoradiography and micro-autoradiography. Macro-autoradiography is the imaging of thin sections obtained from intact organs, organ systems, and/or intact whole-bodies, and micro-autoradiography provides localization of radioactivity at the cellular level in a histological preparation. Macro-autoradiography is typically a lower resolution technique that is most often viewed with the unaided eye and/or hand magnifying glass, while micro-autoradiography is a very high resolution technique that requires the aid of a microscope. Macro-auto-radiography is capable of providing quantitative tissue concentration data when phosphor imaging technology is used to image samples, but micro-autoradiography cannot provide quantitative data, despite numerous claims in the literature. Whole-body autoradiography (WBA) is a general term given to a macro-autoradiography technique where a radioactive compound, which is administered to a lab animal, is visualized by obtaining autoradiographs from thin whole-body sections, which have been obtained from the intact frozen carcass of a lab animal (see Figure 6.1). Micro-autoradiography results are obtained from much thinner cryosections, obtained from individual tissues that have been dissected from an animal and exposed to photographic emulsion on a glass slide, which is used for typical light microscopy (see Figure 6.2).

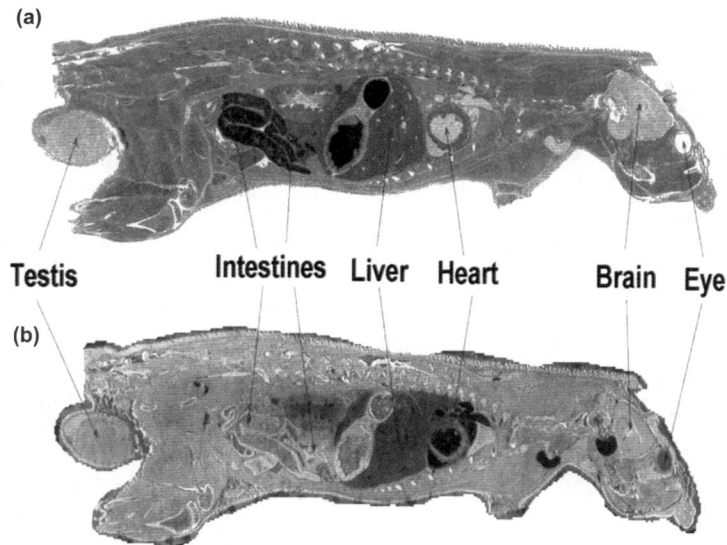

Figure 6.1   (a) A dried whole-body section of a rat obtained after cryosectioning at -20 °C and at a section thickness setting of 40 μm. (b) The resulting auto-radioluminograph of the whole-body section.

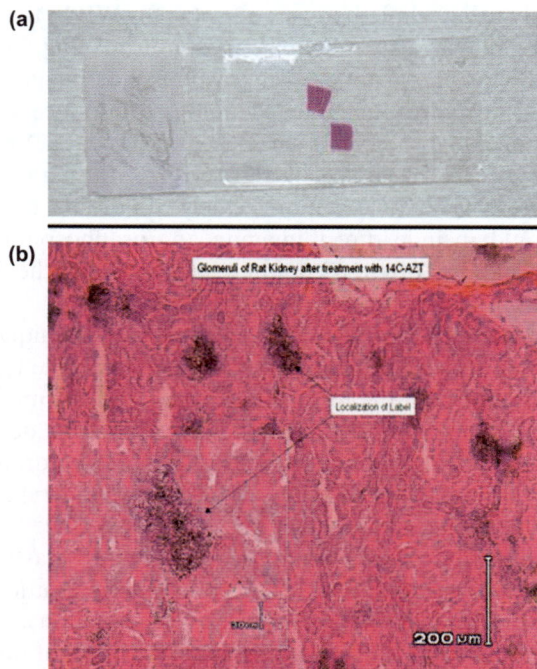

Figure 6.2 (a) A glass microscope slide that has been pre-coated with a photographic emulsion, after overlaying with a thin cryosection (4 μm thick) of kidney. The emulsion was exposed for 2 weeks and then developed and stained with hematoxylin and eosin. (b) A photomicrograph of the kidney showing exposed photographic emulsion from radioactivity (black specks) present throughout the kidney.

As in many areas of science, the terminology used in the literature to describe methods varies and this can be very confusing. This is the case when discussing ARG techniques, owing to the various methods that have been developed to image radioactivity. To that end, it is useful to understand the following definitions to keep them in perspective.

Autoradiography (ARG): any technique whereby radioactivity from within a sample is imaged using a photographic emulsion.

Macro-autoradiography: any technique where radioactivity is imaged in large samples (*e.g.* whole-bodies, organ systems, organs of large animals such as cows) using a photographic emulsion.

Whole-body autoradiography (WBA): a qualitative photographic technique whereby radioactivity administered to an animal is imaged in all tissues of the body using whole-body sections obtained from intact snap-frozen animal carcasses.

Whole-body autoradioluminography (WBAL) or quantitative whole-body autoradiography (QWBA): a qualitative and quantitative digital imaging technique whereby radioactivity administered to an animal is imaged in all tissues of the body using whole-body sections (as for WBA) that are exposed to phosphor imaging plates (also known as phosphor imaging screens), the latter of which are subsequently scanned using a phosphor image scanner, and/or that are imaged using a direct imaging technology to produce an image that can be used to determine quantitative tissue concentration data. Note that the term WBAL refers to QWBA that is specifically performed using phosphor imaging technology, whereas QWBA is a more widely applicable term that includes WBA, which could include direct imaging and/or phosphor imaging technology. Both terms are often used interchangeably in the literature. This chapter will predominantly use the term QWBA.

Micro-autoradiography (MARG): A histological technique where radioactivity from within a small cellular, tissue, or organ sample is imaged using a photographic emulsion and the results are viewed under a microscope.

The roles of WBA and MARG in drug discovery and development have changed over the years, and the use of WBA has increased dramatically due to the development of phosphor imaging technology. Unfortunately, the development of micro-autoradiography has not progressed since its origins, and it remains as a qualitative technique that is also difficult to master. Nevertheless, both techniques have been used extensively in pharmaceutical research and have helped researchers answer many types of questions. Studies can be designed to quantitatively localize varying concentrations of new drugs and/or endogenous compounds in all tissues of lab animals and the literature is full of examples of its utility.

To increase the likelihood of developing a new drug, adequate screening for metabolic and pharmacokinetic liabilities during drug discovery and development has become mandatory, and efforts to evaluate relevant metabolic characteristics of each new chemical entity (NCE) have increased substantially. It is therefore critical for pharmaceutical researchers to re-evaluate new technologies and the role of drug metabolism studies to support the discovery and development of NCEs. The use of more sensitive and novel detection systems has made this process less cumbersome than in the past. However, notably absent for many years were tissue distribution studies, especially those conducted by WBA, which demanded many resources and required a long time to achieve results (4–10 weeks) prior to the implementation of phosphor imaging. Quantitation of the autoradiographic images was also difficult due to the limited linear response of the films used to capture them. However, in recent years, the established methods of WBA have been combined with new phosphor imaging technologies, which now provides reliable, quantitative tissue distribution information for radiolabeled drug discovery compounds. This information can often be obtained within two weeks of dosing and for the last 10–15 years QWBA has been used during drug discovery to answer such pivotal questions regarding tissue pharmacokinetics, routes of elimination, drug–drug interactions, drug localization, clearance, solubility and formulation issues, routes of administration, tumor and brain penetration, inter-species comparisons, and tissue retention. For these reasons, tissue distribution by WBA has also become the preferred technique for demonstrating the definitive tissue distribution of new drugs, information which is required by regulatory agencies for the registration of new drugs. MARG. which is not a technique that is often required by regulatory agencies, provides even higher sensitivity and resolution to enable visualization of drug localization and receptor binding at the cellular level and has also been used to help answer pivotal questions during drug discovery.

This chapter will discuss the history, validation, current applications, instrumentation, and strengths and limitations of WBA and MARG in pharmaceutical discovery and development. It will also present some ideas on how these techniques might be improved in the future to better serve pharmaceutical research.

6.2 Whole-Body Autoradiography

The determination of tissue distribution profiles of new drugs is crucial in understanding the disposition (absorption, distribution, metabolism, and elimination, or ADME, characteristics) and pharmacokinetics of drugs, and these studies are required by worldwide regulatory authorities.

At this point, a detailed description of the current methods used for QWBA is required in order to provide some background for further discussions. The QWBA technique, which is based on the original methods of Ullberg, begins with the administration of a radioactive test substance (beta particle emitter) to a lab animal.[1] Typical doses of radioactivity range from 100–200 microcuries (μCi) kg^{-1}, but this can vary depending on the isotope, dose route, availability of the radiolabeled material, and/or other considerations. Animals are then euthanized at different time points after dosing and their carcasses are snap-frozen by submersion in a dry-ice–hexane bath (Figure 6.3a). Carcasses are left to freeze in the bath for anywhere from 10 min to 2 h depending on the size of the animal, which could be a mouse or a monkey depending on the study design. Frozen carcasses are then freeze-embedded in a block of embedding media (typically 1–5% carboxymethylcelluose) and the blocked animal is then cryosectioned in a large sample cryomicrotome (Figure 6.3b). Whole-body sections are collected onto adhesive tape at section thicknesses of 20 to 50 μm and the sections on tape are dehydrated (usually inside the microtome chamber). The dried sections are then apposed to phosphor imaging plates (or X-ray film), along with a set of radioactive calibration standards (Figure 4.3c). Calibration standards are typically a series of blood aliquots that have been spiked with radioactivity to produce a range of radioactivity concentrations (*e.g.* 0.0001 to 10 μCi g^{-1} blood). It is important to realize that the standards used must match the section thickness of the whole-body sections to be analyzed, otherwise the calibration values will not match the samples and the resulting quantification will be incorrect. Re-usable calibration standards are commercially available but must be validated and usually require re-calibration against standards made under the conditions for which they will be used (*i.e.* using lab-prepared standards that have been sectioned at the same thickness used for QWBA and using standards that reflect a similar tissue matrix, such as blood).

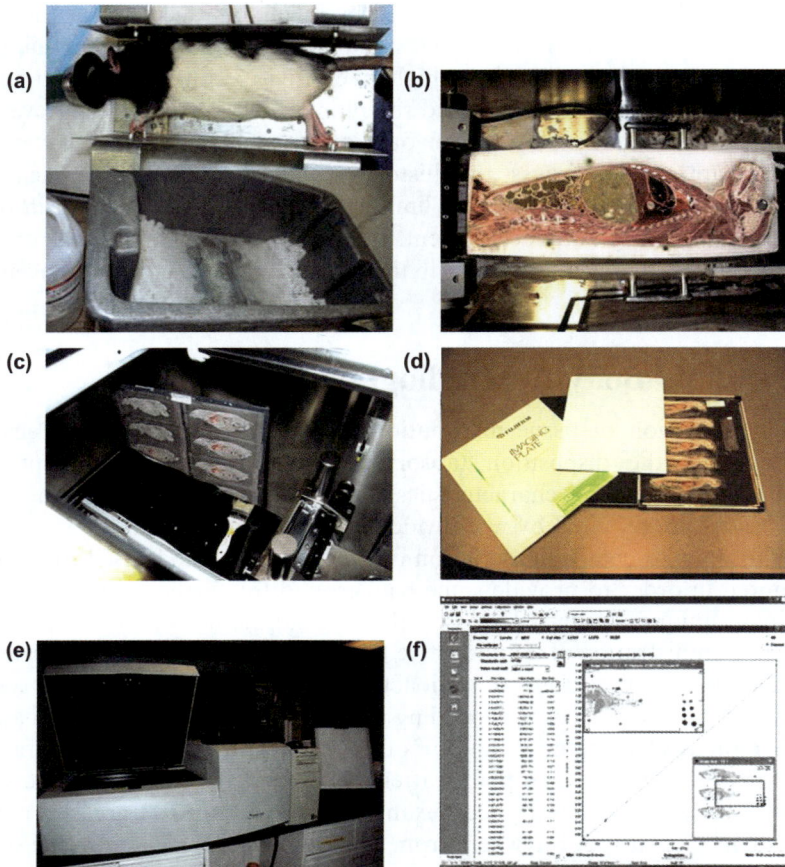

Figure 6.3   (a) A rat carcass being frozen in a hexane-dry ice bath. (b) A partially
cryosectioned monkey carcass inside a Leica CM3600 Cryomacrocut
microtome, co-embedded are ^{14}C quality control standards for verifying
section thickness. (c) A set of rat whole-body sections dehydrating inside a
Leica microtome. (d) Dried, mounted, rat whole-body sections being
apposed to a phosphor imaging plate along with a set of ^{14}C-blood calibration standards inside an exposure cassette. (e) A GE Healthcare
Typhoon 9410 phosphor image scanner. (f) The image calibration function
screen of the MCID image analysis software showing how image density
values produced by the calibration standards are plotted against the known
concentrations of radioactivity in the co-exposed ^{14}C calibration standards.

Typical sample exposure times for phosphor imaging plates are anywhere
from 3 to 7 days for ^{14}C and similar isotopes, but once again this depends on
the nature of the isotope and anticipated amount of radioactivity in the sections. After the appropriate exposure time, the imaging plates are scanned in a
phosphor imager (Figure 6.3d). Modern day phosphor imagers can scan imaging plates at a variety of resolutions (*i.e.* to generate images that have different
pixel sizes) that range from 10 to 1000 μm. Most labs conducting QWBA

studies use a scan resolution of 100 μm, which produces digital image files that are approximately 15 000 megabytes (for an imaging plate that measures approximately 40 × 40 cm). However, the size of the files depends on the size of the imaging plate, which can vary from approximately 20 x 20 cm to 40 x 40 cm. The higher the resolution, the larger the file size, and thus the digital data storage capacity of the laboratory must be considered.

Calibrated digital images enable quantitative determinations of tissue radioactivity concentrations using image density measurements. The phosphor imager used to collect the digital image assigns a grey-scale density value to individual pixels, which is directly related to the radioactive concentration from the region of the sample. For example, a Typhoon 9410 (GE Healthcare) has a linear dynamic range of 100 000:1. The pixel and/or mean pixel density value for a given area of the image can be related to concentrations of radioactivity in tissue by image analysis software that interpolates the sample density reading from a calibration curve of known radioactivity concentrations from a co-exposed calibration (Figure 6.3f).

In addition to phosphor imaging, there are 2 basic types of direct nuclear imaging systems that capture quantitative data by directly detecting and imaging beta disintegrations in real time, although these are less commonly used, owing to limitations on sample analysis time. These will be discussed further in the next section on the history of the development of the WBA technique.

6.2.1 History of Whole-Body Autoradiography

Before discussing the strengths and limitations of WBA it is important to understand how the WBA technique was developed. WBA was a crude method in the beginning and in 1867, Niepse de Saint Victor first described the phenomenon of autoradiography, which he described as the "persistent activity due to an unknown chemical radiation".[2] This observation lead E.S. London to perform an experiment in which an autoradiographic image of a frog treated with radium was first produced in 1904.[3] Fifty years later, Dziewaitkovski used beta radiation to investigate the localization of compounds in biological samples.[4] This was followed by the development of the whole-body autoradiography technique in 1954 by Ullberg, who pioneered the techniques, which are still used today, by administering[35] S-penicillin to mice, followed by freeze-embedding them in water-soaked cotton using dry ice.[1] He then sectioned their entire bodies using a large microtome in a walk-in freezer. The whole-body sections he produced were then apposed to X-ray film, which produced the autoradiographs of tissue distribution. Methods for sectioning the whole-bodies of lab animals were also developed. Methods included the use of dry ice-cooled microtomes;[5] exposing the surface of ground down frozen carcasses to X-ray film;[6] abrasion of resin-embedded carcasses;[7] thick sectioning of frozen carcasses using a circular saw;[8] and finally, the development of a large microtome held inside a chest freezer.[9] Leica Microsystems Inc. (Nussloch, Germany) began manufacturing commercially available large format cryomicrotomes and are the current leading provider of large cryomicrotomes used for WBA today. These new large format microtomes have been improved to tightly control section thicknesses, chamber temperature controls, and section dehydration control. Leica has also developed software that enables documentation of every aspect of cryosectioning for WBA, to aid in regulatory compliance issues, if needed. These instruments are quite large (footprint up to 3 ft wide × 14 ft long × 3–4 ft tall) and expensive (> $250 000), and they are capable of sectioning samples containing hard bone and teeth at section thicknesses as thin as 10 μm.

Quantitation of autoradiographs was the another challenge for pioneering macro-autoradiographers Berlin, Ullberg, and Kutzim who made the first attempts to quantify tissue concentrations in autoradiographs using early image analysis techniques based on grey scale image densities provided by X-ray film.[10,11] Unfortunately, the results achieved were only semi-quantitative, but in the following years (1974–1987), several investigators researched methods to better determine quantitative data from autoradiographs with limited success, owing to the inherent non-linearity of film.[12–17] Also during this time, Schweitzer developed an image calibration method using ^{14}C-spiked blood standards at concentrations bracketing expected tissue concentrations.[18] This robust technique continues to be used in various forms by many investigators today.[19] Luckey revolutionized WBA in 1975 by developing and patenting phosphor imaging technology, or radioluminography, which provided digital images from whole-body sections within days.[20] Most importantly, these images enabled the direct determination of quantitative tissue concentrations that spanned 4–5 orders of magnitude and validation efforts began. A technical validation of the phosphor imaging instrument, and QWBA methods, which more completely described the principles, specifications, and limitations of the instrumentation and QWBA was published in 2000 in a special edition of the Journal of Regulatory Toxicology and Pharmacology.[21] In 1994, a group of autoradiographers in the pharmaceutical industry formed the Society for Whole-Body Autoradiography (SWBA), whose mission was to promote the use of QWBA over traditional organ dissection homogenate methods to determine tissue distribution of new drugs. Further work parameterized quantitative aspects to meet stringent bioanalytical expectations.[22–27] In 1990, a Japanese collaboration of > 20 bioscience companies proposed that QWBA should replace the use of traditional organ dissection and liquid scintillation counting (LSC) assay to determine true tissue distribution during drug development.[28] Dr Yasuo Ohno of the National Institute of Health Science (Tokyo, Japan) concluded his presentation at the 1997 meeting of the SWBA by stating that the Japanese Ministry of Health and Welfare would accept QWBA data in lieu of traditional organ dissection distribution studies for the approval of new drugs as long as the procedures were appropriately validated. Today, pharmaceutical companies have almost entirely eliminated the use of dissection studies to determine the distribution of new pharmaceuticals. Anecdotal information to the author has suggested that the United States Food and Drug Administration (FDA) is now requesting that QWBA studies be provided to determine tissue distribution and, especially, to answer certain questions that arise during drug development.

In addition to autoradiography and autoradioluminography, direct nuclear imaging technologies were also developed. These instruments utilize ionization chambers and different imaging technologies (*e.g.* scintillating sheet, CCD camera) and were developed by Jeavons in 1983.[29] Today's most popular direct imaging instrument utilizes a parallel plaque avalanche chamber, which is based on the 1989 invention by Charpak.[30] These instruments, which are currently sold by Biospace Lab (Paris, France), also image radioactivity in whole-body and smaller tissue sections and have the ability to acquire quantitative images in real-time. These instruments have proved most useful in drug discovery because they are capable of dual isotope analysis and they provide data quickly.

6.2.2 Strengths of Whole-Body Autoradiography

The key strength of the QWBA technique is that it shows true tissue distribution of radiolabeled test articles in a relatively unadulterated, *in situ* sample. Phosphor imaging has also been shown to be a very robust yet sensitive technology for the quantification of radioactivity in whole-body sections. Its wide linear range and sensitivity, which can reliably quantify $\sim$45 dpms distributed over an area of $1/2$ cm^2, exceeds that achieved by LSC for similar sized samples. Phosphor imaging is also able to image the relatively weak energy of ^{14}C and ^3H, which fortunately are also long-lived isotopes, so drugs/metabolites with very long half-lives can be tracked in the body of animals over years. This is not possible using the relatively short-lived isotopes used for *in vivo* positron emission tomography (PET) and single photon emission computer tomography (SPECT) imaging.[31,32] The use of short-lived radioisotopes for tracking xenobiotics *in vivo* is a big limitation when studying drug disposition using *in vivo* techniques, despite their popularity in drug discovery. This is because drugs can remain in tissue for periods of time longer than the half-lives of the isotopes typically used for those techniques. It is not unusual to see drug-related radioactivity in tissues up to 35 days post-administration, and QWBA of ^{14}C-labeled drugs is commonly used to reveal their presence over long periods of time.

Prior to the late 1990s, most drug distribution studies were conducted by organ dissection, homogenization, and LSC ('cut & count' technique), which provided a numerical data set of organ concentrations, which offered no spatial information. Figure 6.4 demonstrates the value of providing detailed image and tissue concentration determination in a rat eye provided by QWBA versus LSC. Organ homogenization in this example resulted in the dilution of drug-derived radioactivity that remained in the pigmented uveal tract at 28 days post-dose and underestimated tissue exposure, whereas QWBA clearly showed the presence of radioactivity. Furthermore, LSC data did not reflect true "tissue" information, nor did it provide data on any tissues that were not specifically collected for analysis.

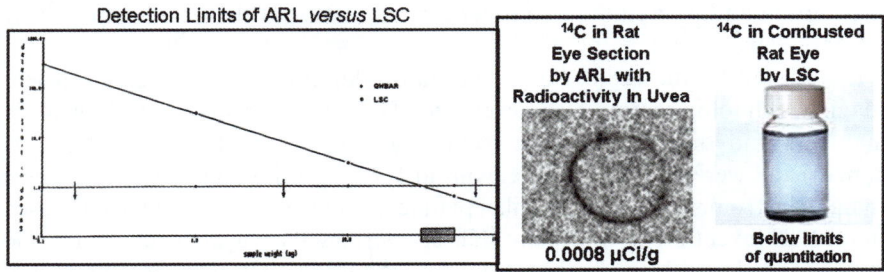

- ARL limits of quantitation ~44 dpm in an area of 0.5 cm² of a 40 µm thick section of tissue and has a lower limit of detection = 0.06 dpm per mm² of ¹⁴C.

- LSC limits of quantitation = 2x background ~ 50-60 dpm per vial.

- For example: if a 0.01 g sample of rat uveal tract has 0.0008 µCi per g, which is approximately 18 dpm, which would be well below background for LSC.

Figure 6.4   A diagram demonstrating the value of having a quantifiable image over a homogenized organ. In this case, the autoradioluminograph revealed specific localization of radioactivity to the uveal tract of the rat eye and a valid tissue concentration. The results of whole-eye combustion and liquid scintillation counting showed a concentration that was below reliable quantitation for the sample size and provides no spatial localization information.

Organ lists for LSC analysis typically included approximately 25 organs and each was homogenized and/or combusted before being assayed for radioactivity in a liquid scintillation counter. Fortunately, several large pharmaceutical companies invested in the new phosphor imaging technology and combined it with established WBA techniques. This quickly became the state-of-the-art technique for studying drug tissue distribution. Furthermore, specific tissue pharmacokinetic parameters for each tissue could be established to produce a detailed description of how drugs were absorbed, distributed, and were eliminated from all tissues over time.[33]

The WBA images can also be reviewed and analyzed at any time so that if an issue arises after the tissue distribution has been completed, investigators can go back and review and quantify the non-routine tissue(s) of interest. QWBA also eliminates the variables of cross-contamination of organs, and the variable effects of organ exsanguination, that inevitably occur during organ removal. The whole-body freezing procedure, which takes approximately 5–15 minutes for rats and mice (depending on body weight), quickly inhibits biological and metabolic processes, which helps to decrease the inter-animal variability of the data, especially at short time points post-dose, and this has been routinely experienced by whole-body autoradiographers.[34] This observation supports the use of a study design where 1 animal per time point and more time points are used, which in turn provides more reliable PK parameters for a more accurate description of tissue compartment PK.[19] This approach also supports many company and national policies for the ethical use of lab animals.

Another strength of the technique, which some may consider to be a weakness, is the flexibility in application of various techniques used to obtain the final result for QWBA. There have been several approaches developed over the years by different laboratories to conduct study designs, euthanasia, carcass freezing, carcass embedding, sectioning, preparing sections for exposure, exposure and imaging of sections, and image calibration and analysis. In 1999, the SWBA conducted a survey which polled 29 scientists from around the world who were conducting QWBA studies.[19] The survey asked approximately 50 questions related to study designs, applications, methods/techniques, tissue quantitation, and regulatory compliance. Results revealed consistencies and inconsistencies among the laboratories that responded. Consistencies were related to: isotope use, doses of radioactivity, number of animals used per time point, exsanguination of animals, freezing methods, section thickness, tissue collection lists, section dehydration, imaging technology, blood and calibration standards, tissues and sections used for quantitation, use of QWBA data for human dosimetry, and QWBA method validation. Inconsistencies were related to: number of time points used, euthanasia methods, carcass freezing times, microtome calibration, section thickness verification, sample collection, validation of commercial calibration standards, use of background measurements during calibration, definition of limits of quantitation, reporting of extrapolated values, re-exposure of sections to determine low concentrations, computer system validation, definitions of raw data, audit trail documentation, and whether or not studies were performed under FDA Good Laboratory Practice guidelines. The survey indicated that many pharmaceutical companies were using QWBA to perform their tissue distribution studies and that they were submitting those studies to regulatory authorities. Although different parts of the QWBA methods vary across different laboratories, laboratories that routinely submit their data to regulatory authorities, and contract research organizations, have a responsibility to validate their systems and procedures to assure quality. The widely accepted use of QWBA for regulated and non-regulated purposes over the past 15 years is a testament to the strengths of QWBA over the outdated and crude organ dissection and homogenization techniques. Furthermore, technological advances in the future will undoubtedly improve the technique and provide results faster and of higher quality, but first the current limitations must be confronted.

6.2.3   Limitations of Whole-Body Autoradiography

This section will discuss limitations of the QWBA process, and will preset some suggestions for improvement. These will begin with the choice of radiolabel used for the test article, and follow through the process of data management.

6.2.3.1   Monitoring Radioactivity Versus Test Molecule

The overall and perhaps most important limitation, which is common to all radiolabeled studies, is that the techniques provide data on the concentration of radioactivity only. That is to say, investigators may not always know the actual molecular identity of the radioactivity they are monitoring. Tissue concentrations determined by QWBA (or LSC) may represent parent drug, plus its metabolites, and/or degradation products, and investigators do not always know if the radioactivity they are measuring reflects the intact molecule and/or those other possible by-products. This is a most important caveat, especially if the investigator doesn't know anything about the metabolism of the test molecule, and the specific radioisotope is located somewhere in the molecule that becomes a metabolite after *in vivo* administration. This is the reason that QWBA tissue concentration data is usually, and correctly, given such units as microgram "equivalents" of test compound per gram of tissue, instead of simply micrograms per gram of tissue.

QWBA has been used to study a variety of compounds that include: small organic molecules, antibodies, peptides, proteins, oligonucleotides, nano-particles, liposomal compounds, and radiopharmaceuticals.[35] Accordingly, the molecular structure influences the choice of isotope that can be used for imaging and reliable quantification. To this end, the degree of analysis depends greatly on the radiopurity and stability of the radiolabel on the molecular entity.

Despite advances in radiochemistry, autoradiographers are still faced with the issue that they are assaying radioactivity and the positive identity of the molecule is always in question, especially at later time points where the radioactivity may reveal the location of an unknown metabolite(s). However, over the last 10–15 years, new imaging instruments that can positively identify and image parent compounds and their metabolites in whole-body and organ sections have been and are being developed. Two such technologies are known as matrix-assisted laser desorption ionization mass spectroscopy imaging (MALDI MSI) and secondary ion mass spectrometric imaging (SIMS-MSI).[43,44] MALDI-MSI is a mass spectrometric imaging technique capable of label-free and simultaneous determination of the identity and distribution of xenobiotics and their metabolites, as well as endogenous substances in biological samples. This makes it an interesting extension to WBA and MARG, eliminating the need for radiochemistry and providing molecular specific information. SIMS-MSI offers a complementary method to MALDI-MS for the acquisition of images with higher spatial resolution directly from biological specimens. Although traditionally used for the analysis of surface films and polymers, SIMS has been used successfully for the study of biological tissues and cell types, thus enabling the acquisition of images at sub-micrometer resolution with a minimum of samples preparation.

6.2.3.2   Radiopurity

Radiopurity is the percentage of actual radiolabeled test molecule and the percentage of degradation products and/or other radioactive contaminants present in a given lot of a radiolabeled test article. The radiopurity of a test compound for any type of study should be 97% or higher, whenever possible, so that the investigator is confident that the results reflect the actual test compound and not other sources of radioactivity (*e.g.*, degradation products). The radiopurity can be affected by the stability of the isotope label in the test molecule and this should be considered when choosing which isotope should be used for labeling. When choosing an isotope, the following factors need to be considered: compound structure (*e.g.* small organic molecule, large oligonu-cleotide, small/large protein), methods of synthesis, cost of synthesis, methods of analysis (WBA, QWBA, MARG), data quality (true quantitation, semi-quantitation, or qualitative), expected image resolution, and the expected *in vivo* stability of the radiolabeled material.

Different isotopes produce images with different resolution due to the energy and types of radioactivity emitted from each isotope. Figure 6.5 shows examples of phosphor images produced by different isotopes and demonstrates that tritium provides images with fine resolution of many small tissues. The resolution provided by ^{14}C may be slightly lower than ^{3}H, especially for micro-autoradiographic purposes, but in most cases ^{14}C is used because it provides a stably labeled test molecule. Conversely, the images produced by ^{125}I may lack the resolution to adequately visualize/localize radioactivity in small tissues.

6.2.3.3 Stability of Radiolabel

The stability of the radiolabel is a key factor in establishing reliable and accurate quantitation. If the radiolabel used does not remain on the chemical entity *in vivo*, then the quantitative results will include radioactive counts from the labeled test article as well as radioactivity from other entities, such as metabolites. This can result in gross over and/or underestimations of the tissue concentrations. Labeling small molecules with ^{14}C in known positions usually results in stably labeled compounds. In contrast, labeling of small molecules with ^{3}H, and large molecules with ^{125}I is often less stable and it is crucial to monitor this *in vivo*, when quantitative results are needed.[35] In general, small organic molecules are labeled with ^{14}C, ^{35}S, or ^{3}H, and larger molecules, such as proteins, and peptides, are labeled using ^{125}I (and sometimes ^{35}S), owing to the related molecular structures.[36] Tritium-labeled xenobiotics are known to undergo hydrogen-exchange with water, and this possibility increases *in vivo*.[37,38] Thus it is important to verify the stability of the ^{3}H label *in vivo* by obtaining urine and/or plasma from the animals being used for QWBA.[37] The extent of *in vivo* stability can be monitored by determining the concentration of radioactivity in fresh (wet) and evaporated samples of plasma and/or urine obtained from the animals to be used for QWBA. The radioactive counts obtained from the wet and dried samples may be compared to see if there is a difference. If radioactivity in the dried samples is less than that of the wet samples, then the exchange of the ^{3}H with water is likely and the results of the QWBA must be interpreted more carefully and/or corrected.

A similar situation exists when using large molecules labeled with ^{125}I. This is becoming more important, as researchers have begun relying on the use of ^{125}I for QWBA to determine the tissue distribution of biotech entities, such as antibodies,

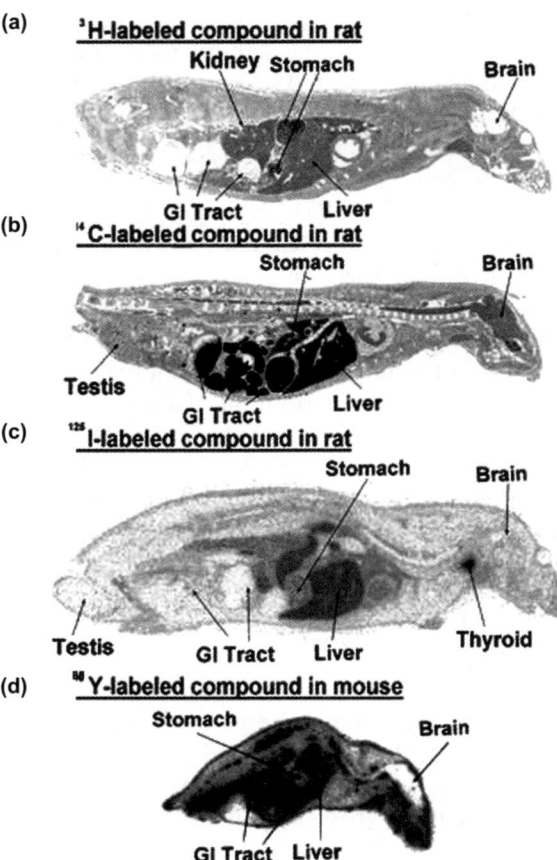

Figure 6.5   Image Resolution Provided by Different Radioisotopes. Figures (a) and (b) are images produced by ³H- and ¹⁴C-labeled compounds, respectively, that were dosed intravenously in the rat and where tissue level detail is evident in heart, bone and testis. Conversely, tissue level image resolution produced by ¹²⁵I- and ⁹⁰Y-labeled compounds (Figures (c) and (d), respectively) in the same tissues is much reduced. Additionally, free ¹²⁵I is evident in the thyroid of the rat treated with ¹²⁵I.

peptides and proteins. Currently, ¹²⁵I QWBA appears to provide the highest resolution and the most quantitative method for determining tissue distribution of large molecules because it is either not possible or extremely expensive to radiolabel these molecules with ¹⁴C or ³H. However, the stability of ¹²⁵I labeling on these large molecules is subject to cleavage *in vivo,* which often results in relatively high concentrations of free ¹²⁵I.[39] Therefore, studying the distribution of ¹²⁵I-labeled biotech molecules requires careful consideration, especially if quantitative data are sought. The determination of drug concentrations in tissues using ¹²⁵I labeled molecules must be considered as semi-quantitative due to the inevitability of measuring free ¹²⁵I along with the test article.

Additionally, care must be taken when interpreting tissue concentrations of ^{125}I-labeled compounds for the thyroid, stomach, kidneys, mammary gland, salivary gland, thymus, epidermis, and choroid plexus. These organs contain a sodium iodide symporter that is involved in the organification and/or elimination of free [125]I.[35] These tissues often have high concentrations of free and/or organified ^{125}I, which is not drug-related and may provide misleading results. Most iodopeptides are absorbed intact through the intestinal tract, but once in the blood stream, de-iodination can occur to varying degrees depending on the labeling.[40] Iodide has a volume of distribution that is about 38% of body weight and is mostly extracellular.[40] Free iodide is eliminated from the plasma primarily by glomerular filtration in the kidney and by organification by the thyroid. In humans, total plasma iodide clearance is about 12% h^{-1} (rate = 45-60 mL min^{-1}). Thyroid clearance varies widely depending on dietary intake, but during periods of high intake, the effect of plasma clearance due to thyroid uptake can be as low as 3–4 mL min^{-1}.[39,40]

Administration of non-radiolabeled sodium iodide to test animals prior to giving the ^{125}I test is one way to reduce the uptake of free ^{125}I and thus reduce the confounding effects of measuring free ^{125}I in the thyroid as well as other tissues.[35] Figure 6.6 shows images resulting from the administration of a ^{125}I-labeled test article alone and then with co-administration of non-radioactive iodine. Note the reduction in radioactivity in the thyroid. This also may act to shunt free ^{125}I to the kidneys, thus facilitating clearance and elimination and reducing the effects of free plasma ^{125}I on tissue quantitation. Although administration of non-radiolabeled iodide can help reduce the background "noise" of free ^{125}I on quantitation, it does not eliminate the effect, and the concentrations of free ^{125}I in the body must be characterized to better

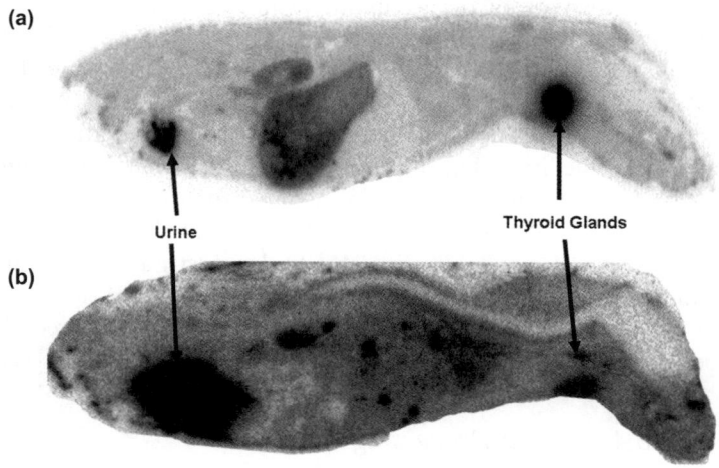

(a)

Urine   Thyroid Glands

(b)

Figure 6.6   Mouse autoradioluminographs resulting from the administration of a ^{125}I-labeled test article alone (a), and with co-administration of non-radioactive iodine (b). Note the lower concentration of radioactivity in the thyroid of the mouse given non-radioactive iodine in its drinking water.

understand the true drug-derived concentrations in tissue. To characterize free *versus* protein-bound (and presumably test article-bound) [125]I, investigators often use trichloroacetic acid (TCA) protein precipitation of plasma (and sometimes tissues) to determine the ratio of protein-precipitable [125]I *versus* free [125]I.[39] This ratio can then be used to correct tissue concentrations obtained using QWBA. This does not account for the further possibility of *in vivo* binding of free ^{125}I to animal proteins and/or distinguish between parent drug, metabolites, and degradation products. To address those questions, further analysis of plasma and/or tissues may be performed using extraction techniques followed by gel-electrophoresis, thin-layer chromatography, ELISA, and/or specialized mass spectroscopy techniques. However, in the experience of the author, these latter techniques are seldom used, and most analysis stops at the TCA precipitation and tissue concentration correction. To these ends, tissue quantitation of drug-derived radioactivity using ^{125}I and QWBA or gamma counting techniques usually carries the caveat of being a semi-quantitative technique, albeit one of the only practical techniques currently available.

6.2.3.4 Short-Lived Isotope Use

Another limitation of QWBA is that it is difficult to evaluate short-lived isotopes, such as those used for radiopharmaceuticals, due to the processing time required, although it is possible to alter the processes to obtain data more quickly, but it requires special attention. For example, when using ^{90}Y, ^{11}C, or ^{18}F, which have half-lives of 2.67 days, 21 minutes, and 60 minutes respectively, as the radiolabel, the frozen 'wet' sections may be exposed to imaging plates immediately and while under freezer conditions.[41] Otherwise, it could take too long to dehydrate the sections and half of the radioactivity could be lost, thus decreasing the sensitivity of the technique. This is one area where improved technology could be very helpful and could increase efficiency by shortening the QWBA process by 2 days, because dehydration of sections within the microtome for 2 days remains a common practice across the industry. Addition of a relatively simple dehydration chamber, which would be attached to the microtome, could facilitate that improvement. In fact, that improvement has been suggested to manufacturers of large microtomes by members of the SWBA but is not readily available at the time of publication. Despite these technical drawbacks, Kaim *et al.* were able to demonstrate that the uptake of ^{18}F-FET in non-neoplastic inflammatory cells in an experimental soft tissue infection model was lower than that of ^{18}F-FDG, thus allowing prediction of a higher specificity for the detection of tumor cells using ^{18}F-FET.[42]

6.2.3.5 WBA Sample Processing Variables

Other factors that may be considered to be a limitation are the variety of ways that the animals are processed for QWBA. The processes of animal anesthesia and euthanasia, animal freezing, carcass embedding, application of standards (section thickness quality control standards and calibration standards), sectioning, section dehydration and handling during exposure, and image analysis, all introduce variables that can affect the end results if special attention is not paid to each step. The most important consideration at all steps of the process is to keep the carcasses frozen and to keep all instrumentation clean so that contamination is minimal and the sample integrity is maintained.

The use of anesthesia and method of euthanasia can affect the tissue distribution of any test article and both should be adjusted in cases where the investigator(s) think it may affect distribution of their test article. For example, it is commonly thought that CO_2 alters the permeability of the blood brain barrier (acidosis) and therefore alters normal brain penetration.[45] Thus euthanasia by CO_2 inhalation may not be a good technique to use if the compound is targeted for the brain.

The most common freezing technique for QWBA is performed by submerging the euthanized or deeply anesthetized animal into a container of hexane and dry ice (see Figure 6.3a), which attains a temperature of approximately $-70°C$.[19] Freezing times may vary from lab to lab however, it is generally agreed that the optimal time for rats and mice (body weights ranging from 30–250 g) is 15–30 min and about an hour for larger animals such as dogs and monkey (body weights ranging from 3–6 kg). Animals may be frozen in a variety of positions ranging from holding the tail of the animal and dipping it into the hexane–dry ice bath, to putting the animal into some sort of positioning frame in an attempt to maintain body positions, which enables uniform presentation for sectioning and image analysis in later steps. Most animals are sectioned in a sagital orientation for whole-body sectioning, but in some cases a cross-sectional orientation may better serve the investigator. Such may be the case where only head structures are being studied and serial frontal sections through the head of a rat are required. In such cases, adjustments may be needed during the embedding procedure to ensure that the head is properly oriented. During embedding there is some risk of thawing, therefore technicians must ensure that the hexane-dry ice bath used for this purpose has been pre-cooled to facilitate quick freezing of the embedding media.

Commonly used section thicknesses range from 20–50 μm.[19] Any of these will work, but any type of calibration standard must be of similar characteristics or validated to work along with the section thickness of the samples to be analyzed. For instance, if whole-body sections were collected at a section thickness of 40 μm, the calibration standards to be co-exposed for image analysis must be of the same section thickness as the whole-body sections and/ or corrected for the difference in thicknesses to ensure that correct quantitative values will be obtained. Furthermore, the concentrations of the calibration standards used for QWBA may also vary. Because image analysis is a quantitative analytical procedure, many laboratories use a range of standards that encompass the possible tissue concentrations and/or detection limits of their instruments.[46] However, owing to the well characterized and robust linear detection of phosphor imaging, the autoradiography laboratories at Novartis Pharmaceutical company (A. Schweitzer, unpublished validation study, Basel, Switzerland) and QPS, LLC (E. Solon and A. Lordi unpublished validation study, Newark, Delaware, USA) have shown that they need only a single calibration standard that is co-embedded, thus included with each section, to reliably calibrate and analyze their whole-body image data.

The use of quality control standards for verifying section thickness uniformity is another consideration that must accounted for, especially when there are several different people performing sectioning for a study (see Figure 6.3b). This is because of the possibility that different people can produce sections of different thickness based on how they pull the sections during sectioning. It is important to control for this because if a section is thicker at one end or another, the resulting quantitation will be compromised and inaccurate. This is most important when the data will be used for submission to regulatory authorities and the quality must be assured. A robotic technology to automatically collect uniformly thick sections would be a valuable asset to an autoradiography lab and undoubtedly increase efficiency and quality. And if that technology could incorporate a morphological identification software, it is possible that the entire sectioning process could be automated. However, it would be critical to assure that the carcasses were frozen and embedded in a uniform position, otherwise robotic sectioning could be more difficult.

Other things to consider when sectioning are the various densities of the tissues being sectioned. For instance, sectioning through the heavy bones and/or teeth of a large animal, such as dog, may require changes in sectioning speeds, knife angle and/or height, and/or a change in the chamber temperature. For those reasons a technician would still be required to monitor the sectioning process.

The handling of dehydrated whole-body sections during the exposure to phosphor imaging plates is another area that requires attention. Care must be used to avoid unnecessary bending and/or contact with the tissue on the tape. A common source of artifacts is bending sections while handling them during exposure to imaging plates, X-ray film, and/or direct detectors. This can result in, for example, the relocation of radioactivity from the sample on the collection tape if the sample undergoes flaking of intestinal contents that may contain particles of highly radioactive material (*e.g.* dosing material and/or ingesta containing radioactive bile from the liver) and/or other tissue. Contamination across the whole-body section can produce images that are difficult to analyze later and may invalidate all previous work if it is extensive. Future technologies that automatically collect sections onto some type of rigid mounting device that can be easily apposed to imaging plates with out further manipulation may be developed. One might envision a microtome fitted with a device that applies tape to the surface of the block, collects the section, and attaches the "wet" section to a conveyor, which goes into a lyophilizer. Lyophilized sections can then be attached to an appropriately sized and pre-labeled cardboard backing (or similar media) coated with glue, which would be ready to be placed in an exposure cassette with an imaging plate.

6.2.3.6 Imaging Technology

Phosphor imaging technology also has its limitations, such as sample exposure time, suitability for ^{3}H or other weak beta emitters, image resolution (pixel size), lower limits of detection (0.06 dpm mm^{-2} of ^{14}C), lower limits of quantitation ($\sim$2220 dpm g^{-1}/0.5 mm^{-2}) image region from sections that are $\sim$40 μm thick), effects and sources of background radiation, image plate scanning time, and digital image file size and storage.[47,33] In drug discovery and development, fast sample analysis is an important key to success and is also one reason why WBA was not as commonly used before phosphor imaging enabled much shorter exposure times. Current typical exposure times still require 3–14 days for ^{14}C and ^{3}H so there is room for improvement. Weak beta emitters like ^{3}H usually require a longer exposure time and 7–14 days is routine. Another limitation of phosphor imaging of ^{3}H is the need to use special imaging plates that can only be used once because they lack the protective coating that is present on plates that image ^{14}C or higher energy isotopes. Currently, ^{3}H imaging plates, which are relatively small (20 × 40 cm), cost approximately $1000–$1500 each. Therefore the cost to image a 10-rat QWBA study can easily exceed $20 000. Direct imagers such as the Beta imager (Biospace Mesures, Paris, France) have improved this by offering real-time imaging, without the need for imaging plates. However, even this can require an overnight acquisition time and often only a few sections from only one small animal can be imaged at a time.[48] As a result, it can still require around10 days of imaging time. Furthermore, the sample holder can accommodate only about 3–4 rat whole-body sections, whereas a large format phosphor imaging plate can image up to 8–10 rat whole-body sections.

Image resolution is another limitation, but one that may not be a big issue for imaging at the tissue level. Current phosphor image scanners (made by Fuji, Kodak, and GE Healthcare) have the capability to scan imaging plates so that an image may have pixel sizes that are anywhere from 10 μm × 10 μm to 1000 μm × 1000 μm. This is acceptable for most tissue distribution studies because tissue morphology obtained from the relatively slow freezing procedure used for QWBA won't support microscopic analysis. Additionally, as one increases the resolution (by decreasing the pixel size during scanning), the scan time greatly increases. For example, a typical scan time of a Typhoon 9410 (GE Healthcare) set at an image resolution of 100 μm × 100 μm, is approximately 20 minutes and produces a digital image file of about 16 000 MB. But when the resolution is increased to produce a resolution of 50 μm × 50 μm, the scan time increases to approximately 1 hour and generates a file size of about 100 000 MB. This can greatly increase the time required to scan dozens of images and also becomes a digital data management issue for a high throughput lab.

6.2.3.7 Image Analysis

Image analysis is the last area where technological limitations and a limited number of qualified individuals are apparent. To date, there are only about 3 major providers of software for industrial pharmaceutical QWBA analysis. The 3 most popular software systems being used for QWBA are AIDA™ (raytest Isotopenmessgeräte GmbH, Straubenhardt, Germany), Seescan 2™ (LabLogic Systems Limited, Sheffield, UK) and MCID™ (InterFocus, Linton, UK). These software packages offer good image sampling tools, region of interest tracking, and powerful algorithmic calibration features that lend themselves to use for regulatory purposes. Image contrasting features also help to identify tissues with very low or very high concentrations during image analysis. One recent improvement made by raytest in their AIDA™ software is that which easily co-registers the autoradioluminograph with a color image of the actual whole-body sections (obtained from a conventional scanner). This enables the user to identify regions to quantify on the color image of the actual section, while the software samples the autoradioluminographic image. This is useful when tissue concentrations are very low, and for beginner autoradiographers who may be less experienced in sampling autoradiographs. However, it can lead the analyst into a false sense of security as artifactual radioactivity can contaminate regions of other tissues and, if careful attention is not paid, the analyst can easily and unknowingly sample regions that have been contaminated with high radioactivity, thus obtaining inaccurate results. Another feature offered by AIDA™ and Seescan2™ is extra features that help address regulatory compliance issues with the capability to add notes and secure audit trails to aid in documentation. All 3 systems have procedures in place to maintain data integrity to match quantitative values to specific regions of the images sampled. The person performing the image analysis needs to have a very good knowledge of animal anatomy, interpreting tissue distribution patterns and artifacts, and possess a complete knowledge of the entire process so that they can recognize such things as improper dosing, freezing, sectioning, section melting or faulty dehydration, and section thickness inconsistencies. Finally, interpreting the data requires years of experience so that QWBA studies can provide the critical information to other pharmaceutical scientists to promote drugs from discovery through development and onto the market to ultimately benefit human health. The next section will provide a few such examples of how QWBA has been applied and made valuable contributions to understanding the characteristics and functions of xenobiotics.

6.2.4 Whole-Body Autoradiography Applications

QWBA offers many applications for researchers in drug discovery and development and can best answer questions related to the biodistribution of new or old drugs. In drug development, QWBA is used most often to perform definitive tissue distribution studies, which are included as part of an ADME package to support new drug applications for regulatory agencies.[35] The QWBA data resulting from those studies are routinely used to determine tissue pharmacokinetics and to predict human radiation dosimetry that might occur during human radiolabeled ADME studies. A routine tissue distribution performed using QWBA evaluates the exposure of 35–40 tissues over various study periods that often go as long as 35 days post-dose. Time points extending to 35 days post-dose or more enable the determination and PK characterization of drug–melanin binding that may occur with some drugs and/or their metabolites.[49] The resulting tissue pharmacokinetic parameters are often used in ^{14}C dosimetry calculations to predict the human individual tissue and whole-body exposure to radioactivity when humans are given the radiolabeled compound to study human drug metabolism. These dosimetry predictions are required by regulatory agencies and drug testing clinics around the world, and are pivotal in proving drug safety for a variety of therapeutic areas.[50] However, variations in the animal study designs and methods used to determine limits of quantitation for QWBA by phosphor imaging can affect this prediction.[33] Furthermore, current regulation and mathematical methods used to conduct human dosimetry estimates are designed to use organ homogenate data and not true tissue concentration data that are provided by QWBA. Thus adjustments to the mathematical methods are necessary to best utilize the higher precision QWBA data to show tissue radioactivity exposure as opposed to organ exposure, which can be very different. For example, QWBA often reveals large differences in radioactive concentration between the cortex and medulla of the kidney and lymph nodes; white and red pulp of the spleen; specific regions of the brain; linings of the intestinal tract; testis and epididymis; and other tissues (see Figure 6.7). These differences in radioconcentrations have been overlooked due to limitations of the organ dissection and LSC analysis methods used before QWBA was available.

QWBA has also been used post-approval to follow up on requests from drug marketing departments and also regulatory requests. For example, Schweitzer *et al.* investigated the distribution of ^{14}C]diclofenac sodium (Voltaren®), which is a drug that has been marketed for years, after a single oral administration in rats.[51] These investigators showed that diclofenac preferentially distributed into the inflamed tissues and achieved exposures that were 26- and 53-fold higher in the inflamed neck and inflamed paws of treated animals than in control rats. All other tissues in treated animals and control animals showed similar distribution and exposure.

Another example, which was conducted to demonstrate a useful drug interaction, was performed by the Bristol Myers Squibb Company. They used QWBA to show how co-administration of ritonavir (RTV, an anti-HIV drug with known enzyme inhibition characteristics) could increase the tissue exposure of a second anti-HIV drug {*N*-[(3-fluorophenyl)methyl]glycyl-*N*-{3-[((3-aminophenyl) sulfonyl)-2-(aminophenyl)amino]-(1S,2S)-2-hydroxy-1-(phenylmethyl)propyl}-3-methyl-L-valinamide, or ^{14}C]L-valinamide} that they were developing.[52] Oral co-administration of ^{14}C]L-valinamide with RTV in rats caused a significant increase in systemic exposure to ^{14}C]L-valinamide. Following a single oral dose of ^{14}C]L-Valinamide, with and without RTV pre-treatment in rats, and subsequent QWBA analysis, which was prepared from samples obtained at 1 and 7 or 8 h post-dose, an increase in

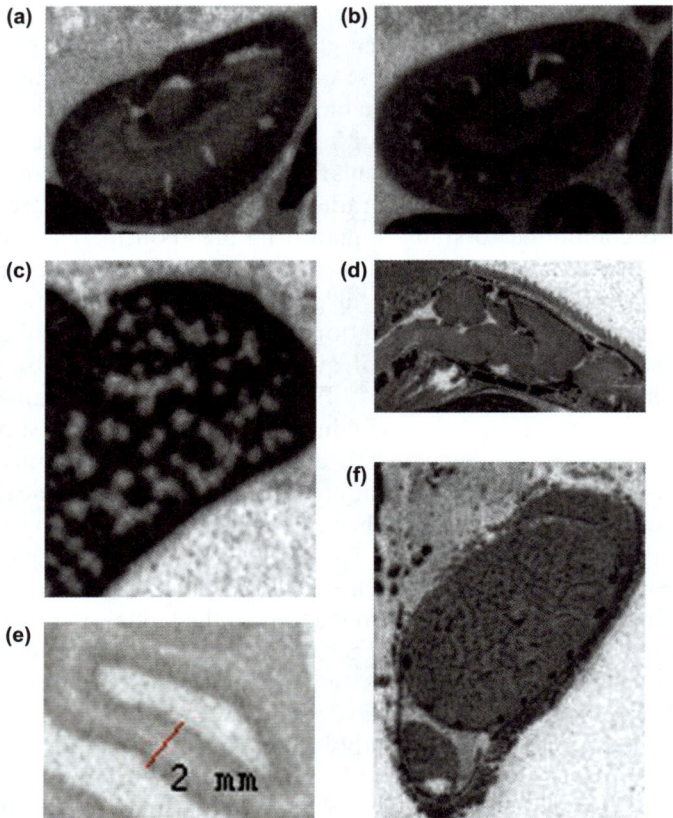

Figure 6.7 QWBA often reveals differences in radioactive concentration between the cortex and medulla of the kidney. Figures (a) and (b) show different distribution of the metabolites of the same drug; white and red pulp of the spleen (c); specific regions of the brain (d); linings of the intestinal tract (e); testis and epididymis (f).

radioactivity in tissues, *e.g.*, brain, and testes was observed upon co-administration. The distribution of radioactivity in the brain parenchyma and ventricles was different, such that the concentration of radioactivity was greater in cerebrospinal fluid (CSF) than in central nervous system (CNS) tissue. Thus the use of CSF concentration of the total radioactivity as a surrogate for brain penetration would result in an overestimation. Valinamide was determined to be metabolized prominently by cytochrome p450 isozme 3A4 (CYP 3A4). The increased tissue exposure to [^{14}C]L-valinamide in rats was largely attributed to inhibition of CYP 3A4 by RTV. [^{14}C]L-Valinamide was also a good substrate for the drug transporter P-glycoprotein (Pgp), with K_m of 4 µM and V_{max} of 13 pmol min^{-1}. The Pgp-mediated transport of [^{14}C]L-Valinamide across Caco-2 cells was readily saturated at

> 10 μM and was inhibited significantly by RTV at 5–10 μM. This data and the reported RTV concentrations suggested that both the Pgp and CYP 3A4 inhibition by RTV played a significant role in enhancing the systemic and tissue exposure to [^{14}C]L-Valinamide in humans, which could aid the treatment of HIV.

QWBA data is also used in drug discovery to support the selection of new drug candidates, based on tissue distribution/uptake/retention, and also to help evaluate and identify potential toxicological issues.[53] In the drug discovery process, pharmacokinetic screening, drug stability studies, evaluation of metabolites, CYP involvement, enzyme induction and inhibition, and excretion studies play a major role. QWBA has considerable merit in identifying "pharmacodeficient" compounds and providing insight on mechanistic questions. QWBA data can provide information related to tissue pharmacokinetics, routes of elimination, CYP or Pgp mediated drug–drug interactions, tissue distribution, site specific drug localization and retention, clearance, compound solubility issues, routes of administration, penetration into specific targets (*e.g.*, tumors), and interspecies kinetics. The timely acquisition and submission of QWBA data relies on a dependable radiochemistry department and a state-of-the-art image acquisition and analysis system. With the support of a radiochemistry laboratory, tritium-labeled compounds can typically be obtained within 10 days of requests, while ^{14}C-labeled compounds usually take longer, but may result in more stable labeling. Typically, labeled compounds are synthesized to have a high specific activity (1–5 mCi mg^{-1}). QWBA has answered a number of specific drug discovery questions that have arisen during the drug selection process and are presented in the next two sections.

6.2.4.1 Brain and Cerebrospinal Fluid Penetration

Currently, the clinical surrogate marker for central nervous system (CNS) penetration of anti-retroviral drugs is the plasma to cerebro-spinal fluid (CSF) concentration ratio.[54] However, unless this model is validated for each compound, it can be misleading when predicting the extent of CNS penetration of a particular compound. Discrepancies between CSF and CNS penetration can result from functional and structural differences between blood/CSF and the blood/brain barriers (*i.e.*, permeability and surface area differences), variable diffusion of compounds from CSF into the CNS, and the rapid rate of bulk CSF flow.[55] Numerous studies performed by pre-clinical pharmacokinetic laboratories have shown that many drugs distribute into CSF and the brain differently.[56,57] Despite the evidence, which has demonstrated clear differences between CSF and CNS drug levels, the HIV literature contains many comparisons of CNS permeability of various therapeutic agents, based on their relative CSF concentrations. QWBA has been useful in this respect to quickly determine whether CSF is a valid clinical end point for determining brain penetration.[53]

6.2.4.2   Melanin Binding

Melanin is a compound responsible for coloration in pigmented animals, but it is absent in albino animals, such as the Sprague Dawley (SD) rat, which is widely used in toxicology studies. Melanin is found in pigmented skin and in the pigmented epithelial layer of the retina, the uveal tract, such as in Long-Evans (LE) rats, and in nerve cells of primates and amphibians.[58–60] Figure 6.8 shows the localization of a ^{14}C-labeled small molecule drug to pigmented structures of the dog eyes. Many lipophilic drugs with pK_a values above 7 bind to melanin and may or may not cause toxicity.[61] Structure activity relationship studies using WBA for melanin binding as a function of various other physicochemical characteristics showed a correlation with volume of distribution, Log P, pK_a and binding energy in pigmented rats. Strongly basic structures, such as piperidine and piperazine moieties and other amines, showed potential for retention in the ocular melanin.[62] Chloroquine, ofloxacin, norephedrine, and diazepam bind to synthetic melanin with varying affinity.[63] Diazepam, flunitrazepam, and Ro-5-4864 are known to induce melanogenesis.[64] Covalent binding of drugs to melanin is rare, however its polyanionic nature and high content of carboxyl and semiquinone subunits facilitates non-covalent binding with many drugs. Although melanin binding of NCEs is not predictive of toxicity, it may prohibit the conduct of human radiolabeled studies where prolonged exposure of pigmented tissues, especially the eye, to radioactivity could pose a health risk.[65]

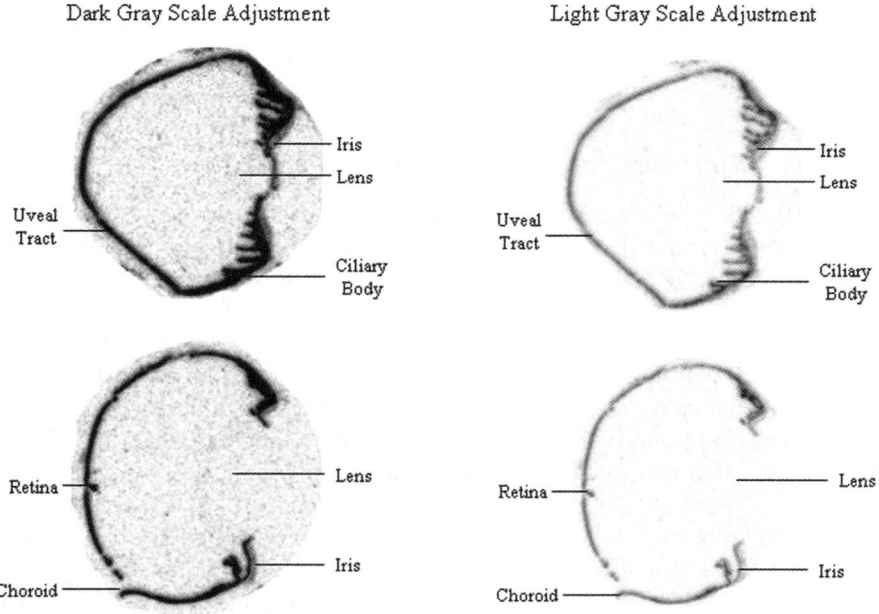

Figure 6.8   Autoradioluminograph showing the localization of a ^{14}C-labeled drug to pigmented structures of the dog eyes.

The neurotoxin 1-methyl-4-phenyl-1,2,3,6-tetrahydropyridine (MPTP) causes selective destruction of melanin containing monoaminergic neurons of the brain in primates and amphibians.[60] MPTP is metabolized by monoamine oxidase type B to 1-methyl-4-phenylpyridine (MPP$^+$), which is known to bind to neuromelanin with high affinity and causes cell death and neurochemical features of Parkinson's disease. The anti-malarial drug, chloroquine, which also has high affinity for melanin, competitively inhibits the binding of MPP$^+$ to neuromelanin in monkeys, providing a protective effect against motor abnormalities.[66] Such distribution studies could be monitored easily by WBA. MPTP related toxicities were not seen in rodents that lack neuromelanin. WBA studies in frogs given single dose of [^{3}H]MPTP showed selective, high accumulation and retention of the radioactivity in melanin containing tissues for up to 15 days. The pigmented cells acted as depots for drug and/or metabolites, resulting in toxic cytoplasmic concentrations causing nerve cell death. Similar melanin binding studies have also been reported for ^{14}C-*N*-*N'*-dicyclopropyl-methyl-piperazine-dichlorohydrate in Cynomolgus monkeys and ^{3}H-aflatoxin in pigs.[59,67]

6.2.5 Whole-Body Autoradiography Conclusion

Although there are many technological improvements that could be made to the current instrumentation, WBA has become a versatile drug discovery and development tool, which provides pharmaceutical scientists with quantifiable high resolution images of the distribution of xenobiotics in practically any biological sample. A wealth of knowledge can be obtained from a single study when the proper study design is applied, and the benefits of WBA and updated study designs over the outdated tissue dissection technique are numerous and substantial. To that end, regulatory authorities should encourage drug developers to use WBA instead of organ dissection and homogenization to conduct tissue distribution studies whenever possible. This is because WBA provides the highest quality quantitative data and a more complete analysis of true tissue distribution, which is especially important when trying to predict human exposure to radiation during human radiolabeled studies.

However, WBA is also limited in that it cannot adequately provide data at the microscopic level due to the freezing technique. Despite the "snap-freezing" techniques used, the freezing of all tissues is too slow to prevent cellular damage due to ice crystal formation. The resulting poor cellular morphology makes cellular identification much more difficult, so that radiolabeled test articles cannot be easily co-localized to specific cellular targets. This is where the application of micro-autoradiography can help.

6.3 Micro-Autoradiography

Micro-autoradiography (MARG) provides pharmaceutical scientists with a high resolution tool to investigate spatial localization of radiolabeled drugs at the tissue and cellular level. MARG is especially good at providing insight regarding *in situ* receptor binding in various cell types and has predictive value for specific drug targeting. In this respect it has been used widely in academic settings, where it can provide important information on cellular mechanisms. MARG has applications in all areas of science, but this report will discuss examples in drug metabolism, pharmacology, toxicology, and molecular biology.

The methods used by the author, which are described in this section, are based on the methods of Appleton and Stumpf, but it is important to realize that there are many variations of the method presented in the literature.[68,69]

To begin, an animal is dosed with a radiolabeled substance (typically ^{3}H, ^{14}C, ^{35}S, or ^{125}I), the animal is exsanguinated and tissues are dissected and snap-frozen in isopentane that is chilled with liquid nitrogen. The tissue is then cryosectioned at $-20\,^{\circ}$C (or the optimal cutting temperature for a given tissue/organ), to obtain 4–5 μm thick sections. Then, under darkroom conditions, sections are thaw-mounted onto dry glass microscope slides that have been pre-coated with nuclear photographic emulsion. The slides are placed into a light-tight box with desiccant and allowed to expose for an appropriate length of time. Figure 6.9 presents a pictorial summary of the methods used for MARG. The collection of cryosections onto dry, pre-coated slides, while under dark-room conditions, is a key step developed by Appleton and it eliminates the possibility of diffusion of soluble compounds, which can happen during slide and section dipping into an aqueous emulsion.[68] In contrast, the original Stumpf and Roth method involved collection of the section into vials for freeze-drying, which required very careful section handling and was very time consuming and prone to sample destruction.[70] Following exposure, the slides are developed in a manner similar to that used for developing photographic film before being stained using conventional histological staining protocols. This may include immunostaining techniques that can provide positive co-localization of drug-derived radioactivity to known cell types, receptors, and/or other structures/markers for which antibody staining protocols exist.[69]

6.3.1 History of Micro-Autoradiography

The first micro-autoradiographic data were produced by Lacassagne in 1924, which lead to further work by Bélanger and Leblond, who poured liquid photographic emulsion onto histological sections to reveal the location of radioactive substances in the tissues.[71,72] Joftes and Warren revised that technique in 1955 and dipped slides into photographic emulsion, which gained wide use due to its ease of manipulation.[73] This technique has survived the years and is sometimes used today. However, if diffusible radiolabeled compound is being imaged, the results could be useless owing to the relocation of the radiolabeled substance, which produces telltale artifacts that can invalidate experiments and discourage investigators. During the 1940s, methods utilizing strips of dried emulsion were developed whereby the histological sample that contained a

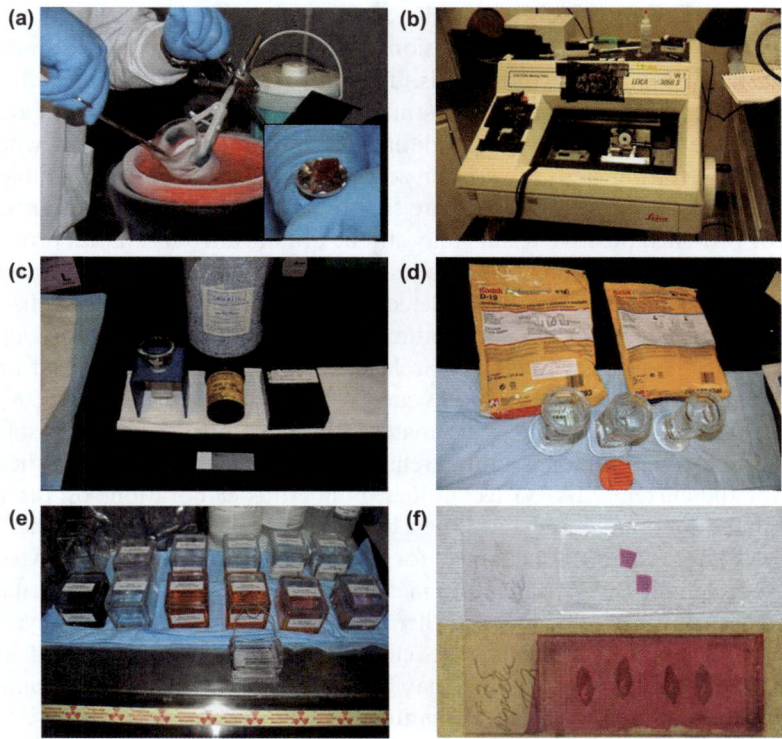

Figure 6.9   A pictorial summary of the methods used for MARG. (a) Shows the snap-freezing apparatus and inset show the fresh tissue sample on a cryostat sample holder for mounting and freezing. (b) A Leica CM3050 Cryostat for cryosectioning of sample and collection onto slides pre-coated with emulsion (note sections must be collected onto glass slides that have been pre-coated with photo-emulsion while under darkroom conditions, thus all equipment lights have been covered with black tape). (c) Dip Miser slide coating cup (Electron Microscopy Sciences, Inc.), Kodak emulsion, black slide exposure box, Drierite desiccant and a coated slide. Slides are dipped into liquid emulsion heated in the Dip Miser and allowed to dry. Sections are collected onto the pre-coated slides and tissue sections are exposed in a sealed black slide box with desiccant at 4 °C. (d) Slides are developed using Kodak Developer and Fixer. (e) Slides may be stained as usual. (f) Examples of slides that have been stained; note the bottom slide shows how emulsion picks up stain, while the upper slide has no emulsion coating.

radioactive substance was placed in direct contact with the dried emulsion strip and allowed to expose it over time. This method proved cumbersome though because the technician would need to maintain the precise position of the strip on the samples during exposure, development, and staining, and often the alignment could not be maintained and the entire process would end in failure. However, in 1964, Appleton first developed the technique of collecting cryo-sections onto slides covered with strips of dried emulsion using a thaw mounting technique.[74] This required sectioning and collection of sections in a darkroom and under safelight conditions, which requires a dedicated staff who can master and maintain their skills. The use of cryopreservation and cryo-sectioning remains critical to the study of diffusible substances because it maintains the spatial locale of the radiolabeled substance in the matrix, whereas liquid tissue fixation steps most often solubilize and relocate the diffusible test article. However, when substances are tightly bound to cellular structures (*e.g.* receptor proteins) positive results may still be obtained from samples processed using conventional histology techniques. MARG procedures are also still used reliably by molecular biologists to detect RNA molecules *in situ* hybridization techniques, and to study the localization of genes and DNA sequences in histological preparations owing to their relatively stable positions in cells. Further refinement of MARG techniques was carried out by Caro during the 1960s, and shortly after that, Stumpf and Roth made additional improvements to establish receptor autoradiography as a more reliable technique.[70,75] This established the basis for the current MARG techniques. Numerous elaborations on the techniques have been presented since then by different investigators, but the basic principles have remained unchanged for > 30 years.[76] Today, as in the past, the MARG technique is very difficult to master, which continues to hamper its use. Researchers must exercise caution when reviewing the literature and relying on articles that report the use of the emulsion dipping technique and claim quantitative data. The conclusions may be questionable and may be disputed in some cases. Validation of the technique is lacking in most laboratories and results can be very subjective. Claims of truly quantifiable results have never been clearly shown by any investigator and they are at best semi-quantitative. This is due to the lack of thickness uniformity of both the tissue sections and the emulsion detection media used. Additionally, there are few, if any, instances where calibration and/or quality control standards have been co-exposed within the samples, which is the only way to clearly prove quantitation of samples.

6.3.2   Micro-Autoradiography Limitations

Several limitations have impeded the progress and wider use of MARG in drug discovery and development. These include: the processing time required to obtain results; the inability of the technique to provide quantitative results, which includes the inability to assure and prove uniformity of tissue and emulsion thicknesses, and lack of internal calibration and/or quality control standards; the high rate and ease of artifact production; and difficulties in collection tissue sections under darkroom conditions. The processing time to obtain results from MARG is a difficult thing to gauge as each tissue must be treated and evaluated differently, depending on how much radioactivity is present. The exposure time can take anywhere from days, to weeks, and even months to obtain optimal results. This often discourages drug discovery scientists, who often work under much shorter time lines. Furthermore, it becomes an overwhelming amount of the work for the scientists working in the drug development area, who may be challenged by regulators to assure high quality results through the use of validated procedures, and quality control and calibration standards for each sample. Technology may help to solve some of these problems if the detection media (*e.g.* emulsions) can be more uniformly produced and made to have inherently linear quantitation. Technology may also help to develop easier methods of collecting uniformly thick tissue sections that can be automatically mounted onto slides for processing, although this would be quite a challenge due to the varying matrices to be sectioned (*e.g.* hard bone, adipose, and eyes). Dependable micro-sized calibration and quality control standards that can be co-exposed with every section would also need to be developed in order to assure reproducibility of quantitation. Finally, the new methods would need to enable a significant reduction in the amount and types of artifacts that are produced. Currently, the following types of artifacts must be controlled: (1) effects on emulsion by slight variations in light, humidity, temperature, tissue characteristics, fixation, freezing, chemicals, pH, developer, fixer, and miscellaneous debris in developer solutions; (2) tissue condition (*e.g.* freezing technique, fixation, autolysis, sectioning temperature, improper section mounting); (3) light leaks; (4) latent image fading; (5) reticulation of emulsion; (6) positive chemography; (7) negative chemography; (8) deviations of pH in processing fluids; (9) pressure artifacts; (10) ice crystals on knife; (11) crystalline deposits from developing process.[69] Some of these are more easily controlled than others, but together they require a high level of skill by the analyst to overcome, and the presence of any can invalidate months of work. Until methods and/or technologies that can better control tissue section and emulsion uniformity and also reduce the sources of and occurrence of artifacts can be developed, the current technique will remain strictly qualitative and will prove to be to daunting for routine use in pharmaceutical discovery and development. The lack of new developments in MARG methods has continued to make MARG an underutilized technique in drug discovery and development, but when performed correctly the results can be of utmost value in promoting a drug candidate and in answering some pivotal questions for pharmaceutical investigators.

6.3.3 Micro-Autoradiography Applications

MARG has made important contributions to drug discovery and development over the years and has provided insight into the localization of pharmaceuticals to support proof of concept studies and studies on: mechanisms of toxicity, efficacy, physiology of hormone action, and cell regulation. For example, MARG is useful for studying skin penetration of various compounds and is routinely used in the cosmetic industry, physiology research, pharmacology, and safety studies. Unilever is a company that makes consumer skin care products and as such is responsible for proving the safety of substances that are applied to skin. They have developed MARG techniques to examine skin penetration in different *in vitro* test models with rat, pig, and human skin samples. They have also coupled MARG with confocal microscopy and other techniques to extend the usefulness of their studies (H. Minter, presented at the 2007 Meeting of the SWBA). Linoleic acid (LA) is commonly used in cosmetics, but its *in vivo* human skin penetration characteristics were not very well demonstrated. However, in 2006, Rauvast and Mavon used a unique, *in situ*, 'virtual' microautoradiographic slide to examine transfollicular delivery of LA in human scalp.[77] They combined their MARG data with an *in vitro* permeation experiment and compartmental analysis to show that most of the LA was localized to the hair sheath, but that none was present in the dermal compartment and that 10% of the total LA recovered was found in the stratum corneum and dermis after 6 h. This supported their notion that the diffusion of LA occurred by a transfollicular route. It also demonstrated the value of high-resolution MARG for providing detailed cellular localization of the molecule. Skin receptor MARG techniques have also been used to study the absorption, penetration, and cellular localization of ^{3}H-Maxacalcitol, which is a vitamin D analog used for the treatment of psoriasis. Hayakawa *et al.* treated the dorsal skin of rats with a ^{3}H-Maxacalcitol ointment and examined skin exposed for periods of 0.5, 2, 8, 24, 48, or 168 h.[78] They discovered two routes of skin penetration: one *via* epidermal cell layers and the other *via* hair follicles. They were also able to distinguish very fine regions of cellular localization, which supported theories on the mechanism of efficacy. Figure 6.10 shows an example of a ^{14}C-labeled test article localized to the sebaceous gland in the skin of a rat after a dermal application (unpublished data from author). This example demonstrates the fine detail of localization that can be obtained with MARG.

Both renal function and localization of various substances in the kidney have been studied by MARG. Young *et al.* used ^{14}C-iodoantipyrine as a tracer to

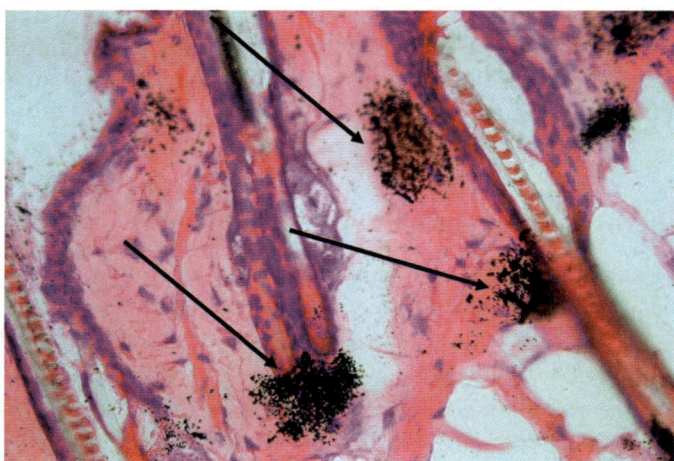

Figure 6.10   An example of a ^{14}C-labeled test article localized to the sebaceous gland in the skin of a rat after a dermal application.

study intra-renal blood flow in nephrectomized rats and they used their microautoradiographs and standards to determine blood flow rates.[79] They noted how MARG was helpful in defining the morphological location of blood flow and they made several conclusions regarding changes in regional renal blood flow: "the interaction between vasoactive mediators and the autonomic nervous system"; and "that medullary blood flow was dependent on local prostaglandin production and is also influenced by sympathetic nervous supply".

Another laboratory used *in vitro* MARG to show localization and density of atrial natriuretic peptide (ANP) in nephrectomy biopsy samples obtained from patients with renal disease.[80] These investigators used [^{125}I]-alpha-human (1-28) ANP and they found localization in the glomerulus and tubular regions in the human biopsy specimens. They also observed that density of ANP binding generally decreased in patients with renal dysfunction and hypertension. Overall though, their study established an *in vitro* MARG method to assess ANP binding in human biopsy specimens.

More recent work by Yamamoto *et al.* used MARG in combination with immunohistochemistry, macro-autoradiography, and positron emission topography to examine intestinal ulceration and healing in the rat.[81] They used ^{18}F-FDG to examine ulcerations in the small intestine of rats, which were induced using indomethacin. MARG combined with immunohistochemistry showed an accumulation of ^{18}F-FDG in inflammatory cells, in granular tissue-forming cells (forming granulation tissue), and around ulcers. ^{18}F-FDG was also found to be present in proliferating intestinal crypt cells and in intact intestinal tissue taken from indomethacin-treated and control animals. They concluded that ulceration could be visualized early by the prominent uptake of ^{18}F-FDG by inflammatory cells, and by the formation of granulation tissue by cells in and around ulcers. This work also demonstrated the value of combining both *in vivo* and *ex vivo* imaging techniques, which provided robust data sets for analysis.

6.4 Conclusions

Autoradiography techniques, which in their basic form have not changed much over the last 60 years, have continued to be used in the pharmaceutical industry to provide valuable information about drug disposition in lab animals. However, the use of WBA techniques have become most useful in the last 15 years, owing to the implementation of phosphor imaging and direct imaging instruments that provide high resolution quantitative and qualitative spatial drug tissue distribution information. QWBA has changed the way investigators evaluate drug distribution and offer much more precise data than has been provided by the former techniques of organ removal, homogenization, and liquid scintillation counting. The obvious benefits of improved technology, especially phosphor imaging and image analysis software, have revolutionized the study of drug distribution by QWBA; however, similar technological improvements to MARG techniques have not been realized. This lack of attention to the MARG techniques has kept its use to a minimum, despite the potential value is offers. Future technological improvements to both techniques will undoubtedly offer valuable contributions to scientists in drug discovery and development and help to bring new drugs to market more quickly for the ultimate benefit of humankind.

References

1. S. Ullberg, *Acta Radiolog. Suppl.*, 1954, **118**, 1–110.
2. V. Niepse de Saint, *Compt. Rend.*, 1867, **65**, 505–507.
3. E. S. London, *Russ. Vrach*, 1904c, **3**, 869–872 .
4. D. D. Dziewaitkovski, *J. Exp. Med.*, 1953, **98**, 119.
5. Y. Cohen and W. Epierre J., *Rapport C.E.A.*, 1961, **2071**.
6. P. Pellerin, *Pathol. Biol. Sent. Hôp.*, 1961, **9**, 233.
7. L. E. Martin, C. Harrison and C. M. Bates, *Biochem. J.*, 1962, **82**, 17.
8. F. Kalberer, *Adv.Tracer Methodol.*, 1966, **3**, 139.
9. S. Ullberg, in *Special Issue on Whole-Body Autoradiography*, ed. O. Elvefeldt, LKB Instr. J., Science Tools, Bromma, Sweden, 1977.
10. M. Berlin and S. Ullberg, *Arch. Environ. Health*, 1963, **6**, 589.
11. H. Kutzim, *Nucl. Med.*, 1962, **15**, 39–50.
12. S. A. M. Cross, A. D. Groves and T. Hesselbo, *Int. J. Appl. Radiot. Isot.*, 1974, **25**, 381–386.
13. S. Longshaw and J. S. L. Fowler, *Xenobiotica*, 1978, **8**, 289–295.
14. R. A. J. Coe, *Int. J. Appl. Radiot. Isot.*, 1982, **36**, 93–96.
15. E. R. Franklin, *Int. J. Appl. Radiot. Isot.*, 1985, **36**, 193–96.
16. W. A. Geary II, A. W. Toga and G. F. Wooten, *Brain Res.*, 1985, **337**, 99–118.
17. W. Steinke, Y. Archimbaud, M. Becka, R. Binder, U. Busch, P. Dupont and J. Maas, *Regul. Toxicol. Pharmacol.*, 2000, **31**, S33–S43.
18. A. Schweitzer, A. Fahr and W. Niederberger, *Appl. Radiat. Isot.*, 1985, **33**, 329–333.
19. E. G. Solon and L. Kraus, *J. Pharmacol. Toxicol. Methods*, 2002, **43**, 73–81.
20. G. Luckey, *US Patent 3,859,527*, 1975.
21. F. Coulson, and C. J. Carr, Eds., *Regul. Toxicol. Pharmacol. (Special Edition)*, 2000, **31**, part 2.
22. M. Sonada, M. Takana, J. Miyahara and H. Kato, *Radiology*, 1987, **148**, 833–838.
23. J. Miyahara, *Chem. Today*, 1989, **223**, 29–36.
24. A. Shigematsu, N. Motoji, A. Hatori and T. Satoh, *Regul. Toxicol. Pharmacol.*, 2000, **22**, 122–142.
25. H. Kolbe and G. Dietzel, *Regul. Toxicol. Pharmacol.*, 2000, **31**, S5–S14.
26. N. Motoji, E. Hayama and A. Shigematsu, *Eur. J. Drug Metab. Pharmacokinet.*, 1995, **20**, 89–105.
27. M. J. Potchioba, T. G. Tensfeldt, M. R. Nocerini and B. M. Silber, *J. Pharmacol. Exp. Ther.*, 1995, **272**, 953–962.
28. M. Tanaka, *Xenobio. Metabol. Dispos.*, 1994, **9**, 393–407.

29. A. P. Jeavons, *IEEE Trans. Nucl. Sci.*, 1983, **30**, 640–645.
30. G. Charpak, D. Imrie, J. Jeanjean, P. Miné, H. Nguyen, D. Scigocki, S. P. K. Tavernier and K. Wells, *Eur. J. Nucl. Med.*, 1989, **15**, 690–693.
31. W. Vaalburg, *Drug Inf. J.*, 1997, **31**, 1015–1018.
32. J. Rao, A. Dragulescu and H. Yao., *Curr. Opin. Biotechnol.*, 2007, **18**, 17–25.
33. E. G. Solon and F. Lee, *J. Pharmacol. Toxicol. Methods*, 2002, **46**, 83–91.
34. R. A. J. Coe, *Regul. Toxicol. and Pharmacol.*, 2000, **31**, S1–S3.
35. E. Solon, *Expert Opin. Drug Discov.*, 2007, **2**, 503–514.
36. E. J. Hahn, *Am. Lab.*, 1983, **15**, 64–71.
37. H. Kim, D. Prelusky, L. Wang, D. Hesk, J. Palamanda and A. Nomeir, *Am. Pharm. Rev.*, 2004, **7**, 44–48.
38. D. Hesk and P. McNamara, *J. Labelled Compd. Radiopharm.*, 2007, **50**, 875–887.
39. M. Motie, K. W. Schaul and L. A. Potempa, *Drug Metab. Dispos.*, 1998, **26**, 977–981.
40. S. Venturi and M. Venturi, *Eur. J. Endocrinol.*, 1999, **140**, 371–372.
41. J. Lazewatsky, Y. Ding, D. Onthank, P. Silva, E. Solon and S. Robinson, *Cancer Biother. Radiopharm.*, 2003, **18**, 413–419.
42. A. Kaim, B. Weber, M. Kurrer, G. Westera, A. Schweitzer, J. Gottschalk, G. von Schulthess and A. Buck, *Eur. J. Nucl. Med.*, 2002, **29**, 648–654.
43. T. C. Rohner, D. And Staab and M. Stoeckli, *Mech. Ageing Dev.*, 2005, **126**, 177–185.
44. L. A. McDonnell, R. M. A. Heeren, R. P. J. de Lange and I. W. Fletcher, *J. Am. Soc. Mass Spectrom.*, 2006, **17**, 1195–1202.
45. V. Fencl, J. R. Vale and J. A. Broch, *J. Appl. Physiol.*, 1969, **27**, 67–76.
46. M. J. Potchoiba, T. G. Tensfeldt, M. R. Nocerini and B. M. Silber, *Pharmacol. Exp. Ther.*, 1995, **272**, 953–962.
47. S. Kanekal, A. Sahai, R. E. Jones and D. Brown, *J. Pharmacol. Toxicol. Methods*, 1995, **33**, 171–178.
48. A. Breskin, *Nucl. Instrum. Methods*, 2000, **A454**, 26–39.
49. B. Leblanc, S. Jezequel, T. Davies, G. Hanton and C. Taradach, *Regul. Toxicol. Pharmacol.*, 1998, **28**, 124–132.
50. Code of Federal Regulations, 21 CFR: Part 361.1, Section b, 3I, April 1, 2004.
51. A. Schweitzer, N. Hasler-Nguyen and J. Zijlstra, *BMC Pharmacol.*, 2009, **9**, 5.
52. E. G. Solon, S. K. Balani, G. Luo, T. J. Yang, P. J. Haines, L. Wang, T. Demond, S. Diamond, D. D. Christ, L.-S. Gan and F. W. Lee, *Drug Metab. Dispos.*, 2002, **30**, 1164–1169.
53. E. Solon, S. K. Balani and F. W. Lee, *Curr. Drug Metab.*, 2002, **3**, 451–462.
54. M. Yazdanian, *J. Pharm. Sci.*, 1999, **88**, 950–954.
55. W. M. Pardridge, *J. Neurochem.*, **70**, 1781–1792.
56. W. P. McNally, P. D. DeHart, C. Lathia and L. R. Whitfield, *Life Sci.*, 2000, **67**, 1847–1857.
57. J. W. Polli, J. L. Jarrett, S. D. Studenberg, J. E. Humphreys, S. W. Dennis, K. R. Brouwer and J. L. Woolley, *Pharmacol. Res.*, 1999, **16**, 1206–1212.
58. C. G. Mason, *J. Toxicol. Environ. Health*, 1977, **2**, 977–995.

59. P. Bernard, Y. Dormard, G. Houin, C. Declume, R. Malbosc, A. E. Tufenkji, P. Galtier and M. Alvinerie, *Arzneimittelforschung*, 1993, **43**, 516–520.

60. A. L. Sokolowski, B.S. Larsson and N.G. Lindquist, *Pharmacol. Toxicol.*, 1990, **66**, 252–258.

61. B. LeBlanc, S. Jezequel, T. Davies, G. Hanton and C. Taradach, *Regul. Toxicol. Pharmacol.*, 1998, **28**, 124–132.

62. P. A. Zane, S. D. Brindle, D. O. Gause, A. J. O'Buck, P. R. Raghavan and S. L. Tripp, *Pharmacol. Res.*, 1990, **7**(9), 935–941.

63. T. Yamada, Y. Okuyama and H. Mukai, *Arzneimittelforschung*, 2001, **51**(4), 299–303.

64. E. Matthew, J. D. Laskin, E. A. Zimmerman, I. B. Weinstein, K. C. Hsu, and D. L. Engelhardt, *Proc. Natl. Acad. Sci. U.S.A.*, 1981, **78**(6), 3935–3939.

65. J. D. Dain, J. M. Collins and W. T. Robinson, *Pharmacol. Res.*, 1994, **11**, 925–928.

66. R. J. D'Amato, G. M. Alexander, R. J. Schwartzman, C. A. Kitt, D. L. Price and S. H. Snyder, *Nature*, 1987, **327**, 324–326.

67. P. Larsson and H. Tjalve, *J. Anim. Sci.*, 1996, **74**, 1672–1680.

68. J. R. J. Baker, *Autoradiography: A Comprehensive Review, Royal Microscopical Society, Microscopy Handbooks 18*, Oxford Science Publications, 1989, 30–32.

69. W. E. Stumpf. Drug Localization in Tissues and Cells. IDDC Press. Library of Congress Control Number 2003105179. 2003.

70. W. E. Stumpf and L. J. Roth, *Stain Technol.*, 1964, **39**, 219–223.

71. A. Lacassagne and J. Lattes, *C. R. Séances Soc. Biol.*, 1924, **90**, 352–353.

72. L. F. Bélanger and C. P. Leblond, *Endocrinol.*, 1946, **39**, 8.

73. D. L. Joftes and S. Warren, *J. Biol. Photogr. Assoc.*, 1955, **23**, 145–150.

74. T. C. Appleton, *J. R. Microsc. Soc.*, 1964, **83**, 277–281.

75. L. G. Caro, *J. Biophys. Biochem. Cytol.*, 1961, **10**, 37.

76. T. Nagata, *Histol. Histopathol.*, 1997, **12**, 1091–1124.

77. V. Rauvast and A. Mavon, *Int. J. Cosmet. Sci.*, 2006, **28**, 117–123.

78. N. Hayakawa, N. Kubota, N. Imai and W. E. Stumpf, *J. Pharmacol. Toxicol. Methods*, 1998, **50**, 131–137.

79. L. S.Young, M. C. Regan, P. Sweeney, K. M. Barry, M. P. Ryan and J. M. Fitzpatrick, *J. Urol.*, 1998, **160**, 926–931.

80. T. Ogura, N. Toshio, N. Asano, E. Katayama, T. Oishi, Y. Mimura, M. D. K. Hironaka, N. Kashihara, H. Makino, Z. Ota and N. Ogawa, *J. Med.*, 1994, **25**, 203–217.

81. M. Yamato, Y. Kataoka, H. Mizuma, Y. Wada and Y. Watanabe, *J. Nucl. Med.*, 2009, **50**, 266–273.

CHAPTER 7.1

In vivo *Fluorescence Optical and Multi-Modal Imaging in Pharmacological Research: From Chemistry to Therapy Monitoring*

RAINER KNEUER,[a] HANS-ULRICH GREMLICH,[a]
NICOLAU BECKMANN,*[a] THOMAS JETZFELLNER[b,c]
AND VASILIS NTZIACHRISTOS[b,c]

[a] Novartis Institutes for BioMedical Research, Global Imaging Group, CH-4056 Basel, Switzerland; [b] Technical University of Munich, Institute for Biological and Medical Imaging, Germany; [c] Helmholtz Center Munich, D-85764 Neuherberg, Germany

7.1.1 Introduction

One of the current strategies adopted by pharmaceutical companies for introducing better (more effective, less side effects) and affordable drugs is to improve the characterization of compounds and their effects in early and relatively non-costly phases, in order to increase their chance of success in late phases of development. Intimately linked to this reasoning stands the knowledge of disease pathophysiology, along with its early diagnosis and characterization. Certainly, the better the mechanism of a disease is known, the

RSC Drug Discovery Series No. 15
Biomedical Imaging: The Chemistry of Labels, Probes and Contrast Agents
Edited by Martin Braddock
© Royal Society of Chemistry 2012
Published by the Royal Society of Chemistry, www.rsc.org

higher the probability to find an appropriate therapy. In addition, the better and earlier a disease can be diagnosed and characterized, the greater will be the chance to interfere in this process with a chemical entity. This reasoning sets the framework for the use of imaging in pharmaceutical research.[1–4]

Methods that allow the visualization and quantification of molecular interactions in the intact organism, *i.e.* molecular imaging techniques, are needed to address important questions in drug discovery and development:

   (i) where do drugs act in the body?
   (ii) do they reach their target?
   (iii) at what dose do side effects or even toxicological effects occur?
   (iv) what organs are affected?
   (v) what are the optimal routes for drug delivery?
   (vi) what is the receptor occupancy at a given dose level?
   (vii) for how long does a compound stay bound?

Broadly speaking, molecular imaging techniques of interest in the context of pharmacological research include nuclear methods such as positron emission tomography (PET) and single photon emission computerized tomography (SPECT), as well as non-nuclear methods such as optical imaging and magnetic resonance imaging (MRI). They exploit either specific molecular probes or intrinsic tissue characteristics as the source of image contrast.

In this chapter, we address the use of *in vivo* optical imaging in pharmaceutical research. Although bioluminescence also plays an important role in drug discovery (see ref. 5 and 6 for reviews), our interest will focus on fluorescence optical imaging. The main assets of the technique are described in the first part of the review. The design of imaging probes is addressed next. Finally, examples of preclinical and clinical imaging activities are presented and selected to illustrate some key points reflecting the advantages, challenges and limitations of *in vivo* optical imaging in pharmaceutical research.

7.1.2   *In Vivo* Optical Fluorescence Imaging

Fluorescence imaging is an attractive tool due to its operational simplicity, safety, and cost-effectiveness. Exogenous fluorochromes (dyes or genetically engineered fluorescent proteins) are excited by *e.g.* laser diodes operating at a frequency close to that of the detected light; the emitted fluorescent light is then detected in a spatially resolved manner by a cooled charge-coupled device (CCD) camera.

The near-infrared (NIR) window is particularly suitable for *in vivo* investigations. For light of wavelengths between 650 and 900 nm, near-infrared fluorescence (NIRF) imaging takes advantage of the low absorbance of tissue chromophores such as oxy- and deoxy-hemoglobin, water, melanin and fat, to study *in vivo* biological processes at the cellular and molecular levels. At these wavelengths, scattering of photons is a more significant attenuation factor than absorption. Currently, many *in vivo* fluorescence imaging approaches are based on planar detection of fluorescent light. Similarly to planar bioluminescence imaging, this approach has obvious restrictions with regard to deriving quantitative information, since the measured fluorescent signal is non-linearly related to the depth at which it was generated. This problem can be overcome by fluorescence molecular tomography, which improves signal quantification.[7–9]

Fluorescence Molecular Tomography (FMT) is a technique capable of determining the location of a fluorescent probe non-invasively and three-dimensionally. Images are obtained by scanning the excitation source over the subject and spatially measuring the emitted fluorescence intensity for each source position. FMT is based on the fact that by changing the position of the excitation source one creates several projections (views) through the medium that, when combined tomographically, offer three-dimensional quantitative images of fluorescence biodistribution.[7,10] The resolution and sensitivity of FMT greatly depend on the dimensions of the subject under study and on the wavelengths used for excitation and emission. For small animal imaging in the NIR, resolution of the order of 1 mm at picomole sensitivity is achievable,[7] worsening as the imaged volume increases.[11] Time-resolved, CCD-based systems have also been proposed for quantitative small animal fluorescence tomography.[8] Additionally, mathematical approaches[12] have enabled full angular measurements both for bioluminescence and fluorescence measurements, greatly improving the accuracy of the reconstructed images.[13–15]

Combining the high versatility of NIRF imaging with high spatial ultrasonic resolution, optoacoustic tomography (OAT) is a recently developed hybrid imaging modality with the potential to visualize optical contrast in tissue in real-time.[16] In contrast to FMT, optoacoustic tomography is sensitive to light absorption, instead of light emitted by a fluorochrome. Imaging is performed by illuminating an object with short laser pulses, in the nano-second range, which generate a thermo-elastic expansion in response to the light absorption. The subsequently generated broadband ultrasound waves in the range of 0.1–100 MHz are recorded by ultrasound transducers in a tomographic detection geometry.[17] OAT is richer in contrast compared to traditional ultrasound imaging but provides the same resolution (20–200 μm). High-resolution optical imaging through several millimeters to centimeters, as enabled by OAT, brings a newfound ability of biological and medical investigations. For example, OAT can image with 20–200 μm resolution the morphology and disease-related vascular changes, oxygenation and blood volume in real-time. It is possible to observe blood oxygenation by resolving the spectral signatures of hemoglobin (Figure 7.1.1(a)).[18] Additionally, markers have been developed to enhance the detection sensitivity and specificity of the method, including dyes,[19] light-absorbing nano-particles,[20] and chromogenic substrates.[21]

In particular, the recent development of multispectral optoacoustic tomography (MSOT) provides the capability to accurately resolve and quantify the volumetric bio-distribution of intrinsic tissue molecules, molecular probes and tissue biomarkers.[22,23] By using several wavelengths the spectral information of molecular probes and tissue can be quantitatively resolved (Figure 7.1.1(b)).[24]

(a)   **(b)**

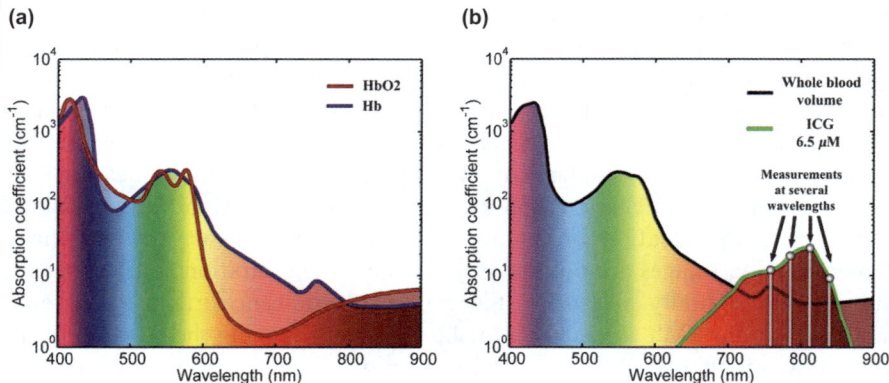

Figure 7.1.1   Absorption Spectrum of Oxygenated and Deoxygenated Blood. **(a)** targeting a molecular probe (Indocyanine Green, ICG), and **(b)** in the whole blood volume by measuring at several wavelengths.

Using multispectral excitation and ultra wide-band acoustic detection, MSOT allows visualization of blood oxygenation levels,[18] and optical probes based on fluorescent dyes or absorbing nanoparticles,[25] up to a depth of several cm. MSOT can differentiate such molecules based on their spectral differences and not only as absorption changes over background. This is of specific interest when studying probes with long distribution times or when performing longitudinal studies of animals where no "baseline" measurements are available, *i.e.* measurements before and after the administration of a probe. Spectral identification can also increase the specificity of molecular detection over single wavelength measurements.[24]

Using intravital microscopy, improved resolution can be achieved at more superficial depths (< 500 µm) compared to the ones in which FMT or MSOT operate (several mm-cm). The method is generally invasive, requiring the preparation of a window in the skin or in the skull (for brain studies), open surgery or endoscopic approaches. Single-cell imaging with intravital microscopy has been used *e.g.* to study cancer cell invasion, seeding in distant organs and dormancy.[26–28] The approach is probably more suited to address fundamental biological questions rather than to be used as a technique in the context of routine drug testing.

7.1.3   Multi-Modal Imaging

In many respects, imaging techniques are complementary; there is no 'all-in-one' imaging modality providing optimal sensitivity, specificity and temporo-spatial resolution. The high spatial resolution achieved with computerized tomography (CT) and magnetic resonance imaging (MRI) provides a good anatomical reference for molecular data obtained with high sensitivity, low-resolution modalities like optical imaging. This might be achieved by post-processing of data obtained in different imaging sessions or by simultaneous multimodality imaging. For such applications, the formats of data acquired sequentially with different instruments need to be compatible among them and–or sophisticated software tools for image co-registration (fusion), visualization and integration across modalities are necessary.

Optical imaging components can be easily arranged in a CT gantry. The development of CT-optical systems is currently pursued in several groups worldwide, and it is almost certain that in the coming years they will become a reality. In this respect, the development of matching fluid-free setups based on non-contact (free space) measurements has been of great importance.[12,13] For combined MRI-optical imaging systems, the main challenge is to develop an optical system that is MRI compatible. So far, concurrent optical tomography and MRI have been combined for breast cancer studies in humans,[29] but in small animals, studies have been relegated to using fiducials and performing sequential imaging sessions see ref. 11). Besides the magnetic field, one reason for this is the small bore needed for high-resolution small animal MRI that severely limits the space available for the optical components.

7.1.4 Molecular Probes and Tracers for Optical Imaging

Optical imaging is based on the detection of light that is emitted by an organic fluorophore, a quantum dot, a fluorescent protein or a molecule in an excited state through fluorescence or chemo-bioluminescence. In other words, for most optical imaging applications a probe is a necessity for signal generation. Reporter molecules for optical imaging consist of a fluorescent dye, which can be coupled to target-specific ligands or carriers such as antibodies, nanoparticles or polymers, proteins, peptides and small molecules, analogously to radiolabeling methods but with certain limitations due to the bulky dye molecules.[30,31] In addition, fluorescence detection allows researchers to design smart sensor reporters based on fluorescence quenching mechanisms, which are not detectable in their native state but are activated by interaction with their target (*e.g.* protease sensors), to increase signal-to-background ratios.[32] In general, optical imaging probes can be divided into three different classes (Table 7.1.1).

Table 7.1.1 Classes of Optical Imaging Probes.

Class	Characteristics
Level of information	Unspecific probes for anatomical and physiological imaging, targeted and activatable probes to report at the functional and molecular level
Anticipated use	Labeled drug analogs, ligands, pathway markers, biomarkers–surrogate markers
Distribution	Vascular, extracellular, intracellular

7.1.4.1 Small Organic Dyes

There are numerous optical imaging probes commercially available to report molecular processes and to highlight sub-cellular compartments for fixed and live cell microscopy (*e.g.* Molecular Probes – Invitrogen, GE Healthcare). In addition, fluorescently labeled primary and secondary antibodies can be purchased from a huge variety of vendors (*e.g.* Abcam, Cedarlane, AbD Serotec, GeneTex, BD Biosciences, Pharmingen, Biolegend) to be used for cell sorting, protein arrays, western blots and immunohistochemistry. Most of these conjugates emit light in the ultraviolet or visible region so that they can be used with standard excitation sources, filters and detection systems. For this reason they are of limited use for whole body *in vivo* imaging applications as they suffer from the same optical limitations as their fluorescent protein counterparts (*e.g.* GFP, RFP) given by the low tissue penetration of visible light and the relatively high scattering in turbid media such as tissues. In addition, these tracers were not designed for *in vivo* use and have very often a poor pharmacokinetic profile, which does not allow a systemic delivery and would require a local administration of the tracer. Nevertheless, imaging technologies such as intravital microscopy,[33] are currently paving the way to use these reporters [*e.g.* Ca^{2+} sensing dyes,[34] rhodamine dyes for leucocyte tracking,[35]] in an *in vivo* imaging setting as well. These minimally invasive techniques allow multiplexing and are not as dependent on tissue penetration since the microscopic lens can be introduced into the tissue close to the region of interest using microsurgery or endoscopic devices.

On the other hand, more and more vendors are offering NIRF (excitation > 650 nm) organic dyes,[36] mainly based on polymethine (*e.g.* cyanine) or oxazine core structures (Figure 7.1.2) with a variety of functional reactive groups and excitation wavelengths. This allows researchers to design and synthesize their own NIRF labeled conjugates particularly suited for whole body *in vivo* imaging applications (see Table 7.1.2 for selected examples of dyes.). Additional work, usually performed in academic centers, is necessary to improve the properties of NIRF dyes as tools for *in vivo* imaging.[37,38]

Even biologically relevant NIR labeled bioconjugates ready to use for small animal optical imaging have been recently marketed and formulated. There are some companies (*e.g.* Caliper, VisEn Medical, Carestream Health, Li-COR Biosciences) dedicated to offer complete solutions for small animal optical imaging, starting from reagents and tracers up to scanners. In addition, they provide support to their customers on how to apply these tools by showing selected *in vivo* applications and by providing protocols (see Table 7.1.3 for selected examples of probes.).

7.1.4.2 Nanoparticles

Nanomaterials constitute another interesting class of optical imaging reporters.[39,40] Dye-doped nanoparticles are designed by either covalent attachment of multiple organic fluorophores to components of the particle [*e.g.* lipophilic

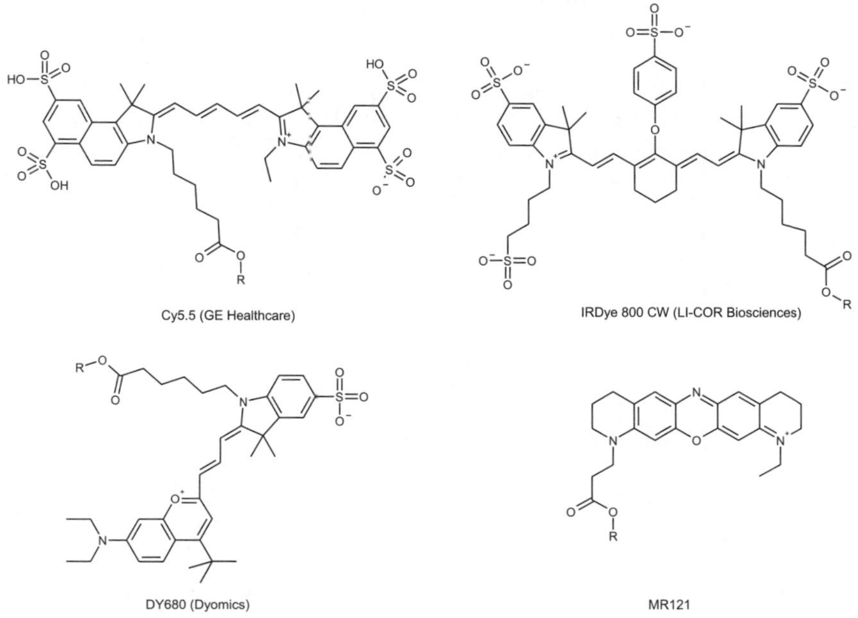

Cy5.5 (GE Healthcare)

IRDye 800 CW (LI-COR Biosciences)

DY680 (Dyomics)

MR121

Figure 7.1.2 Structure of Typical NIRF Dyes Based on Polymethine (*e.g.* Cyanine) or Oxazine Core.

Table 7.1.2 Selected Examples of Commercially Available NIRF Dyes.

Dye	Available chemistries	Vendor	Available emission wavelengths [nm]
IRDyes	COOH, NHS, maleimide	Li-COR Biosciences	687, 698, 786, 794
VivoTag	NHS	VisEn	688, 691, 775
Kodak X-Sight	NHS, TFP	Carestream Health	733, 755
XenoLight	NHS	Caliper	702, 780
CyDye	NHS, COOH, maleimide, hydrazide	GE Healthcare	670, 694, 776
Alexa Fluor	NHS, TFP, maleimide, hydrazide	Invitrogen	668, 690, 702, 723, 775
Atto dyes	COOH, NHS, maleimide, amine, azide, iodoacetamide	Atto-Tec	669, 684, 700, 719, 752, 764
Dyomics dyes	COOH, NHS, maleimide, amine	Dyomics	672–678 (6 dyes), 694–709 (6 dyes), 730–800 (14 dyes)

dye C-18 esters for liposomal formulation].[41] Alternatively, the fluorescent dyes can be trapped in the interior compartment of a hollow particle or can be embedded in a defined shell or matrix of solid particle designs [*e.g.* silica,[42,43] PLGA,[44]]. On the other hand, inorganic fluorophores such as quantum dots

Table 7.1.3 Selected Examples of Biologically Relevant Commercially Available Optical Imaging Tracers.

Tracer	Ligand	Target	Vendor	Potential Application
IRDye® 800CW 2-DG	2-deoxy-D-glucose	GLUT transporter	Li-COR Biosciences	Tumor biology
IRDye® 800CW EGF	recombinant human epidermal growth factor (EGF)	EGF receptor	Li-COR Biosciences	EGFR expressing tumors
IRDye® 800CW RGD, IntegriSense 680/750	peptide with RGD motif, non-peptide small molecule antagonist	$\alpha_v\beta_3$ integrin	Li-COR Biosciences, VisEn	Angiogenesis, metastasis
IRDye® Bonetag™, Osteosense 680/750	Bisphosphonate	Hydroxyapatite	Li-COR Biosciences, VisEn	Bone mineralization, bone metastasis
Superhance 680, Angiosense 680/750, SAIVI™ Alexa Fluor® 680/750 bovine serum albumin	—	Blood pool agents	VisEn, Invitrogen	Vascularity, perfusion and vascular permeability
MMPSense 680/750, ProSense (680/750)	Protease substrates	Proteases (*e.g.* MMP, cathepsins	VisEn	Tumor biology, Inflammation
annexin V, Alexa Fluor® 680	annexin V	phosphatidylserine	Invitrogen	Apoptosis
scVEGF/Cy	scVEGF	VEGF receptor	SibTech	Angiogenesis

consisting of semiconductor materials (*e.g.* CdSe, InP) can also be designed.[45,46] The optical properties of these inorganic nanocrystals may be tuned by adjusting the size of the particles given by the quantum confinement of these materials (size-dependent band gaps). The surface of the nanoparticles has to be coated with amphiphilic polymers to increase their water solubility and to make them biocompatible and to reduce toxicity (Figure 7.1.3). Quantum dots of different colors tuned to target different biological processes by coupling to corresponding carrier molecules will potentially enable multi-plexed imaging *in vivo*. [47]

In general, surface modification is an important task in nanoparticle design not only for imaging applications but also for drug delivery formulations to control the properties of particles in body fluids for *in vivo* applications:

– aggregation
– charge and cellular delivery
– opsonization and recognition by cells of the reticulo-endothelial system
– biological half-life and stability
– toxicity and immunogenicity
– functional groups for further modifications (*e.g.* attachment of targeting molecules)

It is beyond the scope of this chapter to discuss such modifications in detail, but there are excellent reviews with guidelines on how to design particles for a selected application and how to bypass some issues related to the *in vivo* use of these special nanocarriers.[39,48,49] A limitation of nanoparticles is the poor tissue penetration given by the size and slow diffusion of these materials, which restrict their use for deep tissue imaging. Nevertheless, endothelial targets and pathologies with an impaired vasculature such as tumors and sites of

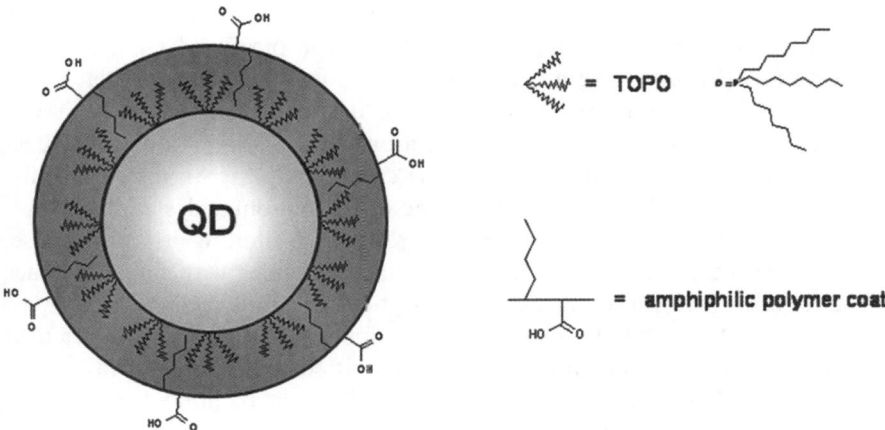

Figure 7.1.3   Schematic Representation of a Typical Biocompatible Quantum Dot Design.

inflammation (enhanced permeability and retention effect) will be accessible. The advantages of such reporters for *in vivo* imaging applications are obvious:

- brightness
- photo stability
- protection of the fluorophore from a degrading environment
- signal amplification
- multiplexing capabilities
- multivalency for targeted derivatives.

In summary, it can be said that the biological question should drive the selection of the label (small organic dye or nanoparticle) and that a rational probe design is the key to success. As an example, one might imagine that it is possible to study the biodistribution of an antibody non-invasively *in vivo* using whole body optical imaging by labeling it with a small organic NIRF dye, but not by using a 20 nm sized quantum dot. The nanocarrier mainly determines the pharmacokinetic properties of such a conjugate. Nevertheless, antibody tagged nanoprobes might be designed as valuable imaging biomarkers to highlight molecular processes *in vivo*.

7.1.4.3   Design of Optical Imaging Probes

Despite the availability of many fluorescent probes on the market, there is a need for the design of further optical imaging tracers given by the specific biological question in a drug discovery environment. There are two main interests driving the development of a novel target-specific contrast agent or tracer in this context:

- *Imaging and measuring the drug biodistribution:* Early information on pharmacokinetic properties, drug biodistribution and off-target accumulation (toxicity alerts) is essential during lead optimization and profiling. Conventionally, such data are obtained in rodents by blood and tissue sampling, or by autoradiography requiring isotopically- (^{14}C, ^{3}H) labeled compounds. More recently, nuclear imaging methods, in particular PET–micro-PET, have been regularly used to derive such information in animals and humans as isotopic substitution with ^{11}C or ^{18}F does not affect the physicochemical properties of the compound.[50,51] An important disadvantage of PET is the short lifetime of the radionuclides and the associated stringent requirements of the technique, limiting its routine use. Alternatively, one could imagine labeling molecules with fluorochromes and using the far more accessible optical imaging techniques as preliminary, fast readouts of drug biodistribution. Compounds selected at this preliminary step would then be submitted to the significantly more involved PET examinations. This approach might be limited to visualize the distribution of large molecular weight compounds such as biopolymers [*e.g.* monoclonal antibodies (mAb), proteins, siRNA] or drug delivery formulations [*e.g.* liposomes, inhalation powders] as the reporter groups for optical imaging are bulky dyes that may affect the properties of the labeled molecule. This influence will be less pronounced on these macromolecules compared to conventional small molecule drugs. The nuclear and optical imaging techniques do not allow a differentiation between the parent compound and a metabolite.

- *Imaging the target distribution–density and pharmacodynamic effects of drugs:* Demand is for specific reporter probes and amplification strategies in order to differentiate target information from non-specific background signal and to cope with the low (sub-nanomolar) target concentrations. Minimization of background signals requires elimination of the unbound and possibly, of the non-specifically bound fraction of the label, which implies a waiting period following injection of the reporter probe. Modulations of the signal from the reporter probe after administration of a drug candidate can be used to assess the compound binding to the target (receptor occupancy or down-regulation) or the effect of the drug on a certain molecular pathway. Reporter probes include targeted agents [*e.g.* small molecules, peptides, metabolites, antibodies or other molecules labeled with fluorochromes] for optical imaging and activatable probes. The latter undergo chemical or physicochemical changes upon interacting with their target. Examples include caged near-infrared fluorochromes,[52,53] protease-activatable dequenching probes,[54] or substrates for reporter genes.[55]

Contrast agent or tracer development is a time consuming project that requires significant resources and commitment. This process can be compared to a full drug discovery program, even for probes applied only in animal models, which should be driven by decision-relevant biological questions. It can take several months to years, from the selection of a suitable ligand until the routine use of a novel tracer. The tasks that might be needed in the tracer–contrast agent development and validation are summarized in Table 7.1.4.

One of the key initial criteria for the success of imaging probe development is the selection of a suitable ligand for further development. You can go through the literature or select analogs of known biochemical entities (*e.g.* fluorodeoxyglucose, fluorodeoxythymidine) as starting points. Big pharmaceutical or biotech companies have a competitive advantage in this field as their small-molecule archives or collections of biologics such as monoclonal antibodies (mAbs), engineered mAb fragments, antibody-like scaffolds or other proteins are obvious and rich sources for novel tracers. In addition, significant amounts of relevant data that could be used to address some of the criteria for the selection of suitable ligands for imaging probe development, such as the site for labeling, high affinity and selectivity for the target, *in vitro/in vivo* pharmacokinetics, structure activity relationship, metabolism, and reasonable clearance of unbound tracer, might be available from the corresponding drug development programs. A suboptimal drug in the compound collection could become a favorable tracer; it is just about digging for the right candidate. In the context of a biomarker strategy, it would be ideal to co-develop an imaging agent

Table 7.1.4   Tasks for Tracer–Contrast Agent Development and Validation.

Steps	*Approaches*
Ligand selection and tracer optimization based on *e.g.* known structure-activity relationship (SAR), lipophilicity, ease of labeling	Phage display, affinity maturation, positional scanning libraries, compound libraries *etc.*
Physicochemical characterization	
In vitro assays	Affinity testing, binding assays, selectivity profiling, metabolism
Cellular assays	Cell uptake, binding kinetics, functional assay, blocking studies
In vivo assays (maybe different species)	Pharmacokinetics (PK) and tissue distribution studies, clearance of unbound tracer, elimination pathways, unspecific (protein) binding, metabolites
Ex vivo analyses	Tissue distribution (organ level), (sub)-cellular distribution, ligand-target co-localization (staining, autoradiography)
In vivo testing (maybe different species)	Validation in disease model, blocking studies, negative control tracers, knock-out models, specific inhibitors

together with a novel drug at the very beginning of the discovery process, as there are synergies in the process and resources can be shared. As a final remark, it is important to keep in mind that most probes synthesized for molecular imaging will be limited to experimental research, since the approval process for human use involves similar hurdles as those for registering drugs.[56]

7.1.4.4   Labeling of Biologics

Biologics are particularly interesting candidates for labeling with fluorescent dyes and subsequent optical imaging studies as their relatively large size allows labeling with minimal changes of their properties. Traditionally small biotech companies developed these highly specific drugs, but recently large pharmaceutical companies have also been complementing their drug development pipelines that consist mainly of small molecule drugs, with biologics. These biopolymers are an obvious source for highly specific tracers. It would either be possible to measure the pharmacokinetics and biodistribution of the biopharmaceutical drug by using labeled derivatives or to study the corresponding therapeutic target non-invasively *in vivo* using biologic tracers, *e.g.* target expression levels or receptor occupancy to prove the mechanism of action or the efficacy of the corresponding drug. In addition, these tracers could be translated into clinically useful probes (*e.g.* for patient selection in clinical trials, for dose selection and scheduling and to highlight toxicity concerns by measuring target expression in non-targeted organs), just by replacing the measurable label into a radioisotope after preclinical validation and optimization in animal models using optical imaging (Figure 7.1.4). In other words, fluorescence imaging might serve as an economical and rapid in-house

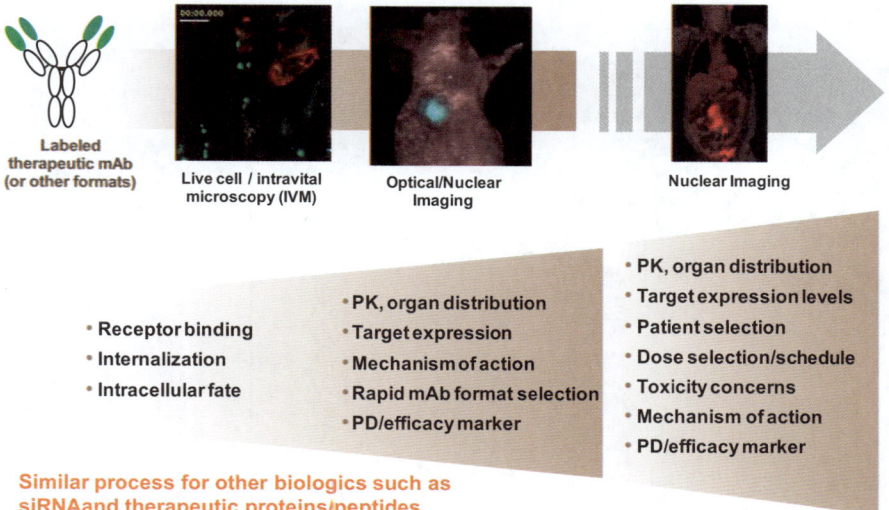

| Labeled therapeutic mAb (or other formats) | Live cell / intravital microscopy (IVM) | Optical/Nuclear Imaging | Nuclear Imaging |

• Receptor binding
• Internalization
• Intracellular fate

• PK, organ distribution
• Target expression
• Mechanism of action
• Rapid mAb format selection
• PD/efficacy marker

• PK, organ distribution
• Target expression levels
• Patient selection
• Dose selection/schedule
• Toxicity concerns
• Mechanism of action
• PD/efficacy marker

Similar process for other biologics such as siRNA and therapeutic proteins/peptides

Figure 7.1.4   Translational Biologics Imaging Platform.

screening tool delivering decision-relevant information to justify a further time consuming and costly PET tracer development:

– for biologics format selection (*e.g.* Fab, minibody, nanobody, PEG conjugates)
– preclinical validation, proof of concept
– isotope selection based on fragment kinetics [*e.g.* 86Y (14.7 h), 89Zr (3.27 d), 64Cu (12.7 h), 68Ga (68 min)].

In this context, the co-development of a biopharmaceutical drug together with a companion molecular imaging diagnostic would save both time and resources in the clinical validation of the biopharmaceutical. There are, however, several limitations using biologics as molecular imaging tracers. Many biopharmaceuticals such as monoclonal antibodies and therapeutic proteins show pharmacokinetics (slow clearance, low target to background ratios) unfavorable for imaging applications. They typically have a low stability or poor tissue penetration due to their large size. Protein engineering and post-translational modifications (*e.g.* PEGylation) set the stage to optimize these biomolecules for imaging applications. Antibody fragments such as scFvs, Fabs, F(ab')2, minibodies or diabodies are more favorable formats due to their enhanced clearance from the circulation compared to their monoclonal antibody (mAb) counterparts.[57,58] Novel antibody-like scaffolds such as affibodies or nanobodies are emerging as innovative therapeutics and diagnostics.[59]

Another restraint is given by the limited chemistry to label proteins with fluorescent dyes (Figure 7.1.5). Most frequently, random labeling technologies are used to couple a fluorescent dye to the protein of interest. *N*-hydroxysuccinimide-esters

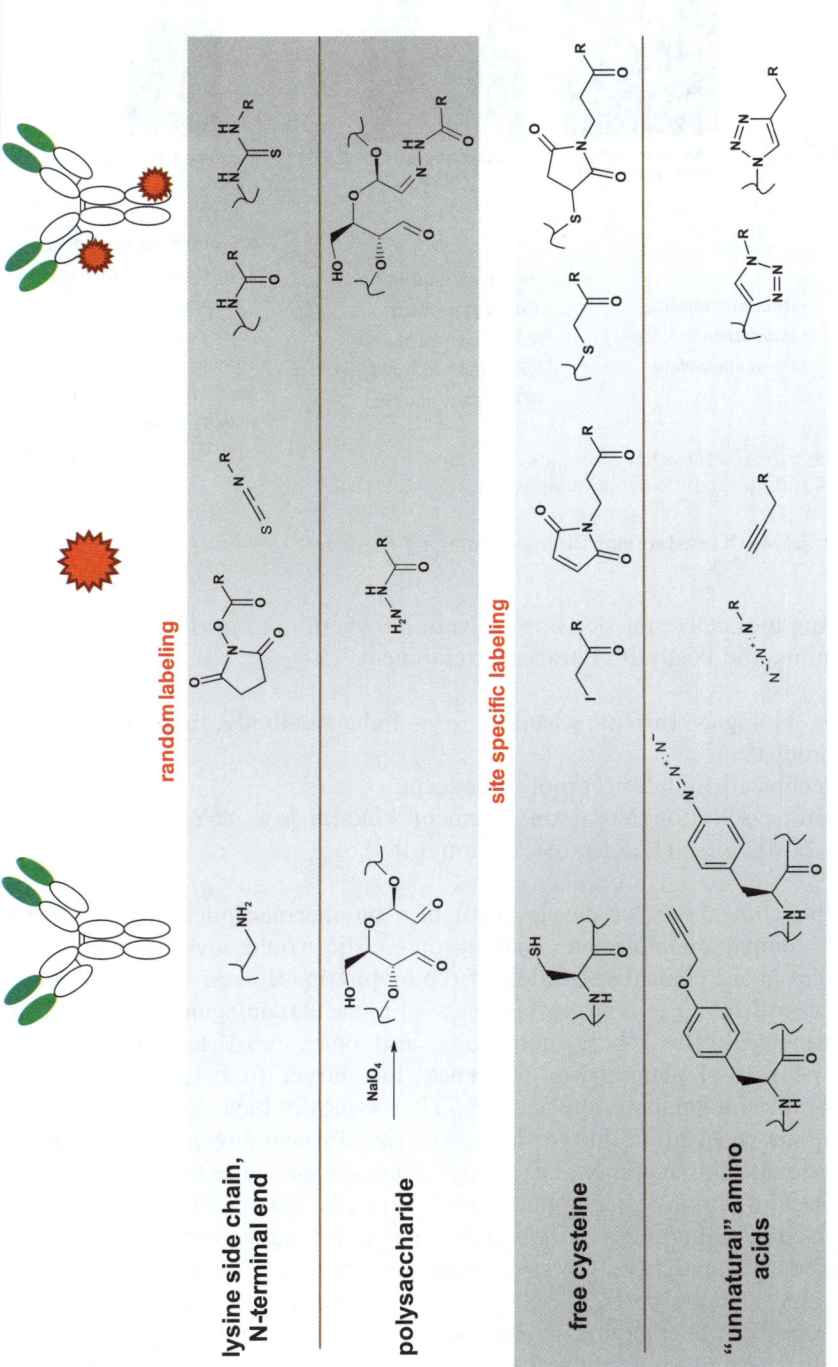

Figure 7.1.5 Labeling Strategies for Biologics.

(or other active esters) and isothiocyanates are available to be attached to lysine side chains or the N-terminal end of the protein.[60] These labeling methods always result in a complex mixture of unlabeled material, different degrees of labeling and regioisomers, as the reaction cannot be directed to certain reactive groups due to the generally equal reactivity of the lysine residues. Separation of the isomers is either labor-intensive or not possible even with sophisticated protein purification methodologies. Labeling of the antigen-binding site might have a strong negative impact on the binding affinity and specificity of the biologics tracer and the degree of labeling needs to be optimized in terms of affinity and brightness of the conjugate. Intensive loading of the protein with fluorochromes might lead to an extensive intramolecular fluorescence quenching of the conjugate by different mechanisms (FRET, electron transfer) and thereby to a reduced brightness.

A further option to label antibodies consists of the oxidation of carbohydrates from the glycosylated part of the protein using periodate and subsequent labeling of the resulting carbonyl groups with hydrazide derivates of the fluorescent dyes (hydrazone formation). These modifications might have an impact on the effector function of *e.g.* an antibody conjugate. Free cysteines (engineered proteins with free cysteines or reduction of disulfide bonds) also show a unique reactivity toward iodoactyl or maleimide reagents at controlled pH conditions and are frequently used to label proteins.[60]

Recently it became possible to incorporate unnatural amino acids into proteins in a site-specific manner by the use of amber codon suppression mutagenesis.[61] For example, unnatural amino acids bearing ketones, azides or alkynes have been incorporated. These modifications allow selective derivatization with labels using hydrazone formation or azide–alkyne cycloaddition (click-chemistry) reactions.[62] The major advantages of unnatural amino acid labeling are the excellent specificity, versatility with respect to small-molecule label, and minimal structural perturbation to the protein of interest. The design of recombinant proteins with engineered site-specific tags offers another option to target the labeling reaction to a specific region of the protein.[63] One approach is to fuse the protein of interest to a peptide or protein recognition sequence, which can then be labeled (covalently or by electrostatic interaction) with the reporter groups (*e.g.* TetraCys, HexaHis, SnapTag, HaloTag). Another methodology is enzyme-mediated protein labeling. A recognition peptide is fused to the protein of interest and a natural or engineered enzyme ligates the reporter group to the recognition peptide (Q-Tag, Biotin-Ligase, SorTag). In general it can be said that there is always a trade-off between tag-size and labeling specificity.[64]

7.1.5 Optical Imaging in Drug Discovery: From Research to the Clinics

In this section, selected examples from different disease areas are provided to illustrate how *in vivo* optical imaging techniques have the ability to provide relevant information in the context of drug discovery. Understanding the role of gene products in the development of disease is crucial for early diagnosis as well as for finding new therapeutic targets and developing novel classes of drugs that interact with them. Based on their sensitivity, optical imaging techniques have the potential to non-invasively visualize specific molecular targets *in vivo*,[53] and thus provide relevant information for target validation. In such applications, it needs to be considered that in addition to target-specific fluorochrome accumulation, non-specific accumulation of the tracer occurs owing to perfusion and vascular permeability effects. Experiments involving the administration of non-target-specific NIR-fluorochromes need to be performed in order to improve the discrimination between both contributions.[65] Optical imaging is used as well to monitor pathology progression and the effects of therapies in animal models of disease.[1,66] In addition, some clinical applications are emerging. As previously mentioned, there is translational potential in the use of optical imaging probes to support the development of selected targeted radiotracers.

7.1.5.1   Cancer

To answer basic questions of *in vivo* tumor development and progression, fluorescence based imaging techniques provide new insights into molecular pathways and targets [see ref. 67–69 for reviews]. For example, von Wallbrunn *et al.*[70] reported on the specific targeting of αvβ3 integrin in human cancer xenografts using a Cy5.5-labeled arginine-glycine-aspartic acid (RGD) motif. The αvβ3 integrin cell adhesion receptor is expressed on proliferating but not quiescent endothelial cells and is involved in tumor progression, angiogenesis, and metastasis.[71,72] The most common integrin binding sequence, the RGD peptide, has been used for endothelial targeting for diagnostic and therapeutic purposes in cancer.[73,74]

Monitoring enzyme activity *in vivo* by NIRF has become routine,[11,75,76] and enzyme-activatable fluorochromes can be detected with high positional accuracy in deep tissues. For instance, FMT has been used to image the activity of cathepsin B, which is consistently over-expressed in human and murine tumors [see ref. 77 for a review], in 9L gliosarcomas stereotactically implanted into unilateral brain hemispheres of nude mice.[11]

Agents allowing the combined use of several imaging modalities are obviously of great advantage. A recent example is the application of AMTA680, a functionally derivatized magneto-fluorescent nanoparticle (MNP), to track tumor-associated macrophages (TAMs) at different resolutions with various imaging modalities, *e.g.*, FMT, MRI, and multiphoton and confocal intravital microscopy.[78] The agent labeled a subset of myeloid cells with an M2 macrophage phenotype, whereas other neighboring cells, including tumor cells and a variety of other leukocytes, remained unlabeled. Quantitative assessment of TAM distribution and activity *in vivo* identified that these cells cluster in delimited foci within tumors, show relatively low motility, and extend cytoplasmic protrusions for prolonged physical interactions with neighboring tumor cells. Noninvasive imaging can also be used to monitor TAM-depleting regimen quantitatively.

Tumor response to chemotherapy has been shown to be accurately resolved *in vivo* by FMT with a phosphatidylserine sensing fluorescent probe based on modified annexins.[79] Following injection of Lewis lung carcinoma cell lines sensitive to and resistant to cyclophosphamide into the left and right mammary pad respectively of female athymic nu/nu mice, animals were treated with cyclophosphamide. Optical imaging revealed a more than 10-fold increase of fluorochrome concentration in cyclophosphamide-sensitive tumors and a 7-fold increase of resistant tumors compared with control studies.

Due to the good penetration depth of NIR light in breast tissue, breast cancer is a principal clinical focus of the bio-optics community. Most studies have used intrinsic absorption contrast based on the different spectral signatures of oxy- and deoxyhemoglobin,[80–82] and a few have made use of indocyanine green (ICG) as a contrast agent to enhance and control the absorption present in the tumor.[29,83]

The detection of tumors *in vivo* with MSOT is not limited to targeting the spectrum of a molecular agent. Instead of resolving the agent, the abnormal vascularization of tumors can be targeted with MSOT. Ku *et al.*[84] imaged the vascular structure of rat brains, where they successfully observed the angiogenesis of brain tumors in rats. These were identified by the distorted vascular architecture of the brain tumorigenesis and related vascular changes, such as hemorrhage. With this approach, tumor treatment can be followed longitudinally even without administering molecular probes. It also provides insight into the vascular structure of tumors, which is especially interesting to assess the effects of anti-angiogenic therapies.

7.1.5.2 Rheumatoid Arthritis

Macrophages possess widespread pro-inflammatory, destructive, and remodeling capabilities that critically contribute to the acute and chronic phases of rheumatoid arthritis (RA).[85,86] Activated macrophages constitute key effector cells in RA, a direct correlation existing between the level of macrophage activity and the observed joint inflammation, articular pain, and bone erosion. Macrophages expressing the F4/80 antigen on their surface accumulate in inflamed joints.[87] This has been used as a strategy to label macrophages for NIRF imaging in a murine arthritis model.[88] Macrophages were targeted by fluorochrome-labeled mAbs directed against the F4/80 antigen. Intravenous injection of Cy5.5-labeled mAb F4/80 against macrophages caused an increased accumulation of fluorochrome probes in arthritic joints of mice. These techniques were used in testing anti-inflammatory compounds.[88] Moreover, imaging of activated macrophages has been achieved by targeting the folate receptor.[89]

Several enzymes of different protease families, such as matrix metalloproteinases (MMPs),[90] and cysteine proteases *e.g.* cathepsin B,[91] produced primarily by synovial fibroblasts and synovial macrophages, are highly upregulated in RA and have a substantial role in the destruction of arthritic joints.[92] *In vivo* imaging of protease activity with NIRF has shown potential to monitor treatment response in experimental arthritis.[93,94] For instance, using a

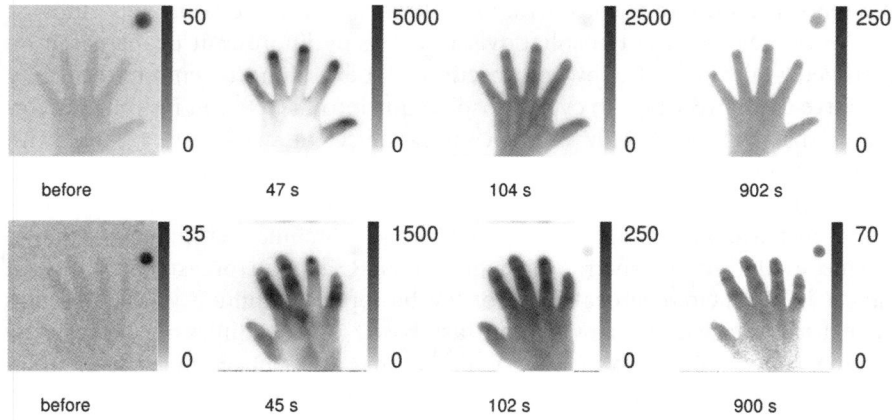

Figure 7.1.6 Fluorescence imaging for the detection of RA in humans. Images of a volunteer (top row) and a patient (bottom row) after the intravenous application of 0.1 mg kg^{-1} of ICG. The time after injection is indicated in each image. Fluorescence intensities are given in counts per second, normalized to the corresponding fluorescence standard in the top right corner. The numbers at the intensity bars indicate the contrast region. Reproduced with permission from Fischer *et al.*[95] © 2010 AUR Published by Elsevier Inc.

protease-activated "smart" probe, Wunder *et al.*[93] examined the presence and distribution of fluorescence in arthritic joints of mice with collagen-induced arthritis by both NIRF imaging and histology. Proteases that target the Lys-Lys cleavage site, including cathepsin B, activate the probe fluorescence. Treatment monitoring data were obtained following methotrexate therapy.[93]

Fluorescent probes still need to be clinically approved in order to transfer protocols from small animal to human imaging systems. The nonspecific NIRF dye, ICG, already approved for cancer examinations, is currently in the approval process for clinical arthritis imaging. A custom-built ICG-enhanced optical imaging device for hand imaging was recently shown by Fischer *et al.* to correlate well with contrast-enhanced MRI for the detection of RA in patients (Figure 7.1.6).[95] This application has been triggered by experimental studies in mice, showing that inflammatory changes in arthritic joints can be detected based on the pharmacokinetic behavior of ICG,[96] characterized by a 98% binding to plasma proteins after intravenous injection and hepatocellular uptake.[97] ICG thus labels the intravascular space; it is removed exclusively by the liver with a half-life of ~150–180 s. Altogether, ICG offers great potential for the early and sensitive detection of inflammatory joint disease in humans.[66,95] Additionally, optical tomography approaches have been proposed for the detection of synovitis in arthritic finger joints.[98]

7.1.5.3 Alzheimer's Disease

At present, the only definitive diagnosis of Alzheimer's disease (AD), the major cause of late-onset dementia, is made post-mortem, based on the verification of the pathological hallmarks: extracellular aggregates of ß-amyloid (Aß) peptide (amyloid plaques) and intracellular neurofibrillary tangles (reviewed by Ittner and Götz).[99] By processes not completely understood, the accumulation of Aß and neurofibrillary tangles produces neurodegeneration, which ultimately accounts for the clinical signs of the disease.

NIRF provides an important and fast alternative to image *in vivo* cerebral plaques in murine models. A NIRF oxazine dye, AOI987, has been demonstrated to readily penetrate the intact blood brain barrier and to bind to amyloid plaques.[100] Using NIRF imaging, a specific interaction of AOI987 with amyloid plaques was shown in APP23 mice *in vivo*, and confirmed by postmortem analysis of brain slices. Quantitative analysis revealed increasing fluorescence signal intensity with increasing plaque load of the animals,[66] and significant binding of AOI987 was observed for APP23 transgenic mice aged 9 months and older.[100] Thus, AOI987 is an attractive probe to monitor disease progression in animal models of AD noninvasively, and to evaluate the effects of potential drugs on the plaque load. In addition, multi-modal FMT-CT imaging allows for accurate signal localization and quantification of amyloid-β plaque burden in APP23 transgenic mice.[101]

Research in the field of AD may significantly benefit from the technical advances provided by MSOT.[102] The technique is able to accurately image brain parameters in small animals with good contrast, high resolution and adequate penetration depth. For example, it can target amyloid plaques labeled Congo-red *in vivo*, allowing longitudinal studies of the same animal.[103] Additionally, MSOT can visualize the blood oxygenation of the brain, which also gives information on the progress of neurodegenerative disease. But MSOT is not only limited to research related to AD. It has a very good performance even for structural imaging of the brain. Figure 7.1.7 depicts images acquired by measuring the head of a 6-day-old mouse. Data related to several slices were acquired and combined to an image stack (Figure 7.1.7(a)). To validate the findings, one slice of the image stack was selected (Figure 7.1.7(b)) and compared to the corresponding cryoslice of the mouse brain (Figure 7.1.7(c)). The results demonstrate the ability of MSOT to provide high-resolution images and good contrast in never before seen performance, compared to other *in vivo* imaging modalities.

7.1.5.4 Inflammation

As central effector cells during allergic airway inflammation, eosinophils are an important clinical therapeutic target. These multifunctional leukocytes degrade and remodel tissue extracellular matrix through production of proteolytic enzymes, release of proinflammatory factors to initiate and propagate inflammatory responses, and direct activation of mucus secretion and smooth muscle cell constriction. Cortez-Remozo *et al.*[104] described the use of an injectable MMP-targeted optical sensor that specifically and quantitatively resolved eosinophil activity in the lungs of mice with experimental allergic

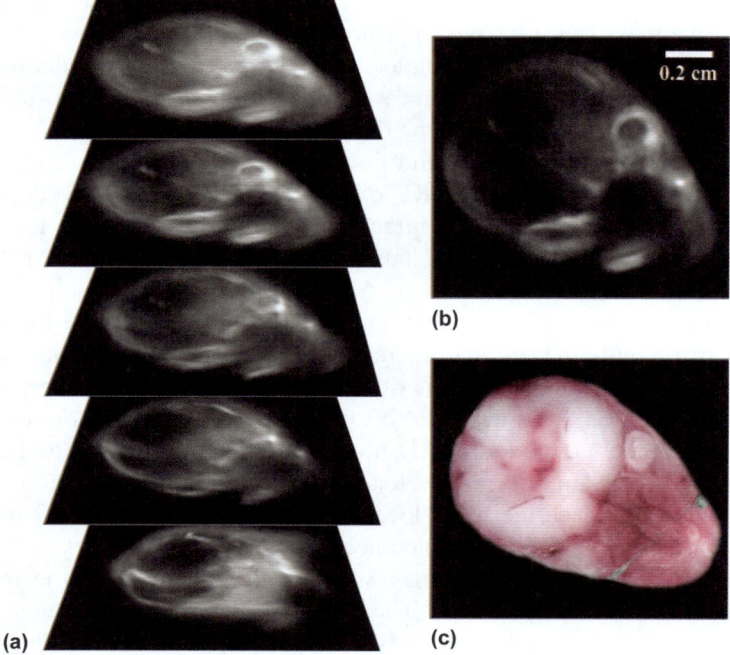

Figure 7.1.7   MSOT Images of the Mouse Brain *Ex Vivo*. **(a)** Three dimensional
image stack acquired in 60 min, **(b)** Single slice of the image stack with
(c) corresponding cryo-slice. Modified with permission from Jetzfellner
et al.[102] © 2011 American Institute of Physics.

airway inflammation. Using real-time molecular imaging methods, visualization of eosinophil responses *in vivo* at different scales was feasible. Eosinophil
responses were seen at single-cell resolution in conducting airways using near-
infrared fluorescence fiberoptic bronchoscopy, in lung parenchyma using
intravital microscopy, and in the whole body using FMT. Using these real-time
imaging methods, the immunosuppressive effects of the glucocorticoid drug
dexamethasone in the mouse model of allergic airway inflammation were
confirmed and a viridin-derived prodrug that potently inhibited the accumulation and enzyme activity of eosinophils in the lungs, has been indentified.[104]
The combination of sensitive enzyme-targeted sensors with noninvasive molecular imaging approaches permitted evaluation of airway inflammation severity
and was used as a model to rapidly screen for new drug effects. Both FMT and
fiberoptic bronchoscopy techniques have the potential to be translated into the
clinic. Non-specific approaches based on MRI have also shown great usefulness
in pharmacological research in the area of respiratory diseases.[105–109]
   Dysfunctions in mucociliary clearance are associated with the accelerated
loss of lung function in several respiratory diseases. Approaches enabling the
in vivo visualization of mucus dynamics in rodents at high resolution and

sensitivity would be beneficial for experimental lung research. Blé *et al.*[110] described the synthesis and characterization of bilabeled aminodextran-based probes binding specifically to mucin. Labeling of secreted mucus and of mucin in goblet cells in the lungs of lipopolysaccharide (LPS)-challenged rats has been demonstrated *in vivo* with NIRF and MRI, and confirmed by histology. The effects of uridine triphosphate (UTP) were then studied in LPS-challenged rats by simultaneously administering the imaging probe and the compound. The data suggest that UTP increased the mucociliary clearance, but at the same time induced a release of mucin from goblet cells, thus not contributing to the overall reduction of mucus in the lung. The outlined approach enables one to derive information on mucus clearance as well as secretion. Such a global view on mucus dynamics may prove invaluable when testing new pharmacological agents aimed at improving mucociliary clearance.

7.1.5.5　Cardiology

Even though fluorescence imaging of the heart is typically limited to invasive *ex vivo* or *in vitro* applications, Sosnovik *et al.*[111] demonstrated that non-invasive FMT images of the heart can be acquired *in vivo* and be coregistered with *in vivo* cardiac MR images. The uptake of the magnetofluorescent nano-particle (MNP) CLIO-Cy5.5 by macrophages was studied in the infarcted myocardium of mice following ligation of the left coronary artery.[111] The MNP CLIO-Cy5.5 was used to provide dual magnetic and fluorescence readouts of postinfarction myocardial macrophage infiltration, and to demonstrate the synergy and congruence of this dual-modality approach, since MNPs are avidly taken up by macrophages. Mice received CLIO-Cy5.5 48 h after surgery, and imaging was performed 48 h later. An increase in MRI contrast-to-noise ratio, indicative of myocardial probe accumulation, was seen in the anterolateral walls of the infarcted mice but not in the sham-operated mice. Fluorescence intensity over the heart was also significantly higher in the FMT images of the infarcted mice, the uptake of CLIO-Cy5.5 by macrophages infiltrating the infarcted myocardium being later confirmed by fluorescence microscopy and immunohistochemistry.[111]

MSOT provides an alternative to FMT images co-registered with MRI, because it allows anatomical imaging at high resolution with good contrast and can also target molecular probes in real-time. Taruttis *et al.*[20] used commercially available gold nanorods with an absorption peak in the NIR spectrum for real-time imaging of cardiovascular dynamics in mice. The absorption peak was at 780 nm and the dimensions of the particles were 10 nm by 38 nm. To resolve the molecular probe, MSOT measurements at five wavelengths were performed (725 nm, 750 nm, 775 nm, 800 nm, 825 nm) to resolve the particles from the background. It was shown that MSOT can resolve cardiovascular dynamics and circulating gold nanorods in real-time (Figure 7.1.8). The carotid arteries, the aorta and the cardiac wall could be imaged as well.[20]

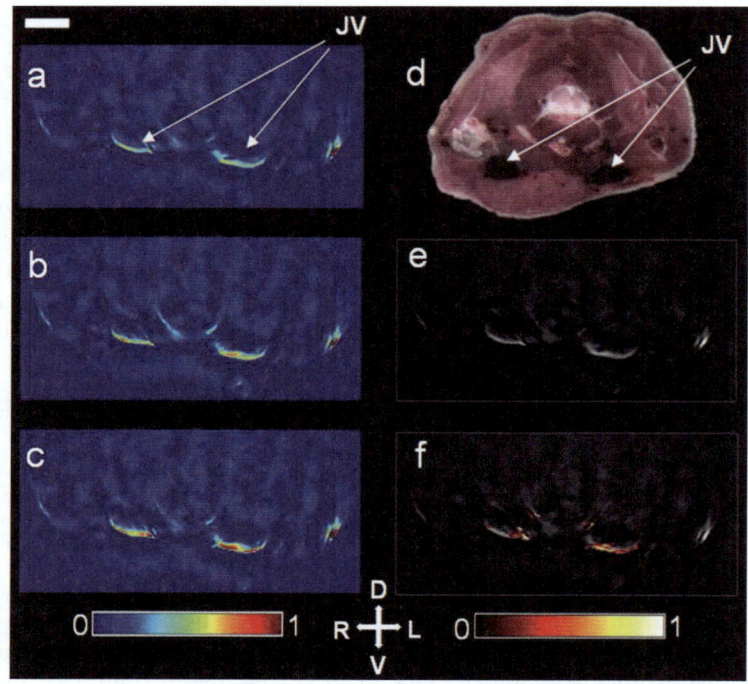

Figure 7.1.8   Optoacoustic Detection of Gold Nanorods Injected Intravenously to a Mouse. Imaging was performed at the level of the jugular veins (JV) in the neck. **(a)** Single-pulse transverse slice through the neck prior to injection. The same slice **(b)** during and **(c)** 10 s after the administration of nanorods, **(d)** Cryosection showing anatomical correspondences, **(e)** MSOT image before injection of nanorods, **(f)** MSOT image post injection showing multispectrally resolved distribution of nanorods overlaid on a single wavelength image. Scale bar 3 mm. Reproduced with permission from Taruttis *et al.*[20] © 2010 The Optical Society of America.

7.1.6   Summary

In vivo molecular imaging has become an integral part of pharmaceutical research and development.[1–4] Specific molecular probes as well as intrinsic tissue characteristics are exploited as the source of image contrast, thereby providing potential for the characterization and quantification of biological processes at the cellular and subcellular levels in intact living subjects, earlier detection and characterization of pathology, and evaluation of treatment. Imaging can provide biomarkers of a disease process and therefore help to define stratified study groups. The non-invasive character of imaging allows for longitudinal assessments in a single individual, of particular relevance in chronic studies. The statistical power of the results is thereby increased. Moreover, in a preclinical setting, more clinically relevant study designs using fewer animals are feasible. Imaging also provides important information on the optimal timing and dosing of drugs.

Among the molecular imaging techniques, optical imaging is of particular interest to pharmaceutical research because of its low cost, versatility and high-throughput capability for experimental studies in small rodents. Many fluorescent probes are already available on the market. Nonetheless, there is a need for the design of further optical imaging tracers because of the demands from specific biological questions in the realm of the drug discovery process. It needs to be kept in mind that the proper validation of such probes may be time consuming. Additionally, we have seen in the present contribution that the technique has translational potential to the clinic. Human examinations are feasible either in areas close to superficial regions like the breast,[81,83] and finger joints,[95,98] or by the combined use of optical imaging instruments and fiberoptic bronchoscopy devices.[104] The greatest translational contributions of optical imaging however, may be in the early support of *in vivo* target validation as well as in the screening and selection of biologic tracers for further PET tracer development. The combined efforts of chemists, biologists and pharmacologists will further expand the use of optical imaging in pharmaceutical research and related disciplines. This will benefit not only the biopharmaceutical sector but also medical practice and public health in general, by speeding up the drug discovery process, thereby resulting in better medicines at lower costs.

Abbreviations

Aß ß-amyloid; AD Alzheimer's disease; CCD charge-coupled device; CLIO cross-linked iron oxide; CT computerized tomography; FMT fluorescence molecular tomography; FRET fluorescence resonance energy transfer; GFP green fluorescent protein; ICG indocyanine green; mAb monoclonal antibody; MMP matrix metalloproteinase; MNP magneto-fluorescent nanoparticle; MR magnetic resonance; MRI magnetic resonance imaging; MSOT multispectral optoacoustic tomography; NIR near-infrared; NIRF near-infrared fluorescence; OAT optoacoustic tomography; PET positron emission tomography; PLGA poly(DL-lactic-co-glycolic acid); RA rheumatoid arthritis; RFP red fluorescent protein; siRNA small interfering RNA; SPECT single photon emission computerized tomography; TAM tumor-associated macrophage.

References

1. M. Rudin and R. Weissleder, *Nat. Rev. Drug Discovery*, 2003, **2**, 123.
2. N. Beckmann, *In Vivo MR Techniques in Drug Discovery and Development*, Taylor & Francis, New York, NY, 2006.
3. T. F. Massoud and S. S. Gambhir, *Trends Mol. Med.*, 2007, **13**, 183.
4. J. K. Willmann, N. van Bruggen, L. M. Dinkelborg and S. S. Gambhir, *Nat. Rev. Drug Discovery*, 2008, **7**, 591.
5. K. Shah and R. Weissleder, *NeuroRx*, 2005, **2**, 215.
6. A. Stell, S. Belcredito, B. Ramachandran, A. Biserni, G. Rando, P. Ciana and A. Maggi, *Q. J. Nucl. Med. Mol. Imaging*, 2007, **51**, 127.

7. V. Ntziachristos, J. Ripoll, L. V. Wang and R. Weissleder, *Nat. Biotechnol.*, 2005, **23**, 313.

8. K. O. Vasquez, C. Casavant and J. D. Peterson, *PLoS One*, 2011, **6**, e20594.

9. G. Zacharakis, H. Kambara, H. Shih, J. Ripoll, J. Grimm, Y. Saeki, R. Weissleder and V. Ntziachristos, *Proc. Natl. Acad. Sci. U.S.A.*, 2005, **102**, 18252.

10. V. Ntziachristos, C. H. Tung, C. Bremer and R. Weissleder, *Nat. Med.*, 2002, **8**, 757.

11. V. Ntziachristos, J. Ripoll and R. Weissleder, *Opt. Lett.*, 2002, **27**, 333.

12. J. Ripoll, R. B. Schulz and V. Ntziachristos, *Phys. Rev. Lett.*, 2003, **91**, 103901.

13. R. B. Schulz, J. Ripoll and V. Ntziachristos, *IEEE Trans. Med. Imaging*, 2004, **23**, 492.

14. A. Garofalakis, G. Zacharakis, H. Meyer, E. N. Economou, C. Mamalaki, J. Papamatheakis, D. Kioussis, V. Ntziachristos and J. Ripoll, *Mol. Imaging*, 2007, **6**, 96.

15. T. Lasser and V. Ntziachristos, *Med. Image Anal.*, 2007, **11**, 389.

16. D. Razansky and V. Ntziachristos, *Med. Phys.*, 2007, **34**, 4293.

17. R. A. Kruger, *Med. Phys.*, 1995, **22**, 1605.

18. J. Laufer, D. Delpy, C. Elwell and P. Beard, *Phys. Med. Biol.*, 2006, **52**, 141.

19. A. Buehler, E. Herzog, D. Razansky and V. Ntziachristos, *Opt. Lett.*, 2010, **35**, 2475.

20. A. Taruttis, E. Herzog, D. Razansky and V. Ntziachristos, *Opt. Express*, 2010, **18**, 19592.

21. L. Li, R. J. Zemp, G. Lungu, G. Stoica and L. V. Wang, *J. Biomed. Opt.*, 2007, **12**, 020504.

22. D. Razansky, C. Vinegoni and V. Ntziachristos, *Opt. Lett.*, 2007, **32**, 2891.

23. D. Razansky, M. Distel, C. Vinegoni, R. Ma, N. Perrimon, R. W. Koster and V. Ntziachristos, *Nat. Photonics*, 2009, **3**, 412.

24. V. Ntziachristos and D. Razansky, *Chem. Rev.*, 2010, **110**, 2783.

25. V. Ntziachristos, *Nat. Methods*, 2010, **7**, 603.

26. R. K. Jain, L. L. Munn and D. Fukumura, *Nat. Rev. Cancer*, 2002, **2**, 266.

27. C. Nombela-Arrieta, R. A. Lacalle, M. C. Montoya, Y. Kunisaki, D. Megías, M. Marqués, A. C. Carrera, S. Mañes, Y. Fukui, C. Martínez-A and J. V. Stein, *Immunity*, 2004, **21**, 429.

28. R. M. Hoffman, *Nat. Rev. Cancer*, 2005, **5**, 796.

29. V. Ntziachristos, A. G. Yodh, M. Schnall and B. Chance, *Proc. Natl. Acad. Sci. U.S.A.*, 2000, **97**, 2767.

30. J. V. Frangioni, *Curr. Opin. Chem. Biol.*, 2003, **7**, 626.

31. K. Licha, *Top. Curr. Chem.*, 2002, **222**, 1.

32. M. Funovics, R. Weissleder and C. H. Tung, *Anal. Bioanal. Chem.*, 2003, **377**, 956.

33. A. Bullen, *Nat. Rev. Drug Discovery*, 2008, **7**, 54.

34. N. L. Rochefort, H. Jia and A. Konnerth, *Trends Mol. Med.*, 2008, **14**, 389.
35. C. Halin, J. R. Mora, C Sumen and U. H. von Andrian, *Annu. Rev. Cell Dev. Biol.*, 2005, **21**, 581.
36. S. Stoyanov in *Practical Spectroscopy, 25 (Near-Infrared Applications in Biotechnology)*, ed. R. Raghavachari, Marcel Dekker Inc., New York, NY, 2001, p. 35.
37. W. Pham, L. Cassell, A. Gillman, D. Koktysh and J. C. Gorea, *Chem. Commun.*, 2008, **16**, 1895.
38. F. Shao, R. Weissleder and S. A. Hilderbrand, *Bioconjug. Chem.*, 2008, **19**, 2487.
39. M. J. Murcia, C. A. Naumann in *Nanotechnologies for the Life Sciences: Biofunctionalization of Nanomaterials*, ed. C. S. S. R. Kumar, Wiley-VCH, Weinheim, 2005, **vol. 1**, p. 1.
40. N. Vijayasree, K. Haritha, V. Subhash and K. R. S. S. Rao, *Res. J. BioTechnol.*, 2009, **4**, 61.
41. V. Deissler, R. Rüger, W. Frank, A. Fahr, W. A. Kaiser and I. Hilger, *Small*, 2008, **4**, 1240.
42. A. Burns, H. Ow and U. Wiesner, *Chem. Soc. Rev.*, 2006, **35**, 1028.
43. A. A. Burns, J. Vider, H. Ow, E. Herz, O. Penate-Medina, M. Baumgart, S. M. Larson, U. Wiesner and M. Bradbury, *Nano Lett.*, 2009, **9**, 442.
44. Y. T. Lim, Y.-W. Noh, J. H. Han, Q.-Y. Cai, K.-H. Yoon and B. H. Chung, *Small*, 2008, **4**, 1640.
45. X. Michalet, F. F. Pinaud, L. A. Bentolila, J. M. Tsay, S. Doose, J. J. Li, G. Sundaresan, A. M. Wu, S. S. Gambhir and S. Weiss, *Science*, 2005, **307**, 538.
46. U. Resch-Genger, M. Grabolle, S. Cavaliere-Jaricot, R. Nitschke and T. Nann, *Nat. Methods*, 2008, **5**, 763.
47. A. M. Smith, S. Dave, S. Nie, L. True and X. Gao, *Expert Rev. Mol. Diagn.*, 2006, **6**, 231.
48. S. M. Moghimi, A. C. Hunter and J. C. Murray, *Pharmacol. Rev.*, 2001, **53**, 283.
49. V. P. Torchilin, *Nat. Rev. Drug Discovery*, 2005, **4**, 145.
50. A. J. Fischman, N. M. Alpert and R. H. Rubin, *Clin. Pharmacokinet.*, 2002, **41**, 581.
51. M. E. Phelps, *Proc. Natl. Acad. Sci. U.S.A.*, 2000, **97**, 9226.
52. D. J. Bornhop, C. H. Contag, K. Licha and C. J. Murphy, *J. Biomed. Opt.*, 2001, **6**, 106.
53. R. Weissleder and V. Ntziachristos, *Nat. Med.*, 2003, **9**, 123.
54. C. H. Tung, U. Mahmood, S. Bredow and R. Weissleder, *Cancer Res.*, 2000, **60**, 4953.
55. V. Josserand, I. Texier-Nogues, P. Huber, M.-C. Favrot and J.-L. Coll, *Gene Ther.*, 2007, **14**, 1587.
56. F. A. Jaffer and R. Weissleder R, *JAMA, J. Am. Med. Assoc.*, 2005, **293**, 855.
57. P. Holliger and P. J. Hudson, *Nat. Biotechnol.*, 2005, **23**, 1126.

58. A. M. Wu and P. D. Senter, *Nat. Biotechnol.*, 2005, **23**, 1137.
59. R. C. Roovers, G. A. van Dongen and P. M. van Bergen en Henegouwen, *Curr. Opin. Mol. Ther.*, 2007, **9**, 327.
60. G. T. Hermanson, *Bioconjugate Techniques*, Academic Press, New York, NY, 1996.
61. L. Wang and P. G. Schultz, *Chem. Commun.*, 2002, **1**, 1.
62. Q. Wang, A. R. Parrish and L. Wang, *Chem. Biol.*, 2009, **16**, 323.
63. N. Johnsson and K. Johnsson, *ChemBioChem.*, 2003, **4**, 803.
64. M. Fernández-Suárez and A. Y. Ting, *Nat. Rev. Mol. Cell. Biol.*, 2008, **9**, 929.
65. K. Licha, B. Riefke, V. Ntziachristos, A. Becker, B. Chance and W. Semmler, *Photochem. Photobiol.*, 2000, **72**, 392.
66. J. Ripoll, V. Ntziachristos, C. Cannet, A. L. Babin, R. Kneuer, H. U. Gremlich and N. Beckmann, *Drugs R&D*, 2008, **9**, 277.
67. W. S. El-Deiry, C. C. Sigman and G. J. Kelloff, *J. Clin. Oncol.*, 2006, **24**, 3261.
68. E. E. Graves, R. Weissleder and V. Ntziachristos, *Curr. Mol. Med.*, 2004, **4**, 419.
69. J. T. Wessels, A. C. Busse, J. Mahrt, C. Dullin, E. Grabbe and G. A. Mueller, *Cytometry Part A*, 2007, **71**, 542.
70. A. von Wallbrunn, C. Höltke, M. Zühlsdorf, W. Heindel, M. Schäfers and C. Bremer, *Eur. J. Nucl. Med. Mol. Imaging*, 2007, **34**, 745.
71. K. M. Hodivala-Dilke, A. R. Reynolds and L. E. Reynolds, *Cell Tissue Res.*, 2003, **314**, 131.
72. H. Jin and J. Varner, *Br. J. Cancer*, 2004, **90**, 561.
73. E. Ruoslahti and M. D. Pierschbacher, *Science*, 1987, **238**, 491.
74. S. E. D'souza, M. H. Ginsberg, G. R. Matsueda and E. F. Plow, *Nature*, 1991, **350**, 66.
75. C. Bremer, C. H. Tung and R. Weissleder, *Nat. Med.*, 2001, **7**, 743.
76. J. W. Chen, M. Querol Sans, A. Bogdanov Jr. and R. Weissleder, *Radiology*, 2006, **240**, 473.
77. I. Podgorski, and B. F. Sloane in *Proteases and the Regulation of Biological Processes: Proceedings of the Biochemical Society Symposium,* ed. J. Saklatvala, H. Nagase and G. Salvesen, Portland Press, London, 2003, vol. 70, p. 263.
78. A. Leimgruber, C. Berger, V. Cortez-Retamozo, M. Etzrodt, A. P. Newton, P. Waterman, J. L. Figueiredo, R. H. Kohler, N. Elpek, T. R. Mempel, F. K. Swirski, M. Nahrendorf, R. Weissleder and M. J. Pittet, *Neoplasia*, 2009, **11**, 459.
79. V. Ntziachristos, E. A. Schellenberger, J. Ripoll, D. Yessayan, E. Graves, A. Bogdanov Jr., L. Josephson and R. Weissleder, *Proc. Natl. Acad. Sci. U.S.A.*, 2004, **101**, 12294.
80. S. Srinivasan, B. W. Pogue, S. Jiang, H. Dehghani, C. Kogel, S. Soho, J. J. Gibson, T. D. Tosteson, S. P. Poplack and K. D. Paulsen, *Acad. Radiol.*, 2006, **13**, 195.
81. R. Choe, S. D. Konecky, A. Corlu, K. Lee, T. Durduran, D. R. Busch, S. Pathak, B. J. Czerniecki, J. Tchou, D. L. Fraker, A. Demichele, B.

Chance, S. R. Arridge, M. Schweiger, J. P. Culver, M. D. Schnall, M. E. Putt, M. A. Rosen and A. G. Yodh, *J. Biomed. Opt.*, 2009, **14**, 024020.

82. S. Srinivasan, B. W. Pogue, C. Carpenter, S. Jiang, W. A. Wells, S. P. Poplack, P. A. Kaufman and K. D. Paulsen, *Antioxid. Redox Signaling*, 2007, **9**, 1143.

83. X. Intes, J. Ripoll, Y. Chen, S. Nioka, A. G. Yodh and B. Chance, *Med. Phys.*, 2003, **30**, 1039.

84. G. Ku, X. D. Wang, X. Y. Xie, G. Stoica and L. H. V. Wang, *Appl. Opt.*, 2005, **44**, 770.

85. M. Cutolo, *Ann. N. Y. Acad. Sci.*, 1999, **876**, 32.

86. R. W. Kinne, R. Brauer, B. Stuhlmuller, E. Palombo-Kinne and G. R. Burmester, *Arthritis Res.*, 2000, **2**, 189.

87. R. J. Bischof, D. Zafiropoulos, J. A. Hamilton and I. K. Campbell, *Clin. Exp. Immunol.*, 2000, **119**, 361.

88. A. Hansch, O. Frey, D. Sauner, I. Hilger, M. Haas, A. Malich, R. Bräuer and W. A. Kaiser, *Arthritis Rheum.*, 2004, **50**, 961.

89. W. T. Chen, U. Mahmcod, R. Weissleder and C. H. Tung, *Arthritis Res. Ther.*, 2005, **7**, R310.

90. G. Murphy, V. Knäuper, S. Atkinson, G. Butler, W. English, M. Hutton, J. Stracke and I. Clark, *Arthritis Res.*, 2002, **4** (Suppl. 3), S39.

91. T. Hansen, P. K. Petrow, A. Gaumann, G. M. Keyszer, P. Eysel, A. Eckardt, R. Bräuer and J. A. Kriegsmann, *J. Rheumatol.*, 2000, **27**, 859.

92. B. Bresnihan, *J. Rheumatol.*, 1999, **26**, 717.

93. A. Wunder, C. H. Tung, U. Müller-Ladner, R. Weissleder and U. Mahmood, *Arthritis Rheum.*, 2004, **50**, 2459.

94. E. S. Izmailova, N. Paz, H. Alencar, M. Chun, L. Schopf, M. Hepperle, J. H. Lane, G. Harrimaa, Y. Xu, T. Ocain, R. Weissleder, U. Mahmood, A. M. Healy and B. Jaffee, *Arthritis Rheum.*, 2007, **56**, 117.

95. T. Fischer, B. Ebert, J. Voigt, R. Macdonald, U. Schneider, A. Thomas, B. Hamm and K. G. Hermann, *Acad. Radiol.*, 2010, **17**, 375.

96. T. Fischer, I. Gemeinhardt, S. Wagner, D. V. Stieglitz, J. Schnorr, K. G. Hermann, B. Ebert, D. Petzelt, R. Macdonald, K. Licha, M. Schirner, V. Krenn, T. Kamradt and M. Taupitz, *Acad. Radiol.*, 2006, **13**, 4.

97. M. J. Luetkemeier and J. A. Fattor, *Clin. Chem.*, 2001, **47**, 1843.

98. A. K. Scheel, M. Backhaus, A. D. Klose, B. Moa-Anderson, U. J. Netz, K. G. Hermann, J. Beuthan, G. A. Müller, G. R. Burmester and A. H. Hielscher, *Ann. Rheum. Dis.*, 2005, **64**, 239.

99. L. M. Ittner and J. Götz, *Nat. Rev. Neurosci.*, 2011, **12**, 65.

100. M. Hintersteiner, A. Enz, P. Frey, A. L. Jaton, W. Kinzy, R. Kneuer, U. Neumann, M. Rudin, M. Staufenbiel, M. Stoeckli, K. H. Wiederhold and H. U. Gremlich, *Nat. Biotechnol.*, 2005, **23**, 577.

101. D. Hyde, R. de Kleine. S. A. MacLaurin, E. Miller, D. H. Brooks, T. Krucker and V. Ntziachristos, *Neuroimage*, 2009, **44**, 1304.

102. T. Jetzfellner, A. Rosenthal, K. H. Englmeier, A. Dima, M. Á. Araque Caballero, D. Razansky, and V. Ntziachristos, *Appl. Phys. Lett.*, 2011, **98**, in press.

103. S. Hu, P. Yan, K. Maslov, J. M. Lee and L. V. Wang, *Opt. Lett.*, 2009, **34**, 3899.

104. V. Cortez-Retamozo, F. K., Swirski, P. Waterman, H. Yuan, J. L. Figueiredo, A. P. Newton, R. Upadhyay, C. Vinegoni, R. Kohler, J. Blois, A. Smith, M. Nahrendorf, L. Josephson, R. Weissleder and M. J. Pittet, *J. Clin. Invest.*, 2008, **118**, 4058.

105. N. Beckmann, C. Cannet, H. Karmouty-Quintana, B. Tigani, S. Zurbruegg, F.-X. Blé, Y. Crémillieux and A. Trifilieff, *Eur. J. Radiol.*, 2007, **64**, 381.

106. F.-X. Blé, C. Cannet, S. Zurbruegg, H. Karmouty-Quintana, N. Frossard, A. Trifilieff and N. Beckmann, *Radiology*, 2008, **248**, 834.

107. F.-X. Blé, C. Cannet, S. Zurbruegg, C. Gérard, N. Frossard, N. Beckmann and A. Trifilieff, *Br. J. Pharmacol.*, 2009, **158**, 1295.

108. H. Karmouty-Quintana, C. Cannet, F.-X. Blé, S. Zurbruegg, J. R. Fozard, C. P. Page and N. Beckmann, *Br. J. Pharmacol.*, 2008, **154**, 1063.

109. L. E. Olsson, A. Smailagic, P. O. Onnervik and P. D. Hockings, *J. Magn. Reson. Imaging*, 2009, **29**, 977.

110. F.-X. Blé, P. Schmidt, R. Kneuer, C. Cannet, C. Gérard, H. Karmouty-Quintana, S. Zurbruegg, K. Coote, H. Danahay, H.-U. Gremlich and N. Beckmann, *Magn. Reson. Med.*, 2009, **62**, 1164.

111. D. E. Sosnovik, M. Nahrendorf, N. Deliolanis, M. Novikov, E. Aikawa, L. Josephson, A. Rosenzweig, R. Weissleder and V. Ntziachristos, *Circulation*, 2007, **115**, 1384.

CHAPTER 7.2

Fluorescence Lifetime Imaging applied to Microviscosity Mapping and Fluorescence Modification Studies in Cells

KLAUS SUHLING,[a] NICHOLAS I. CADE,[†a] JAMES A. LEVITT,[a] MARINA K. KUIMOVA,[b] PEI-HUA CHUNG,[a] GOKHAN YAHIOGLU,[b,c] GILBERT FRUHWIRTH,[d] TONY NG[d] AND DAVID RICHARDS[a]

[a] Department of Physics, King's College London, Strand, London WC2R 2LS, UK; [b] Department of Chemistry, Imperial College London, Exhibition Road, London SW7 2AZ, UK; [c] PhotoBiotics Ltd, 21 Wilson Street, London EC2M 2TD, UK; [d] Randall Division of Cell and Molecular Biophysics and Division of Cancer Studies, Guy's Medical School Campus, King's College London, London SE1 1UL, UK

7.2.1 Introduction

In 1989, the first articles were published describing a fluorescence imaging technique where the contrast in the image is provided by the fluorescence lifetime.[1,2] Since then, fluorescence lifetime imaging (FLIM) has emerged as a powerful technique both for studying the microenvironment of fluorescent dyes

† Present address: Microtubule Cytoskeleton Laboratory, Cancer Research UK, London Research Institute, 44 Lincoln's Inn Fields, London WC2A 3LY, UK

RSC Drug Discovery Series No. 15
Biomedical Imaging: The Chemistry of Labels, Probes and Contrast Agents
Edited by Martin Braddock
© Royal Society of Chemistry 2012
Published by the Royal Society of Chemistry, www.rsc.org

and observing proteins and their interactions in living cells.[3,4] Measurements can be made *in situ,* thus allowing access to biological function within a true physiological context. The vast majority of FLIM applications to date have been in the biomedical and life sciences since FLIM, in common with other optical techniques, is non-destructive, minimally invasive and can be applied to living cells and tissues.[5] The most frequent use of FLIM is to detect Förster Resonance Energy Transfer (FRET) to identify protein interactions or conformational changes of proteins.[6-9] However, applications in diverse areas such as forensic science,[10] combustion research,[11] luminescence mapping in diamond,[12] microfluidic systems,[13,14] art conservation,[15] and lipid order problems in physical chemistry[16] have also been reported. Moreover, efforts are underway to use FLIM, possibly combined with endoscopy, for clinical diagnostics.[17,18]

The power of fluorescence lifetime imaging lies in the ability to remotely monitor the local environment of a molecular probe independent of the fluorescence intensity or local probe concentration.[3,4] The fluorescence lifetime provides an absolute measurement which, compared to fluorescence intensity-based imaging, is also less susceptible to artefacts arising from scattered light, photobleaching, non-uniform illumination of the sample, or light pathlength. The Stokes shift of the fluorescence with respect to the excitation light allows the latter to be entirely eliminated from the image, either by filters or other spectrally selective methods, so only the dyes or proteins of interest are observed. Quantum dots or other nanoparticles have also recently found favour in cell imaging applications due to their high fluorescence quantum yield, low photobleaching susceptibility and narrow, size-dependent emission spectra which can be excited with a single wavelength.[19-22] Fluorescence imaging techniques thus provide a high contrast and allow easy visualisation of quantum dots, dyes and proteins and their environment in cells.

The observation of fluorescence and the use of microscopy are many hundreds of years old (see Figure 7.2.1).[23] However, the understanding of these phenomena and the creation of an appropriate theoretical framework to quantitatively interpret and predict fluorescence and to design a microscope only occurred 100–150 years ago. In particular, over the last 10 or 20 years the field has advanced rapidly and enormously,[5] mainly due to the combination of lasers and beam scanning, powerful computers,[24] and also sensitive detectors and cameras,[25-28] and genetic engineering,[29]—the latter effort being recognised with the award of the Chemistry Nobel Prize in 2008.

Currently the sensitivity of fluorescence detection is at the single molecule level, and point-spread function engineering has allowed fluorescence imaging well below the spatial resolution limit given by classical optical diffraction. Techniques include stimulated emission depletion (STED), structured illumination, saturation microscopy and photoactivated localisation microscopy, as reviewed recently.[30,31] FLIM has been carried out from the UV to the visible and it is not surprising that fluorescence-based imaging is widely used in the biomedical sciences, and that this trend shows no sign of abating.

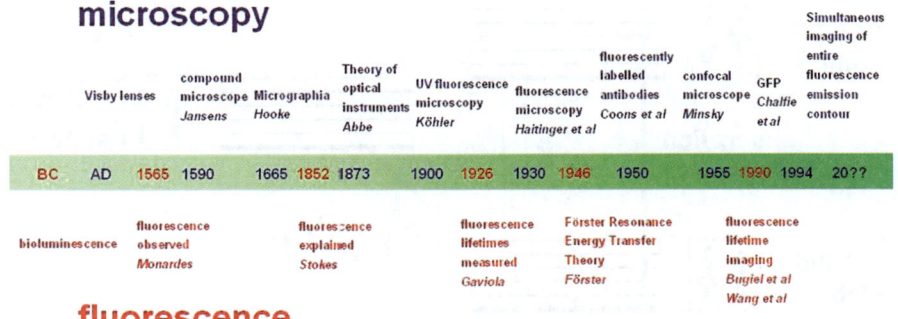

Figure 7.2.1 A Brief History of Fluorescence and Microscopy.

7.2.2 Theoretical Background of Fluorescence

Fluorescence is the radiative de-activation of the first electronically excited singlet state of a fluorescent molecule, or fluorophore.[32,33] It is a multi-parameter signal that can be characterised not only by its intensity and position, as in conventional intensity-based fluorescence imaging, but also by its wavelength, fluorescence lifetime and polarization.[34] Each of these parameters provides an additional spectroscopic dimension, which contains information about the biophysical environment of the fluorescence probe. Indeed, fluorescence characteristics such as wavelength, lifetime, and polarization, can now be imaged relatively easily,[35]—albeit not yet all of them together in a single measurement (see Figure 7.2.1).

A fluorophore can be excited into its first electronically excited singlet state S_1 *via* absorption of a photon (Figure 7.2.2). Other ways of reaching an excited state include heat (thermoluminescence), chemical reactions (chemi- or bioluminescence), an electric current (electroluminescence) or sound (sonoluminescence), thus fluorescence is sometimes also known as photoluminescence. From the excited state a fluorophore can return to its ground state S_0 either radiatively, by emitting a fluorescence photon, or non-radiatively, *e.g.* by dissipating the excited state energy as heat. This depends on the de-excitation pathways available. The fluorescence lifetime τ_f is the average time a fluorophore remains in S_1 after excitation, and is defined as the inverse of the sum of the rate parameters for all depopulation processes,[32,33]

$$\tau_f = \frac{1}{k_r + k_{nr}} \tag{1}$$

where k_r is the radiative rate constant, and the non-radiative rate constant k_{nr} is the sum of the rate constant for internal conversion, k_{ic}, and the rate constant for intersystem crossing to the triplet state, k_{isc}, so that $k_{nr} = k_{ic} + k_{isc}$ (see Figure 7.2.2).

The fluorescence quantum yield Φ_f is the ratio of the number of fluorescence photons emitted to the number of photons absorbed, regardless of their

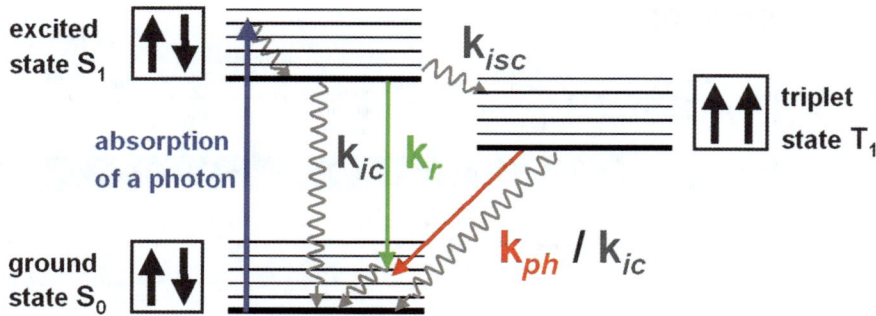

Figure 7.2.2   Schematic energy level diagram of a fluorescent molecule (Jablonski diagram). It depicts the ground state S_0, the first electronically excited singlet state S_1 (both with anti-parallel electron spins), the triplet state T_1 (parallel electron spins), and the transitions between them. The thin lines represent vibrational energy levels. Non-radiative relaxation to S_0 can occur *via* internal conversion k_{ic} and intersystem crossing (from singlet to triplet) k_{isc}. k_r is the radiative rate constant and k_{ph} is the rate constant for phosphorescence.

energy. The fluorescence lifetime is related to the fluorescence quantum yield according to

$$\Phi_f = \frac{k_r}{k_r + k_{nr}} = k_r \tau_f \tag{2}$$

with $0 < \Phi_f < 1$.

Spectrally resolved fluorescence lifetime imaging records the lifetime and spectrum of the fluorescence in each pixel of an image.[36–40] Such multi-dimensional imaging allows for additional contrast in the image and identification of multiple species. Most of these systems work with beam scanning; however, imaging dyes and autofluorescence in cells has been achieved using stage-scanning for multiphoton excitation, and using a streak camera based system with a microlens array for simultaneous acquisition of fluorescence decay and spectrum in 9 or 16 pixels.[41,42]

Time-Resolved Fluorescence Anisotropy

When fluorescence is characterised by polarization-resolved fluorescence life-time measurements, the Brownian rotational mobility of fluorophores can be determined.[43,44] This depends on the viscosity of their surroundings—thus polarization-resolved FLIM, or time-resolved fluorescence anisotropy imaging TR-FAIM allows mapping of viscosity by a fluorescence-based optical method.[45–51] After excitation with linearly polarized light, rotational diffusion of the fluorophore in its excited state results in a depolarization of the fluorescence emission. TR-FAIM measures fluorescence decays at polarizations parallel and perpendicular to that of the exciting light. The time-resolved fluorescence anisotropy $r(t)$ can be defined as,[32,33]

$$r(t) = \frac{I_{\parallel}(t) - GI_{\perp}(t)}{I_{\parallel}(t) + 2GI_{\perp}(t)} \tag{3}$$

where $I_{\parallel}(t)$ and $I_{\perp}(t)$ are the fluorescence intensity decays parallel and perpendicular to the polarization of the exciting light. G accounts for different transmission and detection efficiencies of the imaging system at parallel and perpendicular polarization, and, if necessary, an appropriate background has to be subtracted.[46] For a spherical molecule, $r(t)$ decays as a single exponential and is related to the rotational correlation time θ according to

$$r(t) = (r_0 - r_{\infty})e^{-t/\theta} + r_{\infty} \tag{4}$$

where r_0 is the initial anisotropy and r_{∞} is the limiting anisotropy which accounts for a restricted rotational mobility. For a spherical molecule in an isotropic medium, θ is directly proportional to the viscosity η of the solvent and the volume V of the rotating molecule

$$\theta = \frac{\eta V}{kT} \tag{5}$$

where k is the Boltzmann constant and T the absolute temperature. Thus imaging θ with TR-FAIM can map the rotational mobility of a fluorophore in its environment.[45–51]

Fluorescent Molecular Rotors

Fluorescent molecular rotors are distinctive fluorescent molecules whose fluorescence lifetimes are a function of the viscosity of their microenvironment.[52] Their radiative de-excitation pathway (see Figure 7.2.2) competes with intramolecular twisting, which leads to non-radiative deactivation of the excited state.[53,54] The rate constant for the latter pathway decreases in viscous media, such that the fluorescence lifetime τ_f and quantum yield Φ_f are high in viscous microenvironments and low in non-viscous microenvironments. In contrast to TR-FAIM, no polarization-resolved detection of the fluorescence emission is required—in fact, no polarized excitation is required either. However, polarization-resolved measurements of fluorescent molecular rotors have recently been shown to have an extended dynamic range over which the viscosity can be measured.[55] The fluorescence lifetime can directly be converted into a viscosity value,[56–58] using a calibration based on the Förster Hoffmann model.[59] This model states that the fluorescence lifetime, τ_f, of molecular rotors is a function of the viscosity η of their environment and can be described well by

$$\tau_f = \frac{z}{k_r}\eta^{\alpha} \tag{6}$$

where k_r is the radiative rate constant and z and α are constants.[60]

Metal-Induced Fluorescence Lifetime Modifications

In the presence of a metal, the excited-state molecular dipole can couple with surface plasmon electrons in the metal creating additional radiative k'_r and non-radiative k'_{nr} decay channels.[61-64] The total fluorescence lifetime then becomes

$$\tau'_f = \frac{1}{k'_r + k_r + k'_{nr} + k_{nr}} \tag{7}$$

with the corresponding metal enhanced quantum yield

$$\Phi'_f = \frac{k'_r + k_r}{k'_r + k_r + k'_{nr} + k_{nr}} = (k'_r + k_r)\tau'_f \tag{8}$$

These additional decay channels are strongly dependent on the separation between the emitting fluorophore and the metal;[65-68] hence, eqn (7) and eqn (8) predict that as k'_r increases near a metal surface, the fluorescence quantum yield increases while the fluorescence lifetime decreases. Within ~ 5–10 nm of the metal, the additional non-radiative channel k'_{nr} dominates leading to a strong quenching of the fluorescence.

Fluorescence Decay Analysis

In an ideal case, the decay of the fluorescence intensity follows an exponential decay law

$$I(t) = I_0 e^{-t/\tau_f} + B \tag{9}$$

where t is the time, I_0 represents the fluorescence intensity at $t = 0$, and B is uniform background noise. There are a variety of decay fitting models, and the resulting values for the fitting parameters have to be interpreted in the context of the experimental situation.

A bi-exponential fluorescence decay model has the form

$$I(t) = B + A_1 e^{-t/\tau_1} + A_2 e^{-t/\tau_2} \tag{10}$$

where $I(t)$ is the fluorescence intensity at time t, B is the background, A_1 and A_2 the pre-exponential factors (amplitudes) and τ_1 and τ_2 are the fluorescence lifetimes. For $A_2 = 0$, a single exponential decay is obtained which is an appropriate model for a single fluorophore with a single emitting state. The lifetime may vary with changes in the environment: typically k_{nr} varies, which consequently changes τ_f. The fitting works by choosing a mathematical decay model, and then minimizing the square of the distance between each data point and a point on the curve representing the fitting model for all data points in the decay. The goodness of fit is judged by the distribution of real data points above and below the curve of best fit, which should be random, and the normalized chi squared, which for Poisson noise data should be around 1.[69,70]

A bi-exponential fluorescence decay would be an appropriate model for two different fluorophores with a single emitting state each, or a single fluorophore in two distinct environments. In the latter case, the pre-exponential factors would be a measure of the relative concentration of the fluorophore in each environment.

FLIM Instrumentation

Typically, FLIM measurements are made using either single photon or multiphoton excitation from laser sources, often Ti:Sapphire lasers or diode lasers, or low-cost light-emitting diodes (LEDs). The recent availability of compact broadband supercontinuum sources allows excitation across the wavelength range of the visible region.[71,72] The signal can be detected with a photomultiplier, single photon avalanche diode (SPAD),[25–28] or novel hybrid detectors in scanning systems,[73,74] and with a gated or modulated intensified camera in wide-field imaging systems. While wide-field systems are fast, time-correlated single photon counting (TCSPC)-based FLIM with scanning systems has an excellent signal-to-noise ratio,[75] provides intrinsic optical sectioning, and is easily combined with spectral detection. Recent technical advances have allowed for more rapid acquisition in multiphoton FLIM by the use of multiple scanning beams,[76–78] and spinning (Nipkow) discs.[79,80] The reduction in acquisition times may provide the opportunity to study dynamic processes in cells and is an important step towards high content/throughput screening for specific protein–protein interactions for drug discovery,[81] and for real-time clinical applications.[17,18]

Biological Motivation—Diffusion Studies

Diffusion is often an important rate-determining step in chemical reactions or biological processes, and viscosity is one of the key parameters affecting diffusion of molecules and proteins. In biological specimens, changes in viscosity have been linked to disease and malfunction at the cellular level, and signalling pathways along with protein-protein interactions are dependent on the transport of biomolecules in cells.[4,6–9] Elucidation of intracellular reaction kinetics and mechanisms can thus potentially assist in development and understanding of the mechanisms of targeted therapies for cancer.[82]

While methods to measure the bulk viscosity are well developed, macroscopic sample quantities are required, and mechanical or fluid dynamics approaches are used.[83] However, imaging the microviscosity, for example in single cells, remains a challenge. Indeed, viscosity maps of single cells have until recently been hard to obtain.[84,85]

Among the techniques for measuring diffusion and viscosity based on fluorescence microscopy are single molecule tracking,[86] fluorescence recovery after photobleaching, fluorescence correlation spectroscopy, and raster imaging correlation spectroscopy, which can probe the translational diffusion of

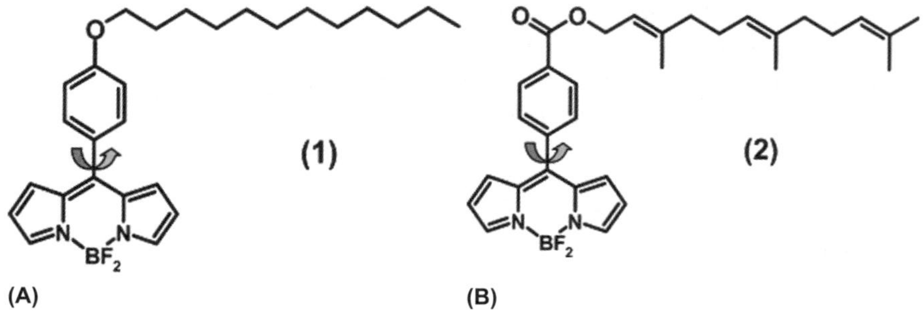

Figure 7.2.3   *Meso*-substituted 4,4'-difluoro-4-bora-3a,4adiaza-*s*-indacene (BODIPY)
fluorescent molecular rotors. Long, hydrophobic tails were designed to
render them membrane-soluble. (A) alkyl chain (B) farnesyl chain.

labelled molecules. The diffusion rate can also be estimated by measuring diffusion-limited processes in cells; for example, formation and decay of cytotoxic singlet molecular oxygen, and viscosity-dependent singlet oxygen quenching reactions within cells. Furthermore, rotational diffusion can be measured and mapped *via* TR-FAIM when combined with imaging.[45-51] Another technique for measuring microviscosity, particularly that of a biological environment, is fluorescence imaging of fluorescent molecular rotors.[52-54,87]

Fluorescent molecular rotors, in which the non-radiative decay of the excited state is influenced by the viscosity of the medium, have recently been highlighted as promising candidates for measurements of local viscosity using the change of fluorescence quantum yield.[52] Indeed, they have been used to measure the microviscosity in polymers, sol–gels, micelles, liposomes and biological structures such as tubulin. However, the main problem with fluorescence intensity-based measurements is distinguishing between viscosity and other factors which affect the fluorescence intensity, in particular the concentration of the fluorophore. A ratiometric approach, using probes that incorporate two linked independent fluorophores, has been suggested to address this problem. In this approach, the fluorescence intensity of one of the two fluorophores is not affected by viscosity and is used to determine the concentration while the other acts as a fluorescent molecular rotor.[88,89] In this way, a calibration of the viscosity in fluorescence intensity measurements is possible and has been demonstrated.[88-90] This is feasible if the rotors are distributed in a homogeneous environment, where the fluorescence decay is monoexponential. However, intensity-based measurements are tricky if the rotors are distributed in a heterogeneous environment which leads to multi-exponential decays. This is because different viscosity and population distributions can result in the same fluorescence intensity, leading to ambiguity.

Alternatively, one can exploit the fluorescence lifetime of fluorescent molecular rotors, which changes along with fluorescence quantum yield when the non-radiative decay contribution changes as a function of viscosity, according to eqn (6). This approach does not require conjugation of the molecular rotor to another fluorescence label, decouples the influence of the viscosity on the

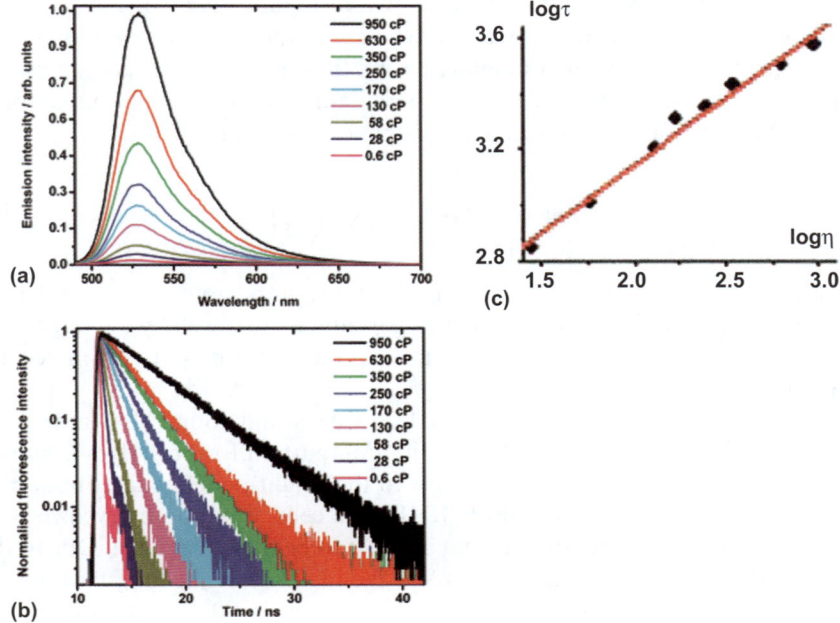

Figure 7.2.4 (a) Fluorescence intensity and (b) fluorescence lifetime of the BODIPY-based molecular rotors as a function of viscosity of the medium. (c) Calibration graph for BODIPY-based molecular rotors showing the logarithm of the fluorescence lifetime versus the logarithm of the viscosity of the medium. This graph allows the conversion of fluorescence lifetime into viscosity according to eqn (6).

fluorescence intensity from that of the probe concentration, and also allows detection of heterogeneous rotor environments *via* multi-exponential fluorescence decays.

Moreover, in combination with FLIM, viscosity maps can be obtained.[56,57] FLIM combines the benefits of measuring the fluorescence lifetime, which is highly sensitive to environmental parameters and independent of fluorophore concentration, with the high spatial and temporal resolution afforded by fluorescence microscopy with ultrafast excitation sources and photon counting detection.[3,4,90]

We recently synthesised a *meso*-substituted 4,4′-difluoro-4-bora-3a,4adiaza-s-indacene (BODIPY, see Figure 7.2.3), demonstrated that it acts as a fluorescent molecular rotor, and showed the feasibility of using it in combination with FLIM to map the microviscosity in living cells.[56,57] Long, hydrophobic tails were designed to render our probes membrane-soluble and, hence, report on the membrane and other hydrophobic domains in living cells. These dyes are well suited for use as probes of biological environments (*e.g.* cells) because their excitation and emission wavelengths are in the visible region.

Fluorescence measurements of the fluorescent molecular rotors made in methanol–glycerol mixtures of different viscosities, show that the fluorescence quantum yield increases dramatically with increasing solvent viscosity, as shown in Figure 7.2.4 (a). The fluorescence lifetime also increases with viscosity, from

0.7 ns in 20 cP to 3.8 ns in 950 cP, as shown in Figure 7.2.4 (b). The observed increase in fluorescence intensity is consistent with the restricted rotation of the phenyl group in the medium of high viscosity, thus preventing relaxation *via* the populating of the dark excited state. The data obtained can be used for a calibration graph according to the logarithmic version of eqn (6),

$$\log \tau_f = \alpha \log \eta + \log \left(\frac{z}{k_r} \right) \tag{11}$$

allowing the conversion of fluorescence lifetime into viscosity, as shown in Figure 7.2.4 (c). A plot of log τ_f versus log η can be fitted well by a straight line with a slope of around 0.5, in agreement with the literature data for molecular rotors in viscous media.

We incubated the BODIPYs in SK-OV-3 cells and found that they were readily taken up. The intracellular distribution pattern for both dyes is shown in Figure 7.2.5 (a) and (c).[56] In addition to the bright punctate distribution of the alkyl chain and farnesyl chain BODIPYs in cells, we also observed regions of lower fluorescence intensity in what appears to be the cell cytosol. Due to the hydrophobic nature of the BODIPYs and the presence of the long tails, which render them membrane-soluble, we expect both BODIPYs to target the membrane domains of intracellular organelles.

FLIM of SKOV-3 cells incubated with both meso-substituted dyes are shown in Figure 7.2.5 (b, d).[56] The FLIM images were obtained using excitation with a pulsed diode laser (467 nm, 20 MHz repetition rate) and monitoring the fluorescence through a bandpass filter (525 ± 25 nm). For both BODIPYs, the fluorescence decays in every pixel of the image can be adequately fitted using a single exponential decay model, eqn (9). By plotting the fluorescence lifetimes extracted from every pixel, we obtain a lifetime histogram, Figure 7.2.5 (e). The histograms are asymmetric for both dyes, although pixel fits are mono-exponential. This is consistent with a bimodal distribution of fluorescence lifetimes across the image. The individual contributions to the histogram (best fit using Gaussian distributions) are at 1.7 and 2.0 ns for the BODIPY with the alkyl chain and at 1.0 and 1.3 ns for the BODIPY with the farnesyl chain. A display of the results by applying a discrete color scale with short (1350–1850 ps, yellow) and long (1850–2200 ps, blue) lifetimes for the BODIPY with the alkyl chain and short (800–1250 ps, yellow) and long (1250–1800 ps, blue) lifetimes for the BODIPY with the farnesyl chain is shown Figure 7.2.5 (b, d).[56]

From these images, it is clear that the measured short and long fluorescence lifetimes are organelle specific; that is, the short lifetimes are predominantly situated in the brighter puncta, and the longer lifetimes are found within what appears to be the cytosol. All the values for the fluorescence lifetimes detected for both the alkyl and the farnesyl BODIPY from cells, lie within the calibrated range of viscosities and, importantly, are within the regime of the good linear fit to the data for the calibration measurements, Figure 7.2.4 (c). According to the calibration curves for the alkyl and the farnesyl BODIPY, the shorter

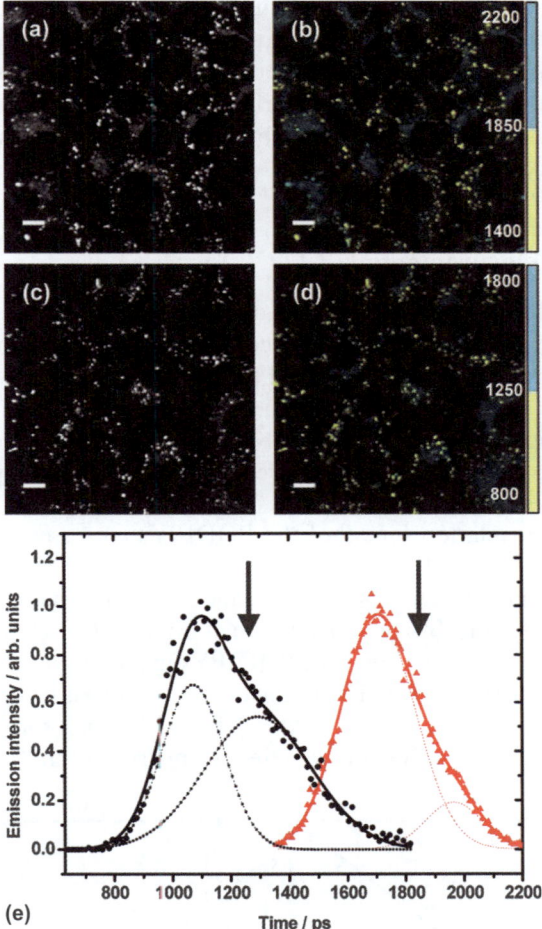

Figure 7.2.5 BODIPY-based molecular rotors in cells. Fluorescence intensity (a, c) and FLIM images (b, d) for live SK-OV-3 human ovarian carcinoma cells incubated with alkyl chain (a, b) and farnesyl chain BODIPY (c, d). The fluorescence intensity images show a punctate and continuous distribution of the rotors. The fluorescence lifetime images show a short lifetime for the punctate distribution (yellow, 1.4–1.85 ns), and a longer lifetime for the continuous distribution (blue, 1.85–2.2 ns). The FLIM images show the differences in fluorescence lifetime for the two dyes, recorded in the different regions of a cell. The discrete colour scale shows shorter lifetimes in yellow and longer lifetimes in blue. (b) for the alkyl chain BODIPY: 1350–1850 ps in yellow and 1850–2200 ps in blue. (d) for the farnesyl chain BODIPY: 800–1250 ps in yellow and 1250–1800 ps in blue. The scale bars are 10 µm. (e) Fluorescence lifetime distribution histograms from FLIM measurements for the alkyl chain BODIPY (filled red triangles) and the farnesyl chain BODIPY (filled black circles) in intracellular environments. The asymmetric distributions can be adequately described and fitted (solid black and red lines) by two contributions (dotted black and red lines) corresponding to a bimodal distribution of the fluorescence lifetimes. The arrows correspond to the cutoff in the discrete color scale bars in (b) and (d) for both BODIPYs with "short" and "long" fluorescence lifetimes arising from the punctate and cytoplasmic regions. respectively.

fluorescence lifetimes correspond to a viscosity value of 160 cP, whereas the longer fluorescence lifetimes correspond to a viscosity of 260 cP. The dashed lines in Figure 7.2.5 (e) serve as indicators for the eye to demonstrate where the intracellular fluorescence lifetime contributions lie on the calibration curves and also demonstrate that the values obtained using both dyes do correspond to the same viscosity values.

Anisotropy Measurements

To ensure that this high viscosity value does not result from the binding of the rotor to the intracellular targets, which could restrict the rotation of the phenyl group, we also performed time-resolved fluorescence anisotropy measurements of the BODIPYs in cells. The rotational diffusion rate can be determined by TR-FAIM using polarization-resolved TCSPC.[35] Time-resolved fluorescence anisotropy decays for both rotors were recorded in methanol–glycerol solutions with viscosities ranging from 28 to 950 cP. The data are plotted in Figure 7.2.6. The rotational correlation time θ of the BODIPYs increases linearly with solvent viscosity.

The slope of this plot allows us to estimate the dimensions of the rotating unit, according to the Stokes-Einstein relation eqn (5). Plots of rotational correlation time *vs.* viscosity for both BODIPYs in methanol–glycerol solutions are shown in Figure 7.2.6 (a). The data can be fitted well using eqn (4) with a zero value for r_∞. The calibration graph for rotational correlation time *vs.* viscosity can be used to calculate the effective microviscosity in cells.

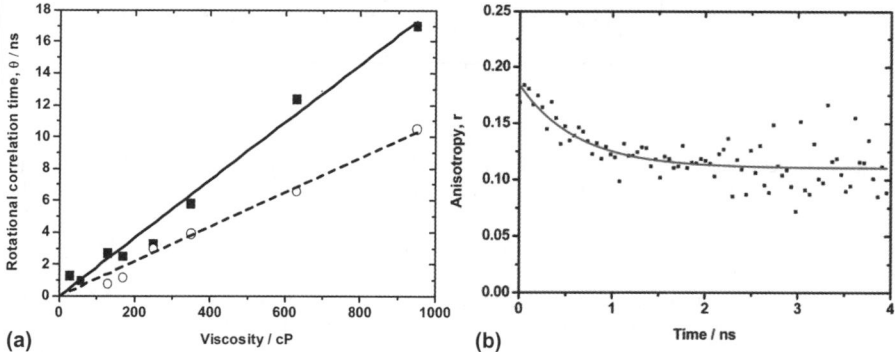

Figure 7.2.6   (a) Plots of rotational correlation time *vs.* viscosity for the BODIPY with the alkyl chain (■) and the BODIPY with the farnesyl chain (○) in methanol–glycerol solutions. The plots show linear fits to the data for both alkyl chain (solid line, —) and farnesyl chain (dashed line, ---) BODIPY, confirming the linear dependence of θ on η. Values of θ at a given η were found by fitting time-resolved anisotropy decays using eqn (4). (b) Representative intracellular time-resolved fluorescence anisotropy decay from SK-OV-3 cells. A fit according to eqn (4) yields a rotational correlation time of 590 ± 110 ps, which corresponds to a microviscosity of 60 cP, in the same order of magnitude of that given by FLIM.

A representative time-resolved fluorescence anisotropy decay for the farnesyl chain BODIPY in SK-OV-3 cells is shown in Figure 7.2.6 (b). Regions of interest in the stained areas of the SK-OV-3 cells were scanned, and fluorescence lifetime measurements were recorded simultaneously for fluorescence polarized parallel and perpendicularly to the polarization of the excitation beam. From these data, the fluorescence anisotropy decay was extracted and fitted with a single exponential decay model according to eqn (4). The measured rotational correlation time was 590 ± 110 ps, which corresponds to a viscosity of 60 cP. This value, along with the value found from the fluorescence lifetime data, is significantly higher than that expected of the aqueous cytoplasmic region.[84] However, they agree well with the values obtained by FLIM.

We note here that we also used a new type of ratiometric fluorescent molecular rotor based on porphyrin dyes.[91,92] It is suitable for imaging intracellular viscosity in live cells, but with the added benefit of inducing cell death by photosensitised singlet oxygen.[82] We demonstrated that the cellular viscosity increases upon cell death, and that such a viscosity increase indeed alters diffusion-dependent kinetics in a cell, illustrated through changes in the photosensitized production and subsequent decay of the cytotoxic species singlet oxygen.[82]

Metal-Modified FLIM for Increased Axial Specificity

In biological studies, high-resolution confocal images are often desirable to give the greatest phenomenological information with a high signal-noise. Although the lateral resolution is generally sufficient to discriminate microscopic features in the cell body, conventional confocal imaging has an *axial* resolution of ~ 500 nm at best; this greatly limits the effectiveness of this technique in detecting subtle protein dynamics in the cellular membrane. Alternative techniques such as total internal reflection microscopy (TIRF), scanning near-field optical microscopy (SNOM)[93] and 4Pi microscopy,[30,31] can give a much greater axial resolution; however, these can be difficult and expensive to implement, and are not in common use.

As discussed previously, the lifetime of fluorescent molecules can be strongly modified by the presence of a metallic structure. This has led to the development of a wide range of biomedical applications,[94] and significant improvements in imaging techniques.[95,96] To quantify this fluorescence lifetime change we have used a calibration system that closely mimics a cell imaging geometry (Figure 7.2.7 (a)); FLIM images were acquired of a microsphere surface-labelled with fluorescein isothiocyanate (FITC) on a 30 nm Au film (Figure 7.2.7 (c)). By analysing the radial lifetime profile and comparing it to the sphere's geometrical profile (Figure 7.2.7 (d)), we find that the FITC lifetime decreases approximately exponentially < 100 nm from the surface of the Au film, but there is no detectable fluorescence lifetime change above a glass surface.

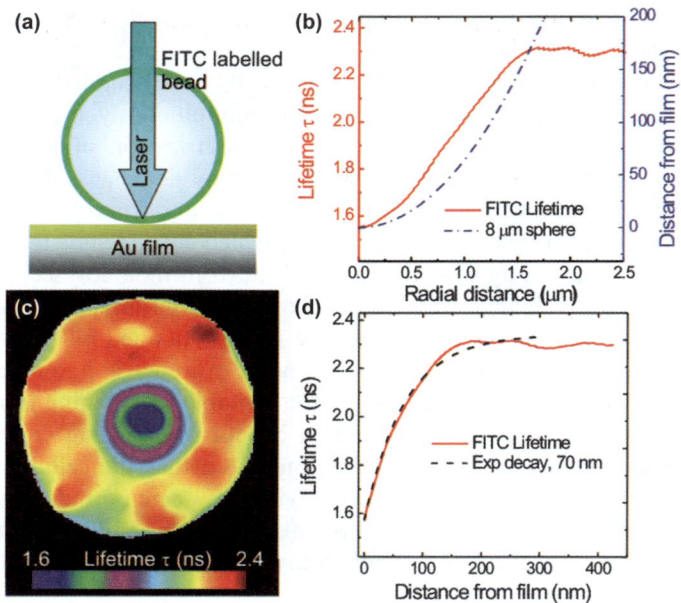

Figure 7.2.7   Distance-dependent Lifetime Calibration Using an 8 μm Diameter FITC Labelled Latex Microsphere on a 30 nm Au Film. (a) Schematic of the experimental geometry: lifetimes were measured in the focal plane of the Au film by imaging through the sphere; 'r' is the radial distance from the point of contact, and 'd' the vertical distance above the film. (b) 10×10 μm FLIM image calculated from a bi-exponential fit with a fixed 170 ps component. (c) Radial lifetime profile obtained from (b, left axis (red)) by averaging 10 line profiles (b, right axis (blue)) Geometrical cross-section of an 8 μm sphere. (d) FITC lifetime *vs*. distance from Au film, calculated from (c). The average unmodified lifetime of conjugated FITC is 2.3 ns and the lifetime decays approximately exponentially near to the film with a constant of 70 nm.

To investigate the increased axial specificity resulting from these metal-induced lifetime modifications, cells expressing a plasma-membrane receptor fused to enhanced Green Fluorescent Protein (eGFP) were grown on a 30 nm Au film. Figure 7.2.8 shows fluorescence intensity and lifetime images of a mammary adenocarcinoma cell; the confocal image shown in Figure 7.2.8 (a) demonstrates the difficulty in identifying the cell morphology without using optical sectioning (Figure 7.2.8 (c)). Even then, spatial information within the confocal volume will be lost. However, the FLIM images in Figure 7.2.8 (b) and (d) show that the eGFP lifetime is significantly reduced in the bottom membrane but is unmodified in parts of the cell further above the Au film, as seen in Figure 7.2.7 (d). We have utilized the sensitivity of this lifetime change to measure small variations in distance in a cellular assay measuring protein redistribution during receptor-mediated endocytosis;[97] this technique can potentially be integrated easily into a high-content screening platform.

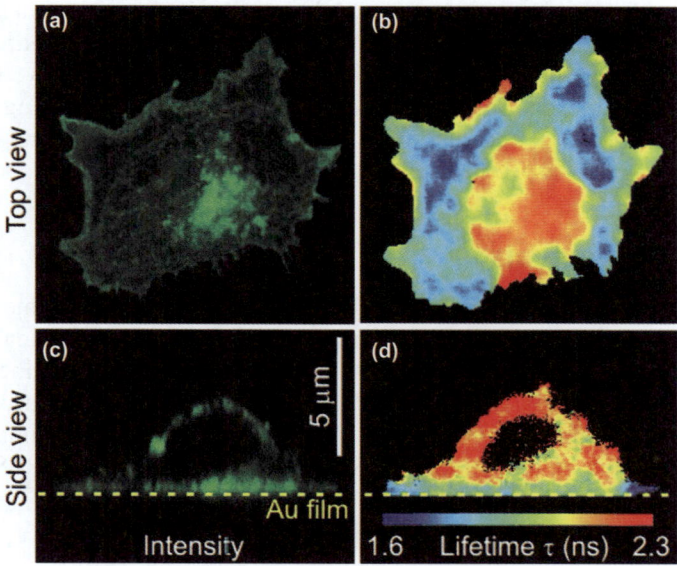

Figure 7.2.8 (a), (c) Fluorescence intensity and (b), (d) FLIM images of a human carcinoma cell expressing a plasma-membrane receptor fused to eGFP on an Au film, imaging from the top and side, respectively. A clear reduction in lifetime occurs only in the bottom membrane, giving a much greater axial specificity than intensity based confocal imaging.

Conclusion

FLIM offers some key advantages over intensity-based measurements of protein interaction and in the measurement of specific analytes, and continues to gain favour as a powerful imaging technique for biological sciences. The information provided by FLIM can give additional degrees of contrast not attainable in conventional microscopy.

The BODIPY fluorescent molecular rotors can report microviscosity *via* variations in their fluorescence lifetimes. Their fluorescence properties and high cellular uptake make them ideal candidates for studies in biological systems. Our measurements using fluorescent molecular rotors confirm the heterogeneity in the viscosity in the stained regions. This result highlights the importance of spatially resolved microscopic scale measurements in biological environments. By tailoring the chemistry and delivery method of the fluorescent molecular rotors, it may be possible to create microviscosity maps on the wide range of intracellular environments and targets on the basis of fluorescence lifetimes measured by FLIM, or ratiometric spectral imaging. In summary, we have developed a practical and versatile approach to measuring the microviscosity of the environment of fluorescent molecular rotors in cells.

Furthermore, by utilising mechanisms that selectively modify the fluorescence lifetime, such as those arising from the presence of a metal surface, much greater specificity can be achieved in measurements. This has important implications for the development of novel biotechnologies such as high-content screening and biosensors. There is a long way to go before we are close to saturating the capabilities of fluorescence imaging for cell biology. Technological advances in excitation sources, detectors,[98–100] data processing and analysis will ensure that it becomes more accessible for researchers, and certain to reveal exciting new aspects in cell biology.

Acknowledgements

We would like to thank our colleagues Dr Fred Festy and Dr Simon Ameer-Beg in the Randall Division of Cell and Molecular Biophysics, King's College London, for stimulating discussions. M.K.K. thanks the UK's Engineering and Physical Science Research Council (EPSRC) Life Sciences Interface program for a personal Fellowship. We would also like to acknowledge funding by the UK's Biotechnology and Biological Sciences Research Council (BBSRC) and the Medical Research Council (MRC).

References

1. I. Bugiel, K. König and H. Wabnitz, *Lasers Life Sci.*, 1989, **3**, 47–53.
2. X. F. Wang, T. Uchida and S. Minami, *Appl. Spectrosc.*, 1989, 840–845.
3. K. Suhling, P. M. W. French and D. Phillips, *Photochem. Photobiol.*, 2005, **4**, 13–22.
4. F. Festy, S. M. Ameer-Beg, T. Ng and K. Suhling, *Molecular BioSyst.*, 2007, **3**, 381–391.
5. F. S. Wouters, *Contemp. Phys.*, 2006, **47**, 239–255.
6. E. A. Jares-Erijman and T. M. Jovin, *Nat. Biotechnol.*, 2003, **21**, 1387–1396.
7. H. Wallrabe and A. Periasamy, *Curr. Opin. Biotechnol.*, 2005, **16**, 19–27.
8. R. R. Duncan, *Biochem. Soc. Trans.*, 2006, **34**, 679–682.
9. M. Peter and S. M. Ameer-Beg, *Biol. Cell*, 2004, **96**, 231–236.
10. D. K. Bird, K. M. Agg, N. W. Barnett and T. A. Smith, *J. Microsc.*, 2007, **226**, 18–25.
11. T. Ni and L. A. Melton, *Appl. Spectrosc.*, 1996, **50**, 1112–1116.
12. G. Liaugaudas, A. T. Collins, K. Suhling, G. Davies and R. Heintzmann, *J. Phys.: Condens. Matter*, 2009, **21**, 364210–364217.
13. R. K. P. Benninger, O. Hofmann, J. McGinty, J. Requejo-Isidro, I. Munro, M. A. A. Neil, A. J. deMello and P. M. W. French, *Opt. Express*, 2005, **13**, 6275–6285.
14. A. D. Elder, S. M. Matthews, J. Swartling, K. Yunus, J. H. Frank, C. M. Brennan, A. C. Fisher and C. F. Kaminski, *Opt. Express*, 2006, **14**, 5456–5467.
15. D. Comelli, C. D'Andrea, G. Valentini, R. Cubeddu, C. Colombo and L. Toniolo, *Appl. Opt.*, 2004, **43**, 2175–2183.

16. D. M. Togashi, R. I. S. Romao, A. M. G. da Silva, A. J. F. N. Sobral and S. M. B. Costa, *Phys. Chem. Chem. Phys.*, 2005, **7**, 3875–3884.
17. J. Requejo-Isidro, J. McGinty, I. Munro, D. S. Elson, N. P. Galletly, M. J. Lever, M. A. A. Neil, G. W. H. Stamp, P. M. W. French and P. A. Kellett, *Opt. Lett.*, 2004, **29**, 2249–2251.
18. G. O. Fruhwirth, S. Ameer-Beg, R. Cook, T. Watson, T. Ng and F. Festy, *Opt. Express*, 2010, **18**, 11148–11158.
19. P. Howes, M. Green, J. Levitt, K. Suhling and M. Hughes, *J. Am. Chem. Soc.*, 2010, **132**, 3989–3996.
20. M. Green, *Ange. Chem*, 2004, **43**, 4129–4131.
21. H. E. Grecco, K. A. Lidke, R. Heintzmann, D. S. Lidke, C. Spagnuolo, O. E. Martinez, E. A. Jares-Erijman and T. M. Jovin, *Microsc. Res. Tech.*, 2004, **65**, 169–179.
22. U. Resch-Genger, M. Grabolle, S. Cavaliere-Jaricot, R. Nitschke and T. Nann, *Nat. Methods*, 2008, **5**, 763–775.
23. K. Suhling, in *Cell Imaging*, ed. D. Stephens, Scion Publishing Ltd., Bloxham, 2006, pp. 219–245.
24. W. B. Amos and J. G. White, *Biol. Cell*, 2003, **95**, 335–342.
25. X. Michalet, O. H. W. Siegmund, J. Vallerga, P. Jelinsky, J. E. Millaud and S. Weiss, *J. Mod. Opt.*, 2007, **54**, 239–281.
26. R. H. Hadfield, *Nat. Photonics*, 2009, **3**, 696–705.
27. G. S. Buller and R. J. Collins, *Meas. Sci. Technol.*, 2010, **21**, 012002.
28. G. Hungerford and D. J. S. Birch, *Meas. Sci. Technol.*, 1996, **7**, 121–135.
29. N. C. Shaner, G. H. Patterson and M. W. Davidson, *J. Cell Sci.*, 2007, **120**, 4247–4260.
30. L. M. Hirvonen and T. A. Smith, *Aust. J. Chem.*, 2011, **64**, 41–45.
31. R. Heintzmann and G. Ficz, *Briefings Funct. Genomics Proteomics*, 2006, **5**, 289–301.
32. B. Valeur, *Molecular Fluorescence*, Wiley-VCH, Weinheim, 2002.
33. J. R. Lakowicz, *Principles of Fluorescence Spectroscopy*, Springer, New York, 3rd edn, 2006.
34. M. Y. Berezin and S. Achilefu, *Chem. Rev.*, 2010, **110**, 2641–2684.
35. J. A. Levitt, D. R. Matthews, S. M. Ameer-Beg and K. Suhling, *Curr. Opin. Biotechnol.*, 2009, **20**, 28–36.
36. Q. S. Hanley, D. J. Arndt-Jovin and T. M. Jovin, *Appl. Spectrosc.*, 2002, **56**, 155–166.
37. D. K. Bird, K. W. Eliceiri, C. H. Fan and J. G. White, *Appl. Opt.*, 2004, **43**, 5173–5182.
38. C. Biskup, T. Zimmer, L. Kelbauskas, B. Hoffmann, N. Klocker, W. Becker, A. Bergmann and K. Benndorf, *Microsc. Res. Tech.*, 2007, **70**, 442–451.
39. A. Rück, C. Hülshoff, I. Kinzler, W. Becker and R. Steiner, *Microsc. Res. Tech.*, 2007, **70**, 485–492.
40. R. R. Duncan, A. Bergmann, M. A. Cousin, D. K. Apps and M. J. Shipston, *J. Microsc.*, 2004, **215**, 1–12.

41. L. Liu, J. Qu, Z. Lin, L. Wang, Z. Fu, B. Guo and H. Niu, *Appl. Phys. B: Lasers Opt.*, 2006, **84**, 379–383.

42. J. L. Qu, L. X. Liu, D. N. Chen, Z. Y. Lin, G. X. Xu, B. P. Guo and H. B. Niu, *Opt. Lett.*, 2006, **31**, 368–370.

43. D. M. Jameson and J. A. Ross, *Chem. Rev.*, 2010, **110**, 2685–2708.

44. C. C. Gradinaru, D. O. Marushchak, M. Samim and U. J. Krull, *Analyst*, 2010, **135**, 452–459.

45. J. Siegel, K. Suhling, S. Lévêque-Fort, S. E. D. Webb, D. M. Davis, D. Phillips, Y. Sabharwal and P. M. W. French, *Rev. Sci. Instrum.*, 2003, **74**, 182–192.

46. K. Suhling, J. Siegel, P. M. P. Lanigan, S. Lévêque-Fort, S. E. D. Webb, D. Phillips, D. M. Davis and P. M. W. French, *Opt. Lett.*, 2004, **29**, 584–586.

47. A. H. A. Clayton, Q. S. Hanley, D. J. Arndt-Jovin, V. Subramaniam and T. M. Jovin, *Biophys. J.*, 2002, **83**, 1631–1649.

48. D. S. Lidke, P. Nagy, B. G. Barisas, R. Heintzmann, J. N. Post, K. A. Lidke, A. H. A. Clayton, D. J. Arndt-Jovin and T. M. Jovin, *Biochem. Soc. Trans.*, 2003, **31**, 1020–1027.

49. A. N. Bader, E. G. Hofman, P. M. P. van Bergen en Henegouwen and H. C. Gerritsen, *Opt. Express*, 2007, **15**, 6934–6945.

50. A. N. Bader, E. G. Hofman, J. Voortman, P. M. P. van Bergen en Henegouwen and H. C. Gerritsen, *Biophys. J.*, 2009, **97**, 2613–2622.

51. A. N. Bader, S. Hoetzl, E. G. Hofman, J. Voortman, P. M. P. van Bergen en Henegouwen, G. van Meer and H. C. Gerritsen, *ChemPhysChem*, 2011, **12**, 475–483.

52. M. A. Haidekker and E. A. Theodorakis, *Org. Biomol. Chem.*, 2007, **5**, 1669–1678.

53. M. A. Haidekker and E. A. Theodorakis, *J. Biol. Eng.*, 2010, **4**, 11.

54. M. A. Haidekker, M. Nipper, A. Mustafic, D. Lichlyter, M. Dakanali and E. A. Theodorakis, in *Advanced Fluorescence Reporters in Chemistry and Biology I. Fundamentals and Molecular Design*, ed. A. P. Demchenko, Springer, Berlin Heidelberg, 2010, vol. 8, pp. 267–308.

55. J. A. Levitt, P. H. Chung, M. K. Kuimova, G. Yahioglu, Y. Wang, J. L. Qu and K. Suhling, *ChemPhysChem*, 2011, **12**, 662–672.

56. J. A. Levitt, M. K. Kuimova, G. Yahioglu, P. H. Chung, K. Suhling and D. Phillips, *J. Phys. Chem. C*, 2009, **113**, 11634–11642.

57. M. K. Kuimova, G. Yahioglu, J. A. Levitt and K. Suhling, *J. Am. Chem. Soc.*, 2008, **130**, 6672–6673.

58. G. Hungerford, A. Allison, D. McLoskey, M. K. Kuimova, G. Yahioglu and K. Suhling, *J. Phys. Chem. B*, 2009, **113**, 12067–12074.

59. T. Förster and G. Hoffmann, *Z. Physik. Chem. NF*, 1971, **75**, 63–76.

60. B. Wilhelmi, *Chem. Phys.*, 1982, **66**, 351–355.

61. W. L. Barnes, *J. Mod. Opt.*, 1998, **45**, 661–699.

62. J. R. Lakowicz, *Anal. Biochem.*, 2005, **337**, 171–194.

63. Y. X. Zhang, K. Aslan, M. J. R. Previte and C. D. Geddes, *Appl. Phys. Lett.*, 2007, **90**, 053107.

64. E. Fort and S. Gresillon, *J. Phys. D: Appl. Phys.*, 2008, **41**, 013001.

65. P. Anger, P. Bharadwaj and L. Novotny, *Phys. Rev. Lett.*, 2006, **96**, 113002.
66. S. Kühn, U. Håkanson, L. Rogobete and V. Sandoghdar, *Phys. Rev. Lett.*, 2006, **97**, 017402.
67. T. Ritman-Meer, N. I. Cade and D. Richards, *Appl. Phys. Lett.*, 2007, **91**, 123122.
68. N. I. Cade, T. Ritman-Meer, K. A. Kwakwa and D. Richards, *Nanotechnology*, 2009, **20**, 285201.
69. D. J. S. Birch and R. E. Imhof, in *Topics in Fluorescence Spectroscopy: Techniques*, ed. J. R. Lakowicz, Plenum Press, New York, 1991, vol. 1.
70. D. V. O'Connor and D. Phillips, *Time-correlated single-photon counting*, Academic Press, New York, 1984.
71. G. McConnell, *Opt. Express*, 2004, **12**, 2844–2850.
72. C. Dunsby, P. M. P. Lanigan, J. McGinty, D. S. Elson, J. Requejo-Isidro, I. Munro, N. Galletly, F. McCann, B. Treanor, B. Önfelt, D. M. Davis, M. A. A. Neil and P. M. W. French, *J. Phys. D: Appl. Phys.*, 2004, **37**, 3296–3303.
73. W. Becker, B. Su, O. Holub and K. Weisshart, *Micros. Res. Tech.*, 2011, DOI: 10.1002/jemt.20959.
74. X. Michalet, A. Cheng, J. Antelman, M. Suyama, K. Arisaka and S. Weiss, *Proc. SPIE 6862*, 2008, p. 68620F.
75. E. Gratton, S. Breusegem, J. Sutin, Q. Ruan and N. Barry, *J. Biomed. Opt.*, 2003, **8**, 381–390.
76. S. Kumar, C. Dunsby, P. A. A. De Beule, D. M. Owen, U. Anand, P. M. P. Lanigan, R. K. P. Benninger, D. M. Davis, M. A. A. Neil, P. Anand, C. Benham, A. Naylor and P. M. W. French, *Opt. Express*, 2007, **15**, 12548–12561.
77. S. Padilla-Parra, N. Auduge, M. Coppey-Moisan and M. Tramier, *Biophys. J.*, 2008, **95**, 2976–2988.
78. M. Straub and S. W. Hell, *Bioimaging*, 1998, **6**, 177–185.
79. E. B. van Munster, J. Goedhart, G. J. Kremers, E. M. M. Manders and T. W. J. Gadella, *Cytometry, Part A*, 2007, **71A**, 207–214.
80. D. M. Grant, J. McGinty, E. J. McGhee, T. D. Bunney, D. M. Owen, C. B. Talbot, W. Zhang, S. Kumar, I. Munro, P. M. P. Lanigan, G. T. Kennedy, C. Dunsby, A. I. Magee, P. Courtney, M. Katan, M. A. A. Neil and P. M. W. French, *Opt. Express*, 2007, **15**, 15656–15673.
81. A. Esposito, C. P. Dohm, M. Bahr and F. S. Wouters, *Mol. Cell. Proteomics*, 2007, **6**, 1446–1454.
82. M. K. Kuimova, S. W. Botchway, A. W. Parker, M. Balaz, H. A. Collins, H. L. Anderson, K. Suhling and P. R. Ogilby, *Nat. Chem.*, 2009, **1**, 69–73.
83. E. Kaliviotis and M. Yianneskis, *Proc. Inst. Mech. Eng. Part H: J. Eng. Med.*, 2007, **221**, 887–898.
84. K. Luby-Phelps, S. Mujumdar, R. Mujumdar, L. A. Ernst, W. Galbraith and A. S. Waggoner, *Biophys. J.*, 1993, **65**, 236–242.
85. J. A. Dix and A. S. Verkman, *Biophys. J.*, 1990, **57**, 231–240.
86. D. Wirtz, *Annu. Rev. Biophys.*, 2009, **38**, 301–326.

87. M. A. Haidekker, T. Ling, M. Anglo, H. Y. Stevens, J. A. Frangos and E. A. Theodorakis, *Chem. Biol.*, 2001, **8**, 123–131.
88. M. E. Nipper, M. Dakanali, E. Theodorakis and M. A. Haidekker, *Biochimie*, 2010, **93**, 988–994.
89. M. Haidekker, T. P. Brady, D. Lichlyter and E. A. Theodorakis, *J. Am. Chem. Soc.*, 2006, **128**, 398–399.
90. X. Peng, Z. Yang, J. Wang, J. Fan, Y. He, F. Song, B. Wang, S. Sun, J. Qu, J. Qi and M. Yan, *J. Am. Chem. Soc.*, 2011, **133**, 6626–6635.
91. M. K. Kuimova, H. Collins, M. Balaz, E. Dahlstedt, J. Levitt, N. Sergent, K. Suhling, M. Drobizhev, N. Makarov and A. Rebane, *Org. Biomol. Chem.*, 2009, **7**, 889–896.
92. K. P. Ghiggino, J. A. Hutchison, S. J. Langford, M. J. Latter, M. A. P. Lee, P. R. Lowenstern, C. Scholes, M. Takezaki and B. E. Wilman, *Adv. Funct. Mater*, 2007, **17**, 805–813.
93. S. Kawata, Y. Inouye and T. Ichimura, *Sci. Prog.*, 2004, **87**, 25–49.
94. J. R. Lakowicz, *Anal. Biochem.*, 2001, **298**, 1–24.
95. E. Le Moal, E. Fort, S. Lévêque-Fort, F. P. Cordelieres, M. P. Fontaine-Aupart and C. Ricolleau, *Biophys. J.*, 2007, **92**, 2150–2161.
96. E. Le Moal, S. Leveque-Fort, M. C. Potier and E. Fort, *Nanotechnology*, 2009, **20**, 225502.
97. N. I. Cade, G. Fruhwirth, S. J. Archibald, T. Ng and D. Richards, *Biophys. J.*, 2010, **98**, 2752–2757.
98. M. J. Stevens, R. H. Hadfield, R. E. Schwall, S. W. Nam, R. P. Mirin and J. A. Gupta, *Appl. Phys. Lett.*, 2006, **89**, 031109.
99. D. U. Li, J. Arlt, J. Richardson, R. Walker, A. Buts, D. Stoppa, E. Charbon and R. Henderson, *Opt. Express*, 2010, **18**, 10257–10269.
100. G. W. Fraser, J. S. Heslop-Harrison, T. Schwarzacher, A. D. Holland, P. Verhoeve and A. Peacock, *Rev. Sci. Instrum.*, 2003, **74**, 4140–4144.

CHAPTER 7.3

Design and Use of Contrast Agents for Ultrasound Imaging

FABIAN KIESSLING,*[a] GEORG SCHMITZ[b] AND
JESSICA GÄTJENS[a]

[a] Institute for Experimental Molecular Imaging, RWTH-Aachen University,
Medical Faculty, Aachen, Germany; [b] Department of Electrical Engineering
and Information Sciences, Bochum, Germany

7.3.1 Indications for Ultrasound Contrast Agents

In the clinical routine B-mode, ultrasound imaging is often used as the diagnostic imaging tool for the first look. Among others, its non-radioactive nature, high mobility and cost effectiveness contribute to the broad acceptance and may explain its preferential use for some indications as compared to CT, MRI and PET. Indications range from screening for bile stones, size measurements of organs (*e.g.* thyroid gland and prostate) and the exclusion of brain disorders in newborn children to the search for neoplastic, infectious and inflammatory lesions. The implementation of non-contrast enhanced Doppler methods extends the application field to cardiology and angiology. The Doppler effect describes the frequency shift of an acoustic (or optic) wave by its reflection and scattering from a moving element. If the object is moving toward the pulse wave emitter (in ultrasound this is the transducer), the frequency increases and if it moves away it decreases. In blood vessels, acoustic pulse waves are mainly scattered by blood cells, in particular from erythrocytes, which form the major blood cell fraction. The pixel-wise colour coding of these frequency shifts and

RSC Drug Discovery Series No. 15
Biomedical Imaging: The Chemistry of Labels, Probes and Contrast Agents
Edited by Martin Braddock
© Royal Society of Chemistry 2012
Published by the Royal Society of Chemistry, www.rsc.org

its overlay on the morphologic B-mode images enables to visualise the vessels by its flow. Also blood flow velocities can be assessed.

Whether a Doppler signal can be detected or not mainly depends on its intensity and on the magnitude of the frequency shift. The intensity depends on the scattering strength of the moving particles. The frequency shift depends on the velocity of the blood cells, on the ultrasound pulse centre frequency and on the angle of insonation. The higher the pulse frequency and the faster the blood flow is, the higher the Doppler shift will be. Therefore, the method favourably works to visualise and characterise large arteries and veins. However, microcirculation within tissues usually is not assessable (Figure 7.3.1).

Therefore, pathologies that can be diagnosed by changes in micro-vascularisation benefit from enhancers of the vascular signal that have high scattering strength. Unclear liver lesions are a prominent example: here contrast enhanced ultrasound clearly improves the differentiation between

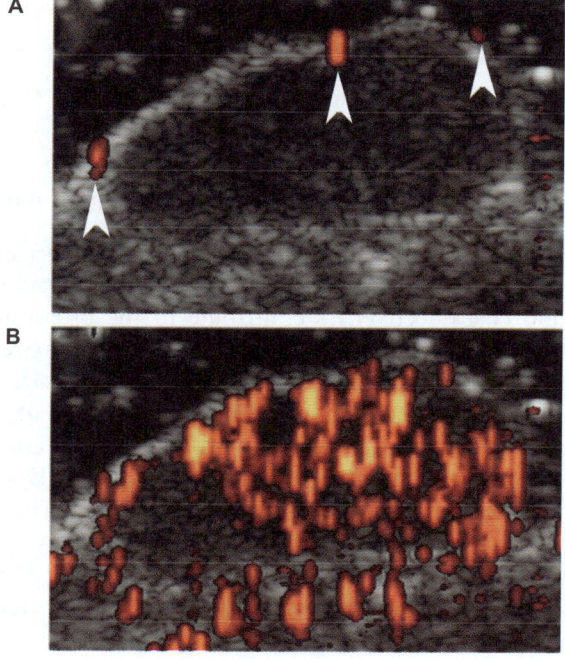

Figure 7.3.1   Doppler images of experimental skin cancers obtained with a clinical ultrasound device (7 MHz) before (A) and after (B) administration of microbubbles. Due to the low sensitivity for blood vessels with slow blood flow, the native Doppler image shows a mostly non vascularised tumour and only few vessels at the tumour periphery (arrowheads). In contrast, the microbubble enhanced image clearly indicates that the tumour is full of microvessels except for a necrotic area at the left side of the image.

focal nodular hyperplasia, hemangioma, metastases or hepatocellular carcinoma.[1] Also in other organs such as prostate, ovary and kidney the use of ultrasound contrast agents can help to identify the solid and vascularised areas that are suspicious for cancer. Contrast enhanced ultrasound is also an interesting option for the assessment of tumour response to therapy. However, it has to be mentioned that ultrasound in its present form is user dependent and has a lower reproducibility than other tomographic imaging modalities. Therefore, for whole body staging and for the monitoring of therapy response of multiple lesions CT, MRI and PET may be better suited. Nevertheless, when using the available non-invasive imaging modalities complementarily, ultrasound may play an essential future role for individualising therapy by indicating the early response of primary tumours and by controlling therapy efficacy in the interval between whole body staging examinations.

Besides the detection of tumour neovascularisation the characterisation of muscle tissue is a clear indication for contrast enhanced ultrasound. Weber and colleagues showed that contrast enhanced ultrasound favourably depicts the enhanced perfusion of peripheral muscle tissue in patients with dermatomyositis and polymyositis.[2] Contrast enhanced ultrasound is also used to characterise the heart muscle and improves the detection of ischaemic lesions.

A more established indication is its use in echocardiography. Here it was shown that by using ultrasound contrast agents the diagnosis of re-flow phenomena and stenosis of the coronary artery are diagnosed more reliably.

The characterisation of vesico-ureteral reflux is an important niche application for ultrasound contrast agents. Here the ultrasound contrast agent is injected into the cystic bladder and a pathologic reflux into the ureter is screened. In the latter case the risk of infectious renal disease and of the development of renal insufficiency are increased.

With the recent development of targeted ultrasound contrast agents a much more detailed characterisation of the vascularisation at molecular level became feasible.[3–7] Although targeted ultrasound contrast agents are not yet in the clinic their indications can be estimated from animal studies. One of them is the detection and characterisation of tumour tissue by its increased angiogenesis. This is particularly interesting in regard to prostate and breast tumours where ultrasound imaging already plays a substantial role in clinical diagnosis. In this context, the longer persisting contrast enhancement as compared with regular ultrasound contrast agents may also facilitate interventions (*e.g.* needle biopsy).

In cardiology-targeted ultrasound, contrast agents may be used to identify re-endothelialisation of implanted grafts and stents and to differentiate stable and vulnerable atherosclerotic plaques. Also the characterisation of the vascular remodelling in ischaemic tissues may be of interest. Preclinical examples that underline the assumptions of the authors about the future of molecular ultrasound imaging are reported in Section 7.3.4.

7.3.2 Microbubbles

The first contrast-enhanced ultrasound experiments were performed by Gramiak and Shah over 40 years ago. They used saline solutions that contained small air bubbles in the echocardiography of the aortic root.[8] Since these free gas bubbles are very short-lived, the search for methods that allow for the stabilisation of the gas–liquid interface was initiated soon thereafter. Nowadays, the most effective ultrasound contrast agents, termed microbubbles, consist of small stabilised gas- or air-filled spheres with different degrees of rigidity of the shell. Their diameters range from several hundred nanometres up to a few micrometres with various shell material and gas cores (*cf.* Table 7.3.1).

Hence, due to their size, microbubbles resemble red blood cells in their behaviour and remain strictly intravascular, circulating in the blood for several minutes.[3] Ultrasound contrast agents improve contrast by acting as highly efficient back-scatterers, *i.e.* they enhance the reflection of the ultrasound waves. They exhibit strong non-linear behaviour upon insonification at higher mechanical indexes, appearing as bright signals on the screen (see also Section 7.3.3.). The contrast-enhancing properties of microbubbles are defined by different physical parameters, of which size and compressibility are the ones that are crucial for their effectiveness and can be easily modified through synthesis. In general, these properties can be influenced through the choice of the shell material, meaning that the softer the shell the higher the elasticity, while more rigid materials provide a substantial gain in stability with a concomitant loss of scattering efficiency.[6,7] Commonly used shell materials consist of phospholipids, albumin, galactose, or polymers such as poly-cyanoacrylates, yielding microbubbles with a narrow size distribution and a shell thickness ranging from approximately 10 to 200 nm. The encapsulation of the gas core with an elastic shell leads to a prolonged blood half-life because of reduced gas diffusion into the blood.[7] Apart from the shell material, the gas core itself is of high

Table 7.3.1   Selected Ultrasound Contrast Agents, Listed According to their Shell Properties.

Name	Shell Material	Gas Core
soft shell microbubbles		
BR14[®]	phospholipid	perfluorocarbon
Definity[®]	phospholipid	perfluorocarbon
EchoGen[®]	surfactant	perfluorocarbon
Echovist[®]	sugar matrix	air
Levovist[®]	fatty acid	air
Sonazoid[TM]	phosphoglyceride	perfluorocarbon
Sonovue[®]	phospholipid	sulfur hexafluoride
hard shell microbubbles		
Albunex[®]	albumin	perfluorocarbon
Imagify[TM]	poly-L-lactide *co*-glycide	perfluorocarbon
Optison[®]	albumin	perfluorocarbon
Quantison[TM]	albumin	air
Sonavist[®]	polycyanoacrylate	air

importance for compressibility and can also easily be modified. It is composed of air, nitrogen, sulfur hexafluoride (SF_6) or perfluorocarbon derivatives (*e.g.* octafluoropropane C_3F_8 or decafluorobutane C_4F_{10}). The use of inert gases with higher molecular mass such as SF_6, C_3F_8 or C_4F_{10} also has an impact on the circulation time due to the low solubility of these gases in aqueous solution or blood, thus slowing diffusion through the shell. Microbubbles are cleared by the reticuloendothelial system after several minutes with the gas being excreted in the expired air.[4]

Soft Shell Microbubbles

Soft shell microbubbles display a high degree of sensitivity when subjected to changes in pressure. These contrast agents display high echogenicity due to their compressibility but lack sufficient stability in order to perform time-consuming ultrasound scans. This makes them highly interesting compounds for contrast-enhanced ultrasound imaging because of their ability to resonate at low ultrasound frequencies. Apart from free air bubbles that are generated by agitation of an injection solution and do not have any coating, they consist of a thin layer of surfactant surrounding the gas core. Commonly used shell materials include sugar matrices (galactose, Echovist®), surfactants (polyoxyethylene propylene, EchoGen®), fatty acids (palmitic acid, Levovist®), or phosphoglycerides (phosphatidyl serine, Sonazoid™). Several commercially available soft shell contrast agents are based on perfluorocarbon or sulfur hexafluoride gas cores encapsulated by phospholipids (*e.g.* dipalmitoylphosphatidylcholine, BR14®, Definity®, Sonovue®). The major advantage of (phospho)lipids as shell material is their ability to spontaneously assemble in mono-, bi-or multi-layers forming micelles or liposomes. Besides, there are an almost unlimited number of different phospholipids available that allow for countless modification of the contrast agents properties, especially with respect to the surface. While their main disadvantage is a rather low stability, introducing polyethylene-glycol (PEG) into the shell of lipid-based contrast agents helps to extend their life-span.[4] Using gas fillings with high-molecular mass also increases the circulation time of the microbubbles. This makes them highly versatile platforms that have proven their efficacy in a number of ultrasound experiments.

Hard Shell Microbubbles

Hard shell microbubbles display a superior stability due to a more rigid shell, which also prevents the loss of gas due to diffusion. They are made of bio-compatible polymers or denatured proteins that provide a gain in stability while the bubbles lose elasticity and thus scattering efficiency. Hence, in order for the hard shell bubbles to resonate or oscillate, ultrasound pulses with higher mechanical indexes are needed (see also Section 7.3.3.). Several hard shell contrast agents are made of denatured human serum albumin as shell material, and have an air or perfluorocarbon-based gas core, *e.g.* Albunex®, Optison®,

Quantison™. Shell thickness is variable, with different degrees of stability. Additional coating allows for further modification of the protein bubbles and might also be used to suppress immunogenicity.[5] More recently, artificial polymers based on *n*-butyl-2-cyanoacrylate as monomer have been used to synthesise hard shell contrast agents (Sonavist®). These air-filled microbubbles with high stability allow for a simple one-step synthesis with easy modification of the surface. Another contrast agent with a biodegradable shell made of poly-[D,L-lactide-*co*-glycolide] (Imagify™) belongs to the same category of polymeric hard shell contrast agents. Although all these approaches seem to be very promising, only very few of these hard shell compounds have found their way into the clinic.

Targeted Microbubbles

In general, microbubbles have been used mostly as blood-pool contrast agents that remain strictly intravascular because of their size (diameter approx. 1–10 μm).[3] There are different strategies to implement targeted (molecular) ultrasound imaging, always keeping in mind that the receptors to molecular targets have to be present in the blood stream or on the vascular endothelial surface. Microbubbles provide the possibility of introducing modifications at the surface, supporting passive targeting through intrinsic properties, such as shell composition and surface charge and architecture. This strategy has been applied in the imaging of inflammation sites, *e.g.* microbubbles with an albumin shell are able to bind to activated adherent leukocytes through integrin receptors.[3] Leukocytes can also be targeted with compounds having negatively charged phospholipid surfaces that promote the attachment of the bubbles.[4] Passive targeting is suggested to be classified as "functional imaging" since it does not strictly display molecular specificity.[4]

Active targeting can be achieved *via* the covalent or non-covalent attachment of targeting moieties that support specific ligand–receptor binding to molecular targets. Appropriate disease-specific ligands are *e.g.* monoclonal antibodies, small peptides, glycoproteins or carbohydrates. Depending on the compound that will be included in the microbubble surface, different coupling techniques can be applied. One possibility is the direct coupling to shell-precursor molecules before the actual microbubble is formed. This coupling can be performed with standard chemical reaction procedures allowing for controlled work-up with conventional purification steps and highly predictable yields. Since the generation of microbubbles usually demands high-shear agitation of the monomer solution, only molecules that can withstand these reaction conditions might be used in this approach. An alternative way is to attach targeting molecules such as antibodies to the surface of preformed bubbles *via* a streptavidin or biotin interaction.[7] The streptavidin and biotin affinity is one of the strongest non-covalent interactions known and biotinylated antibodies are available in large numbers and can be easily tailored. This protein–small molecule interaction is a standard coupling technique in bioconjugate chemistry, with the main advantages being the extremely high stability and a high bioselectivity. In spite of these favourable properties, this system is not likely to find its way into clinical application due to potential immunogenicity. Still, it has already been of great value in preclinical studies. Another coupling strategy is based on the formation of covalent bonds between the microbubble surface and the targeting moieties, also *via* PEG spacers. Several methods are feasible, including amide, ester or thioether formation with activated residues on the microbubble surface. Although these methods have been established in organic chemistry for decades, the transition to the synthesis of specifically tailored contrast agents by chemical coupling is still in its infancy.

One of the main issues that have prevented the use of contrast-enhanced ultrasound in the clinic is associated with the low adhesion efficiency of targeted microbubbles, since only a considerably low fraction of injected contrast agents actually binds to the area of interest.[7] One strategy to overcome this limitation is the development of dual-targeted microbubbles that are expected to show enhanced binding efficiency in diseased or abnormal tissue.

Non Microbubble-based Ultrasound Contrast Agents

Recent work on ultrasound contrast agents has not been limited to micro-bubbles but included also research in other areas. Several new types of contrast agents have been examined for their applicability in ultrasound imaging. Nanoparticles in particular seem to be promising because their small size allows for extravasation. Among others, solid silica and polystyrene nanospheres were tested both *in vitro* and *in vivo*.[9] Interesting results were also presented with superparamagnetic iron oxide nanoparticles that were shown to be detectable by ultrasound, making way for a bimodal application of these particles both in ultrasound and magnetic resonance imaging.[10,11] The use of liquid-filled lipo-somes and perfluorocarbon droplets as *in situ* generated contrast agents, also holds some promise for new-generation ultrasound contrast agents and is the subject of intense research efforts.

Microbubbles as Carriers for Drugs and Genes

At higher ultrasound energy levels microbubbles are prone to disruption of the shell and can temporarily increase vascular leakiness. Both scenarios might be used in an approach to deliver therapeutic payloads such as drugs and genes. Different variants of ultrasound mediated drug-delivery have already been tested and will be presented in the following short summary.[7,12] The easiest way is probably a co-injection of microbubbles and pharmaceuticals that can be distributed through blood circulation. At the site of delivery in the vasculature, a high-energy ultrasound pulse is applied and oscillation (and finally destruc-tion) of the bubbles can lead to opening of endothelial cell membranes (sonoporation) and even rupture of capillaries. This allows the pharmaceuticals to enter endothelial cells through the pores that are created, or through the diseased tissue from capillary ruptures. In a second variant, microbubbles are charged with a drug, *e.g.* by hydrophobic or by electrostatic interaction with the bubble surface. Here, it might also be possible to use hard shell forming polymers and load the interior of the bubble with an appropriate pharmaceutically active compound. Again, the application of an ultrasound pulse leads to the destruction of the bubble with a subsequent release of the drug.

The main difficulty connected with this task is how to deliver a sufficient amount of active compound to the area of interest.

7.3.3   Contrast Enhanced Ultrasound Imaging Methods

If microbubbles are insonated by ultrasound they oscillate in the over-pressure and under-pressure of the sound field and scatter sound energy back to the ultrasound probe. Microbubble behaviour is governed by nonlinear differential equations (for an overview see *e.g.* Ref. 13) and can be divided into three regimes, which can be characterised by the incident sound field's mechanical index

$$\mathrm{MI} = \frac{p_-}{\sqrt{f_0}} \frac{\sqrt{\mathrm{MHz}}}{\mathrm{MPa}} \qquad (1)$$

which is the maximum negative pressure *P-* in Megapascals (MPa) divided by the square root of the probe frequency f_o in Megahertz (MHz). At very low MI levels, scattering by the microbubbles is linear, which means that the amplitude of the scattered echo signal changes proportionally to the amplitude of the incident pulse. Increasing the MI, nonlinear scattering occurs giving the echo a microbubble specific fingerprint.

Whether this so-called "Low Mechanical Index" imaging can be applied successfully, highly depends on the type of microbubble used as discussed below. Eventually, a further increase in ultrasound energy to higher MI will destroy the microbubbles. During their destruction, the gas leaves the cracked microbubbles and creates a very strong nonlinear echo of a free gas bubble, which can be detected by Doppler imaging modes as a "Loss Of Correlation (LOC)" that is interpreted by the ultrasound system as motion. Subsequently, with the collapse of the free gas bubble in the sound field, a shock wave can be emitted. This phenomenon is known as "Stimulated Acoustic Emission (SAE)". Alternatively, the free bubble dissolves.

Based on these properties of ultrasound microbubbles, two major imaging concepts can be distinguished, which will be specified in the following two subsections: the "destructive imaging" and the "non destructive imaging".

Use-oriented Characterisation of Microbubbles

A first important characteristic of microbubbles is the amount of nonlinearity their oscillation shows in the low MI regime. While free gas bubbles show strong nonlinear behaviour, the oscillation of encapsulated gas microbubbles strongly depends on the mechanical properties of the shell. For soft-shelled microbubbles, *e.g.* with shells based on phospholipids like Sonovue®, the bubbles show strong nonlinear behaviour at moderate MI levels while for hard-shelled microbubbles with polymer shells, radius oscillations have typically much lower amplitudes and thus nonlinearity is less pronounced. The nonlinear behaviour of microbubbles can be characterised by the relative amount of power at harmonic, subharmonic and ultraharmonic frequencies when the bubbles are excited by a sine wave pulse at the fundamental frequency. Figure 7.3.2 shows an example of a characterisation measurement of a diluted soft-shelled microbubble suspension at two different energy levels. In this experiment, the spectral response of single microbubbles insonated by sine-bursts at 2.25 MHz centre frequency is plotted over 200 individual experiments.

In all experiments the fundamental is the strongest component. Even at the lower MI the harmonics at two and three times the fundamental frequency are clearly visible. For the higher MI of approximately 0.7 an additional sub-harmonic component at half the insonation frequency and ultraharmonics at *e.g.* 2.5 and 3.5 the fundamental are visible and indicate strong nonlinear oscillation close to instability of the bubbles. Furthermore, irregular broadband SAE signals spanning the whole frequency band emitted from destructed microbubbles can be seen. This measurement demonstrates that the destruction threshold of microbubbles is not fixed but is a random variable and varies due to the random distribution of bubble stability caused by varying shell composition and size. Current developments aim at reducing this variance by well-controlled microbubble production.

The distribution of the destruction threshold of microbubbles can be determined in repeating the experiments shown above at different pressure levels of the ultrasound burst and counting the destruction events relative to the number of microbubbles left intact by the pulse. The result is an S-shaped curve of percentage of destroyed bubbles over pressure running from 0% for low

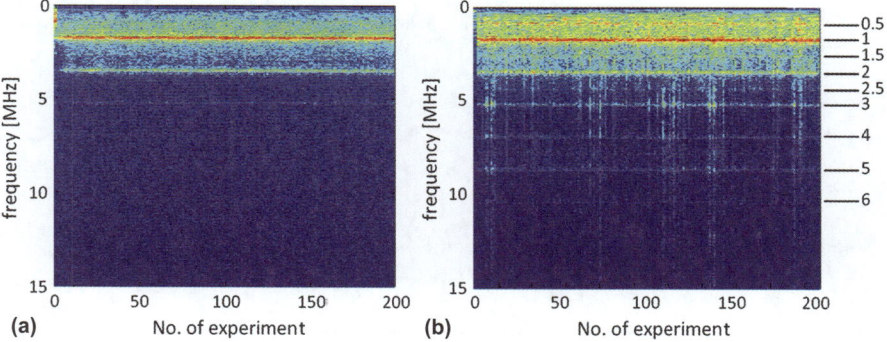

Figure 7.3.2 Spectral representation of scattered echoes from 200 individual experiments of microbubbles insonified with 2.25 MHz with low mechanical index (a), and high mechanical index (b). While in (a) only the fundamental [1] and higher harmonics [2–5] are visible in (b) harmonics [2–6] subharmonics [0.5], ultraharmonics [1.5 and 2.5] and broadband noise from bubble collapse can be observed.

pressure up to 100% at very high pressure. These curves can be used to characterise the stability of the microbubbles and the variance of their properties.

Non destructive Imaging

In non-destructive imaging the nonlinear oscillation of the microbubbles is exploited to separate the microbubble echoes from the linearly scattered signals from tissue. The ultrasound pulses used for imaging are broadband and thus fundamental, harmonic, sub- and ultraharmonics will overlap and cannot be easily separated by filters. However, by pulsing multiple times with different phases or different amplitudes or both together with subsequent signal processing, nonlinear echoes can be separated from linear echoes with high contrast: tissue ratio. The first and very intuitive method is pulse inversion,[14] as illustrated by Figure 7.3.3. When a linear system is excited with pulses of opposite sign, the responses also have opposite sign but identical shape. Thus, they cancel when added and only measurement noise will be left as a result. Microbubbles show different -behaviour in the overpressure and the rarefaction phase making their response nonlinear. Therefore, the echoes from microbubbles do not cancel. This results in a very high contrast : tissue ratio (CTR). In fact, the echo from a single microbubble can be detected as a strong echo although the microbubble cannot be resolved by the imaging device with its resolution in the millimeter to submillimeter range. The CTR is only limited by the fact that the propagation of ultrasound through tissue is also nonlinear, although this effect is much weaker than the bubble nonlinearity and can only be observed at high MI.

There exist further methods following a similar approach, *e.g.* Contrast Pulse Sequencing (CPS),[15] which is a sequence of three pulses with the relative

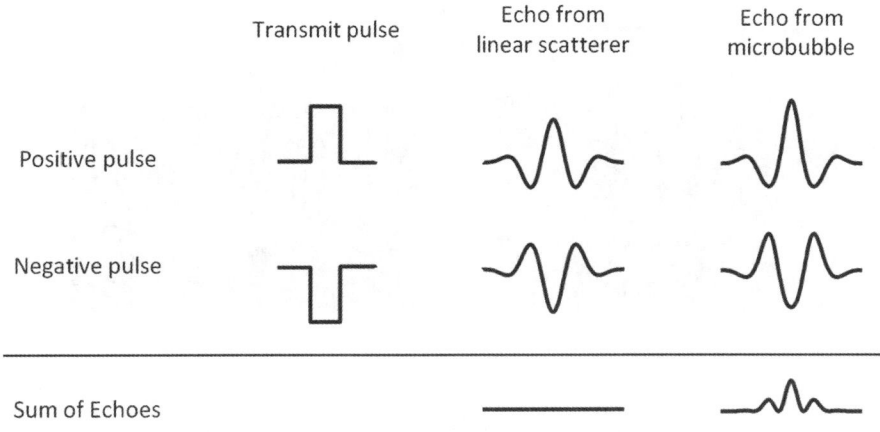

Figure 7.3.3   Principle of Pulse Inversion Imaging: two pulses with opposite sign are used to drive the ultrasound transducer. Echoes from a linear scatterer also have opposite sign and cancel when added, while echoes from nonlinear scatterers like microbubbles do not cancel.

amplitudes of $-1/2$, 1 and $-1/2$. Again, linear echoes of the three pulses will cancel when summed. While for pulse inversion only even-harmonic frequencies are present in the summation signal, in CPS nonlinear contributions in the odd harmonics and the fundamental are also present. Thus, for microbubble echoes the CTR is typically higher for CPS than for pulse inversion. Current research aims at even higher CTR using also subharmonics and ultraharmonics. In general, soft-shelled microbubbles are the better choice for nondestructive imaging because of their strong nonlinear oscillation.

Destructive Imaging

The nonlinearly scattered signals can be used for microbubble detection. For quantification of perfusion or for better estimation of microbubble concentrations it is often necessary to destroy the microbubbles by a very high MI burst and image them before and after the destruction pulse using nondestructive techniques.

For example, different protocols for destruction-reperfusion imaging or replenishment imaging of tissue exist, as first introduced by Wei.[16] Here, microbubbles are destroyed in the whole imaging plane and in subsequent images the nonlinear echoes will increase again due to reperfusion of the tissue. Characteristics of this replenishment curve can be used to characterise tissue perfusion.

Additionally, destruction of microbubbles can be detected by a so-called flash in Doppler ultrasound. Stimulated acoustic emissions and the changing echoes of free gas bubbles after bubble destruction leads to a loss of correlation between Doppler pulses that are interpreted as motion. In clinical systems, the power Doppler mode, which shows the integrated power of the Doppler signal, can sensitively detect the destruction events. However, sequences specialised to detect destruction may be even more sensitive and are a topic of ongoing research.

For quantifying microbubble concentration, it is often necessary to observe a low number of destroyed microbubbles so that single events can be counted. However, for typical concentrations observed in practice very large numbers of microbubbles fall into the destruction range of one image slice. To solve this problem Reinhardt and coworkers proposed a method they named Sensitive Acoustic Particle Quantification (SPAQ).[17] Here, all microbubbles in an image slice are destroyed by a destructive burst. Then the ultrasound probe is moved perpendicular to the image in steps of *e.g.* 10–100 μm and images of bubble destruction are recorded, *e.g.* in power Doppler mode. Consequently, only a small number of microbubbles will be added to the destruction range after movement of the probe. These events can be counted and enable precise quantification of concentrations that can be used in targeted imaging.

For destructive imaging with SPAQ hard-shelled microbubbles are advantageous due to the sudden change of the microbubble echo after cracking, when a free gas bubble is created.

7.3.4   Main Applications of Contrast Enhanced Ultrasound

Oncology

In the abdomen, the use of ultrasound contrast agents is most established for liver imaging. Here it can clearly improve detection and characterisation of pathologies (Figure 7.3.4).

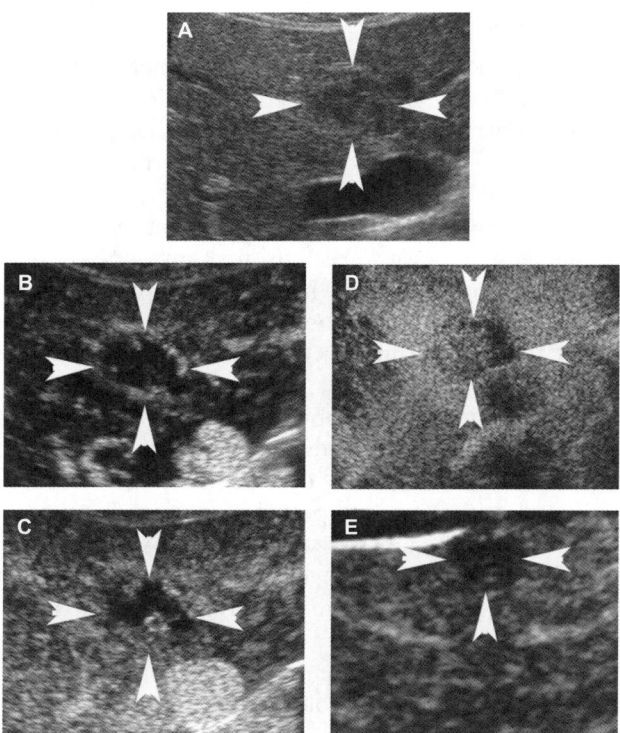

Figure 7.3.4   Ultrasound images of liver lesions. A hypoechogenic liver lesion (arrowheads) in B-mode ultrasound images (as shown in A) often goes in line with a high diagnostic uncertainty. Here microbubble injection can significantly help the clinician to find the correct diagnosis and to decide between malignant and benign processes. After microbubble injection the lesion first enhances at the periphery (arterial phase in B) and then signal intensity starts to increase from the periphery to the center (C) indicating the presence of an atypical hemangioma. A metastasis is shown in D and E. In contrast to the hemangioma, this lesion shows a strong homogeneous early enhancement (arterial phase, D) and later appears less contrasted than the liver tissue (E). Contrast enhanced images were taken with a clinical ultrasound device using a contrast specific inversion pulse sequencing mode. Images were kindly provided by Prof. Dr. Stefan Delorme (German Cancer Research Center, Heidelberg, Germany).

As in CT and MRI three phases of contrast enhancement can be distinguished:

1. The arterial phase describing the first pass of contrast agents. Metastases and hepatocellular carcinomas that are strongly vascularised appear as enhanced lesions. Hemangiomas (and also some metastases) show a strong rim enhancement. Focal nodular hyperplasia – a frequently occurring benign liver tumour – can be diagnosed by vascular assemblies forming a "star sign" or a "spoke-wheel pattern".
2. The venous phase: microbubble inflow *via* the portal vein enhances the liver tissue. The enhancement of hemangiomas and liver tissue becomes comparable. Metastases and hepatocellular lesions become less vascularised.
3. "Liver specific" phase: some microbubbles (*e.g.* Levovist® and Sonazoid®) show a liver specific late enhancement, which most probably derives from microbubbles accumulated in the Disse-spaces and from microbubbles being internalised by cells of the reticulo-endothelial system. This enhancement phase is optimally suited to screening for liver lesions including metastases, which appear less enhanced. However, since benign lesions may appear similar, this phase is sensitive but not very specific. The major advantage of the "liver specific" phase is the persistence of the enhancement for several minutes. This gives the observer time to screen the entire organ carefully.

Also in other organs the visualisation of microvasculature by contrast enhanced ultrasound facilitates the detection of tumour suspicious lesions and the evaluation of their dignity since most malignant tumours go along with increased angiogenesis and vascularisation. This was shown in many clinical studies, *e.g.* on lesions of the breast, pancreas, gynaecologic tumours and for lymph node metastases.

The value of contrast enhanced endorectal ultrasound for prostate cancer detection was evaluated in several large patient trials and superior sensitivity and specificity as compared to non contrast enhanced ultrasound was found. This particularly held true when using microbubble specific non destructive imaging techniques.

Contrast enhanced ultrasound also has the potential to improve the evaluation of tumour response to therapy. In this context it was reported that in animal studies, superior sensitivity of contrast enhanced ultrasound for early therapy effects occurred, prior to tumour volume reduction. By combining non contrast enhanced and contrast enhanced Doppler ultrasound imaging Palmowski and co-workers even reported on the possibility to differentiate mature (less responding) from immature (highly responding) vessel fractions in tumours that were treated with an anti-angiogenic drug.

Also in patients, contrast enhanced ultrasound proved its ability to sensitively capture a patient's response to therapy. In a comparative study Meloni and co-workers showed that contrast-enhanced sonography is an effective alternative to CT and MRI in the follow-up of renal tumours managed with percutaneous radiofrequency ablation.[18] A similar conclusion was drawn by Ricci and co-workers who monitored the success of radiofrequency ablation of liver tumours.[19] Also in this study the accuracy of contrast enhanced ultrasound was comparable to CT. The use of contrast enhanced ultrasound for evaluating the response of stereotactic radiotherapy on liver metastases is another example of its broad applicability. Besides these loco-regional treatments, contrast enhanced ultrasound also sensitively assessed chemotherapy effects on liver metastases, pancreatic and other tumours in patients.

Although not yet in the clinic, it can be assumed that molecular ultrasound imaging of tumour angiogenesis may further improve tumour characterisation and therapy response evaluation in patients in the near future. Several targets have been addressed successfully in tumour bearing mice by molecular ultrasound: $\alpha_v\beta_3$ Integrin has been intensively studied as an indicator of tumour angiogenesis and is highly expressed on activated endothelium. Microbubbles conjugated to cRGD peptides, cRRL peptides or to specific antibodies have demonstrated significant binding capacities to $\alpha_v\beta_3$ Integrin expressing neovasculature (Figure 7.3.5).

Echistatin, a viper venom disintegrin that binds to $\alpha_v\beta_3$ Integrin, is another ligand that was coupled to microbubbles. Using these probes angiogenesis in human gliomas implanted intracerebrally in rats was depicted.[3] Molecular ultrasound imaging of $\alpha_v\beta_3$ Integrin and ICAM-1 (Intercellular Adhesion Molecule 1) was used by Palmowski and colleagues to capture the response of prostate tumours to heavy ion therapy. After 96 hours of irradiation, significant upregulation of both markers was demonstrated indicating vascular activation and early therapy effects.[20] Microbubbles targeting the vascular endothelial growth factor receptor type 2 (VEGFR2) were reported to accumulate within subcutaneously implanted tumours significantly more than unspecific control microbubbles. The "Vascular Endothelial Growth Factor (VEGF)" is the major inductor of tumour angiogenesis, and in particular, by binding to VEGF receptor 2 (VEGFR2) it stimulates endothelial cell proliferation and survival. In this context, it was shown that the retention of VEGFR2-specific microbubbles was higher in "highly invasive and metastatic" than in "non-metastatic" breast cancer xenografts.[3]

Korpanty and co-workers used specific microbubbles against VEGFR2 and Endoglin in two subcutaneous models of pancreatic cancer during anti-angiogenic treatment. Endoglin is a cell membrane glycoprotein and mainly expressed on endothelial cells and over-expressed on tumour-associated endothelium. Using the specific microbubbles, decreasing marker densities after tumour-suppressive therapy could be observed for both targets, a finding that correlated with the treatment effects. In another study, VEGFR2 and $\alpha_v\beta_3$ Integrin expression were imaged by molecular ultrasound. During the growth of untreated tumours the density of both markers increased, whereas reduced binding of targeted microbubbles was found after therapy against MMP 2 and 9. In addition to these examples, there are reports of microbubbles binding to VCAM and P-/E-Selectin, which underline the stability and the potential of the molecular ultrasound approach.[3]

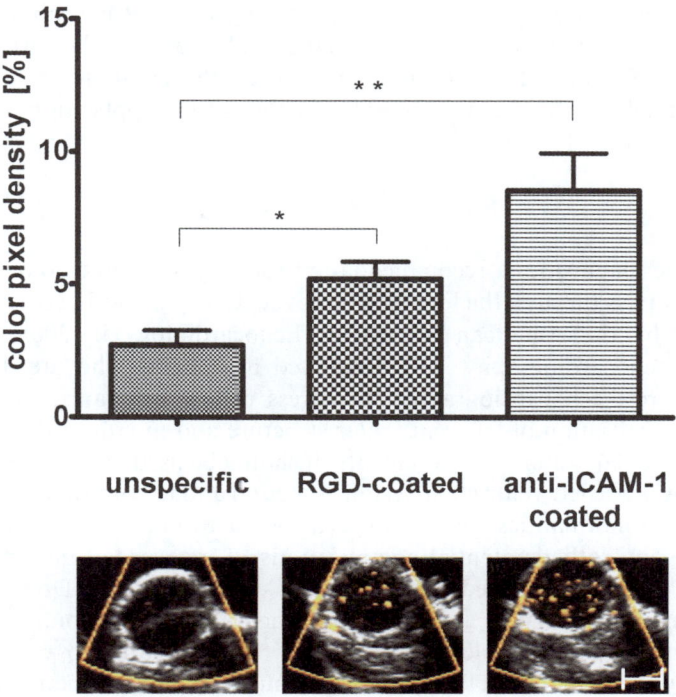

Figure 7.3.5 Volumetric Doppler measurements of experimental prostate carcinomas prove the specificity of targeted microbubbles: after injection of non-targeted microbubbles, no relevant accumulation was observed within tumour vessels. After administration of RGD-coated microbubbles targeted against $\alpha v \beta 3$-integrin, an increased number of Doppler signals (yellow dots) were visible, representing the bursting of stationary microbubbles at their target. Quantitative analysis indicated significant accumulation as compared to non-targeted microbubbles. Accumulation of anti-ICAM-1-antibody-coated microbubbles was even more pronounced (*: $p < 0.05$; **: $p < 0.01$) (n = 8 tumours). Bar: 5 mm. Figure of Palmowski *et al.*, Neoplasia, 2009, 11(9).[20]

Lymph Node Imaging

Active targeting of healthy lymph nodes was reported by Hauff and colleagues,[21] who investigated the feasibility of lymph node imaging. The MECA-79 ligand expressed on the endothelium of peripheral lymph nodes was chosen as a target for MECA-79 antigen-conjugated cyanoacrylate microbubbles. MECA-79 antigen interacts with the L-selectin expressed on circulating lymphocytes and is involved in the homing of lymphocytes to the lymph node. The specific microbubbles accumulated significantly in healthy lymph nodes in mice and dogs and could be detected both *in vitro* (mice) and *in vivo* (dogs). A clinical translation of lymph node targeting is strongly supported by the recently published results of Sever *et al.*[22] who examined 80 consecutive patients with primary breast cancer with a clinically approved "non-targeted" ultrasound contrast agent. After periareolar intradermal injection of microbubbles, they were able to visualise the draining lymphatics and to identify the corresponding sentinel lymph node. These results demonstrate the feasibility of ultrasound based sentinel lymph node imaging and encourage the application of targeted contrast agents in further studies.

Cardiology

For cardiac diagnostics a recommendation for the use of ultrasound contrast agents that pass through the lung like Sonovue, Optison and Levovist has been published by the American Society of Echocardiology in 2008: Contrast enhanced echocardiography should be used in patients who are difficult to image for rest echocardiography and stress echocardiography, in order to improve visualisation of left ventricular structure and to reduce the variability in left ventricular volume measurements. It should be used in all patients for the assessment of the left ventricular systolic function and to confirm or exclude left ventricular abnormalities such as hypertrophic cardiomyopathy, thrombosis, aneurysms and other potential complications of myocardial infarction. Furthermore, contrast enhanced echocardiology should be used if Doppler signals are not sufficiently high to assess diastolic and valvular function.

In addition, microbubbles like Echovist that normally do not cross the pulmonary vascular bed following intravenous injection can be used to diagnose atrial and pulmonary shunts as well as complex congenital heart disease.

All these indications benefit from "microbubble specific" scan techniques such as subharmonic imaging and pulse inversion imaging, which have been implemented in the last generation of high-end ultrasound devices.

Target-specific microbubbles have been used successfully to characterise cardiovascular disease but have not yet been evaluated in patients.[6] In wild-type and ApoE$^{-/-}$ mice fed either with chow diet or hypercholesterimic diet, inflammatory plaques within the aortic arch could be visualised using antibody-conjugated microbubbles targeting VCAM-1. Furthermore, there was a correlation between the different stages of arteriosclerosis and the retention of targeted microbubbles.

In addition, targeted microbubbles have been developed that specifically bind to developing thrombi by binding *e.g.* fibrin or the GPIIb IIIa receptor on immobilized platelets. In successful clinical evaluation, these may act as valuable early indicators of developing vascular occlusion.

Preclinical studies have shown that as well as the imaging of thrombus formation, the post-ischaemic injury can also be characterised by molecular ultrasound imaging,[6,7] which is particularly important for the control of therapy success. This was demonstrated in the myocardium of mice and rats without infarction and in a mouse model of unilateral kidney ischaemia using P/E-selectin-specific microbubbles. Therapy induced re-vascularisation was successfully monitored by Leong-Poi and co-workers whose treated hind-limb ischaemia in mice with FGF (Fibroblast Growth Factor) and used echistatin-conjugated microbubbles targeting $\alpha_v\beta_3$.

Inflammation cells play a major role in the remodelling of ischaemic tissues. Therefore, targeting inflammation cells like leucocytes with microbubbles has been explored as an alternative method ("passive targeting"). After intravenous injection microbubbles bind to inflammation cells which are circulating, but may migrate to the diseased tissue and to those that are already present locally. In this context, it is important to note that even if the cells internalise the microbubbles, they remain intact for a certain period of time and can be detected. The feasibility of this approach was shown in dogs, where the extent of post-ischaemic myocardial inflammation could be evaluated.[3]

Other Applications

Certainly, most indications for the use of ultrasound microbubbles concern cardiovascular and oncological disease. However, to list all other examples where ultrasound microbubbles were applied in experimental or clinical trials would go beyond the scope of this book chapter. The important role of ultrasound microbubbles injected into the urine bladder to depict ureteral reflux was already mentioned in Section 7.3.1. Some groups also use ultrasound microbubbles to diagnose stroke. The characterisation of inflammative, rheumatologic diseases and transplant rejection are further reasonable indications of contrast-enhanced ultrasound imaging since these diseases are accompanied by increased local vascularity (vasodilatation and angiogenesis).

Since ultrasound is moving towards a molecular discipline, it is not surprising that targeted ultrasound has also been evaluated for these indications in small animal studies. For example, specific microbubbles against the gut-specific "mucosal addressing cellular adhesion molecule-1" (MAdCAM-1) were used to characterise inflammatory bowel disease. In other studies, leukocyte-targeting microbubbles (passive targeting) and ICAM-1 targeting microbubbles were used to diagnose acute allograft rejection in rat cardiac transplantation models.

7.3.5   Safety of Ultrasound Contrast Agents

Concerns about the safety of ultrasound contrast agents are the major cause for the lack of clinical use in many countries and for many diagnostic problems. However, since potential risks and safety questions on the organism's general reaction, the induction of an immune response, or ultrasound radiation bioeffects are not completely understood, opinions are highly diverged.

Adverse effects of clinical ultrasound contrast agents are rare and reported in less than 1% of cases. These range from headaches and altered sensation at the site of the injection, nausea, flushing, paraesthesia and taste perversion to abdominal and non specific pain. There are also reports about respiratory problems, pharyngitis, pruritus, rash, abnormal vision, dry mouth, dizziness, insomnia, nervousness, hyperglycaemia, peripheral oedema, ecchymosis and sensory motor paresis.

The explanation for these symptoms may be complex:

With a size of > 1 micrometres, embolisms in the lung or aggravation of tissue ischaemia by temporary occlusion of microvasculature have to be excluded. On the one hand, microbubbles are about half the size of erythrocytes that easily pass microvasculature. On the other hand, microbubbles are less deformable and not discus shaped like erythrocytes. Whether they can cause embolic effects also depends on their blood half life and stability to pressure. Indeed there are more than 11 cases of death following the administration of microbubbles and most of these patients had cardiovascular disease. However, a direct causal connection between these fatal outcomes and vascular occlusions by microbubbles or direct cardiac toxicity was never shown. In addition, all these patients were known to be at high risk of major cardiac complications. Nevertheless, the contrast agent SonoVue, which was administered in most of the cases, was temporarily withdrawn and later reinstated with new limitations on its use in patients with recent acute coronary syndrome or clinically unstable cardiac disease.

The next complication of contrast enhanced ultrasound may derive from the disintegrating microbubbles, inducing a strong local acoustic pressure that can damage tissue and vessel walls. It this context, it was shown that even the blood brain barrier can be opened temporarily using contrast enhanced ultrasound. However, effects on vessel walls and endothelial cells were described to be rapidly reversible. Nevertheless, the intensity of ultrasound radiation described by the mechanical index (MI) and the thermal index should be monitored constantly in clinical practice according to FDA guidelines.[25,26]

In addition to these physical and mechanical side effects, microbubbles may induce allergic reactions like almost every other diagnostic or therapeutic drug. However, as compared with MRI and X-ray contrast agents the incidence of side effects is lower. This may be attributed to the low doses at which ultrasound contrast agents are administered and to their biocompatible chemical composition. Nevertheless, there are few reports on mild to severe hypersensitivity reactions including skin erythema, bradycardia, hypotension or anaphylactic shock.

Allergic reactions more often appear in the presence of electrostatic charges on the particle surface. Therefore, the reduction of surface charges, *e.g.* by using polyethylene glycol spacers on the particle surface, may help to control these problems.

Targeted microbubbles are not yet in the clinics and thus there are no experiences about their safety. On the one hand, it might be possible to use these contrast agents in significantly lower dosages, which reduces the risks mentioned above. On the other hand, the conjugation of bioactive ligands may induce a stronger activation of the immune system and thus increase the number of adverse immunological events. In this context, the biotin – avidinstreptavidin coupling method has to be replaced, which is often used in preclinical studies and known to be immunogenic in humans. First soft shell microbubbles targeting the VEGFR2 that fulfil the criterion of a clinical contrast agent have been developed by Bracco, and tested preclinically for their adhesion in tumour vessels.[23,24] Initial clinical studies are running where these microbubbles are tested for their ability to visualise neovasculature in prostate cancers.

7.3.6 Outlook

Ultrasound contrast agents have already been established in the clinical routine and the number of indications and recommendations for their use is increasing. This particularly holds true for characterising perfusion of organs and tumours. Although side effects have been reported casually, microbubbles are considerably safe as contrast agents in comparison to CT and MRI agents. The diagnostic impact of ultrasound contrast agents clearly improves with the implementation of non destructive "contrast specific" detection techniques. In this context, there is a clear trend towards the use of soft shell microbubbles with pronounced harmonic properties for non targeted applications in the clinics.

Molecular ultrasound imaging is still experimental but may be evaluated in patients for some indications soon. It enables a more specific characterisation of angiogenesis and inflammation and a longer persisting label. The longer persisting label facilitates the investigation of larger body areas (*e.g.* both breasts and axillar lymph nodes) and the localisation of suspect lesions during interventions (*e.g.* biopsy). Furthermore, experimental data suggest that molecular ultrasound may be a sensitive tool for monitoring early therapy effects on tumours and thus may play a future role in personalized therapy approaches. Nevertheless, for the clinical translation of molecular ultrasound there is still significant demand for the development of more quantitative and 3D (and thus better reproducible) imaging techniques.

References

1. R. Lencioni, C. Della Pina, L. Crocetti, E. Bozzi and D. Cioni, *Eur. J. Radiol.*, 2007, **17**(Suppl. 6), 73.
2. M. A. Weber, M. Krix, U. Jappe, H. B. Huttner, M. Hartmann, U. Meyding-Lamadé, M. Essig, C. Fiehn, H. U. Kauczor and S. Delorme., *Radiology*, 2006, **238**, 640.
3. F. Kiessling, J. Huppert and M. Palmowski, *Curr. Med. Chem.*, 2009, **16**, 627.
4. P. A. Dayton and J. J. Rychak, *Front. Biosci.*, 2007, **12**, 5124.
5. K. Ferrara, R. Pollard and M. Borden, *Annu. Rev. Biomed. Eng.*, 2007, **9**, 415.
6. J. R. Lindner, *Nat. Rev. Drug Discovery*, 2004, **3**, 527.
7. A. L. Klibanov, *Invest. Radiol.*, 2006, **41**, 354.
8. R. Gramiak and P. M. Shah, *Invest. Radiol.*, 1968, **3**, 356.
9. J. Liu, A. L. Levine, J. S. Mattoon, M. Yamaguchi, R. J. Lee, X. Pan and T. J. Rosol, *Phys. Med. Biol.*, 2006, **51**, 2179.

10. I. Nolte, G. H. Vince, M. Maurer, C. Herbold, R. Goldbrunner, L. Soly-mosi, G. Stoll and M. Bendszus, *Am. J. Neuroradiol.*, 2005, **26**, 1469.
11. J. Oh, M. D. Feldman, J. Kim, C. Condit, S. Emelianov and T. E. Milner, *Nanotechnology*, 2006, **17**, 4183.
12. E. C. Pua and P. Zhong, *IEEE Eng. Med. Biol.*, 2009, **28**, 64.
13. A. A. Doinikov and P. A. Dayton, *J. Acoust. Soc. Am.*, 2006, **120**, 661.
14. D. Hope-Simpson and C. T. Chin, *IEEE Trans. Ultrason., Ferroelect., Freq. Contr.*, 1999, **46**, 372.
15. H. E. Huppert and P. Phillips, *Proc. – IEEE Ultrason. Symp.*, 2001, , 1739.
16. K. Wei, A. R. Jayaweera, S. Firoozan, A. Linka, D. M. Skyba and S. Kaul, *Circulation*, 1998, **97**, 473.
17. M. Reinhardt, P. Hauff, A. Briel, V. Uhlendorf, R. A. Linker, M. Mäurer and M. Schirner, *Invest. Radiol.*, 2005, **40**, 2.
18. M. F. Meloni, M. Bertolotto, C. Alberzoni, S. Lazzaroni, C. Filice, T. Livraghi and G. Ferraioli, *Am. J. Roentgenol.*, 2008, **191**, 1233.
19. P. Ricci, V. Cantisani, F. Drudi, E. Pagliara, M. Bezzi, F. Meloni, F. Calliada, S. M. Erturk, V. D'Andrea, U. D'Ambrosio and R. Passariello, *Ultraschall Med.,*, 2009, **30**, 252.
20. M. Palmowski, P. Peschke, J. Huppert, P. Hauff., M. Reinhardt, M. Maurer, W. Semmler, P. Huber and F. Kiessling, *Neoplasia*, 2009, **11**, 856.
21. P. Hauff, M. Reinhardt, A. Briel, N. Debus and M. Schirner, *Radiology*, 2004, **231**, 667.
22. A. R. Sever, P. Mills, S. E. Jones, K. Cox, J. Weeks, D. Fish and P. A. Jones, *Am. J. Roentgenol.*, 2011, **196**, 251.
23. M. A. Pysz, K. Foygel, J. Rosenberg, S. S. Gambhir, M. Schneider and J. K. Willmann, *Radiology*, 2010, **256**, 519.
24. S. Pochon, I. Tardy, P. Bussat, T. Bettinger, J. Brochot, M. von Wronski, L. Passantino and M. Schneider, *Invest. Radiol.*, 2010, **45**, 89.
25. N. McDannold, N. Vykhodtseva and K. Hynynen, *Ultrasound Med. Biol.*, 2008, **34**, 930.
26. B. Baseri, J. J. Choi, Y. S. Tung and E. E. Konofagou, *Ultrasound Med. Biol.*, 2010, **36**, 1445.

CHAPTER 8.1

Imaging as a CNS Biomarker

RICHARD HARGREAVES,[1] LINO BECERRA[2] AND
DAVID BORSOOK[2]

[1] Discovery Neuroscience Research, Merck & Co. Inc., West Point, PA
19486, United States; [2] P.A.I.N. Group, Massachusetts General Hospital,
McLean Hospital, and Children's Hospital, Harvard Medical School,
115 Mill Street, Belmont, MA 02478, United States

8.1.1 Introduction

Despite advances in human genetics and drug discovery technologies, the discovery of drugs for the treatment of functional neurological or psychiatric
brain diseases remains largely empirical, with a low probability of success of
drugs for novel targets being registered. Indeed, it is a sobering thought that
from the development of a new medicine to the completion of phase 3 clinical
trials can now take up to 15 years and cost $1.5 billion; yet in areas such a
neuroscience 61% of drugs directed at CNS indications still fail vs. placebo and
11% fail because of lack of efficacy or differentiation.[1] The burden in time and
cost is largely driven by attrition during the drug discovery and development
process, making the key to fast and effective drug discovery one of disciplined
early decision making around the selection and validation of targets, molecules
and therapeutic indications. This realization has given rise to the field of biomarkers in drug discovery and development that can be used to validate targets,
select molecules, specify active dose ranges, stratify patients into subpopula-
tions and provide early evidence for supporting the mechanism, biology and
clinical proof of concept. The issue is central to the discovery of treatments for
CNS disorders since there is often no objective diagnosis, no objective measure

RSC Drug Discovery Series No. 15
Biomedical Imaging: The Chemistry of Labels, Probes and Contrast Agents
Edited by Martin Braddock
© Royal Society of Chemistry 2012
Published by the Royal Society of Chemistry, www.rsc.org

of treatment efficacy and sadly no cure. By better understanding the functional effects of different therapeutic approaches on brain circuits and systems, it may be possible to unravel their impact on CNS disease itself or the co-morbidities that are so frequently yet often heterogeneously manifested as part of the overall condition. Not surprisingly, about 20% of current biomarker research is directed at CNS systems (Frost and Sullivan, http://www.frost.com). As noted in Table 8.1.1, significant numbers of Americans suffer from brain diseases for which there are few effective therapies.

Imaging the brain in conscious subjects has revolutionized our understanding of brain diseases. In CNS drug discovery, imaging is becoming a core technology with which to probe for *Drug Effect* and *Disease State* (Figure 8.1.1). Identifying a brain phenotype defined by 'imaging' has the potential to link wanted and unwanted drug effects to genotype and disease state.

Table 8.1.1   Incidence of Functional Brain Diseases in USA. From: D. Borsook, R. Hargreaves, L. Becerra, *Expert Opin. Drug. Discov.*, 2011 **6**(6).

Disease	*Numbers Affected*	*Reference*
Degenerative Diseases		
Alzheimer's	4.5 million	[108]
	5.3 million	[109]
Parkinson's	> 500 000	www.ninds.nih.gov
	121/100 000	[110]
	3–4 million undiagnosed	www.ninds.nih.gov
System Disorders		
Narcolepsy	25–50/100 000	[111]
		[112]
Psychiatric Disorders		
Mood Disorders	20.9 million	[113]
Depression	17.1 million	www.nimh.nih.gov
Bipolar Disorder	5.7 million	[113]
Dysthymic Disorder	3.3 million	[113]
Anxiety Disorders		
PTSD	7.7 million	[113]
Gen. Anxiety Disorder	40 million	[113]
Panic Disorder	6 million	[113]
Schizophrenia	2.4 million	[114]
Pain Disorders		
Chronic Pain	50 million	
Neuropathic Pain	17%	[115]
Fibromyalgia	0.5% in men 3.4% in women	[116]
Chronic Back Pain	10.2%	[117]
Osteoarthritis	16%	[118]
Migraine	12/100 in adults	[119]
	11.3 million	[120]
Developmental Disorders		
Autism	3.4/1000	[121]
Attention Deficit Hyperactivity Disorder	4.1 million	[113]

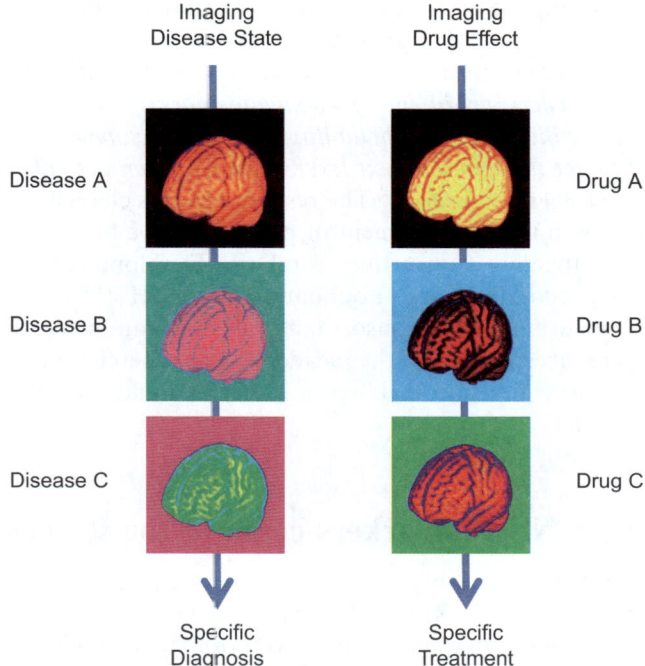

Figure 8.1.1 Biomarkers for Disease State and Drug Effects. The figure shows brains
representing imaging measures for different CNS disease states (A–C,
left panel) as indicated by different brain pseudo-colors. Imaging drug
effects on the brain (A–C) may be specific for either a disease in different
individuals, for subtypes of a particular disease state, or for different
diseases (right panel). Brain images created with FSL (www.fmrib.ox.
ac.uk/fsl/).

8.1.2 Brain Disease and Subjective Measures – In Search of Better Information

Objective biomarkers of CNS function have the potential to supplant subjective assessments that are often highly variable and so inadequate for
guiding many approaches to improving therapy. Indeed, a recent NIH RFA
(Request for Application) commenting upon the situation in the seemingly
easy area of pain highlights this issue: "*Pain research has been greatly ham-
pered by the unreliable nature of self-report based instruments. The establishment of objective, affordable and reliable pain biomarkers and measurements
would advance our understanding of pain mechanisms, provide a basis for
improved clinical management of pain, ... and establish much needed objective
measures of treatment success or failure*" (**RFA-DA-09–017, NIDA**). There is
therefore a clear and unmet medical need to develop better biomarkers for
evaluating drug effects and disease state in order to advance therapy. However, despite the clear acknowledgement that this gap exists, as highlighted by

this commentary on the FDA Critical Path Initiative, "*Biomarker development has languished, probably along with the languishment of clinical pharmacology and experimental medicine around the world. And that needs to be built back up... There are a lot of candidate markers out there, but their usefulness, their utility, their predictability in drug development is not known. And we need to get them to the next level where they can actually be useful in predicting effectiveness or safety*".[2] The response to this challenge is only now gathering pace with the establishment of precompetitive biomarker consortia initiatives (*e.g.*, Imaging Consortium for Drug Development (ICD) (www.imagingpain.org/icdd.html) and Foundation for NIH (fNIH) Biomarkers Consortium (www.biomarkersconsortium.org) that can help validate endpoints that are acceptable to the academic, commercial and regulatory communities to help decision making on safety and efficacy across the drug development cycle.

8.1.3   Can CNS Biomarkers come to the Rescue?

Biomarkers have been variably defined as "*Biological substances or features that can be used to indicate normal biological processes, disease processes, or responses to therapy*" (www.biobankcentral.org/resource/glossary.php) or "*objective, accessible, and easily measurable biologic parameters that correlate either with the presence (trait) or the severity (state) of a disease*".[3] A biomarker usually is an indicator of a biological state of an individual and can be used to define a positive response (*e.g.*, treatment effect), or negative effect (*e.g.*, adverse reaction). As such, the use of biomarkers has been applied to a number of diseases, the most notable of which are cardiac enzymes for myocardial infarction,[4,5] HbA1C, adiponectin, and glucose for diabetes,[6] plasma lipids for atherosclerosis, viral and bacterial loads for infection. With respect to brain disease, defining biomarkers has been a bigger challenge. The early use of neuroimaging as a biomarker in multiple sclerosis has evolved as the relationship between image and disease symptoms has become better understood with its use in the evaluation of effective and ineffective therapies [7, 8]. With the advent of multiple new neuroimaging techniques that can measure changes in function, structure and chemistry in conscious human subjects, a new *era* has opened to the definition, evaluation, and adoption of CNS imaging biomarkers in medicine.

   The challenge of adopting brain imaging biomarkers for disease states such as depression, chronic pain, schizophrenia and other such brain specific diseases is not trivial and a number of issues need to be accounted for such as: (i) sensitivity, specificity and reproducibility; (ii) ability to differentiate brain states (*e.g.*, a disease biomarker from a treatment biomarker or a response biomarker); (iii) understanding how data sets derived from individuals can be interpolated into a group data; and (iv) account for placebo effects that can be pronounced (up to 50%) in patients with Parkinson disease (PD), pain syndromes, and depression.[9] In some neurodegenerative conditions such as

Table 8.1.2 State of the Art – Examples of Current Imaging and Potential CNS Biomarkers.

CNS Disease	Possible CNS Imaging Biomarkers	Main Finding	Reference
Alzheimer's Disease (AD)			
Volumetric	Hippocampal atrophy (HiA)	Regional measures of HiA predicted AD progression	[122]
	Brain atrophy	Volumes of hippocampus, entorhinal cortex, ventricle, and whole brain	[123]
	Hippocampal atrophy	Smaller hippocampal and entorhinal volumes predict conversion to AD	[124]
DTI	White matter integrity (Decreased FA)	Wallerian degeneration secondary to cortical atrophy	[125]
	White matter integrity	Different in aging vs. AD in anterior temporal pole	[29]
Functional	fMRI – default mode network	Distinguishes AD from dementia with Lewy bodies	[126]
MRS	1H Spectroscopy	Metabolic MRS profile in AD differs to Parkinson's dementia	[127]
	1H Spectroscopy	Metabolic changes detectable in presymptomatic mutation carriers of AD	[128]
PET	11C-PIB PET (Amyloid ligand)	Mapping Amyloid Toxicity – temporal pole susceptibility	[129]
	Fluorodexoyglucose PET	Reductions in cerebral metabolic rate for glucose after onset of AD	[130], [131]
Parkinson's Disease (PD)			
Volumetric	Basal ganglia morphometry	Bilateral decrease in *Parkin* mutation carriers > idiopathic PD	[132]
	Putamen and caudate	Increase in gray matter in putamen and globus pallidus	[133]
DTI	DTI substantia nigra	Distinguishes early stage PD	[134]
Functional	Performance task	Lateral prefrontal and caudate activation related to motor disease severity; dopamine shift in this relationship	[95]
	[a]Working Memory and Motor function	Opposite effects of dopamimetic Rx on working memory by different networks	[135]
MRS	1H MRS Glutamate	Decreased glutamate in cerebral cortex in PD	[128]
	MRS	Presupplementary motor area NAA: Cr decreased in PD vs. healthy controls	[136]
	MRS substantia nigra	Healthy subjects, 4 × GABA: Glu ratio vs. cortex correlates with known neurochemistry	[137]
PET	6-[18F]fluorodopa PET	Putamen: caudate and putamen: substantia nigra ratios separate PD from multisystem atrophy	[138]
	Fluorodeoxyglucose PET	Higher metabolic rates for glucose in the pallidum and substantia nigra bilaterally and unilaterally in caudate and pallidum	[139]
Other	Midbrain Iron Content	Increased iron in lateral substantia nigra pars compacta	[140]

Table 8.1.2 (*Continued*)

CNS Disease	Possible CNS Imaging Biomarkers	Main Finding	Reference
Depression (MDD)			
Volumetric	Brain volume (meta analysis)	Reductions in frontal regions (ACC and GOb) and hippocampus putamen and caudate	[141]
	Brain volume 5-HTTLPR gene	MDD homozygous for the L(A) allele show reduced gray matter volumes indicating greater vulnerability for morphological changes	[142]
DTI	White matter	Late life depression, abnormalities in prefrontal regions	[27]
	White matter	MDD showed decreased FA sagittal stratum suggesting dysfunction in limbic-cortical network	[143]
Functional	fMRI NAc and Caudate	Decreased responses to gains *i.e.*, rewarding stimuli	[144]
	fMRI Ventral striatum DMN	Decreased ventral striatum to +ve stimuli	[145]
MRS	¹H MRS GABA	Decreased GABA in treatment- resistant *vs.* controls	[146]
	¹H MRS glutamate GABA	Reduced glutamate: glutamine ratio and GABA in prefrontal regions. Data correlates with postmortem findings	[147]
PET	Glucose metabolism amygdala	*Unmedicated* depressed patients *vs.* controls: increased metabolism in prefrontal cortex (PFC) amygdala, and posterior cingulate cortex; and decreased in the subgenual ACC and PFC. *Medicated vs.* controls: decreased metabolism in amygdala and subgenual cingulated cortex	[148]
Anxiety			
Volumetric	Smaller amygdala	Decreased amygdala volume in pediatric anxiety	[149]
	Putaminal volume	Decreased in panic disorder	[150]
DTI	White matter abnormalities	MDD had lower FA in right fasciculus near Gob, a major tract connecting to the amygdala	[151]
Functional	fMRI emotional images	Enhanced amygdala and insula responses to negative pictures	[152]
	DMN	RSN connectivity of PCC with perigenual ACC and amygdala is associated with symptoms and amygdala predicts future PTSD symptoms	[153]
MRS	¹H MRS	Acetylaspartate: creatine ratio in DLPF	[154]
	¹H MRS	NAA: creatine decrease in ACC PTSD	[155]
PET	[¹¹C]WAY-100635	Decreased binding suggesting 5-HT1A receptor in amygdala and ACC	[156]
	[¹¹C]Flumazenil	Decrease GABA: BZ binding in insular cortex in panic disorder	[157]

Schizophrenia

Category	Method	Description	Ref
Volumetric	Meta-analysis	Reduced bilateral insular cortex, anterior cingulate, parahippocampus, middle frontal gyrus, postcentral gyrus, and thalamus; increased in striatal regions	[158]
	Gray matter density	*Decreased* gray matter in cerebellum thalamus, basal ganglia, middle frontal gyrus, inferior frontal gyrus, precentral gyrus, insula, superior temporal gyrus, fusiform gyrus, parahippocampal gyrus, cuneus, and lingual gyrus; in the left posterior cingulate, superior frontal gyrus, transverse temporal gyrus, and precuneus; and in the right postcentral gyrus. *Increased* gray matter in basal ganglia, anterior cingulate, and medial orbitofrontal cortices. Related to duration of disease.	[159]
DTI	Functional bundles	Altered FA in thalamic – frontal lobe	[160]
	Fiber tractography	Reduced FA in arcuate and inferior longitudinal fasciculus indicated connections between temporal lobe, neocortical regions are abnormal	[161]
Functional	fMRI	Decreased activation in inferior frontal gyrus to semantic encoding	[162]
	Connectivity	Altered functional connectivity in networks involved in auditory processing, executive control and baseline activity.	[163]
MRS	1H MRS	Decrease in GABA	[164]
	^{a}1H MRS	Risperidone treatment produces increases in NAA and mI	[165]
PET	[^{18}F]dopa	Striatal dopamine levels elevated in prodromal symptoms	[166]
	L-[beta-^{11}C]dopa	Increased presynaptic dopamine synthesis in caudate and increased dopaminergic transmission in thalamus and temporal cortex	[167]

Chronic Pain (*e.g.*, Neuropathic Pain)

Category	Method	Description	Ref
Volumetric	Cortical thickness	Differences in cortical thickness in sensory and emotional cortical regions	[168]
	Chronic back pain	Decreased DLPF and thalamic density	[22]
DTI	Probabilistic tractography	Mapping somatosensory pathways will allow for defining pathology	[169]
Functional	fMRI/psychophysical	Altered sensory and emotional networks to evoked stimuli	[170]
	fMRI/psycophysical	Syringomyelia with subtypes of allodynia show different patterns of activation; consistent among these was prefrontal cortical activation	[171]
MRS	1H MRS	Increased NAA concentration in thalamus; NAA decreased in patients who responded to Rx	[172]
PET	^{11}C Diprenorphine	*Central pain* – Decreased opioid binding in midbrain, medial thalamus, insular temporal and contralateral cortices contralateral to pain; *Peripheral pain* – bilateral symmetrical decreases of opioid binding	[173]

Table 8.1.2 (*Continued*)

CNS Disease	Possible CNS Imaging Biomarkers	Main Finding	Reference
Migraine			
Volumetric	Cortical thickness	Cortical thickening SI	[174]
DTI		Altered structural and subjacent white matter in MT+, V3A and superior colliculus and lateral geniculate body	[175]
Functional	fMRI	Interictal altered brainstem function shows abnormal modulatory circuits in migraine patients	[98]
MRS	^{1}H MRS	Interictal MRS metabolites	[72]
PET	^{18}F-fluorodeoxyglucose PET	Interictal metabolic changes show hypometabolism in insula, cinculgate, prefrontal and primary somatosensory cortex.	[176]
Autism			
Volumetric	Gray matter	Reduced gray matter and altered chemistry consistent with decreased neural density	[177]
DTI	FA and fiber length	Altered FA due to white matter organizational abnormalities	[178]
Functional	fMRI	Enhanced visual processing with increased task-related activity in extrastriate areas, and decreased activity in the lateral prefrontal cortex and the medial posterior parietal cortex	[179]
MRS	^{1}H MRS	Decreased NAA in thalamus Correlated with sensory abnormalities	[180]
PET	^{a}^{18}F deoxyglucose PET	Higher metabolic rates in anterior cingulated gyrus and orbitofrontal cortex; correlation of high metabolic rates and response to fluoxetine	[181]
ADHD			
Volumetric	Structural changes	Decreased callosal thickness in anterior and posterior regions (frontal and parietal regions)	[182]
DTI	Meta analysis	Reduced gray matter in right putamen–globus pallidus	[183]
	White matter abnormalities	Distinct clusters of increased FAs in fronto-temporal regions and right parietal-occipital regions	[184]
Functional	fMRI	Ventrolateral prefrontal network dysfunction in ADHD	[185]
	^{a}fMRI	Timing disturbances in prefrontal, cingulate, cerebellar and striatal regions that normalized with methylphenidate	[186]
MRS	^{31}P MRS (assesses overproducing or pruning of synapses)	Lower membrane phospholipid precursor levels in basal ganglia higher levels in inferior parietal	[187]
PET	^{11}C Altropane PET	Increased dopamine transporter binding in caudate in adult ADHD	[188]
	^{11}C Raclopride PET	Decreased dopamine activity in caudate	[189]

^{a}Examples of measures of drug treatments.

Alzheimer's disease a more definitive link between imaging measures and the pathological processes thought to underlie disease progression is being established through initiatives such as Alzheimer's Disease Neuroimaging Initiative (ADNI; http://www.adni-info.org/) by studying the correlation of brain imaging with CSF fluid biomarkers and cognitive performance. In this review we provide a perspective on the promise of appropriately validated imaging biomarkers for brain dysfunction that have the potential to transform our current position of empirically based diagnosis and treatment to rational approaches based on objective markers of brain structure and function.

8.1.4   Criteria for CNS Biomarkers

A number of generalizable criteria need to be agreed upon for disease specific biomarkers,[10] and these also apply to criteria needed for CNS imaging markers. A biomarker should (i) *predict or differentiate* an individuals function having generic features; (ii) be *validated*; (iii) have a *high degree of specificity*; (iv) have a high *degree of reproducibility* with and across subjects; (v) be able to be performed and *measured easily*; (vi) be *defined in the context of the underlying neurobiology* of the specific disease in question; and (vii) have a low *risk* when performing the evaluation.

The brain is however a highly complex network of neurocircuits and diseases affect the integrated biology of these brain systems in a myriad of different ways. While brain function defines behavior (in healthy subjects, in a disease state or in response to drugs) there are a number of factors that make CNS biomarkers a challenge, including static and dynamic changes in CNS function as a result of the disease process or drug effect interweaved with the consequences of aging, gender, genetic background, and environment. Variations in brain function may also reflect subtypes within disorders that are representative of individual patients or different cohorts that change with duration of medication and disease. CNS biomarkers need to be a true measure of the underlying disease process and, by analogy to tumors, defining criteria for structural diseases such as neurodegeneration and strokes may be easier than for functional CNS diseases that will probably require a completely different set of approaches. Lessons could be learnt from cancer research where nearly all biomarkers are characterized through a defined set of criteria collectively known as 'data elements' facilitating standardization and interpretation.[10] Today, many CNS biomarkers are in an evolutionary stage as our understanding of the neurobiology of brain function and disease improves and as a consequence the field needs to be highly adaptive as it iterates towards defining measures that are meaningful and have wide consensus.

8.1.5   CNS Biomarker Technologies

Several technological approaches have potential to become CNS imaging biomarker tools (see Figure 8.1.2). Fusion of these functional, anatomical

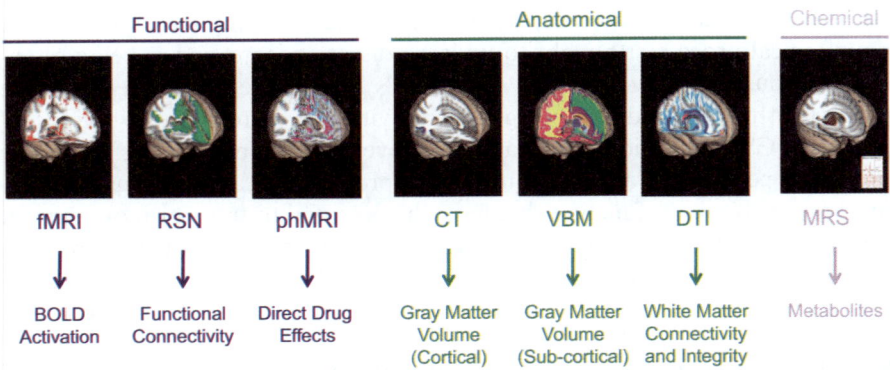

Figure 8.1.2   Functional, Structural and Chemical Approaches to Measures of
Brain Function. The figure shows the different approaches. Examples
of data obtained for functional imaging method for BOLD fMRI,
resting state networks (RSN) and pharmacological MRI (phMRI) are
shown in the first three panels on the left. Morphological or anatomical methods that measure changes in gray matter (voxel based
methods or VBM) include measure of sub-cortical or cortical (cortical
thickness) gray matter changes. Measure of chemical changes using
magnetic resonance spectroscopy (MRS) in neural and glial systems in
the brain is shown in the right panel for changes derived from specific
region of interest. Adapted From: Borsook D, Hargreaves R, Becerra L.
Can Functional Magnetic Resonance Imaging Improve Success Rates in
CNS Drug Discovery? *Expert Opin. Drug Discov.,* 2011 Jun 1;**6**(6).

and neurochemical imaging techniques undoubtedly gives increased power
to evaluate disease and drug effects in brain diseases but the challenge is to
reduce them to practice in their simplest form to enable cost effective and
timely biomarker approaches to diagnosis, treatment and CNS drug
discovery.

8.1.5.1   Anatomical CNS Biomarkers

Volumetric Imaging. Measurements of hippocampal and entorhinal cortex,
ventricular volume and cortical thickness have been central to biomarkers of
neurodegenerative disease and central to the Alzheimer's Disease Neuroimaging Initiative (ADNI; www.adni-info.org). Altered amygdala and hippocampal volumes have been linked to chronic depressive illness.[11,12] This
approach has been mostly used in measures of changes in volume of subcortical
structures.[13] Comparisons of segmentation techniques have been reviewed by
others,[14,15] and high levels of consistency have been shown using this approach
across multiple samples.[16] Regions that have been most reported on include the
hippocampus,[17] basal ganglia,[18] thalamus,[19] and amygdala.[20] These approaches may indicate duration of disease,[21] and provide markers for effective
therapies. Postmortem volumetric and anatomical imaging of the brain will
allow for high resolution of regions and may eventually be a highly useful

Table 8.1.3 Benefits on All Fronts: Example of Chronic Pain. Adapted From: Borsook D, Hargreaves R, Becerra L. Can Functional Magnetic Resonance Imaging Improve Success Rates in CNS Drug Discovery? *Expert Opin. Drug Discov.*, 2011 Jun 1;**6**(6).

Industry	"Due to the continuous increase in time and cost of drug development and the considerable amount of resources required by the traditional approach, companies can no longer afford to continue to late phase 3 with drugs which are unlikely to be therapeutically effective"	• Need to explore imaging as a legitimate tool to enhance decision making in drug development. • Need to decrease costs of evaluating a drug including no-go decisions • Potential to use healthy volunteers in pain imaging studies • Potential to study small numbers of patients for 'go' 'no go' to more extensive clinical trials
Regulatory Agencies	Clinical testing remains the core of registration of new compounds; however, traditional clinical trial methods are facing escalating drug development costs; novel drug targets with unproven therapeutic potential and the need for more compelling/objective methods to evaluate drugs are major hurdles in getting new and better drugs to patients. Objective methods including CNS biomarkers is one approach that may help overcome these hurdles.	• Define imaging biomarkers of the brain in health and disease • Objective approach for evaluation of central drug effects • Supportive evidence for efficacy • Reduced exposure of trial subjects to drugs of unknown value
Patients	Currently there is no objective measure for pain. Clinical treatments and therapeutic modalities are empirical. Subjective reports are highly variable. Treatment efficacy is only around 30% in controlled trials. The result is the use of a 'random' therapeutic approach for treatment of chronic pain.	• Image based patient evaluation • Objective image based evaluation of drug effects • Objective image based treatment decisions • Increased choice of drug therapies • Better treatment of chronic pain

approach since regions can then be a focus of further studies (*e.g.*, staining techniques *etc.*). Most recently more sensitive techniques than standard voxel based approaches have been used to "equilibrate" cortical thickness by ballooning out the brain and these enable the measurement of very small changes in cortical regional volumes (http://surfer.nmr.mgh.harvard.edu) and may have utility in detecting subtle yet significant changes in brain morphology such as those occurring during chronic pain.[22,23]

Diffusion Tensor Imaging: DTI is an MRI technique that measures changes in white matter tracts integrity based on microstructural changes in water diffusion.[24,25] Using this approach, anisotropic differences in normal *vs.*

abnormal tracts can be inferred. The technique has been used in a number of clinical disorders including multiple sclerosis (Roosendaal *et al.*, 2009 19027076[26]), depression,[27] drug abuse,[28] and Alzheimer's Disease,[29] and may offer insights into the underlying changes in brain state. Although there are some limitations to DTI,[30] when combined with fMRI studies, it may help improve our understanding of functional anatomical mapping of processing information.

8.1.5.2 Functional CNS Biomarkers

Evoked Stimuli: BOLD fMRI: This is now a relatively common technique that uses MRI to measure Blood Oxygen Level Dependent (BOLD) contrast changes in the brain as a surrogate for neuronal activity triggered by a wide variety of stimuli.[31,32] Whilst this simple interpretation of BOLD fMRI may have some limitations,[33,34] it has been used extensively to assess brain responses in cognitive, affective and neurological disorders. Its use in pain research has been particularly effective due to the ability to present well characterized discrete objective stimuli that can be cross validated using subjective visual analogue scales.[35]

Resting State Networks (RSN) and Functional Connectivity: The underlying approach here is to use fMRI (BOLD) to evaluate default mode networks (DMN) of brain functioning that takes place during resting states as opposed to functioning that takes place when performing a task. In this default mode, there is an organized baseline level of activity characterized by low frequency BOLD signal fluctuations among specific networks with relevant functions such as self-awareness, novelty of stimuli, *etc.*[36] These fMRI signals correspond to functionally relevant RSNs.[37] A number of studies have shown evidence for cohesive default mode networks,[38–40] that are consistent across healthy subjects,[41] and can be used to differentiate disease states from healthy states,[42,43] and potentially drug effects. Functional connectivity examines temporal correlation of brain response across brain structures.[44] It is believed that in disease and in the presence of a drug, some correlations are either suppressed or enhanced.[45,46]

Pharmacological MRI (phMRI). The evaluation of the functional effects of drugs on the brain through BOLD fMRI allows us to link drug exposure levels to the changes in evoked responses or to assess the dose related effects of drugs on RSN activity as a measure of central activity in healthy individuals or patients. A potential confound in drug studies are systemic changes that influence the BOLD response. An alternative MRI methodology is arterial spin labeling (ASL) where blood flow changes, as a surrogate for neuronal activity, are monitored with improved contrast and signal-to-noise through magnetization of the blood.[47] For both the BOLD and ASL approaches the correlation of brain activity changes with the distribution and occupancy of specific CNS receptor populations by drugs, (defined by molecular PET imaging (see below)), can be used as an efficacy surrogate to monitor the functional and anatomical consequences of drug target engagement and the dose-relationship of central responses. Recent technological advances with the development of combined MRI-PET cameras that will facilitate fusion imaging should pave the way for novel imaging protocols to assess the effects of drugs on brain function in health and disease.[48,49]

8.1.5.3 Chemical CNS Measures for Biomarkers

Magnetic Resonance Spectroscopy (MRS): MRS has the ability to assess neurochemical changes non-invasively.[50] Characterization of metabolic concentration (*e.g.*, glutamate, glutamine and gamma-aminobutyric acid (GABA)) of diverse brain conditions in neurological–psychiatric disorders can be defined.[51,52] Importantly, these changes can be detected before morphological (*i.e.*, tissue damage) changes occur,[53] and may be able to differentiate disease subtypes for example dementia,[54] or other alterations in cognitive function.[55] Most techniques have been focused on examining hydrogen nuclei present in metabolic molecules.[56] Research evaluating MRS measures of other nuclei of biological significance include phosphorous (^{31}P),[57] sodium (^{23}Na),[58] and carbon (^{13}C).[59] Local ^{31}P MRS measures allow for evaluation of high energy phosphates,[60] and fluxes of creatine kinase.[61] Sodium MRS is being used mainly to assess tissue sodium concentration,[62] as it changes significantly in stroke and tumors. ^{13}C measurements, allow for more information on understanding brain metabolism and neural transmission during functional activation in the brain;[63,64] specifically, ^{13}C MRS spectroscopy can be used for detection of metabolism at a high spatial and temporal resolution to measure glutamatergic (excitatory) and GABAergic (inhibitory) neuronal activity.

Receptor Occupancy and CNS Receptor Mapping: The most commonly used translational molecular imaging approach is nuclear imaging that visualizes radiolabeled probes, or radiotracers interacting with protein targets within or on the surface of cells. The two key radionuclide imaging modalities are PET, which uses tracers labeled with positron emitting radioisotopes (most commonly ^{11}C and ^{18}F), and SPECT, which detects tracers labeled with gamma-emitting radioactive isotopes (*e.g.*, ^{123}I). Both can be used to track small molecule and biologic therapeutics. Radiotracers are versatile and sensitive, and can be designed to track the drug itself, image the drug target or monitor key biochemical and physiological processes. Novel molecular tracer probes are the only way to measure receptor populations and pharmacology (at picomolar to nanomolar densities) quantitatively *in vivo* in both animals and humans. The development of small animal tomographic cameras (microPET or microSPECT with computer tomography) has facilitated translational bridging between preclinical and clinical central nervous system (CNS) research. Radionuclide imaging modalities, particularly PET, have become powerful tools for CNS drug discovery and development. The effective use of a PET tracer strategy in a research portfolio relies on parallel discovery efforts in medicinal chemistry and radiochemistry. Early discovery of a PET tracer helps selection of lead molecules in the preclinical lead optimization phase by focusing research on those drug candidates that achieve the highest target engagement with the lowest exposure thereby maximizing the potential therapeutic safety window. To have maximum value, PET tracers need to be clinically validated in advance of phase I clinical studies so that they can be incorporated into early pharmacokinetic, safety and tolerability study paradigms. Confirmation that drugs reach their targets using markers of engagement and pharmacodynamics is central to successful proof-of-concept testing. Data that determine how hard and how long a drug must hit its target to produce the desired pharmacologic effect are critical to later clinical trial designs. Proof-of-concept is established when target engagement can be linked to a meaningful change in a clinical endpoint. If a drug candidate sufficiently engages its target *in vivo* but does not produce the expected biological or clinical effects, the therapeutic concept is flawed and clinical development may be stopped. PET based mapping of neurotransmitter receptors in the brain can provide novel molecular information that can be correlated with functional data and cyto-architectural or anatomical maps,[65] to provide a better understanding of the underlying chemical basis of neurotransmission in health and disease.

8.1.6 Brain State and Biomarker Targets: Hurdles to Navigate

Brain activity is influenced by gender, genetic make-up, and age, in health to disease, in response to therapy and as a result of new adaptive experiences that impact memory, affect and learning.[66–69] Fundamental changes in functional, chemical, and morphological processes in the brain are predicted to define specific phenotypes at two ends of the spectrum – Healthy Brain State *vs.* Diseased Brain State. Using chronic neuropathic pain as an example the brain may be in transitional or "altered modes" that can result from the complex interaction between a number of factors including evolution of the disease, effects of therapy and other factors. The transition from normal to chronic (*e.g.*, neuropathic) pain may be the result of a number of processes that include altered connections based on increased or loss of dendritic sprouting,[70] and alterations in excitatory and inhibitory neurotransmitters.[71,72] These changes are manifest phenotypically as: (i) altered sensation resulting from acute tissue injury (*e.g.*, trauma) or injury to the central nervous system (*e.g.*, stroke); (ii) altered pain sensitivity (gain) to normally noxious and non-noxious stimuli; (iii) altered receptor–gene expression in peripheral and perhaps central nervous systems; (iv) altered brain function. Neuroimaging the sensory, emotional, cognitive, and modulatory changes that can occur in brain function may characterize CNS disorders but to become biomarkers the neuroimaging measures must be highly robust and specific and meet stringent criteria for acceptance. Functional neuroimaging biomarkers that lack sensitivity, specificity and reproducibility will confound the detection and quantification of drug effects on disease.

8.1.7 Biomarkers and Neuro-Psychiatric Clinical Practice

Specific diagnoses of most CNS diseases are relatively straightforward in the *forme fruste* or fulminant condition. However, given the many subtleties of disease subtypes it is unlikely that one approach will alone be useful and a combination of imaging, genetic, and clinical approaches (Figure 8.1.3) as exemplified in the pain field,[73–75] is likely to be needed. As biomarkers evolve, these will show a close association with the disease, correlate with disease change, be sensitive, reproducible and validated across multiple sites.[76] In the domain of personalized or stratified medicine, the development and application of validated biomarkers will allow for the health field to identify patients that will most likely benefit from the treatment or be a candidate for an adverse reaction to treatment.[77] In this way, clinical practice could be radically changed. The impact may also be beneficial in the identification or prediction of

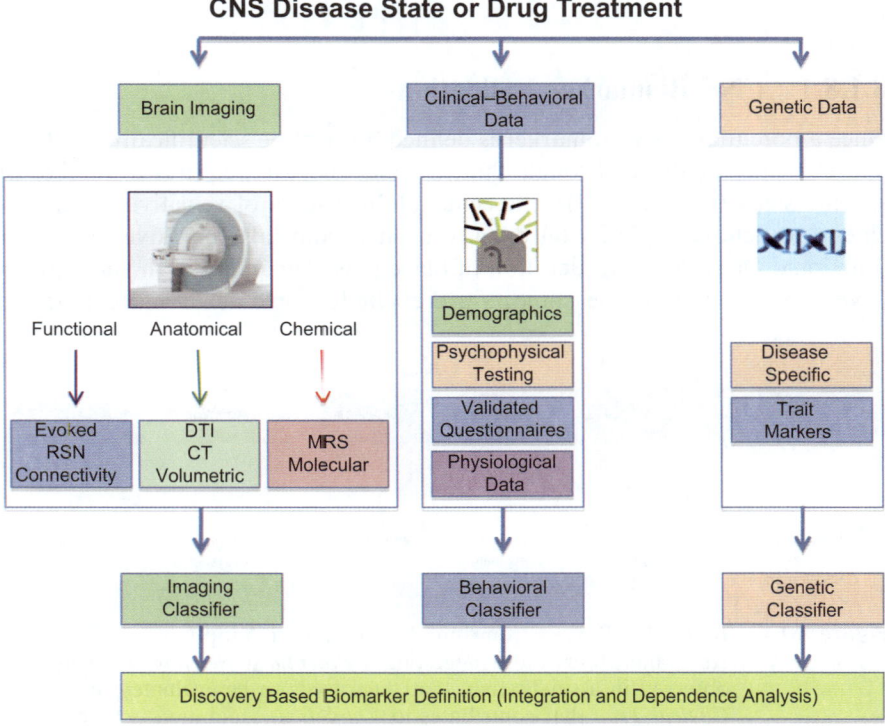

Figure 8.1.3 Discovery Based Biomarker Definition. The figure is a diagrammatic representation of combining imaging, clinical and behavioral data (including that obtained at the time of imaging *e.g.*, physiological data or prior to imaging *e.g.*, clinical profile or other evaluative tools), and genetic data (*e.g.*, trait or specific genetic makers) for the disease. From: D. Borsook, *Brain*, 2011, in press.

responders and non-responders, which may be due to genetic variations, or lack of objective diagnostic processes.

8.1.8   CNS Biomarker Selection and Validation

The use of biomarkers can streamline decision making in the confirmatory stages of clinical development of new CNS drugs by determining whether the drug is reaching and affecting the molecular target in humans, delivering findings that are consistent with preclinical biological hypotheses, and by providing measurable endpoints that predict desired or undesired clinical effects.[78] We now have the ability to measure changes in structure function and chemical changes in the brain and spinal cord using neuroimaging providing potential biomarkers for static or dynamic characterization of disease and drug effects. Current approaches often initially explore multiple imaging modalities grounded in the neuropathology or clinical presentation of the condition to select the best diagnostic to validate as a disease and response biomarker.

8.1.8.1   CNS Biomarker Validation

Once a potential CNS Biomarker is defined, it must be scientifically validated and clinically qualified through clinical trials and this requires a number of specific steps (Figure 8.1.4). These include measures of sensitivity and specificity to determine detection of false positive and false negative rates. For functional measures, consideration of placebo is also an issue. In validating a CNS biomarker, multiple subjects are evaluated on one or more times to

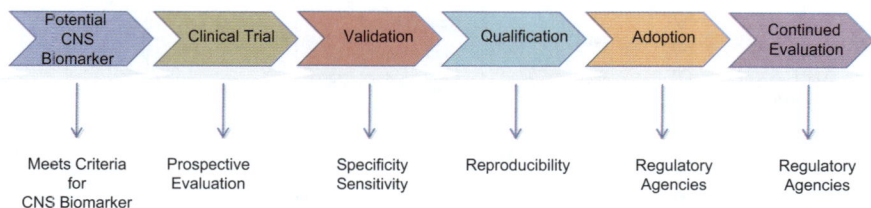

Figure 8.1.4   Biomarker Process from Initial Definition of Adoption.
An initial observation suggesting a *Potential CNS Biomarker* may be observed in a small study that then needs to be evaluated in a larger *Clinical Trial* that contributes to *Validation*. Validation is followed by *Qualification* of the biomaker and then the required *Regulatory Adoption*. Once adopted, the process of continued evaluation of the biomarker defines its continued status (*Continued Evaluation*) that includes potential refinements as technologies and larger clinical datasets become available. From: Borsook D, Hargreaves R, Becerra L. Can Functional Magnetic Resonance Imaging Improve Success Rates in CNS Drug Discovery? *Expert Opin. Drug Discov.*, 2011 Jun 1;**6**(6).

provide both inter-subject as well as intra-subject measure of variability in order to define its robustness and reliability. Longitudinal trials may be needed to evaluate biomarker stability, often an issue in progressive or chronic CNS disease that may change as a result of the disease itself or the drug effect independent of drug effectiveness. Given the plasticity of brain systems and that changes in the brain may precede clinical changes (as exemplified by multiple sclerosis,[79] it is important to understand biomarkers in the context of clinical outcomes. One of the great challenges in validation is the implementation of standards for multi-site collection of CNS imaging data. Integrated research networks are likely to be necessary for wide-scale clinical validation and acceptance of CNS biomarkers and this will require consortia approaches that have well defined research operating plans and shared goals for success.[80] It is fair to say that with perhaps the exception of Alzheimer's disease, qualification of biomarkers for CNS disease and the evaluation of drug candidates has not been as rapid as in other therapeutic areas. It is hoped that increased efforts on CNS biomarker development coordinated through the Neuroscience section of fNIH (http://www.fnih.org/) and active support on their use from regulatory authorities such as FDA (Critical Path Initiative http://www.fda.gov/ScienceResearch/SpecialTopics/CriticalPathInitiative/default.htm) will result in CNS Biomarkers becoming increasingly incorporated into clinical trials.

8.1.9 CNS Biomarkers for Drug Development

In the domain of drug development, CNS biomarkers allow an earlier, more robust assessment to be made of the therapeutic window between drug safety and efficacy.[81–83] Biomarkers may allow for increased success in drug development by eliminating the molecules and therapeutic hypotheses that are most likely to fail. Proof of target engagement is a key step in CNS drug development that provides information on receptor occupancy that is essential to guide clinical efficacy testing. Reliable estimates of dose occupancy relationships requires knowledge of the specificity of human receptor interaction, the CNS receptor distribution of the drug target and the availability of receptor imaging agents. When PET tracers that measure target engagement are lacking, functional pharmacodynamic measures, such as fMRI BOLD and arterial spin labeling (ASL), may provide pharmacodymanic readouts that can be used to guide dosing in clinical proof of concept testing. Occasionally, when CNS receptor markers are lacking, the engagement of central targets that are also expressed in the periphery can be estimated using peripheral surrogate markers when a robust relationship has been established between central and peripheral engagement through preclinical pharmacological investigations. However, many highly receptor targeted drugs (*e.g.*, cannabinoids,[84] or selective serotonin reuptake inhibitors,[85] impact multiple CNS systems, and dissecting the relationship of specific target engagement to effects on brain function and disease can be difficult.[84]

Ideally, a biomarker for a CNS drug candidate would have: (i) biological basis for its effects on the disease; (ii) a reliable or consistent response of the measured response in brain function or behavior to the drug; (iii) a specific response to dosing including dose response. The use of healthy individuals for testing the efficacy of drugs using specific biomarkers is a potentially interesting concept since brain systems affected in disease are essentially intact.[86,87] The concomitant use of genetic or epigenetic measures in volunteers may however help define responder populations that could be the basis of trial enrichment in the early phases of drug discovery, and in the future form the basis for differential diagnosis and personalization of therapy. The judicious use of biomarkers will enhance clinical trials because responsive populations would not get lost when patient selection is based on subjective measures.

8.1.10   Potential CNS Neuroimaging Biomarkers

Various CNS diseases currently have potential brain imaging based biomarker candidates with high face validity for evaluating disease and therapeutic effects.

Parkinson's Disease – Targeting CNS Biomarkers for Dopamine Dysfunction. Like other neurodegenerative diseases, early detection of Parkinson's disease using neuroimaging measures of the key neurotransmitter systems impacted by disease will facilitate diagnosis, tracking disease progression and the efficacy of potential disease modifying treatments.[76,88] Dopamine based imaging biomarkers could help diagnose susceptible individuals (before 70% of dopamine neurons are lost, when clinical signs then begin to become apparent) and also differentiate idiopathic Parkinson's from other forms of movement disorders with different underlying etiology.[89–91] For example positron emission tomography (PET) imaging has been used to assess dopaminergic function,[92,93] and to evaluate the effects of putative neuroprotective agents in clinical trials.[94] fMRI has been used to evaluate the functional abnormalities in cognitive, reward and motor pathways that accompany Parkinson's disease,[95] and this approach may be valuable to assess clinical responses to palliative and putative disease modifying treatments. However, despite well-described neuropathology and the anatomical and neurochemical specificity there are currently no validated CNS biomarkers for Parkinson's disease.

Migraine – Brainstem Abnormalities and Measures of Interictal Dysfunction: Advances in genetics has defined biomarkers for familial migraine.[96] The search for neuroimaging biomarkers for migraine has included imaging of anatomical changes in the brainstem and alterations in chemistry in cortical regions. Brainstem dysfunction has been reported using PET,[97] and fMRI,[98] and findings are suggestive of altered descending modulatory processing that may result in decreased thresholds to migraine trigger factors. Observations of increased iron in the brainstem (and other deep brain nuclei) of migraineurs has been suggested to be associated with dysfunction in anti-antinociceptive processing in these brain regions.[99] Recent MR spectroscopy studies have found changes in excitatory amino acid balance in the brain of migraineurs,[72] that may be linked to the increased cortical excitability that has been well documented in migraine patients particularly those with migraine aura. Finally, several studies have reported morphological alterations in the brains of migraine patients.[100] Each of these traits has been proposed as a potential biomarker for migraine susceptibility but all require clinical qualification.

Alzheimer's Disease – Gray Matter Volume Changes as Targets for Therapeutic Measures: in its early stages Alzheimer's disease can be difficult to separate from normal age related minimal cognitive impairment and other dementias of non-Alzheimer type. The use of neuroimaging markers of Alzheimer pathology such as PET tracers that can identify amyloid burden together with CSF fluid biomarkers that reflect disease seem to hold the most promise in early diagnosis of disease.[101] Anatomical imaging has focused on degeneration in mediotemporal, neocortical, and subcortical areas of the brain, including measures of reduced hippocampal and entorhinal volumes since these regions of the brain are known from post-mortem neuropathology to be focal points for neurodegeneration and to be involved in cognitive brain functions that are impacted by the disease. Brain volumes and shifts in ventricular boundaries have also been shown to be sensitive indices of neurodegeneration in Alzheimer's disease.[102] Today, the Alzheimer's Disease Neuroimaging Initiative (ADNI) leads in defining neuroimaging biomarkers that are becoming scientifically well validated and clinically qualified such that they can now be incorporated as exploratory early outcome measures alongside neurocognitive assessments in phase II and III clinical trials of potential Alzheimer's disease modifying therapies. Paradoxically their definitive validation for drug discovery requires the use of these biomarkers in trials of a successful therapy and, whilst many trials are ongoing, such an advance has yet to be made.

Depression and Treatment Outcomes: In major depressive disorders there are several early indications that neuroimaging biomarkers could have utility to select patients for clinical trials and to evaluate treatment outcomes. For example successful antidepressant therapy has been suggested to be predicted by patterns of pre-treatment brain activity,[103,104] and abnormalities of the hippocampus and basal ganglia in unipolar disorder and amygdala and cerebellum in bipolar disorders,[105] to segregate subtypes of mood disorders.

8.1.10.1 CNS Biomarkers and the FDA

The development of biomarkers including those for CNS disorders require large-scale integrated efforts on the part of government, industry and academia. A key enabling step was taken with the publication of the FDA's critical path initiative which was described as "... the FDA's effort to stimulate and facilitate a national effort to modernize the scientific process through which a potential human drug, biological product, or medical device is transformed from a discovery or "proof of concept" into a medical product" (http://www.fda.gov/ScienceResearch/SpecialTopics/CriticalPathInitiative/default.htm). A direct product of the critical path initiative has been the launch of a Biomarker Consortium at the FNIH (http://www.fnih.org/) that has the goal of developing biomarkers that can be deployed to speed the development of safe and effective targeted drugs for many common diseases including CNS disorders. It brings together NIH, FDA, CMS, PhRMA, BIO, Patient Groups and Academia with the goal of identification and qualification of biomarkers. Once established, these biomarkers could be used to define responsive populations and perhaps to provide an early comparative assessment of the likely effectiveness of new mechanism drugs against established therapies. In the future, objective measures that define the most appropriate patients to treat or those who are most likely to benefit will become an important part of the new drug application, review and approval process.

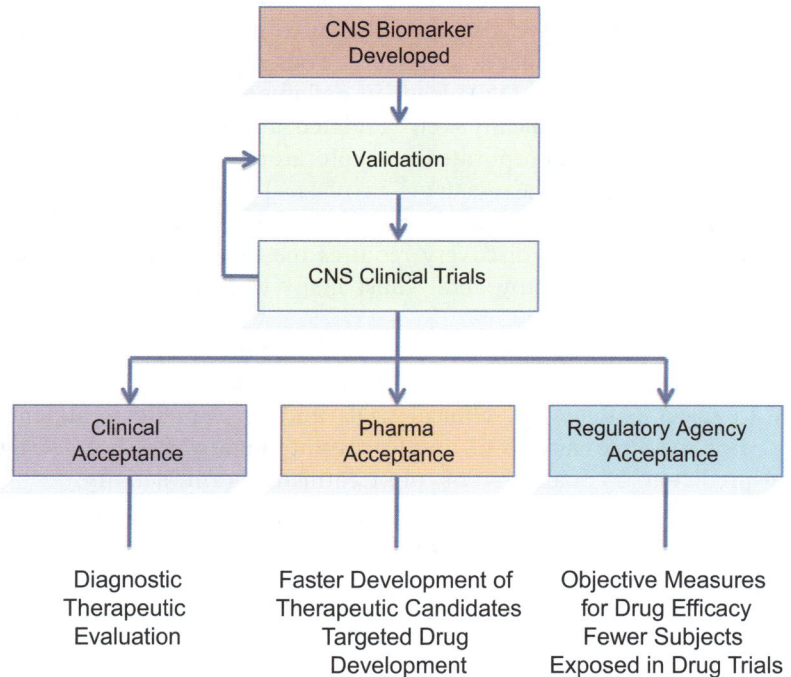

Figure 8.1.5   CNS Biomarker Value. Adoption or acceptance of a CNS biomarker, particularly in difficult to diagnose or evaluate disease conditions may have immediate benefits in the Clinical, Pharmaceutical and Regulatory Agency domains (see also Table 8.1.2).

8.1.11 Conclusions

Whilst the use of CNS biomarkers holds great promise to improve clinical practice, therapy and the discovery of novel CNS drugs, their adoption will depend on a number of issues including their sensitivity, specificity, cost and, as support for drug registration, regulatory acceptance (Figure 8.1.5). During early drug discovery, decision-making based on biomarkers is a development risk that is assumed by the companies to reduce costs and cycle time by getting an early focus on the molecules and clinical hypotheses having the highest probability of success. In clinical practice the use of biomarkers to guide therapy carries different relative risks in terms of defining safety and effectiveness in targeted diseased populations. Current measures of brain function using imaging are undoubtedly complex and require specialized and trained individuals and equipment. However, like many imaging processes, these obstacles can be overcome through purposeful harmonization of data acquisition techniques and the development of "plug and play" applications using standardized tools for data collection and analysis developed to consensus by consortia that involve key stakeholders.

In conclusion, whilst some authors have added notes of caution to the excitement around the use of biomarkers,[83] Biomarkers that are truly *Found in Translation*"[106] have the potential to become central to clinical practice and drug discovery. CNS biomarkers enable the characterization of patient populations and quantitation of the extent to which new drugs reach their intended targets, alter proposed pathophysiological mechanisms and achieve clinical outcome.[107] CNS biomarkers have the potential to inform decision making for pharmaceutical companies and regulatory agencies with respect to candidate drugs and their indications, in order to bring new medicines to the right patients more rapidly and effectively than they are today.[107]

Acknowledgements

Supported by NINDS K24 (NS064050) to DB and the L Herlands Fund (DB,LB).

References

1. M. Gordian, N. Singh, R. Zemmel, T. Elias, *In Vivo*, 2006, **24**, 49–54.
2. J. Woodcock, Update on FDA's Critical Path Initiative, 2004.
3. T. Gasser, *Neurology*, 2009, **72**, S27–31.
4. J. Mair, *Wien. Med. Wochenschr.*, 2007, **157**, 48–56.
5. J. Howie-Esquivel and M. White, *J. Cardiovasc. Nurs.*, 2008, **23**, 124–131.
6. A. Pfutzner, M. M. Weber and T. Forst, *Clin. Lab.*, 2008, **54**, 485–490.
7. D. K. Li, M. J. Li, A. Traboulsee, G. Zhao, A. Riddehough and D. Paty, *Adv. Neurol.*, 2006, **98**, 203–226.

8. M. Rovaris, F. Barkhof, M. Calabrese, N. De Stefano, F. Fazekas, D. H. Miller, X. Montalban, C. Polman, M. A. Rocca, A. J. Thompson, T. A. Yousry and M. Filippi, *Neurology*, 2009, **72**, 1693–1701.

9. N. J. Diederich and C. G. Goetz, *Neurology*, 2008, **71**, 677–684.

10. M. Khatami, *Cell Biochem. Biophys.*, 2007, **47**, 187–198.

11. Y. I. Sheline, M. H. Gado and J. L. Price, *NeuroReport*, 1998, **9**, 2023–2028.

12. Y. I. Sheline, M. Sanghavi, M. A. Mintun and M. H. Gado, *J. Neurosci.*, 1999, **19**, 5034–5043.

13. J. Jovicich, S. Czanner, X. Han, D. Salat, A. van der Kouwe, B. Quinn, J. Pacheco, M. Albert, R. Killiany, D. Blacker D, P. Maguire, D. Rosas, N. Makris, R. Gollub, A. Dale, B. C. Dickerson and B. Fischl, *Neuroimage*, 2009, **46**, 177–192.

14. K. O. Babalola, B. Patenaude, P. Aljabar, J. Schnabel, D. Kennedy, W. Crum, S. Smith, T. F. Cootes, M. Jenkinson and D. Rueckert., *Med. Image Comput. Comput. Assist. Interv. Int. Conf. Med. Image Comput. Comput. Assist. Interv.*, 2008, **11**, 409–416.

15. F. Klauschen, A. Goldman, V. Barra, A. Meyer-Lindenberg and A. Lundervold, *Hum. Brain Mapp.*, 2009, **30**, 1310–1327.

16. A. M. Fjell, L. T. Westlye, I. Amlien, T. Espeseth, I. Reinvang, N. Raz, I. Agartz, D. H. Salat, D. N. Greve, B. Fischl, A. M. Dale and K. B. Walhovd, *Cereb. Cortex*, 2009, **19**, 2001–12.

17. M. C. McKinnon, K. Yucel, A. Nazarov and G. M. MacQueen, *J. Psychiatr. Neurosci.*, 2009, **34**, 41–54.

18. J. C. Looi, O. Lindberg, B. Liberg, V. Tatham, R. Kumar, J. Maller, E. Millard, P. Sachdev, G. Hogberg, M. Pagani, L. Botes, E. L. Engman, Y. Zhang, L. Svensson and L. O. Wahlund, *Psychiatry Res.*, 2008, **163**, 279–288.

19. L. W. de Jong, K. van der Hiele, I. M. Veer, J. J. Houwing, R. G. Westendorp, E. L. Bollen, P. W. de Bruin, H. A. Middelkoop, M. A. van Buchem and J. van der Grond, *Brain*, 2008, **131**, 3277–3285.

20. J. P. Hamilton, M. Siemer and I. H. Gotlib, *Mol. Psychiatry*, 2008, **13**, 993–1000.

21. M. Battaglini, A. Giorgio, M. L. Stromillo, M. L. Bartolozzi, L. Guidi, A. Federico and N. De Stefano, *J. Neurol. Sci.*, 2009, **282**, 55–60.

22. A. V. Apkarian, Y. Sosa, S. Sonty, R. M. Levy, R. N. Harden, T. B. Parrish and D. R. Gitelman, *J. Neurosci.*, 2004, **24**, 10410–10415.

23. A. Kuchinad, P. Schweinhardt, D. A. Seminowicz, P. B. Wood, B. A. Chizh and M. C. Bushnell, *J. Neurosci.*, 2007, **27**, 4004–4007.

24. A. L. Alexander, J. E. Lee, M. Lazar and A. S. Field, *Neurotherapeutics*, 2007, **4**, 316–329.

25. P. Mukherjee, J. I. Berman, S. W. Chung, C. P. Hess and R. G. Henry, *Am. J. Neuroradiol.*, 2008, **29**, 632–641.

26. S. D. Roosendaal, J. J. Geurts, H. Vrenken, H. E. Hulst, K. S. Cover, J. A. Castelijns, P. J. Pouwels, F. Barkhof, *Neuroimage*, 2009, **44**(4), 1397–1403.

27. J. S. Shimony, Y. I. Sheline, G. D'Angelo, A. A. Epstein, T. L. Benzinger, M. A. Mintun, R. C. McKinstry and A. Z. Snyder, *Biol. Psychiatry*, 2009, **66**, 245–252.
28. A. Pfefferbaum, M. Rosenbloom, T. Rohlfing and E. V. Sullivan, *Biol. Psychiatry*, 2009, **65**, 680–690.
29. N. H. Stricker, B. C. Schweinsburg, L. Delano-Wood, C. E. Wierenga, K. J. Bangen, K. Y. Haaland, L. R. Frank, D. P. Salmon and M. W. Bondi, *Neuroimage*, 2009, **45**, 10–16.
30. Y. Assaf and O. Pasternak, *J. Mol. Neurosci.*, 2008, **34**, 51–61.
31. N. K. Logothetis, *Phios. Trans. R. Soc. London, Ser. B*, 2002, **357**, 1003–1037.
32. B. Krekelberg, G. M. Boynton and R. J. van Wezel, *Trends Neurosci.*, 2006, **29**, 250–256.
33. A. Bartels, N. K. Logothetis and K. Moutoussis, *Trends Neurosci.*, 2008, **31**, 444–453.
34. N. K. Logothetis, *Nature*, 2008, **453**, 869–878.
35. R. C. Coghill, C. N. Sang, J. M. Maisog and M. J. Iadarola, *J. Neurophysiol.*, 1999, **82**, 1934–1943.
36. P. Fransson, *Hum. Brain Mapp.*, 2005, **26**, 15–29.
37. M. De Luca, C. F. Beckmann, N. De Stefano, P. M. Matthews and S. M. Smith, *Neuroimage*, 2006, **29**, 1359–1367.
38. M. D. Greicius, B. Krasnow, A. L. Reiss and V. Menon, *Proc. Natl. Acad. Sci. U. S. A.*, 2003, **100**, 253–258.
39. V. G. van de Ven, E. Formisano, D. Prvulovic, C. H. Roeder and D. E. Linden, *Hum. Brain Mapp.*, 2004, **22**, 165–178.
40. P. Bellec, V. Perlbarg, S. Jbabdi, M. Pelegrini-Issac, J. L. Anton, J. Doyon and H. Benali, *Neuroimage*, 2006, **29**, 1231–1243.
41. J. S. Damoiseaux, S. A. Rombouts, F. Barkhof, P. Scheltens, C. J. Stam, S. M. Smith and C. F. Beckmann, *Proc. Natl. Acad. Sci. U. S. A.*, 2006, **103**, 13848–13853.
42. M. D. Greicius, G. Srivastava, A. L. Reiss and V. Menon, *Proc. Natl. Acad. Sci. U. S. A.*, 2004, **101**, 4637–4642.
43. A. G. Garrity, G. D. Pearlson, K. McKiernan, D. Lloyd, K. A. Kiehl and V. D. Calhoun, *Am. J. Psychiatry*, 2007, **164**, 450–457.
44. K. Friston, *PLoS Biol.*, 2009, **7**, e33.
45. C. Kelly, G. de Zubicaray, A. Di Martino, D. A. Copland, P. T. Reiss, D. F. Klein, F. X. Castellanos, M. P. Milham and K. McMahon, *J. Neurosci.*, 2009, **29**, 7364–7378.
46. L. E. Hong, H. Gu, Y. Yang, T. J. Ross, B. J. Salmeron, B. Buchholz, G. K. Thaker and E. A. Stein, *Arch. Gen. Psychiatry*, 2009, **66**, 431–441.
47. J. A. Detre, J. Wang, Z. Wang and H. Rao, *Curr. Opin. Neurol.*, 2009, **22**, 348–355.
48. H. P. Schlemmer, B. J. Pichler, M. Schmand, Z. Burbar, C. Michel, R. Ladebeck, K. Jattke, D. Townsend, C. Nahmias, P. K. Jacob, W. D. Heiss and C. D. Claussen, *Radiology*, 2008, **248**, 1028–1035.

49. T. Beyer and Pichler, *Eur. J. Nucl. Med. Mol. Imaging*, 2009, **36**(Suppl. 1), S1–2.

50. D. P. Soares and M. Law, *Clin. Radiol.*, 2009, **64**, 12–21.

51. A. P. Burlina, T. Aureli, F. Bracco, F. Conti and L. Battistin, *Neurochem. Res.*, 2000, **25**, 1365–1372.

52. I. K. Lyoo and P. F. Renshaw, *Biol. Psychiatry*, 2002, **51**, 195–207.

53. G. Castellino, M. Govoni, M. Padovan, P. Colamussi, M. Borrelli and F. Trotta, *Ann. Rheum. Dis.*, 2005, **64**, 1022–1027.

54. R. S. Jones and A. D. Waldman, *Neurol. Res.*, 2004, **26**, 488–495.

55. A. J. Ross and P. S. Sachdev, *Brain Res. Brain Res. Rev.*, 2004, **44**, 83–102.

56. T. M. Rudkin and D. L. Arnold, *Arch. Neurol.*, 1999, **56**, 919–926.

57. F. Arias-Mendoza and T. R. Brown, *Dis. Markers*, 2003, **19**, 49–68.

58. D. B. Clayton and R. E. Lenkinski, *Acad. Radiol.*, 2003, **10**, 358–365.

59. R. Gruetter, G. Adriany, I. Y. Choi, P. G. Henry, H. Lei and G. Oz, *NMR Biomed.*, 2003, **16**, 313–338.

60. J. J. Ackerman, P. J. Bore, D. G. Gadian, T. H. Grove and G. K. Radda, *Philos. Trans. R. Soc. London, Ser. B*, 1980, **289**, 425–436.

61. F. Du, X. H. Zhu, H. Qiao, X. Zhang X and W. Chen, *Magn. Reson. Med.*, 2007, **57**, 103–114.

62. R. Ouwerkerk, *J. Am. Coll. Radiol.*, 2007, **4**, 739–741.

63. P. G. Henry, I. Tkac and R. Gruetter, *Magn. Reson. Med.*, 2003, **50**, 684–692.

64. R. A. de Graaf, G. F. Mason, A. B. Patel, K. L. Behar KL and D. L. Rothman, *NMR Biomed.*, 2003, **16**, 339–357.

65. K. Zilles and K. Amunts, *Curr. Opin. Neurol.*, 2009, **22**, 331–339.

66. P. M. Rossini and G. Dal Forno, *Phys. Med. Rehabil. Clin. N. Am.*, 2004, **15**, 263–306.

67. D. H. Miller, *NeuroRx*, 2004, **1**, 284–294.

68. T. Endo, C. Spenger, E. Westman, T. Tominaga and L. Olson L, *Exp. Neurol.*, 2008, **209**, 155–160.

69. J. H. Kaas, H. X. Qi, M. J. Burish, O. A. Gharbawie, S. M. Onifer and J. M. Massey, *Exp. Neurol.*, 2008, **209**, 407–416.

70. A. E. Metz, H. J. Yau, M. V. Centeno, A. V. Apkarian and M. Martina, *Proc. Natl. Acad. Sci. U. S. A.*, 2009, **106**, 2423–2428.

71. I. D. Grachev, B. E. Fredrickson and A. V. Apkarian, *Pain*, 2000, **89**, 7–18.

72. A. Prescot, L. Becerra, G. Pendse, S. Tully, E. Jensen, R. Hargreaves, P. Renshaw, R. Burstein and Borsook, *Mol. Pain.*, 2009, **5**, 34.

73. I. Tegeder, M. Costigan, R. S. Griffin, A. Abele, I. Belfer, H. Schmidt, C. Ehnert, J. Nejim, C. Marian, J. Scholz, T. Wu, A. Allchorne, L. Diatchenko, A. M. Binshtok, D. Goldman, J. Adolph, S. Sama, S. J. Atlas, W. A. Carlezon, A. Parsegian, J. Lötsch, R. B. Fillingim, W. Maixner, G. Geisslinger, M. B. Max and C. J. Woolf, *Nat. Med.*, 2006, **12**, 1269–1277.

74. J. Scholz, R. J. Mannion, D. E. Hord, R. S. Griffin, B. Rawal, H. Zheng, D. Scoffings, A. Phillips, J. Guo, R. J. Laing, S. Abdi, I. Decosterd and C. J. Woolf, *PLoS Med.*, 2009, **6**, e1000047.
75. B. Kosarac, A. A. Fox and C. D. Collard, *Curr. Opin. Anaesthesiol.*, 2009, **22**, 476–482.
76. A. W. Michell, S. J. Lewis, T. Foltynie and R. A. Barker, *Brain*, 2004, **127**, 1693–1705.
77. M. R. Trusheim, E. R. Berndt and F. L. Douglas, *Nat. Rev. Drug Discovery*, 2007, **6**, 287–293.
78. J. Kuhlmann, *Ernst Schering Res. Found. Workshop*, 2007, 29–45.
79. P. M. Matthews, *Neurcimag. Clin. N. Am.*, 2009, **19**, 101–112.
80. D. Borsook, D. Bleakman, R. Hargreaves, J. Upadhyay, K. F. Schmidt and L. Becerra, *Neuroimage*, 2008, **42**, 461–466.
81. W. A. Colburn, *J. Clin. Pharmacol.*, 2000, **40**, 1419–1427.
82. C. Carini, *IDrugs*, 2007, **10**, 395–398.
83. E. Marrer and F. Dieterle, *Chem. Bio.l Drug Des.*, 2007, **69**, 381–394.
84. L. Zuurman, A. E. Ippel, E. Moin and J. M. van Gerven, *Br. J. Clin. Pharmacol.*, 2009, **67**, 5–21.
85. G. J. Dumont, S. J. de Visser, A. F. Cohen AF and J. M. van Gerven, *Br. J. Clin. Pharmacol.*, 2005, **59**, 495–510.
86. S. J. de Visser, J. van der Post, M. S. Pieters, A. F. Cohen and J. M. van Gerven, *Br. J. Clin. Pharmacol.*, 2001, **51**, 119–132.
87. S. J. de Visser, J. P. van der Post, P. P. de Waal PP, F. Cornet, A. F. Cohen and J. M. van Gerven, *Br. J. Clin. Pharmacol.*, 2003, **55**, 39–50.
88. D. J. Brooks, *Semin. Neurol.*, 2008, **28**, 435–445.
89. A. J. Hughes, S. E. Daniel and A. J. Lees, *Neurology*, 2001, **57**, 1497–1499.
90. G. M. Knudsen, M. Karlsborg, G. Thomsen, K. Krabbe, L. Regeur, T. Nygaard, C. Videbaek and L. Werdelin, *Eur. J. Nucl. Med. Mol. Imaging*, 2004, **31**, 1631–1638.
91. M. Schreckenberger, S. Hagele, T. Siessmeier, H. G. Buchholz, H. Armbrust-Henrich, F. Rosch, G. Grunder, P. Bartenstein and T. Vogt, *Eur. J. Nucl. Med. Mol. Imaging*, 2004, **31**, 1128–1135.
92. J. Seibyl, D. Jennings, R Tabamo and K. Marek, *Minerva Med.*, 2005, **96**, 353–364.
93. B. Ravina, D. Eidelberg, J. E. Ahlskog, R. L. Albin, D. J. Brooks, M. Carbon, V. Dhawan, A. Feigin, S. Fahn, M. Guttman, K. Gwinn-Hardy, H. McFarland, R. Innis, R. G. Katz, K. Kieburtz, S. J. Kish, N. Lange, J. W. Langston, K. Marek, L. Morin, C. Moy, D. Murphy, W. H. Oertel, G. Oliver, Y. Palesch, W. Powers, J. Seibyl, K. D. Sethi, C. W. Shults, P. Sheehy, A. J. Stoessl and R. Holloway, *Neurology*, 2005, **64**, 208–215.
94. I. Halperin, M. Morelli, A. D. Korczyn, M. B. Youdim and S. A. Mandel, *Neurotherapeutics*, 2009, **6**, 128–140.
95. J. B. Rowe, L. Hughes, B. C. Ghosh, D. Eckstein, C. H. Williams-Gray, S. Fallon, R. A. Barker and A. M. Owen, *Brain*, 2008, **131**, 2094–2105.

96. B. de Vries, R. R. Frants, M. D. Ferrari and A. M. van den Maagdenberg, *Hum. Genet.*, 2009, **126**, 115–132.
97. S. K. Aurora, P. M. Barrodale, R. L. Tipton and A. Khodavirdi, *Headache*, 2007, **47**, 996–1003.
98. E. A. Moulton, R. Burstein, S. Tully, R. Hargreaves, L. Becerra and D. Borsook, *PLoS One*, 2008, **3**, e3799.
99. M. C. Kruit, L. J. Launer, J. Overbosch, M. A. van Buchem and M. D. Ferrari, *Cephalalgia*, 2009, **29**, 351–359.
100. A. May, *Nat. Rev. Neurol.*, 2009, **5**, 199–209.
101. H. Hampel, K. Burger, S. J. Teipel, A. L. Bokde, H. Zetterberg and K. Blennow, *Alzheimers Dement.*, 2008, **4**, 38–48.
102. I. Driscoll, C. Davatzikos, Y. An, X. Wu, D. Shen, M. Kraut M. and S. M. Resnick, *Neurology*, 2009, **72**, 1906–1913.
103. H. S. Mayberg, *Br. Med. Bull.*, 2003, **65**, 193–207.
104. K. C. Evans, D. D. Dougherty, M. H. Pollack and S. L. Rauch, *Ann. Clin. Psychiatry*, 2006, **18**, 33–42.
105. P. Brambilla, F. Barale, E. Caverzasi and J. C. Soares, *Epidemiol Psichiatr Soc*, 2002, **11**, 88–99.
106. B. P. Lockhart and B. Walther, *Med. Sci.*, 2009, **25**, 423–430.
107. R. Frank and R. Hargreaves, *Nat. Rev. Drug Discovery*, 2003, **2**, 566–580.
108. L. E. Hebert, P. A. Scherr, J. L. Bienias, D. A. Bennett and D. A. Evans, *Arch. Neurol.*, 2003, **60**, 1119–1122.
109. Alzheimer's Association, *Alzheimer's Dementia*, 2009, **5**, 234–270.
110. J. A. Driver, G. Logroscino, J. M. Gaziano and T. Kurth, *Neurology*, 2009, **72**, 432–438.
111. W. T. Longstreth, Jr., T. D. Koepsell, T. G. Ton, A. F. Hendrickson and G. van Belle, *Sleep*, 2007, **30**, 13–26.
112. M. M. Ohayon, R. G. Priest, J. Zulley, S. Smirne and T. Paiva, *Neurology*, 2002, **58**, 1826–1833.
113. R. C. Kessler, O. Demler, R. G. Frank, M. Olfson, H. A. Pincus, E. E. Walters, P. Wang, K. B. Wells and A. M. Zaslavsky, *N. Engl. J. Med.*, 2005, **352**, 2515–2523.
114. D. A. Regier, W. E. Narrow, D. S. Rae, R. W. Manderscheid, B. Z. Locke and F. K. Goodwin, *Arch. Gen. Psychiatry*, 1993, **50**, 85–94.
115. C. Toth, J. Lander and S. Wiebe, *Pain. Med.*, 2009.
116. F. Wolfe, K. Ross, J. Anderson, I. J. Russell and L. Hebert, *Arthritis Rheum.*, 1995, **38**, 19–28.
117. J. K. Freburger, G. M. Holmes, R. P. Agans, A. M. Jackman, J. D. Darter, A. S. Wallace, L. D. Castel, W. D. Kalsbeek and T. S. Carey, *Arch. Intern. Med.*, 2009, **169**, 251–258.
118. J. Y. Reginster, *Rheumatology (Oxford)*, 2002, **41**(Supp. 1), 3–6.
119. M. E. Bigal and R. B. Lipton, *Neurol. Clin.*, 2009, **27**, 321–334.
120. S. J. Tepper. *Headache*, 2008, **48**, 730–731.
121. M. Yeargin-Allsopp, C. Rice, T. Karapurkar, N. Doernberg, C. Boyle and C. Murphy, *JAMA, J. Am. Med. Assoc.*, 2003, **289**, 49–55.

122. W. J. Henneman, J. D. Sluimer, J. Barnes, W. M. van der Flier, I. C. Sluimer, N. C. Fox, P. Scheltens, H. Vrenken and F. Barkhof, *Neurology*, 2009, **72**, 999–1007.

123. J. H. Morra, Z. Tu, L. G. Apostolova, A. E. Green, C. Avedissian, S. K. Madsen, N. Parikshak, A. W. Toga, C. R. Jack, Jr., N. Schuff, M. W. Weiner and P. M. Thompson, *Neuroimage*, 2009, **45**, S3–15.

124. C. R. Jack, Jr., M. M. Shiung, J. L. Gunter, P. C. O'Brien, S. D. Weigand, D. S. Knopman, B. F. Boeve, R. J. Ivnik, G. E. Smith, R. H. Cha, E. G. Tangalos and R. C. Petersen, *Neurology*, 2004, **62**, 591–600.

125. D. P. Devanand, G. Pradhaban, X. Liu, A. Khandji, S. De Santi, S. Segal, H. Rusinek, G. H. Pelton, L. S. Honig, R. Mayeux, Y. Stern, M. H. Tabert and M. J. de Leon, *Neurology*, 2007, **68**, 828–836.

126. J. S. Damoiseaux, S. M. Smith, M. P. Witter, E. J. Sanz-Arigita, F. Barkhof, P. Scheltens, C. J. Stam, M. Zarei and S. A. Rombouts, *Hum. Brain Mapp.*, 2009, **30**, 1051–1059.

127. J. Sauer, D. H. ffytche, C. Ballard, R. G. Brown and R. Howard, *Brain*, 2006, **129**, 1780–1788.

128. H. R. Griffith, J. A. den Hollander, O. C. Okonkwo, T. O'Brien, R. L. Watts and D. C. Marson, *Alzheimer's Dementia*, 2008, **4**, 421–427.

129. A. K. Godbolt, A. D. Waldman, D. G. MacManus, J. M. Schott, C. Frost, L. Cipolotti, N. C. Fox and M. N. Rossor, *Neurology*, 2006, **66**, 718–722.

130. G. B. Frisoni, M. Lorenzi, A. Caroli, N. Kemppainen, K. Nagren and J. O. Rinne, *Neurology*, 2009, **72**, 1504–1511.

131. J. B. Langbaum, K. Chen, W. Lee, C. Reschke, D. Bandy, A. S. Fleisher, G. E. Alexander, N. L. Foster, M. W. Weiner, R. A. Koeppe, W. J. Jagust and E. M. Reiman, *Neuroimage*, 2009, **45**, 1107–1116.

132. K. Reetz, C. Gaser, C. Klein, J. Hagenah, C. Buchel, S. Gottschalk, P. P. Pramstaller, H. R. Siebner and F. Binkofski, *Mov. Disord.*, 2009, **24**, 99–103.

133. F. Binkofski, K. Reetz, C. Gaser, R. Hilker, J. Hagenah, K. Hedrich, T. van Eimeren, A. Thiel, C. Buchel, P. P. Pramstaller, H.R. Siebner and C. Klein, *Neurology*, 2007, **69**, 842–850.

134. D. E. Vaillancourt, M. B. Spraker, J. Prodoehl, I. Abraham, D. M. Corcos, X. J. Zhou, C. L. Comella and D. M. Little, *Neurology*, 2009, **72**, 1378–1384.

135. V. S. Mattay, A. Tessitore, J. H. Callicott, A. Bertolino, T. E. Goldberg, T. N. Chase, T. M. Hyde and D. R. Weinberger, *Ann. Neurol.*, 2002, **51**, 156–164.

136. R. M. Camicioli, C. C. Hanstock, T. P. Bouchard, M. Gee, N. J. Fisher and W. R. Martin, *Mov. Disord.*, 2007, **22**, 382–386.

137. G. Oz, M. Terpstra, I. Tkac, P. Aia, J. Lowary, P. J. Tuite and R. Gruetter, *Magn. Reson. Med.*, 2006, **55**, 296–301.

138. D. S. Goldstein, C. Holmes, O. Bentho, T. Sato, J. Moak, Y. Sharabi, R. Imrich, S. Conant and B. A. Eldadah, *Parkinsonism Relat. D.*, 2008, **14**, 600–607.

139. C. Eggers, R. Hilker, L. Burghaus, B. Schumacher and W. D. Heiss, *J. Neurol. Sci.*, 2009, **276**, 27–30.

140. W. R. Martin, M. Wieler and M. Gee, *Neurology*, 2008, **70**, 1411–1417.

141. P. C. Koolschijn, N. E. van Haren, G. J. Lensvelt-Mulders, H. E. Hulshoff Pol and R. S. Kahn, *Hum. Brain Mapp.*, 2009, **30**, 3719–35.

142. T. Frodl, N. Koutsouleris, R. Bottlender, C. Born, M. Jager, M. Morgenthaler, J. Scheuerecker, P. Zill, T. Baghai, C. Schule, R. Rupprecht, B. Bondy, M. Reiser, H. J. Möller and E. M. Meisenzahl, *Mol. Psychiatry*, 2008, **13**, 1093–1101.

143. T. Kieseppa, M. Eerola, R. Mantyla, T. Neuvonen, V. P. Poutanen, K. Luoma, A. Tuulio-Henriksson, P. Jylha, O. Mantere, T. Melartin, H. Rytsälä, M. Vuorilehto and E. Isometsä, *J. Affective Disord.*, 2009, **120**(1–3), 240–4.

144. D. A. Pizzagalli, A. J. Holmes, D. G. Dillon, E. L. Goetz, J. L. Birk, R. Bogdan, D. D. Dougherty, D. V. Iosifescu, S. L. Rauch and M. Fava, *Am. J. Psychiatry*, 2009, **166**, 702–710.

145. J. Epstein, H. Pan, J. H. Kocsis, Y. Yang, T. Butler, J. Chusid, H. Hochberg, J. Murrough, E. Strohmayer, E. Stern and D. A. Silbersweig, *Am. J. Psychiatry*, 2006, **163**, 1784–1790.

146. R. B. Price, D. C. Shungu, X. Mao, P. Nestadt, C. Kelly, K. A. Collins, J. W. Murrough, D. S. Charney and S. J. Mathew, *Biol. Psychiatry*, 2009, **65**, 792–800.

147. G. Hasler, J. W. van der Veen, T. Tumonis, N. Meyers, J. Shen and W. C. Drevets, *Arch. Gen. Psychiatry*, 2007, **64**, 193–200.

148. W. C. Drevets, W. Bogers and M. E. Raichle, *Eur. Neuropsychopharmacol.*, 2002, **12**, 527–544.

149. M. P. Milham, A. C. Nugent, W. C. Drevets, D. P. Dickstein, E. Leibenluft, M. Ernst, D. Charney and D. S. Pine, *Biol. Psychiatry*, 2005, **57**, 961–966.

150. H. K. Yoo, M. J. Kim, S. J. Kim, Y. H. Sung, M. E. Sim, Y. S. Lee, S. Y. Song, B. S. Kee and I. K. Lyoo, *Eur. J. Neurosci.*, 2005, **22**, 2089–2094.

151. K. L. Phan, A. Orlichenko, E. Boyd, M. Angstadt, E. F. Coccaro, I. Liberzon and K. Arfanakis, *Biol. Psychiatry*, 2009, **66**, 691–4.

152. S. G. Shah, H. Klumpp, M. Angstadt, P. J. Nathan and K. L. Phan, *J. Psychiatry Neurosci.*, 2009, **34**, 296–302.

153. R. A. Lanius, R. L. Bluhm, N. J. Coupland, K. M. Hegadoren, B. Rowe, J. Theberge, R. W. Neufeld, P. C. Williamson and M. Brimson, *Acta Psychiatr. Scand.*, 2009, **121**, 33–40.

154. S. J. Mathew, X. Mao, J. D. Coplan, E. L. Smith, H. A. Sackeim, J. M. Gorman and D. C. Shungu, *Am. J. Psychiatry*, 2004, **161**, 1119–1121.

155. M. D. De Bellis, M. S. Keshavan, S. Spencer and J. Hall, *Am. J. Psychiatry*, 2000, **157**, 1175–1177.

156. R. R. Lanzenberger, M. Mitterhauser, C. Spindelegger, W. Wadsak, N. Klein, L. K. Mien, A. Holik, T. Attarbaschi, N. Mossaheb, J. Sacher, T. Geiss-Granadia, K. Kletter, S. Kasper and J. Tauscher, *Biol. Psychiatry*, 2007, **61**, 1081–1089.

157. O. G. Cameron, G. C. Huang, T. Nichols, R. A. Koeppe, S. Minoshima, D. Rose and K. A. Frey, *Arch. Gen. Psychiatry*, 2007, **64**, 793–800.

158. D. C. Glahn, A. R. Laird, I. Ellison-Wright, S. M. Thelen, J. L. Robinson, J. L. Lancaster, E. Bullmore and P. T. Fox, *Biol. Psychiatry*, 2008, **64**, 774–781.

159. P. Tanskanen, K. Ridler, G. K. Murray, M. Haapea, J. M. Veijola, E. Jaaskelainen, J. Miettunen, P. B. Jones, E. T. Bullmore and M. K. Isohanni, *Schizophr. Bull.*, 2008, **36**, 766–77.

160. J. S. Oh, M. Kubicki, G. Rosenberger, S. Bouix, J. J. Levitt, R. W. McCarley, C. F. Westin and M. E. Shenton, *Hum. Brain Mapp.*, 2009, **30**, 3812–25.

161. O. R. Phillips, K. H. Nuechterlein, K. A. Clark, L. S. Hamilton, R. F. Asarnow, N. S. Hageman, A. W. Toga and K. L. Narr, *Schizophr. Res.*, 2009, **107**, 30–38.

162. B. Jeong, C. G. Wible, R. I. Hashimoto and M. Kubicki, *Hum. Brain Mapp.*, 2009, **30**, 4138–51.

163. D. I. Kim, D. H. Mathalon, J. M. Ford, M. Mannell, J. A. Turner, G. G. Brown, A. Belger, R. Gollub, J. Lauriello, C. Wible, D. O'Leary, K. Lim, A. Toga, S. G. Potkin, F. Birn and V. D. Calhoun, *Schizophr. Bull.*, 2009, **35**, 67–81.

164. N. Goto, R. Yoshimura, J. Moriya, S. Kakeda, N. Ueda, A. Ikenouchi-Sugita, W. Umene-Nakano, K. Hayashi, N. Oonari, Y. Korogi and J. Nakamura, *Schizophr. Res.*, 2009, **112**, 192–193.

165. A. Szulc, B. Galinska, E. Tarasow, W. Dzienis, B. Kubas, B. Konarzewska, J. Walecki, A. S. Alathiaki and A. Czernikiewicz, *Pharmacopsychiatry*, 2005, **38**, 214–219.

166. O. D. Howes, A. J. Montgomery, M. C. Asselin, R. M. Murray, I. Valli, P. Tabraham, E. Bramon-Bosch, L. Valmaggia, L. Johns, M. Broome, P. K. McGuire and P. M. Grasby, *Arch. Gen. Psychiatry*, 2009, **66**, 13–20.

167. S. Nozaki, M. Kato, H. Takano, H. Ito, H. Takahashi, R. Arakawa, M. Okumura, Y. Fujimura, R. Matsumoto, M. Ota, A. Takano, A. Otsuka, F. Yasuno, Y. Okubo, H. Kashima and T. Suhara, *Schizophr. Res.*, 2009, **108**, 78–84.

168. A. F. DaSilva, L. Becerra, G. Pendse, B. Chizh, S. Tully and D. Borsook, *PLoS One*, 2008, **3**, e3396.

169. J. Upadhyay, J. Knudsen, J. Anderson, L. Becerra and D. Borsook, *Magn. Reson. Med.*, 2008, **60**, 1037–1046.

170. L. Becerra, S. Morris, S. Bazes, R. Gostic, S. Sherman, J. Gostic, G. Pendse, E. Moulton, S. Scrivani, D. Keith, B. Chizh and D. Borsook, *J. Neurosci.*, 2006, **26**, 10646–10657.

171. D. Ducreux, N. Attal, F. Parker and D. Bouhassira, *Brain*, 2006, **129**, 963–976.
172. S. Fukui, M. Matsuno, T. Inubushi and S. Nosaka, *Magn. Reson. Imaging*, 2006, **24**, 75–79.
173. J. Maarrawi, R. Peyron, P. Mertens, N. Costes, M. Magnin, M. Sindou, B. Laurent and L. Garcia-Larrea, *Pain*, 2007, **127**, 183–194.
174. A. F. DaSilva, C. Granziera, J. Snyder and N. Hadjikhani, *Neurology*, 2007, **69**, 1990–1995.
175. C. Granziera, A. F. DaSilva, J. Snyder, D. S. Tuch and N. Hadjikhani, *PLoS Med.*, 2006, **3**, e402.
176. J. Kim, S. Kim, S. I. Suh, S. B. Koh, K. W. Park and K. Oh, *Cephalalgia*, 2010, **30**, 53–61.
177. S. D. Friedman, D. W. Shaw, A. A. Artru, G. Dawson, H. Petropoulos and S. R. Dager, *Arch. Gen. Psychiatry*, 2006, **63**, 786–794.
178. S. K. Sundaram, A. Kumar, M. I. Makki, M. E. Behen, H. T. Chugani and D. C. Chugani, *Cereb. Cortex*, 2008, **18**, 2659–2665.
179. I. Soulieres, M. Dawson, F. Samson, E. B. Barbeau, C. P. Sahyoun, G. E. Strangman, T. A. Zeffiro and L. Mottron, *Hum. Brain Mapp.*, 2009, **30**, 4082–107.
180. A. Y. Hardan, N. J. Minshew, N. M. Melhem, S. Srihari, B. Jo, R. Bansal, M. S. Keshavan and J. A. Stanley, *Psychiatry Res.*, 2008, **163**, 97–105.
181. M. S. Buchsbaum, E. Hollander, M. M. Haznedar, C. Tang, J. Spiegel-Cohen, T. C. Wei, A. Solimando, B. R. Buchsbaum, D. Robins, C. Bienstock, C. Cartwright and S. Mosovich, *Int. J. Neuropsychopharmacol.*, 2001, **4**, 119–125.
182. E. Luders, K. L. Narr, L. S. Hamilton, O. R. Phillips, P. M. Thompson, J. S. Valle, M. Del'Homme, T. Strickland, J. T. McCracken, A. W. Toga and J. G. Levitt, *Biol. Psychiatry*, 2009, **65**, 84–88.
183. I. Ellison-Wright, Z. Ellison-Wright and E. Bullmore, *BMC Psychiatry*, 2008, **8**, 51.
184. T. J. Silk, A. Vance, N. Rinehart, J. L. Bradshaw and R. Cunnington, *Hum. Brain Mapp.*, 2008, **30**, 2757–65.
185. K. Rubia, A. B. Smith, R. Halari, F. Matsukura, M. Mohammad, E. Taylor and M. J. Brammer, *Am. J. Psychiatry*, 2009, **166**, 83–94.
186. K. Rubia, R. Halari, A. Christakou and E. Taylor, *Philos. Trans. R. Soc. London, Ser. B*, 2009, **364**, 1919–1931.
187. J. A. Stanley, H. Kipp, E. Greisenegger, F. P. MacMaster, K. Panchalingam, M. S. Keshavan, O. G. Bukstein and J. W. Pettegrew, *Arch. Gen. Psychiatry*, 2008, **65**, 1419–1428.
188. T. J. Spencer, J. Biederman, B. K. Madras, D. D. Dougherty, A. A. Bonab, E. Livni, P. C. Meltzer, J. Martin, S. Rauch and A. J. Fischman, *Biol. Psychiatry*, 2007, **62**, 1059–1061.
189. N. D. Volkow, G. J. Wang, J. Newcorn, F. Telang, M. V. Solanto, J. S. Fowler, J. Logan, Y. Ma, K. Schulz, K. Pradhan K, C. Wong and J. M. Swanson, *Arch. Gen. Psychiatry*, 2007, **64**, 932–940.

CHAPTER 8.2

Magnetic Resonance Imaging in Drug Development

JIN XIE AND XIAOYUAN CHEN*

Laboratory of Molecular Imaging and Nanomedicine (LOMIN),
National Institute of Biomedical Imaging and Bioengineering (NIBIB),
National Institutes of Health (NIH), Bethesda, MD 20892, USA

8.2.1 Introduction

Drug development is a lengthy process characteristic of three highs: high expense, high risk, and high attrition rate. On average, it takes about 14.2 years for one agent to be discovered, validated, and finally brought to the market (Figure 8.2.1), and the expense of the whole process is as much as $800 million or more.[1] More critically, according to the statistics, merely 5% of the molecules discovered in the pre-clinical research can make it into the clinical trials, and out of them only 20% can be approved for final marketing.[2] This suggests billions of dollars of inefficient investment on drug R&D every year, and explains the critical significance of finding ways to accelerate the process.

The application of imaging facilities in the drug development cycle is believed to be one possible solution to the problem and is brought to the attention of pharmaceutical companies. Magnetic resonance imaging (MRI), in particular, is of great use in such a context, for its extraordinary capacities in providing anatomical, functional and even molecular information in a non-invasive manner at high spatial resolution. Those features have made it a useful tool to phenotype markers, to characterize lesions, to profile disease progression and to monitor therapeutic response.[3] Compared with other common imaging

RSC Drug Discovery Series No. 15
Biomedical Imaging: The Chemistry of Labels, Probes and Contrast Agents
Edited by Martin Braddock
© Royal Society of Chemistry 2012
Published by the Royal Society of Chemistry, www.rsc.org

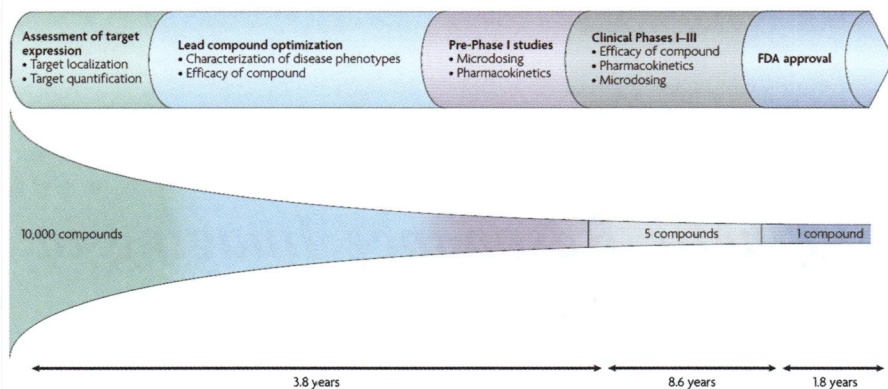

Figure 8.2.1 Schematic Illustration of the Drug Development Cycle. The whole process takes 14.2 years on average, and out of 10 000 compounds that are evaluated in the preclinical studies, only one can finally receive regulatory approval by FDA. MR imaging is found to be of great use at multiple steps of this lengthy process, and its active involvement may help reduce both time and the attrition rates. Adapted with permission from ref. 1.

facilities, MRI has a much longer list of parameters to be tuned, so the tissue contrast can be accordingly altered and enhanced in various ways. Indeed, besides the basic T1-, T2- and T2*- weighted MRI, a number of advanced MRI techniques have recently been developed, such as diffusion weighted MRI, perfusion weighted MRI, functional MRI, *etc.* These, plus the development of various MRI contrast agents, have greatly enriched the capabilities of MRI, and have further strengthened its role in both clinical and preclinical studies. Moreover, in the on-going efforts of building up multimodality imaging systems, MRI is combined with other imaging modalities, such as positron emission tomography (PET), single-photon emission computed tomography (SPECT), and in the preclinical stage, fluorescence and bioluminescence imaging, to construct multimodality imaging systems. Such a combination is of significant importance, for integrating the strengths of each modality and providing a more accurate and comprehensive interrogation.

8.2.2 Preclinical and Clinical Trials

A typical drug development cycle includes a preclinical drug identification step and a subsequent three-phase clinical evaluation step (Figure 8.2.1). In the preclinical stage, a group of molecules with potential therapeutic potency are discovered and are subjected to *in vitro* and small animal tests to arrive at the optimized formulas. Qualified formulas then enter the three-phase clinical trials after regulatory approval. In phase I, the candidate drugs are given to healthy volunteers, with the purpose of establishing safety profiles and understanding the biophysical parameters of the agents. The qualified agents then advance to phase

II, where real patients are dosed with the compounds to evaluate the therapeutic efficacy as well as to further assess its safety profile. And finally, in phase III, the compounds will be tested in large patient cohorts from multiple centers to further confirm the safety and efficacy. The whole process, from the drug discovery to the marketing, takes an average of 14.2 years. And out of ∼10 000 compounds evaluated in preclinical studies, only one will receive regulatory approval by the US Food and Drug Administration (FDA) (Figure 8.2.1).[1,4]

Due to the high expense and risk of the clinical trial process, consolidating the efficacy of a drug candidate before entering clinical trials is of significant importance. It is found that the rate of successful translation from preclinical stage in animal models to phase II on human beings is one critical indicator of the final marketing.[5] Providing improved understanding of disease progression and treatment effects, MRI and other imaging facilities are found to be extremely useful in such a transition, saving tremendous efforts and expense in the late stage development.[6] It also explains the necessity of setting preclinical animal studies in the pipeline, where disease progression can be assessed in a comprehensive and histologically correlated manner, which is hardly affordable in the clinical context. Despite the incongruity, using small animals to model human disease and therapeutics remains irreplaceable in drug development, and MRI is an active player in the process, linking those pre-clinical studies with the later clinical trials.

8.2.3 Basics of MRI

MRI is a non-invasive imaging technique that is widely used in daily medical practice. An MRI map is constructed on the basis of the nuclear magnetic resonance (NMR) signals generated by certain nuclei (*e.g.* ^{1}H, ^{19}F, ^{31}P, and ^{13}C) when irradiated with radio waves in a strong magnetic field.[7] The MRI signals are extremely weak under the field of clinical MRI (*e.g.* 1.5 or 3.0 T), and therefore MRI is only applicable to encode the spatial position of highly concentrated nuclei spins, in most cases ^{1}H of water molecules. During an MRI scan, the position of protons in the body can be determined, and based on which, an image of the body can be constructed.

The MRI signals are from the relaxation, which is essentially a decaying process of protons at excitation state returning back to equilibrium. Such relaxation can be characterized by two parameters, T1 and T2. T1, also called spin-lattice relaxation time, is the recovery of longitudinal magnetization; and T2 is the loss of phase coherence in the transverse plane, also known as spin-spin relaxation time. In real practice, the unavoidable inhomogeneities that exist in the static magnetic field may also contribute to the shortening of relaxation at the transverse plane, and the parameter to depict such observed decay is called T2*, which is always shorter than T2.

Although the intensities of MRI signals are highly dependent upon the local environment, such as the water concentration and molecule mobility, in many cases, the differences may not be sufficient to report a pathological alteration.

Such a drawback has been greatly compensated by the application of contrast agents, such as gadolinium chelates or iron oxide nanoparticles (IONPs), as both can help improve the contrast by fluctuating the local magnetic field and shortening the relaxation rate. The Gd complexes, being paramagnetic ions, are mainly used as T1 agents, which accelerate the longitudinal relaxation rate. And the iron oxide nanoparticles, being ferromagnetic or superparamagnetic, are typically regarded as T2/T2* contrast agents, which function by reducing the transverse relaxation rate.

8.2.4   Advanced MRI Technologies

As just mentioned, the intensities of MRI signals are related to the density of the water protons, and based on that, a density-weighted MRI map can be acquired to depict the anatomical structure of the body. However, this is only a small fraction of the capabilities of MRI. The contrast of MRI is dependent on not only the densities and relaxation times (T1, T2, and T2*), but also water diffusion, water exchange rates and macroscopic motion, *etc.*[8] By choosing an appropriate imaging pulse and sequence, only one mechanism can be emphasized, which gives MRI tremendous flexibility and allows a far more comprehensive scope than simply providing structural information.

For instance, MRI can be weighted to reveal the water diffusion (diffusion weighted imaging, DWI). This is mostly conducted with RF spin echo sequences, where a pulsed field gradient is first applied to induce a position dependent signal phase shift, followed by a second field gradient that restores the phase shift of the retained molecules. However, for the part of molecules that have moved during the intervals, typically caused by Brownian motion or flow, the signal re-phasing is incomplete, and the mismatch can be used to depict the diffusion of the water molecules on the MR images.[7] Several MRI techniques based on diffusion theory have been developed, such as apparent diffusion coefficient (ADC) mapping and diffusion tensor imaging (DTI). These techniques are now utilized in characterizing a wide range of pathological alternation characterizations, such as in cerebral ischemia,[9–11] tumors,[12,13] and sclerotic metastases.[14]

Perfusion-weighted imaging (PWI) provides the hemodynamic information of the tissues by recording the dynamic changes caused by the passage of the labeled agents. The labeled agents could be simply endogenous water protons. But more efficiently, exogenous tracers are employed, which are typically contrast agents like gadolinium-diethylene-triamine-pentaacetic acid (Gd-DTPA) and albumin-(Gd-DTPA). Those contrast agents are administrated *via* bolus injection, and the induced intensity changes encode the information of the tissue microstructure. PWI is widely used to evaluate angiogenesis or angiogenesis targeted treatment.[15,16] Also, it is the technique of choice to assess myocardial ischemia and blood flow.[17,18] Although both T1 and T2* weighted sequences can be used in PWI, T1 weighted sequences are mainly used to measure the permeability of the tissue, and T2* weighted sequences, with better sensitivity and contrast,[19] is mainly used to measure capillary density, such as in assessing the relative cerebral blood volume (rCBV).[14]

Recently, functional MRI (fMRI) has emerged as an important tool, especially in the assessment of brain functions. It works on the basis that the brain activities are highly oxygen-consuming processes, accompanied with an elevated level of oxygenated hemoglobin and increased MRI signals.[20] By comparing the images of subjects in thinking and in rest, it is possible to relate brain regions with specific functionalities.[21] Such a feature is widely used to measure the response of the brain to pharmacological stimuli in CNS drug development. Especially, in stroke therapy, fMRI provides a solitary approach to measure the therapeutic responses, which is conventionally assessed subjectively by behavior.

8.2.5 MRI in Drug Development

With a high spatial resolution, 3D imaging nature, and superior soft tissue contrast, MRI is highly useful in characterizing many types of diseases. Compared with the destructive histological analysis, MRI is a noninvasive method, which offers the opportunity of assessing the therapeutic intervention on the same living subject in a repetitive and consistent manner. In the following section, by the category of diseases, we will give some cases and prospects of using MRI to facilitate the drug development process.

8.2.5.1 Degenerative Joint Diseases

MRI remains the premier tool in the anatomical characterization of the complex joint architecture, and is the method of choice in the studies of degenerative disorders of the joints.[22-24] In rheumatoid arthritis (RA) diagnosis, MRI is used widely to characterize the bone spacing and synovial hyperplasia,[25] which are regarded as biomarkers in RA profiling. For example, high resolution 3D MRI assisted in the efficacy evaluation of Sandimmune and Neoral on collagen-induced arthritis rat models.[26] The pathomorphological changes, such as the increase of the joint space, as well as the cartilage and bone erosion, were longitudinally monitored with MRI, and the observation was found to be well-correlated with the histological results. Compared with the control group, strong protective effects against cartilage and bone destruction were observed for those administrated with Neoral.[26] In the assessment of synovitis, DCE-MRI is becoming a powerful tool, which measures the leakiness of vessels. Clinical trials using DCE-MRI to investigate the therapeutic effects of leflunomide and methotrexate have been reported (Figure 8.2.2).[6] Also, being an inflammatory process, rheumatoid arthritis is usually accompanied with macrophage infiltration. By labeling the macrophages with IONPs, the *in vivo* migration of macrophage, such as the accumulation at the arthritis sites, can be tracked by MRI. In an antigen induced arthritis model, iron-laden macrophages were administrated and significant signal reduction was observed

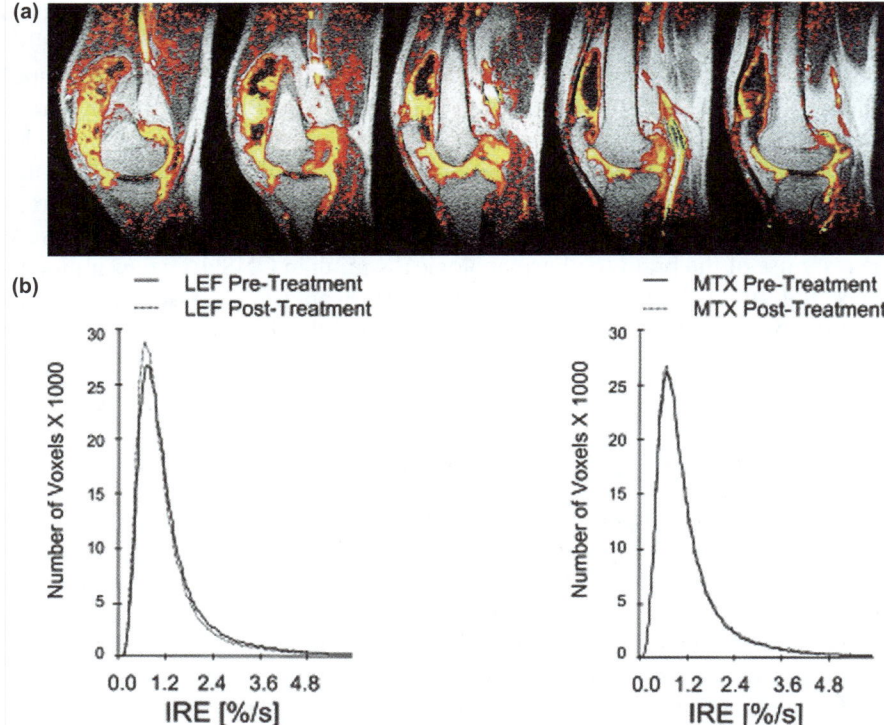

Figure 8.2.2 (a) Five characteristic dynamic gadolinium-enhanced magnetic resonance imaging sagittal scans across synovial space. (b) Pooled initial rate
of enhancement (IRE) histograms for patients with leflunomide (LEF)
and methotrexate (MTX) treatment (n = 17 in each group), at baseline
and after 4 months of therapy. The curve approximating leflunomide
treatment for 4 months is shifted to the left, showing an improvement in
the IRE and in synovial inflammation. The histogram of IRE values for
methotrexate-treated patients shows virtually no change other than a
slight shift to the right of the tail portion of the curve. s = seconds;
%/s = percent signal intensity change per second. Adapted from ref. 6
with permission.

in the synovium of arthritic knees.[27] Such signal attenuation was attributed to
the accumulation of iron contents in the synovium and was validated on histological analysis. This may become another possible biomarker of early stage
RA progression and is of potential application in the therapeutic studies. Also,
IONPs alone can be used to detect arthritis-associated macrophages.[28]

Unlike RA, osteoarthritis (OA), also known as degenerative arthritis, is a
non-inflammatory process. It is initiated with cartilage damage, which in turn
causes damage to subchondral bone and muscle atrophy. MRI is found to be a
useful tool in monitoring the OA progression for its capability to perceive the
changes occurring to the collagen and the proteoglycans. Such characterization
is usually achieved with dynamic MRI using Gd-DTPA as the contrast agent.

The idea is that the equilibrium concentration of Gd-DTPA^{2-} (hence the proton T1) in the tissue is inversely correlated with the proteoglycan concentration, hence its fluctuation is reflective of the proteoglycan changes.[29] In addition, MRI, in combination with ultrasonography imaging, has been used in hand and wrist osteoarthritis and rheumatoid arthritis.[30]

8.2.5.2 Stroke

Cerebral infarction remains one of the leading causes of mortality and morbidity in industrialized countries, and despite the continuous efforts, no apparent therapy is currently available.[8] Due to the noninvasiveness and the functional imaging capacity, MRI has long served as one major facility in stroke characterizations, at both preclinical and clinical levels.

The most commonly used animal stroke model is the focal ischemia model induced by unilateral middle cerebral artery occlusion (MCAO) in rats. The vasogenic edema formed in a stroke is exhibited as hyperintense under T2 weighted maps, which was previously confirmed to be well-correlated with the histological results and is long adopted as a measure of the infarction.[31] A number of therapeutic agents, including calcium antagonists, *N*-methyl-D-aspartate (NMDA) antagonists, AMPA antagonists, ion channel modulators, free radical scavengers, anti-inflammatory drugs, *etc.*,[8] have been evaluated on this model. Despite the variation in mechanisms, all these agents are essentially cytoprotective compounds which aim to reduce the extent of cellular damage in the acute phase of stroke. MRI, as an efficient tool in monitoring the edema volume change, has been actively anticipated in those assessments.

In the context of clinical translation, the transient MCAO model (rather than the permanent MCAO model) is more relevant, and DWI, being able to assess early stage ischemic damage, has been proven to be a powerful tool for such a model.[32] ADC signal reduction, in particular, is found to be closely associated with the ischemia and is regarded as a characteristic mark of stroke onset and progression. It is also very interesting that the loss of ADC can be regained during pharmacological intervention, making it an ideal tool to measure the therapeutic responses. For example, it was demonstrated in an NMDA induced neurotoxicity model that MK 801 can attenuate the damage, and a regain of ADC signal was accompanied with that process.[33] Besides DWI, other MRI techniques are also contributive to stroke characterizations in various aspects. For example, PWI is actively employed to assess the changes of cerebral blood flow after stroke onset. Also, magnetic resonance angiography, capable of detecting a slight reduction of blood flow after reperfusion, was used to visualize the cerebral vasculature on both permanent and temporal MCAO models.[34]

However, the endeavor for acute stroke therapy has so far led to an unsatisfactory outcome, and there is a growing interest in therapeutics and assessments that target the subchronic and chronic stages of a stroke. Anti-inflammation therapy could be a topic that falls under the umbrella of those studies, because inflammation is involved in the stroke progression and is a cause of the delayed increase of infarct volume.[3] Using MRI to track the migrations of leukocytes *in vivo* may provide a unique means to look into the inflammation process. In one proof-of-concept demonstration, IONPs were administrated into a permanent MCAO model, and the process of particles migrating and concentrating at the lesion site was successfully observed by MRI. The subsequent histological analysis confirmed the observation by finding large populations of iron containing macrophages in the infracted tissue.[35]

Besides physical characterizations, MRI is also playing a unique role in evaluating the recovery progression of patient–animal models, such as deciding whether an improved performance is caused by the cytoprotective therapy or by the functional reorganization of the brain.[3,36] Sauter *et al.* used fMRI to investigate the therapeutic response of isradipine on a permanent MCAO model *via* electrical stimulation of both forepaws. While vehicle-treated rats did not show fMRI responses in the infarcted somatosensory cortex throughout the study, several of the isradipine-treated animals displayed functional recovery in the cytoprotected cortex at days 5 (3 out of 5 rats) and 12 days (5 out of 10 rats).[37] As addressed earlier, this role of performing function evaluations makes MRI the method of choice in stroke studies to monitor the therapeutic responses.

8.2.5.3   Oncology

In the early days, MRI's application in cancer research was limited to differentiating malignant masses from benign tissues by the intrinsic contrast.[38] Such contrast, however, is in many cases inefficient, leading to a poor prognosis, especially for early stage cancerous cell detection. To improve the sensitivity and to expand the applicability, various types of contrast agents and MRI techniques have since been developed. Now, MRI is playing a leading role in cancer staging and therapeutic response monitoring.

Using MRI to assess physiological parameters of tumor, including tumor volume, perfusion, permeability, and neovascularization, is now a common technique in the drug development cycle. The changes of these parameters, acquired in a noninvasive manner, are tightly associated with the therapeutic interventions hence are valuable in optimizing the treatment regimen. Using albumin-Gd-DTPA as the contrast agent, it was demonstrated on a rat R3230 mammary carcinoma model that, the MR-derived microvascular characteristics (plasma volume and capillary permeability) were well correlated with histological capillary density ($r^2 = 0.85$).[39] In a clinical study conducted with 20 breast cancer patients, it was found that there was a good correlation between tumor angiogenesis and Gd-DTPA induced contrast enhancement ($r^2 = 0.59$).[40] Blood-oxygen-level-dependent (BOLD) MRI, which measures the relative tissue oxygenation, is becoming an important technique in tumor identification.[41] In a clinical trial, fourteen volunteers with meningiomas, gliomas, and metastatic tumors were scanned by BOLD MRI before surgery (Figure 8.2.3). Significant differences were found between tumors and surrounding normal brain in all the subjects, and in meningiomas and gliomas, selection of a voxel

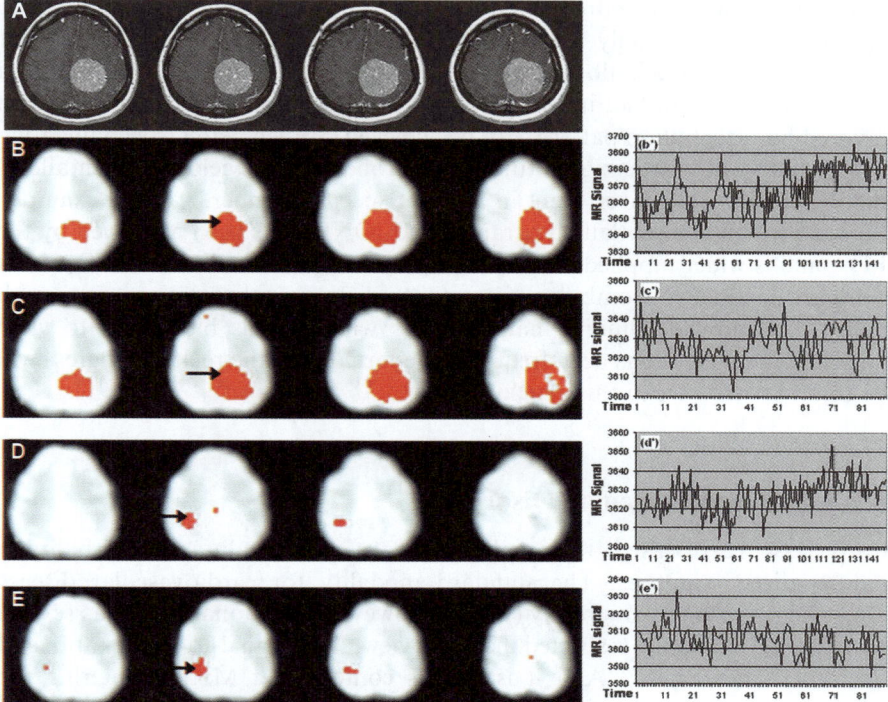

Figure 8.2.3 Representative Patient With a Grade I Meningioma. Images show the effect of resting state *versus* motor activity on the BOLD signal intensity. (B–E), seed points in the tumor (B and C) or motor area (D and E) are analyzed during resting or motor activity. There is no difference in the highlighted areas and their signals, whether the VOIs are selected while the patient is at rest or finger tapping. (A), Anatomic map. (B and C), Correlation map and time series for a tumor seed (arrow) in the resting paradigm (B, B') and in the bilateral tapping paradigm (C, C'). (D and E), Correlation map and time series for a normal VOI (arrow) in the resting paradigm (D, D'), and the bilateral tapping paradigm (E, E'). Arrows indicate VOI seed points. Adapted from ref. 42 with permission.

in the tumor for signal-intensity analysis highlighted the entire tumor mass while excluding the normal tissue.[42] On top of that, by quantifying the BOLD signal's temporal complexity using a fractal dimension index, Wardlaw *et al.* developed a novel approach to evaluate tumor microvasculature.[43]

There are quite a few cases of using MRI to assist the clinical/preclinical evaluations of therapeutic agents. For instance, MRI was involved in a comparison study on a canine prostate model to evaluate the effects of several *aza*steroid inhibitors of steroid 5 alpha-reductase.[44] It was also used in determining the effects of dexamethasone in murine pancreatic adenocarcinoma (Panc02) model.[45] Moreover, it has been used to monitor the treatment effects of ZD6474,[46] and ZD4190,[47] both of which are VEGF receptor-2 (KDR) tyrosine kinase inhibitors, on murine xenograft models.

There is an on-going effort of harnessing the migrations of contrast agents to achieve probes that only highlight the areas of interest. Taking IONPs for example, by carefully tailoring the particle size, charge, stability *etc.*, a desired pharmacokinetic profile, including an appropriate circulation half-life and a favored biodistribution pattern, can be achieved.[48] Even better, by conjugating the particles with targeting motifs, smart probes can be yielded, which allows the dissection of a cancer event in a target specific manner.[49,50] For example, Feridex, which is essentially dextran coated IONPs, has been approved by the FDA and is widely applied to assist tumor detection in the liver and spleen. Its compact analogue, Combidex, is used to help visualize lesions in lymph nodes.[48] Meanwhile, linking nanoparticles with various biovectors to yield targeting conjugates has been frequently reported, and the related progress will be summarized later in this article.

8.2.5.4 Cardiovascular Disorders

Cardiovascular disease is the number one cause of mortality and morbidity in industrialized countries. The standard modality for cardiovascular disease assessment is echocardiography (ECG). However, traditional ECG has several limitations, including restricted field of view, low signal-to-noise ratio and operator dependence.[51] And it is in this context that MRI is entering the mainstream of the imaging practice, which provides both structural and functional assessments of the cardiovascular system. One critical issue in small animal cardiovascular MRI is the animals' fast heart rates. The heart rates are as high as 300 beats per min for rats and 600 beats per min for mice, respectively. This makes ECG-gating a mandatory technique in small MRI data acquisition, which synchronizes imaging with a phase of the cardiac cycle to minimize motion artifacts.

MRI is a proven useful tool to study the left ventricular (LV) hypertrophy development in both pre-clinical and clinical settings. Laurent *et al.* used MRI to monitor the changes of LV mass, wall thickness and volume change on hypertensive (SHR) and renal hypertensive (RHR) in rat models.[52] And in a clinical trial with 1100 asymptomatic individuals, Rosen and his colleagues confirmed with MRI observation that age, increased LV mass, and decreased myocardial perfusion are related to delayed myocardial contraction and greater extent of dyssynchrony.[53] Moreover, MRI is used to characterize myocardial infarction (also known as heart attack), which is one major form of cardiovascular disorders. Myocardial infarction is induced by insufficient blood supply, which in turn causes damage to the heart muscle tissue. Owing to the formation of *edema*, the infarcted myocardium regions are usually displayed as hyperintensities on T2 weighted MRI maps, and this character has been used widely to delineate the lesions.[54] More impressively, when complemented with contrast agents, PWI can be performed to assess the perfusion status of the myocardium. For example, using this method, Wilke *et al.* demonstrated a good correlation between regional perfusion reserve and the coronary flow reserve, suggesting the feasibility to detect reduced coronary flow reserve with MRI.[55] Later, Al-Saadi *et al.* and Schwitter *et al.* performed separate clinical studies, and both confirmed the high accuracy of using MR first-pass perfusion to measure coronary artery stenosis.[56] This technique also holds the potential to allow early stage identification of myocardial perfusion deficits, which is of significant diagnosis meaning.

Most of the pharmacological efforts are focused on angiotensin-converting enzyme (ACE) inhibitor or anti-anginal drug development, and MRI has proved to be a powerful tool in assisting the evaluation. Umemura *et al.* performed a longitudinal study, in which MRI was used to evaluate the effects of spirapril (an ACE inhibitor). Specifically, rats with LV hypertrophy were fed with spirapril (10 mg kg^{-1} per day), and LV parameters, both structural (LV mass and LV wall thickness) and functional (LV end-systolic and end-diastolic volumes, stroke volume (SV), ejection fraction (EF)), were assessed with MRI and were compared with the control group. It was found that, while the untreated group showed a significant increase in heart weight, the LV weight of the spirapril group remained stable; moreover, the LV end-diastolic volume, LV end-systolic volume, and SV were found smaller in the spirapril treated group.[57] Lund *et al.* used MRI to assess the acute effects of nicorandil on ischemic injury of myocardium (Figure 8.2.4). The animal model was established on rats by inducing 30 min coronary occlusion followed by 24 h

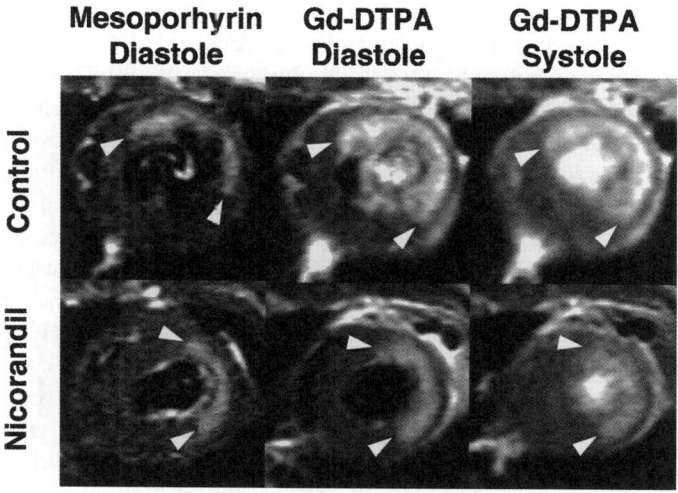

Figure 8.2.4 Functional MR images in short-axis view show the mesoporphyrin- and gadolinium-enhanced regions during diastole and systole in control and nicorandil-treated rats. In the nicorandil-treated rat, a reduced LV dilatation and improved LV wall thickening was observed, as compared with the control rat. Moreover, the gadolinium-enhanced region (arrowheads) is substantially larger than the mesoporphyrin-enhanced region (arrowheads) in the nicorandil-treated animal. Adapted with permission from ref. 58.

reperfusion. In the therapy group, nicorandil was fused during the occlusion and early reperfusion. Mesoporphyrin- and gadopentetate dimeglumine-enhanced MRI was performed, and a reduced infarction size as well as an improved systolic reduction was detected.[58]

8.2.5.5   Respiratory Diseases

MRI was previously considered to be non-favorable in lung disease imaging. This is not only because of the low proton density and the high level of paramagnetic oxygen in the lung, but also due to the blood flow and physiological motion from cardiac pulsation and respiration.[21] However, the advances of the MR techniques and the application of ECG-gating in MRI data acquisition have changed the status. For example, due to the lack of lung parenchyma, some lesions, such as non-cardiogenic pulmonary *edema*, can be detected by MRI with high sensitivity.[59–61] On the other hand, by applying some new MRI techniques, such as back-projection (BP), the lung parenchyma can be delineated, and using this method to assess lung pathologies, including ozone-, hyperoxia- and paraquat-induced lung injury has been reported.[62]

One case of using MRI to assist drug development was given by Beckmann *et al.* in a rodent allergic pulmonary inflammation model. In this study, both MRI and bronchoalveolar lavage (BAL) fluid analysis were performed to monitor the edema formation, and in the group treated with budesonide (which is a glucocorticosteroid), a good correlation between MR signals and BAL parameters (such as inflammatory cell numbers, eosinophil peroxidase and myeloperoxidase activities) was found.[63] Later, Tigani *et al.* used MRI to study the effects of both budesonide and 4-(8-benzo[1,2,5]oxadiazol-5-yl-[1,7]naphthyridin-6-yl)-benzoic acid (NVP-ABE171, a selective phosphodiesterase 4 inhibitor) on ovalbumin induced inflammation in actively sensitized rats. In both groups, increased rates of resolution of established *edema*tous signals were detected by MRI. Interestingly, although such rapid resolution was confirmed with the findings of perivascular *edema* in the histological analysis, it was not detected by BAL fluid analysis, implying a special role that MRI may play in rapid anti-inflammatory effect assessment.[64]

Rather than imaging the lung parenchyma, the anatomical structure of the airways can also be obtained by inhaling hyperpolarized noble gases such as ^{3}He and ^{129}Xe. Freely distributing in the airways, these hyperpolarized gases work as contrast agents to facilitate the acquisition of high quality images and help the detection of defects in the ventilation. Chen *et al.* demonstrated on the rodent panacinar emphysema model that early stage emphysema could be detected by measuring the ADC of hyperpolarized ^{3}He.[65] Later, on the same animal model, Peces-Barba *et al.* confirmed a good correlation between ADC and the corresponding morphometric parameters in mild emphysema.[66] However, the application of this technique in a drug development setting has not been fully exploited yet.

8.2.6 Cell Trafficking

Cell therapy is a rapidly growing area of great potential. With the Obama administration increasing the relative funding support and lifting the ban on the embryonic stem cell research, a new cycle of breakthroughs in this realm is highly expected. MRI is playing a critical role in the related studies for providing an unprecedented cell trafficking capacity. Working in conjugation with *ex vivo* cell labeling techniques, MRI is capable of tracking the *in vivo* migration and proliferation of the implanted cells in a noninvasive and longitudinal manner, making it the method of choice in assessing the dynamics of the engrafted cells.

Gd complexes, as the most common T1 contrast agents, have been employed for cell labeling. For example, neural stem cells (NSCs) were pre-labeled with gadolinium-rhodamine dextran (GRID) and were implanted into the contralateral hemisphere of the rats with unilateral stroke damage. The cells were visible under MRI due to the Gd induced T1 relaxation time changes. Also, a lesion-directed trans-hemispherical cell migration was successfully tracked by MRI, and was subsequently corroborated by histological analyses.[67] Crich *et al.* reported using Gd-HPDO3A to label endothelial progenitor cells (EPCs). The labeled cells preserved their viability and pro-angiogenesis capacity, and the hyperintensity could be tracked over 14 days when injected under the mouse kidney capsule or grafted subcutaneously on a Matrigel plug.[68]

Recently, paramagnetic chemical exchange saturation transfer (PARA-CEST) agents, which are mostly some lanthanide ion complexes, are emerging as a new class of MRI contrast agents. It works on the basis of magnetization transfer, which occurs upon irradiation between lanthanide-bound water and bulk water. However, unlike the traditional MRI contrast, the radio frequency is unique to each PARACEST agent, which makes multiplexing MR imaging possible. For example, Aime *et al.* labeled HTC (rat hepatoma) cells with Tb(dotamGly) and Eu(dotamGly), respectively, and performed an MRI phantom study on the labeled cells. It was shown that, when irradiated at 600 ppm, Tb^{III} labeled cells were exclusively detected; whereas 50 ppm irradiation only highlighted the Eu^{III} containing cells.[69]

Despite the versatility of the paramagnetic contrast agents, due to a relatively low contrast, their applications in cell labeling are limited. According to Daldrup-Link *et al.*, who labeled hematopoietic progenitor cells were with a number of contrast agents (including ferumoxides, ferumoxtran, magnetic polysaccharide nanoparticles-transferrin (MPNT), P7228 liposomes, and gadopentetate dimeglumine liposomes (GDL)), at least twice as many GDL labeled cells are needed to reach the minimum detection limit of the ferumoxides, P7228 liposomes or MPNT labeled ones. So in practice, the IONP based cell labeling is still dominantly used.[70,71]

One iron oxide formula that has proved to be efficient in cell labeling is magnetodendrimer or MD-100.[72] These agents are essentially iron oxide nanocrystals coated with a generation 4.5 carboxylated dendrimer,[67] which is highly charged and can facilitate the cell translocation. A long list of cell lines, including rat oligodendroglial progenitors, rat neural stem cells, human mesenchymal stem cells (MSCs), human neural stem cells (NSCs), human embryoid body derived pluripotent stem cells,[73] has been labeled by MD-100 with satisfactory efficacy. However, MD-100 is not yet marketed, so the availability is a concern. Recently, a convenient cell labeling method using commercial Feridex particles as the iron source has been developed. Feridex particles were found inefficient to label non-phagocytic cells and is not necessarily a good labeling agent by itself. However, when used in combination with transfection agents (*e.g.* poly-L-lysine (PLL), Lipofectamin, protamine), it complexes with the highly positively charged compounds, forming adducts that can be internalized by a wide range of cell lines.[70,74–77] Taking PLL-Feridex for example, when incubated at a concentration of 25 μg Fe ml^{-1}, can result in an iron uptake of 10–20 pg Fe per cell,[78] which is comparable to MD-100.[72] According to Heyn *et al.*, as little as 1.3–3.0 pg Fe per cell is sufficient to be detected by MRI.[79] If this is true, 20 pg Fe per cell means a tracking period of no less than 6 cell divisions (Figure 8.2.5).[80] To achieve even higher iron uptake, micron-sized particles (MPIOs) can be used. Even though the internalization rate (in terms of particle number) is low for MPIOs, just a few of them could supply enough Fe to be sufficient for even single cell detection.[81]

8.2.7   Target Specific Molecular Imaging

Molecular imaging is a relatively new concept, which emphasizes metabolic and functional information acquisition, rather than just structural and anatomic. Compared with traditional imaging methods, which differentiate lesions by simple physical properties, molecular imaging uses smart probes, *i.e.*, probes with targeting features, to help highlight a specific biological event. In MRI molecular imaging, in particular, such smart probes are constructed by coupling the contrast agents with biovectors, after which the migration of the probes can be governed in a target specific manner. Affording a unique, subcellular viewpoint, molecular MRI is believed to play an important role in the future drug development pipeline, especially for dose-response profiling and marker phenotype.[5]

Many examples of proof-of-concept research using target-specific MRI probes for disease delineation have been reported. For example, Gd-DTPA has been conjugated with polyglucose associated macrocomplex (PGM), and was used to image lymph nodes in both rats and rabbits of lymph node metastases.[82,83] Also, conjugating Gd-DTPA with an antibody to achieve specific targeting and imaging has been assessed.[84] Recently, paramagnetic high-density lipoprotein (HDL)-like naoparticles were developed by Frias *et al.*, which contain 15–20 gadolinium agents on the surface of each particle. The r_1 relaxivity of the HDL nanoparticles was evaluated to be 10 s^{-1} mM^{-1},[85] compared to that of 3.8 s^{-1} mM^{-1} for Gd-DTPA alone.[86] A demonstration of using such particles for atherosclerosis imaging was also given.[85] However,

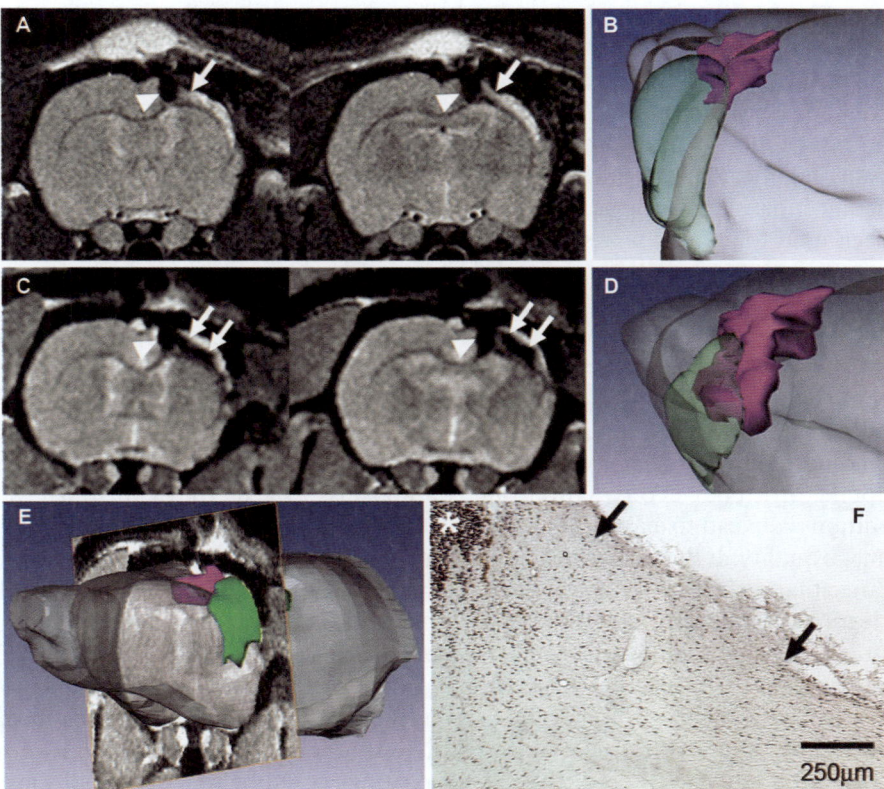

Figure 8.2.5 About 3×10^5 iron-laden hCNS-SCns cells (human central nervous system stem cells grown as neurospheres cells) were implanted in rats after distal middle cerebral artery occlusion. The MRI was able to detect the intraparenchymal-targeted migration of hCNS-SCns toward the ischemic brain. (A and C) two consecutive coronal sections showing the bolus of hCNS-SCns (arrowhead) medial to the hyperintense stroke area in T2-weighted images 1 week (A) and 5 weeks (C) after transplantation. Three-dimensional reconstruction and surface rendering of the rat brain based on high resolution T2-MRI as illustrated in (E). Posterior view of the rat brain 1 (B) and 5 (D) weeks after transplantation showing the graft (pink) and the stroked area (green). Note the broad migration of hCNS-SCns along the medial border of the stroke in the anterior posterior and craniocaudal direction, resulting in a significant increase in graft volume. (E) Three-dimensional reconstruction of the rat brain illustrating the segmentation process of the stroked area (green) and the graft (pink) based on coronal T2-MRI. (F) Histological section corresponding to the MRI (C) at 5 weeks stained with the human-specific nuclear marker SC101, showing the lateral bolus edge on the left side (asterisk) and robust migration (arrows) toward the infarct. Adapted from ref. 80 with permission.

mainly due to the intrinsic low sensitivity, the successful cases of using Gd based smart probes for *in vivo* targeting and imaging are limited.

On the contrary, using IONPs as nanoplatforms to build targeting probes could be a more promising approach. The r2 relaxivity of Feridex nanoparticles is about 150 s^{-1} mM^{-1},[87] and this number for Zn and Mn doped IONPs could be as high as 860 s^{-1} mM^{-1}.[88] IONPs conjugated with various kinds of targeting molecules, such as peptides,[87,89] antibodies,[90–92] folate acid,[93,94] have been used for tumor targeting and imaging. Also, using IONP-biovector conjugates to image cardiovascular diseases, such as atherosclerosis,[95,96] thrombosis,[97,98] and myocardial infarcts,[99] is reported.

One of the greatest concerns in smart probe engineering is biophysical property changes caused by the signal reporter docking. Either paramagnetic complexes or IONPs, with a relatively large size, can alter the pharmacokinetics of the parent molecules in a non-trivial way. As a consequence, a compromised binding affinity, a decreased circulation half-life, an increased acute hepatic uptake or a poor extravasation rate is always associated. All of these degenerations can lead to a lower than expected targeting profile and compromise the image quality. IONPs, specifically, have a diameter of tens of nanometers, a size range that tends to provoke acute reticuloendothelial system (RES) uptake and clearance. Such an issue can be partially compensated with appropriate coating strategies, which confine the overall size of particles and help reduce the rates of aggregation and the serum adsorption.[48] In the past two decades, in parallel to the advances of IONP preparation, many surface modification methods have been developed, with a special emphasis on good *in vivo* stability and long circulation half life. Indeed, the surface coatings could be of such importance that they themselves can decide the targeting features. IONPs, for instance, are generally regarded as short-lived vasculature contrast agents marked with high acute hepatic clearance rate. But when appropriately engineered, it is possible to endow them with a long circulation half life; and even better, they may accumulate at the sites of microvascular leakage, where pathological processes are usually associated. Differing from those whose migrations are governed by biovectors, such targeting genre is called passive targeting. Using such features to image cancer,[100,101] diabetes,[102,103] and atherosclerosis,[104] has been reported with various levels of success.

8.2.8 Multimodality

Each imaging modality has its specialties and limitations, and so far no single modality can provide information of all aspects with satisfactory spatial resolution, temporal resolution, detection sensitivity, tissue penetration, signal-to-noise and quantitative accuracy.[105] One of the current efforts is to combine two (or more) modalities with the hope of integrating the strengths of both compliments, while compensating the weaknesses of each individual. Such a marriage, especially between those with anatomical specialties and those with functional merits, could greatly benefit the drug development process by providing more accurate and comprehensive observation over a biological event.

One of the most successful examples is PET–CT. It adds precision of anatomic localization (by CT) to the PET functional imaging, making it possible to perform simultaneous structure-function interrogation therefore dramatically improving the efficiency and accuracy of diagnostics or therapeutic monitoring. The combination of MRI and PET is another prospective combination along the way. Such integration will benefit from the great spatial resolution and excellent morphological discrimination of MRI, as well as the high sensitivity and significant signal-to-noise ratio of PET. And even better, the versatile advanced MRI technologies, such as the blood-oxygenation-level-dependant (BOLD) imaging, can be readily incorporated into the system, to make it a multi-functional imaging platform. Compared to PET–CT which is now commercially available and deployed across the nation in clinical institutes, the development of PET–MRI is still in progress. Several major options were considered, such as to minimize the size of the PET detectors in order to maintain the MRI gantry opening, to develop a PET detector which is insensitive to high magnetic fields, and to avoid any magnetic PET detector parts that might cause heterogeneities of the magnet's B_0 field.[106] Regardless, the attractive features over the current PET–CT system, including a much better soft tissue contrast and simultaneous PET and MRI scans (as compared to the sequential scanning in PET–CT), have consistently driven the PET–MRI system development, and have lead to some promising results.[105] Based on some recent progress, it is reasonable to expect nationwide implementation of PET–MRI systems soon.[107]

Compared to the hardware development, the progress of making PET–MRI dual functional probes is relatively smoother. Choi *et al.* labeled manganese doped IONPs with ^{124}I and used the probes for sentinel lymph node imaging.[108] Jarrett *et al.* labeled dextran sulfate coated IONPs with ^{64}Cu and demonstrated their feasibility for atherosclerotic plaque imaging.[109] However, in these two cases, the probes are without biovectors and the targeting was achieved through a passive route. Recently, the Chen group developed a synthetic approach of making polyaspartic acid polymer (PASP) coated IONPs. Since both amine and carboxyl groups are available from PASP, the particles were dually labeled with ^{64}Cu chelates and RGD peptides to be rendered PET–MRI dual positive while possessing integrin $\alpha_v\beta_3$ targeting specificity (Figure 8.2.6).[87]

Due to the relatively limited accessibility to PET facilities in the pre-clinical studies, the combination of MRI and optical imaging has become a popular substitute, where optical imaging, instead of PET, is used to complement the low sensitivity of MRI. Dye labeled IONPs were conjugated with antivascular cell adhesion molecule antibodies,[95] E-selectin-binding peptides,[101] RGD peptide,[110,111] and annexin V *etc.*[112] were used in preclinical studies.[113] On the other hand, optical–MRI dual functional probes derived from Gd complexes,[114–116] Mn nanoparticles,[117] and Gd particles[118] have also been reported. However, due to the limited tissue penetration depth of light, optical–MRI is limited in small animal imaging, and its clinical relevance should be not over-exaggerated.

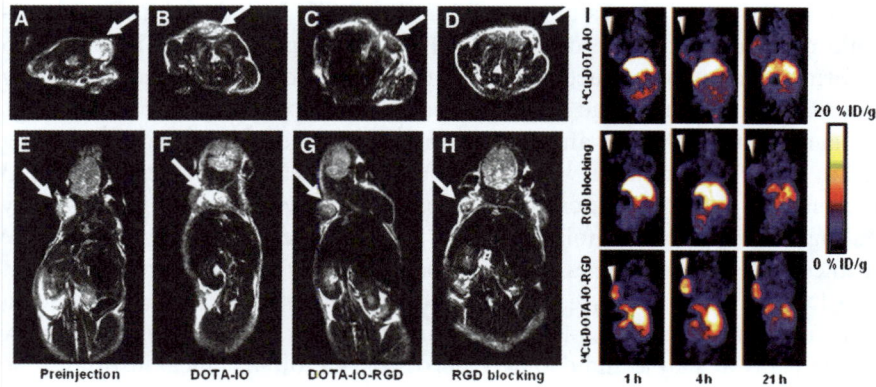

Figure 8.2.6 (A-H), sagittal and axial MR imaging on U87MG xenograft model. (A and E) are the MRI images before particle administrations. (B and F) are mice dosed with DOTA-IONPs. (C and G) were taken on mice administrated with IONPs that were pre-labeled with both DOTA chelator and RGD peptide. (D and H) were taken on mice administrated with both DOTA-IO-RGD particles and blocking dosage of free RGD peptides. Note that only (C and G) showed significant signal attenuation at the tumor sites. I, the PET results of mice administrated with IONPs. Similar to the MRI results, the tumor homing was observed when dosed with ^{64}Cu-DOTA-IO-RGD but not when dosed with ^{64}Cu-DOTA-IO. Also, co-administration of RGD would block the tumor homing, suggesting that the particle accumulation at the tumor sites was mediated via RGD-integrin interaction. Adapted with permission from ref. 87.

8.2.9   Conclusion

MRI is actively involved nowadays in the drug development process and is highly contributive at both clinic and pre-clinic levels. Unlike the traditional destructive histological methods, MRI is a noninvasive technique, which is highly favored in clinics. In preclinical research, such non-invasiveness is also of critical significance, which, besides adding statistical strengths to the disease progression monitoring, also saving the number of animals used in the studies. This is especially true for the research on chronic animal models, where repetitive and longitudinal monitoring is needed, and on higher species, where economics becomes a critical concern. Moreover, many MRI characteristics are now serving in clinics as important surrogate markers for disease identification and staging. Using such markers to guide the animal model establishment is of critical importance to the successful clinic translation, or in other words, MRI is playing a unique role in linking the human and animal pharmacology.

Compared with other imaging facilities, such as CT, PET and SPECT, *etc.*, MRI has the strength of high spatial resolution and no need of ionization radiation. More impressively, it alone provides a variety of data acquisition methods and even allows manifold data collection in the same experimental setting. This has been demonstrated, for instance, in stroke studies, where information on local brain perfusion, local cerebral blood volume, oxygen deficit, cytotoxic and vasogenic edema, and functional responsiveness can be obtained on the same living subject in one MRI session.[8] Similar comprehensive characterizations can be made on oncology and cardiovascular disorder studies; however, no other imaging facility can afford the equivalent power.

As a relatively young modality in the imaging family, there is much potential of MRI that has not been fully exploited. For instance, the on-going efforts of developing IONP or Gd complex based targeting probes have resulted in some promising results, and may lead to even more exciting achievements. Also, MRI is working with other imaging facilities, either in a spatially separated setting or potentially as an integrated system, to provide a more accurate and comprehensive assessment over a pathological process. All these findings and successes, in one way or another, will ultimately benefit the pharmaceutical industry.

References

1. J. K. Willmann, N. van Bruggen, L. M. Dinkelborg and S. S. Gambhir, *Nat. Rev. Drug Discovery*, 2008, **7**(7), 591–607.
2. B. Booth and R. Zemmel, *Nat. Rev. Drug Discovery*, 2004, **3**(5), 451–456.
3. N. Beckmann, D. Laurent, B. Tigani, R. Panizzutti and M. Rudin, *Drug Discovery Today*, 2004, **9**(1), 35–42.
4. B. P. Zambrowicz and A. T. Sands, *Nat. Rev. Drug Discovery*, 2003, **2**(1), 38–51.
5. P. V. Prasad, *Magnetic resonance imaging: methods and biological applications*, Humana Press, Totowa, N. J., 2006.
6. R. J. Reece, M. C. Kraan, A. Radjenovic, D. J. Veale, P. J. O'Connor, J. P. Ridgway, W. W. Gibbon, F. C. Breedveld, P. P. Tak and P. Emery, *Arthritis Rheum.*, 2002, **46**(2), 366–372.
7. J. C. Richardson, R. W. Bowtell, K. Mader and C. D. Melia, *Adv. Drug Delivery Rev.*, 2005, **57**(8), 1191–1209.
8. M. Rudin, N. Beckmann, R. Porszasz, T. Reese, D. Bochelen and A. Sauter, *NMR Biomed.*, 1999, **12**(2), 69–97.
9. M. E. Moseley, Y. Cohen, J. Mintorovitch, L. Chileuitt, H. Shimizu, J. Kucharczyk, M. F. Wendland and P. R. Weinstein, *Magn. Reson. Med.*, 1990, **14**(2), 330–346.
10. K. Kohno, M. Hoehn-Berlage, G. Mies, T. Back and K. A. Hossmann, *Magn. Reson. Imaging*, 1995, **13**(1), 73–80.
11. S. A. Roussel, N. van Bruggen, M. D. King and D. G. Gadian, *J. Cereb. Blood Flow Metab.*, 1995, **15**(4), 578–586.
12. Y. Mazaheri, H. Hricak, S. W. Fine, O. Akin, A. Shukla-Dave, N. M. Ishill, C. S. Moskowitz, J. E. Grater, V. E. Reuter, K. L. Zakian, K. A. Touijer and J. A. Koutcher, *Radiology*, 2009, **252**(2), 449–457.
13. S. H. Kim, E. S. Cha, H. S. Kim, B. J. Kang, J. J. Choi, J. H. Jung, Y. G. Park and Y. J. Suh, *J. Magn. Reson. Imaging*, 2009, **30**(3), 615–620.

14. L. Fass, *Mol. Oncol.*, 2008, **2**(2), 115–152.
15. M. Y. Su, J. A. Taylor, L. P. Villarreal and O. Nalcioglu, *Magn. Reson. Imaging*, 2000, **18**(3), 311–317.
16. C. D. Pham, T. P. Roberts, N. van Bruggen, O. Melnyk, J. Mann, N. Ferrara, R. L. Cohen and R. C. Brasch, *Cancer Invest.*, 1998, **16**(4), 225–230.
17. O. Muhling, M. Jerosch-Herold, M. Nabauer and N. Wilke, *Herz*, 2003, **28**(2), 82–89.
18. M. Jerosch-Herold, C. Swingen and R. T. Seethamraju, *Med. Phys.*, 2002, **29**(5), 886–897.
19. A. R. Padhani, *J. Magn. Reson. Imaging*, 2002, **16**(4), 407–422.
20. D. Borsook, L. Becerra and R. Hargreaves, *Nat. Rev. Drug Discovery*, 2006, **5**(5), 411–424.
21. I. Rodriguez, S. Perez-Rial, J. Gonzalez-Jimenez, J. Perez-Sanchez, F. Herranz, N. Beckmann and J. Ruíz-Cabello, *J. Pharm. Sci.*, 2008, **97**(9), 3637–3665.
22. T. M. Link, *Radiol. Clin. N. Am.*, 2009, **47**(4), 617–632.
23. J. Dawson, S. Gustard and N. Beckmann, *Arthritis Rheum.*, 1999, **42**(1), 119–128.
24. C. G. Peterfy, *J. Rheumatol.*, 2001, **28**(5), 1134–1142.
25. P. W. Hellings, E. M. Hessel, J. J. Van Den Oord, A. Kasran, P. Van Hecke and J. L. Ceuppens, *Clin. Exp. Allergy*, 2001, **31**(5), 782–790.
26. N. Beckmann, K. Bruttel, H. Schuurman and A. Mir, *J. Magn. Reson.*, 1998, **131**(1), 8–16.
27. N. Beckmann, R. Falk, S. Zurbrugg, J. Dawson and P. Engelhardt, *Magn. Reson. Med.*, 2003, **49**(6), 1047–1055.
28. S. Lefevre, D. Ruimy, F. Jehl, A. Neuville, P. Robert, C. Sordet, M. Ehlinger, J. L. Dietemann and G. Bierry, *Radiology*, 2011, **258**(3), 722–728.
29. A. Bashir, M. L. Gray and D. Burstein, *Magn. Reson. Med.*, 1996, **36**(5), 665–673.
30. A. Iagnocco, C. Perella, M. A. D'Agostino, E. Sabatini, G. Valesini, and P. G. Conaghan, *Rheumatology*, 2011.
31. R. N. Bryan, M. R. Willcott, N. J. Schneiders, J. J. Ford and H. S. Derman, *Radiology*, 1983, **149**(1), 189–192.
32. M. E. Moseley, M. F. Wendland and J. Kucharczyk, *Top. Magn. Reson. Imaging*, 1991, **3**(3), 50–67.
33. H. B. Verheul, R. Balazs, J. W. Berkelbach van der Sprenkel, C. A. Tulleken, K. Nicolay and M. van Lookeren Campagne, *Brain Res.*, 1993, **618**(2), 203–212.
34. T. Reese T, D. Bochelen, A. Sauter, N. Beckmann and M. Rudin, *NMR Biomed.*, 1999, **12**(4), 189–196.
35. M. Rausch, A. Sauter, J. Frohlich, U. Neubacher, E. W. Radu and M. Rudin, *Magn. Reson. Med.*, 2001, **46**(5), 1018–1022.
36. R. M. Dijkhuizen, J. Ren, J. B. Mandeville, O. Wu, F. M. Ozdag, M. A. Moskowitz, B. R. Rosen and S. P. Finkelstein, *Proc. Natl. Acad. Sci. U.S.A.*, 2001, **98**(22), 12766–12771.

37. A. Sauter, T. Reese, R. Porszasz, D. Baumann, M. Rausch and M. Rudin, *Magn. Reson. Med.*, 2002, **47**(4), 759–765.
38. R. Damadian, *Science*, 1971, **171**(976), 1151–1153.
39. C. F. van Dijke, R. C. Brasch, T. P. Roberts, N. Weidner, A. Mathur, D. M. Shames, J. S. Mann, F. Demsar, P. Lang and H. C. Schwickert, *Radiology*, 1996, **198**(3), 813–818.
40. C. Frouge C, J. M. Guinebretiere, G. Contesso, R. Di Paola and M. Blery, *Invest. Radiol.*, 1994, **29**(12), 1043–1049.
41. F. A. Howe, S. P. Robinson, D. J. McIntyre, M. Stubbs and J. R. Griffiths, *NMR Biomed.*, 2001, **14**(7–8), 497–506.
42. S. C. Feldman, D. Chu, M. Schulder, M. Barry, E. S. Cho and W. C. Liu, *Am. J. Neuroradiol.*, 2009, **30**(2), 389–395.
43. G. Wardlaw, R. Wong and M. D. Noseworthy, *Phys. Med.*, 2008, **24**(2), 87–91.
44. S. M. Cohen, J. G. Werrmann, G. H. Rasmusson, W. K. Tanaka, P. F. Malatesta, S. Prahalada, J. G. Jacobs, G. Harris and T. M. Nett, *Prostate*, 1995, **26**(2), 55–71.
45. P. G. Braunschweiger, K. Reynolds, T. R. Nelson and E. Maring, *Magn. Reson. Imaging*, 1987, **5**(6), 483–492.
46. D. Checkley, J. J. Tessier, J. Kendrew, J. C. Waterton and S. R. Wedge, *Br. J. Cancer*, 2003, **89**(10), 1889–1895.
47. D. Checkley, J. J. Tessier, S. R. Wedge, M. Dukes, J. Kendrew, B. Curry, B. Middleton and J. C. Waterton, *Magn. Reson. Imaging*, 2003, **21**(5), 475–482.
48. J. Xie, J. Huang, X. Li, S. Sun and X. Chen, *Curr. Med. Chem.*, 2009, **16**(10), 1278–1294.
49. J. Xie, S. Lee and X. Chen, *Adv. Drug Delivery Rev.*, 2010, **62**(11), 1064–1079.
50. S. Lee, J. Xie and X. Chen, *Chem. Rev.*, 2010, **110**(5), 3087–3111.
51. R. Maksimovic, T. Dill, P. M. Seferovic, A. D. Ristic, P. Alter and D. S. Simeunovic, *Herz*, 2006, **31**(7), 708–714.
52. D. Laurent D, P. R. Allegrini and W. Zierhut, *J. Hypertens.*, 1995, **13**(6), 693–700.
53. B. D. Rosen, V. R. Fernandes, K. Nasir, T. Helle-Valle, M. Jerosch-Herold and D. A. Bluemke, *Circulation*, 2009, **120**(10), 859–866.
54. M. T. McNamara, C. B. Higgins, N. Schechtmann, E. Botvinick, M. J. Lipton and K. Chatterjee, *Circulation*, 1985, **71**(4), 717–724.
55. N. Wilke, M. Jerosch-Herold, Y. Wang, Y. Huang, B. V. Christensen and A. E. Stillman, *Radiology*, 1997, **204**(2), 373–384.
56. N. Al-Saadi, E. Nagel, M Gross, A. Bornstedt, B. Schnackenburg and C. Klein, *Circulation*, 2000, **101**(12), 1379–1383.
57. K. Umemura, W. Zierhut, M. Rudin, D. Novosel, E. Robertson and B. Pedersen, *J. Cardiovasc. Pharmacol.*, 1992, **19**(3), 375–381.
58. G. K. Lund, C. B. Higgins, M. F. Wendland, N. Watzinger, H. J. Weinmann and M. Saeed, *Radiology*, 2001, **221**(3), 676–682.
59. J. R. Mayo, N. L. Muller, B. B. Forster, M. Okazawa and P. D. Pare, *J. Can. Assoc. Radiol.*, 1990, **41**(5), 281–286.

60. S. D. Caruthers, C. B. Paschal, N. A. Pou and T. R. Harris, *J. Magn. Reson. Imaging*, 1997, **7**(3), 544–550.
61. G. M. Glazer, *Chest*, 1989, **96**(Suppl. 1), 44S–47S.
62. A. G. Cutillo, *Application of magnetic resonance to the study of lung,* NY Futura Publishing Company, Armonk, 1996.
63. N. Beckmann, B. Tigani, D. Ekatodramis, R. Borer, L. Mazzoni and J. R. Fozard, *Magn. Reson. Med.*, 2001, **45**(1), 88–95.
64. B. Tigani, C. Cannet, S. Zurbrugg, E. Schaeublin, L. Mazzoni and J. R. Fozard, *et al. Br. J. Pharmacol.*, 2003, **140**(2), 239–246.
65. X. J. Chen, L. W. Hedlund, H. E. Moller, M. S. Chawla, R. R. Maronpot and G. A. Johnson, *Proc. Natl. Acad. Sci. U. S. A.*, 2000, **97**(21), 11478–11481.
66. G. Peces-Barba, J. Ruíz-Cabello, Y. Cremillieux, I. Rodriguez, D. Dupuich and V. Callot, *Eur. Respir. J.*, 2003, **22**(1), 14–19.
67. M. Modo, K. Mellodew, D. Cash, S. E. Fraser, T. J. Meade and J. Price, *Neuroimage*, 2004, **21**(1), 311–317.
68. S. G. Crich, L. Biancone, V. Cantaluppi, D. Duo, G. Esposito and S. Russo, *Magn. Reson. Med.*, 2004, **51**(5), 938–944.
69. S. Aime, C. Carrera, D. Delli Castelli, S. Geninatti Crich and E. Terreno, *Angew. Chem. Int. Ed. Engl.*, 2005, **44**(12), 1813–1815.
70. H. E. Daldrup-Link, M. Rudelius, R. A. Oostendorp, M. Settles, G. Piontek and S. Metz, *Radiology*, 2003, **228**(3), 760–767.
71. A. Bhirde, J. Xie, M. Swierczewska and X. Chen, *Nanoscale*, 2010, **3**(1), 142–153.
72. J. W. Bulte, T. Douglas, B. Witwer, S. C. Zhang, E. Strable and B. K. Lewis, *Nat. Biotechnol.*, 2001, **19**(12), 1141–1147.
73. M. Modo, M. Hoehn and J. W. Bulte, *Mol. Imaging*, 2005, **4**(3), 143–164.
74. D. L. Kraitchman, A. W. Heldman, E. Atalar, L. C. Amado, B. J. Martin and M. F. Pittenger, *Circulation*, 2003, **107**(18), 2290–2293.
75. S. A. Anderson, J. Glod, A. S. Arbab, M. Noel, P. Ashari and H. A. Fine, *Blood*, 2005, **105**(1), 420–425.
76. L. Kostura, D. L. Kraitchman, A. M. Mackay, M. F. Pittenger, and J. W. Bulte, *NMR Biomed.*, 2004, **17**(7), 513–517.
77. A. S. Arbab, G. T. Yocum, H. Kalish, E. K. Jordan, S. A. Anderson and A. Y. Khakoo, *Blood*, 2004, **104**(4), 1217–1223.
78. J. A. Frank, B. R. Miller, A. S. Arbab, H. A. Zywicke, E. K. Jordan and B. K. Lewis BK, *Radiology*, 2003, **228**(2), 480–487.
79. C. Heyn, C. V. Bowen, B. K. Rutt and P. J. Foster, *Magn. Reson. Med.*, 2005, **53**(2), 312–320.
80. R. Guzman, N. Uchida, T. M. Bliss, D. He, K. K. Christopherson and D. Stellwagen, *Proc. Natl. Acad. Sci. U. S. A.*, 2007, **104**(24), 10211–10216.
81. K. A. Hinds, J. M. Hill, E. M. Shapiro, M. O. Laukkanen, A. C. Silva and C. A. Combs, *Blood*, 2003, **102**(3), 867–872.
82. L. Harika, R. Weissleder, K. Poss and M. I. Papisov, *Radiology*, 1996, **198**(2), 365–370.
83. L. Harika, R. Weissleder, K. Poss, C. Zimmer, M. I. Papisov and T. J. Brady, *Magn. Reson. Med.*, 1995, **33**(1), 88–92.

84. C. Curtet, C. Tellier, J. Bohy, M. L. Conti, J. C. Saccavini and P. Thedrez, *Proc. Natl. Acad. Sci. U. S. A.*, 1986, **83**(12), 4277–4281.
85. J. C. Frias, K. J. Williams, E. A. Fisher and A. A. Fayad, *J. Am. Chem. Soc.*, 2004, **126**(50), 16316–16317.
86. M. Mikawa, H. Kato, M. Okumura, M. Narazaki, Y. Kanazawa and N. Miwa N, *Bioconjug. Chem.*, 2001, **12**(4), 510–514.
87. H. Y. Lee, Z. Li, K. Chen, A. R. Hsu, C. Xu and J. Xie, *J. Nucl. Med.*, 2008, **49**(8), 1371–1379.
88. J. T. Jang, H. Nah, J. H. Lee, S. H. Moon, M.G. Kim and J. Cheon, *Angew. Chem. Int. Ed. Engl.*, 2009, **48**(7), 1234–1238.
89. J. Xie, K. Chen, H. Y. Lee, C. Xu, A. R. Hsu and S. Peng, *J. Am. Chem. Soc.*, 2008, **130**(24), 7542–7543.
90. J. H. Lee, Y. M. Huh, Y. Jun, J. Seo, J. Jang and H. T. Song, *Nat. Med.*, 2007, **13**(1), 95–99.
91. D. Artemov, N. Mori, R. Ravi and Z. M. Bhujwalla, *Cancer Res.*, 2003, **63**(11), 2723–2727.
92. M. A. Funovics, B. Kapeller, C. Hoeller, H. S. Su, R. Kunstfeld and S. Puig, *Magn. Reson. Imaging*, 2004, **22**(6), 843–850.
93. J. Sudimack and R. J. Lee, *Adv. Drug Delivery Rev.*, 2000, **41**(2), 147–162.
94. H. Choi, S. R. Choi, R. Zhou, H. F. Kung and I. W. Chen, *Acad. Radiol.*, 2004, **11**(9), 996–1004.
95. A. Tsourkas, V. R. Shinde-Patil, K. A. Kelly, P. Patel, A. Wolley and J. R. Allport, *Bioconjug. Chem.*, 2005, **16**(3), 576–581.
96. K. A. Kelly, J. R. Allport, A. Tsourkas, V. R. Shinde-Patil, L. Josephson and R. Weissleder, *Circ. Res.*, 2005, **96**(3), 327–336.
97. F. A. Jaffer and R. Weissleder, *Circ. Res.*, 2004, **94**(4), 433–445.
98. L. O. Johansson, A. Bjornerud, H. K. Ahlstrom, D. L. Ladd and D. K. Fujii, *J. Magn. Reson. Imaging*, 2001, **13**(4), 615–618.
99. R. Weissleder, A. S. Lee, B. A. Khaw, T. Shen and T. J. Brady, *Radiology*, 1992, **182**(2), 381–385.
100. M. K. Yu, Y. Y. Jeong, J. Park, S. Park, J. W. Kim and J. J. Min, *Angew. Chem. Int. Ed. Engl.*, 2008, **47**(29), 5362–5365.
101. H. Lee, M. K. Yu, S. Park, S. Moon, J. J. Min and Y. Y. Jeong, *J. Am. Chem. Soc.*, 2007, **129**(42), 12739–12745.
102. M. C. Denis, U. Mahmood, C. Benoist, D. Mathis and R. Weissleder, *Proc. Natl. Acad. Sci. U.S.A.*, 2004, **101**(34), 12634–12639.
103. A. Moore, S. Bonner-Weir and R. Weissleder, *Diabetes*, 2001, **50**(10), 2231–2236.
104. F. A. Jaffer, P. Libby and R. Weissleder, *J. Am. Coll. Cardiol.*, 2006, **47**(7), 1328–1338.
105. S. R. Cherry, *Annu. Rev. Biomed. Eng.*, 2006, **8**, 35–62.
106. B. J. Pichler, M. S. Judenhofer and H. F. Wehrl, *Eur. J. Radiol.*, 2008, **18**(6), 1077–1086.
107. O. Ratib, and T. Beyer, *Eur. J. Nucl. Med. Mol. Imaging*, 2011.
108. J. S. Choi, J. C. Park, H. Nah, S. Woo, J. Oh and K. M. Kim, *Angew. Chem. Int. Ed. Engl.*, 2008, **47**(33), 6259–6262.

109. B. R. Jarrett, B. Gustafsson, D. L. Kukis and A. Y. Louie, *Bioconjug. Chem.*, 2008, **19**(7), 1496–1504.
110. W. J. Mulder, G. J. Strijkers, J. W. Habets, E. J. Bleeker, D. W. van der Schaft and G. Storm, *FASEB J.*, 2005, **19**(14), 2008–2010.
111. K. Chen, J. Xie, H. Xu, D. Behera, M. H. Michalski and S. Biswal, *Biomaterials*, 2009, **30**(36), 6912–6919.
112. D. E. Sosnovik, E. A. Schellenberger, M. Nahrendorf, M. S. Novikov, T. Matsui and G. Dai, *Magn. Reson. Med.*, 2005, **54**(3), 718–724.
113. S. Lee and X. Chen, *Mol. Imaging*, 2009, **8**(2), 87–100.
114. V. S. Talanov, C. A. Regino, H. Kobayashi, M. Bernardo, P. L. Choyke and M. W. Brechbiel, *Nano Lett.*, 2006, **6**(7), 1459–1463.
115. J. C. Frias, Y. Ma, K. J. Williams, Z. A. Fayad and E. A. Fisher, *Nano Lett.*, 2006, **6**(10), 2220–2224.
116. W. J. Mulder, R. Koole, R. J. Brandwijk, G. Storm, P. T. Chin and G. J. Strijkers, *Nano Lett.*, 2006, **6**(1), 1–6.
117. J. Huang, J. Xie, K. Chen, L. Bu, S. Lee and Z. Cheng, *Chem. Commun.*, 2010, **46**(36), 6684–6686.
118. Bridot JL, Faure AC, Laurent S, Riviere C, Billotey C, Hiba B, M. Janier, V. Josserand, J. L. Coll, L. V. Elst, R. Muller, S. Roux, P. Perriat and O. Tillement, *J. Am. Chem. Soc.*, 2007, **129**(16), 5076–5084.

CHAPTER 8.3

MRI in Practical Drug Discovery

K. K. CHANGANI,* M. V. FACHIRI AND S. HOTEE

MRI Centre, GlaxoSmithKline Medicines Research Centre, Gunnels Wood Road, Stevenage Hertfordshire, SG1 2NY, UK

8.3.1 Introduction

Drug discovery and development is a time-consuming, expensive and risk-intensive process. The industry is marked by long development cycles and soaring research and development (R&D) costs, which have risen by 152% since 1991.[1] Moreover, whilst R&D technological developments have increased the number of potential drug targets entering the developmental process, since 1995 there has been a reduced number of drug approvals. The highest compound failure rate is in the late clinical phases of drug development at 90%, and is principally due to problems with efficacy and safety.[2] Late-stage drug failures are particularly expensive as clinical trials are resource intensive, and therefore accurate early assessment of the potential of a drug candidate is crucial. There is therefore a growing demand for new technologies to provide more reliable predictive data on which early cost saving Go–No Go decisions may be based.[3]

In vivo imaging technologies, including magnetic resonance imaging (MRI) could provide the solution. Imaging methods provide anatomical, functional and molecular quantitative data, allowing the real-time assessment of pharmacokinetic and pharmacodynamic properties of compound candidates in both animal and man.[4] Moreover, imaging techniques are non-invasive, enabling longitudinal studies to be conducted for full characterisation of

RSC Drug Discovery Series No. 15
Biomedical Imaging: The Chemistry of Labels, Probes and Contrast Agents
Edited by Martin Braddock
© Royal Society of Chemistry 2012
Published by the Royal Society of Chemistry, www.rsc.org

disease processes. Imaging studies also enable development of predictive disease-based biomarkers and animal models that can be translated from early to late phases in drug development. In particular, magnetic resonance imaging (MRI) is a non-invasive and non-ionising technology that produces images with excellent soft-tissue contrast, high spatial and temporal resolution.[5] MRI has therefore become a useful tool in preclinical and clinical investigations. Investigations into the efficacy and safety of new drugs performed at earlier stages increases the efficiency of the drug development process and hence, reduces the cost in time and money.

This chapter will commence by providing an overview of the drug discovery process and will follow by highlighting the value of applying imaging technology. The use of MRI in pharmaceutical research will be discussed in more detail, including a description of advances in MRI technology, illustrative examples of MRI in R&D and consideration of the technique's advantages and limitations.

8.3.2 The Drug Development Process

8.3.2.1 Target Identification and Validation

Drug discovery and development consists of several stages beginning with the identification, prioritisation and validation of a biological target that is important in human disease (proof of concept testing). Significant investment in new technologies and advances in biomedical technology over the past ten years has led to an exponential increase in the number of therapeutic targets.[5] Target identification and validation is therefore a highly time consuming process, requiring considerable prioritisation of many possible targets and confirmation of the target's role in disease development.

Target identification may be either physiology-based or target-based, or a combination of the two paradigms. Physiology-based drug discovery uses physiological readouts of disease to screen and profile compounds, with later stages investigating drug target function. Alternatively, target-based drug discovery commences with understanding drug target function and its contribution to disease.[6] As knowledge of the underlying mechanisms involved in disease grows, new therapies are increasingly being targeted at pathways of multiple proteins, rather than at individual proteins.

Target validation involves demonstration of a therapeutic target's crucial involvement in the process of a disease and that target modulation will have therapeutic effect.[6] Validation of potential targets has been greatly helped by the use of transgenic or genetic knock out animals.[5]

8.3.2.2 Screening and Hits to Leads

Subsequently, high throughput screening and start of chemistry (SoC) are used to identify a short-list of lead candidate compounds and to validate their involvement in cellular and tissue-isolated processes using both *in vitro* and *in vivo* assays. Chemical entities that display activity in the screening assays are chemically modified to optimise drug like properties such as bioavailability and stability, with the aim of generating preclinical drug candidates.[6] Lead optimisation addresses issues such as drug-target binding affinity, drug-target specificity, pharmacokinetic properties (*e.g.* rate and efficiency of absorption), and toxicological profile.[5] The ability to manufacture the compounds on a large scale is also investigated and formulation is assessed to ensure the compounds can be produced in a marketable form, such as a tablet.

8.3.2.3 Preclinical Studies

Compounds are then progressed to preclinical investigations involving extensive *in vitro* and *in vivo* investigations into the pharmacological and toxic characteristics of the drug. These studies involve characterisation of several biophysical parameters (drug absorption, metabolism, metabolite toxicity, speed of drug and metabolite excretion).[5,6] Animal models are crucial at this stage for providing an intact disease-mimicking biological system, in which the drug's involvement in the disease process may be further validated and potential safety concerns identified. This process can take more than 6 years to complete.[1]

8.3.2.4 Clinical Trials (Phase I-III)

Following sufficient preclinical safety evaluation and initial regulatory approval, a lead compound is selected and advanced to three successive stages of clinical investigations. Phase I or 'first time in human' (FTIH) studies are conducted on a small selection of healthy human volunteers. These studies determine the metabolic and pharmacological drug actions, provide an initial safety assessment and establish side effects with increasing doses in man.[6] Phase II tests the compound in diseased patients to provide preliminary data on drug efficacy and to further identify adverse effects of the drug. Next, phase III trials are conducted on large patient cohorts, in order to establish further data on safety and effectiveness, to compare the drug against other currently available treatments, and therefore to identify the benefit-risk relationship of the compound.[1,6] During Phase I-III, the FDA can prevent clinical studies being performed if they are deemed unsafe or ineffective in achieving the desired objectives.[6]

8.3.2.5 Regulator Review, Market Approval and Monitoring

Finally, data from preclinical and clinical studies are collected and reviewed by regulatory authorities who decide whether the drug is of sufficient quality, efficacy and safety for market approval.[6] Once the drug is authorised and commercially available, the effectiveness of the drug when used more extensively will be determined and side effects continuously monitored.

8.3.2.6   Attrition Rates in Pharma

Pharmaceutical drug discovery and development is a resource intensive process and costs have been gradually rising, for example from ~ $20 billion in 1990 to ~ $50 billion in 2000.[7] The most costly stages are the clinical phases, representing 50–70% of the R&D cost.[6] Moreover, 90% of new chemical entities (NCEs) progressing through clinical trials fail mostly due to lack of efficacy (27%), commercial and market considerations (21%) and safety (20%). Interestingly, different therapeutic areas experience different rates of attrition, with cardiovascular (20%) and arthritis pain (17%) more successful than CNS disorders (8%) and oncology (5%).[6] As the distribution of cost across the development cycle is so heavily weighted at the clinical study stages, a reduction of 10% of attrition here would give a saving of 40% to the cost of development of the entire cycle.[1] Additionally, costs of R&D are only recouped at the end of the development cycle after the drug makes it to market. Each molecule has a finite patent life (approximately 17–20 years) after which generic alternatives can be produced and marketed without the extensive R&D costs, often meaning that they are cheaper.[1] It is for this reason that reducing the time taken for a drug to be developed, and as a result extending the marketing time that is patent protected, is of great importance.

8.3.2.7   Technological Advances in Pharma

As part of the industry's efforts to both reducing development time and unnecessary expenditure, investments have been made in new technologies, including high-throughput chemistry, modern lead optimisation techniques and pharmacokinetic modelling.[1] Despite these improvements attrition rates remain high.

Closer examination reveals that most advances have been made on *in vitro* or virtual technologies, with few parallel advances in animal model technologies over the past 120 years.[2] Paradoxically, preclinical animal data is deemed the most reliable predictive data on drug efficacy and safety in clinical trials used by regulators, such as the US Food and Drug Administration (FDA) and hence is crucial to the drug discovery process.[2] Therefore improvements in animal models of disease and advances in the development of bench to bedside readouts are required to improve attrition rates in pharmaceutical research, to provide more reliable predictions of drug success in the clinic.

8.3.3   Imaging Technology

Imaging technology provides a powerful means of assessing disease development in the intact, living organism, therefore improving understanding of disease pathology and crucially enabling earlier and more accurate selection of drug candidates. Currently, the major application of imaging technologies to the R&D process is in late preclinical phases through to the clinical phases, but their use in drug discovery is increasing.[3] Such is the perceived importance and potential of *in vivo* imaging techniques in R&D, that most of the major pharmaceutical companies are making considerable investments in both preclinical and clinical imaging centres.[8] Preclinical imaging modalities include magnetic resonance imaging (MRI), computerised tomography (CT), positron emission tomography (PET) and bioluminescent imaging (BLI).

8.3.3.1 Advantages of Imaging Technology

Imaging techniques provide high-resolution anatomical, functional, cellular and molecular data that is derived non-invasively. The non-invasive nature of imaging is especially important in the study of chronic diseases such as neurodegenerative disorders (Alzheimer's and Parkinson's disease) and cardiovascular diseases (artherosclerosis). This is because they enable long term monitoring of disease progression and response to therapy.[9] Non-invasive imaging studies allow efficacy of drug candidates to be assessed in subjects over relatively long time periods, enabling anatomical and physiological alterations to be compared with the original pre-treatment condition.[5] Additionally, longitudinal studies may be conducted on a single animal or patient, such that each subject can act as its own statistical control. This reduces intra-individual variability, increases the statistical quality of the data and subsequently means that fewer animals and patients are required to achieve statistically significant data.[5,10] For example, in rat models of asthma it is estimated that MRI techniques use ~80–90% fewer animals than would histological or bronchoalveolar lavage fluid (BLF) techniques.[11] Imaging also helps to sort patient and animal populations into 'homogenous' study groups prior to administration of candidate compounds, which improves the statistical relevance of data.[5]

The non-invasive nature of imaging studies means that animals are alive following experiments and can therefore be examined repeatedly, for example using behavioural tests.[5] Imaging data may be used to complement information derived from traditional analytical techniques, such as histological methods, both of which have shown to correlate well. For example in rat models of allergic asthma, histological derived measurements and proton MRI determined signals of a key inflammatory factor, perivascular oedema, were significantly correlated.[11] The efficacy of anti-asthma drugs can therefore be rapidly assessed *in vivo* using measurements of the rate of oedematous signal resolution.[11] Maximum data collection from all possible sources will improve the power and accuracy of diagnosis, prognosis and treatment management.[5]

Imaging techniques offer several advantages over traditional analytical techniques. Imaging data enables the collection of 3-dimensional images, which differs from the more site specific nature of histological techniques. This makes imaging studies more suited to the study of heterogeneous tissues than traditional techniques are, as possible sampling errors may be avoided.[5] Imaging data can be complemented with histological techniques, as it can be used to direct histological tissue sampling for increased specificity.[5,11] Additionally the non-invasive quality of imaging methods allows tissues to be examined in the host setting.[5] This reduces artefacts caused by traditional methods of tissue collection and analysis, such as global tissue ischaemia induced during tissue extraction. Morphological changes can be induced by histological analysis, which introduces inaccuracies into measurements of morphology.[5] Non-invasive anatomical imaging is particularly useful in animal models where tissue is difficult to access by conventional methods, such as in transgenic tumour models where tumours lie in deep tissues.[3] Advances in imaging technology have also enabled earlier and more subtle detection of biological changes than was previously feasible using traditional techniques.[10]

Imaging methods have a high temporal resolution (sub-seconds to seconds), which permits real-time monitoring of dynamic data that may not be possible using other methodologies.[5] This is useful in fMRI studies, where the functional response of the brain to stimulation, can be assessed to the temporal resolution of seconds.[5] Imaging also permits the fast, computerized collection of *in vivo* data which quickens evaluation of drug candidates and decreases the possibility of human error.[2] This is in contrast to the labour-intensive methods of traditional analysis methods, which can take considerably longer to perform.[5] In addition, semi-automatic image analysis allows data to be evaluated relatively quickly, so that results can often be compiled on the same day as the experiment was conducted.[11]

8.3.3.2 Combined Imaging Technology

Imaging modalities vary in their respective strengths and weaknesses, and a full understanding of these differences is required for their effective employment. The properties of small animal imaging modalities are compared in Table 8.3.1. Imaging modalities are complementary techniques that provide distinct information and can be combined to benefit from their strengths. MRI and CT are anatomical imaging technologies, and the anatomical data can be enhanced by fusion with higher sensitivity functional data, such as that of the nuclear technologies PET or SPECT.[8] Combined imaging simultaneously benefit from sensitivity and resolution of different imaging modalities.[1] However, potential movement of patients and the time period between scans makes fusion of data a complicated and inexact process.[12] In order to overcome these problems, combined scanners have been introduced, such as PET-MRI, PET-CT, and SPECT-CT.[4,13] The increased availability of small-animal combined imaging technologies will allow drug design to be assisted by MRI.[1]

In view of the many advantages of imaging technology, the appropriate employment of imaging techniques will have a great impact on the drug discovery and development process. Imaging studies permit disease phenotyping of patients and animal models of human disease, which allows preclinical identification and refinement of markers of disease severity and progression (biomarkers), bench to bedside translation of study protocols and *in vivo* evaluation of drug candidates. The improvements offered by imaging technology will help to streamline developmental phases of new drugs, allow more rapid proof-of-concept assessment in both preclinical and clinical stages, and reduce attrition rates.[10]

Table 8.3.1 Comparison of Small Animal Imaging Technologies. BLI, bioluminescence imaging; CT, X-ray computed tomography; MRI, magnetic resonance imaging; PET, positron emission tomography; SPECT, single photo emission computed tomography.[2,4,10]

Imaging modality	Resolution	Depth	Time	Cost	Advantages and applications	Disadvantages
MRI	10–100 μM	No limit	Min – h	>$300 000	• Excellent soft-tissue contrast • Non ionising radiation • Anatomical physiological and molecular imaging	• Expensive • Requires high technical expertise • Relatively poor sensitivity
CT	50 μM	No limit	Min	$100–300 000	• Intermediate cost • Anatomical and physiological imaging	• X-Ray • Low sensitivity for molecular imaging
PET	1–2 mm	No limit	Min	>$300 000	• Best for pharmacokinetics and pharmacodynamics • Highest sensitivity for imaging labelled probes • Can quantify regional kinetic parameters • Physiological and molecular imaging	• Expensive • Requires technical expertise • Ionising radiation • Limited spatial resolution • Synthesis of a radionucleotide
SPECT	1–2 mm	No limit	Min	$100–300 000	• Intermediate cost • Relatively simple, although expertise required for optimal data • Image labelled peptides and molecular probes • Physiological and molecular imaging	• Lower sensitivity than PET • Less suited to imaging labelled drugs than PET and MRI • Ionising radiation
BLI	Several mm	Centimetres	Min	$100–300 000	• Intermediate cost • Relative high throughput • Relatively simple • Gene expression, cell and bacterial tracking	• Low resolution • Limited depth

8.3.4 MRI

8.3.4.1 Theory and Technological Advances

MRI technology focuses a powerful magnetic field on the body area under study, to align the nuclei of hydrogen atoms.[2] This is subsequently disrupted by a radiowave pulse transmitted through the tissue, leading to the emission of electromagnetic energy as nuclei return to their original positions. The energy emitted varies according to the type of tissue exposed, and this is detected by the MRI scanner which produces a cross-sectional image of the anatomy.[2] The MRI image is controlled by various parameters including proton density, relaxation times T_1, T_2, T_2^*, water diffusion and water exchange rates.[9] These parameters can be modified to generate the optimal contrast for detailed differentiation of soft tissue pathologies and normal tissue structures.[9]

MRI is able to employ a broad range of endogenous contrast techniques that permit imaging of morphological, physiological, cellular and molecular information *in vivo,* making it a successful imaging modality for both clinical and preclinical studies.[5] MRI contrast derives from variable proton density and heterogeneity in relaxation times of distinct proton populations. In some cases, the intrinsic contrast of tissue is insufficient for differentiation of diseased and normal tissue. Image contrast can therefore be enhanced by optimisation of pulse sequences and the administration of contrast agents, such as the paramagnetic ions gadolinium (Gd) and manganese.[14] For example gadolinium-based probes have been generated to detect fibrin in thrombi.[15,16]

Cellular and metabolic imaging has become possible with the development of various MR-based techniques. Firstly, the high spatial resolution of MRI makes it an attractive technology for cell tracking, as it enables a quantitative measurement of cell migration within intact organisms.[5] There is considerable interest in stem cell transplantation for regeneration of lost nervous tissue, myocardium and pancreatic beta cells.[5] The development of effective cell therapies requires non-invasive characterisation of the location, distribution and viability of these cells which can be performed using MRI techniques.[17] Serial MRI studies over time track migration of stem cells labelled with iron oxide nanoparticles, to target tissues.[18,19] Diffusion-weighted MRI is another important MRI imaging technique. It is sufficiently sensitive to identify degradation of the cell membrane following stroke, and has therefore become the preferred method of identifying cerebro-vascular pathology.[5]

Molecular imaging uses para- or superparamagnetic contrast agents, such as gadolinium or ultra-small superparamagnetic iron oxides, as MR probes to specifically target a particular molecular target of interest.[5] For example, MR probes have been developed that can detect factors involved in cardiovascular disease including fibrin (thrombosis),[15,16] and integrins (angiogenesis).[20–22] This enables the *in vivo* visualisation of molecular and cellular interactions of normal and pathophysiological processes, which will greatly facilitate understanding of disease mechanisms and evaluation of drug-host interactions.[5] The main limitations of molecular MRI are the low sensitivity of the approach and the need for large para- or superparamagnetic reporter groups.[23]

MR spectroscopy (MRS) is a forerunner to MRI and has recently enjoyed a resurgence both clinically and experimentally as the *in vivo* method of choice to track disease progression and monitor therapeutic approaches. Several key metabolites can be probed using phosphorus (ATP, phosphocreatine, inorganic phosphate, pH, lipid precursors and glycolytic intermediates), proton (lipids, lactate, amino acids, choline containing compounds *etc.*) and carbon (glucose, glycogen, glycolytic intermediates, long chain fatty acids *etc.*) magnetic resonance. For example, MRS with probes for ATP and phosphocreatine, was used to identify abnormal myocardial energetics in heart failure.[24] The high spatiotemporal resolution provided by MRS means that metabolite information can be gained not only from the complete organ, but also from discrete regions of interest ($1mm^3$) which can be subsequently serially re-scanned.

Despite the many advances in MRI technology and their potential applicability to R&D, the number of publications applying MRI to the drug discovery process is relatively small. This is however growing in number, and reflects the considerable potential MRI based studies have, for improving efficiency in the drug discovery and development process.[14]

8.3.4.2 Advantages of MRI Technology

Like other imaging modalities, MRI enables 2 and 3 dimensional, high quality, dynamic real-time images to be obtained in any orientation and from any organ. The principal advantages of MRI over other imaging modalities are the excellent soft-tissue contrast capabilities, the control and flexibility over the chosen image contrast and the absence of ionising radiation.[12]

MRI detects the absorption and emission of energy from tissues, meaning that the type of MR signal received is dependent on the intrinsic properties of the experimental tissue.[5] This is in contrast to other imaging techniques which detect transmission of signals through tissues, and involve administration of extrinsic substances such as labelled particles, into the study tissue.[5] Thus the MRI signal emitted depends on the type of tissue examined, for example tissue with high fat or water content will have different properties.[5] The dependence of MRI on multiparametric signals gives MRI excellent soft-tissue contrast properties, enabling the differentiation of healthy and diseased tissues and the evaluation of candidate drug effects on experimental tissue.[24]

Moreover unlike PET and CT techniques, MRI uses radio-frequency (RF) pulses that are non-ionising and non-destructive, and it is therefore possible for multiple measurements of each patient or animal to be taken over long periods of time.[5] The ability for multiple scans to be performed and averaged also helps to overcome the inherent low sensitivity of the MRI technique, although this extends measurement times.[5] A further advantage of MRI is that RF signals are able to penetrate deep into tissue with little energy loss, and therefore MRI resolution is maintained in scans of high tissue depth.[5]

MRI technological advances now enable anatomical, functional, cellular and molecular imaging to be performed, which can be further complemented with data from other imaging methods to provide a comprehensive analysis of the experimental subject.[5] The advantages of MRI means that the technique is increasingly contributing to many stages of the drug discovery process; for study of disease mechanisms, identification and development of animal models and biomarkers, evaluation of candidate drugs, and as a bench to bedside technology.

8.3.5 MRI Applications

Since the 1970s MRI has been a vital technology for clinical diagnosis, providing information on anatomy, composition and function of almost every tissue.[2] Recent technological advances, such as those derived from functional imaging (fMRI),[25] imaging of the heart and open MRI,[9] have facilitated many clinical studies to be performed using MRI. MRI is extensively used to diagnose visceral pathologies such as cardiovascular disease, or musculoskeletal diseases such as rheumatoid arthritis.[10,26] Indeed MRI is well established as a method of assessing lesion progression in multiple sclerosis and stroke, and is the preferred method for diagnosis of joint diseases.[3,27] Radiological studies have greatly benefited from the introduction of high-field whole body MRI scanners (over 4 Tesla), which have high soft tissue contrast properties and sub-millimetre image resolution.[12] The non-invasive, excellent soft-tissue contrasting ability, multi-dimensional imaging capability of MRI makes it a useful methodology for diagnostic purposes.

MRI was originally developed to provide high spatial resolution (on the order of 10 μm given sufficient signal-to-noise ratio) morphological images for clinical diagnosis.[5] Over the past decade, MRI has become a technique that enables imaging of functional and physiological parameters such as degree of oxygenation.[24] Physiological and functional MRI are important for identifying pathophysiological changes that often occur before observable structural changes.[24] MRI is now evolving towards molecular imaging, which uses labelled molecules to probe biological targets.[3,24] Molecular imaging has much to offer the pharmaceutical process, as it permits measurement of drug absorption, distribution and binding, in addition to determination of pharmacodynamic properties.[5] The versatility of MRI is advantageous, for example MRI enables morphological and molecular images to be compared in the same subject.[24]

MRI is increasingly being applied to pharmaceutical research as it provides the means of establishing the relationships between three key components of drug discovery; target, disease and therapy.[8] It is important for example, to validate the strength of the relationship between novel targets and disease, and also to confirm that targets are indeed modifiable by an external drug. MRI is an attractive technology for comprehensively characterising animal models of disease, and subsequently enabling development and translation of biomarkers and study protocols, and thorough evaluation of drug candidates in the drug discovery and development process.

8.3.5.1 Animal Models

After the completion of the human genome sequence in the 1990s and the production of many animal model genome sequences, the utility of animal models in pharmaceutical research is growing.[5] Knockout and transgenic technology permit examination of the effects of gene products on disease processes, through modification of specific genes and their downstream products.[24] Transgenic animals are useful both as disease models and for target validation.[24] Importantly it is believed that selecting compounds that are likely to demonstrate benefit in man rely essentially on the predictability of preclinical animal models of human disease. It is therefore necessary to undergo an iterative process of animal model development and refinement which involves full characterisation and benchmarking of the animal model.

MRI small animal systems have now been developed that can be operated in basic laboratory settings, are actively shielded, and do not need the radio frequency shielded screen that is required in clinical set-ups.[5] These relatively high throughput scanners are usually cheaper than their clinical counterparts, making them suitable tools for development and validation of NCEs.[5] MRI animal scanners operate up to 17 Tesla (T) and obtain *in vivo* images with sub 50 μm in-plane resolution.[12] Moreover the non-invasive nature of the MRI technique provides the possibility of reducing animal numbers, increasing statistical power through longitudinal studies, improving subtlety over traditional techniques such as histology, and translating developments from preclinical to clinical stages.[5]

8.3.5.2 Dissecting Disease Mechanisms

The versatility of the MRI approach is such that it enables *in vivo* comparison of anatomy and disease morphology, analysis of physiological and functional properties such as blood flow and oxygenation, and also assessment of molecular and cellular parameters, including cellular proliferation and metabolism.[28] MRI can therefore be used to assess the phenotype of selected tissues and organs in established animal models and clinical studies, and provide non-invasive assessment of disease mechanism and progression. Knowledge gained from investigations into the mechanisms underlying disease development, can subsequently be used to refine animal models of disease, to develop new biomarkers, and to identify new therapeutic targets of disease.

8.3.5.3 Specific Disease Areas

MRI is well suited for investigating disease pathways and progression as it provides direct visualisation and quantification of disease mechanisms over time. These qualities make MRI an effective tool for understanding the pathophysiology of a variety of diseases, including those of the central nervous system (CNS), cardiovascular system and the musculoskeletal system.

Indeed, MRI is the preferred imaging modality for longitudinal investigations into diseases of the central nervous system, including stroke, multiple sclerosis and neurodegenerative disorders.[5] For example, MRI can be used to image successive phases of stroke pathophysiology in both animal and man. MRI angiography is used to provide high resolution images of initial vascular occlusion,[29] and diffusion-weighted MRI is used to assess early cerebral infarction.[30] An illustrative example of disease phenotyping in animal models is in the mouse model for the neurodegenerative disorder, Alzheimer's disease (AD).[27] AD mice exhibit neuroanatomical alterations linked to AD, including plaques containing amyloid-β and deposition of amyloid-β in cerebral wall vessels.[31] MRI techniques have been introduced to image amyloid-β plaques,[32] but acquisition times remain over 10 hours long, even with the use of contrast enhancing molecular probes specifically targeted against amyloid-β plaques.[33] A different approach is to use fMRI for assessment of brain function in transgenic mice overexpressing the amyloid precursor protein.[34] For example, fMRI was used in a study that recorded a significantly reduced cerebral haemodynamic response to administration of bicuculline (GABA$_A$ antagonist), compared to age-matched controls.[34]

Moreover, MRI is now a routine technique for assessment of cardiovascular function, including measurement of ventricular function and myocardial viability, and detection of ischemia.[5] The high temporal resolution of dynamic MRI cardiac imaging enables data collection of functional parameters that would otherwise be difficult to measure.[5] The heart, however, provides specific challenges for imaging, including cardiac and respiratory motion, and magnetohydrodynamic effects caused by flow. These problems are more pronounced in rodents, which have higher cardiac and respiratory rates.[24] However, they can be overcome by application of different acquisition techniques or adjustment of MRI sequences for cardiac MRI imaging.[35] For example, heart contraction causes considerable motion artefacts which greatly reduce the readability of images produced by MRI. This problem is resolved by synchronisation of MRI acquisition to electrical activity of the heart, for the purpose of 'freezing' cardiac motion.[35]

MRI is also the preferred technique for analysing animal models of joint diseases, including rheumatoid arthritis and osteoporosis.[27] This is because MRI is able to generate reproducible, 3-dimensional images with high spatial resolution; differentiating bone, cartilage, tendons, synovium, muscle and adipose tissue.[27] MRI permits longitudinal studies to be conducted that monitor changes in the intricate joint architecture.[27] MRI is also commonly used for assessment of both musculoskeletal and neoplastic disease.[5]

8.3.5.4   Comparison of MRI Technology with Conventional Analytical Techniques

Non-invasive investigations such as MRI provide advantages over conventional techniques of disease analysis, such as tissue sampling and excision. Firstly, MRI studies are effective where it is not possible to perform invasive traditional techniques on the target organ of study. For example, it is not possible to perform biopsies on the living brain, and therefore traditional techniques have relied on secondary read-outs such as behavioural abnormalities, to diagnose and characterise cerebral disorders.[5] MRI provides structural, functional and molecular data that can be used in the phenotyping of cerebral disorders.[5] For example, the main goal of stroke therapy is to improve functional brain recovery.[27] This can be assessed by behavioural studies, or by functional imaging techniques such as fMRI or positron emission tomography (PET).[27] fMRI is particularly a useful tool for studying brain function in rodents due to its high spatial and temporal resolution.[24]

Secondly, MRI enables longitudinal studies of disease progression to be performed, in contrast to traditional techniques, which are known to provide less relevant 'snapshots' of disease processes. Moreover, MRI is better able to detect more subtle changes in disease compared to traditional invasive techniques. For example, airway diseases are characterised by inflammation in the lung which are generally measured in animal models using brochoalveolar lavage (BAL) fluid analysis or histology. These are invasive and less sensitive to subtle disease changes,[36] whereas MRI can be used to detect allergen-induced oedematous signals that correlate with inflammatory factors in BAL fluid.[37] Moreover pretreatment with the glucosteroid budesonide caused a rapid resolution of the MRI signal, and this is not reflected in analysis of BAL fluid.[38] Compared to other established imaging techniques, MRI is better able to identify constituents of high risk plaques (lipid, fibrous, calcium) in manifestations of atherosclerosis.[39] Such measures would facilitate selection of high-risk people to take part in clinical trials. Therefore MRI can provide more information on disease mechanisms than traditional invasive techniques and other imaging modalities.[27]

Advances in MRI technologies have also meant that earlier pathophysiological changes can be identified *in vivo*, increasing our understanding of disease mechanisms. For example, MRI was able to detect early changes associated with new vessel growth, such as imaging of blood supply to myocardium, and slight changes to regional myocardial contractile ability.[40]

MRI has therefore become an indispensible tool for the qualitative and quantification analysis of disease processes. The improvements to disease analysis permitted by MRI, will lead to a better understanding of disease mechanisms in animals and man, and will allow more subtle models to be developed that better reflect human disease. This will reduce the severity level of disease state required within the animal models for disease studies, and benefit the translation of scientific knowledge across species because observations in animal models are more relevant to human studies. Enhanced subtlety and earlier detection of disease will also enable greater efficacy based differentiation of NCEs and accelerated decision making potential for the drug discovery process.

8.3.5.5 Target Validation and Candidate Drug Evaluation

Once underlying disease mechanisms are identified, key targets in disease pathways are examined for possible therapeutic modulation. Large compound libraries are screened for the identification of chemical lead structures, which are then validated in secondary assays and optimised.[5] Evaluation of lead compounds involves assessment of drug efficacy and safety.[5]

Firstly, novel targets of disease can be assessed by precise *in vivo* manipulation of targets using compounds of known effect, such as target antagonists. For example, if *in vivo* administration of target antagonists significantly alleviates disease symptoms in an animal model, then this would confirm the up regulation of the target in the disease setting. This would provide essential target validation and indicate its potential as a target for therapeutic modulation.

Like other imaging technologies MRI is not a suitable tool for general screening, as it is not a high throughput approach (~ 50 animals scanned per day). In screening of compound libraries, compounds are screened in numbers to the order of 10^5 or 10^6.[8] Limitations on MRI throughput are imposed by a necessity for animal handling, image collecting times and image processing; making imaging technologies such as MRI an ineffective general screening technology.[8]

Despite having insufficient throughput for general screening, MRI is still able to play an important role in the drug discovery stages by helping to validate potential targets and to profile drug candidates. MRI measures physiological mechanisms in an intact biological system over extended time periods, enabling comprehensive investigations of the target, its modulation and effect on disease.[8] During target validation, compounds are selected based on criteria which include their ability to manipulate the disease state and their safety profile in animal subjects.[9]

Non-invasive and longitudinal studies enable the real-time evaluation of labelled drugs (pharmacokinetic) and pharmacodynamic drug properties, which includes assessment of drug absorption, distribution, therapeutic action, metabolism and elimination aspects.[9] For example, it is possible to evaluate pharmacodynamics and therapeutic responses through changes in physiology and metabolism using dynamic contrast enhanced MRI.[4] MRI can also be used to measure morphological alterations induced by candidate drugs. For example, the efficacy of anti-tumour drugs is now routinely assessed using morphological MRI to quantify tumour incidence and growth in animal models.[3]

MRI is limited by its poor sensitivity due to the low quantum energy involved and the spectral overlap of metabolites.[9] MRI is therefore unable to detect micromolar tissue concentrations of metabolites required for pharmacokinetic studies which is possible with other nuclear techniques such as positron emission tomography.[9,12] Despite its low sensitivity, MRI has much higher spatial resolution (in the order of 100 μm for rodent studies) than other imaging approaches (PET resolution ≥ 1 mm), such that detailed anatomical visualisation can be obtained in small animals in any orientation.[24] Indeed, both the high spatial contrast and high temporal resolution provided by MRI allows detection of precisely where and when drug candidates have an effect on, and are influenced by, normal and pathological biological processes.[2] For example, the temporal resolution provided by MRI allows the behaviour of compounds over time to be assessed in rat models of lung inflammation MRI. This can be used to complement histological techniques, as it enables a choice of time points, at which histological analysis will be conducted, in order for more detailed data on drug mechanism to be collected.[11]

The principal role of MRI in drug discovery and development is to evaluate the effect of drug candidates on tissue morphology, physiology and biochemistry. Earlier use of imaging the pharmaceutical process will enable earlier assessment of drug efficacy and safety, and therefore earlier GO/NO GO decisions to be made, which will reduce attrition and unnecessary costs.

8.3.5.6 Imaging Endpoints

There is considerable focus on the development and validation of image-based biomarkers, which indicate disease presence or severity at both preclinical and clinical stages.[3] Biomarkers are important for providing preclinical and clinical readouts for evaluation of target effects of drug candidates. This enables target validity to be assessed in normal and diseased animal models, in different species and if it is fully translatable, in early clinical trials.[8]

Identification of reliable biomarkers is especially important in late phases of drug discovery for studies of chronic disease with late clinical end points, where biomarkers are required to assess drug efficacy.[5] For example in animal models of neurodegenerative disease such as stroke, MRI studies provide imaging end points such as degree of haemorrhage, for the evaluation of neuroprotective therapeutic approaches.[3]

One of the greatest contributions MRI is believed to have on the drug discovery and development process is in the biomarker field. The flexibility of MRI enables investigation of different types of biomarkers, from anatomical (cardiac ventricular wall thickness, pulmonary oedema volume, fat: muscle ratio) or biochemical (quantification of phosphorous in specific muscle types) or functional changes (brain activity). Further examples of potential biomarkers that may be assessed by MRI include bone erosion to establish presence of rheumatoid arthritis, or measurement of proteoglycan depletion in cartilage for assessment of osteoarthritis.[27]

In addition it is possible to measure the concentration of endogenous metabolite biomarkers using MRS. MRS is used for example, to measure up regulated choline levels in several cancers including breast, prostate, colon and brain.[28] Choline levels are now used as a biomarker for response to anticancer drug candidates.[28]

Potential biomarkers for assessment of drug efficacy and safety include measurements of gene transcription and translation levels, which frequently require sampling of tissue and body fluid.[41] Whilst these biomarkers have high sensitivity, they require complicated biostatistical methods to interpret, are prone to sampling errors in heterogenous tissue, and provide data from one time point only.[41] Contrast imaging technology such as *in vivo* MRI enables both spatial and temporal determination of biomarkers. An illustrative example is that of rheumatoid arthritis (RA). Disease severity in RA patients is determined using assessments such as clinical status and markers of inflammation in the serum. These methods are limited due to their subjectivity, the indirect nature of their measurements of synovial inflammation, and their inability for standardisation.[27] Synovitis is an early pathophysiological feature of RA that can also be measured by synovial biopsy, but this technique is too invasive for application to longitudinal studies.[27] Noninvasive assessment of synovitis is possible using dynamic gadolinium-enhanced MRI, which uses vessel permeability to detect synovitis in clinical trials.[42] Other MRI biomarkers include bone marrow oedema and synovial membrane volumes.[43]

Functional imaging biomarkers therefore offer considerable advantages over conventional biomarkers due to their non-invasive, quantitative, simple and direct nature. There is a gradual shift underway from the use of imaging biomarkers in the clinic, to their additional application in preclinical studies where they can make early contributions to the GO/NO GO decision making process, and consequently help make cost and time savings.[3]

8.3.5.7 Analysis of Drug-Release Mechanisms

MRI provides 2 and 3 dimensional images taken from any organ, in any orientation and on multiple occasions. These traits make MRI an effective tool for studying mechanisms of drug-release in drug development. For example in rats, controlled release of the antibiotic gentamicin from a hydrogel was monitored using 3-dimensional MRI assessment of hydrogel volume.[44] Over 7–10 days MRI derived images demonstrated a strong relationship between decreasing volume of hydrogel and antibiotic release. A further MRI study on subcutaneous implants in rats, demonstrated a difference in polymer degradation and release processes between poly(hydroxyl esters) and poly(anhydride).[45,46] Moreover, when MRI results were compared to *in vitro* data, it was shown that whilst a poly(hydroxyl ester) swells *in vitro*, it only deforms *in vivo*.[45,46] In addition to the capacity of MRI to identify differences between *in vitro* and *in vivo* studies, an important advantage of the technique is its ability to simultaneously detect the disappearance of subcutaneous implants and the biological response, such as oedema.[14]

An important site in which monitoring candidate compounds and dosage forms using MRI would be useful, is in the gastro-intestinal (GI) tract. Knowledge of the effect of food on drug pharmacokinetic and pharmacodynamic properties would be greatly increased by MRI-mediated description of drug distribution and GI removal processes. Monitoring these factors in the GI tract is challenging however due to motion artefacts induced by breathing and peristaltic movements, and the inherent poor signal intensities emitted by these organs.[14] These difficulties have been overcome by the development of fast imaging sequences that provide real-time images of human antrum and peristaltic movements.[47]

MRI provides spatiotemporally resolved images of the mechanisms of drug release and delivery systems, in both animal and man. It enables simultaneous visualisation of drug release mechanisms and biological responses, and provides novel information to that provided by *in vitro* analysis of drug release. MRI will therefore prove an effective tool for studying drug release mechanisms.

8.3.5.8 Toxicology

Unpredicted toxicity is a major cause of late-stage candidate failure, and frequently leads to the expensive termination of drugs in late stage clinical development. Therefore, there remains a significant unmet need for novel techniques to be introduced that confirm drug safety and improve the reliability of drug toxicity predictions.[3]

Where longitudinal assessment of a candidate compound is required, MRI allows repeat measurements from the same test subject. Due to each subject acting as its own control greater statistical power of the data can be achieved through reduced variability. This improves our confidence in toxicity measures and produces an early idea of dose values for more expensive clinical trials.[5] MRI application to safety testing will facilitate faster production of drug toxicity profiles, informing decision making on candidate drug potential at an earlier stage, and subsequently helping to reduce attrition rates. However the acceptance and implementation of imaging techniques for the purposes of safety testing has been slow, as application of new technologies to evaluate drug safety is deemed risky. Academia, the FDA, and industry have recently begun collaborations in order to identify the roles that imaging technology could play in safety testing, and for the development of image-based safety biomarkers.[3] MRI technology will increase the ability of industry to identify and manage safety risks associated with liver, kidney, brain and heart in drug development.[5]

8.3.5.8.1 Liver Toxicity

The most common reason for withdrawal of drug candidates is because of liver toxicity.[5] The effect of drug candidates on liver anatomy and function is therefore a major consideration for safety testing. The liver has the critical role of filtering and detoxifying blood that leaves the stomach and intestine for the portal vein.[5] Moreover, the liver is vital for drug metabolism, clearance or the potential production of toxic drug derivatives that cause secondary safety issues.[5] Toxicity in the liver most often manifests as fatty liver, glycogen deposition and hepatocyte necrosis.[5] Advances in knowledge of liver processing of NCEs will increase the accuracy of identifying and controlling liver safety risks.

MRI provides a valuable method for identifying liver toxicology issues in safety studies. MRI is particularly useful where usual serum chemistry measurements such as enzyme levels are undetectable, safety biomarkers are unavailable and histological techniques lack sufficient sensitivity.[5] For example, a common drug effect in preclinical studies is the induction of hepatic steatosis, a form of fatty liver disease.[5] Presence of this condition is often uncorrelated to increases in hepatic serum enzymes.[48] Moreover in preclinical studies, whilst elevated serum transaminase levels show strong correlation to histological changes in the liver, they are often not correlated at the drug dose levels administered. MRS is also a useful tool for probing liver metabolism and connected toxicity.[5]

Imaging approaches are limited by the confounding effects of respiratory activity and organ movements, but sampling errors caused by these factors are minimal relative to those linked to conventional biopsy techniques.[5]

8.3.5.8.2   Kidney Toxicity and Nephrotoxicity

Kidney toxicity is a rare cause of attrition, and is mainly observed in chronic diseases at late stages of development.[5] The kidney filters blood components, including drugs and its metabolites, in order to control blood homeostasis. Unlike the liver, a kidney is not a solid organ but consists of ~million individual nephrons packed together.[5] Compared to traditional methods, such as insulin agents which measure global level renal perfusion, MRI enables production of region-specific images, facilitating comparison with histological data.[5] Contrast enhanced MRI can be used to identify nephrotoxicity following therapy. For example, nephrotoxicity was identified in a study on a mouse model treated with the drug Cisplatin and a gadolinium dendrimer MRI contrast agent.[49] Nephrotoxicity can also be assessed using high resolution MRS complemented with pattern recognition techniques for spectral analysis of body fluids, such as urine.[50,51]

8.3.5.8.3   Cardiotoxicity

MRI has emerged over the past two decades as the preferred imaging modality for studies of cardiac function.[35] No other non-invasive techniques exist for assessment of heart anatomy and function, making MRI an ideal tool for cardiac studies.[27] MRI properties of high spatial contrast and temporal resolution, combined with anatomical imaging of the heart during the cardiac cycle, permits quantification of cardiac volume and regional specific functional analysis.[35] MR myocardial tagging methods can also be used to evaluate myocardial wall deformation and myocardial contraction processes, whilst velocity-encoded MRI provides dynamic measurements of flow through cardiac valves and assessments of myocardial motion.[35]

Conventional invasive techniques, such as structural histopathological techniques measure toxicity indicators that are released from heart cells, too late for effective evaluation of candidate drugs. Clinical follow up studies of patients receiving highly cytotoxic anthracycline chemotherapy is normally performed using repeated echocardiographic studies of left ventricle function. However this approach may not detect early changes that often have considerable clinical importance.[35] Other approaches including biopsy are also limited, for example in their ability to predict clinical outcome.[52] In contrast, MRI is able to precisely quantify and assess ventricular function whilst also imaging myocardial damage in a non-invasive manner.[35] In addition to providing an accurate indication of cardiac function, MRI has the potential to image early tissue changes that often occur before function changes.[52] For example, MRI was used to detect early changes in myocardial contrast and slight decreases in cardiac function in patients administered with cytotoxic anthracyclines.[52] MRI can therefore function as a straightforward, non invasive and sensitive tool for the follow-up of patients receiving treatment in clinical trials and of commercially available products.

8.3.5.8.4 Brain Toxicity

Many diseases of the central nervous system, including cancer and neurodegeneration are chronic conditions requiring longitudinal studies. The non-invasive property of MRI therefore makes it a useful tool for studying the safety profile of CNS therapies. Moreover, MRI sensitivity is sufficient to image white-matter related pathologies such as white matter lesions of the cerebellum.[5] For example, rats treated with the CNS drug vigabatrin, were shown by MRI to exhibit changes in the morphology and relaxation time constants of cerebellar white matter lesions, and this correlated with histopathological findings.[53] The non-invasive property of MRI is also advantageous for elucidating the reversibility of tissue pathologies identified by histological techniques.[5] For example, the reversibility of drug-induced microvacuolation (intramyelin oedema) was demonstrated in rats and dogs through comparison of changes in MRI T_2 signal intensities with histopathology.[54]

8.3.6 Translational Applicability

MRI has been developed for use in both animals and man, and is therefore positioned to provide translational information between the two.

Firstly, imaging techniques can be used to assess the degree of cross species similarity, and therefore to assess the value of animal models. For example, MRI and MRS technology was applied to mice to give quantitative *in vivo* readings of cardiac anatomy, function, infarction, perfusion, metabolism and calcium ion influx.[55] These studies illustrated that the normal mouse heart is a good model for human cardiac disease.[55] Other questions that may be addressed include the similarity of dose-response curves in man and preclinical model species, and whether side effects observed in preclinical models are likely to be seen in the clinic.[8]

Moreover the translatability of MRI technology enables development and refinement of biomarkers at preclinical stages rather than at clinical stages where it may be impossible due to strict regulation or expense. For example, biomarkers measure the same biological phenomenon in both preclinical and clinical settings. However clinical trials have very highly regulated study designs, in which it is not possible to develop biomarkers. In contrast, early application of imaging technologies in the preclinical stages allows the most appropriate imaging biomarkers to be identified and validated. For example at

GlaxoSmithKline, MRI mediated characterisation of mechanisms associated with both allergic and non allergic rhinitis led to defining the biomarkers of vasodilation and oedema, which are associated with rhinorrhoea and nasal mucosa swelling. The same biomarkers were then used to develop new targets and therapies in early stages of drug discovery, and also in the clinic to assess drug efficacy.

Moreover, translation of information is bidirectional, as biomarkers identified in human MRI studies of disease mechanisms can help to identify new biomarkers and develop more representative disease models. Preclinical imaging studies also enable the decision of where to apply biomarkers (in which model choice or disease area) and at what time point to measure them between treatment and disease progression. Such early decision making will result in enhanced efficiency and reduced costs for later stage clinical trials.[28]

In addition to disease biomarker equivalence, the techniques of acquiring data and undertaking image analysis can also be aligned across the species, building confidence in interpretation of the data. In preclinical studies it is possible to select the most appropriate MRI parameters, and to optimise other physical experimental protocols, such as the introduction of exogenous contrast agents. Moreover, identification of the optimal stage of disease to administer therapy, the decision to use active *versus* inactive dosing concentrations and imaging time points for a specific candidate drug, can be well established during preclinical phases and subsequently translated to clinical protocols.[3,8] The heterogeneity of the disease population can also be assessed during preclinical stages, enabling appropriate stratification of the clinical patient cohort.[8] For example, the location and severity of cerebral infarction varies considerably in stroke patients, and therefore MRI data could guide the sorting of such patients for clinical trials.[27]

It is therefore possible with an effectively developed preclinical imaging protocol to establish solutions to key questions concerning design of experimental protocols and biomarkers. The translatability of MRI permits bidirectional translation of information between preclinical and clinical stages, leading to efficiency and cost savings in the drug development process.

8.3.7   Limitations of MRI Technology

Whilst MRI technology has already greatly contributed to the drug discovery and development process, various improvements are required in order to expand its applicability.

A key disadvantage of MRI technology is the expense of establishing and operating MRI facilities. Firstly there is a requirement for specialised equipment and purpose-built facilities to accommodate the MRI scanner. MRI is now routinely used in hospitals and the demand for MRI technology in pharmaceutical research is growing. The costs involved are rapidly decreasing, for example allowing for Health Service inflation the total real costs of running MRI at a Coventry hospital between 1989 and 1996–1997 fell by 30%.[56] MRI also requires trained staff to develop and optimise MRI protocols, including physicists, scan programmer support staff and trained biologists. The overall staffing costs of clinical MRI in Coventry increased by 150% between 1989 and 1996–1997 due to the employment of more experienced and higher salaried staff, extension of working hours and higher numbers of clerical staff.[56] However this considerable rise in staffing costs is offset by the large reduction in the capital costs of MRI.[56]

Another key limitation of MRI is the relatively low throughput that it provides. There is always a trade-off between high throughput demands and image quality, as reduced acquisition time will invariably result in reduced image resolution. The acquisition time is determined however, by the image resolution that needs to be achieved, which need only be sufficient for the particular type of analysis that needs to be conducted. The degree of resolution required will depend on the degree of biological change that is expected to occur in the disease model. For example, in studies where large biological differences are expected, it may be possible to afford lower resolution scans with shorter acquisition times. In contrast, in studies with small expected biological differences, higher resolution images are required to reduce the margin of error associated with each measurement. It is therefore important at a very early stage for image analysts to be involved in critically assessing the adequacy of data for the endpoint required. Additionally, throughput has been improved by the advent of fast imaging protocols that has reduced total scanning time, whilst maintaining sufficient image contrast for accurate analysis. It is also now possible to parallelize data collection, which reduces acquisition time or permits data acquisition from multiple animals simultaneously.[57,58]

A key problem with imaging technologies is heterogeneity across distinct imaging centres. Different imaging facilities use distinct scanning protocols and a range of expertise, which limits inter-centre comparisons and makes validation of results difficult.[59] A future concern is therefore standardisation across imaging centres to promote widespread validation and adoption of imaging techniques.[5]

Finally, the MRI field is experiencing development of exciting new technologies that permits *in vivo* imaging of cellular and molecular phenomenon. However, MRI remains limited by its relatively poor sensitivity, which may be improved by developments in RF coil design with enhanced signal-to-noise properties.[12] Moreover, future challenges for the routine use of molecular MRI in pharmaceutical research include improvement of cell penetrating strategies by development of new generation small nanoparticles.[12] This will enable imaging of intracellular targets and permit imaging across the blood-brain barrier.[12]

8.3.8 Conclusion

It is becoming increasingly costly to bring a medicine to market, and the pharmaceutical industry is constantly looking for new ways to streamline drug development, build confidence in the data, and improve efficiencies around all processes. As a consequence, techniques such as MRI are being invested in to bring only the most promising molecules to the clinic and thus reduce attrition rates, particularly at later developmental stages. This is being achieved by characterising animal models of disease, early application of imaging biomarkers, and bidirectional translation of information between preclinical and clinical stages.

Imaging technologies have become valuable tools in the drug discovery and development process. They have wide applicability, from non-invasive measurements of tissue morphology and function, to detailed analysis of physiological and molecular processes. Despite the significant costs of establishing and operating imaging centres, imaging applications offer considerable advantages over existing highly developed technologies that will make the drug discovery and development process both more efficient and economic.

Imaging techniques, such as MRI share the common advantages of longitudinal study design; internal control; molecular information and quantitative data.[2] They allow non-invasive, longitudinal study designs to be performed that allow long term monitoring of disease progression and response to therapy in a single animal. This reduces animal study numbers and increases statistical power, which is particularly important for studies on higher species, such as non-human primates.[24] Imaging permits earlier and more subtle detection of disease states than traditional techniques, generates clinically relevant disease endpoints, and facilitates the sorting of patient cohorts for clinical trials. Fast, computerised collection of *in vivo* data combined with semi automatic methods of analysis, quickens the assessment of efficacy and safety of drug candidates. The translational aspect of imaging methodology allows the use of the same hardware, analysis and most importantly, biomarkers of disease in early and late stages of drug development. Careful consideration and complementation of the distinct strengths and weaknesses of different imaging modalities is essential for their successful application to areas of pharmaceutical research.

The principal assets of MRI are the exceptional soft-tissue contrast abilities, non-ionising nature, high temporal resolution, multi-dimensional imaging capacity and flexibility of the approach. The versatility of MRI enables comprehensive evaluation of the morphological, functional, physiological, cellular and molecular nature of normal and diseased tissue, with high resolution in the same animal.[24] One of the greatest benefits of MRI technology is the ability to directly connect preclinical and clinical data through imaging biomarkers.[24] MRI has already significantly contributed to many areas of the drug discovery process; the characterisation of disease mechanisms in animal models, evaluation drug candidates in both preclinical and clinical stages, production of biomarkers, and provision of directly transferable data and protocols. Moreover, advances in the throughput capabilities and in the ability of MRI to generate analytically robust molecular data will enable MRI to have a greater impact on more areas of the pharmaceutical process.

References

1. M. D Silva and S. Chandra, *Methods Mol. Med.*, 2006, **24**, 299.
2. P. R. Contag, *Drug Discovery Today*, 2002, **7**, 555.
3. http://www.contractpharma.com/articles/2010/05/imaging-in-drug-research (last accessed June 2010)

4. Z. Lu, *Pharm. Res.*, 2007, **24**(6), 1170.
5. M. Rudin, N. Beckmann and M. Rausch, *Imaging in Drug Discovery and Early Clinical Trials (Progress in Drug Research)* Foreword Birkhauser, Switzerland, 2005, **62**, 135.
6. J. Eckstein, *ISOA/ARF drug development tutorial*, http://www.alzforum.org/drg/tut/ISOATutorial.pdf (last accessed Aug 2011).
7. H. Kubinyi, *Drug Discovery Today*, 2002, **7**, 707.
8. http://www.ngpharma.com/article/Noninvasive-Imaging-in-Drug-Discovery-and-Development-Where-to-Begin/ (last accessed Aug 2011).
9. M. Rudin, N. Beckmann, R. Porszasz, T. Reese, D. Bochelen and A. Sauter, *NMR Biomed.*, 1999, **12**, 69.
10. M. Rudin and R. Weissleder, *Nat. Rev. Drug Discovery*, 2003, **2**, 123.
11. N. Beckmann, C. Cannet, H. Karmouty-Quintana, B. Tigani, S. Zurbruegg and F.X. Blé, *Eur. J. Radiol.*, 2007, **64**, 381.
12. P. G. Kluetz, C. C. Meltzer, V. L. Villemagne, P. E. Kinahan, S. Chander, M. A. Martinelli and D. W. Townsend, *Clin Positron Imaging*, 2000, **3**, 223.
13. S. R. Meikle, F. J. Beekman and S. E. Rose, *Drug Discovery Today: Technol.*, 2006, **3**(2), 187.
14. J.C. Richardson, R.W. Bowtell, K. Mader and C.D. Melia, *Adv. Drug Delivery Rev.*, 2005, **57**(8), 1191.
15. G. Lanza, C. Lorenz, S. Fischer, M. J. Scott, W. P. Cacheris, R. J. Kaufmann, P. J. Gaffney and S. A. Wickline, *Acad. Radiol.*, 1998, **5**(1), 173.
16. R. M. Botnar, A. S. Perez, S. Witte, A. J. Wiethoff, J. Laredo, J. Hamilton, W. Quist, E. C. Parsons Jr., A. Vaidya, A. Kolodziej, J. A. Barrett, P. B. Graham, R. M. Weisskoff, W. J. Manning and M. T. Johnstone, *Circulation*, 2004, **109**, 2023.
17. W. J. Rogers, C.H. Meyer and C.M. Kramer, *Nat. Clin. Pract. Cardiovasc. Med.*, 2006, **3**, 554.
18. R. Weissleder, H. Cheng, A. Bogdanova and A. J. Bogdanov, *J. Magn. Reson. Imaging.*, 1997, **7**, 258.
19. J. M. Hill, A. J. Dick, V. K. Raman, R. B. Thompson, Z. X. Yu, K. A. Hinds, B. S. Pessanha, M. A. Guttman, T. R. Varney, B. J. Martin, C. E. Dunbar, E. R. McVeigh and R. J. Lederman, *Circulation*, 2003, **108**, 1009.
20. P. M. Winter, A. M. Morawski, S. D. Caruthers, R. W. Fuhrhop, H. Zhang, T. A. Williams, J. S. Allen, E. K. Lacy, J. D Robertson, G. M. Lanza and S. A. Wickline, *Circulation*, 2003, **108**, 2270.
21. D. A. Sipkins, D. A. Cheresh, M. R. Kazemi, L. M. Nevin, M. D. Bednarski and K. C. Li, *Nat. Med.*, 1998, **4**, 623.
22. S. A. Anderson, S. A. Wickline and J. J. Kotyk, *Magn. Reson. Med.*, 2000, **44**, 433.
23. N. Beckmann, T. Mueggler, P. R. Allegrini, D. Laurent and M. Rudin, *Anat. Rec.*, 2001, **265**, 85.
24. S. Neubauer, M. Beer, W. Landschutz, J. Sandstede, T. Seyfarth, C. Lipke, H. Kostler, T. W. Pabst, M. Meininger, M. von Kienlin, M. Horn, K. Harre and D. Hahn, *MAGMA*, 2000, **11**, 73.
25. G. Honey and E. Bullmore, *Trends Pharmacol. Sci.*, 2004, **25**, 366.

26. J. C. Waterton, V. Rajanayagam, B. D. Ross, A. Whittemore and D. Johnstone, *Magn. Reson. Imaging.*, 1993, **11**, 1033.
27. N. Beckmann, D. Laurent, B. Tigani, R Panizzutti and M Rudin, *Drug Discovery Today.*, 2004, **9**(1), 35.
28. www.criver.com/SiteCollectionDocuments/DIS_I_Imaging.pdf (last accessed June 2010)
29. T. Reese, D. Bochelen, A. Sauter, N. Beckmann and M. Rudin, *NMR Biomed.*, 1999, **12**, 189.
30. J. Mintorovitch, G. Y. Yang, H. Shimizu, J. Kucharczyk, P. H. Chan and P. R. Weinstein, *J. Cereb. Blood Flow Metab.*, 1994, **14**, 332.
31. J. C. De la Torre and T. Mussivand, *Neurol. Res.*, 1993, **15**, 146.
32. H. Benveniste, G. Einstein, K. R. Kim, C. Hulette and G. A. Johnson, *Proc. Natl. Acad. Sci. U.S.A.*, 1999, **96**, 14079.
33. J. F. Poduslo, T. M. Wengenack, G. L. Curran, T. Wisniewski, E. M. Sigurdsson, S. I. Macura, B. J. Borowski and C. R. Jack Jr, *Neurobiol. Dis.*, 2002, **11**, 315.
34. C. Sturchler-Pierrat, D. Abramowski, M. Duke, K. Wiederhold, C. Mistl, S. Rothacher, B. Ledermann, K. Bürki, P. Frey, P. A. Paganetti, C. Waridel, M. E. Calhoun, M. Jucker, A. Probst, M. Staufenbiel and B. Sommer, *Proc. Natl. Acad. Sci. U S.A.*, 1997, **94**, 13287.
35. A. L. Baert and K. Sartor, in *Clinical Cardiac MRI Series: Medical Radiology Subseries: Diagnostic Imaging*, ed. J. Bogaert, S. Dymarkowski and A. M. Taylor, Springer, Germany, 2005, XII, 556 p.
36. Beckmann, N. B. Tigani, L. Mazzoni and J. R. Fozard, *Trends Pharmacol. Sci.*, 2003, **24**, 550.
37. N. Beckmann, B. Tigani, D. Ekatodramis, R. Borer, L. Mazzoni and J. R. Fozard, *Magn. Reson. Med.*, 2001, **45**, 88.
38. B. Tigani, E. Schaeublin, R. Sugar, A. D. Jackson, J. R. Fozard and N. Beckmann, *Biochem. Biophys. Res. Commun.*, 2002, **292**, 216.
39. S. B. Feinstein, P. Voci and F. Pizzuto, *Am. J. Cardiol.*, 2002, **89**, 31C.
40. J. D Pearlman, R. J. Laham, M. Post, T. Leiner and M. Simons, *Curr. Pharm. Des.*, 2002, **8**, 1467.
41. M. Rudin, M. Rausch and M. Stoeckli, *Mol. Imaging Biol.*, 2005, **7**(1), 5.
42. R. J Reece, M. C. Kraan, A. Radjenovic, D. J. Veale, P. J. O'Connor, J. P. Ridgway, W. W. Gibbon, F. C. Breedveld, P. P. Tak and P. Emery, *Arthritis Rheum.*, 2002, **46**, 366.
43. M. Østergaard, M. Hansen, M. Stoltenberg, P. Gideon, M. Klarlund, K. E. Jensen and I. Lorenzen, *Arthritis Rheum.*, 1999, **42**, 918.
44. R. Weissleder, K. Poss, R. Wilkinson, C. Zhou and A. Bogdanov, *Antimicrob. Agents Chemother.*, 1995, **39**, 839.
45. K. Mäder, G. Bacic, A. Domb, O. Elmalak, R. Langer and H. M. Swartz, *J. Pharm. Sci.*, 1997, **86**, 126.
46. K. Mäder, Y. Crémmilleux, A. J. Domb, J. F. Dunn and H. M. Swartz, *Pharm. Res.*, 1997, **14**, 820.
47. M. Shapiro, M. A. Jarema and S. Gravina, *J. Control. Release.*, 1996, **38**, 123.
48. K. J. Mortele and P. R. Ros, *Semin Liver Dis.*, 2001, **21**, 195.

49. H. Kobayashi, S. Kawamoto, S. K. Jo, N. Sato, T. Saga, A. Hiraga, J. Konishi, S. Hu, K. M. Togashi, M. Brechbiel and R. Star, *Kidney Int.*, 2002, **61**, 1980.

50. M. L. Anthony, B. C. Sweatman, C. R Beddell, J. C. Lindon and J. K Nicholson, *Mol. Pharmacol.*, 1994, **46**, 199.

51. E. Holmes, B. C. Sweatman, M. E. Bollard, C. A. Blackledge, C. R. Beddell, I. D. Wilson, J. C. Lindon and J. K. Nicholson, *Xenobiotica.*, 1995, **25**, 1269.

52. R. Wassmuth, S. Lentzsch, U. Erdbruegger, J. Schulz-Menger, B. Doerken, R. Dietz and M. G Friedrich, *Am. Heart J.*, 2001, **141**, 1007.

53. G. D. Jackson, S. R. Williams, R. O. Weller, N. van Bruggen, N. E. Preece, S. C. Williams, W. H. Butler and J. S. Duncan, *Epilepsy Res.*, 1994, **18**, 57.

54. R. G. Peyster, N. M. Sussman, B. L. Hershey, W. E. Heydorn, L. R. Meyerson, J. T. Yarrington and J. P. Gibson, *Epilepsia.*, 1995, **36**, 93.

55. F. H. Epstein, *NMR Biomed.*, 2007, **20**, 238.

56. J. Fletcher, M. D. Clark, F. A. Sutton, R. Wellings and K. Garas, *Br. J. Radiol.*, 1999, **72**(857), 432.

57. K. P. Pruessmann, *Top. Magn. Reson. Imaging*, 2004, **15**(4), 237.

58. N. A. Bock, N. B. Konyer and R. M. Henkelman, *Magn. Reson. Med.*, 2003, **49**, 158.

59. D. S. Tan, G. V. Thomas, M. D. Garrett, U. Banerji, J. S. de Bono, S. B Kaye and P. Workman, *Cancer J.*, 2009, **15**, 406.

CHAPTER 8.4

Peering Into the Future of MRI Contrast Agents

DARREN K. MACFARLAND

Luna nanoWorks, a Division of Luna Innovations, Incorporated 521 Bridge St., Danville, VA 24541, United States

8.4.1 Introduction

Medical imaging is a critical part of any medical practitioner's toolbox, providing a non-invasive look at a patient's internal workings. Dating back to the first medical use of X-ray photography, the ability to see inside a patient has had a widely understood value. Imaging has gone through many stages and taken many forms, but the basic goal, to penetrate through the body to an image capturing device, has remained essentially unchanged. The method by which the body is penetrated has varied remarkably from the use of radiation (X-rays, radioisotopes) to radiowaves.

Early X-ray photography represented a huge advance in medical technology, providing the first images of the inside of the body without surgery. However, this technology has limitations. First, it is most effective for hard masses such as bone; this is why anyone who has a suspected broken bone has likely had an X-ray (to decide if it is broken). Soft tissues are not well defined by X-ray photography. Second, X-rays themselves are high energy and are not biologically innocent. Berrington de Gonzalez reported that up to 8% of cancers were caused by exposure to X-rays.[1]

RSC Drug Discovery Series No. 15
Biomedical Imaging: The Chemistry of Labels, Probes and Contrast Agents
Edited by Martin Braddock
© Royal Society of Chemistry 2012
Published by the Royal Society of Chemistry, www.rsc.org

As a result, researchers developed new technologies that would give improved images of soft tissue and remove or reduce the risks inherent in high-energy imaging. The current choice of most hospitals is MRI, Magnetic Resonance Imaging. An estimated 60 million MRI examinations were given worldwide in 2007. MRI is a specific application of the laboratory technology known as NMR (Nuclear Magnetic Resonance), but the use of the word "nuclear" in the medical application was deemed too frightening for patients for whom the term brings to mind Chernobyl and Three Mile Island. MRI instruments are now widely available and provide soft tissue structure while exposing patients to only radio frequency (RF) energy. An estimated 60 million MRI procedures are performed every year.

Improvements in MRI have taken the form of improved hardware (*e.g.* better, stronger magnets) and computer software, but also with the addition of compounds administered to patients that enhance signal. These compounds as a class are termed *contrast agents* because they change the signal in areas in which they are concentrated, while unaffected areas remain at the original signal. This improvement in signal in one area but not another increases the observed contrast.

Why is improving signal contrast important? Perhaps the best answer is enhanced ability to observe small structures. Whether the small structure is the start of a ligament tear in a baseball pitcher's arm, or the early stages of a cancerous tumor, diagnoses are improved by increased detail in the image. Tumors are of particular interest, as detection of cancer in its early stages greatly increases the odds of survival, and are the subject of the most intense area of research. Other uses include visualizing damage in the blood-brain barrier,[2] assessing the success of a treatment by visualizing necrotic tumor tissue, and imaging the heart, both for accumulation of plaque and for damage after heart attacks.[3–7] Broadly, contrast agents allow doctors to see things they would not otherwise be able to see.

Exactly how MRI works, and the physics behind it, is outside the scope of this chapter, and numerous reviews may be consulted for details.[8–13] However, a few terms need to be defined, so a short discussion is warranted.

In essence, MRI (and NMR) works by inserting a patient into a strong magnetic field. The magnetic field causes some of the nuclear spins to align with the magnetic field. RF energy is applied, and some of the aligned nuclei are knocked out of alignment with the applied field. With time, the nuclei "relax" (return to alignment with the magnetic field), so termed because the nuclei are returning to a lower energy state. The exact frequency of energy required to knock a nucleus out of alignment is dependent on the isotope of the nucleus; as such, many different elements (those with odd numbers of protons or neutrons) may be selectively observed. In MRI applications, only ^1H is important, and specifically the ^1H nuclei of water. Conveniently water is quite abundant in the body and in different concentrations in different locations; for example, blood vessels have a higher concentration of water than say, the liver. Differences in water concentration lead to differences in brightness in the MRI image. Through these differences in brightness, different biological structures can be identified.

After RF excitation, nuclear spin relaxation takes place through two different pathways, termed R1 and R2. R1 and R2 are derived from the time required for ($x\%$) of ^{1}H nuclei to relax. Relaxivity is the primary metric by which contrast agents are judged. It may be reported on a per atom or per molecule basis, but in either case, the higher the relaxivity, the better the contrast agent. The metric that is measured in the laboratory, from which relaxivity is calculated, is the relaxation time. Depending on the pathway, these times are termed T1 and T2.

MRI contrast agents are unlike contrast agents used in any other technique because the agents themselves are not what are observed. Contrast agents work by changing the relaxation time (T1 or T2) of ^{1}H nuclei in water.[9,15–17] The physics of the interaction between the contrast agents and water is beyond the scope of this work, but suffice it to say that the magnetic field of the contrast agent creates a pathway for water molecules to relax. MRI instruments measure the water relaxation time, so changing T1 or T2 changes the observed image. Contrast agents typically dominate one pathway or the other and are thus divided into "T1 agents" or "T2 agents". T1 agents make an image brighter (signal enhancement or positive contrast) where they are concentrated (see Figure 8.4.1); T2 agents make the image darker (signal loss or negative contrast)(see Figure 8.4.2). Generally, T1 agents are preferred as the dark locales created by T2 agents are easily confused with void volume images.

The primary limitation of MRI is resolution, as the image is brightest where blood flow is highest; without contrast agents some structures are not visibly defined. Even with current commercially available contrast agents, small tumors are not visualized.

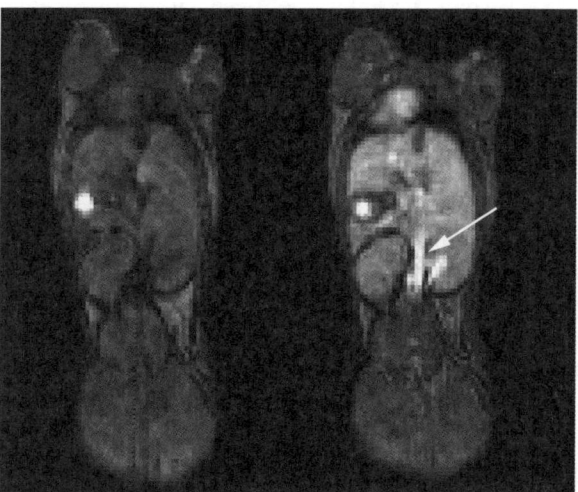

Figure 8.4.1   Mouse MRI image before (left) and after (right) injection with non-targeted T1 agent Hydrochalarone-6. Arrow notes the aorta. From ref. 14.

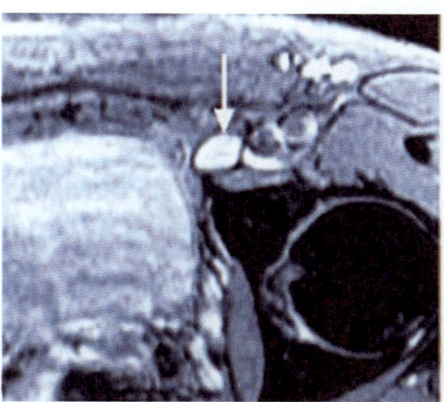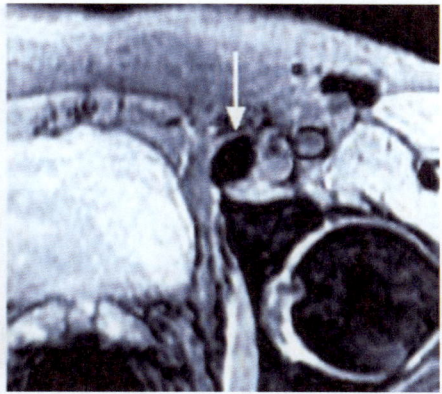

Figure 8.4.2   Image Metastatic Lymph Node Before SPIO Injection (left) and After (right). Arrow points to the lymph node; note the darkening in the right image, indicating accumulation of SPIO. From ref. 18.

8.4.2   Current Commercial Agents

Currently marketed agents fall into two main categories: Gadolinium^{III} (Gd^{3+}) chelates,[19,20] that are the most commonly used, and iron oxide clusters.[21] A much smaller third group is Mn chelates.[21] The market leader is Magnevist, with 2007 sales of over \$400 million.

Gd^{III} is used because it offers an unusual seven unpaired electrons in its valence shell (4f). These unpaired electrons create a large magnetic moment for the complex; this magnetism enhances relaxation of protons that interact with it. For chelate complexes, interaction between the paramagnetic Gd^{3+} and water can occur in two ways, termed inner-sphere and outer-sphere. Inner-sphere interactions involve brief, reversible bonding of water to the Gd^{3+} center. Outer-sphere interactions are through space, not direct bonding. Outer-sphere effects are generally much smaller and decrease rapidly with increasing distance from the Gd atom (effect $\sim 1/r^6$). This proposed mechanism of action is described mathematically by the Solomon-Bloembergen-Morgan (SBM) equation.[22,23]

An interesting prediction of the SBM analysis is that T1 relaxivity is enhanced by slowing the tumbling rate of the chelate complex. This prediction has been confirmed in numerous Gd^{3+}-chelate studies.

Gd^{3+} chelates can be separated into two general subcategories based on whether the chelating ligand is open-chain or cyclic (Figure 8.4.3). Linear chelate products include Magnevist (Gd(DTPA)), Omniscan (Gd(DTPA-BMA)(H$_2$O)), Optimark (Gd(DTPA-BMEA)(H$_2$O)), and Multihance (Gd(BOPTA)(H$_2$O)$^{2-}$) while macrocyclic products include Prohance (Gd(HP-DO3A)(H$_2$O)), Gadovist (Gd(DO3A-butrol)(H$_2$O)), and Dotarem (Gd(DOTA)(H$_2$O)$^-$). In all cases the Gd^{3+} ion is bound by at least 4 atoms and typically 6. This high degree of coordination was expected to prevent Gd^{3+} escape, but as will be discussed later, biologically relevant mechanisms for freeing the Gd^{3+} exist.

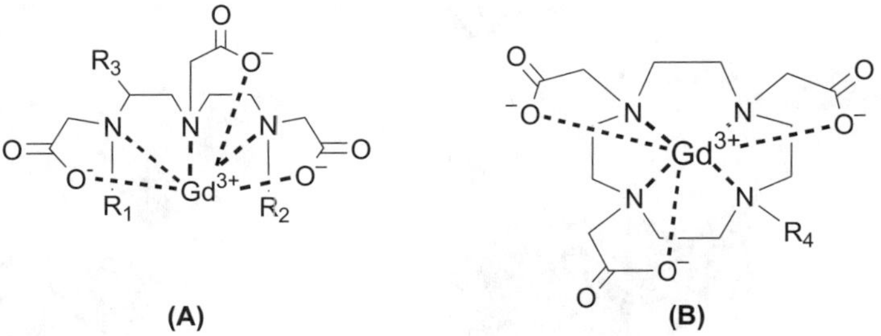

(A)   **(B)**

Figure 8.4.3 Generic Classes of Commercial Gd^{3+} Chelate MRI Contrast Agents. (A) Linear or open-chain chelates. (B) Cyclic chelates. From ref. 24.

Several commercial agents are currently marketed with relatively subtle structural differences within the open-chain or cyclic categories. Subtle chelate differences do influence important factors (such as water–Gd^{3+} exchange rates); however among commercial agents the most important metrics, relaxivity (R1) and clearance, are quite similar. Relaxivities are approximately 4 mM^{-1} s^{-1} for all commercial agents. All agents are cleared by healthy patients within 60 minutes by the renal system. It should be noted that the cyclic structures as a group are more kinetically stable than the open-chain structures.

An interesting observation is that relaxivity of Gd^{3+} chelates is enhanced by slowing the translational rotation (tumbling) rate of the chelate complex. Tumbling is related to size – larger molecules tumble more slowly. Much work has been reported in grafting Gd^{3+} chelates onto macromolecules in an effort to make the size of the molecule larger. Work in progress on this avenue will be discussed below.

Iron oxide nanoparticles are the other commercially important contrast agent group. These are represented by the formula Fe_3O_4, though this is somewhat misleading, as they exist as nanoparticles of many thousands of atoms and are not discreet molecules as with the Gd chelates. Other iron-containing nanoparticles have also been investigated, such as FeCo, FePt, and $MnFe_2O_4$, but the Fe_3O_4 particles are the most economically important. Commercially available agents include Cliavist or Resovist, Combidex or Sinerem, and Endorem or Feridex (trade names vary by country).

Iron oxide nanoparticles can be produced in different sizes depending on the method of production.[18,25–27] For the sake of categorization, they are generally divided into three groups based on their hydrodynamic radius:[28,29] super-paramagnetic particles of iron oxide (SPIO), > 50 nm; ultrasmall super-paramagnetic particles of iron oxide (USPIO),[29] 20–50 nm; monocrystalline iron oxide nanoparticles (MION), < 20 nm. It should be noted that the particles are virtually always coated with some biologically acceptable group that serves to increases circulation time of the particle and to increase hydro-philicity. As a result the iron particles are a relatively small portion of the total volume of the particle in water (4–7 nm diameter of a 30 nm particle, for example); also, the dry radius of the particles is often very different from the particle size in water, when the hydrophilic coatings absorb water and swell.

The size of the particles is critical for their biodispersity. Particles in the > 50 nm range are taken up by the reticuloendothelial system (RES). These can be used to image the liver (and this is their primary clinical usage) where RES cells are concentrated. However, RES absorbed particles tend to be cleared quickly, giving a very short window for observation. In addition, as the liver serves as a kind of dumping ground for the body, it is a particularly easy target. USPIO are small enough to avoid uptake in the RES, and serve as blood pool agents; that is, they remain in the circulatory system and allow imaging of areas in which blood is concentrated.

An interesting physical property of the Fe_3O_4 nanoparticles is super-paramagnetism. That is, under an external magnetic field (such as the MRI instrument) these particles become strongly magnetized through ordering of the magnetic fields of the thousands of iron atoms in the cluster. The magnetic field thus produced influences water molecules without requiring bonding, in contrast with the major pathway (inner-sphere) for Gd chelate-water interaction.

A SPIO is typically coated with a cover that inhibits biological recognition, and therefore clearance; slower clearance equates to longer circulation times that are desirable, up to a limit of a couple of days. The coating also serves to prevent aggregation, which would increase particle size in an uncontrolled manner. Various synthetic methods have been developed to coat iron nano-particles, and a number of different coatings have been tested. Most common is the use of poly-dextrans. Encasing the SPIO has great potential for targeting applications because it allows covalent surface attachment of functional groups that can be used to attach other molecules to the SPIO. One can imagine that this may include a wide variety of targeting agents (antibodies, proteins, receptor-active small molecules, etc), other imaging agents (for multimodal imaging), or even drugs to combine imaging and treatment. These possibilities will be discussed further later.

While clearance is an issue (too fast or too slow) for iron nanoparticles, the fact that the metal involved is a biologically common one and that degradation of the particles releases only small amounts of a relatively innocuous metal greatly decreases the potential for toxicity. In fact, typically particles that are not cleared from the body are metabolized into the body's iron pool. This is in stark contrast to the Gd^{3+} ions used in the most common commercial agents; Gd^{3+} is highly toxic if released into the biosystem. The case for a switch to iron-based systems based on safety concerns has been made.[30]

Problems with SPIO typically include 1) polydispersity of particle size within a sample, 2) uniformity of encapsulating layer, 3) clearance.

8.4.3 Design Criteria Moving Forward: What Will Make a Good Contrast Agent?

To consider where the contrast agent field is going, one must first establish the metrics by which any new agent must be judged. In other words, as one sifts through the sea of literature discussing new agents, how can one find the new design ideas or new agents that may ultimately be commercially important, and which ones will remain an academic curiosity? Given that a good number of compounds have already made their way into the market, where is the room for improvement?

The most important criterion for success is also one of the last to be tested: biocompatibility. This takes different forms. Is the compound toxic? Can it be biodegraded into products that are toxic? How long does it circulate; by another phrasing, how quickly is it eliminated? How is it eliminated (kidney (desirable), liver (acceptable), or RES (not usually acceptable, though used in liver studies)? Is it eliminated or does it accumulate? If it accumulates, where does it accumulate? Agents that accumulate indefinitely face a steep path to FDA approval. Elimination in the first pass through the renal system is also less than desirable, as the window for acquisition of the MRI is small; also, for targeted agents, quick elimination hinders chances for finding the binding target. The acid test for all these concerns is animal testing, and ultimately human testing, but not all compounds can be animal tested, so some screening must be done based on less important criteria. Included in the *in vitro* metrics are relaxivity, solubility, characterization, and particularly for nanoparticles, uniformity–polydispersity and particle size.

When considering the Gd^{3+} chelates, two issues arise that leave room for improvement. First, though a hexa- or octa-dentate ligand would intuitively seem to have an irreversible grip on the Gd^{3+} ion, in fact transmetallation takes place with Zn^{2+}, a bioavailable metal.[31,32] Gd^{3+} is quite toxic when freed of its chelate, so this represents an important problem, as recently recognized by the US FDA.[33] Gd^{3+}-chelate contrast agents have been linked to nephrogenic systemic fibrosis (NSF).[34–36] This risk is associated with patients' compromised renal systems that lead to longer than usual circulation times. This implies that any Gd^{3+} chelate agent must be cleared from the body quickly. Targeting agents that both take time to accumulate at a target and remain at the target for an extended period are thus handicapped because they must ultimately fail the quick-clearance test. Second, the relaxivity (R1) of commercial Gd^{3+} chelates is quite modest at around 4 $(mM^{-1} s^{-1})$. As such, large doses of the agent are required; market leader Magnevist, for example, is sold as a 500 mM solution. Improvement in R1 would allow lower Gd^{3+} dosing into the patient, as well as potentially greater signal enhancement.

When considering the SPIOs, nanoparticle issues are key. Is a new agent highly uniform with regard to both composition and size? Will the particle be able to escape the vasculature and localize at a target? Can the coating be modified to get desirable biological interactions? How well characterized is the compound (mixture)?

The clear direction of contrast agents is toward targeted species.[13] This is true for both Gd^{3+}T1 agents and iron oxide T2 agents. Targeting offers many potential benefits. First, if the agent is highly specific, it will accumulate at the targeted site and provide greatly enhanced imaging at that site and only that site. In cancer, for example, much smaller tumors than are currently visible would be detectable, allowing for earlier diagnosis and treatment. Second, much lower dosing would be needed, because the agent would not be diluted through the entire vasculature (or body) but concentrated only at the target; thus only enough agent to distinguish the target would be needed. Third, by proper choice of targeting group, the nature of a tumor could be determined. Is a tumor metastatic? Add a targeting group specific for metastases and one could find out. Thus, when evaluating new agents, the specificity of the targeting agent is an important criterion. The more localized the enhancement is to the target, the better the agent.

8.4.4 Targeting Groups

As the use of targeting groups is a primary theme of current research in contrast agents, a discussion of targeting groups is appropriate here. What is a targeting agent? What makes a good targeting agent? Of all the millions of known compounds, biological and otherwise, how is a contrast agent designer to know what to use to direct the contrast agent to a specific site? How should the contrast agent be attached to a targeting agent once one is chosen? How can success be determined?

A targeting agent is a moiety, chemical or, more commonly, biochemical, that delivers the contrast agent preferentially to a site of interest, such as a tumor. An ideal targeting agent will bind to the targeted site with 100% specificity, delivering the contrast exclusively to the site. It accumulates quickly, remains at the site for a period of hours to a few days, but then detaches to allow safe removal of the contrast agent from the body. It enhances the inherent relaxivity of the contrast agent, and is entirely unaffected by the binding of the contrast agent. It may act in either of two ways (if the contrast agent is good enough): cellular uptake (and therefore concentration) or cell membrane receptor binding.

Selection of targeting agents is more art than science. Chemists depend on biologists and research medicine to provide insight about what makes a particular target different from its surroundings. In some cases, it is the over-expression of certain receptors on malignant cells. For example, cancer cells have an overabundance of folic acid or folate receptors, so a common targeting agent for cancerous tumors is folic acid.[37–40] Folic acid is good because it is relatively inexpensive and stable, but suffers from non-specificity.[37] Both the kidney and liver have a large number of folate receptors naturally, and all cells have some receptors. More specific targeting agents are usually antibodies,[41–44] or peptides.[45–47] Both are much more expensive, and usually less shelf-stable; however, once a disease research group has found a particular antibody receptor or protein sequence that is unique to the disease, very high specificity can be achieved in contrast agent binding. Unfortunately, the contrast agent itself may interfere with binding if it is placed in a location critical to host-guest recognition. These are problems that can be overcome through iterative studies by chemists; the truly difficult part is finding a handle on the target that is unique to that target.

An additional challenge for targeting groups is the relatively low number of receptors per cell (nM-pM g^{-1} tissue).[48] Even with precise target saturation, the scarcity of receptors and the low relaxivities of current commercial agents combine to prevent imaging through this pathway. Receptors with higher concentrations than others may be chosen, but even at their most common, their number will remain quite low. Thus, for targeting applications, contrast agents with higher molecular relaxivities must be developed.

In some cases, an untargeted agent may be used to target an organ simply because it accumulates there through natural processes. This is most common for the liver, for which there are many examples using iron-oxide based agents, but also for lymph node imaging with iron oxide particles with smaller hydrodynamic radii.

8.4.5   Directions in MRI Contrast Agent Design

The next generation of contrast agents must address the limitations of current agents. Truly disruptive technologies (technology that is entirely different from existing commercial products and offering improvements large enough to capture a majority of market share) are challenging because they cannot rely on existing comparative data for understanding bioactivity, or even mechanism of action. Our inherent skepticism of new ideas raises the burden of proof higher.

A survey of current literature shows several areas of development.

8.4.5.1   Alternative Gadolinium-chelate delivery systems

As mentioned above, slowing the tumbling rate of Gd^{3+} chelates improves their relaxivity. Many research groups are pursuing methods to take advantage of this property to increase the effectiveness of the existing Gd^{3+} chelates without changing the basic chemistry of the chelate or its mode of action. Most commonly this means adding the chelate used in Magnevist, DTPA, though other commercial chelates are used as well. The unifying theme of these approaches is incorporating a Gd^{3+} chelate within a large, biologically acceptable molecule whose size will slow tumbling. Many examples multiply the molecular relaxivity by binding multiple Gd^{3+} ions to the large molecule.

8.4.5.1.1   Dendrimers

Dendrimers—highly branched spherical polymers—offer an intriguing mix of molecular and nanoparticle properties.[49] They have well-defined structures, like small molecules, and their size is typically uniform and well defined. Particle size is determined by the number of "generations," or iterations of the monomer from the core: they are essentially monodisperse, unlike polymers. Dendrimers are nanoparticles in that their size is in the 1–100 nm range (typically 2–10 nm) and they have a large number of reactive sites, making them ideal for functionalization. A generation 5 PAMAM dendrimer, for example, has 128 reactive sites.

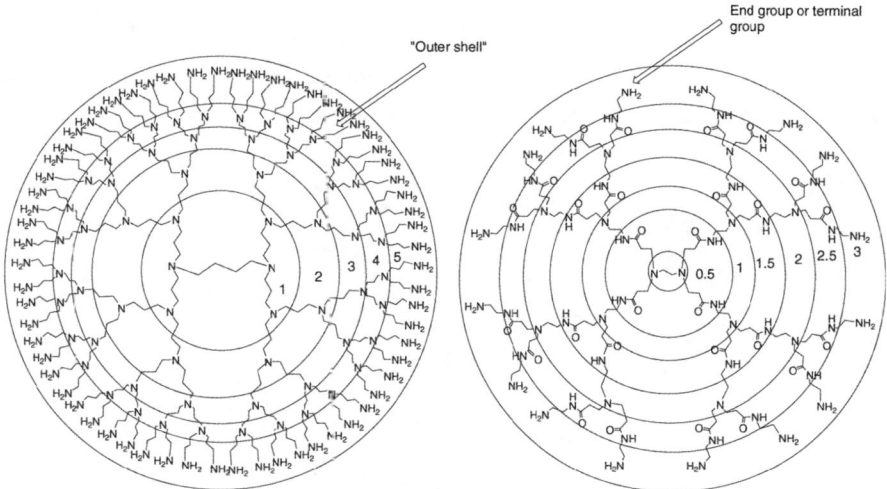

Figure 8.4.4 Example of a Generation 5 PAMAM Dendrimer. Reproduced from ref. 50.

The dendrimer most commonly used is PAMAM (polyamidoamine), trademarked Starburst™ (Figure 8.4.4). PAMAM is highly water soluble and is excreted renally. As large molecules (*e.g.* PAMAM generation 5 is ~30 kD) their tumbling rate is much slower than small molecules. In addition, the large number of reactive groups (amines) on the exterior of the dendrimer allows for binding of numerous chelating groups and therefore Gd^{3+} ions. With many Gd^{3+} ions attached, the contrast agents can be considered pre-concentrated; one target binding event delivers many Gd^{3+} contrast agents. The first dendrimer–Gd^{3+}–chelate MRI contrast agent was reported in 1994. Numerous reports have since described derivatives with different dendrimers, chelating groups and with different targeting groups.[51–63] The variety of targeting groups includes peptides, antibodies, and, most commonly, folic acid. Let's look at a couple of specific examples that highlight the potential of dendritic agents.

Weiner reported the first dendrimer–Gd^{3+}–chelate MRI contrast agent in 1994, using a PAMAM dendrimer.[64] As has become typical procedure, he attached a chelating moiety to the terminal amino groups on the surface of the dendrimer, then introduced a Gd^{3+} ion solution and simply allowed the chelates to soak up the Gd^{3+}. Weiner reported a remarkable 170 Gd^{3+} ions per molecule, and a maximum molecular relaxivity of 5800 $mM^{-1} s^{-1}$, or 34 $mM^{-1} s^{-1}$ per Gd ion, itself nearly 7 times higher than the equivalent R1–Gd^{3+} atom of a small molecule chelate complex.

Using a different dendrimer, poly(propylimine) and achieving a high level of Gd^{3+}–DTPA loading, Meijer reported very high relaxivities at both the per Gd and per dendrimer levels.[66,67] With 64 Gd^{3+} ions per dendrimer, it was recorded that per Gd^{3+}, R1 = 19.7 $mM^{-1} s^{-1}$ and per dendrimer, R1 = 1248 $mM^{-1} s^{-1}$.

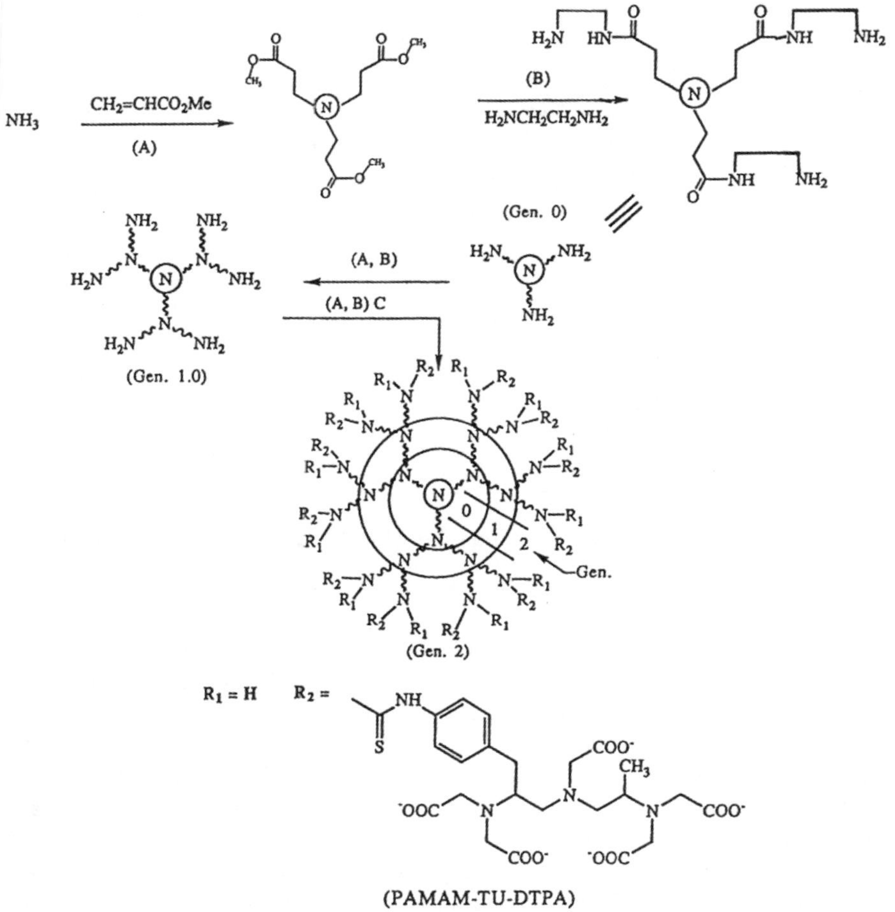

Figure 8.4.5   Preparation of Example Chelate-Covered Dendrimer. From ref. 65.

Incorporating targeting agents with the contrast agent is a recurring theme in this paper, and dendrimers lend themselves to this because of the large number of reactive groups.[68–73] Though precise control of the number of attached targeting agents remains a challenge, simply adding a known stoichiometry of targeting groups (and making the number sufficiently high to assure that essentially all dendrimers will receive at least one targeting group) gives a very testable mixture.

Baker has reported a PAMAM generation 5–Gd^{3+} (DOTA) dendrimer with ~60 Gd^{3+} ions attached, with and without folic acid as a tumor targeting agent.[73] An $R1 = 19–26$ per Gd^{3+} ion, and 532–598 mM^{-1} s^{-1} per dendrimer molecule is reported. The ranges given above reflect the R1 values for the un-folate and the folate dendrimers. In addition, they report a statistically significant uptake of the dendrimer contrast agent in human epithelial cancer cells, showing the targeting groups to be effective in concentrating the contrast agent at the desired site. Wiener reported a folate–dendrimer–Gd^{3+} (DTPA) chelate that showed selective binding to ovarian tumors, and a remarkable molecular R1 of 1646 mM^{-1} s^{-1} per dendrimer.[70]

Brechbiel has reported a PAMAM–generation 3–Gd^{3+}–DTPA dendrimer with peptide targeting groups, and 18–23 Gd ions per molecule.[68] Consistent with the smaller size of the dendrimer, relaxivities (R1) were reported between 8.9–10.1 mM^{-1} s^{-1} per Gd^{3+} atom. As predicted, the smaller dendrimer has both a tumbling rate and R1 intermediate between the larger PAMAM generation 5 dendrimer and the smaller free chelates.

Pseudo-dendrimers have been reported in which the terminal amines of the chelating agent DTPA have been functionalized with acetylated sugars.[74] This created an intermediate-sized species that would be expected to tumble somewhat more slowly than the free chelate complexes, but also be small enough to allow for the potential of escaping the vasculature. However, R1 values were quite low, between 4–5 mM^{-1} s^{-1}.

8.4.5.1.2 Biopolymers: DNA and Proteins

Both DNA and proteins have been conjugated to Gd^{3+}–chelates to improve R1 by two pathways: slowing down tumbling by using a large macromolecule, and concentrating the effective concentration of Gd^{3+} by binding multiple Gd^{3+} ions in each molecule. Shapiro and Hamilton report using oligonucleotides to self-assemble into DNA quadruplexes with cyclic chelates attached to the terminus.[75] Again, the size of the assembled quadruplex makes a marked difference in the R1 value. In the single-DNA species, R1 is 6.4 mM^{-1} s^{-1} per Gd^{3+} atom, while the quadruplex increases to 11.7 mM^{-1} s^{-1} per Gd^{3+} atom and 46.8 mM^{-1} s^{-1} per molecule.

Proteins have been used in a number of different ways. Epix patented a Gd^{3+}–chelate with a group that binds albumin once it is in the blood stream. This *in vivo* assembly dramatically increased its effective size and slowed tumbling.[76] R1 values were enhanced up to 9–fold on albumin binding. Brasch and Oog reported attachment of 30 Gd(DTPA) complexes on albumin,[77,78] before injection, effectively pre-concentrating the local Gd^{3+} concentration as with the dendrimer approach. Barron has engineered a protein with lysine groups at regular intervals to bind Gd^{3+} ions; on average their protein bound 8–9 Gd^{3+} ions, though R1 was only 7.3 mM^{-1} s^{-1} per Gd and 62.6 mM^{-1} s^{-1} per molecule.[79] Yang and Liu adapted an existing protein with additional binding sites to aid Gd^{3+} coordination.[80] Different R1 values were reported for Gd^{3+} ions in different binding locations, but with a high R1 of 117 mM^{-1} s^{-1}. Short peptides have been attached to chelating groups as well.[81,82] Lu reported a cyclic decapeptide bound to DTPA that selectively bound fibrin-fibronectin in tumors.[83] Though the R1 was a low 4.2 mM^{-1} s^{-1}, a good mouse image of a tumor was obtained because the peptide concentrated the contrast agent at the target site. Peptides have been derivatized at both ends to make them amphiphilic. This allows for self-assembly into nanostructures.[84–86] In various states of assembly, R1 was reported between 14.7 mM^{-1} s^{-1} and 22.8 mM^{-1} s^{-1}.

8.4.5.1.3 Polymers

While less common than dendrimers, nanoparticles of linear polymers have been used to create "nanocomplexes" with Gd^{3+} chelates.[87] Several different polymers have been tested, including poly(glutamic acid),[88,89] poly-ethyleneglycol (PEG),[90,91] a polymeric form of the chelate DTPA,[90,92] and a polylactic acid (PLA)–PEG mixture.[87] In the last example, the polymers were combined to create blank nanoparticles into which the Gd chelates were incorporated. The PLA–PEG matrix gave better images of rat livers than commercial agents; this is largely accomplished by greatly reducing the renal clearance rate by increasing particle size, reported to be 188 nm. Poly(L-glu-tamic acid) Gd^{3+} (DTPA) gave a low relaxivity ($R1 = 8.1 mM^{-1} s^{-1}$), but had the interesting effect of accumulating in the necrotic area around treated tumors, providing an effective targeting method.[88]

8.4.5.1.4 Liposomes

Self-assembling macrostructures such as liposomes have been used effectively to concentrate Gd^{3+} ions.[91,93,94] Liposomes are related to micelles, but are kine-tically stable. They are formed by mixing a natural lipid (usually a phospho-diester), and the inclusion group that has been modified to be amphiphilic; different ratios, phosphodiesters, and additives (*e.g.* cholesterol) give the lipo-some its particular properties.[95–98] By attaching a fatty acid phosphate to the cyclic chelate DOTA, Nicolay has created a complex that is liposome-ready.[91] After liposome formation, particle size is 125 nm, and though the relaxivity is a modest 11 $mM^{-1} s^{-1}$ per Gd^{3+} ion, an average of 40 000 Gd^{3+} ions are loca-lized in each liposome. Thus, directing even a few liposomes to cellular receptors has the multiplicative effect of very large concentrations of contrast agent. Further study of this system showed that the cellular localization of the contrast agent had a marked effect on the relaxivity;[99] the reason for this is not clear, but with the concept holding so much promise, further study will undoubtedly follow.

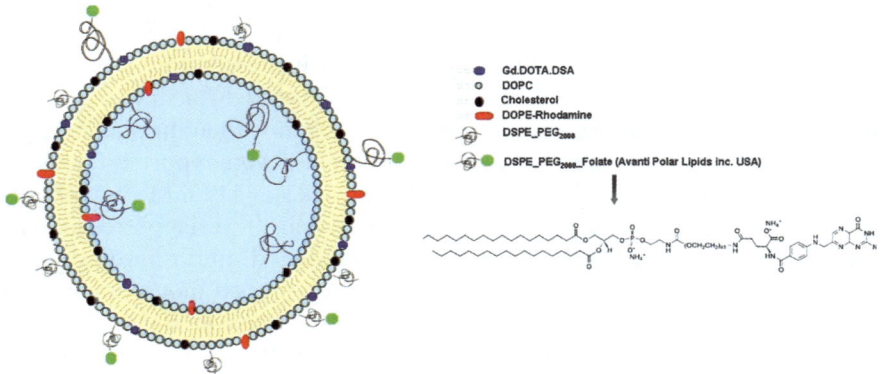

Figure 8.4.6   Gd(DOTA) Containing Liposome. From ref. 94.

Miller has reported a different liposome formulation that has the added benefit of targeting. Liposomal targeting can be done differently than standard approaches. Rather than bind the targeting agent directly to the contrast agent, the targeting agent is incorporated into the lipid formulation. In this example, a folic acid–PEG2000–lipid unit was added during liposome formulation. The contrast agent itself was similar to the phospholipid-bound DOTA described above. While the R1 values are quite small (R1 $= 1.3$ mM^{-1} s^{-1}), the large number of Gd^{3+} ions in the liposome, and the delivery of the liposome to a tumor site demonstrates the promise of the liposome-based targeting approach.

8.4.5.2 Gadolinium Encapsulation in Nanoparticles

8.4.5.2.1 Fullerenes

Gadolinium-containing metallofullerenes (termed gadofullerenes) have tantalized researchers for over a decade.[100] Shortly after fullerenes were described by Smalley and Kroto,[101] their void interior drew speculation about the properties of compounds that could be trapped within the interior of the fullerene sphere.[102–105] By the 1990's researchers had succeeded in encapsulating atoms and small clusters within fullerenes of various sizes.[106] These "filled" fullerenes were termed endohedral fullerenes; in cases where the trapped atom was a metal, the compounds were termed metalloendohedral fullerenes or endohedral metallofullerenes (EMFs). Gadolinium, already in use as a commercial contrast agent in chelate form, was a natural element to work on encapsulating.

Gadofullerenes permanently enclose Gd^{3+} ions within the interior of the spherical fullerene. Because fullerenes are very stable, and do not allow exchange between their interior and the exterior environment, internalized Gd^{3+} ions are prevented from escaping into the biosystem. Transmetallation, the primary mechanism of Gd^{3+} escape from chelates, is not possible with gadofullerenes. This eliminates the primary concern with Gd^{3+} chelates, namely release of Gd^{3+} ions and subsequent toxicity. In addition, the fullerene surface provides many reactive sites that may be used to attach bioactive groups. In this sense they are modular building blocks.

Gadofullerenes include several variations, with 1–3 Gd ions per fullerene cage and different cage sizes. The species used in the majority of research are Gd@C60, Gd@C$_{82}$, Gd$_2$@C$_{82}$, and Gd$_3$N@C$_{80}$. In metalloendohedral notation, the species to the left of the @ is inside the fullerene cage whose size is defined by the term to the right of the symbol. Thus, Gd@C60 is a single Gd^{3+} ion inside a 60-carbon fullerene cage.

Complete encapsulation of the Gd^{3+} ion comes with a price. Isolated from the biosystem by the carbon cage, no direct bonding between Gd^{3+} and water can occur. As the dominant mechanism for enhancing relaxivity in Gd^{3+} chelates is inner-sphere (rapid exchange of free and bound water molecules), gadofullerenes were not predicted to give good relaxivities (R1) if governed by the same rules as chelates. However, several examples of gadofullerenes gave very high relaxivities. As with any surprising result, this stimulated more

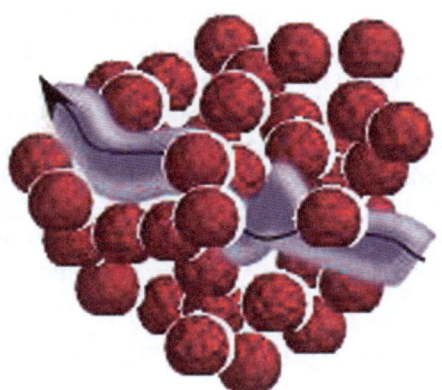

Figure 8.4.7   Restricted Water Passage Through Fullerene Aggregates. From ref. 107.

research and interest from both theoretical and practical perspectives. Mechanisms for the enhanced relaxivity have been proposed. Restricted water passage through large clusters of hydroxy-gadofullerenes was championed by Merbach,[107] and indeed when disaggregation is achieved through high salt concentrations, relaxivity drops.[107,108] However, this is not universally true. Results reported by Luna nanoWorks describe a gadofullerene class, trade-named Hydrochalarone that has high relaxivity with particles at or near the single-molecule size.[14,109] Here a different mechanism must be at work; how the combination of individual particles and high relaxivities is achieved has not been reported.

Primary among the challenges of making fullerenes biocompatible is solubility. Fullerenes are essentially insoluble in water. Thus, to be medically viable, fullerenes must be derivatized to become water soluble. Further complications are raised when one considers the nature of the derivative. How does derivatization influence R1? What is to be considered water-soluble?

Water solubility becomes a trickier definition than it first appears. Most gadofullerene derivatives that appear water soluble are in fact particulate clusters, with particle sizes ranging up to several hundred nanometers. As with SPIOs *vide infra*, particle size becomes an important parameter, as particles that are in the hundreds of nanometers may cause problems while circulating. Additionally, while larger particles tend to have higher relaxivities and remain in the circulation for a longer time than their smaller counterparts, they tend to be unable to escape the vasculature. This limits their use to blood pool imaging and inhibits targeting. To escape the vasculature and penetrate to target sites (other than targets in which the vasculature is compromised), particles should generally be <10 nm, a challenging goal. Most research in the nanoparticle field, and virtually all of the gadofullerene work, has centered on decorating the exterior of the particle with groups that render it hydrophilic while not aggregating the individual particles into untenable clusters.

Early attempts at water-solubilization were through the addition of many hydroxyl groups to the fullerene cage.[110–114] Depending on reaction conditions,

20–40 hydroxyl groups were added to the cage, with other oxygenated functional groups as well. Shinohara hydroxylated Gd@C_{82} and found an average of 40 hydroxyl groups.[115] Relaxivities of these species were surprisingly high, with R1 = 20 mM^{-1} s^{-1} to 83 mM^{-1} s^{-1} reported by different groups, though the range of reported R1 values draws attention to the lack of reproducibility in the procedure. High relaxivity was not predicted, given the lack of direct water-Gd binding, the purported main avenue for relaxivity in Gd^{3+}–chelates. Such high R1 values and the apparent contradiction to accepted wisdom spurred both a burst of new synthetic derivatives and more study on the mechanism of relaxivity. Unfortunately, as discussed, fullerene hydroxylation yielded clusters with particle sizes in the 100 s of nanometers. In addition, Xing found that the large degree of derivatization in early hydroxylated species compromised the integrity of the fullerene cage;[114] limiting hydroxylation to 16–22 left the cage intact, but relaxivity suffered at R1 = 19 mM^{-1} s^{-1}. Finally, it was found that hydroxygadofullerene aggregates cause thrombosis,[116] eliminating them from further consideration.

Gd@C_{60} was solubilized by Bolskar as Gd@$C_{60}[C(COOH)_2]_{10}$;[117] a modest R1 = 4.6 mM^{-1} s^{-1} was reported. However, mouse studies (at 35 mg kg^{-1}) showed the compound to be well tolerated and excreted through the kidneys with minimal retention.[117]

Gd@C82 reappeared with organophosphate decorations in work by Dorn.[118] In their structural formulation, both hydroxyl and phosphonate groups decorate the cage, and are given the formula Gd@$C_{82}O_2(OH)_{16}$ $(C(PO_3Et_2)_2)_{10}$, with the understanding that this represents average numbers of substituents. R1 values of 19.9–38.9 mM^{-1} s^{-1} are reported, depending on magnetic field strength. Aggregates are formed as with earlier hydroxylated derivatives, with particles ranging into hundreds of nanometers. Still, organophophosphonates represent a new derivative class.

Another example of mixed derivatization of fullerenes was reported by Wang.[119,120] Using Gd@C_{82}, he attached both β-alanine and hydroxyl groups in a one-pot reaction, giving the average formula Gd@$C_{82}O_m(OH)_n$ $(NHCH_2CH_2COOH)_l$ ($m \sim 6$; $n \sim 16$; $l \sim 8$). The relaxivity of this derivative (9.1 mM^{-1} s^{-1}) is quite low by gadofullerene standards, but the introduction of terminal functional groups that may be used to attach targeting agents makes this an important step in the development of gadofullerene contrast agents.

A different gadofullerene is the trimetalnitride fullerene Gd$_3$N@C_{80}, discovered by Dorn and Stevenson,[121] and trademarked Trimetasphere™ by Luna Innovations. Trimetaspheres have a cluster of a central nitrogen atom bound to three Gd atoms inside the unusual C_{80} cage. This class has been recently reviewed.[122] The presence of three Gd ions offers the potential of tripling the relaxativity, with the highly paramagnetic state of 21 unpaired electrons. Interestingly, both the Gd$_3$N unit, and the C_{80} cage are unstable in the absence of the other. Other size cages (C_{82}–C_{96}) with the Gd$_3$N unit have been reported but the C_{80} species is the most abundant and most studied. Like other gadofullerenes, Gd atoms are entirely contained, but the increased size of the cluster requires Gd atoms to be closer to the fullerene cage than in the single Gd

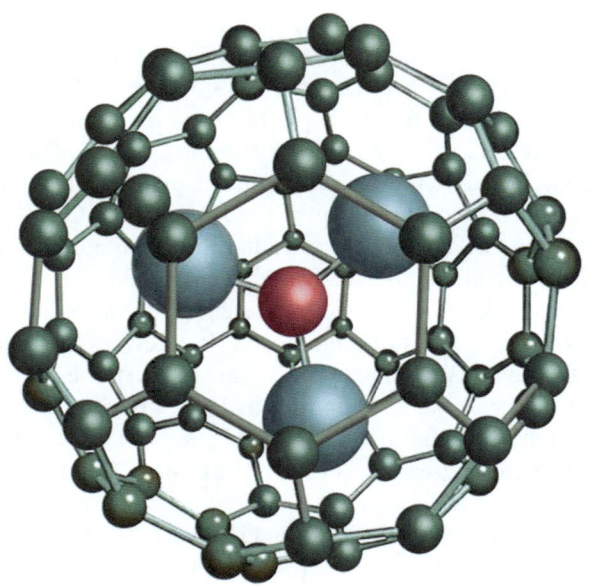

Figure 8.4.8 Gd₃N@C₈₀ "Trimetasphere". From ref. 109.

gadofullerenes.[123] As with earlier gadofullerenes, the first soluble derivative was a polyhydroxylated form,[124] but as with earlier cases, aggregation remained a problem.

Dorn reported a combination of polyhydroxylation and two poly-ethyleneglycol (PEG) 5000 groups.[125] High R1 values of 32–143 mM^{-1} s^{-1} were reported at different field strengths. Mouse studies gave difficult to interpret results because the sample was injected directly into the brain rather than intravenously, and no biodistribution or toxicity data were reported.

Fatouros and Dorn recently reported another Gd₃N@C₈₀ derivative with mixed derivatization,[126] this time using a combination of hydroxyl groups and carboxyl groups on a two carbon chain. A high degree of substitution was achieved, with an average formula of Gd₃N@C₈₀(OH)$_{\sim 26}$(CH2CH2 COOM)$_{\sim 16}$. High relaxivities were reported, at R1 = 207 mM^{-1} s^{-1} at 2.4 T. Interestingly, in the presence of phosphate in PBS, R1 drops dramatically to 35 mM^{-1} s^{-1}. The authors ascribe this to disaggregation in PBS, as the particle size in water is 70–80 nm.

A different approach is the Hydrochalarone series,[24,109,127] in which several short-chain oligo-glycol methyl ether groups are added to Gd₃N@C₈₀ (Tri-metaspheres™) and the cage is charged. By varying the length of the oligoglycol the particle size can also be varied. Particles at the single-particle level (<2 nm) by DLS were reported. With molecular R1 values over 200 mM^{-1}s^{-1} (roughly 50x higher than commercial agents) Hydrochalarones may be used at much lower concentrations than Gd chelates. The greatly increased R1 makes these compounds promising for targeting applications, as smaller quantities of

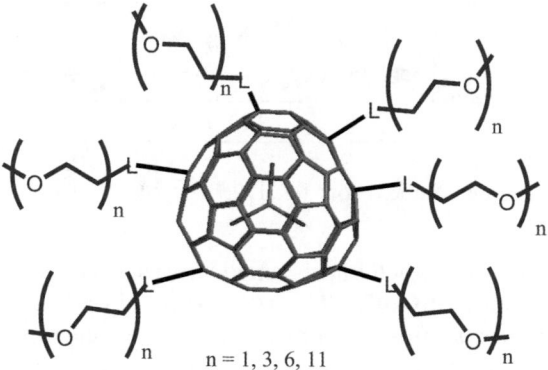

Figure 8.4.9 Hydrochalarone Series. From ref. 14.

contrast agent would need to be delivered to generate good images of the target. In mouse studies, less than 10% concentration of Hydrochlarone was needed to provide the image intensity of commercial Magnevist, and the mice survived 5 days before being sacrificed.

Mixed metal species of the Trimetasphere type have also been reported as contrast agents.[113] Incorporating a mix of Sc^{3+} and Gd^{3+}, and established hydroxylation procedures, Wang reports $Sc_xGd_{3-x}@C_{80}O_m(OH)_n$ ($x = 1,2$; $m \sim 12$; $n \sim 26$). Relaxivities are substantially lower than the pure Gd_3N species, at R1 = 17–20. This is not unexpected, as the Sc does not contribute to the magnetic moment, but does establish the principle that the Gd^{3+} spins are additive for the three atoms.

As mentioned at the start of this section, fullerenes have the potential for attaching targeting groups because of the large number of reactive sites. Empty-cage derivatization with a wide variety of groups is well established,[128] but the gadofullerene field has largely focused on the more immediate need for water solubility. Signalling the progress of the field beyond the solubilization stage is a recent report describing the first targeted gadofullerene.[43] Using an antibody as the targeting agent, and $Gd@C_{82} O_6(OH)_{16}(NHCH_2CH_2COOH)_8$, successful binding to a protein was established. Additionally, the antibody-fulleride complex was shown to have a higher R1 (12 mM^{-1} s^{-1}) than the unconjugated fulleride (8 mM^{-1} s^{-1}). Though this particular example has R1 values too low to be useful for targeted imaging, it succeeded as a "proof of principle" experiment.

8.4.5.2.2 Gadonanotubes

A related nanoscale approach, Gd^{3+} ions in ultra-short nanotubes, $Gd^{3+}_n@USTube$, has only recently been reported by Wilson.[129–133] In this case nanotubes – long hollow tubes with near infinite aspect ratio and walls made of pure graphene carbon – are cut with molecular fluorine and pyrolysis into very

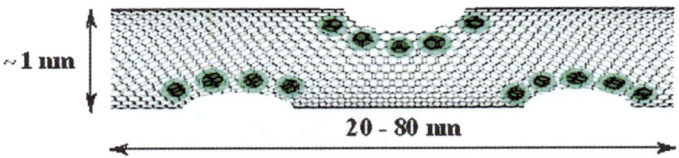

Figure 8.4.10   Gadonanotube. From ref. 129.

short pieces (20–100 nm), then impregnated with Gd^{3+} ions simply by soaking in an aqueous solution of $GdCl_3$. This remarkably simple approach allows for production of large quantities of contrast agent, a problem that has yet to be resolved with gadofullerenes. Extremely high relaxivities are reported, with $R1 = 180 \text{ mM}^{-1} \text{ s}^{-1}$ per Gd^{3+} ion. Within each tube the Gd^{3+} ions are found in clusters with chloride ions, of up to 10 Gd^{3+} atoms. How this clustering might contribute to the record high R1 values is not clear, though the authors suggest a superparamagnetic effect. Given that each tube is loaded with a relatively large number of Gd^{3+} (though the number is not uniform) molecular relaxivities may be calculated in the thousands. Additionally, unlike many gadofullerenes, minimal aggregation is observed, with particle sizes in the 55–75 nm range across a range of pH conditions.[129] Nanotubes offer the possibility of functionalization,[134,135] as with fullerenes, so these may eventually be used with targeting agents. Recently Wilson reported adding amino acids and peptides to gadonanotubes in particular.[136]

An important question about such an open-ended structure is the potential for escape of the Gd^{3+} ion. If the Gd^{3+} ions leaked out the open ends, or through defects in the walls, of the nanotubes, presumably by the same route by which they entered, then safety concerns would arise. To date, Wilson has reported that Gd does not escape when $Gd^{3+}{}_n$@USTube are challenged with PBS, heat, or a range of pH conditions.[129] Biological studies have not yet been reported, but the promise of $Gd^{3+}{}_n$@USTube-type contrast agents seems unlimited.

8.4.5.3   SPIO

Many iron oxide-based agents have been reported, with hundreds of articles in medical journals reporting their various efficacies. Its use in imaging specific sites has been clinically tested with brain injury,[137,138] central nervous system,[139] breast cancer,[78,140,141] lymph nodes,[142–144] spleen,[145] prostate cancer,[141,146–148] cardiovascular,[149] pelvic tumors,[150] joints,[151] ovarian tumors[38] and atherosclerotic plaque.[152–156]

Most work with SPIOs has been in developing coatings that can be chemically bound to targeting groups and exploring new targeting groups. While coatings like dextran (and chemical derivatives of dextran) are the most established, other coatings such as PVP,[157] polyethyleneimine,[158] and PEG have been reported. Critical to the success of the coating are: hydrophilicity, type and number of pendant functional groups, and effect on biodistribution. Coatings may be charged or neutral, a feature that strongly influences the way the particles interact with cells. Coatings largely determine the overall size of the particle in water, which in turn influences circulation time. For example, the USPIO ferumoxtran-10 has a particle size of 30 nm, compared with other approved USPIOs that are 25 nm. The change in size leads to an increased circulation time from 3–6 hours to 24–36 hours.[29,30] The most common pendant groups are carboxylates and amines, as these can be attached through peptide bonds to many biochemicals, the class from which most targeting groups are drawn.

Targeted imaging with SPIOs is done in a similar manner to Gd^{3+} compounds. SPIO surface modification allows for well-established chemistry to be used to attach peptides, antibodies, aptamers, *etc.*

8.4.5.3.1 Coatings

Typically coatings are applied as the particles are formed; that is, particles are formed from soluble iron precursors (usually Fe^{2+} and Fe^{3+} salts) in a solution that contains the coating, whether it is dextran (most common), a polymer, or starches, *etc.*[18] Bare SPIOs can be synthesized and coated after isolation, but this is less common. The coatings influence retention times, and importantly they allow for derivatization.

8.4.5.3.2 Liposomes

Liposomes have been used for USPIO and SPIO in much the same way as for Gd chelates.[159,160] Micelles, relatives of liposomes have also been reported with SPIOs.[161] As with Gd-based agents, the advantages of liposomes are pre-concentration of the contrast agent and delivery of agents to targets *via* targeting groups in the liposome rather than the contrast agent.

8.4.6 Multi-use MRI Contrast Agents

8.4.6.1 Multimodal Agents

Multimodal agents contain imaging agents that allow for imaging by multiple techniques.[160,162–164] This approach gathers more information and the complementary information can cover up limitations in either technique. In principle any two (or more) techniques can be combined by synthesis of the appropriate compound that contains the active agents for the techniques. In the simplest example of this approach, an existing MRI contrast agent is chemically bound to an existing contrast agent for a different analytical method. For example, the combination of a single-crystal iron oxide nanoparticle (MION) with a fluorescence-active porphyrin[165] yields an agent that uses two tools with a single injection. Another example using iron oxide nanoparticles not only attached a fluorescent marker, but also peptide targeting sequences, and observed binding to thrombi with both methods.[166]

An alternative to direct chemical bonding is to combine agents in a single delivery system, such as a liposome. Miller has reported a bimodal liposome that is active in both MRI and fluorescence assays.[93,94]

8.4.6.2   Theranostics

Theranostics is a term for an area that was until recently only speculative.[165,167–169] Simply put, a theranostic would combine an imaging agent and a therapy agent. Two methods are envisioned: one in which a single targeting agent is used to direct the imaging agent, and then if needed, a therapeutic agent. The second, grander concept is a single agent that contains both imaging and therapeutic functionalities.

An example of the first approach is provided by Brechbiel.[170] His group synthesized a chelating group, C-NETA, that successfully binds both Gd^{3+} for MRI imaging and radioisotopes such as ^{177}Lu, ^{90}Y, and ^{212}Bi. C-NETA was bound to an antibody (Trastuzumab) that delivered the entire complex to a tumor site, whether it carried Gd for diagnosis or a radioisotope for treatment.

Nanotechnology has provided an avenue to realize the second approach. When one considers the number of potential attachment sites on the surface of a coated SPIO, or the 128 sites on a generation 5 PAMAM dendrimer, one can see that by partial reaction of the surface just three times, one could add a targeting group, and imaging group (unnecessary for SPIO cases), and a therapeutic.

Yet another approach is to use different properties of the same agent. Recent publications describe thermotherapy, or hyperthermia for cancer treatment.[147,171,172] In one example, a USPIO was directly injected into a tumor to image it, then the iron atoms were thermally excited through an alternating magnetic field. The Fe becomes hot and kills tumor cells where it is localized. A second example developed iron-based nanoparticles (FeNPs),[172] that were

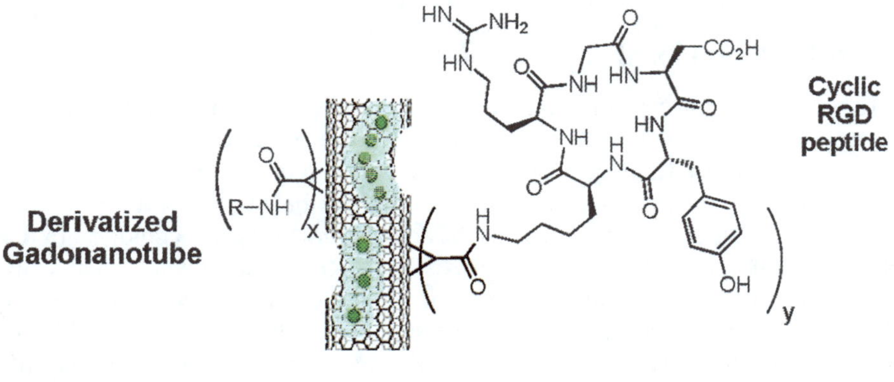

Figure 8.4.11   Derivatized Gadonanotube. From ref. 136.

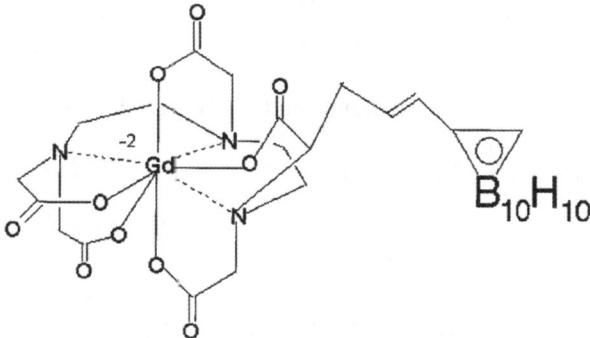

Figure 8.4.12   Mixed Gd–Chelate–Borohydride for Imaging and Neutron Capture Therapy (NCT). From ref. 173.

excited the same way. The FeNP's were synthesized by reduction of $FeCl_2$ in the presence of carboxyl-terminated PEG, providing peripheral carboxyl groups for attachment of directing groups in future work.

A traditional chemistry approach is reported by Wiener who covalently bound a DTPA chelate with a borohydride used in neutron capture therapy (NCT). Wiener also proposed the use of [157]Gd that could be activated by neutron bombardment and act as an NCT agent as well as an MRI contrast agent.[173]

8.4.7   Conclusion

MRI imaging is experiencing an exciting growth phase, with success being reported using numerous innovative approaches. Many of these involve nanoparticles, from dendrimers to polymers to the various iron oxide particles. As nanotechnology continues to develop as a fundamental science, contrast agents will improve by incorporating the new technology. Biotechnology grows ever more sophisticated in both identification of specific markers for disease and production methods of biomolecules. As antibodies and peptides become more available, their incorporation into contrast agents will follow. One can imagine prepared nanoparticles with binding-ready functional groups being mixed at the MRI site with a targeting agent peptide, antibody, *etc.* for the patient's particular needs. Though not all of the exciting laboratory ideas will make it through the rigorous testing procedures, the wealth of sturdy concepts in the field promises a bright future.

References

1. A. Berrington de Gonzalez and S. Darby., *Lancet*, 2004, **363**, 345–351.
2. T. N. Nagaraja, K. Karki, J. R. Ewing, R. L. Croxen and R. A. Knight, *Stroke*, 2008, **39**, 427–432.

3. M. Nahrendorf, D. Sosnovik, J. W. Chen, P. Panizzi, J. L. Figueiredo, E. Aikawa, P. Libby, F. K. Swirski and R. Weissleder, *Circulation*, 2008, **117**, 1153–1160.

4. M. Nahrendorf, D. E. Sosnovik and R. Weissleder, *Basic Res. Cardiol.*, 2008, **103**, 87–94.

5. D. E. Sosnovik, M. Nahrendorf and R. Weissleder, *Circulation*, 2007, **115**, 2076–2086.

6. D. E. Sosnovik, M. Nahrendorf and R. Weissleder, *Nat. Clin. Pract. Cardiovascular Medicine*, 2008, **5**, S63–S70.

7. G. Korosoglou, R. G. Weiss, D. A. Kedziorek, P. Walczak, W. D. Gilson, M. Schar, D. E. Sosnovik, D. L. Kraitchman, R. C. Boston, J. W. M. Bulte, R. Weissleder and M. Stuber, *J. Am. Coll. Cardiol.*, 2008, **52**, 483–491.

8. L. B. Andersen and R. Frayne, in *Advanced Imaging in Biology and Medicine*; ed. C. W. Sensen and B. Hallgrímsson, Springer, Berlin, Heidelberg, 1st Edn, 2009, pp. 363–393.

9. A. S. R. Guimaraes and R. Weissleder, in *In vivo Imaging of Cancer Therapy*, ed. A. F. Shields and, P. Price, Springer, Netherlands, 1st edn, 2007, pp. 259–280.

10. T. Persigehl, W. Heindel and C. Bremer, *Abdominal Imaging*, 2005, **30**, 342–354.

11. T. Persigehl, W. Heindel and C. Bremer, *Radiologe*, 2008, **48**, 863–870.

12. D. Sosnovik and R. Weissleder, in *Progress in Drug Research*; ed M. Rudin; Birkhauser Verlag, Basel, 1st Edn, 2005, **vol. 62**.

13. D. E. Sosnovik and R. Weissleder, *Curr. Opin. Biotechnol.*, 2007, **18**, 4–10.

14. D. K. MacFarland, K. L. Walker, R. P. Lenk, S. R. Wilson, K. Kumar, C. L. Kepley and J. R. Garbow, *J. Med. Chem.*, 2008, **51**, 3681–3683.

15. J. W. M. Bulte and L. H. Bryant Jr., in *Physics and Chemistry Basis of Biotechnology*; ed. M. d. Cuyper and J. W. M. Bulte, Springer, Netherlands, 1st Edn, 2002, **vol. 7**, pp. 191–221.

16. D. E. Sosnovik, M. Nahrendorf and R. Weissleder, *Bas. Res. Cardiol.*, 2008, **103**, 122–130.

17. E. Tóth, L. Helm, A. E. Merbach, J. A. McCleverty and T. J. Meyer, in *Comprehensive Coordination Chemistry II*; Pergamon, Oxford, 1st Edn, 2003, **vol. 9**, pp. 841–881.

18. D. L. J. Thorek, A. Chen, J. Czupryna and A. Tsourkas, *Ann. Biomed. Eng.*, 2006, **4**, 23–38.

19. M. F. Bellin, *Eur. J. Radiol.*, 2006, **60**, 314–323.

20. M. F. Bellin and A. J. Van Der Molen, *Eur. J. Radiol.*, 2008, **66**, 160–167.

21. M. F. Bellin, in *Contrast Media*; ed. H. S. Thomsen, J. A. W. Webb, Springer, Berlin Heidelberg, 2nd edn, 2009, pp. 205–212.

22. N. Bloembergen and L. O. Morgan, *J. Chem. Phys.*, 1961, **34**, 842–850.

23. I. Solomon, *Phys. Rev.*, 1955, **99**, 559–565.

24. K. Kumar, D. K. MacFarland, Z. Zhou, C. L. Kepley, K. L. Walker, S. R. Wilson and R. P. Lenk, in *Drug Delivery Nanoparticles Formulation and Characterization*, ed. Y. Pathik, Informa Healthcare, New York, 1st Edn, 2009, pp. 330–348.

25. B. A. Larsen, M. A. Haag, N. J. Serkova, K. R. Shroyer and C. R. Stoldt, *Nanotechnology*, 2008, **19**, 265–271.
26. B. A. Larsen, M. A. Haag, M. Stowell, D. C. Walther, A. P. Pisano and C. R. Stoldt, *Controlling nanoparticle aggregation in colloidal microwave absorbers via interface chemistry, Proc. SPIE,* 2007, **6525**, 652519.
27. C. Sun, J. S. H. Lee and M. Q. Zhang, *Adv. Drug Delivery Rev.*, 2008, **60**, 1252–1265.
28. A. Bjornerud and L. Johansson, *NMR Biomed.*, 2004, **17**, 465–477.
29. M. E. Kooi, S. Heeneman, M. J. A. P. Daemen, J. M. A. v. Engelshoven, K. B. J. M. Cleutjens, in *Nanoparticles in Biomedical Imaging*; J. W. M. B. Modo. and M. M. J. , Springer, New York, 1st Edn, 2008, **vol. 102**, pp. 63–90.
30. E. A. Neuwelt, B. E. Hamilton, C. G. Varallyay, W. R. Rooney, R. D. Edelman, P. M. Jacobs and S. G. Watnick, *Kidney Int.*, 2009, **75**, 465–474.
31. P. Wedeking, K. Kumar and M. F. Tweedle, *Magn. Reson. Imaging*, 1992, **10**, 641–648.
32. M. F. Tweedle, J. J. Hagan, K. Kumar, S. Mantha and C. A. Chang, *Magn. Reson. Imaging*, 1991, **9**, 409–415.
33. *FDA Requests Boxed Warning for Contrast Agents Used to Improve MRI Images*, FDA, May 23, 2007.
34. T. Grobner, *Nephrol., Dial., Transplant.*, 2006, **21**, 1104–1108.
35. P. Marckmann, L. Skov, K. Rossen, A. Dupont, M. B. Damholt, J. G. Heaf and H. S. Thomsen, *J. Am. Soc. Nephrol.*, 1996, **17**, 2359–2362.
36. C. Tharal, J. Alhiri and J. L. Abraham, *Contrast Media Mol. Imaging*, 2007, **2**, 199–205.
37. A. Hagooly, R. Rossin and M. J. Welch, *Handb. Exp. Pharmacol.*, 2008, **185**(2), 93–129.
38. Z. J. Wang, S. Boddington, M. Wendland, R. Meier, C. Corot and H. Daldrup-Link, *Pediatr. Radiol.*, 2008, **38**, 529–537.
39. M. D. Salazar and M. Ratnam, *Cancer Metastasis Rev.*, 2007, **26**, 141–152.
40. E. I. Sega and P. S., *Cancer Metastasis Rev.*, 2008, **27**, 655–664.
41. H. Zhao, K. M. Cui, A. Muschenborn and S. T. C. Wong, *Mol. Med. Rep.*, 2008, **1**, 131–134.
42. G. H. Chen, W. J. Chen, Z. Wu, R. X. Yuan, H. Li, J. M. Gao and X. T. Shuai, *Biomaterials*, 2009, **30**, 1962–1970.
43. C. Y. Shu, X. Y. Ma, J. F. Zhang, F. D. Corwin, J. H. Sim, E. Y. Zhang, H. C. Dorn, H. W. Gibson, P. P. Fatouros, C. R. Wang and X. H. Fang, *Bioconjugate Chem.*, 2008, **19**, 651–655.
44. L. L. Yang, H. Mao, Y. A. Wang, Z. H. Cao, X. H. Peng, X. X. Wang, H. W. Duan, C. C. Ni, Q. G. Yuan, G. Adams, M. Q. Smith, W. C. Wood, X. H. Gao and S. M. Nie, *Small*, 2009, **5**, 235–243.
45. R. Bieker, T. Kessler, C. Schwoppe, T. Padro, T. Persigehl, C. Bremer, J. Dreischaluck, A. Kolkmeyer, W. Heindel, R. M. Mesters and W. E. Berdel, *Blood*, 2009, **113**, 5019–5027.

46. P. Antunes, M. Ginj, M. A. Walter, J. H. Chen, J. C. Reubi and H. R. Maecke, *Bioconjugate Chem.*, 2007, **18**, 84–92.
47. P. Antunes, M. Ginj, H. Zhang, B. Waser, R. P. Baum, J. C. Reubi and H. Maecke, *Eur. J. Nucl. Med. Mol. Imaging*, 2007, **34**, 982–993.
48. A. D. Nunn, K. E. Linder and M. F. Tweedle, *Quart. J. Nucl. Med.*, 1997, **41**, 155–162.
49. U. Boas and P. M. H. Heegaard, *Chem. Soc. Rev.*, 2004, **33**, 43–63.
50. U. Boas, J. B. Christensen and P. M. H. Heegaard, *J. Mater. Chem.*, 2006, **16**, 3786–3798.
51. H. Kobayashi, S. Kawamoto, S. K. Jo, N. Sato, T. Saga, A. Hiraga, J. Konishi, S. Hu, K. Togashi, M. W. Brechbiel and R. A. Star, *Kidney Int.*, 2002, **61**, 1980–1985.
52. H. Kobayashi, S. Kawamoto, T. Saga, N. Sato, A. Hiraga, T. Ishimori, Y. Akita, M. H. Mamede, J. Konishi and K. Togashi, *Magn. Reson. Med.*, 2001, **46**, 795–802.
53. H. Kobayashi, S. Kawamoto, R. A. Star, T. A. Waldmann, Y. Tagaya and M. W. Brechbiel, *Cancer Res.*, 2003, **63**, 271–276.
54. H. Kobayashi, T. Saga, S. Kawamoto, N. Sato, A. Hiraga, T. Ishimori, J. Konishi, K. Togashi and M. W. Brechbiel, *Cancer Res.*, 2001, **61**, 4966–4970.
55. H. Kobayashi, N. Sato, S. Kawamoto, T. Saga, A. Hiraga, T. Ishimori, J. Konishi, K. Togashi and M. W. Brechbiel, *Magn. Reson. Med.*, 2001, **46**, 457–464.
56. H. Kobayashi, N. Sato, S. Kawamoto, T. Saga, A. Hiraga, T. Ishimori, J. Konishi, K. Togashi and M. W. Brechbiel, *Magn. Reson. Med.*, 2001, **46**, 579–585.
57. B. Misselwitz, H. Schmitt-Willich, M. Michaelis and J. J. Oellinger, *Invest. Radiol.*, 2002, **37**, 146–151.
58. G. M. Nicolle, E. Toth, H. Schmitt-Willich, B. Raduchel and A. E. Merbach, *Chem.-Eur. J.*, 2002, **8**, 1040–1048.
59. H. Kobayashi, N. Sato, T. Saga, J. Konishi, K. Togashi and M. W. Brechbiel, *Radiology*, 2000, **217**, 140.
60. M. M. Ali, M. Woods, P. Caravan, A. C. L. Opina, M. Spiller, J. C. Fettinger and A. D. Sherry, *Chem.- Eur. J.*, 2008, **14**, 7250–7258.
61. Y. Koyama, V. S. Talanov, M. Bernardo, Y. Hama, C. A. S. Regino, M. W. Brechbiel, P. L. Choyke and Kobayashi, *J. Magn. Reson. Imaging*, 2007, **25**, 866–871.
62. C. A. S. Regino, S. Walbridge, M. Bernardo, K. J. Wong, D. Johnson, R. Lonser, E. H. Oldfield, P. L. Choyke and M. W. Brechbiel, *Contrast Media Mol. Imaging*, 2008, **3**, 2–8.
63. K. Vetterlein, U. Bergmann, K. Buche, M. Walker, J. Lehmann, M. W. Linscheid, G. K. E. Scriba and M. Hildebrand, *Electrophoresis*, 2007, **28**, 3088–3099.
64. E. C. Wiener, M. W. Brechbiel, H. Brothers, R. L. Magin, O. A. Gansow, D. A. Tomalia and P. C. Lauterbur, *Magn. Reson. Med.*, 1994, **31**, 1–8.

65. E. C. Wiener, F. P. Auteri, J. W. Chen, M. W. Brechbiel, O. A. Gansow, D. S. Schneider, R. L. Belford, R. B. Clarkson and P. C. Lauterbur, *J. Am. Chem. Soc.*, 1996, **118**, 7774–7782.

66. S. Langereis, Q. G. de Lussanet, M. H. P. van Genderen, W. H. Backes and E. W. Meijer, *Macromolecules*, 2004, **37**, 3084–3091.

67. S. Langereis, Q. G. de Lussanet, M. H. P. van Genderen, E. W. Meijer, R. G. H. Beets-Tan, A. W. Griffioen, J. M. A. van Engelshoven and W. H. Backes, *NMR Biomed.*, 2006, **19**, 133–141.

68. C. A. Boswell, P. K. Eck, C. A. S. Regino, M. Bernardo, K. J. Wong, D. E. Milenic, P. L. Choyke and M. W. Brechbiel, *Mol. Pharmaceutics*, 2008, **5**, 527–539.

69. S. D. Konda, M. Aref, M. Brechbiel and E. C. Wiener, *Invest. Radiol.*, 2000, **35**, 50–57.

70. S. D. Konda, M. Aref, M. S. Wang, M. Brechbiel and E. C. Wiener, *Magn. Reson. Mater. Phys., Biol. Med.*, 2001, **12**, 104–113.

71. S. D. Konda, S. Wang, M. Brechbiel and E. C. Wiener, *Invest. Radiol.*, 2002, **37**, 199–204.

72. Q. G. de Lussanet, S. Langereis, R. G. H. Beets-Tan, M. H. P. van Genderen, A. W. Griffioen, J. M. A. van Engelshoven and W. H. Backes, *Radiology*, 2005, **235**, 65–72.

73. S. D. Swanson, J. F. Kukowska-Latallo, A. K. Patri, C. Y. Chen, S. Ge, Z. Y. Cao, A. Kotlyar, A. T. East and J. R. Baker, *Int. J. Nanomed.*, 2008, **3**, 201–210.

74. G. Yu, M. Yamashita, K. Aoshima, M. Takahashi, T. Oshikawa, H. Takayanagi, S. Laurent, C. Burtea, L. Vander Elst and R. N. Muller, *Bioorg. Med. Chem. Lett.*, 2007, **17**, 2246–2249.

75. J. F. Cai, E. M. Shapiro and A. D. Hamilton, *Bioconjugate Chem.*, 2009, **20**, 205–208.

76. P. Caravan, G. Parigi, J. M. Chasse, N. J. Cloutier, J. J. Ellison, R. B. Lauffer, C. Luchinat, S. A. McDermid, M. Spiller and T. J. McMurry, *Inorg. Chem.*, 2007, **46**, 6632–6639.

77. M. D. Ogan, U. Schmiedl, M. E. Moseley, W. Grodd, H. Paajanen and R. C. Brasch, *Invest. Radiol.*, 1987, **22**, 665–671.

78. K. Turetschek, A. Preda, E. Floyd, D. M. Shames, V. Novikov, T. P. L. Roberts, J. M. Wood, Y. J. Fu, W. O. Carter and R. C. Brasch, *Eur. J. Nucl. Med. Mol. Imaging*, 2003, **30**, 448–455.

79. L. S. Karfeld, S. R. Bull, N. E. Davis, T. J. Meade and A. E. Barron, *Bioconjugate Chem.*, 2007, **18**, 1697–1700.

80. J. J. Yang, J. H. Yang, L. X. Wei, O. Zurkiya, W. Yang, S. Y. Li, J. Zou, Y. B. Zhou, A. L. W. Maniccia, H. Mao, F. Q. Zhao, R. Malchow, S. M. Zhao, J. Johnson, X. P. Hu, E. Krogstad and Z. R. Liu, *J. Am. Chem. Soc.*, 2008, **130**, 9260–9267.

81. P. Caravan, B. Das, Q. Deng, S. Dumas, V. Jacques, S. K. Koerner, A. Kolodziej, R. J. Looby, W. C. Sun and Z. D. Zhang, *Chem. Commun.*, 2009, **4**, 430–432.

82. F. Ye, X. Wu, E.-K. Jeong, Z. Jia, T. Yang, D. Parker and Z.-R. Lu, *Bioconjugate Chem.*, 2009, **20**, 402.
83. F. R. Ye, X. M. Wu, E. K. Jeong, Z. J. Jia, T. X. Yang, D. Parker and Z. R. Lu, *Bioconjugate Chem.*, 2009, **20**, 402–402.
84. S. R. Bull, M. O. Guler, R. E. Bras, T. J. Meade and S. I. Stupp, *Nano Lett.*, 2005, **5**, 1–4.
85. S. R. Bull, M. O. Guler, R. E. Bras, P. N. Venkatasubramanian, S. I. Stupp and T. J. Meade, *Bioconjugate Chem.*, 2005, **16**, 1343–1348.
86. S. R. Bull, L. C. Palmer, N. J. Fry, M. A. Greenfield, B. W. Messmore, T. J. Meade and S. I. Stupp., *J. Am. Chem. Soc.*, 2008, **130**, 2742.
87. Z. J. Chen, D. X. Yu, S. J. Wang, N. Zhang, C. H. Ma and Z. J. Lu, *Nanoscale Res. Lett.*, 2009, **4**, 618–626.
88. E. F. Jackson, E. Esparza-Coss, X. X. Wen, C. S. Ng, S. L. Daniel, R. E. Price, B. Rivera, C. Charnsangavej, J. G. Gelovani and C. Li, *Int. J. Radiat. Oncol., Biol., Phys.*, 2007, **68**, 830–838.
89. X. X. Wen, E. F. Jackson, R. E. Price, E. E. Kim, Q. P. Wu, S. Wallace, C. Charnsangavej, J. G. Gelovani and C. Li, *Bioconjugate Chem.*, 2004, **15**, 1408–1415.
90. E. Toth, I. van Uffelen, L. Helm, A. E. Merbach, D. Ladd, K. Briley-Saebo and K. E. Kellar, *Magn. Reson. Chem.*, 1998, **36**, S125–S134.
91. S. Hak, H. Sanders, P. Agrawal, S. Langereis, H. Grull, H. M. Keizer, H. M. F. Arena, E. Terreno, G. J. Strijkers and K. Nicolay, *Eur. J. Pharm. Biopharm.*, 2009, **72**, 397–404.
92. E. Toth, L. Helm, K. E. Kellar and A. E. Merbach, *Chem.-Eur. J.*, 1999, **5**, 1202–1211.
93. N. Kamaly, T. Kalber, A. Ahmad, M. H. Oliver, P. W. So, A. H. Herlihy, J. D. Bell, M. R. Jorgensen and A. D. Miller, *Bioconjugate Chem.*, 2008, **19**, 118–129.
94. N. Kamaly, T. Kalber, M. Thanou, J. D. Bell and A. D. Miller, *Bioconjugate Chem.*, 2009, **20**, 648–655.
95. W. T. Al-Jamal and K. Kostarelos, *Nanomedicine*, 2007, **2**, 85–98.
96. Y. Weili and H. Leaf, *Polym. Rev.*, 2007, **47**, 329–344.
97. G. P. van Balen, C. A. M. Martinet, G. Caron, G. Bouchard, M. Reist, P. A. Carrupt, R. Fruttero, A. Gasco and B. Testa., *Med. Res. Rev.*, 2004, **24**, 299–324.
98. V. Weissig, S. V. Boddapati, S. M. Cheng and G. G. M. D'Souza, *J. Liposome Res.*, 2006, **16**, 249–264.
99. M. B. Kok, S. Hak, W. J. M. Mulder, D. W. J. van der Schaft, G. J. Strijklers and K. Nicolay, *Magn. Reson. Med.*, 2009, **61**, 1022–1032.
100. R. D. Bolskar, *Nanomedicine*, 2008, **3**, 201–213.
101. H. W. Kroto, J. R. Heath, S. C. Obrien, R. F. Curl and R. E. Smalley, *Nature*, 1985, **318**, 162–163.
102. J. Cioslowski, *J. Am. Chem. Soc.*, 1991, **113**, 4139–4141.
103. J. Cioslowski and E. D. Fleischmann, *J. Chem. Phys.*, 1991, **94**, 3730–3734.

104. J. Cioslowski and A. Nanayakkara, *J. Chem. Phys.*, 1992, **96**, 8354–8362.
105. T. Weiske, D. K. Bohme, J. Hrusak, W. Kratschmer and H. Schwarz, *Angew. Chem., Int. Ed. Engl.*, 1991, **30**, 884–886.
106. H. Shinohara, *Rep. Prog. Phys.*, 2000, **63**, 843–892.
107. S. Laus, B. Sitharaman, E. Toth, R. D. Bolskar, L. Helm, L. J. Wilson and A. E. Merbach, *J. Phys. Chem. C*, 2007, **111**, 5633–5639.
108. S. Laus, B. Sitharaman, E. Toth, R. D. Bolskar, L. Helm, S. Asokan, M. S. Wong, L. J. Wilson and A. E. Merbach, *J. Am. Chem. Soc.*, 2005, **127**, 9368–9369.
109. D. K. MacFarland, *ECS Trans.*, 2008, **13**, 117.
110. S. R. Zhang, D. Y. Sun, X. Y. Li, F. K. Pei and S. Y. Liu, *Fullerene Sci. Technol.*, 1997, **5**, 1635–1643.
111. H. Kato, Y. Kanazawa, M. Okumura, A. Taninaka, T. Yokawa and H. Shinohara, *J. Am. Chem. Soc.*, 2003, **125**, 4391–4397.
112. M. Okumura, M. Mikawa, T. Yokawa, Y. Kanazawa, H. Kato and H. Shinohara, *Acad. Radiol.*, 2002, **9**, S495–S497.
113. E. Y. Zhang, C. Y. Shu, L. Feng and C. R. Wang, *J. Phys. Chem. B*, 2007, **111**, 14223–14226.
114. J. Zhang, K. M. Liu, G. M. Xing, T. X. Ren and S. K. Wang, *J. Radioanal. Nucl. Chem.*, 2007, **272**, 605–609.
115. M. Mikawa, H. Kato, M. Okumura, Y. Narazaki, Y. Kanazawa, N. Miwa and H. Shinohara, *Bioconjugate Chem.*, 2001, **12**, 510–514.
116. A. Radomski, P. Jurasz, D. Alonso-Escolano, M. Drews, M. Morandi, T. Malinski and M. W. Radomski, *Br. J. Pharmacol.*, 2005, **6**, 146.
117. R. D. Bolskar, A. F. Benedetto, L. O. Husebo, R. E. Price, E. F. Jackson, S. Wallace, L. J. Wilson, J. M. Alford and J. M , *J. Am. Chem. Soc.*, 2003, **125**, 5471–5478.
118. C. Y. Shu, C. R. Wang, J. F. Zhang, H. W. Gibson, H. C. Dorn, F. D. Corwin, P. P. Fatouros and T. J. S. Dennis, *Chem. Mater.*, 2008, **20**, 2106–2109.
119. C. Y. Shu, L. H. Gan, C. R. Wang, X. L. Pei and H. B. Han, *Carbon*, 2006, **44**, 496–500.
120. C. Y. Shu, E. Y. Zhang, J. F. Xiang, C. F. Zhu, C. R. Wang, X. L. Pei and H. B. Han, *J. Phys. Chem. B*, 2006, **110**, 15597–15601.
121. S. Stevenson, G. Rice, T. Glass, K. Harich, F. Cromer, M. R. Jordan, J. Craft, E. Hadju, R. Bible, M. M. Olmstead, K. Maitra, A. J. Fisher, A. L. Balch and H. C. Dorn, *Nature*, 1999, **401**, 55–57.
122. L. Dunsch and S. Yang, *Small*, 2007, **3**, 1298–1320.
123. S. Stevenson, J. P. Philips, J. E. Reid, M. M. Olmstead, S. P. Rath and A. L. Balch, *Chem. Commun.*, 2004, **24**, 2814–2815.
124. E. B. Iezzi, F. Cromer, P. Stevenson and H. C. Dorn, *Synth. Met.*, 2002, **128**, 289–291.
125. P. P. Fatouros, F. D. Corwin, Z. Chen, W. C. Broaddus, J. L. Taturn, B. Kettenmann, Z. Ge, H. W. Gibson, J. L. Russ, A. P. Leonard, J. C. Duchamp and H. C. Dorn, *Radiology*, 2006, **240**, 756–764.

126. C. Y. Shu, F. D. Corwin, J. F. Zhang, Z. J. Chen, J. E. Reid, M. H. Sun, W. Xu, J. H. Sim, C. R. Wang, P. P. Fatouros, A. R. Esker, H. W. Gibson and H. C. Dorn, *Bioconjugate Chem.*, 2009, **20**, 1186–1193.

127. D. K. MacFarland, K. W. Walker, R. P. Lenk, S. R. Wilson, K. Kumar, C. L. Kepley and J. R. Garbow, *J. Med. Chem.*, 2008, **51**, 3681–3683.

128. A. Hirsch and M. Brettreich, in *Fullerenes Chemistry and Reactions*; Wiley-VCH, Weinheim, 1st Edn, 2005.

129. K. B. Hartman, S. Laus, R. D. Bolskar, R. Muthupillai, L. Helm, E. Toth, A. E. Merbach and L. J. Wilson, *Nano Lett.*, 2008, **8**, 415–419.

130. K. B. Hartman and L. J. Wilson, *Carbon Nanostructures as a New High-performance Platform for MR Molecular Imaging*, in *Nanostructures for Biomedical Applications*, ed. Warren Chan, Landes Bioscience and Eurekah, Georgetown, Texas, 2007, pp. 74–84.

131. B. Sitharaman, K. R. Kissell, K. B. Hartman, L. A. Tran, A. Baikalov, I. Rusakova, Y. Sun, H. A. Khant, S. J. Ludtke, W. Chiu, S. Laus, E. Toth, L. Helm, A. E. Merbach and L. J. Wilson, *Chem. Commun.*, 2005, **31**, 3915–3917.

132. B. Sitharaman, X. F. Shi, X. F. Walboomers, H. B. Liao, V. Cuijpers, L. J. Wilson, A. G. Mikos and J. A. Jansen, *Bone*, 2008, **43**, 362–370.

133. B. Sithararnan and L. J. Wilson, *Int. J. Nanomed.*, 2006, **1**, 291–295.

134. J. M. Ashcroft, K. B. Hartman, K. R. Kissell, Y. Mackeyev, S. Pheasant, S. Young, P. A. W. Van der Heide, A. G. Mikos and L. J. Wilson, *Adv. Mater.*, 2007, **19**, 573.

135. J. M. Ashcroft, K. B. Hartman, Y. Mackeyev, C. Hofmann, S. Pheasant, L. B. Alemany and L. J. Wilson, *Nanotechnology*, 2006, **17**, 5033–5037.

136. Y. Mackeyev, K. B. Hartman, J. S. Ananta, A. V. Lee and L. J. Wilson, *J. Am. Chem. Soc.*, 2009, **131**, 8342.

137. C. Chapon, F. Franconi, F. Lacoeuille, F. Hindre, P. Saulnier, J. Benoit, J. J. Le Jeune and L. Lemaire, *Magn. Reson. Mater. Phys., Biol. Med.*, 2009, **22**, 167–174.

138. A. Pasco, A. Ter Minassian, C. Chapon, L. Lemaire, F. Franconi, D. Darabi, C. Caron, J. P. Benoit and J. J. Le Jeune, *Eur. Radiol.*, 2006, **16**, 1501–1508.

139. V. Dousse, C. Gomez, K. G. Petry, C. Delalande and J. M. Caille, *Magn. Reson. Mater. Phys., Biol. Med.*, 1999, **8**, 185–189.

140. J. Meng, J. Fan, G. Galiana, R. T. Branca, P. L. Clasen, S. Ma, J. Zhou, C. Leuschner, C. S. S. R. Kumar, J. Hormes, T. Otiti, A. C. Beye, M. P. Harmer, C. J. Kiely, W. Warren, M. P. Haataja and W. O. Soboyejo, *Mater. Sci. Eng., C*, 2009, **29**, 1467–79.

141. O. Rodriguez, S. Fricke, C. Chien, L. Dettin, J. VanMeter, E. Shapiro, H. N. Dai, M. Casimiro, L. Ileva., J. Dagata, M. D. Johnson, M. P. Lisanti, A. Koretsky and C. Albanese, *Cell Cycle*, 2006, **5**, 113–119.

142. R. A. M. Heesakkers, A. M. Hovels, G. J. Jager, H. C. M. van den Bosch, J. A. Witjes, H. P. J. Raat, J. L. Severens, E. M. M. Adang, C. H. van der Kaa, J. J. Futterer and J. Barentsz, *Lancet Oncol.*, 2008, **9**, 850–856.

143. A. M. Hovels, R. A. M. Heesakkers, E. M. Adang, G. J. Jager, S. Strum, Y. L. Hoogeveen, J. L. Severens and J. O. Barentsz, *Clin. Radiol.*, 2008, **63**, 387–395.

144. J. O. Barentsz and P. P. Tekkis, in *Nanoparticles in Biomedical Imaging Emerging Technologies and Applications*; ed. J. W. M. Bulte and M. M. J. Modo, Springer, New York, 2008; **vol. 102**, pp. 25–40.

145. S. H. Kim, J. M. Lee, J. K. Han, J. Y. Lee, W. J. Kang, J. Y. Jang, K. S. Shin, K. C. Cho and B. I. Choi, *Eur. Radiol.*, 2006, **16**, 1887–1897.

146. K. A. Kelly, S. R. Setlur, R. Ross, R. Anbazhagan, P. Waterman, M. A. Rubin and R. Weissleder, *Cancer Res.*, 2008, **68**, 2286–2291.

147. A. Z. Wang, V. Bagalkot, F. Gu, F. Alexis, C. Vasilliou, M. Cima, S. Jon and O. Farokhzad, *Int. J. Radiat. Oncol., Biol., Phys.*, 2007, **69**, S110–S111.

148. A. Z. Wang, V. Bagalkot, C. C. Vasilliou, F. Gu, F. Alexis, L. Zhang, M. Shaikh, K. Yuet, M. J. Cima, R. Langer, P. W. Kantoff, N. H. Bander, S. Y. Jon and O. C. Farokhzad, *ChemMedChem*, 2008, **3**, 1311–1315.

149. A. J. Nijdam, T. R. Nicholson, J. P. Shapiro, B. R. Smith, J. T. Heverhagen, P. Schmalbrock, M. V. Knopp, A. Kebbel, D. Wang and S. C. Lee, *Curr. Nanosci.*, 2009, **5**, 88–102.

150. T. M. Keller, S. C. A. Michel, J. Frohlich, D. Fink, R. Caduff, B. Marincek and R. A. Kubik-Huch, *Eur. Radiol.*, 2004, **14**, 937–944.

151. C. S. Reiner, A. M. Lutz, F. Tschirch, J. M. Froehlich, S. Gaillard, B. Marincek and D. Weishaupt, *Eur. Radiol.*, 2009, **19**, 1715–1722.

152. R. Chamberlain, D. Reyes, G. L. Curran, M. Marjanska, T. M. Wengenack, J. F. Poduslo, M. Garwood and C. R. Jack, *Magn. Reson. Med.*, 2009, **61**, 1158–1164.

153. H. Gao, Q. Long, M. Graves, J. H. Gillard and Z. Y. Li, *J. Magn. Reson. Imaging*, 2009, **30**, 85–93.

154. A. Lombardo, V. Rizzello, L. Natale, M. Lombardi, S. Coli, F. Snider, L. Bonomo and F. Crea, *Int. J. Cardiol.*, 2009, **136**, 103–105.

155. Y. Kurosaki, K. Yoshida, H. Endo, N. Sadamasa, O. Narumi, M. Chin and S. Yamagata, *Stroke*, 2009, **40**, E200–E200.

156. H. Alsaid, G. De Souza, M. C. Bourdillon, F. Chaubet, A. Sulaiman, C. Desbleds-Mansard, L. Chaabane, C. Zahir, E. Lancelot, O. Rousseaux, C. Corot, P. Douek, A. Briguet, D. Letourneur and E. Canet-Soulas, *Invest. Radiol.*, 2009, **44**, 151–158.

157. H. Y. Lee, S. H. Lee, C. J. Xu, J. Xie, J. H. Lee, B. Wu, A. L. Koh, X. Y. Wang, R. Sinclair, S. Xwang, D. G. Nishimura, S. Biswal, S. H. Sun, S. H. Cho and X. Y. Chen, *Nanotechnology*, 2008, **19**, 65101–65107.

158. A. Masotti, A. Pitta, G. Ortaggi, M. Corti, C. Innocenti, A. Lascialfari, M. Marinone, P. Marzola, A. Daducci, A. Sbarbati, E. Micotti, F. Orsini, G. Poletti and C. Sangregorio, *Magn. Reson. Mater. Phys., Biol. Med.*, 2009, **22**, 77–87.

159. J. A. Dagata, N. Farkas, C. L. Dennis, R. D. Shull, V. A. Hackley, C. Yang, K. F. Pirollo and E. H. Chang, *Nanotechnology*, 2008, **19**, 35101–35115.

160. J. Zheng, C. Allen and D. Jaffray, *Med. Phys.*, 2005, **32**, 2054–2054.

161. H. Ai, C. Flask, B. Weinberg, X. T. Shuai, M. D. Pagel, D. Farrell, J. Duerk and J. Gao, *Adv. Mater.*, 2005, **17**, 1949–1952.

162. J. A. Kaufman, E. T. Ahrens, D. H. Laidlaw, S. Zhang and J. M. Allman, *Am. J. Phys. Anthropol.*, 2006, 112.

163. A. Obenaus, M. Robbins, G. Blanco, N. R. Galloway, E. Snissarenko, E. Gillard, S. Lee and M. Curras-Collazo, *J. Neurotraum.*, 2007, **24**, 1147–1160.

164. A. M. Tang, D. F. Kacher, E. Y. Lam, M. Brodsky, F. A. Jolesz and E. S. Yang, *Multi-modal imaging: Simultaneous MRI and ultrasound imaging for carotid arteries visualization*, in *Engineering in Medicine and Biology Society, EMBS 2007: 29th Annual International Conference of the IEEE Lyon, France,* August 23-26, 2007, pp. 2603–2606.

165. J. R. McCarthy, F. A. Jaffer and R. Weissleder, *Small*, 2006, **2**, 983–987.

166. J. R. McCarthy, P. Patel, I. Botnaru, P. Haghayeghi, R. Weissleder and F. A. Jaffer, *Bioconjugate Chem.*, 2009, **20**, 1251–1255.

167. D. Pan, S. D. Caruthers, G. Hu, A. Senpan, M. J. Scott, P. J. Gaffney, S. A. Wickline and G. M. Lanza, *J. Am. Chem. Soc.*, 2008, **130**, 9186.

168. A. H. Schmieder, S. D. Caruthers, H. Zhang, T. A. Williams, J. D. Robertson, S. A. Wickline and G. M. Lanza, *FASEB J.*, 2008, **22**, 4179–4189.

169. B. Sumer and J. M. Gao, *Nanomedicine*, 2008, **3**, 137–140.

170. H. S. Chong, H. A. Song, X. Ma, D. E. Milenic, E. D. Brady, S. Lim, H. Lee, K. Baidoo, D. Cheng and M. W. Brechbiel, *Bioconjugate Chem.*, 2008, **19**, 1439–1447.

171. M. Johannsen, U. Gneueckow, B. Thiesen, K. Taymoorian, C. H. Cho, N. Waldofner, R. Scholz, A. Jordan, S. A. Loening and P. Wust, *Eur. Urol.*, 2007, **52**, 1653–1662.

172. C. G. Hadjipanayis, M. J. Bonder, S. Balakrishnan, X. Wang, H. Mao and G. C. Hadjipanayis, *Small*, 2008, **4**, 1925–9.

173. A. T. Tatham, H. Nakamura, E. C. Wiener and Y. Yamamoto, *Magn. Reson. Med.*, 1999, **42**, 32–36.

Subject Index

As ligands and contrast agents have very long, complex chemical names they are indexed under their standard abbreviated form. Elements are indexed under their full name.